Lecture Notes in Mechanical Engineering

Lecture Notes in Mechanical Engineering (LNME) publishes the latest developments in Mechanical Engineering—quickly, informally and with high quality. Original research reported in proceedings and post-proceedings represents the core of LNME. Volumes published in LNME embrace all aspects, subfields and new challenges of mechanical engineering. Topics in the series include:

- Engineering Design
- Machinery and Machine Elements
- Mechanical Structures and Stress Analysis
- Automotive Engineering
- Engine Technology
- Aerospace Technology and Astronautics
- Nanotechnology and Microengineering
- Control, Robotics, Mechatronics
- MEMS
- Theoretical and Applied Mechanics
- Dynamical Systems, Control
- Fluid Mechanics
- Engineering Thermodynamics, Heat and Mass Transfer
- Manufacturing
- Precision Engineering, Instrumentation, Measurement
- Materials Engineering
- Tribology and Surface Technology

To submit a proposal or request further information, please contact the Springer Editor of your location:

China: Ms. Ella Zhang at ella.zhang@springer.com
India: Priya Vyas at priya.vyas@springer.com
Rest of Asia, Australia, New Zealand: Swati Meherishi at swati.meherishi@springer.com
All other countries: Dr. Leontina Di Cecco at Leontina.dicecco@springer.com

To submit a proposal for a monograph, please check our Springer Tracts in Mechanical Engineering at https://link.springer.com/bookseries/11693 or contact Leontina.dicecco@springer.com

Indexed by SCOPUS. All books published in the series are submitted for consideration in Web of Science.

More information about this series at https://link.springer.com/bookseries/11236

Volodymyr Tonkonogyi ·
Vitalii Ivanov · Justyna Trojanowska ·
Gennadii Oborskyi · Ivan Pavlenko
Editors

Advanced Manufacturing Processes IV

Selected Papers from the 4th Grabchenko's International Conference on Advanced Manufacturing Processes (InterPartner-2022), September 6–9, 2022, Odessa, Ukraine

Springer

Editors
Volodymyr Tonkonogyi
Odessa Polytechnic National University
Odessa, Ukraine

Vitalii Ivanov
Sumy State University
Sumy, Ukraine

Justyna Trojanowska
Poznan University of Technology
Poznan, Poland

Gennadii Oborskyi
Odessa Polytechnic National University
Odessa, Ukraine

Ivan Pavlenko
Sumy State University
Sumy, Ukraine

ISSN 2195-4356 ISSN 2195-4364 (electronic)
Lecture Notes in Mechanical Engineering
ISBN 978-3-031-16650-1 ISBN 978-3-031-16651-8 (eBook)
https://doi.org/10.1007/978-3-031-16651-8

This Springer imprint is published by the registered company Springer Nature Switzerland AG
The registered company address is: Gewerbestrasse 11, 6330 Cham, Switzerland

Preface

This volume of Lecture Notes in Mechanical Engineering contains selected papers presented at the 4th Grabchenko's International Conference on Advanced Manufacturing Processes (InterPartner-2022), held in Odessa, Ukraine, on September 6–9, 2022. The conference was organized by Odessa Polytechnic National University, National Technical University "Kharkiv Polytechnic Institute", Sumy State University, and the International Association for Technological Development and Innovations in partnership with Poznan University of Technology (Poland).

InterPartner Conference Series promotes research and developmental activities, intensifying scientific information interchange between researchers, developers, and engineers.

InterPartner-2022 received 91 contributions from 15 countries around the world. After a thorough peer-review process, the Program Committee accepted 55 papers written by authors from 13 countries. Thank you very much to the authors for their contribution. These papers are published in the present book, achieving an acceptance rate of about 60%.

We want to take this opportunity to thank members of the Program Committee and invited external reviewers for their efforts and expertise in contributing to reviewing, without which it would be impossible to maintain the high standards of peer-reviewed papers.

Thank you very much to keynote speakers Dr. Slawomir Luscinski (Kielce University of Technology, Poland), Dr. Justyna Trojanowska (Poznan University of Technology, Poland), and Prof. Oleh Onysko (Ivano-Frankivsk National Technical University of Oil and Gas, Ukraine).

The book "Advanced Manufacturing Processes IV" was organized into seven parts according to the main conference topics: Part 1 – Design Engineering, Part 2 – Manufacturing Technology, Part 3 – Machining Processes, Part 4 – Advanced Materials, Part 5 – Quality Assurance, Part 6 – Mechanical Engineering, Part 7 – Process Engineering.

The first part "Design Engineering" includes recent advancements in design processes and adaptive control systems. It presents studies in designing mechatronic systems, spatial elements of wheels, and automated control of gear profiles for hydraulic machines. This part includes studies in design measures to ensure the reliability of bearings, pneumatic cylinders, and various technological equipment. Ways to design self-adjusting systems are also presented in this part.

The second part "Manufacturing Technology" includes ways to implement intelligent solutions for integrated management systems. It presents recent developments in the manufacturing of parts using additive manufacturing technologies. This part contains studies on controlling system parameters, ensuring the accuracy of manufactured products, and increasing the efficiency of adaptive groups in layered manufacturing. Finally, optimal conditions for the deformation of the stamping-drawing process from aviation materials, ways to increase the efficiency of dynamic characteristics, and control of thermomechanical conditions for heterogeneous materials at finishing operations are also presented in this part.

The third part "Machining Processes" consists of numerical simulations of the diamond grinding process, vibration processing of parts, and evaluation of operating conditions upon intermittent grinding. Studies on deformation mechanics during broaching of cast iron workpieces and finite element analysis of the cutting forces in face milling of gray cast iron are also included in this part. Additionally, the third part deals with optimizing the cutting process based on thermophysical characteristics and evaluating the influence of the back rake angle of a threading cutter on the drill-string tool-joint pitch diameter.

The fourth part "Advanced Materials" is devoted to applying new complex treatments to ensure the functional properties of the surface layers of machine parts and the method for evaluating the resource of diffusion coatings under fatigue conditions. Investigation of the corrosion resistance of permeable porous materials with protective coatings and the effect of ligatures on microstructure and mechanical properties of automotive materials are presented in this part. The fourth part also aims to synthesize copper nanoparticles on graphite using transient glow-to-arc discharge plasma and evaluate surface texture in laser selective melted alloy parts processed by shot peening. It finally includes a comprehensive analysis of the fatigue strength of steel samples after friction treatment.

The fifth part "Quality Assurance" presents a general approach for tolerance control in quality assessment for technology quality analysis, a universal quality control system for industrial enterprises, and modernization of the internal audit process using a risk-based approach. The advanced technology of economic efficiency evaluation and the taxonomy approach for engineering education outcomes assessment are also included in this part. The fifth part also describes problems in the modernization of packaging technologies and improving operational parameters for high-precision tribosystems.

The sixth part "Mechanical Engineering" is based on recent developments in vibration damping of lifting mechanisms, the dynamic behavior of a vibratory plate compactor on the elastic–viscous–plastic surface, and stabilization of natural frequency oscillation equipment. It includes theoretical studies in particle dynamics under oscillating and rotary movements and Lyapunov function-based approach to estimate attractors for dynamic systems. Moreover, the sixth part is based on recent advancements in wave propagation, tube buckling analysis, strength evaluation for cast parts, and contact between elements of hydrovolumetric transmissions.

The seventh part "Process Engineering" presents research studies on temperature distribution in vehicle disk brakes, the efficiency of convective heat exchange in friction elements, and fluid cooling of friction couples. Research work on the changes in the output parameters of hydraulic machines and numerical modeling of point defect formation at nuclear power plants are also presented in this part. The seventh part also includes failure analysis of refractories in rotary kilns and ways to improve the performance of vortex superchargers.

The editors appreciate the outstanding contribution of all the authors. We are deeply convinced that the research papers presented in the book will be helpful to scientists, industrial engineers, and highly qualified practitioners worldwide.

We appreciate a reliable partnership with Springer Nature, iThenticate, and EasyChair for their support during the preparation of InterPartner-2022.

Thank you very much to InterPartner Team. Their involvement, devotion, and hardwork were crucial to the success of the conference.

InterPartner's motto is "Science unites people together".

September 2022

Volodymyr Tonkonogyi
Vitalii Ivanov
Justyna Trojanowska
Gennadii Oborskyi
Ivan Pavlenko

Organization

Honorary Chair

Anatolii Grabchenko	National Technical University "Kharkiv Polytechnic Institute", Ukraine

General Chair

Volodymyr Tonkonogyi	Odessa Polytechnic National University, Ukraine

Co-chair

Vitalii Ivanov	Sumy State University, Ukraine

Steering Committee

Vitalii Ivanov	Sumy State University, Ukraine
Gennadii Oborskyi	Odessa Polytechnic National University, Ukraine
Ievhen Ostroverkh	National Technical University "Kharkiv Polytechnic Institute", Ukraine
Ivan Pavlenko	Sumy State University, Ukraine
Volodymyr Tonkonogyi	Odessa Polytechnic National University, Ukraine
Justyna Trojanowska	Poznan University of Technology, Poland

Program Committee

Jean-Francois Antoine	University of Lorraine, France
Katarzyna Antosz	Rzeszow University of Technology, Poland
Michal Balog	Technical University of Kosice, Slovak Republic
Yevheniia Basova	National Technical University "Kharkiv Polytechnic Institute", Ukraine
Kristina Berladir	Sumy State University, Ukraine

Cristian Barz	Technical University of Cluj-Napoca, Romania
Ricardo Branco	University of Coimbra, Portugal
Dagmar Caganova	Slovak University of Technology, Slovak Republic
Emilia Campean	Technical University of Cluj-Napoca, Romania
Olivian Chiver	Technical University of Cluj-Napoca, Romania
Vasile George Cioata	Polytechnic University of Timisoara, Romania
Olaf Ciszak	Poznan University of Technology, Poland
Nadezda Cubonova	University of Zilina, Slovak Republic
Ivan Cvitic	University of Zagreb, Croatia
Radu Cotetiu	Technical University of Cluj-Napoca, Romania
Predrag Dasic	High Technical Mechanical School of Professional Studies Trstenik, Serbia
Yuliia Denysenko	Sumy State University, Ukraine
Oleksandr Derevianchenko	Odessa Polytechnic National University, Ukraine
Sergey Dobrotvorskiy	National Technical University "Kharkiv Polytechnic Institute", Ukraine
Mihai Dragomir	Technical University of Cluj-Napoca, Romania
Tygran Dzhuguryan	Maritime University of Szczecin, Poland
Milan Edl	University of West Bohemia, Czech Republic
Sulaymon Eshkabilov	North Dakota State University, USA
Mathieu Gautier	University Lyon, France
Renata Gnatowska	Czestochowa University of Technology, Poland
Marta Grabowska	Poznan University of Technology, Poland
Jakub Grabski	Poznan University of Technology, Poland
Michal Hatala	Technical University of Kosice, Slovak Republic
Ihor Hrytsay	Lviv Polytechnic National University, Ukraine
Ihor Hurey	Lviv Polytechnic National University, Ukraine
Jozef Husar	Technical University of Kosice, Slovak Republic
Vitalii Ivanov	Sumy State University, Ukraine
Maryna Ivanova	National Technical University "Kharkiv Polytechnic Institute", Ukraine
Malgorzata Jasiulewicz-Kaczmarek	Poznan University of Technology, Poland
Lydmila Kalafatova	Donetsk National Technical University, Ukraine
Isak Karabegovic	University of Bihac, Bosnia, and Herzegovina
Serhii Khovanskyi	Sumy State University, Ukraine
Lucia Knapcikova	Technical University of Kosice, Slovak Republic
Dmytro Konovalov	Admiral Makarov National University of Shipbuilding, Ukraine
Kateryna Kostyk	National Technical University "Kharkiv Polytechnic Institute", Ukraine
Jan Krmela	Alexander Dubcek University of Trencin, Slovak Republic
Ivan Kuric	University of Zilina, Slovak Republic

Jozef Zajac	Technical University of Kosice, Slovak Republic
Predrag Zivkovic	University of Nis, Serbia

Invited External Reviewers

Olena Avdieieva	National Technical University "Kharkiv Polytechnic Institute", Ukraine
Vladimir Bulej	University of Zilina, Slovak Republic
Nikolaos Karkalos	National Technical University of Athens, Greece
Alexey Kotliar	National Technical University "Kharkiv Polytechnic Institute", Ukraine
Panagiotis Karmiris Obratanski	National Technical University of Athens, Greece
Emmanouil Papazoglou	National Technical University of Athens, Greece
Svetlana Radchenko	Technical University of Kosice, Slovak Republic
Michal Sajgalik	University of Zilina, Slovak Republic
Serhii Sharapov	Sumy State University, Ukraine
Dimitrios Skondras-Giousios	National Technical University of Athens, Greece
Ioan Radu Sugar	Technical University of Cluj-Napoca, Romania
Valentyna Tkachuk	Lutsk National Technical University, Ukraine
Tatiana Volina	National University of Life and Environmental Sciences of Ukraine, Ukraine
Ihor Yakovenko	National Technical University "Kharkiv Polytechnic Institute", Ukraine
Dmytro Zhyhylii	Sumy State University, Ukraine
Tetiana Zhylenko	University of Cologne, Germany

InterPartner Team

Anna Balaniuk	Odessa Polytechnic National University, Ukraine
Kristina Berladir	Sumy State University, Ukraine
Vitalii Ivanov	Sumy State University, Ukraine
Gennadii Oborskyi	Odessa Polytechnic National University, Ukraine
Ievhen Ostroverkh	National Technical University "Kharkiv Polytechnic Institute", Ukraine
Ivan Pavlenko	Sumy State University, Ukraine
Andrey Pavlyshko	Odessa Polytechnic National University, Ukraine
Volodymyr Tonkonogyi	Odessa Polytechnic National University, Ukraine
Justyna Trojanowska	Poznan University of Technology, Poland

Contents

Manufacturing Technology

Machining Processes

Process Engineering

Design Engineering

Mechatronic Transducer for Two-Level Adaptive Control of CNC Machines

Anatoly Gushchin[1], Vasily Larshin[1](✉), Oleksandr Lysyi[2], and Victor Marchuk[3]

[1] Odessa Polytechnic National University, 1, Shevchenko Avenue, Odessa 65044, Ukraine
vasilylarshin@gmail.com

[2] Odessa Military Academy, 10, Fontanskaya Doroga Street, Odessa 65009, Ukraine

[3] Lutsk National Technical University, 75, Lvivska Street, Lutsk 43018, Ukraine

Abstract. A new direction has been formulated in developing hierarchical adaptive control of machine tools based on using a mechatronic module (MM) mounted in the electro-spindle of a modern CNC metal cutting machine. The MM operation principle is due to its chosen method: either a device for generating a control servo signal for the machine CNC system or a mechatronic power converter. In the first case (generating a control servo signal), the MM generates an unbalance servo signal which represents the required (according to the machining technology of the part) dependence of the force-torque parameters on the path traveled (along the CNC program trajectory) with adjusting the feed from the CNC machine system. In the second case (power converter), the MM works independently of the machine CNC system, maintaining communication with it. An example of a one-dimensional servo automatic control system is considered for the operations of drilling holes and forming grooves when diamond grinding inscriptions and patterns on the surface of parts, including parts made of superhard materials. During multidimensional machining of complex-profile surfaces (turbine blades, impellers, implants, etc.), the MM is embedded in CNC machines' corresponding coordinate electric drives. In addition to adaptive control of the axial force and cutting torque, the developed MM design can take into account other signals (other than axial force and torque) generated by the monitoring system of the CNC machine for intelligent control.

Keywords: Mechatronic module · Equilibrium state · Hierarchical control · Compensating link · Tracking system · Cutting torque · Superhard material · Artificial intelligence · Product innovation

1 Introduction

Scientists of the past and present century have done a lot of work on introducing adaptive control systems for the mechanical machining of parts made of materials with high strength characteristics and special anisotropic properties into production. However, the results obtained in practice, in terms of the effectiveness of such systems (excess of

V. Tonkonogyi et al. (Eds.): InterPartner 2022, LNME, pp. 3–12, 2023.
https://doi.org/10.1007/978-3-031-16651-8_1

benefits over costs), turned out to be inadequate to the expected results. It is caused by the belated detection of an imbalance of forces arising in the cutting zone.

The paper will show how to immediately direct the identified imbalance to eliminate this imbalance. To do this, you must first create a balanced system of cutting forces (and/or torques) in a closed elastic system of the machine. In case of violation of the force balance, it is necessary to restore it automatically (by returning to the force balance) by changing the size of a special "compensating link" (in the closed dimensional chain). In machine assembly technology, this method has been called the "regulation one". However, until now, it has not been used for adaptive control to stabilize the size of the so-called "closing link". In other words, by feeding excessive force-moment effects into the machining zone, it is necessary not only to fix them but also to simultaneously use the resulting reactions to restore the equilibrium state of the closed elastic system of the machine. In this regard, the following three theses have been formulated and implemented.

1. A mechanical system that implements any tool movements (affecting the resulting size) in the cutting zone from the machine side must allow reciprocal (i.e., directed in the opposite direction) movements of the tool, the power source of which is excess energy $\Delta E = T \cdot \Delta\varphi$, where T is the torque (N·m); $\Delta\varphi$ is the increment of the angle of rotation of the tool (rad). This excess energy can be compensated by changing the linear size Δz of the "compensating link". For example, when drilling $\Delta E = F_{thrust} \cdot \Delta z$, where F_{thrust} is the increment of the axial force, N. Therefore, to compensate for the emerging imbalance, the "compensating link" of a closed dimensional chain must change its size by the following amount proportional to the increment of the angle of rotation of the axial tool (or spindle) $\Delta z = \frac{T}{F_{thrust}} \Delta\varphi$. Thus, a mechanism is needed to convert the excess energy ΔE into the increment of the tool rotation angle $\Delta\varphi$. For example, cam mechanism, screw, or friction ones are suitable for this.
2. The response movements of the tool (because of their smallness) can be "absorbed" by the elastic displacements of the tool cutting edge from the workpiece. Hence, for the mechatronic module (hereinafter MM), the operation of the criterion of constancy of elastic displacements that absorb the increment Δz can be used. A change in the size of the "compensating link" of the dimensional chain by an amount Δz occurs at a constant cutting force (and/or cutting torque).
3. Adaptive control systems on CNC machines must have "reasonable behavior", which is inherent in living organisms. To do this, their design should include "sensory organs" and elements of artificial intelligence. In this case, the property of a living organism is used to get rid of irritation (and to protect itself), preventing its development over time. In other words, the MM either immediately generates a control signal for the machine's CNC system (monitoring mode) or autonomously (i.e., by its own means) eliminates the irritation (and protects itself) that has arisen (adaptive control mode). All three theses can be implemented using the MM in a tracking servo drive for the feeding tool on a CNC machine.

That is why the purpose of the paper is to develop a mechatronic transducer based on the MM mentioned that generates a control signal for interaction with a higher-level system, which is the CNC system of a mechatronic CNC machine.

2 Literature Review

Mechatronics is a new scientific discipline that has taken the high position that previously belonged to cybernetics over the past few decades. Today, this synergetics integrates different disciplines: electronics, information technology, and mechanics [1]. Key elements of mechatronics – physical system modeling, signals and systems, computers and logic systems, software and data acquisition, sensors, and actuators – are naturally accompanied by automatic control. That is why system interfacing, instrumentation, and control systems themselves are the subject of studies [2].

Modern control theory's well-known control principles (by deviation and disturbance) are accompanied by a new direction: distributed and hierarchical control systems. According to hierarchical control, there are four system levels (from bottom to top): a component level, and information preprocessor level, an intellectual preprocessor level, and a top-level [3, 4]. There are three feedback kinds from the lower levels to the upper ones: (1) low-level sensory feedback, (2) intermediate-level feedback, and (3) top-level feedback [3]. So we have the so-called distribution of control available geographically and functionally in a complex control servo system. The computers in the hierarchical system communicate using a suitable communication network. Information transfer in both directions (up and down) should be possible for best performance and flexibility. However, there is no concrete example of the implementation of such hierarchical control.

The two-level industrial servo system has the same shortage [5]. For modeling a multi-jointed robot, this system contains a mechanism part, an actuator at the lower level, and an upper-level controller at the upper level. The servo system of each axis is constructed by the motor part, power amplifier part, current control part, velocity control part, position control part, and sensor (position detector, velocity detector, current detector). But nothing is said about the mechatronic servo actuator, which generates a position signal.

Essential control theory, as well as transfer functions and state space approaches in this theory, are the base for developing the related systems mentioned above [6]. But the professional society literature ignores electromechanical devices other than, e.g., generators and motors. Because: (1) their designs are diverse and may have strange-looking structures, (2) their engineering is based largely on judgment, inventiveness, and experimentation as well as on mathematical analysis [7]. At the same time, in the real world, mechanical engineering and electrical engineering are inextricably entwined. Every electrical device is a mechanical device designed for its electrical properties and manufactured in a factory of mechanical machines. Many mechanical devices are partly electrical, and most are made by machines that are electrically powered and electrically controlled [7]. Moreover, machine tools are mechatronic systems themselves [8]. Hence, they are open to the inclusion of mechatronic mechanisms in them, but there are no hints of it in the literature.

Permanent magnetic force is equivalent to buoyant force and is used to reduce bearing friction in watt-hour meters and other instruments. Besides, electro-magnetic force levitation (MAGLEV) is a technique to reduce friction in experimental high-speed trains [7]. But there is no information on the details of the MAGLEV device. There is no information on the idea that many electro-magnetic transducers generate an intermediate parameter, such as mechanical displacement [7]. This information is exactly what this paper is about. But there is only an idea on a mechatronic transducer.

There is a design of a mechatronic electro-spindle containing a mechatronic actuator [9, 10]. But the hierarchy of the mechatronic servo system containing the machine CNC at the upper control level is not disclosed. The same applies to the so-called intelligent automated drilling in laminate composites and hybrid materials [11].

At the end of the review, it is necessary to mention works on a study of the quality of machined parts made of carbon fiber [12], CFRP [13], and composites [14, 15], as well as works on the physics of electromagnetism [16, 17]. The first case (quality of machined parts) determines the relevance of mechatronic technological systems, including the mechatronic servo system. The second one (physics of electromagnetism) allows the development of such systems.

Monitoring CNC machining applications allows the production to be analysed and improved, improving part quality [18, 19]. Online process monitoring can allow the machine to become more intelligent and adapt to its conditions internally. Machine learning and adaptive machining could further help improve manufacturing efficiency. Therefore there is a need for a low-cost, flexible sensor system that is easy to apply to currently existing CNC technologies [20].

3 Research Methodology

3.1 Conclusions from the Review

Literature analysis allows the formulation of two possible modes of MM operation as part of hierarchical mechatronic systems: (1) the use of MM as a transducer (source of information); (2) the use of MM as a power mechanism (convertor) for adaptive control. In the first case, MM detects a deviation, for example, based on the current position of the working body (or tool). The machine's CNC system eliminates this deviation detected by the appropriate feed selection (deviation control). In the second case, the MM is used directly for adaptive control by disturbance.

Literary analysis has shown that when developing control systems, such features of the control object as anisotropy and high hardness of materials to be machined are not taken into account. In this regard, the proposed methodology contains the following four principles (see also three theses in the paper's introduction).

1. "Intervention" in the cutting process should begin at the stage of its preparation (pre-production stage). In this case, it will be controlled based on reference and empirical data.
2. Possible deviations during the cutting process, e.g., the cutting torque increment, must be used either as a deviation signal for a higher level system, for example, CNC level. This signal is used for direct control "by disturbance".

3. The force-torque parameters of the interaction of a cutting tool with a workpiece – parameters that are formed in the cutting zone – must be included in the servo system as an independent cutting parameter set by the MM, just as speed or displacement (path) are independent parameters in the conventional CNC machine system.
4. Control over the progress of the cutting process must be carried out according to the totality of many parameters (taking into account their weighting significance), including indirect signs that precede the occurrence of undesirable phenomena in the cutting zone (increasing noise, vibration, temperature).

All four above principles have been verified by many years of practice in introducing fundamentally new technologies for cutting hard-to-machine materials on a CNC machine into industrial production (Fig. 1).

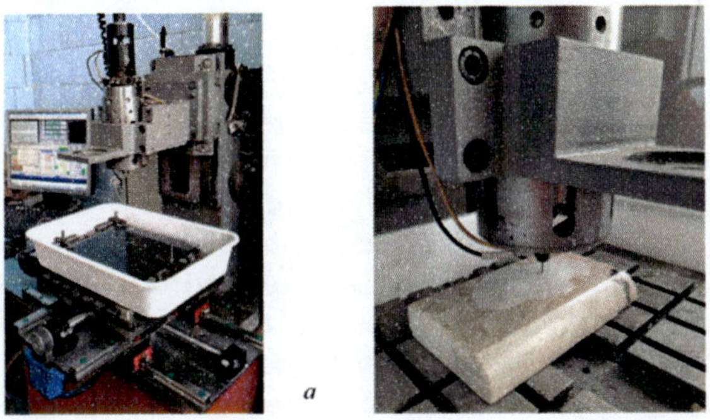

a *b*

Fig. 1. Mechatronic technological system (MTS) based on a 3D CNC machine when processing glass (*a*) and stone (*b*).

3.2 Multi-level Control

Two-level control at the pre-production and production stages allows the creation of a single integrated design and production automation system. In such a system, optimization plays the part of both a design method – at the upper level of hierarchical management – and a control method – at the lower level.

Let us explain the above in greater detail. In the language of control system theory, on the upper control level, a relative decision is made which is not based on some control (reference) points depending on the machine shop's actual conditions. Consequently, it will be some formal decision. On the contrary, a robust control system on the lower control level considers the shop's existing conditions. Sometimes, it will be with the help of the so-called RDT and E laboratory (an intermediary between the upper and lower levels of control) (Fig. 2).

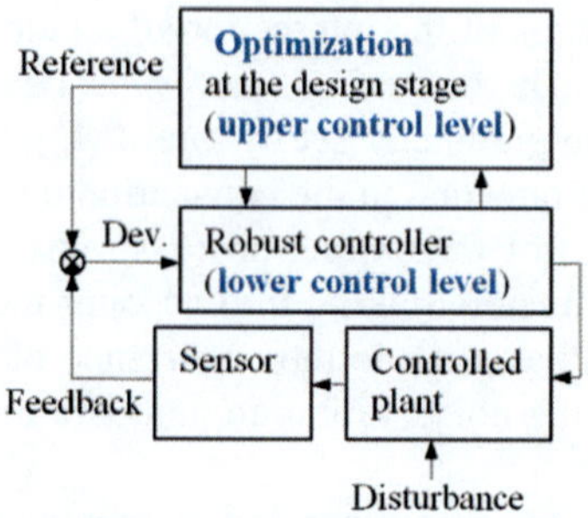

Fig. 2. General scheme of the two-level hierarchical automatic control system.

3.3 Methodology for Programming Systems of the Upper and Lower Levels

MM is intended mainly for CNC machines to ensure maximum productivity with guaranteed quality due to the automatic control servo system operation. This servo system is got (in each tool axial movement coordinate) the required value of the cutting torque, e.g. $T_i(z_i)$, found at the pre-production stage. For example, for the one-dimensional case (hole drilling), this servo system, due to the MM, implements the functional dependence $T_i(z_i)$ in the tracking servo mode. With an increase in the number of controlled coordinates, e.g., up to three, the tracking servo control is provided for each of these coordinates by the corresponding electric drives for each point of the tool trajectory in space. In other words, in the servo mode, three functional dependencies are simultaneously performed: $T_i(x_i)$, $T_j(y_j)$ and $T_k(z_k)$, where x_i, y_j, z_k are the coordinate values at the points with ordinal numbers i, j and k.

For example, the coordinate z_k (instantaneous scalar value of the vector z at the point k) is set by the machine CNC system (the upper control level system). The corresponding torque values are set by the controlling computer that sets the torque parameters of the cutting operation (lower level control system), i.e., robust servo system of automatic direct control "by disturbance". The controlling computer has a vector $\boldsymbol{T}(z)$ generator. Thus, when compiling a program for the tool spatial movement, predetermined force-torque characteristics of the machining process are introduced simultaneously into the upper and lower control systems, i.e., both the machine CNC and the tracking servo system, respectively. The tracking servo system fulfills this instruction by maintaining a machining process's force and torque characteristics at a given level (the machine CNC system sets the level).

4 Results and Discussion

The MM includes a base 1, on which mounting elements 2 are installed, carrying a ferromagnetic housing 3 of a linear DC actuator (Fig. 3). Housing 3 also acts as an external magnetic circuit of the same actuator. In housing 3, coaxially to it, a cylindrical internal magnetic core 4 is rigidly fixed, having a field winding divided into two Sects. 5 and 6, creating a total magnetic flux (not shown) in the above-mentioned magnetic circuits 3 and 4. The magnetic flux has the desired both size and direction. In the working gap between the magnetic circuits 3 and 4, there is a hollow cylindrical non-ferromagnetic armature 7, which has an armature winding 8 on its left side. A ring-shaped (annular) permanent magnet 9 is fixed on the right end of armature 7, which moves together with armature 7 as a single solid body.

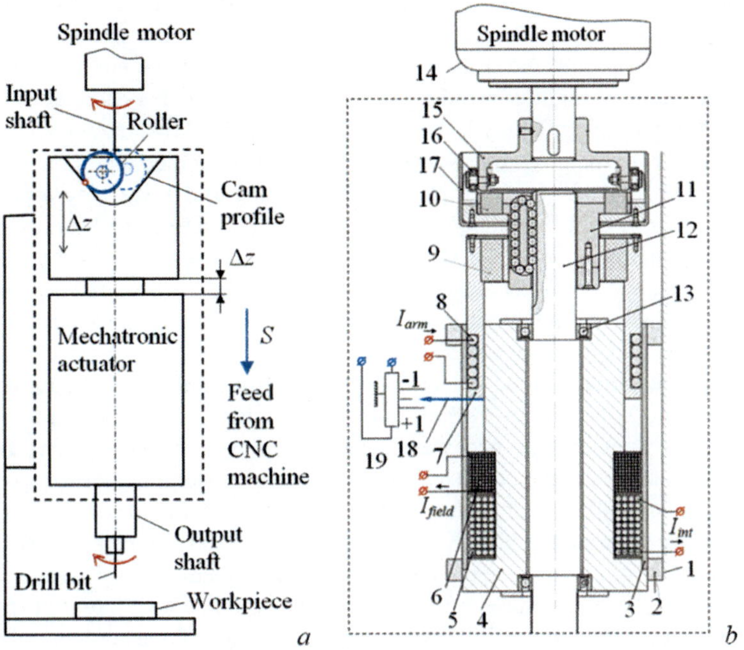

Fig. 3. General scheme of a mechatronic module (*a*) and the same in details (*b*).

Cylindrical magnets 10 (at least two pieces) are installed on the peripheral part of the end face of the flange-ball-spline assembly 11, opposite the magnet 9, so that only repulsive forces arise between the ring-shaped magnet 9 and cylindrical magnets 10. The same name poles of magnets 9 and 10 should be located opposite each other. The overall dimensions and the number of cylindrical magnets 10 are selected depending on the overall design features of the MM.

The ball spline assembly 11 is mounted on shaft 12 (with ball spline grooves). It can transmit torque from the right half coupling with cutouts (cam internal surface) and reciprocate movements along the longitudinal axis of shaft 12.

The shaft 12 is rigidly fixed in the magnetic circuit 4 using the lower (not indicated in Fig. 3) and the upper 13 ball bearings, resulting in which it can rotate and transmit torque.

The torque to the shaft 12 is transmitted from the motor 14 utilizing a driving coupling 15, on the outer cylindrical surface of which two ball bearings 16 are symmetrically fixed so that their axes are perpendicular to the longitudinal axis of the driving coupling 15, and the outer rings have the possibility of rotation.

Ball bearings 16 enter slots (the working surface of the inner cam) made on the lateral cylindrical surface of the driven coupling 17. These slots have the shape of a triangle with a semicircular vertex. The coupling half 17, in turn, is fixed to the ball slot connection 11.

To fix the linear movements of the armature 7, a movable element 18 of the linear displacement sensor 19 is fixed on armature 7, which is informatively connected to the controlling computer of the lower-level tracking servo system (not shown in Fig. 3). The armature winding 8 is powered from a DC source through a unit controlled by the same computer. Section 6 of the field winding is also controlled by this computer.

Section 5 of the field winding is an element of the lower level of control, which can make adjustments to the program of actions of the CNC machine based on information coming from the zone of force interaction of the working tool of the same machine with the external environment (by analogy with the peripheral nervous system of man). To perform such functions, Sect. 5 of the field winding receives signals generated by the control-and-measuring equipment from the zone of force interaction. These signals characterize the adequacy of the ongoing machining process according to the programmed power parameter (axis force or torque).

To fix any deviations in the machining process from the norm (the upper control level sets the norm), the MM as a whole is equipped with various kinds of control-and-measuring equipment and the necessary sensors: video surveillance cameras, microphones, temperature sensors, strain gauges, vibration acceleration sensors, etc. These sensors should be regulated for each specific machine operating in certain conditions.

Special control programs have been developed for an experimental CNC machine containing MM (Fig. 1) to realize the methodology described above. The programs take into account the specific features of the material to be machined: strength (for glass), thermo-physical properties (for CFRP, syntegran), and others. For example, experimental studies have been carried out on contour 2D grinding of inscriptions and patterns with a grinding depth of 0.5 mm on glass with a thickness of 3 mm with diamond cylindrical tools with a diameter of 1.1–4.0 mm. The power parameters of the process – axial force F and torque T– for contour grinding were selected by adjusting the current in Sect. 6 of the field winding (I_{field} in Fig. 3, *b*) and the armature winding 8 (I_{arm} in Fig. 3, *b*) in the following ranges: $0.8 \leq F \leq 1.2$ N and $0.04 \leq T \leq 0.06$ N·m. As a result, exclusive products were obtained (Fig. 4) from ordinary technical glass (*a*), from a mammoth tusk (*b*, *c*), and special optical glass (*d*).

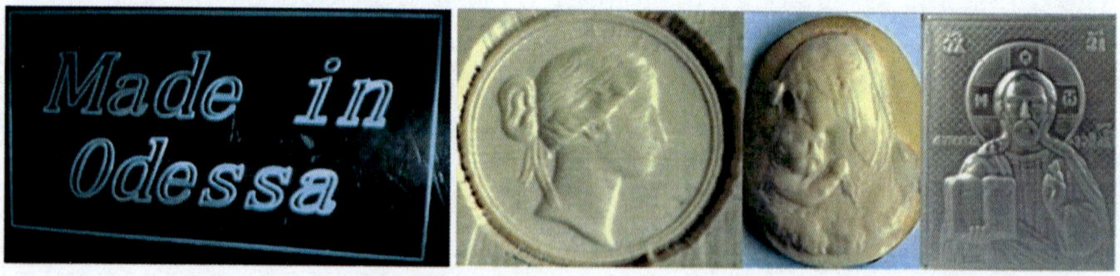

a *b* *c* *d*

Fig. 4. Exclusive products made of hard-to-machine materials.

5 Conclusions

A new scientific and technical direction has been formulated in developing hierarchical adaptive control of machine tools based on the use of a mechatronic module (MM) mounted, as an example, directly on the carriage of vertical feeds of a CNC metal cutting machine.

The principle of operation of a mechatronic servo system based on the MM is determined by the chosen method of its implementation: either by the method of generating a control servo signal for the machine CNC system (transducer and monitoring function) or by the method of automatic control with the power mechatronic converter. In the first case (formation of the control signal), the MM generates a control signal proportional to the amount of feed the machine CNC system sets according to this signal to ensure the tracking servo mode of forming the cutting torque. This case is considered in the paper. In the second case (power converter), the MM has the necessary autonomy and maneuverability for stabilizing the cutting torque independently of the CNC machine system but when working in conjunction with it.

The paper considers an example of a one-dimensional servo automatic control system on the example of drilling and diamond grooving when grinding inscriptions and patterns on the surface of superhard and other hard-to-machine materials. In multidimensional machining of complex-shaped surfaces (turbine blades, impellers, implants, etc.), the MM is built into the corresponding coordinate electric drives of CNC machines and works according to the methods described above.

In addition to adaptive control for cutting axial torque, the developed MM design can take into account other (except for force and torque) signals generated by modern monitoring systems on CNC machines: acoustic emission (in the frequency range of 0.2–2.0 MHz, sound signal (0.1–20.0 kHz), Barkhausen noise (0.06–1.0 MHz), temperature, electromagnetic field, radioactive radiation, etc.

References

1. Bishop, R.H.: Mechatronics Handbook. CRC Press, Bosa Raton, London, Washington (2002)
2. Bishop, R.H.: Mechatronics: an introduction. CRC Press, Taylor & Francis Group University of Texas at Austin, USA, Boca Raton (2006). https://doi.org/10.1201/9781420037241
3. De Silva, C.W.: Mechatronics: A Foundation Course. CRC Press, Taylor & Francis, Boca Raton (2010)
4. Mitrishkin, Y., Rodolfo, E.: Hierarchical control system for complex dynamical plants. In: ICINCO 2009. Proceedings of the 6th International Conference on Informatics in Control, Automation and Robotics, Intelligent Control Systems and Optimization, pp. 56–65. Italy, Milan (2009)
5. Nakamura, M., Goto, S., Kyura, N.: Mechatronic servo system control: problems in industries and their theoretical solutions. Springer-Verlag, Berlin Heidelberg (2004)
6. Billingsle, J.: Essential of Mechatronics. John Wiley & Sons, Inc., Hoboken, New Jersey (2006)
7. Kamm, L.J.: Understanding electro-mechanical engineering: an introduction to mechatronics. The Institute of Electrical and Electronics Engineers Inc., New York (1996)
8. De Silva, C.W.: Mechatronic Systems: Devices, Design, Control, Operation, and Monitoring. CRC Press, Taylor & Francis group, Boca Raton (2008). https://doi.org/10.1201/9780849307768

9. Larshin, V.P., Gushchin, A.M.: Mechatronic technological system information support. Appl. Aspects Inform. Technol. **4**(2), 153–167 (2021). https://doi.org/10.15276/aait.02.2021
10. Dobrinski, A., Möhring, H.-C., Stehle, T.: Development of an adaptronic spindle for a faultless machining of homogeneous and inhomogeneous materials. J. Mach. Eng. **21**(2), 5–23 (2021). https://doi.org/10.36897/jme/136277
11. Dobrinski, A., Dudarev, A.: Intelligent automated drilling in the laminate composites and hybrid materials. Mater. Today: Proc. **38**(4), 1980–1983 (2021). https://doi.org/10.1016/j.matpr.2020.09.723
12. Duntschew, J., Eschelbacher, S., Schluchter, I., Möhring, H.-C.: Discrete wavelet transformation as a tool for analysing the borehole quality when drilling carbon fibre reinforced plastic aluminium stack material. J. Mach. Eng. **21**(1), 78–88 (2021). https://doi.org/10.36897/jme/132930
13. Shimana, K., et al.: Surface integrity of machined surface in simultaneous cutting of CFRP. J. Mach. Eng. **20**(1), 98–106 (2020). https://doi.org/10.36897/jme/117781
14. Sultana, I., Shi, Z., Attia, M.H., Thomson, V.: Surface integrity of holes machined by orbital drilling of composites with single layer diamond tools. Procedia CIRP **45**, 23–26 (2016). https://doi.org/10.1016/j.procir.2016.02.067
15. Fleischer, J., Teti, R., Lanza, G., Mativenga, P., et al.: Composite materials parts manufacturing. CIRP Annals Manuf. Technol. **67**(2), 603–626 (2018). https://doi.org/10.1016/j.cirp.2018.05.005
16. Pawlak, A.M.: Sensors and actuators in mechatronics: design and application. CRC Press, Taylor & Fransis Group, Bosa Raton (2007). DOI: https://doi.org/10.1201/9781315221632
17. De Silva, C.W.: Sensors and actuators: engineering system instrumentation. 2nd edn. CRC Press, Boca Raton (2015). DOI: https://doi.org/10.1201/b18739
18. Karpuschewski, B., Inasaki, I.: Monitoring Systems for Grinding Processes, in Condition Monitoring and Control for Intelligent Manufacturing. Springer, London (2006)
19. Tonshoff, H.K., Friemuth, T., Becker, J.C.: Process Monitoring in Grinding. CIRP Annals – Manufacturing Technology 51, 551−571 (2002). DOI: https://doi.org/10.1016/S0007-8506(07)61700-4
20. Machine Monitoring-Stoney CNC, https://stoneycnc.co.uk/cnc/machine-monitoring, last accessed 2022/07/01

Reverse Engineering and Design Process as Set of Procedures

Viktor Ivanov[1](✉), Liubov Bovnegra[1], Svitlana Ivanova[2], Galyna Naleva[3], and Olha Kononova[4]

[1] Odessa National Polytechnic University, 1, Shevchenko Avenue, Odesa 65044, Ukraine
ivv@op.edu.ua

[2] South Ukrainian National Pedagogical University named after K. D. Ushynsky, 26, Staroportofrankivs'ka Street, Odesa 65020, Ukraine

[3] National University "Odesa Maritime Academy", 34, Mechnikova Street, Odesa 65029, Ukraine

[4] Odessa National Maritime University, 8, Didrikhsona Street, Odesa 65052, Ukraine

Abstract. Equipment that has failed can be repaired, redesigned, or an alternative method of use with limited functionality can be found. Reverse engineering of equipment includes all or part of the stages of the design process. The design process was explored in more detail than the relatively new direction of reverse engineering. Using the experience gained in structuring and automating the design process, it is advisable to improve the reverse engineering methodology. It was proposed to present the design process and reverse engineering procedures. The designing and reverse engineering processes were identified. The procedures were divided into heuristic, design techniques calculation, and metrological procedures. The reverse engineering model shows the place and interaction of these groups. The procedures are considered a symbolic designation, allowing us to present the reverse engineering process as a symbolic sequence.

Keywords: Design process · Reverse engineering · Morphological map · Sustainable manufacturing

1 Introduction

Most researchers narrowly understand reverse engineering: "reconstruction of CAD models from measured data" [1]. A similar idea is at the heart of reverse engineering of software, based on the analysis of dirty source code and finding the initial specification [2], discovering initial models from the legacy artifacts composing a given system, in other words, "discover initial models from the legacy artifacts composing a given system" [3]. With this approach, reverse engineering of equipment involves the following steps: determination of product parameters based on measurements of a real object; the use of special equipment for scanning surfaces; the use of software for the building of 3D surfaces. Researchers are looking for ways to describe the surface of a part as accurately as possible, such as using "tensor-product surfaces" [4] or making the surface "topologically consistent, and it is flexibly editable" [5].

V. Tonkonogyi et al. (Eds.): InterPartner 2022, LNME, pp. 13–22, 2023.
https://doi.org/10.1007/978-3-031-16651-8_2

Simultaneously, the main attention is paid to the accuracy of the original product reproduction. Concerning mechanical engineering problems, such an approach is not always justified. To establish the shape of the bolt surface, we only need the outer diameter of the thread. For the shaft, only the diameters of the seating surfaces are important. The surface shape of the cast casing is not important if a welded casing replaces it.

2 Literature Review

Analysis of publications devoted to reverse engineering shows a wide range of this term applications. These are:

- Military or commercial espionage;
- Compatibility of third-party assemblies with the original product design;
- Analysis of the conformity of the goods to the declared certificates and standards;
- Redesign of obsolete products in the absence of documentation on them;
- Creation of unlicensed and unauthorized copies;
- Analysis of hidden features not declared by the manufacturer;
- Search for alternatives to use part of the damaged product functions [6, 7]

Since this list includes military and commercial espionage and the creation of unlicensed and unauthorized copies, the issue of legal restrictions for reverse engineering is essential. As a result of many years of litigation, it is recognized that legally purchased product, the owner has the right to research, repair, and redesign [8, 9].

The following definition of reverse engineering is proposed: reproducing a device, object, or system by analyzing its structure, function, and action [8]. Reverse engineering, as applied to mechanical engineering, includes the following stages:

- Analysis of equipment damage;
- Design analysis to fill gaps in technical documentation;
- Deciding on options for the further use of equipment;
- Repair or purchase of components, replacing part of equipment units with compatible units of a different design is possible.

Engineering companies working in the field of mechanical engineering are engaged in: the development of compatible devices, analysis of the conformity of products, overwhelmed with technical parameters, as well as investigation of accidents. Rcpurposing and redesigning require a complete design process that is no less difficult than creating an original product.

The solution to the problem of reverse engineering is based on three components. The first one is metrological equipment. This is not only equipment for scanning and measuring complex surfaces but in general, all types of metrological equipment, including equipment for chemical analysis. The second component is software. These are 3D modelling software that is used to analyze scan results. And also CAD/CAM/CAE software is required for analysis using CAE calculations and comparison with standards and databases of unified CAD elements. The software is required for CAD design and for organizing the CAM workflow. The third component is heuristic methods. The tasks of investigating accidents, searching for an object with an alternative purpose, and redesigning are exclusively heuristic.

Consider the heuristic aspect of reverse engineering and the design process. Heuristic methods are widely used in the design process, especially in the early stages of design. Due to the heuristic nature of the design process, the term design thinking has even appeared. A parametric design thinking (PDT) model is proposed. This model includes cognitive models, parametric design, and a digital design process computer model [10].

Heuristic or cognitive models are a typical design and reverse engineering feature. Parametric synthesis (design) is presented as a choice of design options for a gearbox, considering the restrictions imposed by the general design rules and specific restrictions of the automotive industry [11]. Investigation of the relationships between elements in an existing product is the opposite of parametric design. This task can be solved using the DSM matrix (design structure matrix). The matrix's interconnected elements form a cluster corresponding to a machine node or unit [12].

The "interaction matrix" solves similar problems. This method's main task is to establish the relationship in each pair of elements [13]. The task of the "matrix diagram" is not only to establish the fact of the relationship between elements, but also to quantitatively or qualitatively assess the degree of interconnection. Moreover, the elements can be not only parts of the structure but also the functions of elements and assemblies [14].

During the product repair, minor changes may be made to its design. Analysis of permissible changes taking into account the interrelation of design parts is considered using the CICO method (change impacts on the complex product). The method includes a graph of product nodes and a matrix similar to the "matrix diagram" [15]. The Change Prediction Method (CPM) consists of the simultaneous use of DSM and risk matrices. It is not the relationship of elements that is evaluated but their possible interaction, for example, when one is damaged [16]. Without technical documentation for the equipment, its design can be considered a black box. The relationship between the possible nodes of the product and its functions is assessed using a morphological map [17].

3 Research Methodology

Reverse engineering and product design have a number of similar aspects. It is advisable to divide the design process and reverse engineering process into elements to establish which elements match or are similar. This makes it possible to apply the knowledge gained in the design experience to develop methods and techniques for reverse engineering.

The first obvious similarity between reverse engineering and design processes is the widespread use of heuristic methods. Heuristic methods can be represented as a set of heuristic techniques. Let's give the following definition of a heuristic technique - a procedure containing operations that require a designer's participation. The procedure concept is convenient for formalizing the description of the design process and the reverse engineering process.

We will accept the following definitions of design procedures and operations:

- "Design procedure is a sequence of actions formalized in a certain way, as a result of which a design solution is developed" [18];
- "Design operation is an action or a sequence of actions that are part of a design procedure; the algorithm of such actions is unchanged for a set of design procedures" [19].

Thus, our task is to divide the design process into procedures and operations, establish reverse engineering procedures, coincide with the design process procedures, and establish the order of procedures for the design and reverse engineering stages. According to the definition of heuristic technique, this is a procedure. According to the definition of the procedure - the operation is an integral part. Thus, the designer executes a set of procedures. Some of them are heuristic techniques and the procedures may consist of operations. A particular sequence of actions is usually called an algorithm. However, this sequence cannot be implemented as a program, because it contains procedures - heuristic techniques performed directly by the designer. Therefore, calling such a set of procedures a human algorithm is more accurate.

Table 1 summarizes procedures that are part of the overwhelming majority of heuristic methods.

For example, the procedure of collective discussion uses methods of brainstorming, S.C.A.M.P.E.R, Brainsketching, and Synectics. A random association uses the method of focal objects, garlands of accidents and associations, Affinity diagram [20]. The expert evaluation uses in Brainstorming, Synectics, C-Sketch, etc. [20]. Here is just a subset of the methods as an example. There are actually many more methods using the same procedures.

Along with heuristic procedures, design techniques are common to the design process and reverse engineering (Table 1). The designer, whenever possible, uses standard units and elements. These are threaded connections, gear trains, bearings, etc. When analyzing an existing damaged product, measurements are taken of these particular elements to restore the features of its design. For the subsequent possible product redesign, it is essential to understand the existing design principles. For example, design decisions made for the possibility of self-installation of bearings must be implemented in the redesigned product. Calculations performed during the design process may be required for reverse engineering (Table 2). These are calculations related to investigating the causes of the accident or determining permissible loads during operation, including considering the wear of units and elements that cannot be replaced. The scanning of the elements ends with the creation of 3D models, which are then transformed into CAD models.

Table 1. Heuristic and design techniques.

№	Content of heuristic procedures	Designation	Content of design techniques	Designation
1	Collective discussion	p_{H1}	Search and selection of analogs	p_{D1}
2	Using random associations	p_{H2}	Using the principle of unification	p_{D2}
3	Using the four types of analogy	p_{H3}	Using the principles of standardization and manufacturability. CAM modules	p_{D3}
4	Use of expert evaluations	p_{H4}	Using libraries of standard elements and parts of elements	p_{D4}
5	Selection of elements. Establishing relationships between elements	p_{H5}	Design using CAD modules	p_{D5}
6	Identification of interrelated and independent groups of elements	p_{H6}	Using the principle of equal strength	p_{D6}
7	Identify elements that the designer can vary and unchanging elements	p_{H7}	Using the principle of uniform load distribution	p_{D7}
8	Using matrix and graph models	p_{H8}	Using the principle of mechanical power split	p_{D8}
9	Assign grades to each of the item combinations	p_{H9}	Using the principle of self-installation	p_{D9}

Reverse engineering also contains procedures that are not inherent in the design. These are metrological measurements and studies of material properties (Table 2).

Table 2. Calculation and metrological procedures

№	Content of metrological procedures	Designation	Content of calculation procedures	Designation
1	Measurement of gears, splines, cams and other special profiles	p_{M1}	Use of CAE modules. Standard methods for calculating elements	P_{c1}
2	Macro and microanalysis of metals and alloys	p_{M2}	Using CAE modules for calculating dynamics	p_{C2}
3	Testing the mechanical properties of materials	p_{M3}	Using CAE modules for calculating FEM	p_{C3}
4	Vibroacoustic measurements	p_{M4}	Creation of a 3D model of complex surfaces	p_{C4}
5	Measurement of housing center distances	p_{M5}	Determination of design loads for units	p_{C5}
6	Measurement of seating surfaces	p_{M6}		
7	Photo fixation	p_{M7}		

In mechanical engineering, mechanisms have been in operation for years, such as a floating crane, a car dumper, and a drawbridge mechanism can serve for decades. During this time, fatigue damage has accumulated; the moving parts of the mechanisms are worn out. The technical documentation is incomplete and partially lost, and the manufacturing companies no longer exist. The task of reverse engineering is to restore the functionality of the existing equipment. The first step is to investigate damage to parts and units (Fig. 1). The second step is to study the product design. For a possible replacement of a unit, you need to know precisely the connecting dimensions, permissible loads and rotation speeds, temperature conditions, etc. Based on the data obtained in the first two phases, a decision is made on the further use of the equipment and ways to restore its functionality. Metrological procedures are used for both damage investigation and structural analysis. Calculations are necessary both for analyzing damage causes and assessing the load capacity of parts by their dimensions. Heuristic procedures play an essential role. The heuristic aspect is always present when determining the causes of an accident, searching for an original design idea, and making a decision. The bottom of the reverse engineering model represents a typical design process. It is important to note that heuristic procedures and calculation and design procedures are completely identical both in reverse engineering and in the design process.

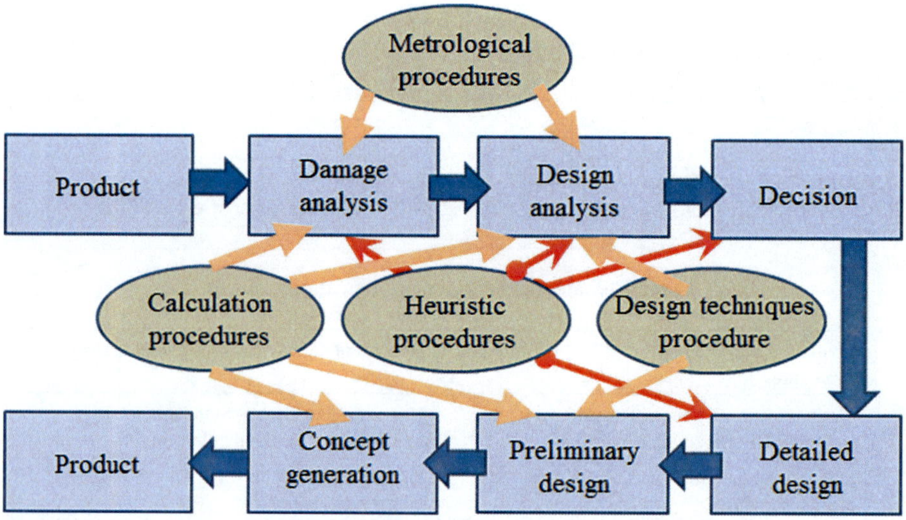

Fig. 1. Reverse engineering model.

4 Results and Discussion

The simplest version of the procedures set performed by the designer at the preliminary design stage will present. First, the designer must decide which units and elements are included in the product. This is the procedure p_{H5} "Selecting elements. Establishing relationships between elements". The structure of the designed product is initially presented as a diagram: this is the procedure p_{H8} "Using models in the form of matrices and graphs". Establishing which units can be purchased and which units are produced at the enterprise for similar products is necessary. Procedure p_{D2} - "Using the principle of unification". When considering the design of the original units, based on the nomenclature of production, it is important to "Identify the elements with which the designer can vary and unchanged elements" - p_{H7}. Using standard elements, the number of original elements should be minimized - this is the procedure "Using libraries of standard elements and part elements" - p_{D4}. Elements calculations are performed with the use of CAE modules. Standard methods for calculating elements "- p_{C1}. Next, the sketch drawings of the original units and the layout of the product is created: "Design using CAD modules" - p_{D5}. When developing the design of units and parts, it is imperative to take into account the possibility of their manufacture "Using the principles of standardization and manufacturability. CAM modules "- p_{D3}. Then, the load capacity, durability and heat resistance of the units are checked: "Determination of the design loads for the units" - p_{C5}. And, finally, the drawings of the conceptual design of the product are created. The procedure p_{D5} - "Design with CAD modules" is used again.

$$S_{PREL.DES} = \{p_{H5}, p_{H8}, p_{D2}, p_{H7}, p_{D4}, p_{C1}, p_{D5}, p_{D3}, p_{C5}, p_{D5}\}.$$

Developing conceptual design can be recurrent, and part of the procedure sequence can be repeated. For example, it has been found that using some standard elements greatly increases the machine's dimensions. In this case, original elements are designed to replace the standard ones, and the sequence of procedures starting with procedure p_{D4} must be repeated.

$$S_{PREL.DES} = \{p_{H5}, p_{H8}, p_{D2}, p_{H7}, p_{D4}, p_{C1}, p_{D5}, p_{D3}, p_{C5}, p_{D5}, p_{D4}, p_{C1}, p_{D5}, p_{D3}, p_{C5}, p_{D5}\}.$$

For the design of some units and elements, additional calculations by the finite element method "Using CAE modules for calculating FEM" (p_{C2}) or analysis of the dynamics of the unit "Using CAE modules for calculating dynamics" (p_{C3}) may be required.

$$S_{PREL.DES} = \{p_{H5}, p_{H8}, p_{D2}, p_{H7}, p_{D4}, p_{C1}, p_{D5}, p_{C3},$$

$$p_{D3}, p_{C5}, p_{D5}, p_{D4}, p_{C1}, p_{C2}, p_{D5}, p_{D3}, p_{C5}, p_{D5}\}.$$

For a particular type of product, for example, an engine, a minimum invariant sequence of procedures can be developed, which is always performed. This sequence can be extended depending on the specifics of a certain engine. Similarly, sequences for the stages Concept generation - $S_{CONC.DES}$ and Detailed design - $S_{DETAL.DES}$ can be developed. Reverse engineering of damaged equipment often used the morphological matrix of damages [21] (Fig. 2). And for the analysis of the design, the identification of interrelated and independent groups of elements, the identification of units - as part of their own method, most researchers use DSM matrix. [13, 16, 18, 21, 22]. Depending on the damage's degree, the solution may be to purchase new equipment or components from the manufacturer. We do not consider these options since there is no subject to the design process here. The solutions that contain the design process are repair, redesign, or simplification. Simplification use is possible when not all functions are needed. For example, a radial drilling machine converts to a drilling machine.

Inspection of the product usually begins with photo fixation since, in some cases, this is also necessary from a legal point of view - the procedure is p_{M7}. This is followed by studying surfaces with wear, spalling, and cracks using the procedure "Macro- and microanalysis of metals and alloys" - p_{M2}. The analysis of the accident causes is carried out using the morphological matrix of damage - p_{H8}.

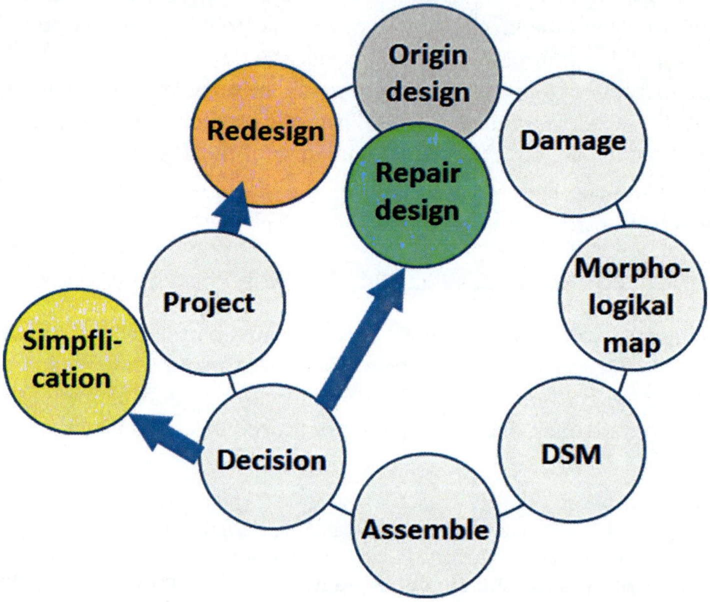

Fig. 2. Reverse engineering options containing the design process.

To analyze the design, the procedure p_{H5} is performed "Selection of elements. Establishing relationships between elements". The DSM matrix - pH8 identifies interrelated and independent groups of pH6 elements. After it has been established which parts are mating, the center distances of the housing p_{M5} and the seating surfaces p_{M6} are measured. Once this has been determined which units are included in the product, unified units are identified "Using the principle of unification" - p_{D2}. Determine which elements are standard "Using standard elements libraries and part of elements" p_{D4}. They finally form an idea of the product design in the form of diagrams and drawings: "The use of models in the form of matrices and graphs - p_{H8}", "Design with using CAD modules - p_{D5}". The decision is made using the methods of brainstorming "Collective discussion p_{H1}" and "Using of expert evaluations p_{H4}". This part of the work inherent in reverse engineering can be represented as a sequence

$$R = \{ p_{M7}, p_{M2}, p_{H8}, p_{H5}, p_{M5}, p_{M6}, p_{D2}, p_{D4}, p_{H8}, p_{D5}, p_{H1}, p_{H4} \} .$$

Then there are the procedures of the design process. Moreover, all stages of the design process are carried out during redesign. The sequence of procedures for redesign is

$$R_{redesign} = R \cap S_{CONC.DES} \cap D_{PREL.DES} \cap D_{DETAIL.DES}.$$

5 Conclusions

Based on a detailed analysis of the structure of the reverse engineering process and the design process structure, it was possible to present these processes as a set of procedures. It has been established that many procedures are standard for these two scientific directions. This, in turn, opens up the opportunity to use a massive array of accumulated information which refers to the structuring and automation of the design process. This contributes to the improvement of the description of the reverse engineering process.

It is established that the main part of the design procedures is used for certain types of reverse engineering - repair and simplification. It was found that in the case of a redesign, all the design procedures are repeated as in developing a new product. The importance of the wide use of heuristic methods at various stages of the design process and phases of reverse engineering is emphasized. The most commonly used heuristic methods are morphological map and DSM matrix, which are very suitable for damage analysis and product design. We state that a distinctive feature of reverse engineering is the presence of a large number of various metrological procedures.

Revealed: the description of the processes of design and reverse engineering as a sequence of procedure sets allows applying some of the functions of the decision support system of the design process for reverse engineering. It is advisable to formalize the experience gained while implementing a specific reverse engineering project and, by analogy, use it in new projects. Prerequisites for creating a decision support system for reverse engineering have also been developed.

References

1. Buonamici, F., Carfagni, M., Furferi, R., Governi, L., Lapini, A., Volpe, Y.: Reverse engineering modeling methods and tools: a survey. Comput.-Aided Des. Appl. **15**(3), 443–464 (2018). https://doi.org/10.1080/16864360.2017.1397894
2. Noor, N.M.M., Ghazali, A.F., Saman, M.Y.M., Zafarina, Z.: Reverse engineering approach in a development of a decision support system for forensic DNA analysis. Appl. Math. Sci. **6**(108), 5369–5382 (2012)
3. Bruneliere, H., Cabot, J., Dupé, G., Madiot, F.: Modisco: a model driven reverse engineering framework. Inf. Softw. Technol. **56**(8), 1012–1032 (2014). https://doi.org/10.1016/j.infsof.2014.04.007
4. Vaitkus, M., Várady, T.: Parameterizing and extending trimmed regions for tensor-product surface fitting. Comput. Aided Des. **104**, 125–140 (2018). https://doi.org/10.1016/j.cad.2017.11.008
5. Wang, J., Zhou, L., An, L., Tan, C.: Reconstruction of solid model with complex profile features. Chin. Mech. Eng. **11**(17), 1157–1161 (2006)

6. Kent Weaver, R.: Reverse engineering and policy design. In: Capano, G., Michael Howlett, M., Ramesh, A.V. (eds.) Making Policies Work, pp. 173–190. Edward Elgar Publishing (2019). https://doi.org/10.4337/9781788118194.00020
7. Andrea, G., Mauro, G.: Why china has not caught up yet: military-technological superiority and the limits of imitation. Reverse engineering and cyber espionage. Int. Secur. **43**(3), 141–189 (2019). https://doi.org/10.1162/isec_a_00337
8. Elsan, M.: Legal aspects of reverse engineering. The Judiciarys Law J. **79**(89), 31–54 (2015)
9. Orozco, D.: Amending the economic espionage act to require the disclosure of national security-related technology thefts. Cath. UL Rev. **62**, 877 (2012)
10. Oxman, R.: Thinking difference: Theories and models of parametric design thinking. Des. Stud. **52**, 4–39 (2017). https://doi.org/10.1016/j.destud.2017.06.001
11. Münzer, C., Shea, K., Helms, B.: Automated parametric design synthesis using graph grammars and constraint solving. In: Proceedings of the ASME 2012 International Design Engineering Technical Conferences and Computers and Information in Engineering Conference, pp. 517–528. Chicago, Illinois, USA (2012). https://doi.org/10.1115/DETC2012-70313
12. Yassine, A.: An introduction to modeling and analyzing complex product development processes using the design structure matrix (DSM) method. Urbana **51**(9), 1–17 (2004)
13. Vezzetti, E., Marcolin, F., Guerra, A.L.: QFD 3D: a new C-shaped matrix diagram quality approach. Int. Qual. Reliab. Manag. **33**(2), 178–196 (2016). https://doi.org/10.1108/IJQRM-07-2013-0112
14. Harrington, H.J.: Matrix Diagram/Decision Matrix. The Innovation Tools Handbook, vol. 3. Tailor & Francis Group (2016)
15. Wynn, D.C., Caldwell, N.H., Clarkson, P.J.: Predicting change propagation in complex design workflows. J. Mech. Des. **136**(8), 081009 (2014). https://doi.org/10.1115/1.4027495
16. Day, R., Stone, R.B., Lough, K.G.: Validating module heuristics on large scale products. In: Proceedings of the ASME 2009 International Design Engineering Technical Conferences & Computers and Information in Engineering Conference IDETC/CIE 2009, pp. 1079–1087. San Diego, California, USA (2009)
17. Ivanov, V., Urum, G., Ivanova, S., Volkova, M.: Development of the positive engagement continuously variable transmission design with the application of graph theory. Eastern-Eur. J. Enterp. Technol. **3**(1 (93)), 43–50 (2018). https://doi.org/10.15587/1729-4061.2018.131101
18. Ulrich, K.T.: Design: creation of artifacts in society. SSRN (2011). https://doi.org/10.2139/ssrn.1951106
19. Funge, J.D.: Artificial Intelligence for Computer Games: An Introduction. A K Peters/CRC Press (2004). https://doi.org/10.1201/9781439864807
20. Ivanov, V., Dimitrov, L., Ivanova, S., Olefir, O.: Heuristic techniques as part of heuristic methods and interaction of personality types in their application. Adv. Sci. Technol. Eng. Syst. J. **6**(1), 208–217 (2021). https://doi.org/10.25046/aj060123
21. Ma, H., Chu, X., Xue, D., Chen, D.: A systematic decision making approach for product conceptual design based on fuzzy morphological matrix. Expert Syst. Appl. **81**, 444–456 (2017). https://doi.org/10.1016/j.eswa.2017.03.074
22. Masmoudi, M., Leclaire, P., Zolghadri, M., Haddar, M.: Dependency identification for engineering change management (ECM): an example of computer-aided design (CAD) -based approach. In: the 20th International Conference on Engineering Design (ICED 15), vol. 3, pp. 199–208. Université de Sfax, Milan, Italy (2015)

Design Measures to Reduce Specific Loads on Support Surfaces of Slide Bearings

Mykola Kiyanovsky[1], Natalia Tsyvinda[1], Vasyl Nechayev[1], Dariya Kravtsova[1], and Yurii Yarovyi[2(✉)]

[1] Kryvyi Rih National University, 11, Vitalii Matusevych St., Kryvyi Rih 50027, Ukraine
[2] Odessa Polytechnic National University, 1, Shevchenko Ave., Odesa 65044, Ukraine
yurovoy.yu@gmail.com

Abstract. The article deals with problems of minimizing wear of friction surfaces in rotor mechanisms, namely surfaces of support rotor journals and slide bearing supports working in the hydrodynamic mode. Characteristic features of the design of a slide bearing, load modes, regularities of length- and crosswise pressure distribution on support surfaces of hydrodynamic slide bearings were analyzed considering specifics of their work in the mechanism. Theoretical and experimental results of studying design and operational factors that influence the stability of the support lubricant film at the stages of acceleration and working speed of the rotor, duration of the boundary friction period, regularities of the location of support surface wear areas, unevenness of geometry of the support surface of the bearing were analyzed for reducing the rate of support surface wear. The determined design and operational factors influence the friction surface wear rate. There were also determined regularities of the trajectory of the shaft entering the position of relative equilibrium of the rotational motion, achievement of stable parameters of the support lubricant film, and their influence on the duration of boundary friction on supporting surfaces of the bearing. To solve the friction parameter stabilization and shaft-bearing wear problems, a particular shape of the slide-bearing bushing is suggested to be adaptive to manual and automatic loads to achieve an evener load diagram.

Keywords: Rotor · Stabilization · Eccentricity · Load · Reliability · R&D investment

1 Introduction

In most cases, rotor machine life is conditioned by the durability of friction surfaces of the operating element supports (slide bearings). Minimizing the work of friction force on friction surfaces primarily depends on the friction coefficient and factors that influence the level of friction forces. On the one hand, reducing costs for overcoming friction forces depends on the design perfection of rotor mechanisms and machines used in engineering systems. On the other hand, it is conditioned by personnel's continuous attention paid to preventing the influence of the factors that can change the type and mode of friction and wear of the working mechanism irrespective of its perfection [1, 2].

V. Tonkonogyi et al. (Eds.): InterPartner 2022, LNME, pp. 23–31, 2023.
https://doi.org/10.1007/978-3-031-16651-8_3

These facts show that accelerated wear of bearings depends significantly on impaired stability of parameters of friction processes on the friction surface caused by the typical hydrodynamic bearing design. With the modern design of bearings, there are certain objective regularities in overcoming friction forces and loss of friction surface durability, but the mechanism of such loss should become clear and be subject to conscious design correction of bearing elements [3, 4].

2 Literature Review

According to the tasks of the work, it is necessary to identify regularities of working surface wear [5, 6].

The prevailing influence of mechanical wear processes on wear rates and summarization of the following research results should be considered: O. Reynolds, D. Tabor – regularities of rolling friction and influence of slipping and the support internal material of the support; J. F. Archard – the equation to determine the influence of the ratio of mechanical properties and the stress level on the amount of wear; A.Yu. Ishlinskyi – regularities of friction at relaxing elastic-viscous deformation of supports; I.V. Krahelskyi – a study of wear at discrete contact of solids, friction fatigue of friction pairs; M.M. Khrushchov, M.A. Babichev – a study of relative wear resistance of materials; B.I. Kostetskyi – determination of critical rates of corrosion wear of surfaces), the following relation seems to be the most unified method of determining the wear rate [7]:

$$\gamma = kp^m v^n \tag{1}$$

where k is the constant that depends on the material and wears conditions; p is pressure on friction surfaces; v is the relative slip rate of friction bodies.

This dependence quite clearly outlines directions for minimizing the part wear rate.

This paper [8] summarizes the conventional texture shape and new shape design of surface texture of radial and thrust sliding bearings and puts forward ideas for the extensive application of composite texture and bionic texture in sliding bearings.

This paper [9] analyzes the changing loading of sliding bearing under different rotating speeds to control the vibration within the allowable range.

This work [10] presents the investigation results of a journal bearing lubricated with magnetorheological fluid activated by a local constant magnetic field to vary the local flow resistance and pressure. Thicker fluid films at low speeds, beneficial pressure distribution, and higher friction losses under all operating conditions are observed in the bearing with magnetorheological fluid compared to the oil lubrication.

3 Research Methodology

The article is aimed to study the regularities of the interaction of parts of a slide bearing in rotor machines at different stages of its use, the character, size, and distribution of friction forces on surfaces depending on the design and operational factors to increase the durability of the functional element supports. The problem's study consists of a detailed analysis of slide bearings, and their inherent dynamic processes that condition loads on support surfaces [11, 12].

Slide bearings are an integral part of many large and critical units; they are widely used in power equipment, electric motors, powerful pumps, and compressors [13]. Physical processes that occur in slide bearings are quite complex and depend on relations of many external and internal factors [14, 15].

During the operation of machines and equipment with bearing units, failures and accelerated uneven wear are caused by uneven load distribution along the length and contact area of the shaft and the bearing unit. These phenomena lead to overheating, accelerated wear, vibration, seizure, and, consequently, the loss of equipment efficiency [13, 16].

Basically, all the operational problems of the slide bearing state are divided into three groups: problems related to the condition of bearing working surfaces, problems related to the clearance between the rotor support surface and the anti-friction half-liner, and problems related to the bearing capacity of lubricant film [17].

The conditions for creating a bearing lubricant film are similar to hydrodynamic processes arising between the plane and the plate, moving at some angle to the plane.

When the bearing clearance is filled with lubricant and the force P is applied to the shaft, the latter shifts from the central position and forms the eccentricity ε between the shaft and the orifice, and thus it forms a V-shaped clearance. When the shaft rotates, the lubricant stuck to it is captured in the V-shaped clearance, compressed in a narrow throat of the clearance, and creates a lifting force that keeps the shaft from contacting the bearing. Determination of hydrodynamic pressure in the lubricant film in the bearing is one of the tasks of the hydrodynamic theory of lubrication. Hydrodynamic pressure p develops in the V-shaped lubricant clearance and is balanced by the load P acting on the shaft.

Figure 1 presents the shaft position in the bearing: at the beginning of its work.

(Fig. 1a) when the angular velocity of the shaft ω is still close to zero, and during the period of steady movement, when the angular velocity ω exceeds the critical value, which corresponds to transition to the fluid friction mode (Fig. 1b).

When machines start operating, their mechanisms work with acceleration and, as a result, dynamic forces (inertia forces) occur and cause great, sometimes significant, loads and additional stresses in rotor-slide bearing kinematic pairs and increase friction and wear of their elements (Fig. 1) [18]. When overloading or reaching the boundary wear of the slide bearing's friction surfaces, the rotor axis shifts from the bearing axis. Thus, causing an increase in self-vibration amplitudes, impairing the normal lubricating mode and turning fluid friction into a boundary or even dry one. This fact leads to a sharp increase in temperature and a decrease in the lubricant viscosity. Favorable conditions for intensive wear of various types arise in the unit, mainly for the wear caused by a mechanical seizure.

When changing the angular velocity, the center of the shaft changes its position accordingly; the trajectory of its movement in the bearing is approximate to the circular arc. When $\omega \to \infty$, the cylindrical surfaces of the shaft are exposed to friction, and the bearing becomes almost concentric, forming a constant circular clearance that is equal to δ. The minimum clearance occurs at the intersection, which is shifted to the angle β between the shaft, the bearing center line, and the load action line P. In addition, the angle of the shaft position in the bearing in the hydrodynamic mode completely determines eccentricity ε between the shifted centers O and O_1. If the stability of the clearance is not maintained artificially, the shaft descends under the action of gravity. Here, the hydrodynamic effect is activated because the shaft is lubricated and forces the lubricant into a V-shaped clearance which narrows in the direction of movement. In this case, additional pressure occurs inside the lubricant wedge (Fig. 1).

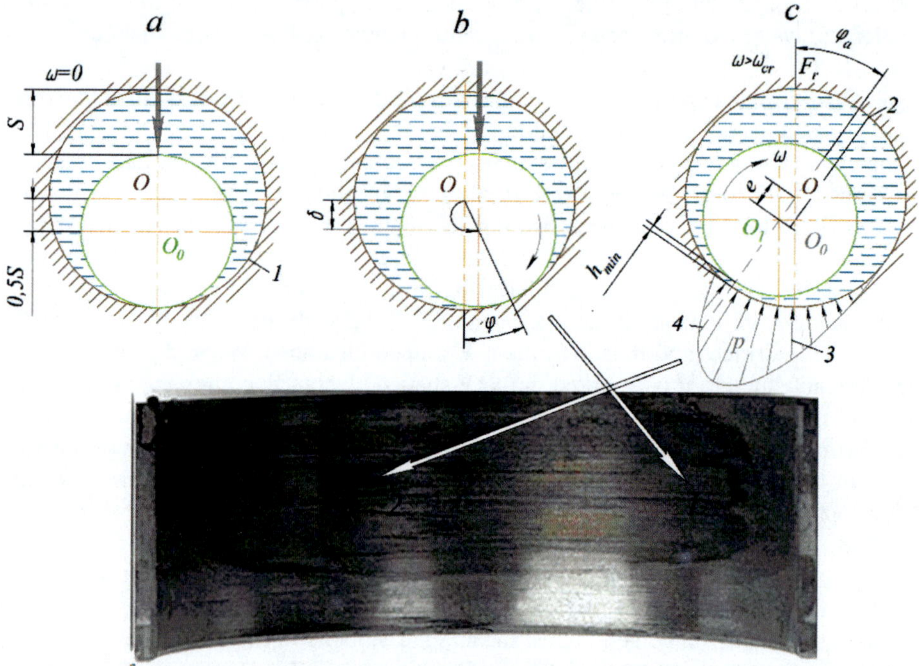

Fig. 1. Areas of intensive wear on the support surface of the half-liner: 1 – the area of wear in boundary friction conditions at the transition from $\omega = 0$ to ω_W; 2 – the area of wear p_{Max} (maximum specific pressure).

The bearing capacity of the lubricant wedge of the slide bearing is the main operational parameter of its condition. It is a complex nonlinear function of clearance size between the shaft and the anti-friction half-liner. The thinner the lubricant film is, the higher the bearing capacity becomes. On the other hand, reduction of the lubricant film decreases resistance of the bearing to dynamic loads, and the shaft is more likely to mechanically touch the half-liner.

The wedge is the thickest at the entry point of the working surface of the rotating shaft into the bearing zone and the thinnest at the exit from it. The greater the load on the bearing is, the thinner the lubricant film carrying the radial load becomes.

Under certain conditions, rotors of units supported by slide bearings may lose stability and switch to the mode of radial self-vibrations. This most often happens at a significant reduction in the rotor shaft load on the bearing for many reasons. The complexity of the hydrodynamic operation of the bearing creates a dynamically unstable mechanism of working surface wear due to fluctuations in the level and nature of the load and shifts of the load spot. This results in uneven and accelerated wear. The location of intensive wear areas is shown in Fig. 1 (lower half-liner).

Attention should be paid to the axial shift of the intensive wear area (Fig. 1). This is explained by the diagram of the axial distribution of pressure on the support surface (Fig. 2). This diagram is representative of hydrodynamic slide bearings [19]. A comparison of wear of the support surface of the bearing half-liner and regularities of the wear rate resulting from the action of design and operational factors during significant service life (Fig. 2) shows considerable unevenness of wear on the area of friction contact, which fully corresponds to regularities of wear. It is this wear pattern of the friction surface that conditions the need for improvement of the bearing design to stabilize the field of working pressure on the support surface.

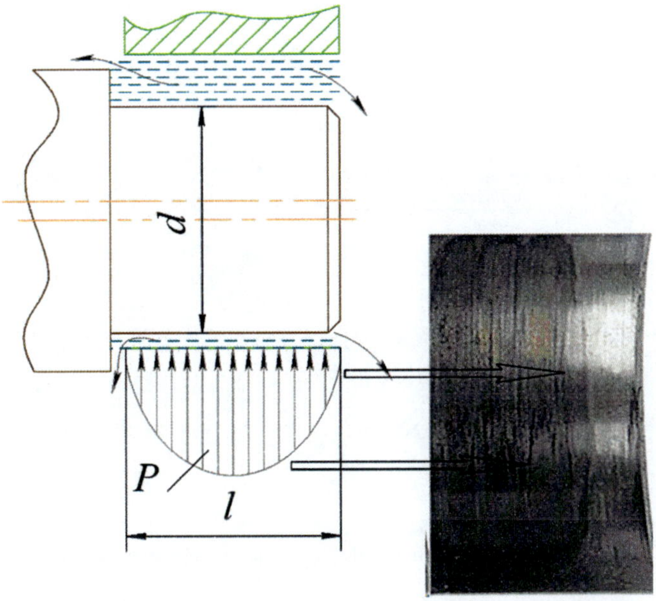

Fig. 2. Location of the half-liner's intensive axial wear areas depending on the pressure distribution diagram in the support lubricant film.

4 Results and Discussion

Researches of friction surface wear and methods of bearing calculation pay little attention to the fact that technological equipment of industrial production mainly consists of stationary machines with rotating working motion. Thus, according to d'Alembert's principle, the forces arising in the unit are cyclical, and because of this, lubricant dripping from the working space of the bearing is not laminar but is accelerated by impacts of working load forces. Due to this, mechanical wear of the support surfaces is combined with fatigue wear [20].

A significant increase in clearances in service leads to impact loads. They also contribute to an increase in the wear rate or even to a breakdown of individual parts or failure of the entire machine. The expediency of changing the slide bearing design to reduce the wear rate is substantiated by solving the task of increasing the support contact spot on friction surfaces. Design improvements of the existing bearing model aim to ensure an evener distribution of lubricant pressure on the shaft surface. The solution to the task involves using blind and circular grooves in the bearing bushing to create a hydroplastic effect to reduce radial clearance in the bearing end.

The idea behind the solution consists of the following. In the slide bearing, even lubricant pressure in the lubricant film over the bushing surface is created because in the support surface of the bushing, there are special circular grooves filled with hydroplastic that enable controlling elastic deformation of the groove wall [21].

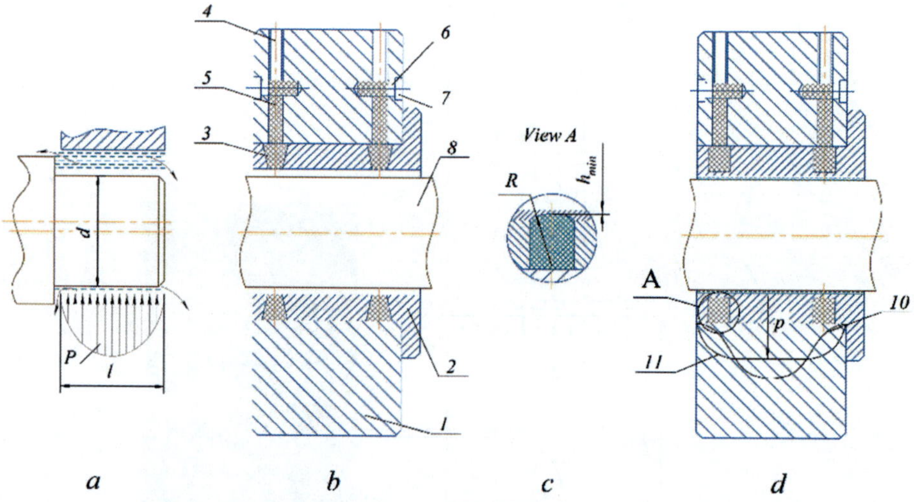

Fig. 3. The slide bearing with the pressure control mechanism.

Figure 3 explains the design improvement where b shows the cross-section of the bearing; c – the detail (View A); d – the diagram of lubricant pressure distribution on the shaft surface.

A hydrodynamic slide bearing with the pressure control mechanism (Fig. 3b) consists of the bearing body 1, structural elements of the bushing 2, the shaft 8, special chambers in the bushing body filled with hydroplastic 3, the orifice 4 for the power plunger 6 which is connected with the pressure screw 7 to create pressure in the cavity, and adjusting screws 5. The principle of operation of the slide bearing unit is as follows: the bushing 2 with its outer surface is pressed in the slide bearing 1, and the inner surface of the bushing 2 and the cylindrical surface of the shaft 8 form a friction pair. Hydroplastic 3 in a liquid form is fed through the orifice 4 for the power plunger 6 into the cavity between the bushing and the bearing body. The power plunger 6 is kinematically connected with the pressure screw 7. Control screws 4 are provided to remove free air from the body's cavity and relieve pressure in the bearing unit.

Figure 3d presents a loaded area in the bearing, which is shifted due to creating a lubricant wedge. As a result, uneven lubricant pressure distribution 10 occurs on the shaft surface. However, considering the structural elements of the bushing (circular grooves filled with hydroplastic), lubricant pressure 11 is made even on the shaft surface.

The slide bearing with the pressure control mechanism is used to create a hydroplastic effect to reduce radial clearance in the bearing end and distribute lubricant pressure evenly on the shaft surface. The solution to the task involves using blind and circular grooves in the bearing bushing to create a hydroplastic effect to reduce radial clearance in the bearing end.

The regularities of sliding bearing support surface wear in the hydrodynamic friction mode are first investigated. The presence of local zones of intense wear on the bearing support surface is revealed. It has been determined that the project durability of the bearing is declining in proportion to the actual support surface.

Investigating the relationships between the sliding bearing support surface wear and the design and operational factors (radial clearance, the size of support surface, oil feed area location, load vector location, speed pattern, and oil viscosity) is conducted. The factors that significantly influence the speed and irregularity of friction surface wear have been identified. The regularities of pressure distribution on the support surfaces of hydrodynamic sliding bearings are set out in the longitudinal and transverse directions.

The analysis enables concluding that engineering and design methods do not consider the availability of uneven axial pressure of the shaft on support surfaces of the bearing. General parameters of formation and influence of a lubricant film in the hydrodynamic mode and its state during loading a slide bearing, complexity of conditions of maintaining fluid friction on peripheral parts of the support surface, and consequent uneven wear of the bearing support surface are determined as well.

Regularities of the trajectory of shaft output into the position of relative equilibrium during rotation have been investigated theoretically and experimentally.

Conditions for stable parameters of support oil layer achievement have been established. The influence of support oil layer parameters on the duration of limit friction on bearing support surfaces has been studied.

A new sleeve design of bearing support is proposed to solve the problem of shaft stability and bearing friction and wear.

5 Conclusions

The inventive structural design of the slide-bearing unit with a pressure control mechanism makes it possible to evenly distribute the pressure field along the entire length of the shaft-bearing pair. The developed bearing construction decreases the specific pressure on the friction surface by up to 30% through the distribution of the pressure field along the entire support surface of the bearing. As a result, the stability of the shaft axis at operating speeds is ensured. Consequently, the operational reliability is improved, and the durability of rotor machines is increased.

The wear rate of the bearing support surface is proportional to the specific pressure. In this way, the wear rate of the bearing support surface can be reduced by up to 50% in the newly designed slide bearing unit. In that connection, its overhaul interval doubles, and the number of repairs halves.

A promising direction for further research is the study of the impact of peripheral radial clearance on pressure area distribution along the bearing, paying special attention to rotor run-up and coast-down stages. The upgrade of the unit design by using other technical media instead of hydroelastic is also an area of interest.

References

1. Pavlenko, I., Simonovskiy, V., Ivanov, V., Zajac, J., Pitel, J.: Application of artificial neural network for identification of bearing stiffness characteristics in rotor dynamics analysis. In: Ivanov, V., et al. (eds.) DSMIE 2018. LNME, pp. 325–335. Springer, Cham (2019). https://doi.org/10.1007/978-3-319-93587-4_34
2. Makarenko, V.D., Maksimov, S., Makarenko, V.V., Panchenko, O.S.: Study of durable strength of steel mining and metallurgical equipment. Solid State Phenom. **332**, 111–121 (2022). https://doi.org/10.4028/p-295y1h
3. Rednikov, S.N., Akhmedyanova, E.N., Zakirov, D.M.: Comprehensive diagnostics of the state of metallurgical equipment. In: Radionov, A.A., Gasiyarov, V.R. (eds.) ICIE 2021. LNME, pp. 1205–1211. Springer, Cham (2021). https://doi.org/10.1007/978-3-030-54817-9_140
4. Kondrachuk, M., Khabutel, M., Pashechko, M., Korbut, Ye.: Tribology. "NAU-druk", Kyiv (2009)
5. Sosnovskiy, L.: Tribo-fatigue: wear-fatigue damageandits prediction. Springer-Verlag, Berlin and Heidelberg Gmb & Co, KG, Berlin (2005)
6. Matsushita, O., Tanaka, M., Kobayashi, M., Keogh, P., Kanki, H.: Basics of plain bearings. In: Matsushita, O., Tanaka, M., Kobayashi, M., Keogh, P., Kanki, H. (eds.) Vibrations of Rotating Machinery: Volume 2. Advanced Rotordynamics: Applications of Analysis, Troubleshooting and Diagnosis, pp. 19–40. Springer Japan, Tokyo (2019). https://doi.org/10.1007/978-4-431-55453-0_2
7. Menezes, P.L., Ingole, S.P., Nosonovsky, M., Kailas, S.V., Lovell, M.R. (eds.): Tribology for scientists and engineers. Springer, New York (2013). https://doi.org/10.1007/978-1-4614-1945-7
8. Song, F., Yang, X., Dong, W., Zhu, Y., Wang, Z., Min, W.: Research and prospect of textured sliding bearing. The Int. J. Adv. Manufact. Technol. **121**(1–2), 1–25 (2022). https://doi.org/10.1007/s00170-022-09281-2
9. Ding, K., et al.: Measurement and analysis of sliding bearing vibration. J. Phys: Conf. Ser. **1965**, 012117 (2021). https://doi.org/10.1088/1742-6596/1965/1/012117
10. Quinci, F., Litwin, W., Wodtke, M., van den Nieuwendijk, R.: A comparative performance assessment of a hydrodynamic journal bearing lubricated with oil and magnetorheological fluid. Tribol. Int. **162**, 107143 (2021). https://doi.org/10.1016/j.triboint.2021.107143
11. Rud, Yu.: Fundamentals of machine design: a textbook for students of engineering of higher educational institutions. 2nd ed., revised. FOP Cherniavskyi D.O., Kryvyi Rih (2015)
12. Piterska, V., et al.: Diagnostics of the strength and stiffness of the loader carrier system structural elements in terms of thinning of walls by numerical methods. Diagnostyka **22**(3), 73–81 (2021). https://doi.org/10.29354/diag/141313
13. Vaneev, S.M., Martsynkovskyy, V.S., Kulikov, A., Miroshnichenko, D.V., Bilyk, Y., Smolenko, D.V., Lazarenko, A.D.: Investigation of a turbogenerator based on the vortex expansion machine with a peripheral side channel. J. Eng. Sci. **8**(1), F11–F18 (2021). https://doi.org/10.21272/jes.2021.8(1).f2

14. Arif, M., Kango, S., Shukla, D.K.: Analysis of textured journal bearing with slip boundary condition and pseudoplastic lubricants. Int. J. Mech. Sci. **228**, 107458 (2022). https://doi.org/10.1016/j.ijmecsci.2022.107458
15. Zhang, H., Liu, Y., Dai, S., Li, F., Dong, G.: Optimization of boundary slip region on bearing sliders to improve tribological performance. Tribol. Int. **168**, 107446 (2022). https://doi.org/10.1016/j.triboint.2022.107446
16. Fedorynenko, D., Boyko, S., Sapon, S.: The search of the spatial functions of pressure in adjustable hydrostatic radial bearing. Acta Mechanica et Automatica **9**(1), 23–26 (2015). https://doi.org/10.1515/ama-2015-0005
17. Kuric, I., Kandera, M., Klarák, J., Ivanov, V., Więcek, D.: Visual product inspection based on deep learning methods. In: Tonkonogyi, V., et al. (eds.) InterPartner 2019. LNME, pp. 148–156. Springer, Cham (2020). https://doi.org/10.1007/978-3-030-40724-7_15
18. Kolobov, A.: Vibrodiagnostics: Theory and Practice. Infra-Inzheneria, Moscow (2020)
19. Kiyanovsky, M., Bondar, O.: Design provision of sustainability of the balanced position of rotor machine shafts in sliding supports. J. Kryvyi Rih Natl Univ. **47**, 112–115 (2018)
20. Komanduri, R., Hou, Z.B.: Analysis of heat partition and temperature distribution in sliding systems. Wear **251**(1–12), 925–938 (2001)
21. Kiyanovsky, M., Demida N.: The Slide bearing with the pressure control mechanism. Patent of Ukraine no. 145421 (2020)

Automated Control of the Gear Profile for the Gerotor Hydraulic Machine

Sergey Kiurchev[1](✉), Mamadamon A. Abdullo[2], Tetiana Vlasenko[3], Svitlana Prasol[3], and Valentyna Verkholantseva[1]

[1] Dmytro Motornyi Tavria State Agrotechnological University, 18, B. Khmelnitsky Avenue, Melitopol 72310, Ukraine
tia_tgatu@ukr.net

[2] Tajik Technical University named after Academician M. Osimi, 10, Academicians Rajabov's Avenue, Dushanbe 734042, Tajikistan

[3] State University of Biotechnology, 44, Alchevskikh Street, Kharkiv 61002, Ukraine

Abstract. In hydraulic drives for construction, railway, drilling, and other self-propelled equipment, gerotor-type hydraulic machines are widely used. One of the main components of these hydraulic machines is the rotors block, which is a precision pair of rotors with toothed surfaces. Moreover, the performance of a gerotor hydraulic machine is largely determined by the accuracy of manufacturing the toothed surface of the inner and outer rotors. It has been established that the external rotor of the displacement unit, being an assembly unit, is an integral structure. The paper substantiates the methodological possibility of automating the check of the toothed surface of the rotors of gerotor hydraulic machines. To determine the checked parameters, a design scheme of the rotor block of a gerotor hydraulic machine was developed and mathematical dependencies were obtained that describe the design elements of the measuring devices. A production check of the developed automated check devices for rotors showed that the proposed check method makes it possible to carry out precision assembly of the rotor block of gerotor hydraulic machines by selectively selecting the appropriate pairs.

Keywords: Rotors block · Sealing of working chambers · Precision assembly · Selective selection · Quality · Energy efficiency

1 Introduction

Currently, hydraulic machines (both pumps and hydraulic motors) of the gerotor type are increasingly used in hydraulic drives for construction, railway, agricultural, drilling, communal, logging, and other self-propelled equipment [1]. The main unit of gerotor-type hydraulic machines is a working fluid displacement unit, represented by internal and external rotors [2].

Structurally, the inner and outer rotors are a toothed pair with an internal hypocycloidal engagement, the equidistant contour approximated by circular arcs [3]. During operation, the inner rotor of the hydraulic machine, rolling around inside the outer

V. Tonkonogyi et al. (Eds.): InterPartner 2022, LNME, pp. 32–43, 2023.
https://doi.org/10.1007/978-3-031-16651-8_4

rotor, displaces the working fluid and simultaneously separates (seals) the high- and low-pressure zones [4]. Thus, the block of rotors is a movable precision pair with a complex toothed surface, and the manufacturing accuracy conditions the performance of the gerotor hydraulic machine.

The provision of the required sealing in the rotor block to separate the high- and low-pressure zones is carried out by a selective selection of the rotors of the gerotor hydraulic machines. To carry out the selective assembly of the rotor unit, it is necessary to have appropriate methods and means of automated check of the deviations of the measured parameters determined by the gear contours of the inner and outer rotors.

Therefore, the issues related to the development of methods and means of automated check of the toothed contours of the rotors of a gerotor hydraulic machine in order to ensure the precision accuracy of the assembly of the displacement unit are relevant.

2 Literature Review

A mathematical model [5] for the kinematic description of the working process of pump-turbines, a numerical model for determining the operating characteristics [6] and calculating the efficiency [7] were proposed. The characteristics of the flow [8] of pump-turbines were investigated. The flow was simulated, and the changes were determined in [9]. Dynamic response [10] and hydrodynamic conditions causing vibrations in a turbine [11] were evaluated. The geometric parameters of the flow parts of vortex-chamber [12] and labyrinth-screw [13] pumps have been optimized, the performance ranges of vortex-chamber blowers have been determined [14], a computational model [15], and an experimental system [16] of energy losses in hydraulic circuits are proposed to save energy in vortex devices proposed to use a confuser [17]. An assessment of the influence of the gas content of the working fluid [18, 19] and vapor cavitation [20, 21] on the parameters of axial piston machines has been carried out. An analysis of the literature shows [22] that there is very little published literature on the methodology for designing and manufacturing gerotor pumps and orbital engines.

The ways of rational design of distribution systems of planetary hydraulic machines are proposed [23], justified rational kinematic diagrams [24] and geometric parameters of elements of liquid distribution systems [25, 26] using distribution windows of various shapes [27], represented by a segment [28], a circle [29] and a groove [30], have been substantiated. Parametric [31, 32], dynamic [33, 34] and experimental [35] studies of the processes that occur in the distribution systems of hydraulic machines are presented, and regression approaches for evaluating the operating parameters [36, 37] are proposed, and optimization parameters are justified [38]. The work processes that take place in orbital hydromachines have not been investigated. Also, approaches for ensuring the capacity of hydromachines are proposed in [39, 40].

A mathematical model was developed [41], the forces and moments acting in the gearing of gerotor pumps [42] were considered, and the influence of the geometry of the curved parts of machines on their functional parameters was studied in [43, 44].

A kinematic scheme of the movement of the rotors [4] and a method for modeling changes in the technical condition of the rotors [2] were proposed in order to determine the reliability of the orbital hydraulic motor. Parametric [3], dynamic [34, 45] and experimental [46] studies of the processes that occur in the system of rotors of an orbital hydraulic motor are presented. Ways for improving the strength parameters of shaft material were proposed in [47, 48].

The influence of technological gaps on the hydrodynamic characteristics of the pump was studied [49]. The checked parameters determine the material selection [50] and form error evaluation during manufacture of these rotors [51]. The disadvantage of methods and means for monitoring the accuracy of manufacturing rotors is that when measuring the deviations of the parameters of the toothed surface of the rotors, it must be set to the measuring position as many times as there are teeth. It should be noted that the considered schemes and devices for checking the accuracy of rotor manufacturing are quite well implemented in their selective check (at the time of manufacture), but, unfortunately, cannot be used in automated rotor check.

An analytical review of well-known literary sources shows that the issue of manufacturing gear surfaces of rotors of gerotor hydraulic machines was practically not considered in the open press [22]. The issue of checking the manufacture of these surfaces (the necessary, especially its automation) has not been considered at all. This can be explained by the commercial interests of the manufacturer of orbital hydraulic motors (Danfoss), which has practically not changed the design of these hydraulic machines over the past 20 years. At the same time, a group of Ukrainian researchers A. Panchenko, A. Voloshina, A. Rogovoy, et al. (analysis of their research is given above) successfully solve issues related to the calculation, design, manufacture and testing of orbital hydraulic motors. Therefore, their research experience is very important in developing methods and means of automated check of manufacturing gear surfaces of rotors of gerotor hydraulic machines. An automated check provides the possibility of precision assembly by selecting appropriate pairs of rotors, which is an urgent task today.

3 Research Methodology

For the development of methods and means of automated check of the toothed surface of the rotors of gerotor hydraulic machines, in order to ensure their precision assembly by a selective selection of the appropriate pairs, it is necessary:

- to develop a design scheme of the rotor block of a gerotor hydraulic machine to determine the checked parameters;
- to substantiate the methodological possibility of automating the check of the toothed surface of the rotors of gerotor hydraulic machines;
- to develop methods and means of automated checking the toothed surface of rotors of gerotor hydraulic machines and to analyze the measurement results.

The main working unit of a gerotor hydraulic machine is its block of rotors (Fig. 1), the toothed surfaces forming working chambers interacting with the liquid [4]. Working chambers (Fig. 1) are located symmetrically about the vertical axis and form high- and low-pressure zones. The sealing of these zones is carried out at the contact points of the teeth of the inner and outer rotors located on the normals formed by the angle of "contact" k and at the points lying on the line of the radius of the "contact" R_0 [51].

The outer profile of the inner rotor 1 (Fig. 1) is a toothed surface with the number of teeth z_1, and the contour of the teeth z_1 is approximated by circular arcs of radius r_1.

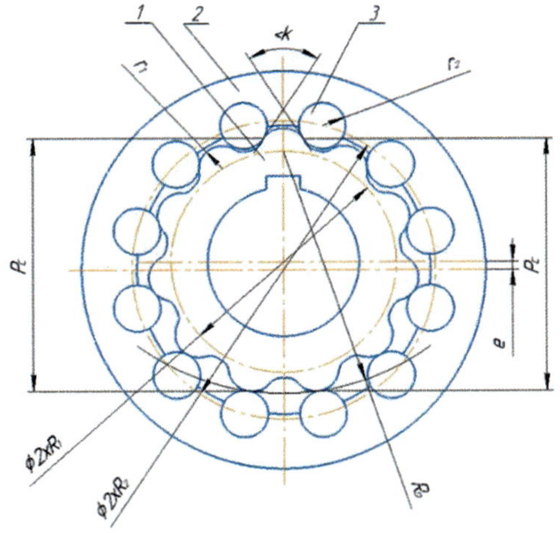

Fig. 1. Design diagram of the rotor block of a gerotor hydraulic machine: 1 – inner rotor; 2 – outer rotor; 3 – tooth of the outer rotor (roller); R_1 and R_2 – radii of the location of the centers of the teeth of the inner and outer rotors, respectively; r_1 and r_2 – radii of the teeth of the inner and outer rotors, respectively; e – eccentricity; $<k$ – "contact" angle; R_0 – the "contact" radius; P_c – monitored parameter.

The centers of the circles of the teeth z_1 of the inner rotor 1 are located on a circle with a radius R_1. The teeth z_2 of the outer rotor 2 (Fig. 1) are made in the form of insert rollers 3 with a radius r_2. Their centers are located on a circle with a radius of R_2. It is known [3] that in hypocycloidal gear drives with internal gearing, the connection between the teeth z_2 of the outer rotor and the teeth z_1 of the inner rotor is determined by the relationship $z_2 = z_1 + 1$. The center of the inner rotor 1 is displaced relative to the center of the outer rotor 2 by the amount of eccentricity *e*.

An outer rotor is an assembly unit in which the inner toothed surface (Fig. 2) is formed by the insert rollers 3. Such a design of the outer rotor 2 allows roller 3 to rotate in hole 1 during the operation of the hydraulic machine. In this case, the existing sliding friction between the toothed surfaces of the rotors is replaced by rolling friction. To ensure the integrity of such a (assembled) design of the outer rotor 2 (Fig. 2), its holes 1 ("nests") are made with a diameter D_{2h} and are "open" by the value *L*.

The nominal values of the diameter D_{2r} of the roller 3 and the diameter D_{2h} of the hole 1 for the rollers are numerically equal (Fig. 2). The ability to rotate the roller 3 in the hole 1 under load is provided by the fit *H7/g6*.

According to the design documentation, the difference between the hole diameter D2h and the size *L* is $D_{2h} - L = 0.2\ldots\ 0.3$ mm. The edges of hole 1, located at a height *h*, keep roller 3 in the working position, preventing it from "falling out". This (integral) design of the outer rotor 2 with the roller 3 makes it possible to automate its parameter (P_c) check operations. The running-in of one rotor inside the other with the sealing of the working zones (operability) of the considered hypocycloidal gearing is ensured by the corresponding fit *H7/f7*.

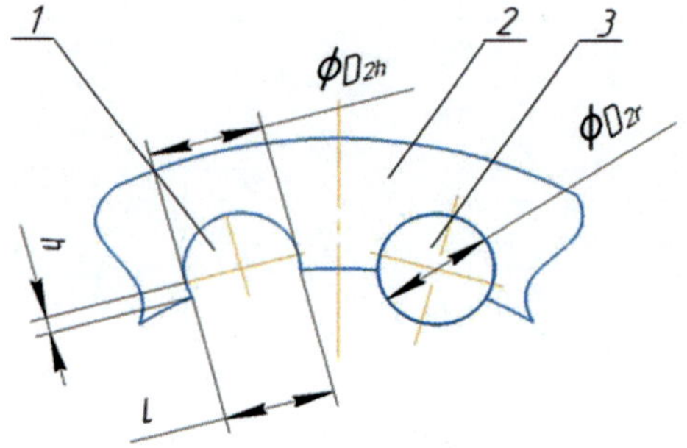

Fig. 2. Formation of the toothed surface of the outer rotor: 1 – hole ("socket") for the roller; 2 – outer rotor; 3 – roller (tooth of the outer rotor).

The nominal value of the checked parameters Pc for the inner and outer rotors (Fig. 2) of the gerotor hydraulic machine is equal and is determined from the design diagram according to the formula:

$$P_c = (2R_1 - r_1) \cdot \cos\frac{\pi}{2z_1} + r_1 \cdot \sin\left(\arccos\frac{R_2 \cdot \sin\frac{\pi}{z_2}}{r_1 + r_2}\right) \qquad (1)$$

The value of the technological gap G. It ensures the tightness of the rotor block during assembly. It is set by the design documentation and is determined by the difference in the values of the dimensions of the checked parameters P2c of the outer rotor and P1c of the inner rotor (i.e., G = P2c – P1c).

Depending on the geometrical dimensions of the rotors, the technological gap G between the toothed surfaces of the rotors of the gerotoron hydraulic machine is in the range of 0.02…0.04 mm. It is quite problematic to provide such a (specified) value of the gap technologically in the manufacture of rotors. This is because the limiting deviations of the nominal dimensions of the checked parameters Pc of each rotor are practically commensurate with the size of the technological gap G. Therefore, the required clearance between the toothed surfaces of the rotors of the displacement unit is provided by a selective selection of the appropriate pairs, which indicates the need to automate the process of measuring the checked parameters Pc for each rotor.

The imitation of the rotor system's operation principle in the gerotor hydraulic machine when measuring the checked parameters P_c for each rotor (Fig. 3) made it possible to implement the developed method for the automated process of monitoring the rotors.

Since the check conditions are as close as possible to the exploitation conditions of the rotors, the pitch between the teeth of the base gear 2 should be equal to the pitch between the teeth of the mating rotor (Fig. 3).

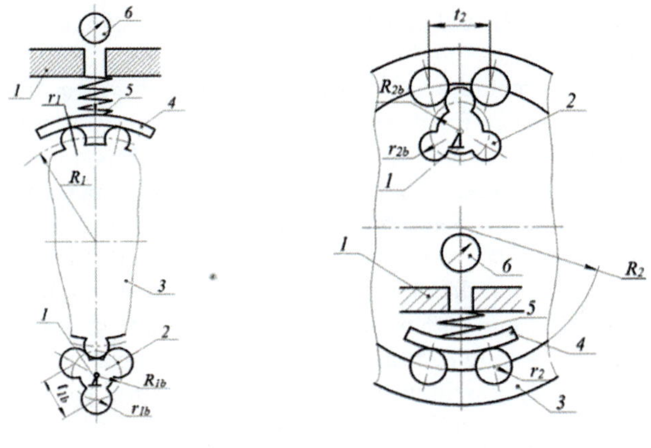

Fig. 3. Schemes of automated rotor check devices: a – internal; b – external; 1 – rack; 2 – base gear; 3 – checked rotor; 4 – measuring tip; 5 – spring; 6 – indicator.

Therefore, when checking the inner rotor to determine the parameters of the base gear 2 (Fig. 3, a), taking into account that t1b = t2, the geometric parameters of the outer rotor 3 (Fig. 3, b) are used as conjugate. Then, the radius of the tooth of the base gear r1p is equal to the radius of the tooth r2 of the outer rotor (roller) r1b = r2, and the radius of the pitch circle of the base gear 2 is determined by the expression

$$R_{1b} = R_2 \cdot \sin\frac{\pi}{z_2} / \sin\frac{\pi}{z_{1b}}, \tag{2}$$

where R_2 is the pitch circle radius of the outer rotor.

Similarly, the calculation of the parameters of the base gear 2 is carried out when the external rotor is checked (Fig. 3, b), taking into account that $t_{2b} = t_1$, and $r_{2b} = r_1$

$$R_{2b} = R_1 \cdot \sin\frac{\pi}{z_1} / \sin\frac{\pi}{z_{2b}}, \tag{3}$$

where R_1 is the pitch circle radius of the inner rotor.

4 Results and Discussion

Measuring devices for an automated check of the inner and outer rotors are made according to the same kinematic scheme, which implements the proposed method (Fig. 3) and includes the same basic elements.

The basis of the measuring devices (Fig. 3) are their racks 1, on which the drives are rigidly fixed, rotating around the axis of the base gears 2. When measuring, the base gears 2 are kinematically connected with the checked rotors 3.

Therefore, during the check process, rotor 3 rotates. The rotor 3, when checked, makes at least one complete revolution, which makes it possible to measure along the entire gear contour in one installation. Movable tips 4 have measuring surfaces described by the radius of "contact" R_0 and under the action of springs 5 are in constant contact with the checked rotors 3.

Deviations from the nominal size of the checked parameter Pc are determined using indicators 6 installed on racks 1 and kinematically connected with measuring tips 4.

For the production verification of prototypes of the developed means of automated check of gear surfaces of rotors, a gear pair with internal hypocycloidal gearing was selected. The kinematic and geometric parameters of the inner and outer rotors are respectively equal: the number of teeth – $z_1 = 11$ and $z_2 = 12$; radius of curvature of the teeth – $r_1 = r_2 = 7.0$ mm; the radii of the centers of the teeth – $R_1 = 33.35$ and $R_2 = 45.0$ mm; eccentricity – e = 2.35 mm. The nominal size of the checked parameter P_c for both rotors is $P_c = 76.22$ mm, and the selected fit H7/f7 guarantees the operability of the gear pair under study, providing a minimum technological gap G = 0.03 mm. The maximum clearance for the selected fit is G = 0.09 mm.

Analysis of the measurement results (Fig. 4) of the deviations of the inner rotor shows that the deviations of its toothed surface from the nominal size have minor fluctuations caused by the error in the shape of the toothed contour [3]. Fluctuations in the error of the shape of the toothed contour are within the limits specified by the corresponding quality $f7^{-0.03}_{-0.06}$, providing a dispersion of sizes in the range of 0.03 mm. The scatter of the measured values of deviations from the nominal size of the monitored parameter P_c is 0.009 mm and is in the range −0.048…−0.039 mm, which corresponds to the specified quality.

Analysis of the measurements of deviations of the outer rotor shows (Fig. 4) that deviations of its toothed surface from the nominal size have more significant fluctuations caused by the error in the shape of the toothed contour [3]. Fluctuations of the form error of the gear contour of the outer rotor are also within limits specified by the corresponding quality $H7^{+0.03}_{0}$, providing a dispersion of dimensions in the range of 0.03 mm. For the external rotor, the scatter of measured values of deviations from the nominal size of the checked parameter P_c is 0.019 mm and in the range of 0.006…0.025 mm, which corresponds to the specified quality.

The actual minimum gap for the test check pair is Gmin = 0.045 mm, and the maximum Gmax = 0.073 mm, which corresponds to the selected fit H7/f7 and, therefore, ensures the operability of the rotor unit (running in and sealing).

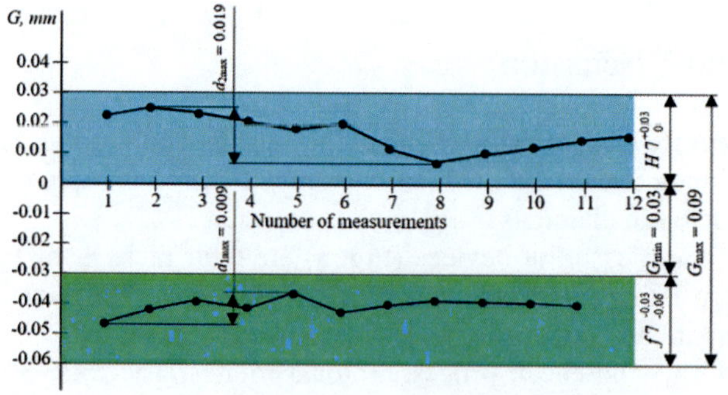

Fig. 4. Measured deviations of the checked parameters P_c for the external and internal rotors.

The production check showed that the proposed method and means of automated check of the deviations of the toothed profiles of the rotors of gerotor hydraulic machines allow precision assembly of their block of rotors by selecting the appropriate pairs.

5 Conclusions

The studies established that the outer rotor of the displacement unit of a gerotor hydraulic machine, being an assembly unit, is an integral structure similar to the inner rotor.

The paper substantiates the methodological possibility of automating the check of the toothed profiles of the rotors of gerotor hydraulic machines. For determining the checked parameters, a design diagram of the rotor block of a gerotor hydraulic machine was developed, and mathematical dependencies were obtained that describe the structural elements of the check devices.

The production check of prototypes of the developed automated check tools for the toothed surfaces of the rotors showed that the proposed check method allows precision assembly of the displacement units of gerotor hydraulic machines by a selective choosselecting the corresponding pairs of rotors.

References

1. Panchenko, A., Voloshina, A., Milaeva, I., Luzan, P.: Operating conditions' influence on the change of functional characteristics for mechatronic systems with orbital hydraulic motors. In: Nadykto, V. (ed.) Modern Development Paths of Agricultural Production, pp. 169–176. Springer, Cham (2019). https://doi.org/10.1007/978-3-030-14918-5_18
2. Panchenko, A., Voloshina, A., Panchenko, I., Titova, O., Pastushenko, A.: Reliability design of rotors for orbital hydraulic motors. IOP Conf. Series: Mater. Sci. Eng. **708**(1), 012017 (2019). https://doi.org/10.1088/1757-899X/708/1/012017
3. Panchenko, A., Voloshina, A., Milaeva, I., Panchenko, I., Titova, O.: The influence of the form error after rotor manufacturing on the output characteristics of an orbital hydraulic motor. Int. J. Eng. Technol. **7**(43), 1–5 (2018). https://doi.org/10.14419/ijet.v7i4.3.19542
4. Panchenko, A., Voloshina, A., Luzan, P., Panchenko, I., Volkov, S.: Kinematics of motion of rotors of an orbital hydraulic machine. IOP Conf. Series: Mater. Sci. Eng. **1021**, 012045 (2021). https://doi.org/10.1088/1757-899X/1021/1/012045
5. Makarov, V., Rezvaya, K., Drankovskiy, V., Mikhaylo, C.: Determination of the main parameters of the pump-turbine using the block-hierarchical approach. In: 2020 IEEE KhPI Week on Advanced Technology (KhPIWeek), pp. 536–540 (2020). https://doi.org/10.1109/KhPIWeek51551.2020.9250164
6. Barrio, R., Fernández, J., Blanco, E., Parrondo, J., Marcos, A.: Performance characteristics and internal flow patterns in a reverse-running pump-turbine. J. Mech. Eng. Sci. **226**(3), 695–708 (2012). https://doi.org/10.1177/0954406211416304
7. Amelio, M., Barbarelli, S.: A one-dimensional numerical model for calculating the efficiency of pumps as turbines for implementation in micro-hydro power plants. In: Proceedings of the 7th Biennial Conference on Engineering Systems Design and Analysis, vol. 1, pp. 65–72 (2004). https://doi.org/10.1115/esda2004-58080
8. González, J., Oro, J., Argüelles-Díaz, K.: Flow analysis for a double suction centrifugal machine in the pump and turbine operation modes. Int. J. Numer. Meth. Fluids **61**(2), 220–236 (2009). https://doi.org/10.1002/fld.1951

9. Gentner, C., Sallaberger, M., Widmer, C., Braun, O., Staubli, T.: Numerical and experimental analysis of instability phenomena in pump turbines. IOP Conf. Ser.: Earth and Environ. Sci. **15**(3), 0320422 (2012). https://doi.org/10.1088/1755-1315/15/3/032042
10. Egusquiza, E., Valero, C., Presas, A., Huang, X., Guardo, A., Seidel, U.: Analysis of the dynamic response of pump-turbine impellers Influence of the rotor. Mech. Syst. Signal Process. **68–69**, 330–341 (2016). https://doi.org/10.1016/j.ymssp.2015.05.034
11. Mao, X., Dal Monte, A., Benini, E., Zheng, Y.: Numerical study on the internal flow field of a reversible turbine during continuous guide vane closing. Energies **10**(7), 988 (2017). https://doi.org/10.3390/en10070988
12. Rogovyi, A., Korohodskyi, V., Khovanskyi, S., Hrechka, I., Medvediev, Y.: Optimal design of vortex chamber pump. J. Phys: Conf. Ser. **1741**, 012018 (2021). https://doi.org/10.1088/1742-6596/1741/1/012018
13. Andrenko, P., Rogovyi, A., Hrechka, I., Khovanskyi, S., Svynarenko, M.: Characteristics improvement of labyrinth screw pump using design modification in screw. J. Phys: Conf. Ser. **1741**, 012024 (2021). https://doi.org/10.1088/1742-6596/1741/1/012024
14. Rogovyi, A., Korohodskyi, V., Medvediev, Y.: Influence of Bingham fluid viscosity on energy performances of a vortex chamber pump. Energy **218**, 119432 (2021). https://doi.org/10.1016/j.energy.2020.119432
15. Van de Ven, J.: On fluid compressibility in switch-mode hydraulic circuits – Part I: Modeling and analysis. J. Dyn. Syst. Measur. Control **135**(2), 021013 (2012). https://doi.org/10.1115/1.4023062
16. Van de Ven, J.D.: On fluid compressibility in switch-mode hydraulic circuits—Part II: experimental results. J. Dyn. Syst. Meas. Control **135**(2), 021013 (2012). https://doi.org/10.1115/1.4023063
17. Rogovyi, A., Khovanskyi, S., Hrechka, I., Gaydamaka, A.: Studies of the swirling submerged flow through a confuser. In: Ivanov, V., Pavlenko, I., Liaposhchenko, O., Machado, J., Edl, M. (eds.) DSMIE 2020. LNME, pp. 85–94. Springer, Cham (2020). https://doi.org/10.1007/978-3-030-50491-5_9
18. Vaneev, S.M., et al.: Investigation of a turbogenerator based on the vortex expansion machine with a peripheral side channel. J. Eng. Sci. **8**(1), F11–F18 (2021). https://doi.org/10.21272/jes.2021.8(1).f2
19. Andrenko, P., Rogovyi, A., Hrechka, I., Khovanskyi, S., Svynarenko, M.: The influence of the gas content in the working fluid on parameters of the the hydraulic motor's axial piston. In: Ivanov, V., Pavlenko, I., Liaposhchenko, O., Machado, J., Edl, M. (eds.) DSMIE 2021. LNME, pp. 97–106. Springer, Cham (2021). https://doi.org/10.1007/978-3-030-77823-1_10
20. Vacca, A., Klop, R., Ivantysynova, M.: A numerical approach for the evaluation of the effects of air release and vapour cavitation on effective flow rate of axial piston machines. Int. J. Fluid Power **11**(1), 33–45 (2010). https://doi.org/10.1080/14399776.2010.10780996
21. Moloshnyi, O.M., Szulc, P., Sotnyk, M.I.: Influence of an inlet rotating axial device on the cavitation processes in a low specific speed centrifugal pump. J. Eng. Sci. **6**(1), E25–E32 (2019). https://doi.org/10.21272/jes.2019.6(1).e5
22. Gamez-Montero, P., Codina, E., Castilla, R.: A review of gerotor technology in hydraulic machines. Energies **12**, 2423 (2019). https://doi.org/10.3390/en12122423
23. Yeremenko, O., Abdullo, M.A., Boltianska, N., Mikhalchenko, S., Verkholantseva, V.: Reducing working fluid pulsations in planetary hydraulic machines by rational design of the distribution systems. In: Ivanov, V., Pavlenko, I., Liaposhchenko, O., Machado, J., Edl, M. (eds.) Advances in Design, Simulation and Manufacturing V: Proceedings of the 5th International Conference on Design, Simulation, Manufacturing: The Innovation Exchange, DSMIE-2022, June 7–10, 2022, Poznan, Poland – Volume 2: Mechanical and Chemical Engineering, pp. 133–143. Springer International Publishing, Cham (2022). https://doi.org/10.1007/978-3-031-06044-1_13

24. Voloshina, A., Panchenko, A., Boltyansky, O., Panchenko, I., Titova, O.: Justification of the kinematic diagrams for the distribution system of a planetary hydraulic motor. Int. J. Eng. Technol. **7**(4.3), 6–11 (2018). https://doi.org/10.14419/ijet.v7i4.3.19544
25. Voloshina, A., Panchenko, A., Panchenko, I., Titova, O., Zasiadko, A.: Improving the output characteristics of planetary hydraulic machines. IOP Conf. Ser.: Mater. Sci. Eng. **708**(1), 012038 (2019). https://doi.org/10.1088/1757-899X/708/1/012038
26. Pavlenko, I., Liaposhchenko, A., Ochowiak, M., Demyanenko, M.: Solving the stationary hydroaeroelasticity problem for dynamic deflection elements of separation devices. Vibr. Phys. Syst. **29**, 2018026 (2018)
27. Panchenko, A., Voloshina, A., Panchenko, I., Pashchenko, V., Zasiadko, A.: Influence of the shape of windows on the throughput of the planetary hydraulic motor's distribution system. In: Ivanov, V., Pavlenko, I., Liaposhchenko, O., Machado, J., Edl, M. (eds.) DSMIE 2021. LNME, pp. 146–155. Springer, Cham (2021). https://doi.org/10.1007/978-3-030-77823-1_15
28. Kiurchev, S., Luzan, P., Zasiadko, A., Radionov, H., Boltianska, N.: Influence of the flow area of distribution systems on changing the operating parameters of planetary hydraulic motors. IOP Conf. Ser.: Mater. Sci. Eng. **1021**, 012037 (2021). https://doi.org/10.1088/1757-899X/1021/1/012037
29. Voloshina, A., Panchenko, A., Boltyansky, O., Titova, O.: Improvement of manufacture workability for distribution systems of planetary hydraulic machines. In: Ivanov, V., et al. (eds.) DSMIE 2019. LNME, pp. 732–741. Springer, Cham (2020). https://doi.org/10.1007/978-3-030-22365-6_73
30. Voloshina, A., Panchenko, A., Panchenko, I., Zasiadko, A.: Geometrical parameters for distribution systems of hydraulic machines. In: Nadykto, V. (ed.) Modern Development Paths of Agricultural Production, pp. 323–336. Springer, Cham (2019). https://doi.org/10.1007/978-3-030-14918-5_34
31. Voloshina, A., Panchenko, A., Boltyansky, O., Zasiadko, A., Verkholantseva, V.: Improvement of the angular arrangement of distribution system windows when designing planetary hydraulic machines. In: Tonkonogyi, V., Ivanov, V., Trojanowska, J., Oborskyi, G., Pavlenko, I. (eds.) InterPartner 2021. LNME, pp. 53–63. Springer, Cham (2022). https://doi.org/10.1007/978-3-030-91327-4_6
32. Pylypaka, S., Volina, T., Nesvidomin, A., Zakharova, I., Rebrii, A.: Particle movement in a centrifugal device with vertical blades. In: Ivanov, V., Pavlenko, I., Liaposhchenko, O., Machado, J., Edl, M. (eds.) DSMIE 2021. LNME, pp. 156–165. Springer, Cham (2021). https://doi.org/10.1007/978-3-030-77823-1_16
33. Voloshina, A., Panchenko, A., Titova, O., Panchenko, I.: Changes in the dynamics of the output characteristics of mechatronic systems with planetary hydraulic motors. J. Phys: Conf. Ser. **1741**, 012045 (2021). https://doi.org/10.1088/1742-6596/1741/1/012045
34. Panchenko, A., Voloshina, A., Sadullozoda, S.S., Boltyansky, O., Panina, V.: Influence of the design features of orbital hydraulic motors on the change in the dynamic characteristics of hydraulic drives. In: Ivanov, V., Pavlenko, I., Liaposhchenko, O., Machado, J., Edl, M. (eds.) Advances in Design, Simulation and Manufacturing V: Proceedings of the 5th International Conference on Design, Simulation, Manufacturing: The Innovation Exchange, DSMIE-2022, June 7–10, 2022, Poznan, Poland – Volume 2: Mechanical and Chemical Engineering, pp. 101–111. Springer International Publishing, Cham (2022). https://doi.org/10.1007/978-3-031-06044-1_10
35. Voloshina, A., Panchenko, A., Titova, O., Pashchenko, V., Zasiadko, A.: Experimental studies of a throughput of the distribution systems of planetary hydraulic motors. IOP Conf. Ser.: Mater. Sci. Eng. **1021**, 012054 (2021). https://doi.org/10.1088/1757-899X/1021/1/012054

36. Merzliakov, I., Pavlenko, I., Chekh, O., Sharapov, S., Ivanov, V.: Mathematical modeling of operating process and technological features for designing the vortex type liquid-vapor jet apparatus. In: Ivanov, V., et al. (eds.) DSMIE 2019. LNME, pp. 613–622. Springer, Cham (2020). https://doi.org/10.1007/978-3-030-22365-6_61
37. Voloshina, A., Panchenko, A., Titova, O., Milaeva, I., Pastushenko, A.: Prediction of changes in the output characteristics of the planetary hydraulic motor. In: Tonkonogyi, V., et al. (eds.) InterPartner 2020. LNME, pp. 744–754. Springer, Cham (2021). https://doi.org/10.1007/978-3-030-68014-5_72
38. Panchenko, A., Voloshina, A., Titova, O., Panchenko, I., Caldare, A.: Design of hydraulic mechatronic systems with specified output characteristics. In: Ivanov, V., Pavlenko, I., Liaposhchenko, O., Machado, J., Edl, M. (eds.) DSMIE 2020. LNME, pp. 42–51. Springer, Cham (2020). https://doi.org/10.1007/978-3-030-50491-5_5
39. Kondratiev, A., Prontsevych, O., Nabokina, T.: Analysis of the bearing capacity of adhesive joint of honeycomb cores of sandwich structures with the continuous adhesive layer. Key Eng. Mater. **864**, 228–240 (2020). https://doi.org/10.4028/www.scientific.net/KEM.864.228
40. Kondratiev, A., Prontsevych, O.: Analysis of the bearing capacity of an adhesive connection between a cellular filler and sheathing at the addressed application of the adhesive onto the ends of honeycombs. Eastern-Eur. J. Enterp. Technol. **4**(7–106), 24–32 (2020). https://doi.org/10.15587/1729-4061.2020.208940
41. Stryczek, J., Bednarczyk, S., Biernacki, K.: Strength analysis of the polyoxymethylene cycloidal gears of the gerotor pump. Arch. Civ. Mech. Eng. **14**(4), 647–660 (2014). https://doi.org/10.1016/j.acme.2013.12.005
42. Ivanovic, L., Miric, N., Devedzic, G., Ćuković, S.: Analysis of forces and moments in gerotor pumps. J. Mech. Eng. Sci. **224**(10), 2257–2269 (2010). https://doi.org/10.1243/09544062JMES2041
43. Chiu-Fan, H.: Flow Characteristics of gerotor pumps with novel variable clearance designs. J. Fluids Eng. **137**(4), 041107 (2015). https://doi.org/10.1115/1.4029274
44. Pylypaka, S.F., Klendii, M.B., Nesvidomin, V.M., Trokhaniak, V.I.: Particle motion over the edge of an inclined plane that performs axial movement in a vertical limiting cylinder. Acta Polytech. **59**(1), 67–76 (2019). https://doi.org/10.14311/AP.2019.59.0067
45. Panchenko, A., Voloshina, A., Titova, O., Panchenko, I., Zasiadko, A.: The study of dynamic processes of mechatronic systems with planetary hydraulic motors. In: Tonkonogyi, V., et al. (eds.) InterPartner 2020. LNME, pp. 704–713. Springer, Cham (2021). https://doi.org/10.1007/978-3-030-68014-5_68
46. Panchenko, A., Voloshina, A., Titova, O., Panchenko, I.: The influence of the design parameters of the rotors of the planetary hydraulic motor on the change in the output characteristics of the mechatronic system. J. Phys.: Conf. Ser. **1741**, 012027 (2021). https://doi.org/10.1088/1742-6596/1741/1/012027
47. Sychuk, V., Zabolotnyi, O., Somov, D.: Technology of effective abrasive jet machining of parts surfaces. In: Ivanov, V., et al. (eds.) DSMIE 2018. LNME, pp. 166–176. Springer, Cham (2019). https://doi.org/10.1007/978-3-319-93587-4_18
48. Kostyk, K., et al.: Simulation of diffusion processes in chemical and thermal processing of machine parts. Processes **9**(4), 698 (2021). https://doi.org/10.3390/pr9040698
49. Pellegri, M., Vacca, A.: Numerical simulation of Gerotor pumps considering rotor micro-motions. Meccanica **52**(8), 1851–1870 (2016). https://doi.org/10.1007/s11012-016-0536-6

50. Knapčíková, L., Trojanowska, J., Martiček, M., Husár, J.: Composite materials based on GPPS/PE-HD matrix used in automotive industry. In: Machado, J., et al. (eds.) Innovations in Mechanical Engineering II, pp. 329–338. Springer International Publishing, Cham (2023). https://doi.org/10.1007/978-3-031-09382-1_29
51. Panchenko, A., Voloshina, A., Boltianska, N., Pashchenko, V., Volkov, S.: Manufacturing error of the toothed profile of rotors for an orbital hydraulic motor. In: Tonkonogyi, V., Ivanov, V., Trojanowska, J., Oborskyi, G., Pavlenko, I. (eds.) InterPartner 2021. LNME, pp. 22–32. Springer, Cham (2022). https://doi.org/10.1007/978-3-030-91327-4_3

Non-circular Wheels from Congruent Arcs

Tetiana Kresan[1], Serhii Pylypaka[1], Tatiana Volina[1,2(✉)], Iryna Rybenko[2], and Oleksandr Tatsenko[2]

[1] National University of Life and Environmental Sciences of Ukraine, 15, Heroyiv Oborony Street, Kyiv 03041, Ukraine
t.n.zaharova@ukr.net

[2] Sumy National Agrarian University, 160, Kondratieva Street, Sumy 40021, Ukraine

Abstract. Closed flat curves are designed based on which toothed gears are called centroids. Their characteristic feature is the continuity of the transfer function. However, numerous engineering tasks require centroids with different transfer functions. Non-circular wheels can consist of any number of symmetrical arcs that intersect in pairs at right angles. Determined that the right angle is the minimum value of the angle at which the designed non-circular wheels can roll without jamming. Pairs of wheels have a characteristic feature: there is no sliding of surfaces when functioning. The result is the absence of friction forces, which leads to minimizing working surface wearing. In addition, there are devices for which transfer function type is not essential, unlike the number of complete revolutions of the wheel. Non-circular wheels are a pair of closed curves that rotate around fixed centers and simultaneously roll one by one without sliding. Non-circular wheels serve as centroids in designing cylindrical toothed gears with variable gear ratios. The method of construction of pairs of non-circular wheels, which consist of individual symmetrical arcs intersecting at right angles, is developed in the article. A quadratic polynomial is used to form the corresponding curves in the polar coordinate system. The construction of the wheels needs the initial data: the number of elements of both wheels (the drive one and the driven one).

Keywords: Industrial growth · Toothed gear · Rolling without sliding · Transfer function · Drive and driven wheel · Quadratic polynomial

1 Introduction

Firstly, gears that transmit rotational motion during mutual rolling of surfaces one by one on friction between them are called frictional. They have a significant disadvantage: they cannot transmit high power torque but are basic for designing powerful toothed gears. Secondly, non-circular wheels are a pair of closed curves that rotate around fixed centers and simultaneously roll one by one without sliding. Non-circular wheels serve as centroids in designing cylindrical toothed gears with variable gear ratios. Closed flat curves are designed based on which toothed gears are called centroids. The centroids of some non-circular toothed gears are quite well studied. As a rule, they are constructed not from individual congruent arcs but a continuous closed curve. Their characteristic feature

V. Tonkonogyi et al. (Eds.): InterPartner 2022, LNME, pp. 44–53, 2023.
https://doi.org/10.1007/978-3-031-16651-8_5

is the continuity of the transfer function, while for composite centroids, the transfer function is not continuous. However, the mechanisms require centroids with different transfer functions. In addition, there are devices (for example, counting devices) for which transfer function type is not essential, unlike the number of complete revolutions of the wheel.

It is determined that the right angle is the minimum value of the angle at which the designed non-circular wheels can roll without jamming. Non-circular wheels can consist of any number of symmetrical arcs that intersect in pairs at right angles. Against this background, this work aims to develop a method for constructing pairs of non-circular wheels from symmetrical arcs of congruent curves that intersect in pairs at right angles. A quadratic polynomial is used to form the corresponding curves in the polar coordinate system. The construction of the wheels needs the initial data: the number of elements of both wheels (the drive one and the driven one). A characteristic feature of pairs of wheels is the absence of sliding between surfaces during their functioning. The result is the absence of friction forces, which leads to minimizing working surface wearing.

2 Literature Review

Firstly, the task of constructing geometric objects is widely spread in engineering practice and is proposed to be solved in various methods. For example, in the article [1] the geometric design of multidimensional space objects by approximating the differential equations solution is presented. Furthermore, geometric simulation of multilateral processes is carried out by multidimensional parabolic interpolation [2]. In [3], the space geometric interpolants are taken as a basis. So, the range of problems that require geometric methods for solving is quite wide. Besides, such methods are quite diverse. Without any doubt, methods of differential geometry can take their place in this row.

Secondly, many works are dedicated to reliability and durability improvement of details. In [4], a new method of sulfidation of surfaces of parts by electroerosion alloying was proposed. The articles [5, 6] presented ways to strengthen heat-treated steel parts. It includes cementation and nitriding. At the same time, the scientists developed a method to increase the durability of surfaces of details by using a multiple layers coating by electroerosion alloying [7]. However, as noted in [8], the current recommendations proposed by scientists are insufficient. In these works, it is proposed to increase the wear resistance using a new method of strengthening surfaces, which is quite valuable and time-consuming. Instead, the cause of wear can be eliminated by geometric methods.

There are works devoted to the methods of designing non-circular wheels, which testifies to the attention to this direction of research among scientists. There are well-known examples of rolling a curvilinear profile on a straight line and vice versa – rolling a straight-line segment along a curve. The subsequent movements are classical for the first case: the rolling of a circle on a straight line, during which the point of the circle describes the cycloid, and the rolling of the line on a circle, as a result of which the point of the line describes the involute of the circle. The construction of non-circular wheels through the flowing curve along the straight one is shown in [9]. Besides, in [10], each section's module during the tooth's construction was calculated. The expression of angular velocity was analyzed theoretically according to the radius. Hao Xu [11] studied a general profile of non-circular gears, which can transmit periodic movement with simultaneous provision of speed variability. It is provided by an irregular shape of the gear and eccentricity of centers of rotation. Mircea Niculescu [12] applied a non-circular gear in the drive mechanism by improvement of the billet reheating furnace unloading door. The article [13] presents the methodology of non-circular gear construction with some modifications.

Moreover, elliptical gears were considered in [14]. Moreover, other curvilinear profiles are widely used in designing cylindrical gears (e.g., cycloid [15], improved cycloid [16], and involute ellipse [17]), modular dynamic separators [18, 19], prilling equipment [20], vortex-type apparatuses [21], and so on. The characteristic feature of toothed gears is the continuity of the transfer function. However, numerous engineering tasks require centroids with different transfer functions. Because of these findings, it can be hypothesized that constructing non-circular wheels consisting of any symmetrical arcs that intersect in pairs at right angles can be applied to solve this problem.

3 Research Methodology

Let us start with the image of a pair of non-circular wheels, which should be the result. An example of such a pair is shown in Fig. 1,a.

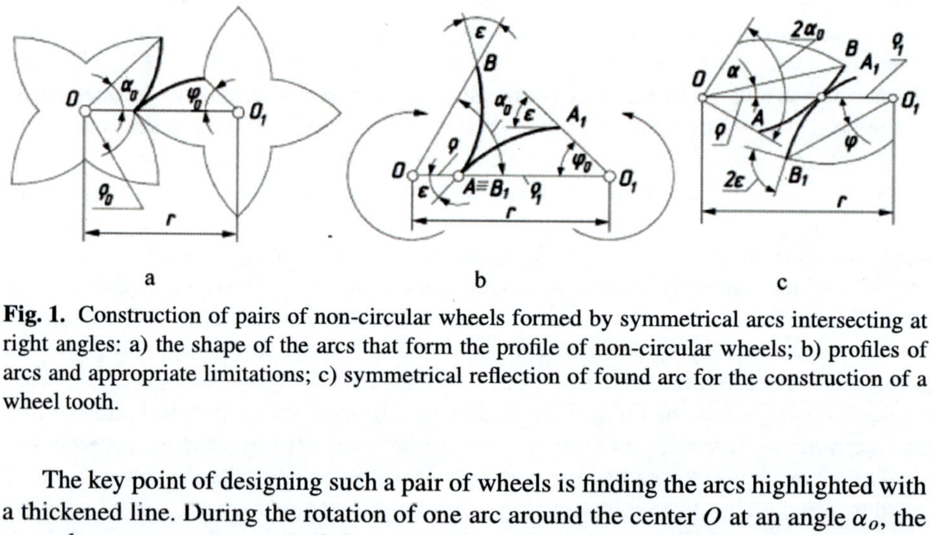

Fig. 1. Construction of pairs of non-circular wheels formed by symmetrical arcs intersecting at right angles: a) the shape of the arcs that form the profile of non-circular wheels; b) profiles of arcs and appropriate limitations; c) symmetrical reflection of found arc for the construction of a wheel tooth.

The key point of designing such a pair of wheels is finding the arcs highlighted with a thickened line. During the rotation of one arc around the center O at an angle α_o, the second arc must rotate around the center O_1 at an angle φ_o so that the curves have a common point of contact and roll without sliding. If these arcs are mapped symmetrically regarding the straight lines connecting their ends with the points O and O_1, then a part of a non-circular wheel can be obtained as a tooth. The minimum value of the angle at the tooth top should be 90°, so with a smaller value, the rolling of wheels becomes impossible due to jamming. At the corresponding angles α_o and φ_o, the number of teeth must correspond to an integer that divides the circle to the specified number of parts. For the case in Fig. 1,a, the number of teeth for both wheels is 4, so the angles α_o and

φ_o are equal and are $\pi/4$. During the rotation of the teeth and rolling their profiles one by one, the top of the tooth of one wheel enters the hollow of the other. The angle in the hollows of the teeth must also be straight.

The arcs *AB* of the driving wheel and A_1B_1 of the driven wheel (Fig. 1,b) can be found in the polar coordinate system. For the angles in the vertices and hollows of the teeth of non-circular wheels to be straight, the angle ε between the tangents to the curves and the polar radius at the beginning and end of the arcs *AB* and A_1B_1 should be equal to 45°. The adding of polar radiuses ρ and ρ_1 (the poles are in the points *O* and O_1) to the contact point should be equal to the distance from center to center: $\rho + \rho_1 = r$ (Fig. 1,b,c).

The rotation of one wheel at an angle α_o must be accompanied by rotation of the second one at an angle φ_o (Fig. 1,a). These angles are $\alpha_o = \pi/n$, $\varphi_o = \pi/n_1$. Here *n* and n_1 are the teeth quantity. When the tooth of the drive wheel rotates at the current angle α, the tooth of the driven wheel rotates at the angle φ (Fig. 1,c), provided that the lengths of the arcs of the curves are equal. Based on this, let us assume that the angle φ depends on α: $\varphi = \varphi(\alpha)$.

The expression of the drive wheel tooth profile is given in the polar coordinate system: $\rho = \rho(\alpha)$. Here α is the polar angle. The parametrical equations of *AB*:

$$x = \rho(\alpha)\cos\alpha;\ y = \rho(\alpha)\sin\alpha. \tag{1}$$

At the current angle α, Eq. (1) sets the value and direction of the polar radius ρ. The vector of the tangent to the curve (1) at the current point can be determined by its derivatives regarding the variable α:

$$\dot{x} = \dot{\rho}\cos\alpha - \rho\sin\alpha;\ \dot{y} = \dot{\rho}\sin\alpha + \rho\cos\alpha. \tag{2}$$

According to the known formula of vector algebra, the angle ε between (1) and (2) is:

$$\cos\varepsilon = \frac{\dot{\rho}}{\sqrt{\rho^2 + \dot{\rho}^2}}. \tag{3}$$

According to our conditions, the angle ε should be equal to 45° at point *A* (Fig. 1,b), namely at $\alpha = 0$, and at point *B* at $\alpha = \alpha_o$. The initial value of the polar radius ρ_o at $\alpha = 0$ is still essential (Fig. 1,a). The dependence $\rho = \rho(\alpha)$, which would satisfy these requirements, will be found as a quadratic polynomial:

$$\rho = a + b\alpha + c\alpha^2. \tag{4}$$

For this, it is necessary to find the expressions of the constants a, b, c. The derivative of expression (4):

$$\dot{\rho} = b + 2c\alpha. \tag{5}$$

Substitution (5) and (4) in (3) at $\alpha = 0$ and $\varepsilon = 45°$ gives the following dependence:

$$\frac{a^2}{\sqrt{a^2 + b^2}} = \frac{\sqrt{2}}{2}. \tag{6}$$

From (6): $a = b$. For $\alpha = 0$ from (4): $\rho_o = a$. Subsequent studies have shown that the constant a plays the role of a scaling coefficient that does not affect the shape of the curve. Therefore, it can be assumed that $a = b = 1$. Based on the previous, dependence (4) takes the form:

$$\rho = 1 + \alpha + c\alpha^2. \tag{7}$$

The constant c can be found by substitution (7) in (3), and its derivative $\dot{\rho} = 1 + 2c\alpha$ at $\varepsilon = 45°$ $\alpha = \alpha_o$, the equation is obtained:

$$\frac{1 + 2c\alpha_0}{\sqrt{(1 + \alpha_0 + c\alpha_0^2)^2 + (1 + 2c\alpha_0)^2}} = \frac{\sqrt{2}}{2}. \tag{8}$$

Equation (8) gives two solutions:

$$c = \frac{1}{2 - \alpha_0};\ c = -\frac{1}{\alpha_0}. \tag{9}$$

By substituting in turn (9) in (7), one can obtain two polar equations of the arc, which are constructed by parametrical Eqs. (1) when the angle α changes within limits $\alpha = 0 \ldots \alpha_o$:

$$\rho = 1 + \alpha + \frac{\alpha^2}{2 - \alpha_0}. \tag{10}$$

$$\rho = 1 + \alpha - \frac{\alpha^2}{\alpha_0}. \tag{11}$$

In Fig. 2,a an arc is constructed for $\alpha_o = 45°$. The number 1 indicates the arc corresponding to (10), and the number 2 – the arc corresponding to (11).

By successive rotation of the tooth (Fig. 2,b) around the origin, a non-circular wheel, shown in Fig. 1,a, can be obtained (it is shaped like a gear). Similarly, a non-circular wheel from convex elements can be obtained by successive rotation of the arcs (Fig. 2,c). For the construction of a toothed gear with n teeth, the value of the angle α_o should be $\alpha_o = \pi/n$, and for a non-circular wheel with the same number of convex elements $\alpha_o = 2\pi/n$.

Let begin with the construction of the arc A_1B_1 (Fig. 1,b). Considering $\rho_1 = r - \rho$ (Fig. 1,b,c) and the fact, that the radius vector ρ_1 oppositely rotates at an angle φ from the angle α, the parametrical equations of the sought curve on which A_1B_1 is located:

$$x_1 = (r - \rho)\cos\phi + r; \quad y_1 = -(r - \rho)\sin\phi, \tag{12}$$

where the expression for ρ is given in (10) or (11) depending on the type of non-circular wheel. The dependence between the angles $\varphi = \varphi(\alpha)$ is determined based on the equality of the lengths of the arcs that pass the wheels during mutual rolling [8]:

$$\phi = \int \frac{\rho}{\rho - r} d\alpha. \tag{13}$$

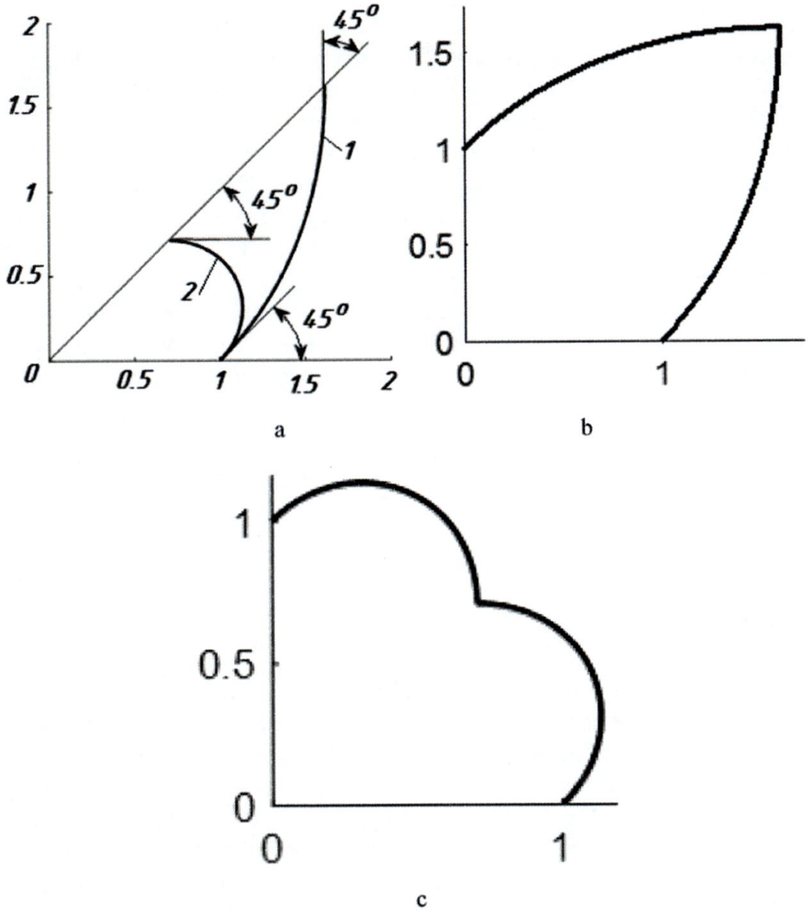

Fig. 2. Graphic illustrations for the profile designing of the driving non-circular wheel: a) two variants of the arc corresponding to polar Eqs. (10) and (11); b) the creation of a tooth by a symmetrical reflection of the arc 1; c) the creation of part of the profile of the wheel by a symmetrical reflection of the arc 2.

The integration of (3) after substitution into it (7), considering the required value of the integration constant (at $\alpha = 0$ $\varphi = -\varphi_o$) gives the following result:

$$\phi = \frac{2r}{\sqrt{4c(1-r)-1}}\left(\text{Arctg}\frac{1}{\sqrt{4c(1-r)-1}} - \text{Arctg}\frac{1+2c\alpha}{\sqrt{4c(1-r)-1}}\right) - \alpha - \phi_0. \tag{14}$$

The parametrical equations of the curve on which A_1B_1 is located can be found by substituting the expression ρ from (7) and the expression of the angle φ from (14) in (12). However, if an arc A_1B_1 is constructed according to these equations at the angle α changes within limits $\alpha = 0\ldots\alpha_o$ and an arbitrary value of the center-to-center distance r, then the final value of the angle φ (i.e., φ_o) will be different with the given because it depends on r. To obtain the sought value of the angle φ_o, the value of which is due to the number of teeth or convex elements of the driven wheel, it is essential to choose the center distance r properly. For this, it is necessary to substitute their respective values instead of the angles α and φ in (14). One should not forget that after the radius vector ρ rotates at an angle α_o, the radius vector ρ_1 will acquire a final position at which $\varphi = 0$ because this angle was negative ($-\varphi_o$) at the beginning of the rotation (Fig. 1,a). After substitution in (14), the values of the angles $\alpha = \alpha_o$ and $\varphi = 0$, the obtained equation should be solved regarding r. However, it cannot be done analytically, so numerical methods must be used.

If construct an arc A_1B_1 after that, it will be constructed correctly, but it will not be in contact with the arc AB. It was found that the arc A_1B_1 should be additionally rotated at an angle $\pi + \varphi_o$. The final expression of the angle φ can be obtained after summing its previous value (14) and the angle $\pi + \varphi_o$:

$$\phi = \frac{2r}{\sqrt{4c(1-r)-1}}\left(\mathrm{Arctg}\frac{1}{\sqrt{4c(1-r)-1}} - \mathrm{Arctg}\frac{1+2c\alpha}{\sqrt{4c(1-r)-1}}\right) - \alpha + \pi. \tag{15}$$

4 Results and Discussion

Let us consider examples of the designing of non-circular wheels: wheels from convex elements. Let the drive wheel has 2 convex elements. According to the expression $\alpha_o = 2\pi/n$ for $n = 2$: $\alpha_o = \pi$. Choose the polar Eq. (11), which corresponds to the specified type of wheel. By Eq. (1), taking into account (11), AB if α changes within limits $\alpha = 0 \ldots \pi$ can be constructed (Fig. 3,a). Let the driven wheel also has 2 elements. Therefore, $\varphi_o = \pi$. The center-to-center distance r should be derived. The values $\varphi_o = \pi$, $c = -1/\alpha_o = -1/\pi$ should be substituted in (14). According to the second expression (9): $\alpha = \alpha_o = \pi$ and $\varphi = 0$. After solving (14) by numerical methods with respect to r: $r = 3{,}11$.

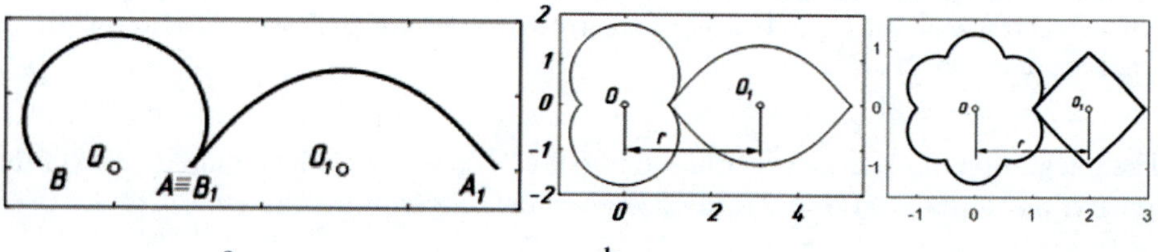

a b c

Fig. 3. Image of non-circular wheels with convex elements: a) initial arcs for constructing non-circular wheels; b) non-circular toothed gears if $n = 2$ and $n_1 = 2$; c) non-circular toothed gears if $n = 6$ and $n_1 = 4$.

The arc A_1B_1 is constructed by Eqs. (12) at the angle α changes within limits $\alpha = 0 \ldots \alpha_o$. In this case, the dependencies $\rho = \rho(\alpha)$ from (11) and $\varphi = \varphi(\alpha)$ from (15) at $c = -1/\pi$, $r = 3.11$ should be substituted there. The received arcs must be sequentially rotated around their centers O and O_1 at the required angles. In this case, they are rotated at 180° (Fig. 3,b). A pair of non-circular wheels are similarly constructed in Fig. 3,c. The initial data is the number of convex elements of the drive wheel $n = 6$ and the number of elements of the driven wheel $n_1 = 4$. The values of the necessary parameters for further calculations: $\alpha_o = 2\pi/n = \pi/3$, $\varphi_o = 2\pi/n_1 = \pi/2$, $c = -3/\pi$, $r = 1.97$.

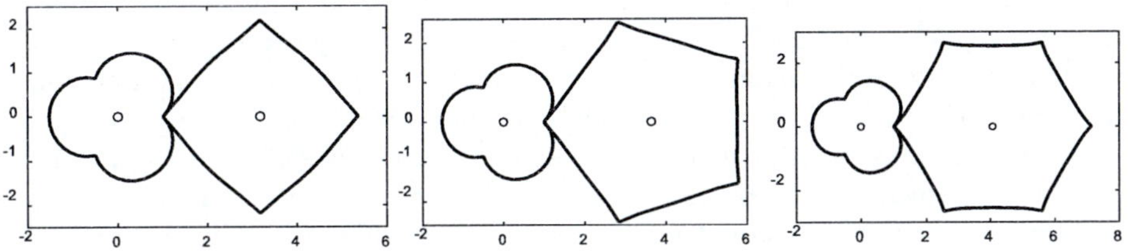

a b c

Fig. 4. Pairs of non-circular wheels in which the drive wheel has three convex elements $n = 3$ and a different number of elements in the driven wheels: a) $n_1 = 4$, $r = 3.18$; b) $n_1 = 5$, $r = 3.64$; c) $n_1 = 6$, $r = 4.07$.

In Fig. 4, pairs of non-circular wheels with the same number of convex elements of the drive wheel $n = 3$ and a different number of elements of the driven wheel are constructed. The transformation of the driven wheel with an increment quantity of its elements can be seen further (Fig. 4).

The construction of non-circular toothed gears can be carried out similarly. It is essential to apply the dependence (10) and the corresponding value from (9) to find the constant c. In addition, the values of the angles α_o and φ_o are twice smaller: $\alpha_o = \pi/n$, $\varphi_o = \pi/n_1$. In Fig. 5, pairs of wheels with different ratios of teeth are constructed. The number of teeth n and n_1 are the initial data.

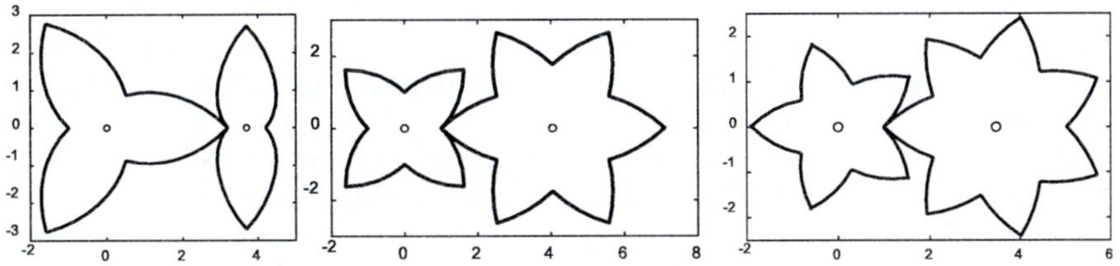

a b c

Fig. 5. Pairs of non-circular wheels, which are similar in shape to toothed gears: a) $n = 3$, $n_1 = 2$, $r = 3.7$; b) $n = 4$, $n_1 = 6$, $r = 4.05$; c) $n = 5$, $n_1 = 7$, $r = 3.47$.

Such pairs of wheels can function like gears. There is no friction between the surfaces of the teeth, which provides for the absence of wear. The disadvantage is that the gear ratio is not constant, and the center-to-center distance cannot be set, but it is calculated depending on the number of wheel teeth.

5 Conclusions

Thus, applying a quadratic polynomial makes it possible to create composite non-circular wheels of two types. In one case, it consists of convex elements, in the second – elements similar to toothed gear. Pairs of wheels have a characteristic feature: there is no sliding of surfaces when functioning. The result is the absence of friction forces, which leads to minimizing working surface wearing. The curves' arcs that delineate the wheel elements' profile intersect at right angles. The construction of the wheels needs the initial data: the number of elements of both wheels (the drive one and the driven one). The center-to-center distance depends on their number so that it can be calculated. Images of non-circular wheels with convex elements with teeth numbers 2/2 and 6/4 are presented.

Moreover, pairs of non-circular wheels in which the drive wheel has 3 convex elements and different elements in the driven wheels (4, 5, 6) are shown. Also given as an example are pairs of non-circular wheels, similar in shape to toothed gears, with the number of teeth 3/2, 4/6, 5/7.

References

1. Konopatskiy, E., Voronova, O., Shevchuk, O., Bezditnyi, A.: About one method of numeral decision of differential equalizations in partials using geometric interpolants. In: CEUR Workshop Proceedings, vol. 2763, pp. 213–219 (2020). https://doi.org/10.30987/conferencearticle_5fce27708eb353.92843700
2. Konopatskiy, E., Bezditnyi, A.: Geometric modeling of multifactor processes and phenomena by the multidimensional parabolic interpolation method. J. Phys.: Conf. Ser. **1441**, 012063 (2020). https://doi.org/10.1088/1742-6596/1441/1/012063
3. Konopatskiy, E., Bezditnyi, A., Shevchuk, O.: Modeling geometric varieties with given differential characteristics and its application. In: CEUR Workshop Proceedings, vol. 2744 (2020). https://doi.org/10.51130/graphicon-2020-2-4-31
4. Tarelnyk, V., et al.: New sulphiding method for steel and cast iron parts. IOP Conf. Ser.: Mater. Sci. Eng. **233**, 012049 (2017). https://doi.org/10.1088/1757-899X/233/1/012049
5. Tarelnyk, V., et al.: New method for strengthening surfaces of heat treated steel parts. IOP Conf. Ser.: Mater. Sci. Eng. **233**, 012048 (2017). https://doi.org/10.1088/1757-899X/233/1/012048
6. Kostyk, K., et al.: Simulation of diffusion processes in chemical and thermal processing of machine parts. Processes **9**(4), 698 (2021). https://doi.org/10.3390/pr9040698
7. Tarel'nik, V.B., Martsinkovskii, V.S., Zhukov, A.N.: Increase in the reliability and durability of metal impulse end seals. Part 1. Chem. Pet. Eng. **53**(1–2), 114–120 (2017). https://doi.org/10.1007/s10556-017-0305-y
8. Martsinkovsky, V., Yurko, V., Tarelnik, V., Filonenko, Y.: Designing thrust sliding bearings of high bearing capacity. Proc. Eng. **39**, 148–156 (2012). https://doi.org/10.1016/j.proeng.2012.07.019
9. Kresan, T., Pylypaka, S., Ruzhylo, Z., Rogovskii, I., Trokhaniak, O.: External rolling of a polygon on closed curvilinear profile. Acta Polytech. **60**(4), 313–317 (2020). https://doi.org/10.14311/AP.2020.60.0313
10. Jang, H.-S., Lee, C.-H., Park, G.-Y., Kim, C.: Study on design of non-circular gears for speed control of the squid belly opening and gutting machine (SBOGM). Appl. Sci. **11**, 3268 (2021). https://doi.org/10.3390/app11073268

11. Xu, H., Fu, T., Song, P., Zhou, M., Fu, G., Mitra, N.J.: Computational design and optimization of non-circular gears. Comput. Graph. Forum **39**, 399–409 (2020). https://doi.org/10.1111/cgf.13939
12. Niculescu, M., Andrei, L.: Using noncircular gears for the unloading door kinematics modification. In: Proceedings of the MATEC Web of Conferences, vol. 112, 6013. Iasi, Romania (2017). 0.1051/matecconf/201711206013
13. Zhou, M., Yang, Y., Wei, M., Yin, D.: Method for generating non-circular gear with addendum modification and its application in transplanting mechanism. Int. J. Agric. Biol. Eng. **13**(6), 68–75 (2020). https://doi.org/10.25165/j.ijabe.20201306.565
14. Nezir, E., Nihat, Y., Fatih, E., Bahadır, K., Burak, Ş.: Non-circular gear design, elliptical gears as example. In: Ejons VI – International conference on mathematics – Engineering – Natural & Medical Sciences, pp. 624–641. Adana, Turkey (2019)
15. Thai, N., Giang, T.: Influence of geometrical dimensions on the profile slippage in the hypogerotor pump. Vietnam, J. Sci. Technol. **50**(4), 482–491 (2018). https://doi.org/10.15625/2525-2518/56/4/9625
16. Thai, N., Tien, T., Dung, P., Huy, N.: Influence of the designing parameters on flow fluctuation and pressure of the improved roots blower. In: International Conference of Fluid Machinery and Automation Systems – ICFMAS2018, pp. 196–203 (2018)
17. Thai, N.: Shaping the tooth profile of elliptical gear with the involute ellipse curve. Sci. Technol. Dev. J. Eng. Technol. **4**(3), 1048–1056. https://doi.org/10.32508/stdjet.v4i3.820
18. Pavlenko, I., Liaposhchenko, A., Ochowiak, M., Demyanenko, M.: Solving the stationary hydroaeroelasticity problem for dynamic deflection elements of separation devices. Vibr. Phys. Syst. **29**, 2018026 (2018)
19. Liaposhchenko, O., Pavlenko, I., Monkova, K., Demianenko, M, Starynskyi, O.: Numerical simulation of aeroelastic interaction between gas-liquid flow and deformable elements in modular separation devices. In: Ivanov, V., et al. (eds.) DSMIE 2019. LNME, pp. 765–774. Springer, Cham (2020). https://doi.org/10.1007/978-3-030-22365-6_76
20. Yurchenko, O., Sklabinskyi, V., Ochowiak, M., Ostroha, R., Gusak, O.: Rational choice of a basket for the rotational vibropriller. J. Eng. Sci. **9**(1), F16–F20 (2022). https://doi.org/10.21272/jes.2022.9(1).f3
21. Merzliakov, I., Pavlenko, I., Chekh, O., Sharapov, S., Ivanov, V.: Mathematical modeling of operating process and technological features for designing the vortex type liquid-vapor jet apparatus. In: Ivanov, V., et al. (eds.) DSMIE 2019. LNME, pp. 613–622. Springer, Cham (2020). https://doi.org/10.1007/978-3-030-22365-6_61

Dynamics of Clamping Pneumatic Cylinder for Technological Equipment

Volodymyr Sokolov[1(✉)], Oleg Krol[1], Oleksandr Golubenko[1], Petko Tsankov[2], and Dmytro Marchenko[1]

[1] Volodymyr Dahl East Ukrainian National University, 59-a, Tsentralnyi Avenue, Severodonetsk 93400, Ukraine
sokolov.snu.edu@gmail.com

[2] Trakia University – Stara Zagora, 38, Graf Ignatiev Street, Yambol 8600, Bulgaria

Abstract. The question of mathematical modeling of the dynamic characteristics for a single-acting pneumatic drive of the clamping device for technological equipment, as well as research on its working process in the clamping pneumatic cylinder, is considered. The pneumatic settlement scheme for a drive is presented. The working stroke of the clamping pneumatic cylinder is performed when the compressed air is supplied to the piston cavity; the return stroke occurs under the action of the built-in spring. The mathematical model of the dynamic characteristics of the pneumatic drive has been developed. The mathematical model is based on the pneumatic cylinder piston's motion equations, the piston cavity's continuity equation, and the mass flow rate for air entering the piston cavity. The piston motion equations separately consider the positional load of the built-in spring, the resistance force proportional to the movement velocity, and the dry friction force. The thermodynamic process in the piston cavity of the pneumatic cylinder is adiabatic. The continuity equation considers the variable volume of the piston cavity and the volume of the connected pneumatic line. The equation for the mass flow rate of compressed air entering the pressure chamber of the pneumatic cylinder considers the subcritical and supercritical gas flow regimes. The example of calculating the dynamic characteristics for the clamping pneumatic cylinder is given. The presence of characteristic stages of the working process is shown.

Keywords: Dynamic characteristics · Mathematical model · Thermodynamic process · Adiabatic exponent · Gas constant · R&D investment

1 Introduction

Pneumatic clamping devices have been widely applied in technological equipment of machine-building industries [1, 2]. These devices are used for fixing parts in metal-cutting machine tools and systems, equipment for plastic shaping, and other equipment for mechanical processing of materials [3, 4].

Pneumatic clamping devices, the action of which is based on the conversion of compressed air energy into mechanical work, have a number of advantages over both hydraulic drives and mechanical devices [5, 6]. The main advantages of pneumatic drives

V. Tonkonogyi et al. (Eds.): InterPartner 2022, LNME, pp. 54–63, 2023.
https://doi.org/10.1007/978-3-031-16651-8_6

are simple and cheap drive design and ease of use. Few wear parts, no leakage typically found in hydraulic systems, and quick response. Essential advantages of pneumatic clamping devices are providing constant clamping force for the workpiece, the ability to adjust the clamping force, and the ability to clamp the workpiece quickly and release the resulting part [7, 8]. Significant disadvantages include only specific cumbersomeness of the structures of clamping devices with pneumatic drive, as well as limitations in the creation of heavy forces due to the use of pneumatic cylinders of large diameters in this case. Since the clamping pressures in such devices are relatively low, using pneumatic drives is rational, except for securing large-sized parts [9, 10].

2 Literature Review

It is quite convenient to design a clamping device using the single-acting pneumatic cylinder (Fig. 1) [1]. The working stroke in such cylinders is carried out under the action of compressed air, and the output link is returned to its home position by the built-in spring or from an external load.

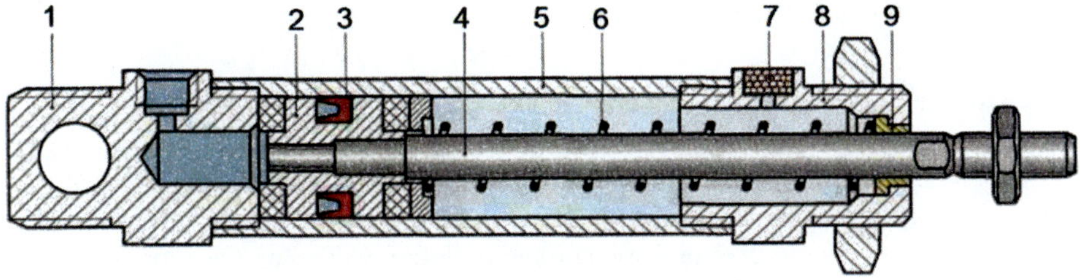

Fig. 1. The single-acting pneumatic cylinder.

In the design under consideration, the liner 5 (cylindrical body) of the pneumatic cylinder is closed on both sides by covers 1 and 8, and in the rear cover 1, there is a hole for supplying compressed air, and the front cover 8 has decompression hole with built-in filter element 7. Piston 2 divides the internal space of liner 5 into two cavities: the rod cavity, in which there is rod 4 rigidly connected to piston 2, and the piston cavity. The cavities are delimited by the hermetic seal 3 (for example, the cup) located in an annular groove on the outer cylindrical surface of the piston 2. The front (through) cover 8 is equipped with the guide bushing 9, which supports sliding rod 4, which transfers the force from piston 2 to the external object. The return spring 6 is mounted inside the cylinder and covers rod 4.

The working stroke of the pneumatic cylinder is performed when compressed air is supplied to the piston cavity; the return stroke is carried out by the built-in spring, which results in less air consumption compared to the double-acting pneumatic cylinders of the same size [11]. In addition, the single-acting pneumatic cylinders are somewhat simpler in design since they do not require complete sealing of the rod cavity, which is constantly connected to the atmosphere, and the absence of additional seals reduces friction losses [12, 13].

The application area of epy single-acting pneumatic cylinders is limited by the disadvantages inherent in this design: the working force is reduced due to the counteraction of the spring (by about 10%); the low effort during the return stroke (about 10% of the working); the limited movement of the rod (usually up to 100 mm); the increased longitudinal dimensions (the length of the compressed spring is added) [14, 15]. Thus, the single-acting pneumatic cylinders are widely used in cases where the transfer of force is required in only one direction, and the return occurs unimpeded (i.e., in the absence of opposing load), as well as safety reasons. It is necessary to ensure the retracted position of the rod when the power is turned off (when the pressure of compressed air drops in the pneumatic network) [16, 17].

Increasing the efficiency of technological equipment and improving its operational characteristics requires the study of working processes occurring in the clamping devices with the single-acting pneumatic cylinders [18, 19]. Especially important is the evaluation of the dynamic characteristics of the pneumatic clamping device since its response time determines the operating time of technological equipment [20, 21] and hence its performance [22, 23].

This article aims to develop the mathematical model of the dynamic characteristics for single acting pneumatic drive of the clamping device for technological equipment, to study the working process in the clamping pneumatic cylinder.

3 Research Methodology

The settlement scheme for the single-acting pneumatic drive of the clamping device for technological equipment is shown in Fig. 1. The drive's main elements are marked on the scheme: PC – the clamping pneumatic cylinder, PD – the two-position three-way pneumatic distributor. For the position of the pneumatic distributor PD, shown in the scheme, the working stroke of the clamping pneumatic cylinder PC is carried out, and when switching the position of the distributor PD, the piston of the cylinder PC is returned.

It is quite appropriate to set the absolute pressure in the rod cavity equal to the atmospheric pressure. Let's separate the dry friction force and the drag force proportional to the speed (usually, this is the viscous friction force). Then, the dynamic characteristics of the pneumatic cylinder piston can be described by the following differential equations [24, 25]

$$\frac{dy}{dt} = V; \tag{1}$$

$$m\frac{dV}{dt} = (p - p_a)S - F - F_s - cy - F_f - F_d signV, \tag{2}$$

where y, V – displacement and velocity of the PC piston; t – time; m – a reduced mass of moving parts; p – absolute pressure in the piston cavity of the PC; p_a – atmosphere pressure; S – piston area; F – rod load; F_s – spring preload force; c – spring rate; F_d – dry friction force; F_f – viscous friction force

$$F_f = k_f V, \tag{3}$$

k_f – viscous friction force coefficient (Fig. 2).

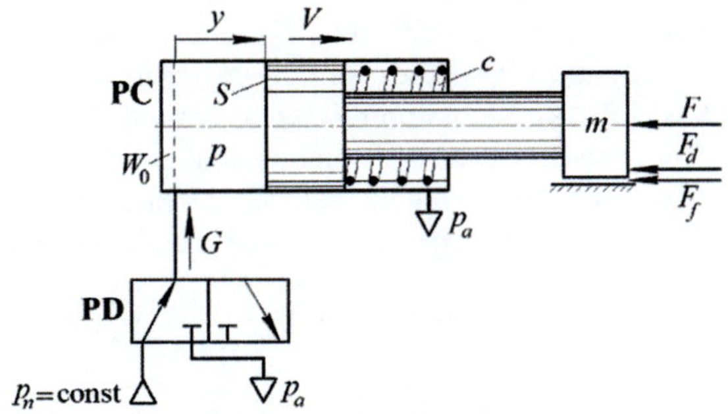

Fig. 2. The settlement scheme.

For further numerical integration of the mathematical model, it is necessary to note the restrictions [26] for the displacement and velocity of the piston

$$0 \leq y \leq H; \tag{4}$$

$$V \geq 0 \;\; and \;\; dV/dt \geq 0, \;\; if \;\; y = 0; \tag{5}$$

$$V \leq 0 \;\; and \;\; dV/dt \leq 0, \;\; if \;\; y = H. \tag{6}$$

In Eqs. (4), (6) H – piston stroke.

The continuity equation for the piston cavity of the clamping pneumatic cylinder, considering the compressibility of air in the adiabatic thermodynamic process [27, 28], can be represented as

$$\frac{W_0 + Sy}{kRT_0}\frac{dp}{dt} = G - \frac{p}{RT_0}SV, \tag{7}$$

where W_0 – "dead" volume for the piston cavity taking into account the volume of the connected pneumatic line; k – adiabatic exponent for air; R – gas constant for air; T_0 – the average absolute temperature in the piston cavity; G – mass flow rate of compressed air entering the piston cavity.

The equation for the mass flow rate of compressed air entering the piston cavity of the pneumatic cylinder taking into account the subcritical and supercritical gas flow regime

$$G = \begin{cases} f_e p_n \sqrt{\frac{2k}{(k-1)RT_n}\left[\left(\frac{p}{p_n}\right)^{\frac{2}{k}} - \left(\frac{p}{p_n}\right)^{\frac{k+1}{k}}\right]}, & if \;\; \frac{p}{p_n} > \varepsilon^*; \\ f_e\left(\frac{2}{k+1}\right)^{\frac{1}{k-1}} p_n \sqrt{\frac{2k}{(k+1)RT_n}}, & if \;\; \frac{p}{p_n} \leq \varepsilon^*, \end{cases} \tag{8}$$

where f_e – effective area for the pneumatic distributor (product of area and flow rate coefficient); P_n – absolute pressure in the pneumatic network; T_n – the absolute temperature in the pneumatic network; ε^* – critical pressure ratio

$$\varepsilon^* = \left(\frac{2}{k+1}\right)^{\frac{k}{k-1}}. \tag{9}$$

The system of Eqs. (1)–(9) forms the mathematical model of the dynamic characteristics for the single-acting pneumatic drive of the clamping device for technological equipment. For air, $k = 1.4$ and $R = 287$ J/(kg·K), assuming that the normal atmospheric pressure is p_a = 101325 Pa, and also representing the differential equations in the Cauchy form, based on the Eqs. (1)–(9), we finally obtain the mathematical model for practical engineering calculations of the dynamics for the working stroke of the clamping pneumatic cylinder

$$0 \leq y \leq H; \tag{10}$$

$$V \geq 0, \ \ if \ \ y = 0; \tag{11}$$

$$V \leq 0, \ \ if \ \ y = H; \tag{12}$$

$$\frac{dy}{dt} = V; \tag{13}$$

$$\frac{dV}{dt} = \frac{(p - 101325)S}{m} - \frac{F - F_s}{m} - \frac{c}{m}y - \frac{k_f}{m}V - \frac{F_d}{m} signV; \tag{14}$$

$$dV/dt \geq 0, \ \ if \ \ y = 0; \tag{15}$$

$$dV/dt \leq 0, \ \ if \ \ y = H; \tag{16}$$

$$G = \begin{cases} 0,1562 f_e p_n \sqrt{\frac{1}{T_n}\left[\left(\frac{p}{p_n}\right)^{1,428} - \left(\frac{p}{p_n}\right)^{1,714}\right]}, & if \ \frac{p}{p_n} > 0,528; \\ 0,0404 f_e \frac{p_n}{\sqrt{T_n}}, & if \ \frac{p}{p_n} \leq 0,528; \end{cases} \tag{17}$$

$$\frac{dp}{dt} = \frac{401,8T_0}{W_0 + Sy}G - \frac{1,4p}{W_0 + Sy}SV. \tag{18}$$

The initial conditions for the working stroke

$$y(t) = 0; \ \ V(t) = 0; \ \ p(t) = p_a. \tag{19}$$

During the return stroke in the clamping pneumatic cylinder, the piston cavity is connected to the atmosphere through the pneumatic distributor and is emptied. Then, the mathematical model will consider expressions for the mass flow rate G_{out} flowing from the piston cavity with absolute pressure p and absolute temperature T_0 into the atmosphere

$$G_{out} = \begin{cases} 0,1562 f_e p \sqrt{\frac{1}{T_0}\left[\left(\frac{101325}{p}\right)^{1.428} - \left(\frac{101325}{p}\right)^{1.714}\right]}, & if \ \frac{101325}{p} > 0,528; \\ 0,0404 f_e \frac{p}{\sqrt{T_0}}, & if \ \frac{101325}{p} \leq 0,528. \end{cases} \tag{20}$$

Also, the continuity equation will change

$$\frac{dp}{dt} = -\frac{401,8T_0}{W_0 + Sy}G_{out} - \frac{1,4p}{W_0 + Sy}SV. \tag{21}$$

Thus, the mathematical model of the dynamics for the return stroke of the clamping pneumatic cylinder has the form

$$0 \le y \le H; \tag{22}$$

$$V \ge 0, \ \ if \ y = 0; \tag{23}$$

$$V \le 0, \ \ if \ y = H; \tag{24}$$

$$\frac{dy}{dt} = V; \tag{25}$$

$$\frac{dV}{dt} = \frac{(p - 101325)S}{m} - \frac{F - F_s}{m} - \frac{c}{m}y - \frac{k_f}{m}V - \frac{F_d}{m}signV; \tag{26}$$

$$dV/dt \ge 0, \ \ if \ y = 0; \tag{27}$$

$$dV/dt \le 0, \ \ if \ y = H; \tag{28}$$

$$G_{out} = \begin{cases} 0,1562f_e p\sqrt{\frac{1}{T_0}\left[\left(\frac{101325}{p}\right)^{1,428} - \left(\frac{101325}{p}\right)^{1,714}\right]}, \ if \ \frac{101325}{p} > 0,528; \\ 0,0404f_e\frac{p}{\sqrt{T_0}}, \ if \ \frac{101325}{p} \le 0,528; \end{cases} \tag{29}$$

$$\frac{dp}{dt} = -\frac{401,8T_0}{W_0 + Sy}G_{out} - \frac{1,4p}{W_0 + Sy}SV. \tag{30}$$

It is quite appropriate to set the following initial conditions for the return stroke

$$y(t) = H; \ V(t) = 0; \ p(t) = p_n. \tag{31}$$

4 Results and Discussion

On the basis of the presented mathematical models, the example of calculating the dynamic characteristics of the clamping pneumatic cylinder for technological equipment was made for the following basic initial data: p_n = 6,01325·10^5 Pa (corresponds to the overpressure in the pneumatic network 5 Bar); T_n = 300 °K; S = 1,257·10^{-3} m^2 (corresponds to the piston diameter 40 mm); H = 0.1 m; F = 150 N; F_s = 50 N; F_d = 5 N; c = 2000 N/m; k_f = 120 kg/s; m = 5 kg; f_e = 1,948·10^{-6} m^2; W_0 = 1,2·10^{-3} m^3.

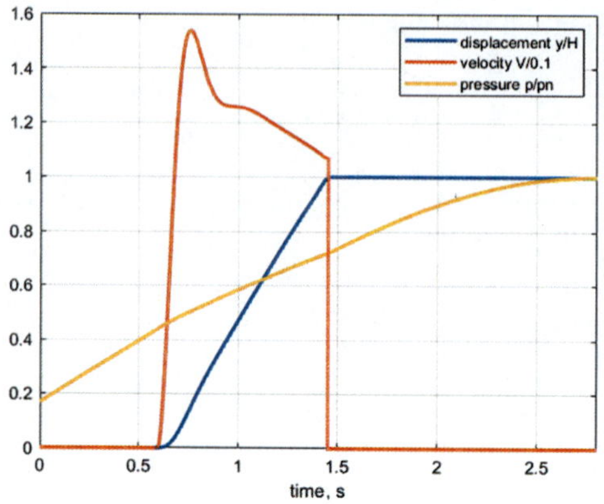

Fig. 3. The dynamic characteristics of the clamping pneumatic cylinder at the working stroke.

The calculations were carried out in the environment of the MATLAB software package [29, 30].

The results of calculating the pressure in the piston cavity of the pneumatic cylinder, displacement, and velocity of the piston during the working stroke are shown in Fig. 3.

As can see, the working process in the pneumatic clamping cylinder has four characteristic time stages. The first one is filling the piston cavity with compressed air up to the pressure at which the piston of the cylinder will move. The second stage is the acceleration process of the piston to its conditionally stable movement. Further, this is the movement of the piston to the extreme position or the position determined by the dimensions of the part or workpiece being clamped. In the last stage, after the piston reaches the extreme right position, the pressure in the piston cavity rises to the pressure in the pneumatic network.

The working process in the clamping pneumatic cylinder and during the piston's return stroke has similar stages (Fig. 4).

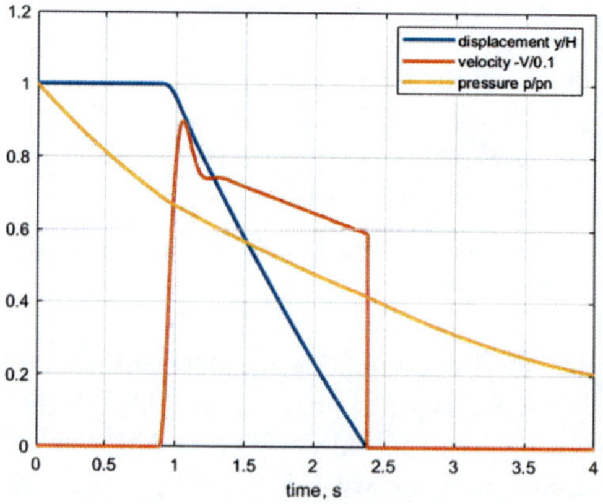

Fig. 4. The dynamic characteristics of the clamping pneumatic cylinder at the return stroke.

For ease of understanding of the material, the figure shows the piston velocity with the opposite sign since the direction of movement during the working stroke is taken as positive. The difference between the piston's return stroke lies in the fact that here the presence of the first stage is due to the emptying of the piston cavity to the pressure at which the return movement of the piston will begin. Also, at the last stage, further emptying of the piston cavity is carried out, where the pressure can drop to the limit of atmospheric pressure.

Evaluating the dynamic characteristics of a pneumatic cylinder allows you to analyze the design of equipment clamping devices to optimize their design parameters according to specified criteria. It should be noted that the time of the working and return stroke for the piston of the clamping pneumatic cylinder are the components of the operating time for technological equipment, which determines its performance [31, 32].

5 Conclusions

Thus, the mathematical modeling of the dynamic characteristics of a single-acting pneumatic drive of the clamping device for technological equipment is considered, and the working process in the clamping pneumatic cylinder is investigated.

The pneumatic settlement scheme for a drive is presented according to the working stroke of the clamping pneumatic cylinder is performed when the compressed air is supplied to the piston cavity, and the return stroke occurs under the action of the built-in spring. The mathematical model of the dynamic characteristics for the pneumatic drive has been developed. The mathematical model is based on the pneumatic cylinder piston's motion equations, the piston cavity's continuity equation, and the mass flow rate for air entering the piston cavity. The piston motion equations separately consider the positional load of the built-in spring, the resistance force proportional to the movement velocity, and the dry friction force. The thermodynamic process in the piston cavity of the pneumatic cylinder is considered adiabatic. The continuity equation takes into account the variable volume of the piston cavity and the volume of the connected pneumatic line. The equation for the mass flow rate of compressed air entering the pressure chamber of the pneumatic cylinder takes into account the subcritical and supercritical gas flow regimes. The example of calculating the dynamic characteristics for the clamping pneumatic cylinder is given, and the presence of characteristic stages of the processes during the working and return strokes is shown.

Evaluating the dynamic characteristics of the pneumatic cylinder allows you to analyze the design of equipment clamping devices and optimize their design parameters according to specified criteria. The piston's working and return stroke time are the components of the operating time for technological equipment, which determines its performance. The need to evaluate the performance of technological equipment determines the relevance of studying the dynamics of the clamping pneumatic cylinder.

References

1. Beater, P.: Pneumatic Drives. System Design, Modelling and Control. Springer, Berlin, Heidelberg (2007)

2. Fomin, O., Lovska, A.: Establishing patterns in determining the dynamics and strength of a covered freight car, which exhausted its resource. East. Eur. J. Enterprise Technol. **6**(7(108)), 21–29 (2020). https://doi.org/10.15587/1729-4061.2020.217162
3. Abrahamova, T., Bushuyev, V., Gilova, L.: Metal-Cutting Machine Tools. Machinery Engineering, Moscow (2012)
4. Shevchenko, S., Mukhovaty, A., Krol, O.: Gear transmission with conic axoid on parallel axes. In: Radionov, A.A., Kravchenko, O.A., Guzeev, V.I., Rozhdestvenskiy, Y.V. (eds.) ICIE 2019. LNME, pp. 1–10. Springer, Cham (2020). https://doi.org/10.1007/978-3-030-22041-9_1
5. Karpus, V.E., Ivanov, V.A.: Locating accuracy of shafts in V-blocks. Russ. Eng. Res. **32**(2), 144–150 (2012). https://doi.org/10.3103/S1068798X1202013X
6. Ivanov, V., Vashchenko, S., Rong, Y.: Information support of the computer-aided fixture design system. In: CEUR Workshop Proceedings 1614, pp. 73–86 (2016)
7. Sokolova, Y.: The synthesis of system of automatic control of equipment for machining materials with hydraulic drive. East. Eur. J. Enterprise Technol. **2**(2(68)), 56–68 (2014). https://doi.org/10.15587/1729-4061.2014.23396
8. Merzliakov, I., Pavlenko, I., Chekh, O., Sharapov, S., Ivanov, V.: Mathematical modeling of operating process and technological features for designing the vortex type liquid-vapor jet apparatus. In: Ivanov, V., et al. (eds.) DSMIE 2019. LNME, pp. 613–622. Springer, Cham (2020). https://doi.org/10.1007/978-3-030-22365-6_61
9. Lovska, A., Fomin, O.: A new fastener to ensure the reliability of a passenger car body on a train ferry. Acta Polytech. **60**(6), 478–485 (2020). https://doi.org/10.14311/AP.2020.60.0478
10. Pavlenko, I., Liaposhchenko, A., Ochowiak, M., Demyanenko, M.: Solving the stationary hydroaeroelasticity problem for dynamic deflection elements of separation devices. Vibr. Phys. Syst. **29**, 2018026 (2018)
11. Schön, H.: Compressed gas cylinders. In: Handbook of Purified Gases, pp. 143–153. Springer, Berlin, Heidelberg (2015). https://doi.org/10.1007/978-3-540-32599-4_6
12. Tan, K.K., Putra, A.S.: Servo hydraulic and pneumatic drive. In: Tan, K.K., Putra, A.S. (eds.) Drives and Control for Industrial Automation, pp. 9–44. Springer, London (2011). https://doi.org/10.1007/978-1-84882-425-6_2
13. Kundrák, J., Mitsyk, A., Fedorovich, V., Markopoulos, A., Grabchenko, A.: Simulation of the circulating motion of the working medium and metal removal during multi-energy processing under the action of vibration and centrifugal forces. Machines **9**(6), 118 (2021). https://doi.org/10.3390/machines9060118
14. Klinzing, G., Rizk, F., Marcus, R., Leung, L.: Control of pneumatic transport. In: Pneumatic Conveying of Solids. Particle Technology Series, vol. 8, pp. 459–474. Springer, Dordrecht (2010)
15. Kundrák, J., Mitsyk, A., Fedorovich, V., Markopoulos, A., Grabchenko, A.: Modeling the energy action of vibration and centrifugal forces on the working medium and parts in a vibration machine oscillating reservoir with an impeller. Manuf. Technol. **21**(3), 364–372 (2021). https://doi.org/10.21062/mft.2021.042
16. Chung, J.: Coordinated control of a new pneumatic gripper. In: Kim, J.-H., Yang, W., Jo, J., Sincak, P., Myung, H. (eds.) Robot Intelligence Technology and Applications 3. AISC, vol. 345, pp. 561–570. Springer, Cham (2015). https://doi.org/10.1007/978-3-319-16841-8_50
17. Rasskazova, Y.: Automation of control processes of technological equipment with rotary hydraulic drive. East. Eur. J. Enterprise Technol. **2**(2(80)), 44–50 (2016). https://doi.org/10.15587/1729-4061.2016.63711
18. Andrenko, P., Rogovyi, A., Hrechka, I., Khovanskyi, S., Svynarenko, M.: The influence of the gas content in the working fluid on parameters of the the hydraulic motor's axial piston. In: Ivanov, V., Pavlenko, I., Liaposhchenko, O., Machado, J., Edl, M. (eds.) DSMIE 2021. LNME, pp. 97–106. Springer, Cham (2021). https://doi.org/10.1007/978-3-030-77823-1_10

19. Sokolov, V.: Transfer functions for shearing stress in nonstationary fluid friction. In: Radionov, A.A., Kravchenko, O.A., Guzeev, V.I., Rozhdestvenskiy, Y.V. (eds.) ICIE 2019. LNME, pp. 707–715. Springer, Cham (2020). https://doi.org/10.1007/978-3-030-22041-9_76
20. Pavlenko, I.: Static and dynamic analysis of the closing rotor balancing device of the multistage centrifugal pump. Appl. Mech. Mater. **630**, 248–254 (2014). https://doi.org/10.4028/www.scientific.net/AMM.630.248
21. Prydalnyi, B.I., Sulym, H.T.: Mathematical model of the tensioning in the collet clamping mechanism with the rotary movable input link on spindle units. J. Eng. Sci. **8**(1), E23–E28 (2021). https://doi.org/10.21272/jes.2021.8(1).e4
22. Rogovyi, A., Korohodskyi, V., Khovanskyi, S., Hrechka, I., Medvediev, Y.: Optimal design of vortex chamber pump. J. Phys. Conf. Ser. **1741**, 012018 (2021)
23. Rogovyi, A., Korohodskyi, V., Medvediev, Y.: Influence of Bingham fluid viscosity on energy performances of a vortex chamber pump. Energy **218**, 119432 (2021). https://doi.org/10.1016/j.energy.2020.119432
24. Popov, D., Panaiotti, S., Ryabinin, M.: Hydromechanics. MSTU, Moscow (2014)
25. Sveshnikov, V.: Hydrodrives of Tools. Machinery Engineering, Moscow (2008)
26. Kovalevskyy, S., Kovalevska, O., Koshevoy, A., Tasić, I.: Using wave signatures for identifying mechanical objects. IOP Conf. Ser. Mater. Sci. Eng. **568**, 012117 (2019)
27. Popov, D.: Mechanics of Hydro- and Pneumodrives. MSTU, Moscow (2001)
28. Loytsyanskiy, L.: Mechanics of Liquid and Gas. Drofa, Moscow (2003)
29. Nuruzzaman, M.: Modeling and Simulating in Simulink for Engineers and Scientists. AuthorHouse, Bloomington (2004)
30. Tewari, A.: Modern Control Design with MATLAB and Simulink. John Wiley & Sons Ltd., Weinheim (2002)
31. Krol, O., Sokolov, V.: Modelling of spindle nodes for machining centers. J. Phys: Conf. Ser. **1084**, 012007 (2018)
32. Sokolov, V., Porkuian, O., Krol, O., Baturin, Y.: Design calculation of electrohydraulic servo drive for technological equipment. In: Ivanov, V., Trojanowska, J., Pavlenko, I., Zajac, J., Peraković, D. (eds.) DSMIE 2020. LNME, pp. 75–84. Springer, Cham (2020). https://doi.org/10.1007/978-3-030-50794-7_8

Check for updates

Synthesis Structural Scheme Self-adjusting Floating Bollard Ship Gateway

Ihor Sydorenko[1] (✉), Predrag Dasic[2], Vladimir Semenyuk[1], Valeriy Lingur[1], and Vera Salii[1]

[1] Odessa National Polytechnic University, 1, Shevchenko Ave., Odesa 65044, Ukraine
igs.ods@gmail.com

[2] Academy of Professional Studies Sumadija, 8, Kosovska St., 34000 Kragujevac, Serbia

Abstract. Floating bollards are one of the essential elements that determine the operability of a ship's lock. The growing demand for increasing the operability of this type of device under the action of unpredictable loads exceeding the nominal requires the introduction of new design solutions. At the same time, it is rational to present the bollard in the form of a passive self-regulating mechanical device. The article presents a synthesis of the structural diagram of an automated floating bollard based on the theory of modified kinematic graphs. The analysis of the obtained solutions to the synthesis problem is carried out. As a result, n optimal variant of the structural diagram of such a device is presented. Possible variants of constructive solutions for some units of the device corresponding to the synthesized structural diagram are shown. It was found that the presence of the proposed method for evaluating the input signal and its transformation for control corresponds to a self-adjusting system with a mechanical control system.

Keywords: Floating bollard · Structural diagram · Control system · Graph theory · Graph model · Industrial investment

1 Introduction

A modern shipping lock is a highly mechanized hydraulic structure with automatic control of the process of mooring ships. However, in this automated chain of passage, there are still a number of problems associated with mooring a vessel in an airlock. One of these problems is as follows.

When mooring large-capacity vessels, rigid and heavy mooring ropes are used. It is practically impossible to moor a ship with such tight ropes. Sagging ropes and stretching lead to significant movements of ships along and across the lock chamber. Such movements cause inertial jerks, which negatively affect not only the ropes but also the ship's mooring mechanisms and the mooring equipment of the lock itself. In addition, the very locking of heavy-duty vessels during the design mode of filling (emptying) the lock chamber often causes a sharp and significant excess of the permissible standard hydrodynamic force, which also leads to inertial jerks.

V. Tonkonogyi et al. (Eds.): InterPartner 2022, LNME, pp. 64–74, 2023.
https://doi.org/10.1007/978-3-031-16651-8_7

Notably, this negative phenomenon is significantly intensified in winter when the vessels themselves change their weight due to freezing. As a result, the kinetic energy of jerks, not absorbed by the mooring ties, leads to the breakage of the mooring lines and the destruction of both the hooks of the floating bollards (floating eyelets) and the floating bollards themselves. Moreover, the destruction of the structure of the bollard is often associated with the destruction of its metal structure and the destruction of the building structures of the gate itself, which are in direct contact with it.

In this regard, the design of new structures of mooring ties that have a new structure that can be adapted to changes in the magnitude of the action of external loads, preventing possible destruction, is a prerequisite that determines the efficiency and reliability of these devices.

2 Literature Review

One of the main elements of the lock's mooring ties is a floating bollard (Fig. 1a). To assess its possible main damage under the action of inertial jerks and associated overloads, it is advisable to consider it in more detail. Bollard consists of a float 1, interconnected and structurally identical to the lower (under water) and upper 2 carriages [1, 2].

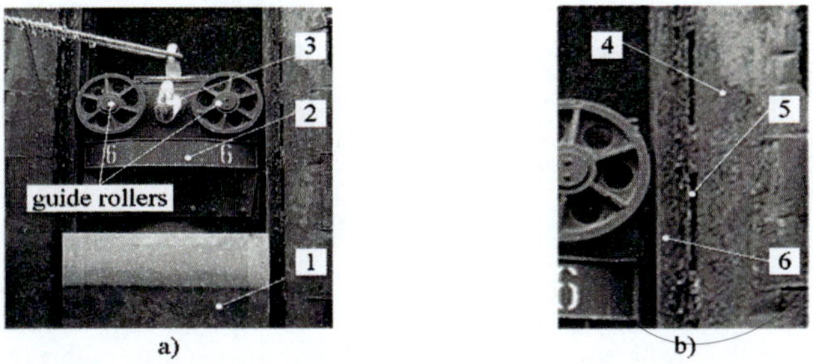

Fig. 1. Floating bollard: general view of the device (a); bollard guide system (shaft) in the airlock (b).

The upper bogie 2 has an eye (hook) 3 for attaching the mooring cable. The structure moves along the guide system (shaft) 4 of the airlock. The guiding system consists of embedded elements 5 in the reinforced concrete structure of the sluice and external guides of the metal structures 6 (Fig. 1b).

There are two main reasons for the failure of floating bollards at the place of their occurrence [3, 4]. One of the reasons is the destruction in the bollard structure itself, which manifests itself in the destruction (critical plastic deformation) of both the eye itself and the metal structures of the bollard carriages 2, especially in the places where the guide rollers are attached (Fig. 1a). Such destruction of a floating bollard accounts for 53% of the recorded cases [5]. Equivalent in terms of negative is another reason for the failure of the bollard, namely, the destruction of the bollard shaft, which was recorded in 42% of cases [4, 5]. Such destruction manifests itself in the form of destruction of concrete 4 in the places where the guides of the shaft of the bollard 5 are attached or destruction (critical plastic deformation) of the guides 6, which causes the bollard to jam in the shaft (Fig. 1b). In 5% of cases, simultaneous destruction of both the bollard itself and the elements of its shaft was recorded [4, 6].

It has been established that the leading cause of the presented destructions is the inertial jerks of the moored vessel [7]. Based on this, it follows that fighting against the negative manifestations of inertial jerks in bollard structures is an urgent task [8].

Several design schemes are known that determine the design of floating bollards. In some of them, to reduce inertial jerks, the bollard eye (hook) is spring-loaded [9], in another design – bollard rollers are assembled in the form of units including elastic scattering elements [10]. There are bollards in which the indicated technical solutions are implemented in a single design [11].

Differences in the available structural diagrams of bollards determine the differences in the loading of the bollard's supporting elements in the form of its guide rollers (Fig. 2). In the case of a conventional bollard, which is called a "classic" structure bollard, the inertial jerk on the eyelet is directly transmitted to the guide rollers of the bollard and the guides in contact with them in the shaft (Fig. 2a).

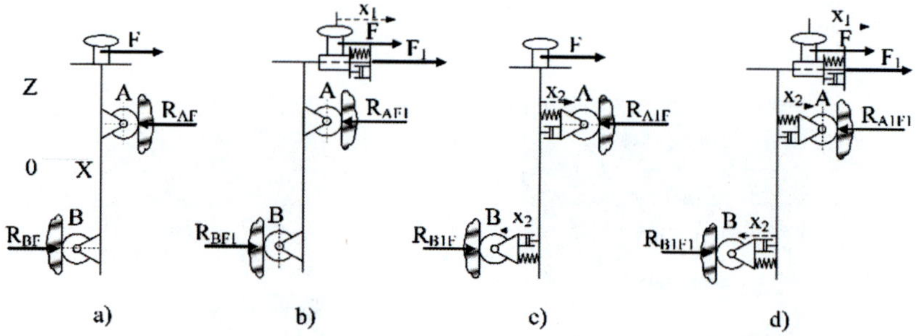

Fig. 2. Structural schemes of bollards: "classical" scheme (a); the eye (hook) of the bollard is spring-loaded (b); bollard rollers with elastic-dissipative elements (c); combined scheme (d).

When the bollard eye is spring-loaded, the inertial jerk applied to it will be slightly reduced from the value F to F1 due to the presence of an elastic-dissipative element, which determines a slower increase in its value and partial damping (Fig. 2b). In this case, the reactions on the bollard rollers R_{AF1}, R_{BF1} and the shaft guides in contact with them will be slightly lower than the R_{AF} R_{BF} in the case of the "classic" structure.

Considering the structural diagram, when the bollard rollers are assembled in the form of assemblies including elastic scattering elements, it should be noted that the reactions on the R_{A1F} R_{B1F} bollard rollers and the shaft guides in contact with them will also be lower than in the case of the "classical structure" (Fig. 2c).

A decrease in reactions in bollard rollers $R_{A1}F_1$ $R_{B1}F_1$ and guide joints in contact with them in the case of a combined bollard structure – also takes place (Fig. 2d).

However, the last two of the presented schemes have not found widespread use because the compliance of the roller units determines the linear displacements x_2, which negatively affects the device's operability due to its distortions in the mine. In addition, the presence of a large number of flexible bollard elements, especially in the case of its combined structure, adds other negative dynamic processes to the oscillatory system of the ship-rope-bollard-pier. Nevertheless, there is a positive experience of using bollards with a spring-loaded eyelet [12].

But let's try to approach the work of a spring-loaded hinge not just as a bollard element that dampens inertial jerks but somewhat from the other side. The amount of deformation of the elastic element of the eyelet $\times_1$ is related to the force causing this deformation, e.g., proportionally, in the case of using a compression spring as an elastic element (Fig. 2b). In other words, in this case, we have a device for monitoring the external situation, in which the amount of deformation of the elastic element of the eyelet can be considered as a controlled signal. The eyelet itself will be in some way a sensor of this signal. In this regard, having a controlled signal and its defining element, we received a basis for creating the structure of a particular controlled device.

In the structures of controlled mechanisms, a direct or inverse relationship exists between the controlled signal and some corrective effect associated with this signal [13]. In the case of a bollard, the corrective effect should be aimed at reducing the negative manifestation of inertial jerks, which are manifested over the permissible design loads on the bollard rollers and the shaft guides in contact with them. According to the authors, one solution can be considered a corrective effect. The resistance of materials and construction mechanics shows that the beam's reaction depends on the support's location relative to the applied force [14]. For example, consider a statically definable beam as the simplest bollard model (Fig. 3a). Let us take its basic geometric parameters, which determine the position of the supports and the point of application of the force, close to real bollards a = 0.2 m, b = 0.6 m, c = 0.9 m.

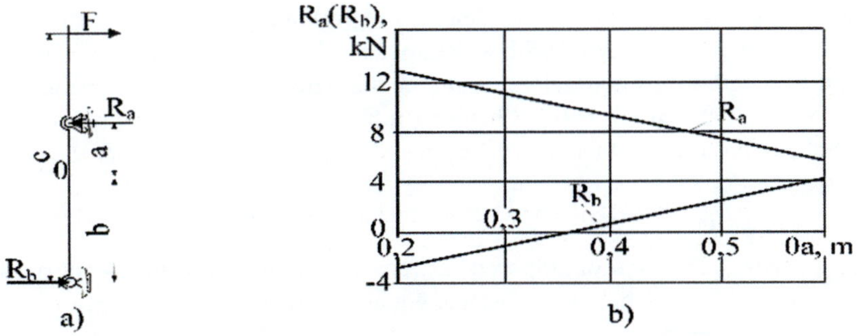

Fig. 3. Reaction in beam supports with variations in their location: calculation scheme (a); graphical interpretation of calculation results (b).

Graphical interpretation of the results of calculating the reactions R_A and R_B in the case of loading the model with a constant force F = 10 kN and variation with the parameter. a = 0.2… 0.6 m indicates their linear change (Fig. 3b). Moreover, it was found that the change in reactions occurs in a wide range, for example, at 0a = 0.2 m corresponding to R_A = 12.89 kN and at 0a = 0.6 m – R_A = 5.67 kN.

Based on this, it can be concluded that creating a bollard structure, in which the supports would automatically occupy the optimal position, which determines the minimum value of reactions under the action of some external load, is a possible solution to a rather significant scientific and applied problem.

3 Research Methodology

The type of controlled signal predetermines the study methodology and its defining element mentioned above.

There is quite a lot of experience in using the deformation value of a particular elastic element (kinematic signal) as a control signal for automatic (including self-adjusting) devices [15]. Typically, this type of signal is converted into an electrical signal with subsequent conversion into a control signal using a control system. This is one of the main features of many active electro-mechanical self-controlled systems, which have advantages and disadvantages. In some cases, the main disadvantages of such systems include their connection to energy sources, response inertia, dimensions, complexity, and operation cost.

However, a scientific direction has recently been developed on the synthesis and analysis of the structures of passive self-adjusting devices in which a controlled mechanical signal is converted into a mechanical corrective action without changing its nature [16, 17]. For this, a technique for analyzing and synthesizing structures of self-adjusting mechanical devices was developed and tested on the basis of the theory of modification of kinematic graphs [17]. The main provisions of this methodology determine three main conditions that the model of a self-regulating mechanism represented by a modified kinematic graph must meet.

The first condition states that the idealized representation of the structure of a conventional planar passive device with a parameter subject to controlled change (which is caused by a self-adjusting device) cannot be a bipartite modified kinematic graph or a modified kinematic graph with a "hanging top". In this case, the degree of mobility calculated based on the elements of the corresponding graph must satisfy the equality

$$W = 3(p_\Sigma - 1) - 2q_5 - q_4 - q_c = 1 - q_a \tag{1}$$

where p_Σ – the total number of graph vertices corresponding to the number of rigid links in the device; q_5 – the number of graph edges equal to the number of kinematic pairs of the 5th class; q_4 – the number of graph edges equal to the number of kinematic pairs of the 5th class; q_c – the number of graph edges equal to the number of controlled parameters of a non-mechanical nature (elastic, dissipative, etc.) associated with a particular mechanical movement; q_t – the number of graph edges equal to the number of tuning mechanical movements associated with a change in specific parameters of both mechanical and non-mechanical nature.

The presence in expression (1) of the connection q_c determines the modification of the usual kinematic graph. According to the applied method, this connection between the rigid links and the device is considered not to affect the kinematic characteristics determining its performance and can be considered as a virtual kinematic pair of the 4th class. It should be noted that although we are talking about controlled and variable parameters of a non-mechanical nature, these changes are always associated with some mechanical movement. So, for example, the controlled overlap of the throttle hole of the shock absorber changes its dissipative properties, and a change in the position of the elastic element with respect to the line of action of the force causing its deformation changes the reduced stiffness of such an element.

According to the second condition, the nature of self-control by the properties of a device, when simulating it with a modified kinematic graph, can be judged using an indicator in the form of a cyclomatic number, which must satisfy the inequality

$$\sigma = q - p + 1 \geq 2 \tag{2}$$

where q is the number of graph edges regardless of their type; p is the number of vertices in the graph.

Remarkably, the cycles due to the presence of bipartite parts of the graph are not considered when checking by condition (2). The cyclomatic number of the mechanism model, represented by the graph, shows the corresponding number of closed cycles of the relationship between the elements of the mechanism. Moreover, one of the cycles corresponds to the interaction that determines the main functional value of the device (main cycle), and the rest are interactions of the tuning nature (tuning cycles).

Given the importance of this circumstance, the third condition interprets the following. The formed cycles, both the main and the tuning character, must have a standard edge that defines the parameter or properties to be changed.

The fulfillment of these conditions determines the graphic model of the self-adjusting structure, which can be further transformed into a kinematic diagram with subsequent constructive implementation in the prototype of the device.

For example, consider a bollard with a spring-loaded ring, which can be adopted as a prototype for the subsequent synthesis of a self-controlled device (Fig. 4a).

The apparent uncontrolled state of this mechanical system is entirely consistent with the accepted methodology and is confirmed by the presence in the graph model of the device under consideration of both the hanging vertex p2, which determines the bipartite graph p2 – qc – p1 – q5, and the non-fulfillment of condition (1), which, with an appropriate structure of terms, takes the form

$$W = 3(5-1) - 2 \cdot 5 - 0 - 1 = 1 \neq [1 - q_t] = 1 - 1 = 0 \tag{3}$$

The state of the mechanism in terms of manageability will change if explicit controls are added to its structure. So, the use of a lead screw p_5 and two kinematic pairs of the 4th class q_4 to implement the previously adopted control method in the form of moving the supports (transformation of φ to x_2), as already mentioned above, will cause changes in the graphic model of the device and according to certain conditions (Fig. 4b).

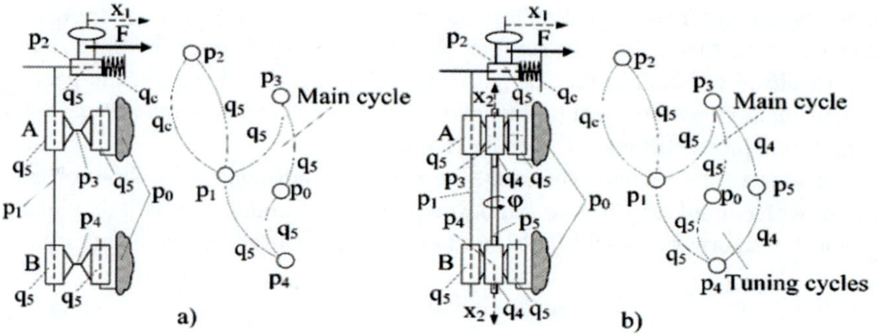

Fig. 4. Bollard modeling modified kinematic graphs: bollard with a spring-loaded eye (a); bollard with a spring-loaded eyelet and an element for adjusting the position of the supports (b)

In this case, a "hanging" vertex of the graph p2 indicates the absence of self-government in the mechanism. However, condition (1) which, with an appropriate structure of terms, takes the form

$$W = 3(6-1) - 27 - 0 - 1 = 1 \neq [1 - q_t] = 1 - 1 = 0 \quad (4)$$

In addition, the calculation of the cyclomatic number by expression (2) shows that if we exclude the "hanging" vertex, then the model will contain two cycles

$$\sigma = 6 - 5 + 1 = 2 \geq 2. \quad (5)$$

These results can be taken as reference information when synthesizing a new device's structure, which fully meets the three main conditions of the adopted technique.

4 Results and Discussion

To carry out the synthesis, let us consider as a basic block diagram and a model of a device adopted as a prototype (Fig. 4b). According to the accepted methodology, the modified kinematic graph of the synthesized device should not have a "hanging" vertex, in our case the vertex p_2.

This can be achieved by including this pole in some of the existing cycles, or by creating a new cycle if the number of "hanging" vertices is more than two. In our case, we have one hanging vertex, so we will include this vertex in the existing cycle. The condition for including the vertex in the cycle will be regulated by expression (1) and is multivariate.

Let us consider the first option, suppose that kinematic pairs of the 5th class will expand the structural diagram of the device. Then, from (1), it follows that the number of corresponding arcs q_5 determining the increment of the graph model by the poles n has the following relationship

$$q_5 = 0.5(3(p_{f\Sigma} + n - 1) - q_4 - q_c - 1 + q_a), \quad (6)$$

where $p_{f\Sigma}$ is the number of poles in the base model.

From expression (6), it follows that the addition of one pole (n = 1) to the model, corresponding to the expansion of the structure of the basic device by one link, will require the presence of two arcs q5, corresponding to the expansion of the structure of the basic one by two kinematic pairs of the 5th class. In this case, the problem of the "hanging" vertex of the graph can be solved (Fig. 5a).

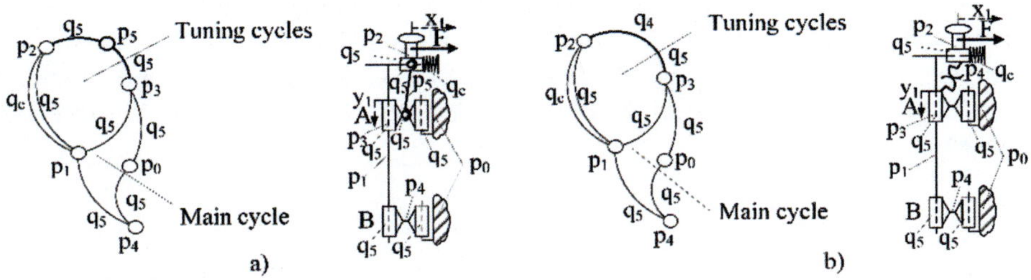

Fig. 5. Modified kinematic graph and bollard block diagram for synthesis: variation by the poles of the graph (a); variations by the edges of the graph (b).

In this case, the result of this synthesis will correspond to three main conditions that determine the self-regulation mechanism according to the adopted method, namely, mobility and cyclomatic number.

$$W = \left[3(\underline{\underline{6}} - 1) - 2 \cdot \underline{\underline{7}} - 0 - 1\right] = 0 = \left[1 - q_t\right]; \sigma = [5 - 6 + 1] = 2 \geq 2, \quad (7)$$

as well as the formed cycles, both basic and tuning, have a common edge that determines the parameter or properties to be changed (Fig. 5a).

The second synthesis variant can be built on the graph model's variation of the number and type of arcs. So, if we put n = 0 and p_5 = const as the initial condition, then according to expression (6), the problem of the "hanging" vertex can also be class 4 kinematic pair (Fig. 5b).

Moreover, the obtained synthesis result will also meet three accepted basic conditions that determine the self-adjusting mechanism in terms of mobility and cyclomatic number

$$W = \left[3(5 - 1) - 2 \cdot 5 - 0 - \underline{\underline{2}}\right] = 0 = \left[1 - q_t\right]; \sigma = [6 - 5 + 1] = 2 \geq 2, \quad (8)$$

as well as the formed cycles, both basic and tuning, have a common edge that determines the parameter or properties to be changed (Fig. 5b).

Although the solutions obtained in the process of the carried out synthesis determine a passive self-adjusting mechanism. However, such a mechanism cannot be called optimal from the point of view of its structure. So, the proposed solutions determine the mobility of only one support A, and support B, which has the same structural purpose, remains uncontrollable. In this regard, it is possible to somewhat extend the synthesis problem by including the pole p_4, corresponding to support B, in a certain control cycle. According to the indicators of the graph model, this action corresponds to an increase in the number of existing cycles by one.

The joint solution of the system of equations is composed of expressions (1) and (2) under the condition: variation by the edges of the graph, $\sigma = 2 + 1 = 3$; $n_{min} = 1$; expansion of the structure only due to kinematic pairs of class 4 – allows you to install an additional number of elements of the graph model. From the solution, a variant of the graph model +1p and +q4 was obtained, corresponding to the block diagram's extension with one more link and a kinematic pair of the 4th class. And compliance with the third condition of the methodology used and the synthesis condition for increasing the cyclomatic number – allows you to get the relationship between the existing and emerging poles of the graph model. At the same time, the graph model and structural diagram of the synthesized device fully correspond to the accepted synthesis problem (Fig. 6).

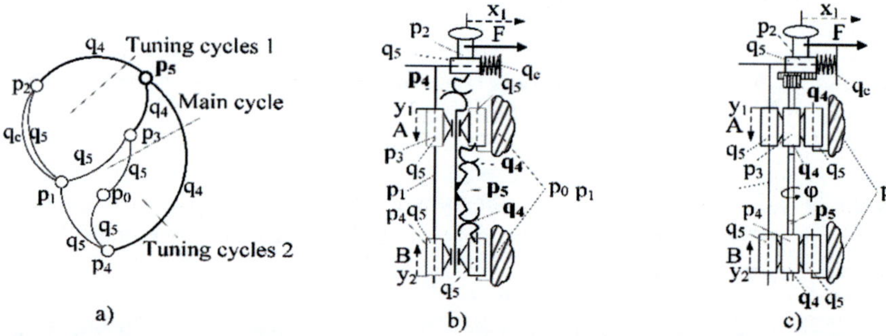

Fig. 6. Synthesis of a block diagram of a self-adjusting floating bollard: a model in the form of a modified kinematic graph (a); block diagram (b); block diagram with some design solutions (c).

After analyzing the obtained solution, some similarities in its block diagram can be noticed, including elements of kinematics (Fig. 6c) with the previously considered scheme of an uncontrolled device (Fig. 4b). This similarity makes it possible to describe the functional interaction between the poles of the graph model and also to determine their implementation by some constructive solutions in an actual structure. So, the external load leads to deformation of the elastic element, which determines the linear and proportional to the load controlled movement x_1 of the element with the hook (pole p_2) and the rope attached to it relative to the general frame of the device (pole p_2).

The arc q_4 between the poles p_2 and p_5 corresponds to the connection between the elements included in the control cycles and determines the transformation of the translational motion x_1 into rotational φ. In this case, the arc q_4 in an actual design can be implemented as a toothed rack – a toothed gear. The presence of the p_5 pole, included in the control cycles with the p_2 and p_3 poles and based on the conditions of its further intended use, characterizes it as a lead screw. Then the problem of two arcs q_4 in control cycles consists of the inverse transformation of rotational motion φ into translational motion y_1 и y_2. This, in turn, determines their possible implementation in an actual design. So, if these arcs are implemented as running nuts, then, depending on the direction of the thread and its pitch, this approach allows you to obtain different patterns of relationship between the poles p_5, p_2, and p_3 (displacements y_1 and y_2 for supports A and B, respectively).

In the authors' opinion, it is possible to significantly expand the possibilities of the relationship under consideration in the case of implementing these arcs in the form of elements of cam rotational mechanisms. In this case, variations are possible in the magnitude of the displacement and its direction. Moreover, the function describing the curvature of the cam guides of such a mechanism can be considered a function of the control algorithm.

After considering the modern development of technologies, the proposed design solutions may well be implemented in the actual design of such a device with a slight increase in its cost concerning the existing ones. However, given its apparent advantages, it can be argued that the decision to create such a device would be justified.

5 Conclusions

At the same time, the following was established. It was found that when synthesizing a self-adjusting device with a mechanical control system, it is necessary to have an element whose deformation does not affect the device's performance and is part of the functional interaction between its elements.

Also, the most optimal element, the deformation of which can be taken as a controlled signal, should be recognized as coiled springs since the magnitude of their deformation is proportional to the load applied to them.

It is rational to solve the synthesis problem by expanding the structure of the mechanism by kinematic pairs of the 4th class since the number of additional links required to solve such a problem will be minimal.

When solving the synthesis problem, it is necessary to have all the elements that determine the setting of the mechanism in the corresponding tuning cycles of its model in the form of a modified kinematic graph.

The constructive implementation of kinematic pairs of the 4th class in the synthesized mechanism depends on the type and required regularity of the implemented corrective movement. It is rational to perform such pairs as gear mechanisms to solve some problems. When the control algorithm is rather complicated, then, as an alternative, it is possible to use cam mechanisms with kinematic closure.

References

1. Felski, A., Zwolak, K.: The ocean-going ships-challenges and threats. J. Marine Sci. Eng. **8**, 41–50 (2020)
2. Yang, S., Ringsberg, J.: Towards the assessment of impact of unmanned vessels on maritime transportation safety. Realiab. Eng. Syst. Saf. **165**, 155–169 (2017)
3. Bergdahl, L., Palm, J., Eskilsson, C., Lindahl, J.: Dynamically scaled model experiment of a mooring cable. J. Mar. Sci. Eng. **4**, 5–12 (2016)
4. Das, S.N., Kulkarni, S., Kudale, M.D.: Design of safe mooring arrangement for large oil tankers. Proc. Eng. **116**, 528–534 (2014)
5. Hsu, W.K.: Assessing the safety factors of ship berthing operations. J. Navig. **68**, 576–588 (2015)

6. Johanning, L., Smith, G., Wolfram, J.: Measurements of static and dynamic mooring line dampingand their importance for floating WEC devices. Ocean Eng. **34**, 1918–1934 (2007)
7. Martinelli, L., Ruol, P., Cortellazzo, G.: On mooring design of wave energy converters: the Seabreath application. Coast. Eng. Proc. **1**, 3–18 (2012)
8. Spanos, P.D., Arena, F., Richichi, A., Malara, G.: Efficient dynamic analysis of a nonlinear wave energy harvester model. J. Offshore Mech. Arctic Eng. **138**, 40–49 (2016)
9. Tsukrov, I., Eroshkin, O., Paul, W., Celikkol, B.: Numerical modeling of nonlinear elastic components of mooring systems. IEEE J. Ocean. Eng. **30**, 37–46 (2005)
10. Xu, Z., Huang, S.: Numerical investigation of mooring line damping and the drag coefficients of studless chain links. J. Mar. Sci. Appl. **13**(1), 76–84 (2014). https://doi.org/10.1007/s11804-014-1235-0
11. Chai, Y., Varyani, K., Barltrop, N.: Semi-analytical quasi-static formulation for three-dimensional partially grounded mooring system problems. Ocean Eng. **29**, 627–649 (2002)
12. Pascoal, R., Huang, S., Barltrop, N., Soares, C.G.: Equivalent force model for the effect of mooring systems onthe horizontal motions. Appl. Ocean Res. **27**, 165–172 (2005)
13. Kraynev, A.F.: Mechanics (the Art of Building) Machines. Fundamental Dictionary Engineering, Moscow (2014)
14. Ambrose, J.: Simplified Mechanics and Strength of Materials. John Wiley & Sons, New York (2012)
15. Chiu, T.: An improved algorithm for the construction of generalized kinematic chains. Mechanism Mach. Theory **78**, 229–247 (2014)
16. Zhang, Y., Sydorenko, I., Tonkonogyi, V., Bovnegra, L., Dašić, P.: Structural analysis of direct passive pressure reducing valves using modified kinematic graphs. In: Karabegović, I. (ed.) NT 2020. LNNS, vol. 128, pp. 114–121. Springer, Cham (2020). https://doi.org/10.1007/978-3-030-46817-0_13
17. Sydorenko, I.: Passive Vibration Isolation Device with Elements of Active Systems. Palmarium Academic Publishing, Saarbrücken, Germany (2011)

Manufacturing Technology

An Increase in Heavy Machines' Accuracy by Controlling the Carrier System Parameters

Yana Antonenko(✉), Viktor Kovalov, Yana Vasylchenko, Maxim Shapovalov, and Nikolay Malyhin

Donbass State Engineering Academy, 72, Akademichna Street, Kramatorsk 84313, Ukraine
yana.s.antonenko@gmail.com

Abstract. The main objective of this work is to solve the actual scientific and technical problem of increasing the heavy machine's accuracy by providing the rigidity of the supporting system at the minimum possible weight. There were investigated the parameters of heavy lathe frames on full-scale samples. The displacements of frames along coordinate axes were analyzed as a result of their loading. The results of specific torsional resistance of the tested frames were obtained. A method of investigating the composite frames' accuracy using mathematical modeling has been developed. The technology of designing the carrier systems for heavy lathes using the results of preliminary calculation on contact deformation (internal forces, displacements) as boundary conditions, allowing to obtain the design geometry of section with the minimum possible mass while maintaining the given norms of productivity and accuracy of machining has been developed. The recommendations on the design and modernization of heavy lathe frames were given.

Keywords: Heavy lathes · Welded frame · Cast frame · Carrier system · Torque · Industrial growth

1 Introduction

A feature of modern mechanical engineering is a significant increase in the nomenclature of manufactured products, reducing their seriality and the production cycle period. The variations of loads on the components of a heavy machine tool during roughing and finishing [1] and temperature changes cause changes in the conditions of interaction between the tool and the workpiece during machining [2]. Generally, they characterize the instability of the process in time, reducing the system accuracy [3], which directly depends on the dynamic properties of the heavy metal-working equipment.

The variety of processes that occur in such a complex system requires serious scientific research in solving problems of improving machining accuracy [4].

V. Tonkonogyi et al. (Eds.): InterPartner 2022, LNME, pp. 77–89, 2023.
https://doi.org/10.1007/978-3-031-16651-8_8

The traditional approach in the study of machining accuracy usually concentrates on different aspects of the accuracy problem, that is, to limit the research to individual issues of the technological and design aspects [5], rather than a comprehensive solution to the problem.

In this regard, the analysis and development of theoretical methods that simulate machining processes on heavy machines, taking into account the dynamic errors introduced by the carrier system [6], working elements, and the cutting process, are directed to reduce the deviation of the machined part, as well as improving accuracy by rationalizing the design of elements of the carrier system while reducing metal intensity [7] is an actual problem of mechanical engineering.

2 Literature Review

Based on the analysis of works of national and foreign scientists and a detailed study of approaches to the design of heavy machines, reliability, and accuracy of technological systems, the following conclusions can be made:

1. With decreasing tolerance for machining parts on a CNC lathe, the proportion of error due to geometric inaccuracy of the machine tool and its kinematic errors becomes determinative [7].
2. The geometrical inaccuracy of the machine tool, as well as its kinematic errors, whose components are errors of the position encoder [8], errors caused by zero drift of elements of electronic circuits, torque error; non-linearities included in the closed loop of the servo drive [9]; accuracy of the loop processing at sharp changes of input influences [10]; errors caused by the control system transfer function [11]; CNC system errors [12] and others influence the accuracy of processing of parts on a CNC lathe.
3. Increasing the accuracy of machining parts on a CNC lathe is possible through the use of such methods as adjustment of machine parameters of the CNC system, rationalization of the design of the machine's carrier system, use of combined systems of feed drives [13], creation of an adaptive system [14], etc.

There are three main directions among the existing ways to improve the accuracy of machine tools. The first one, the classical one, consists of the fact that to obtain the required accuracy of machining on the machine tool, it is necessary to increase to the required level the accuracy of the machine units themselves, and as a rule, it must be an order of magnitude higher than the required machining accuracy.

The second direction is a set of various original design solutions and technological methods, allowing under certain conditions to achieve high accuracy of machining at a low accuracy of the equipment.

The most promising at present is the third direction - the creation of adaptive automatic control systems [9]. The principle of adaptivity consists in obtaining information about the parameters of the technological process and external disturbing factors through a set of sensors and the subsequent application of this information for adequate intervention during the technological process.

In terms of implementing adaptive control systems, there are two ways to improve the accuracy characteristics:

- automatic control of the elements of elastic systems of metal-cutting machines that is, the adaptation of carrier systems to changing operating conditions;
- automatic control of the machining process, taking into account the error of the machined product.

This paper proposes a methodology to improve the accuracy of heavy lathes, considering both directions.

3 Research Methodology

The methodological basis of the work is a comprehensive approach to studying the process of machining parts on heavy lathes, their conditions and features, and the laws of the processes [10].

Reduced rigidity [3] and considerable masses of moving assemblies [2], making it difficult to measure some values, a complex pattern of the irregular distribution of rigidity and temperature of carrying frames in different directions, the domination of weight loads over cutting forces, features of the tool used and cutting modes - all these determine the specifics of the dynamic characteristics of heavy lathes [13].

The support system is the most specific system of the machine, and its characteristics determine the interaction of all its components. The main criteria of the load-bearing system performance are stiffness and vibration resistance in the sense of providing the possibility of stable operation of the machine at the specified modes and limiting the amplitudes of forced vibrations to the permissible limits [15].

Considering that in the machining process, the geometric errors of the frame guides are copied onto the finished workpiece, making a significant contribution to the form error, the frame is of particular interest to us (Fig. 1).

Fig. 1. The frame of the heavy-duty lathe.

In modern automated production, various nomenclature and complex configuration items are produced during machining on CNC lathes [6]. Estimating the stress-strain state of such products is quite important, and for large-size non-rigid parts, it becomes the most relevant [8].

To comprehensively consider the deformation of workpiece elements, it is necessary to involve modern methods for analyzing the stress-strain state based on representations of solid media mechanics. The most suitable method for this purpose is the finite element method (FEM), which has been taken as the basis for mathematical modeling and calculating distributed stiffness matrices at the contact point of the tool and the workpiece.

By knowing the distribution of stiffness characteristics along the machining axis, it is possible, using the developed algorithm [9], to calculate the cutting system alignment trajectory variations during the machining process using the developed algorithm [16].

The most characteristic errors in machining on heavy machines have been considered [13]. The size error is identical in its properties to the dimensioning of the technological system, is caused by its change, and can therefore be compensated by its change. In contrast, the surface shape error can be completely independent of the dimensional configuration of the technological system [17]. For this reason, surface shape error is a much more serious problem than size error, especially for heavy machines [8]. This point is also supported by an analysis of experience with heavy machinery in production environments. This analysis identified and classified the factors that significantly influence each particular type of error and found that the most significant factor in assessing the balance of accuracy is the accuracy of shape in the longitudinal section. Heavy machine frames are connected to several sections to form a "solid" supporting structure [9].

4 Results and Discussion

A welded frame of a heavy lathe was designed. The carrying system of the heavy lathe consists of two stands – the frame of the support and the frame of the workpiece. Each of the frames consists of two sections connected by bolts. The length of each section is 7.8 m. Based on calculating the limits of the distributed loads acting on the frames, the simulation of the deformation strength by the finite element method using the Cosmos Works tool package was conducted. The simulation was performed for the cases that have the most significant impact on the deformation value of the load-bearing system: the position of the support in the middle of the frame section (modeling results are shown in Fig. 2a), when the load at the joints of the sections (Fig. 2b) and at the edge of the frame (Fig. 2c).

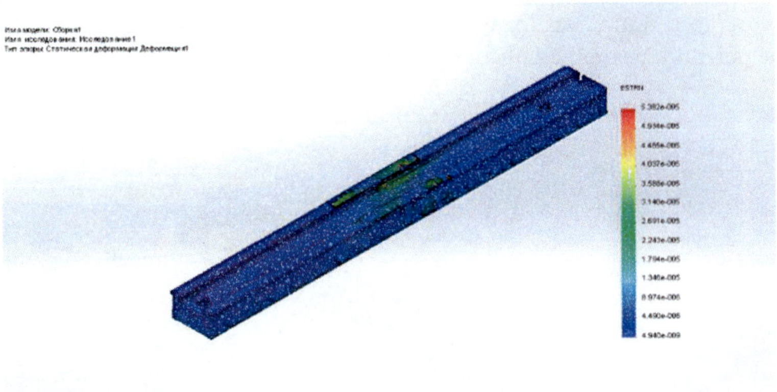

a)

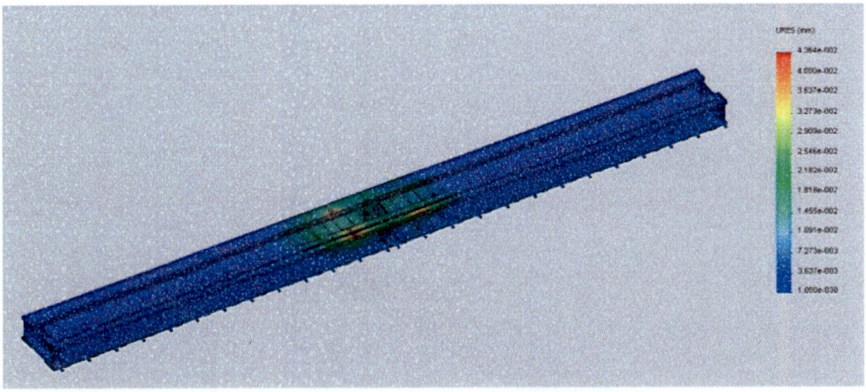

b)

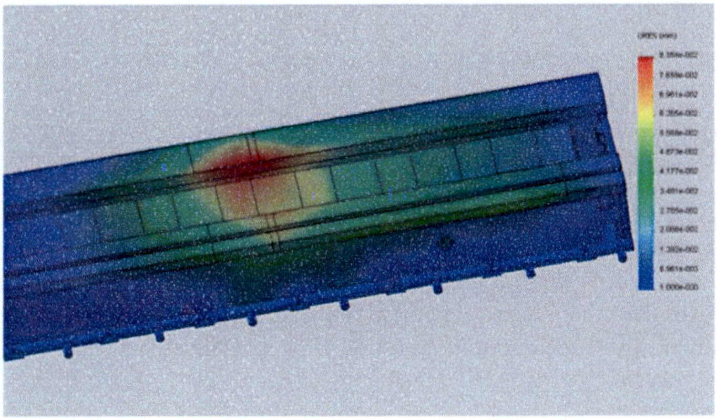

c)

Fig. 2. Tensed-deformed condition of the frame (a) in the middle of the section, (b) at the junction of sections, and (c) at the edge of the section.

The workpiece frame was loading in two places: on the main section of the frame and on the position where the tailstock was placed (Fig. 3).

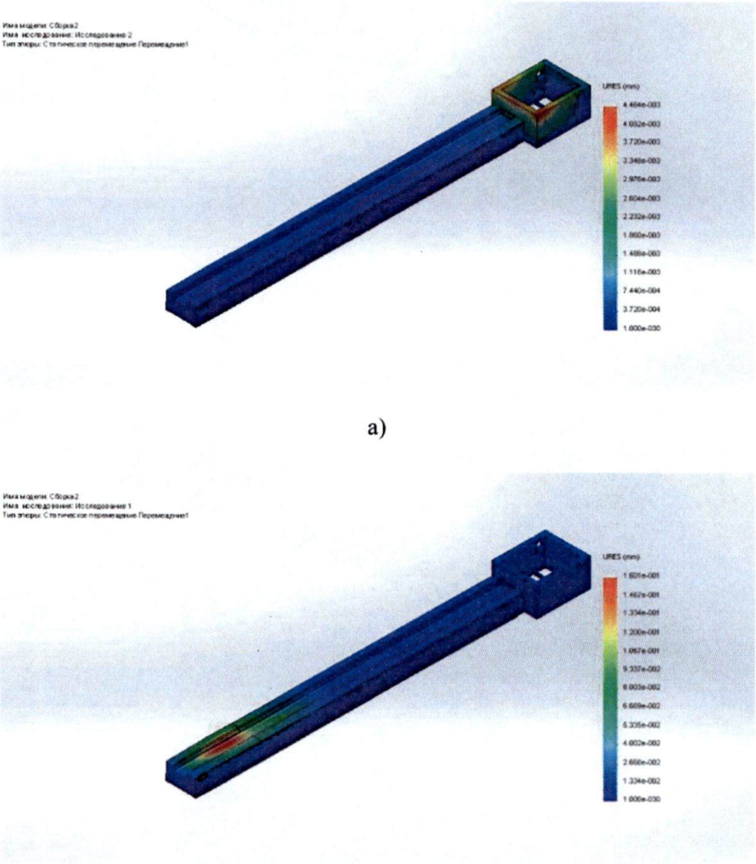

a)

b)

Fig. 3. Tensed-deformed condition of the frame of the workpiece (a) of the main section, (b) in the location of the tailstock.

Analyzing the diagrams, we can conclude that the entire support group of components passed the test, as the maximum stress state is in the green zone of the scale, corresponding to the nominal strain gauge. The state of a welded frame of a heavy lathe under the influence of force and temperature loads was simulated (Fig. 4).

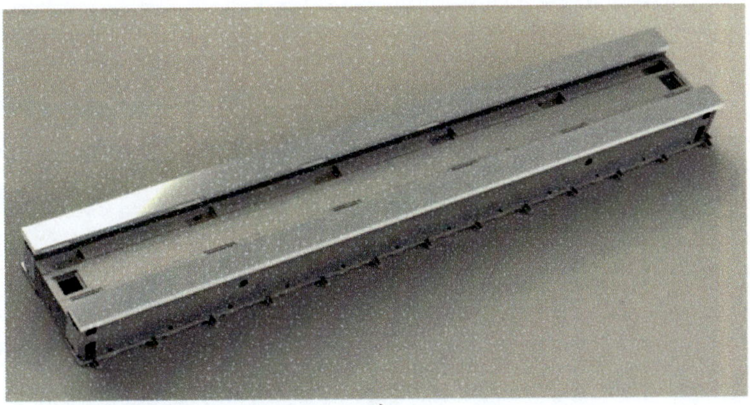

a)

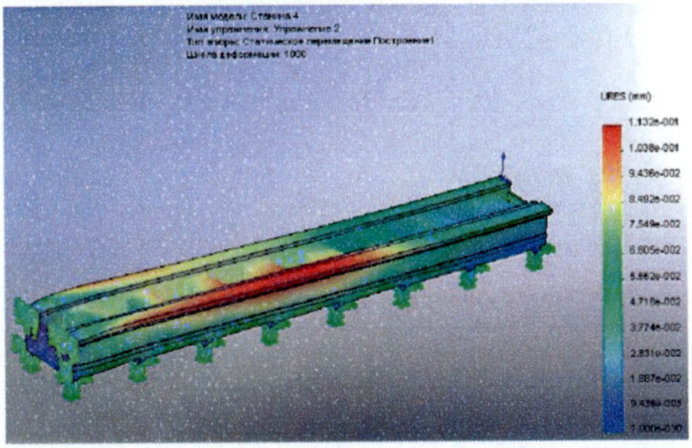

b)

Fig. 4. 3D model (a) and diagram (b) of the welded section of the frame.

To determine the optimal design of a 100-ton heavy-duty lathe, a comparative analysis of the results of computer simulation of the welded frame and full-scale tests of the cast frame of heavy-duty lathes was carried out (Fig. 5).

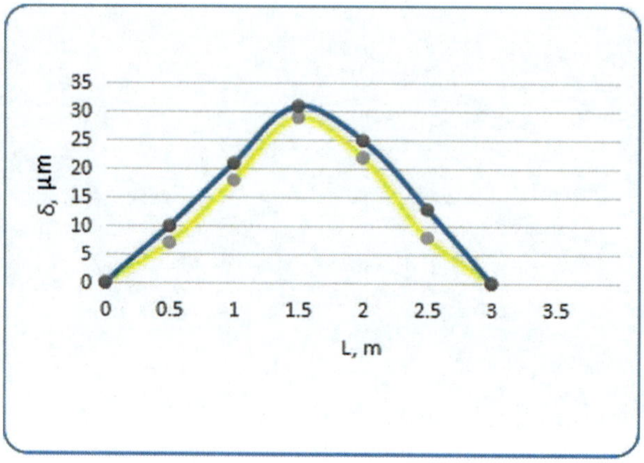

Fig. 5. Comparison of the results of field tests and computer simulation.

The tests of the cast frame were carried out in the laboratory at PJSC KZTS, in the full-size frame section.

Tests and modeling were carried out in two directions: the frame deformation in the "spacer" was displaced, and the frame deflection under its loading was determined (Fig. 6).

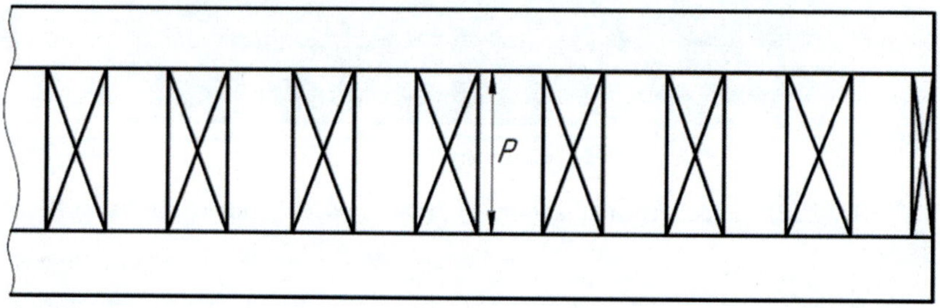

a)

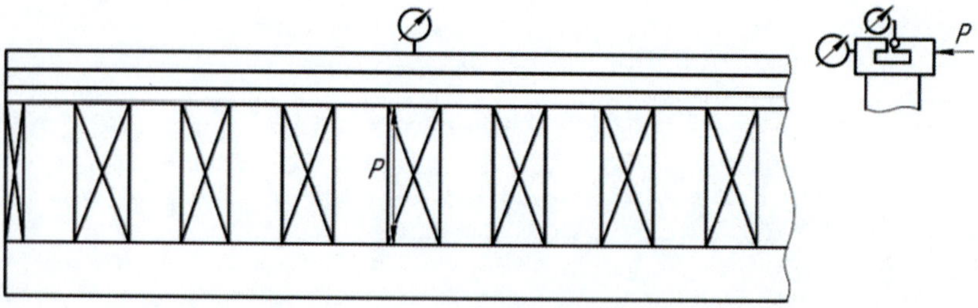

b)

Fig. 6. Scheme of welded frame extension on transverse braces (a), between transverse braces (b).

To confirm the accuracy of the results obtained by simulation of the welded frame, a comparative analysis of the simulation results with the results of full-scale tests of heavy model 1K670 lathes with a welded frame of similar design was carried out (Figs. 7 and 8).

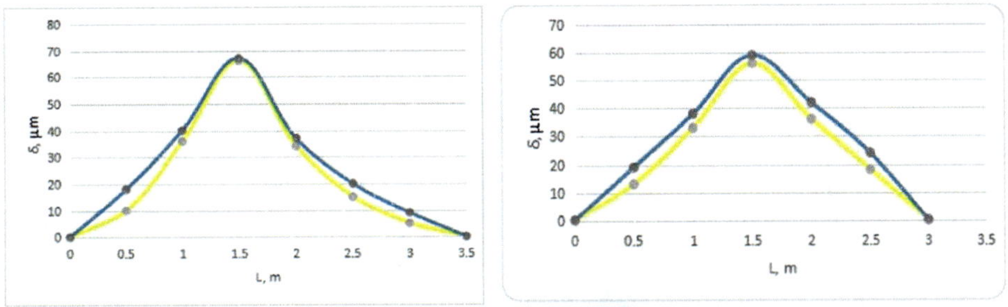

a) b)

Fig. 7. Comparison of the results of field tests and computer simulation extension on transverse braces (a); between transverse braces (b).

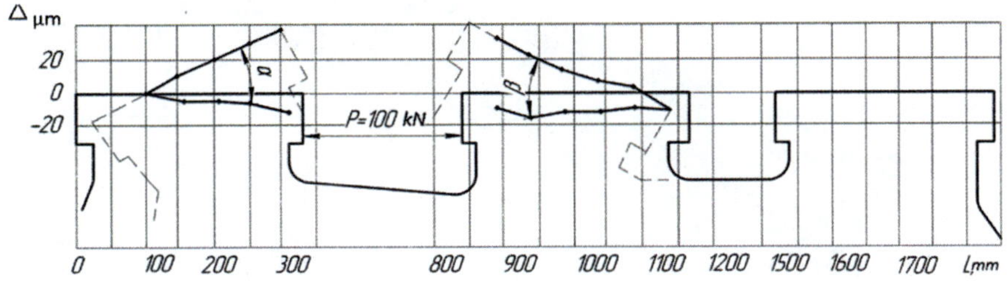

Fig. 8. Scheme for measuring the deformation at the flange distance of the cast frame of the 1A665 machine with an indicator.

Investigation of the deformation of the tailstock flange of a cast frame and the tailstock flange of a designed welded frame (Figs. 9, 10 and 11). A 3D model of the section of the support frame is presented in Fig. 9.

Fig. 9. Diagram of displacement of the welded support frame.

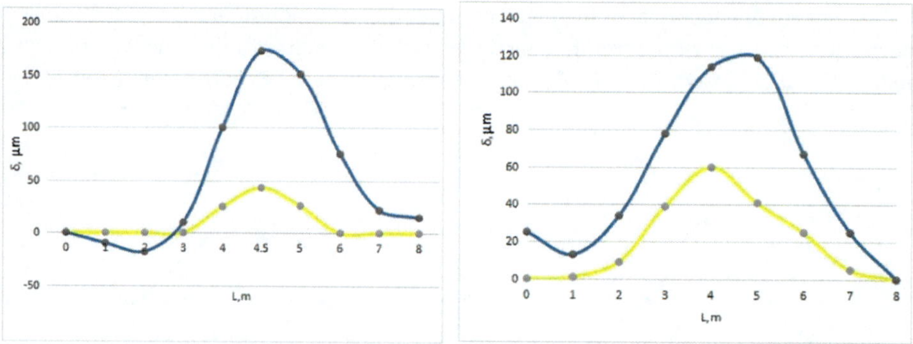

Fig. 10. Comparison of the results of deformation of the tailstock flange of a cast frame and the tailstock flange of a designed welded frame (a), deformation of support flange of a cast frame, and designed support flange welded frame (b).

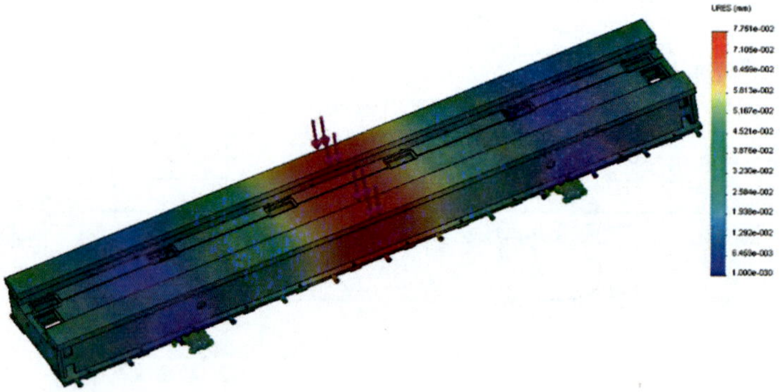

Fig. 11. Diagram of displacement of the welded support frame.

A comparative assessment of the deformation of the cast and welded frames' supporting flanges under a load of 7.2 tons was carried out. (Fig. 12).

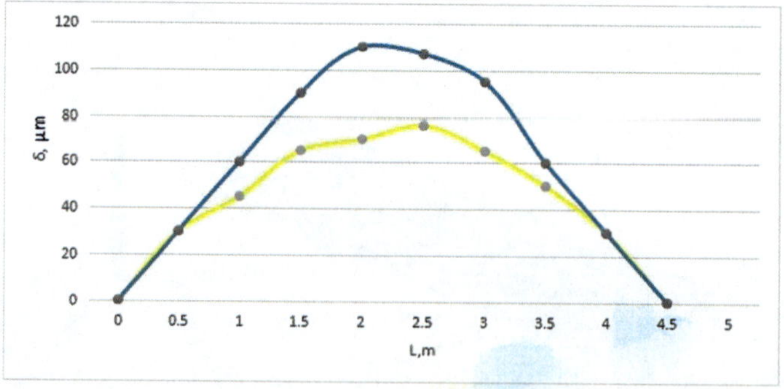

Fig. 12. Deformation of the cast frame and the designed welded frame during loading.

Full-scale production tests of the stands in the laboratory of PJSC "KZTS", at the frame section of full-scale dimensions - determination of the frame deflection under its load, torsional stiffness tests, bending of the frame on one and two stands. In work, there are schemes of loading the frame of the support and the frame of the workpiece by the torque (Fig. 13).

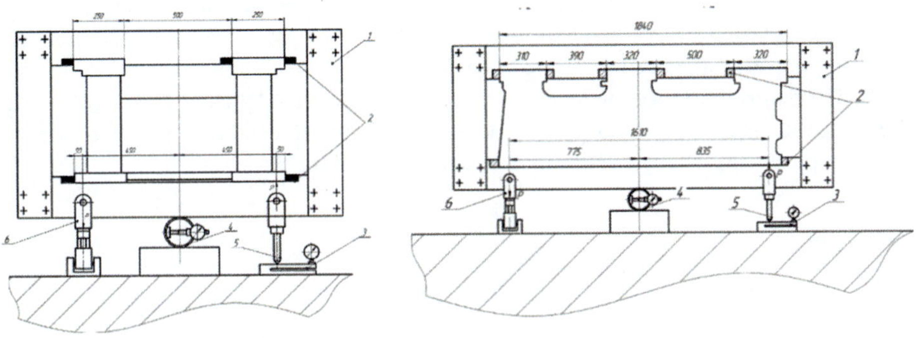

a) b)

Fig. 13. Scheme of loading the welded frame of the workpiece with torque (a) scheme of loading the welded frame of the support with torque (b).

One of the most promising ways to further improve the accuracy of machines is to equip them with adaptive systems. In particular, adapting machine tool carrying systems to changing operating conditions improves their accuracy.

The optimum conditions, stiffness, mass, shape, and design can be established by analyzing the influence of individual parameters on the flexibility of a heavy lathe's load-bearing system. Rational areas of application of cast and welded base parts in heavy machine tools are determined depending on their mass, scope of production, and labor intensity of manufacturing. Welded construction increases the structure's rigidity by 2–3 times.

Control of elastic displacements of the frame allows practically excluding the component of total error due to frame bending (about 60% loss of accuracy) results, which allows using the developed method for research and real design of new generation machines.

A method for investigating the accuracy of composite frames using mathematical modeling has been developed.

5 Conclusions

A satisfactory agreement of theoretical and experimental results has been achieved, as a result of production tests, it has been found that the coincidence of the results of mathematical modeling with field tests is 6–25%.

The technology for designing the load-bearing structures of heavy machines has been developed by using the results of preliminary calculation by contact deformation (internal forces, displacements) as boundary conditions for designing individual load-bearing structures, allowing to obtain the design geometry of the cross-section having the minimum possible weight while maintaining the specified norms of serviceability and processing accuracy.

A three-dimensional model of the heavy machine bed was developed, and test results under high loads allow to conclude about its operability. Design of high-precision heavy-duty lathe frame with a load-carrying capacity of 100 tons, machining capacity up to 12.5 m, diameter of 2.5 m and maximum cutting force of 200 kN. The recommendations for designing the carrying systems of heavy CNC high-precision machine tools were given. The deformations of the load-bearing frame at extreme loads in the range from 29 mcm to 83 mcm are presented. The work results are introduced at the polygon of PJSC KZTS in producing heavy CNC lathes of a new generation.

References

1. Urbikain, G., Campa, F.-J., Zulaika, J.-J., López de Lacalle, L.-N., Alonso, M.-A., Collado, V.: Preventing chatter vibrations in heavy-duty turning operations in large horizontal lathes. J. Sound Vib. **340**, 317–330 (2015). https://doi.org/10.1016/j.jsv.2014.12.002
2. Liu, L.L., et al.: Modeling and analysis of the crossfeed servo system of a heavy-duty lathe with friction. Mech. Based Des. Struct. Mach. **41**, 1–20 (2013). https://doi.org/10.1080/15397734.2012.675873
3. Li, H., Li, Y., Wang, W.: Feature based machine tool accuracy analysis method. Procedia CIRP **27**, 216–222 (2015). https://doi.org/10.1016/j.procir.2015.04.069
4. Lee, K.-I., Yang, S.-H.: Accuracy evaluation of machine tools by modelling spherical deviation based on double ball-bar measurements. Int. J. Mach. Tools Manuf. **75**, 46–54 (2013). https://doi.org/10.1016/j.ijmachtools.2013.09.001
5. Möhring, H.-C., et al.: Materials in machine tool structures. CIRP Ann. Manuf. Technol. **64**(2), 725–748 (2015). https://doi.org/10.1016/j.cirp.2015.05.005
6. Castro, H.F.F., Burdekin, M.: Dynamic calibration of the positioning accuracy of machine tools and coordinate measuring machines using a laser interferometer. Int. J. Mach. Tools Manuf. **43**(9), 947–954 (2003). https://doi.org/10.1016/S0890-6955(03)00083-X
7. Mekid, S., Ogedengbe, T.: A review of machine tool accuracy enhancement through error compensation in serial and parallel kinematic machines. Int. J. Precis. Technol. **1**(3/4), 251–286 (2010). https://doi.org/10.1504/IJPTECH.2010.031657
8. Kovalev, V.D., Vasilchenko, Y.V., Dašić, P.: Adaptive optimal control of a heavy lathe operation. Journal of Mechanics Engineering and Automation **4**(4), 269–275 (2014)
9. Kovalov, V., Antonenko, Y., Dašić, P.: Method of Structural Design of Heavy Machine Tools. Procedia Technol. **22**, 146–152 (2016). https://doi.org/10.1016/j.protcy.2016.01.023
10. Luciv, I., Leshuk, R.: Dynamic characteristics of adaptive type machine tool subsystems. Visnik TDTU **14**(4), 99–107 (2009). [in Ukrainian]
11. Burggräfa, P., Wagnera, J., Steinberga, F., Heinbacha, B., Wiggera, M., Saßmannshausen, T.: Life cycle assessment for adaptive remanufacturing: incorporating ecological considerations into the planning of maintenance activities – a case study in the German heavy machinery industry. Procedia CIRP **105**, 320–325 (2022). https://doi.org/10.1016/j.procir.2022.02.053
12. Chen, Y.-L., et al.: Implementation and verification of a four-probe motion error measurement system for a large-scale roll lathe used in hybrid manufacturing. Meas. Sci. Technol. **28**(10), 105004 (2017). https://doi.org/10.1088/1361-6501/aa7d33

13. Lutsiv, I., Voloshyn, V., Buhovets, V.: Definition of component elements position errors of integrated self–adjusting equipment for turning. Technological Complexes **1**(13), 98–105 (2016)
14. Cai, L., Tian, Y., Liu, Z., Cheng, Q., Xu, J., Ning, Y.: Application of cloud computing to simulation of a heavy-duty machine tool. Int. J. Adv. Manuf. Technol. **84**(1–4), 291–303 (2015). https://doi.org/10.1007/s00170-015-7916-2
15. Archenti, A., Nicolescu, M.: Accuracy analysis of machine tools using elastically linked systems. CIRP Ann. Manuf. Technol. **62**(1), 503–506 (2013). https://doi.org/10.1016/j.cirp.2013.03.100
16. Berselli, G., Gadaleta, M., Genovesi, A., Pellicciari, M., Peruzzini, M., Razzoli, R.: Engineering methods and tools enabling reconfigurable and adaptive robotic deburring. In: Eynard, B., Nigrelli, V., Oliveri, S.M., Peris-Fajarnes, G., Rizzuti, S. (eds.) Advances on Mechanics, Design Engineering and Manufacturing, pp. 655–664. Springer International Publishing, Cham (2017). https://doi.org/10.1007/978-3-319-45781-9_66
17. Han, Z.Y., et al.: Finite difference method-based calculation of gravity deformation curve for the large-span beam of heavy-duty vertical lathe. Adv. Mech. Eng. **8**(4), 1–8 (2016). https://doi.org/10.1177/1687814016646072

Dimensional Accuracy of Porous Structures Manufactured Using Air Controller

Ender Emir[1], Erkan Bahçe[2], Alper Uysal[3](✉), and Eshreb Dzhemilov[4]

[1] Elbistan Vocational School, Istiklal University, 5, Omer Halisdemir Street, 46300 Kahramanmaras,, Turkey
[2] Inonu University, 44280, Malatya-Elazig Way, 44000 Malatya, Turkey
[3] Yildiz Technical University, Besiktas, 34349 Istanbul, Turkey
auysal@yildiz.edu.tr
[4] Crimean Engineering and Pedagogical University, 8, Uchebniy side Street, Simferopol 29501, Russia

Abstract. Today, the additive manufacturing (AM) method is used in aerospace, defense, and biomedical fields due to its advantages. However, it is important to be able to control the production environment in these production methods. This study investigates the effects of the air control unit (ACU) in production with the fused deposition modeling (FDM) method on dimensional accuracy. Gyroid and body-centered cubic (BCC) porous structures were produced in an AC and non-AC environment. After production, the dimensional accuracy and surface quality of the porous structure structures were investigated. Significant surface defects were observed in the porous structure structures produced in a non-AC environment. Better quality surfaces were obtained in the productions carried out by providing AC. In addition, in the results obtained, it was observed that there was an average of 2% deviation in terms of dimensions in the productions carried out using the ACU and an average of 2.5% deviation in the productions without ACU.

Keywords: Additive manufacturing · Air controller · FDM · Dimensional accuracy · Process innovation

1 Introduction

AM technologies have received positive responses from various industries around the globe, such as automobile, aerospace, medical, etc. [1, 2]. AM is an innovative production method based on the production of 3D-designed parts by building layer by layer [3, 4]. It covers seven main methods according to the AM ASTM F42 standard, where different production materials and methods are used. Although it provides convenience with intervenable production processes in all AM methods, some problems are encountered. At the beginning of these problems, production parameters are unsuitable for the production material, the heat generated, and the support structures used to produce inclined surfaces. All these problems cause dimensional deviations between the designed and produced geometries. Especially due to the inability to control the high temperature

V. Tonkonogyi et al. (Eds.): InterPartner 2022, LNME, pp. 90–97, 2023.
https://doi.org/10.1007/978-3-031-16651-8_9

and particle dynamics that occur during production, the formation of deviations in the dimensions of the final products draws attention.

Today, AM is preferred to produce porous structures that provide high energy absorption, good heat transfer, and light [5]. Since porous structures have complex geometries, different production techniques are used during their production. The FDM method is primarily preferred in investigating different porous structures' mechanical and dimensional properties because it is both easy to access and cheaper. In studies, porous structures such as diamond, gyroid [6], BCC [7], and hexagonal [8] are generally preferred. However, the surface quality and dimensional accuracy are adversely affected because a clean production environment cannot be provided in the FDM method. Especially during production, ultrafine particles must adhere to the part at high temperatures. For this reason, more precise printing results and a more flexible layout should be preferred in additive manufacturing. With the ACU to be used, minimum adhered particle formation and better surface quality can be achieved.

2 Literature Review

It has been stated that production parameters, environment, and materials affect the final product in production with the FDM technique. The pre-processing factors, layer thickness, build orientation, road width, air gap [9], temperature, etc., directly affect the surface quality of FDM modeled parts [10]. In the literature, studies have been carried out to determine the dimensional deviations that occur in the productions carried out using AM methods. Nancharaiah et al. [11] examined the surface quality and dimensional accuracy of the samples they produced using the FDM method using ANOVA analysis. Their results found that the layer thickness and path width greatly affect the surface quality and part accuracy. Zharylkassyn et al. [12] investigated the effect of process parameters and materials on the dimensional accuracy of FDM parts. Akbas et al. [13] the effects of nozzle temperature and feed rate on the dimensions of the parts in production with the FDM method were analyzed experimentally and numerically. Their results stated that nozzle temperature and feed rate are effective parameters on layer widths. Garg et al. [14] investigated the effect of part placement orientation on the surface quality and dimensional accuracy of parts in production with the FDM method. They also investigated the effects of the change in surface quality due to the cold steam treatment of dimethylketone (acetone). In their results, they found a significant improvement in the surface quality of the components due to the steam treatment process. Ceretti et al. [15] worked on optimizing the process parameters for the dimensional accuracy and design repeatability of the multilayer scaffold structures produced using the FDM method. Alexandru-Victor et al. [16] used two different production methods, FDM and stereolithography (SLA), for three-dimensional accuracy evaluation in the production of dental models. Their results measured an average deviation of 0.207 mm in the production using the SLA method and 0.156 mm in the production with FDM. Islam et al. [17] examined the dimensional accuracy of the parts produced by three-dimensional printing. Their results said that the dimensional deviation in the production direction (z) is larger than in the other directions (x, y). They also stated that the hole sizes are smaller than the nominal diameter value. Demircioglu et al. [18] aimed to find the relationship between fusion temperature and dimensional accuracy of 3D printed parts. Three sphere samples were produced at five different extruder temperatures (185, 195, 205, 215, and 220 °C). The dimensions of the samples produced after the design and manufacturing processes were measured by image processing techniques. Their results said that the minimum dimensional error occurred at the fusion temperature of 185 °C at a value of 0.290797 mm and at a rate of 3%.

In the literature studies, parameters such as layer thickness and nozzle diameter were used to increase the production quality in studies carried out using the FDM technique. However, there has been a lack of studies on the control of airflow, which is an important parameter in production with the FDM method. This study produced porous structures produced using an ACU and porous structures produced without using a non-ACU. It is aimed to examine the effects of the ACU on the dimensional accuracy and surface quality of the porous structure structures produced by both methods.

3 Research Methodology

TPMS-based gyroid and BCC porous structures were used in the study. The porous structure structures were created with the help of the level set equations given in Eqs. 1 and Eqs. 2 defined in the literature. In addition, x, y, z $\in$ [0, 20] were determined to generate the overall dimensions of the porous structure.

$$\mathrm{Sin}[x]^*\mathrm{Cos}\left[y\right] + \mathrm{Sin}\left[y\right]^*\mathrm{Cos}[z] + \mathrm{Sin}[z]^*\mathrm{Cos}[x] \tag{1}$$

$$\begin{gathered}\mathrm{Cos}\left[2^*x\right] + \mathrm{Cos}\left[2^*y\right] + \mathrm{Cos}\left[2^*z\right] \\ -2^*\mathrm{Cos}[x]^*\mathrm{Cos}\left[y\right] - 2^*\mathrm{Cos}[z]^*\mathrm{Cos}\left[y\right] - 2^*\mathrm{Cos}[z]^*\mathrm{Cos}[x]\end{gathered} \tag{2}$$

In the study, porous structure structures were designed to investigate the effect of air control on dimensional accuracy in production with the FDM method. 20 $\times$ 20 $\times$ 20 mm sized gyroid and BCC porous structure was produced using the Ultimaker 2+ Connect with ACU and the same model and brand 3D printer without ACU (see Fig. 1). The parameters used during production are given in Table. 1. To evaluate the dimensional accuracy after production, measurements were taken from each sample using a light microscope. In the last stage, the dimensional differences in the porous structure structures produced using two different methods and the differences in surface quality were examined. As given in Fig. 2, the porous structure's height, depth, and width were measured.

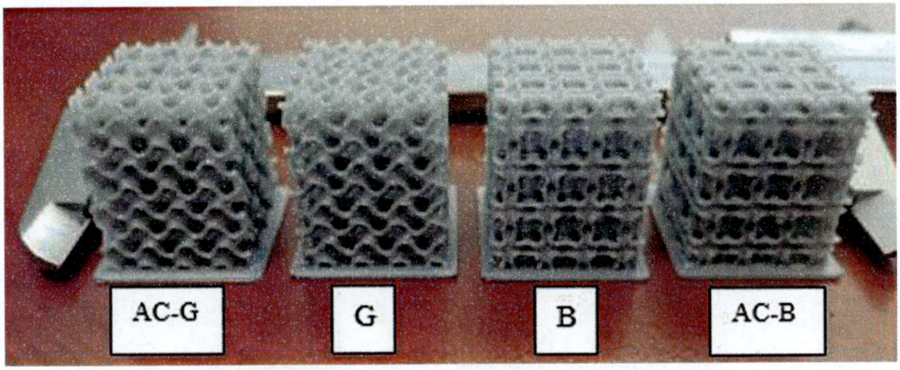

Fig. 1. Manufacturing samples.

Table 1. Manufacturing parameters.

Properties	Value
Layer thickness	0.15
Infill ratio (%)	30
Print speed	40
Nozzle diameter (mm)	0.4
Layer width (mm)	0.15
Nozzle temperature (°C)	210

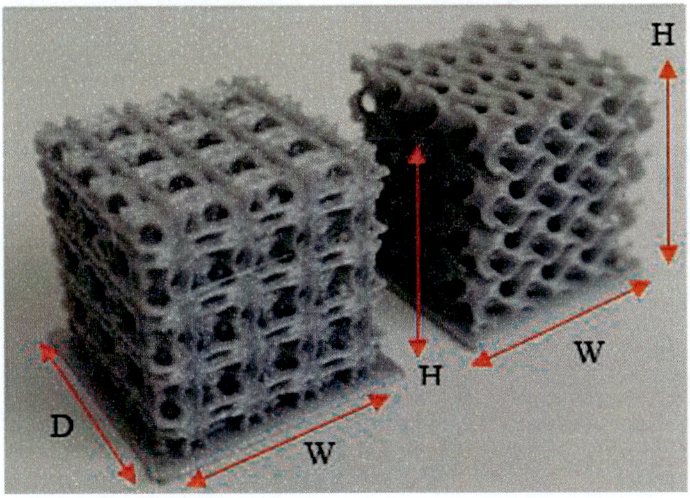

Fig. 2. Dimensional measurements of porous structure structures.

4 Results and Discussion

4.1 Surface Quality

Figure 3 shows the images taken from the upper and lateral regions of the gyroid and BCC porous structure.

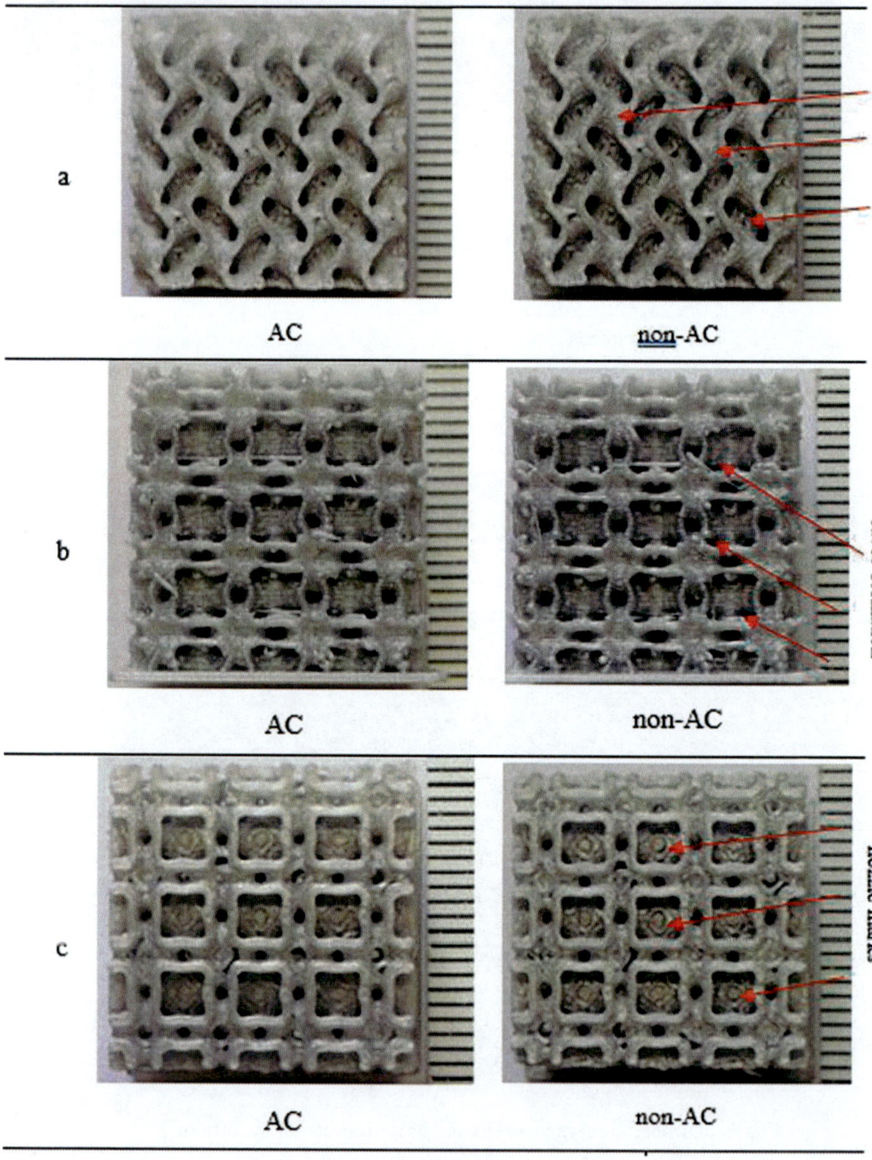

Fig. 3. Microscope images of porous structure structures.

In the images obtained, it has been determined that the upper layer gaps are formed in the productions using ACU compared to the productions without AC (see Fig. 3a). It was observed that fringe and yarn formation occurred in the production of BCC porous structure (see Fig. 3b). It is a problem that occurs as a result of the plastic remaining at the tip of the nozzle creeping and sticking to the other area during the transition of the nozzle from one region to another in a 3D model. Images taken from the upper regions of the BCC porous structure are given in Fig. 3c. In the images obtained, it has been seen that there are obvious nozzle marks in the production direction without AC. Many of the surface defects displayed are due to the heat occurring in the production environment and the inability to provide controlled airflow in the production environment. It has been confirmed that there are significant differences in production quality thanks to the ACU.

4.2 Dimensional Accuracy

In Table 2, the dimensional measurement results of gyroid and BCC porous structure produced using ACU and gyroid and BCC porous structure produced without non-ACU are given. In the measurement results, it has been determined that the dimensional deviations that occur in the productions carried out in an AC environment are less.

Table 2. Dimensional measurement values.

	Non-AC		AC	
	Gyroid	BCC	Gyroid	BCC
Width (mm)	19.25	19.3	19.35	19.45
Depth (mm)	19.45	19.35	19.5	19.4
Height (mm)	19.35	19.5	19.75	19.65

In the results obtained, it was observed that the gyroid porous structure's width, depth, and height produced in a non-AC environment, deviated from the nominal dimensions of 3.75%, 2.75%, and 3.25% in the negative direction. In the productions carried out using the AC, it was determined that the gyroid porous structure's width, depth, and height deviated from their nominal dimensions by 3.25%, 2.5%, and 1.25%. In the BCC porous structure, the deviations in width, depth, and height values were 3.5%, 3.25%, and 2.5% in the productions made without the use of a non-ACU, while the deviations of 2.75%, 3%, and 1.75% were measured in the productions carried out using the ACU. The deviations that occur in AC productions are lower because the heat generated during the creation of the 3D model is distributed to the whole part in a more controlled manner, and more controlled cooling is provided.

5 Conclusions

The study investigated the effects on the dimensional accuracy and surface quality of 3D productions using an ACU. It has been observed that the deterioration in surface quality and deviations in dimensional accuracy in productions carried out utilizing an ACU occur at lower levels than in production without a non-ACU. It has been determined that the samples produced using ACU show better surface quality than the production with the traditional FDM method. Significant differences were measured in the porous structure structures' width, depth, and length measurements. According to the nominal dimensions of the porous structure structures in the productions carried out without a non-ACU, %3.25 deviation has occurred in the gyroid porous structure, and %2.33 deviation has occurred in the productions made using an ACU. In the BCC porous structure, %3.08 deviation in production without non-ACU and %2.5 deviation in production with ACU was measured.

References

1. Guo, N., Leu, M.C.: Additive manufacturing: technology, applications and research needs. Front. Mech. Eng. **8**, 215–243 (2013). https://doi.org/10.1007/s11465-013-0248-8
2. Jain, P.K., Pandey, P.M., Rao, P.V.M.: Tailoring material properties in layered manufacturing. Mater. Design. **31**(7), 3490–3498 (2010). https://doi.org/10.1016/j.matdes.2010.02.029
3. Gibson, I., Rosen, D., Stucker, B.: Additive Manufacturing Technologies. Springer, New York, NY (2015). https://doi.org/10.1007/978-1-4939-2113-3
4. Urhal, P., Weightman, A., Diver, C., Bartolo, P.: Robot assisted additive manufacturing: a review. Robot. Comput. Integr. Manuf. **59**, 335–345 (2019). https://doi.org/10.1016/j.rcim.2019.05.005
5. Li, C., et al.: Crushing behavior of multi-layer metal porous structure panel fabricated by selective laser melting. Int. J. Mech. Sci. **145**, 389–399 (2018). https://doi.org/10.1016/j.ijmecsci.2018.07.029
6. Karimipour-Fard, P., Behravesh, A.H., Jones-Taggart, H., Pop-Iliev, R., Rizvi, G.: Effects of design, porosity and biodegradation on mechanical and morphological properties of additive-manufactured triply periodic minimal surface scaffolds. J. Mech. Behav. Biomed. Mater. **112**, 104064 (2020). https://doi.org/10.1016/j.jmbbm.2020.104064
7. Dar, U.A., Mian, H.H., Abid, M., Topa, A., Sheikh, M.Z., Bilal, M.: Experimental and numerical investigation of compressive behavior of porous structure structures manufactured through projection micro stereolithography. Mater. Today Commun. **25**, 101563 (2020). https://doi.org/10.1016/j.mtcomm.2020.101563
8. McGregor, D.J., Tawfick, S., King, W.P.: Mechanical properties of hexagonal porous structure structures fabricated using continuous liquid interface production additive manufacturing. Addit. Manuf. **25**, 10–18 (2019). https://doi.org/10.1016/j.addma.2018.11.002
9. Taufik, M., Jain, P.K.: Laser assisted finishing process for improved surface finish of fused deposition modelled parts. J. Manuf. Process. **30**, 161–177 (2017). https://doi.org/10.1016/j.jmapro.2017.09.020
10. Taufik, M., Jain, P.K.: Role of build orientation in layered manufacturing: a review. Int. J. Manuf. Technol. Manag. **27**, 47–73 (2013). https://doi.org/10.1504/IJMTM.2013.058637
11. Nancharaiah, T., Ranga Raju, D., Ramachandra Raju, V.: An experimental investigation on surface quality and dimensional accuracy of FDM components. Int. J. Emerg. Technol. **1**(2), 106–111 (2010)
12. Zharylkassyn, B., Perveen, A., Talamona, D.: Effect of process parameters and materials on the dimensional accuracy of FDM parts. Mater. Today: Proc. **44**, 1307–1311 (2021). https://doi.org/10.1016/j.matpr.2020.11.332

13. Akbaş, O.E., Hıra, O., Hervan, S.Z., Samankan, S., Altınkaynak, A.: Dimensional accuracy of FDM-printed polymer parts. Rapid Prototyp. J. **26**(2), 288–298 (2019). https://doi.org/10.1108/RPJ-04-2019-0115
14. Garg, A., Bhattacharya, A., Batish, A.: On surface finish and dimensional accuracy of FDM parts after cold vapor treatment. Mater. Manuf. Process. **31**(4), 522–529 (2015). https://doi.org/10.1080/10426914.2015.1070425
15. Ceretti, E., Ginestra, P., Neto, P.I., Fiorentino, A., Da Silva, J.V.L.: Multi-layered scaffolds production via fused deposition modeling (FDM) using an open source 3D printer: process parameters optimization for dimensional accuracy and design reproducibility. Procedia CIRP **65**, 13–18 (2017). https://doi.org/10.1016/j.procir.2017.04.042
16. Burde, A.V., Gasparik, C., Baciu, S., Manole, M., Dudea, D., Câmpian, R.S.: Three-dimensional accuracy evaluation of two additive manufacturing processes in the production of dental models. Key Eng. Mater. **752**, 119–125 (2017). https://doi.org/10.4028/www.scientific.net/KEM.752.119
17. Islam, M.N., Boswell, B., Pramanik, A.: An investigation of dimensional accuracy of parts produced by three-dimensional printing. In: Proceedings of the World Congress on Engineering, pp. 522−525, London, U.K. (2013)
18. Demircioglu, P., Bogrekci, I., Sucuoglu, S., Guven, E.: The relationship between the fusion temperature and dimensional accuracy of 3d printed parts. Sigma J. Eng. Nat. Sci. **38**(1), 21–28 (2020)

The Efficiency of Adaptive Slicing Group of Rationally Oriented Products for Layered Manufacturing

Yaroslav Garashchenko[1](✉) and Predrag Dasic[2]

[1] National Technical University "Kharkiv Polytechnic Institute", 2, Kyrpychova Street, Kharkiv 61002, Ukraine
yaroslav.garashchenko@gmail.com

[2] Academy of Professional Studies Sumadija, 8, Kosovska Street, 34000 Kragujevac, Serbia

Abstract. The article presents the results of further research on developed algorithm possibilities for adaptive slicing using the example of a 3D models group located in the workspace of the additive machine. In this algorithm for adaptive slicing, each build step is calculated considering of distribution density of angles between the build direction vector and normals of product group surfaces that fell into the current layer. The developed algorithm provides for some truncation of this distribution, which makes it possible to additionally increase the build step and reduce the number of layers. Thus, conditions are created for rational support and reduced build time. At the same time, a decrease in build time in comparison with the existing strategies of slicing is obtained. The usefulness of the adaptive slicing was evaluated based on a comparative analysis concerning the results obtained in previous studies, subject to the manufacture of individual products. The main compared characteristics are the number of layers and predicted deviations from the correct shape of surfaces concerning 3D models of industrial products. The products are placed in the workspace with different numbers and orientations. An increase in efficiency of the layered process with an increase in density of products placed in the workspace and setting of their rational orientation by minimizing the surface area with predicted most considerable deviations from the correct shape is revealed. The research was carried out using the developed system "Technological preparation of materialization of complex products by additive manufacturing".

Keywords: Additive manufacturing · Technological preparation · Layered slicing · Shaping accuracy · Sustainable manufacturing

1 Introduction

One of the main tasks of planning additive manufacturing processes is slicing 3D models located in the layered building workspace [1]. The problem of layered slicing significantly determines the efficiency of using additive machines. The number of layers significantly affects the build time of products. Accuracy of shaping the surfaces of product is mainly determined by building layers thickness and orientation of a surface relative to

V. Tonkonogyi et al. (Eds.): InterPartner 2022, LNME, pp. 98–108, 2023.
https://doi.org/10.1007/978-3-031-16651-8_10

the coordinate axis O_Z, i.e., build directions [2]. For example, inclined surfaces, which normal is inclined at a small angle to build a direction vector (it doesn't matter which direction, both positive and negative), have "stepped" appearance.

The result of the procedure for 3D model splitting is a set of cross-sectional contours and layer thicknesses (build steps). A single slice of a polygonal (triangulation) 3D model (usually described in STL-file) is formed by intersecting with the *XY* plane. A solution to this problem has two problems: the first one is a determination of layer thickness; the second is associated with the determination of closed contour geometry or several contours without self-intersections.

A typical slicing strategy is performed for a given constant build step [3]. This strategy does not consider changes in the product geometry in the building direction (along the Z-axis), which leads to a decrease in the accuracy of the resulting surfaces or productiveness. In some cases, ensuring a rational ratio of productivity of layered manufacturing and accuracy of product surfaces is possible when using adaptive slicing (with variable build step). In this case, the build step is determined based on the analysis of the 3D model surface in the current layer according to the given criterion [4].

2 Literature Review

In work [5] comparative analysis of known methods of adaptive slicing of products 3D-models in terms of building time and indicators of surface shaping error was carried out.

There are approaches to the 3D model: uniform slicing by a set of planes [6], adaptive slicing with section contour slicing with piecewise linear curves [7], piecewise linear approximation [8], and curved lines [9, 10].

With adaptive slicing, build step hi is determined taking into account the following characteristics: peak height or valley depth of steps formed on the product surface during layering [4, 11, 12]; a relative difference in adjacent slices areas [13]; parameter of surface roughness Ra [14, 15]; arithmetic mean error in its vertical and horizontal components within the surface layer [16]; a volumetric error of product building [17, 18]; octree structure on product material distribution in space [19]; building time [18].

Calculating adaptive build step can be performed jointly with correction of slicing contour to minimize building error [17]. Article [20] presents an adaptive slicing based on extracting the number of candidate feature points for the different areas of the 3D models.

There is an approach for data-processing of product 3D model by Boolean operations of polygons before adaptive slicing [21]. This provided a reasonable balance of time and built accuracy.

For Robotic AM, multi-direction and curved layer slicing are used for each sub-volume according to geometrical part features with the realization of multi-mode 3D printing [22].

Evaluation of shaped surface characteristics is carried out based on two profile descriptions in steps [2, 23] or radius sections [24].

The main problem is that in existing works with adaptive slicing, the build step is selected based on limiting values of the selected parameter (manufacturing error or surface quality). This problem can be corrected in whole or partly by considering the distribution nature of selected parameter values. Regardless of the selected parameter as a criterion, it will depend on the build step and φ_{NZ} angle between the build direction vector and normals of surfaces that fall into the layer cutting plane. Therefore, this problem should be considered based on the distribution density of φ_{NZ} angles over their relative area [4]. Considering that usually, products are not made individually but in a group for one or several loads of machines, it is necessary to consider possibilities of adaptive slicing, taking into account orientation and relative placement of products in the workspace of layered building.

In this work, we consider the scientific hypothesis that the efficiency of adaptive slicing of the 3D models' group located in the workspace can be increased using statistical analysis of φ_{NZ} angles distribution. Such analysis should take into account relative surface area. This will make it possible to determine build steps to ensure specified accuracy and reduce build time for the products group.

This article aims to reveal adaptive slicing possibilities for a group of product 3D models based on a statistical analysis of the distribution of angles between the vector of build direction and surface normals. This will provide desired shaping accuracy with minimal build time, considering orientation and placement in a layered build workspace.

Tasks that were solved to achieve the aim:

- adaptive slicing of 3D models group depending on the orientation and relative placement according to limiting values of φ_{NZ} angles and taking into account φ_{NZ} angles distribution truncation;
- comparative analysis of options for slicing of 3D-models group, depending on the option of orientation and placement in the workspace, as also the value of φ_{NZ} angles distribution density;
- effectiveness evaluation of proposed slicing strategy concerning possibilities offered in the Ultimaker Cura program (from version 3.2.0, adaptive slicing is supported);
- statistical analysis of deviations distribution from the correct shape for various options for slicing 3D models with constant and variable build steps.

3 Research Methodology

The research was carried out in the system of technological preparation of materialization of complex products by additive manufacturing, developed at the Department of Integrated Technologies of Mechanical Engineering of the National Technical University "Kharkiv Polytechnic Institute" [4]. This system makes it possible to assess the efficiency of solving optimization problems of additive manufacturing processes planning based on a statistical analysis of researched features of a polygonal, voxel, and layered 3D model of workspace with products.

The developed subsystem of layered statistical analysis has the following main capabilities:

- a creation of layers set with 2D-sections based on polygonal product 3D-model (STL-file) according to slicing strategy with constant or variable build step, developed taking into account the works [2, 4];
- statistical analysis of researched features distribution (in this work, build step h_i and deviations from correct surface shape Δ_S);
- conclusion of the main statistical characteristics in a single table for all layers.

Earlier in [4], the adaptive strategy for 3D model slicing was proposed. This research of advanced algorithm capabilities for adaptive slicing was carried out on examples of loading 1–3 pcs test 3D models of simple and complex products shown in Fig. 1.

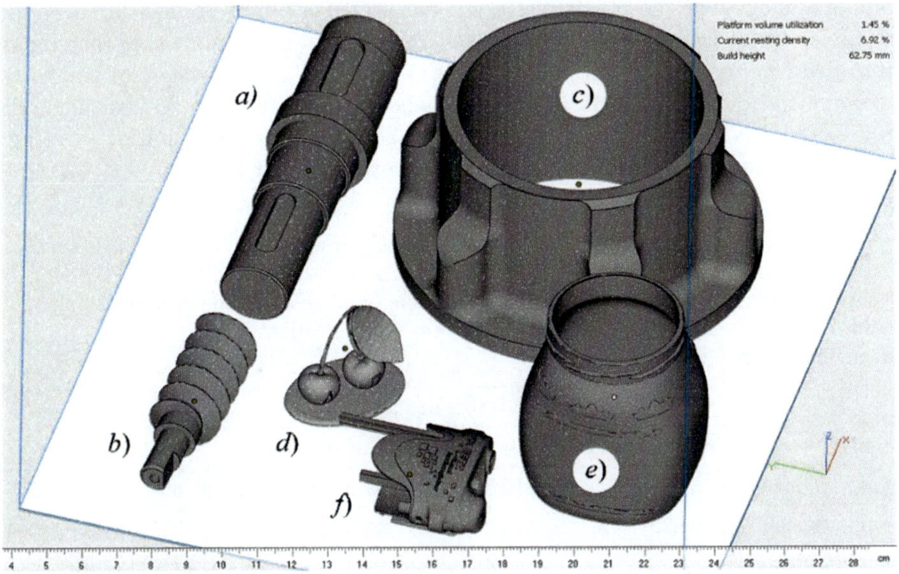

Fig. 1. Test 3D models (overall dimensions, mm): *a* – shaft (30 × 108 × 30); *b* – auger (20 × 20 × 72); *c* – case (105 × 105 × 63); *d* – souvenir (37 × 26 × 35); *e* – container (51 × 47 × 63); *f* – lid (42 × 51 × 22).

One of the most common approaches to determining the build step h_i is to perform a calculation based on the given limitation Δ_{Limit} on deviations from the correct shape of surfaces specified by the 3D model (maximum permissible shaping error) [2, 4, 24]:

$$h_i = \frac{\Delta_{Limit}}{\cos \varphi_{NZ\,\min}},$$

where φ_{NZmin} is the minimum value of angles between the Z-axis and faces normals that fall into the current layer.

4 Results and Discussion

4.1 Research of the Influence of 3D Model Orientation and Adaptive Slicing Parameters on the Number of Build Layers

To provide comparative analysis, slicing of test 3D models, previously oriented in the workspace, was carried out according to strategies with constant and variable build steps. The research considered two options for orienting 3D models: initial (Fig. 1) and rational (ensuring minimization of surface area with most significant deviations from correct shape, determined by the method described in [25]). Versions for placing products in the workspace differ in the number of loaded product groups from 1 to 3. 3D-models placement was carried out in Materialize Magics using EOSPACE automatic placement module. Strategy with the constant step was performed at $h_i = 0.06$ mm. Strategy with a variable step at $\{h_i\}_{\min} = 0.06$ mm, $\{h_i\}_{\max} = 0.2$ mm, and allowable (maximum) error of surface formation $\Delta_{Limit} = \{0.06; 0.1\}$ mm. The selected range of build steps is recommended for Ultimaker 3D printers when using a 0.4 mm AA extruder. The proposed adaptive slicing strategy was performed at 0–20% truncation of φ_{NZ} angle distribution. Results of model calculations are shown in Table 1.

Table 1. Results of layered slicing of industrial products test models by the number of layers.

Quantity of each product, orientation option	Build height H_B, mm	Constant step, h_i = 0.06 mm	Variable step when truncating of distribution φ_{NZ}, %				
			0	5	10	15	20
		Number of slices, N_L					
Permissible deviation from correct surface shape $\Delta_{Limit} = 0.06$ mm							
1 pcs., original	62,75	1046	900	724	651	603	546
1 pcs., rational	108,25	1805	1230	917	813	726	664
2 pcs., original	135,00	2250	1933	1602	1355	1240	1143
2 pcs., rational	132,50	2209	1989	1425	1139	1026	945
3 pcs., original	202,50	3375	2806	2386	2145	1934	1759
3 pcs., rational	202,50	3375	2928	2127	1745	1533	1385
Permissible deviation from the correct surface shape $\Delta_{Limit} = 0.1$ mm							
1 pcs., original	62,75	1046	540	454	418	403	381
1 pcs., rational	108,25	1805	802	661	621	594	574
2 pcs., original	135,00	2250	1172	999	893	851	813
2 pcs., rational	132,50	2209	1172	915	804	765	740
3 pcs., original	202,50	3375	1717	1494	1378	1293	1229
3 pcs., rational	202,50	3375	1745	1378	1234	1168	1121

Comparative analysis of build layers number for all researched variants of 3D models placed in the workspace (Table 1) confirms the well-known [4] advantage of strategies with variable steps compared to constant steps. Adaptive slicing performed at $\Delta\varphi_{NZ} = 0\%$ (such adaptive strategy is known from [26]) makes it possible to sufficiently reduce layers number compared to slicing with constant step $h_i = 0.06$ mm. Parameters choice significantly affects the effectiveness of the adaptive strategy. With permissible deviation from correct surface shape $\Delta_{Limit} = 0.06$ mm, it is possible to reduce layers number by 10.0–31.9%. At $\Delta_{Limit} = 0.1$ mm, we observe a decrease in layers number by 46.9–55.6%. The proposed approach in [4] to truncate the distribution density of angles made it possible to reduce further the number of build layers for all researched options for placing 3D models. When truncated by 5% - by 28.8–63.4% compared with the strategy with constant build step (comparatively to truncated option $\Delta\varphi_{NZ} = 0\%$ by 13.0–28.4%). When truncated by 20% - by 47.8–68.2% (comparatively to slicing at $\Delta\varphi_{NZ} = 0\%$ by 28.4–52.7%). Results of model calculations using the example of different options for orientation and placement of 3D models in the workspace made it possible to identify some tendencies. Figure 2 presents in the form of graphs the results obtained by building layers numbers.

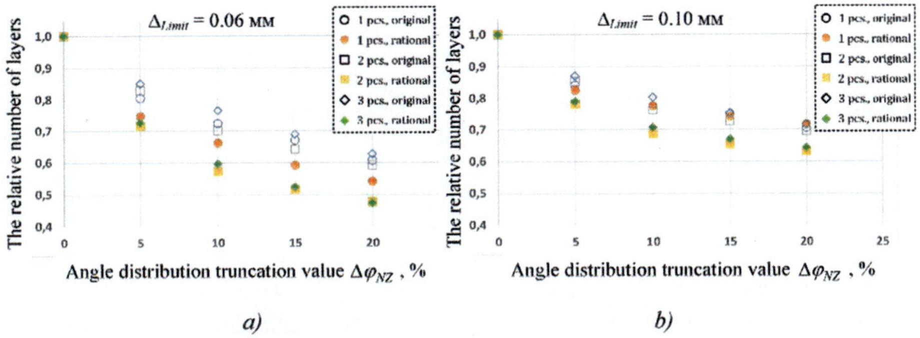

Fig. 2. The relationship between the relative number of layers for building test 3D models on the truncation value of φ_{NZ} angle distribution.

The more significant relative decrease in build layers number n_L can be achieved at $\Delta\varphi_{NZ} \in [0; 10]\%$. With a permissible value of $\Delta_{Limit} = 0.06$ mm as compared to $\Delta_{Limit} = 0.10$ mm, greater efficiency of the proposed adaptive strategy is observed (Fig. 2). Rational orientation compared to the original allows the production of products group in fewer layers. Therefore, it can be concluded that provided advantage of rational orientation both for obtaining products with fewer deviations from the correct shape and for reducing layers number for their layered formation. This is especially true with increased requirements for shaping accuracy, i.e., for a smaller value Δ_{Limit}.

4.2 Analysis of Effectiveness of Adaptive Slicing

For methodological reasons, the developed slicing strategy was evaluated using the example of the test 3D models group (Fig. 1) based on a comparison with adaptive slicing offered in Ultimaker Cura for the Ultimaker 3 Extended printer.

While exploring the adaptive slicing capabilities of Ultimaker Cura, various combinations of the following parameters were considered: adaptive layers minimum variation, $R_h = 0.14$ mm; layers variation step size (difference in h_i for adjacent layers), $\Delta_h =$ 0.01–0.1 mm; adaptive layers threshold (probability of setting h_i lower in value), $p_h =$ 100–200.

Table 2 shows the data obtained from previously created 3D models slicing with groups of test products (Fig. 1) according to strategies with constant and variable build steps.

Table 2. Results of layered slicing of various orientation and placement options for test 3D models group in Ultimaker Cura.

Quantity of each product, orientation option	Constant step, $h = 0{,}06$ MM		Variable step	
	Number of layers N_L	Build time t_b, h	Number of layers N_L	Build time t_b, h
1 pcs., original	1042	59,1	561 ÷ 2445	39,1 ÷ 168,7
1 each, rational	1800	63,1	833 ÷ 2568	39,4 ÷ 126,6
2 each, original	2246	212,7	1245–5892	133,4–436,1
2 each, rational	2204	197,1	1240–4791	120,4–423,5
3 each, original	3371	358,7	1020–5278	71,1–847,3
3 each, rational	3371	326,0	1844–3939	195,6–420,7

Data analysis (Table 2) revealed a clear advantage of developed adaptive slicing with truncation φ_{NZ} (calculation data are in Table 1) concerning that used in Ultimaker Cura for all placement options for the test of the 3D models' group. But this advantage is not apparent. Adaptive slicing in Ultimaker Cura was found to be unusable in all cases. In those cases when it was possible to reduce the build time of the 3D models' group, this happened at overestimated values of Δ_h. Overestimation of Δ_h value leads to less algorithm flexibility for selection of build step, taking into account the product's geometry. This means these situations will be overestimated deviations from the correct surface shape. As applied to adaptive slicing of separately manufactured test 3D models, it made it possible, with a certain reduction in build time, to provide a sufficiently adequate selection of build steps (analysis results are presented in [4]).

Calculated data on build time of test 3D models groups (Table 2) indicate high productivity of layered building for options 1 and 2, i.e., to load a 3D printer one batch of products. This conclusion is valid only for the considered 3D printer model.

Adaptive slicing proposed in [26] (approximately satisfy developed slicing at $\Delta\varphi_{NZ} = 0$) is inferior in the number of layers for all variants of test 3D models placement (it is possible to reduce the number of build layers by 28.4–52.7%).

4.3 Statistical Layered Analysis of Deviations from Correct Shape

Using adaptive layered slicing means the guaranteed provision of a given level of quality and resulting product surfaces accuracy. Therefore, a layered assessment of deviations from the correct Δ_S shape by the arithmetic mean was also carried out. The peak height of the step by slicing was taken as predicted deviations from the correct shape [2, 4, 23].

Figure 3 shows "Box Whiskers" amplitudes distribution $\overline{\Delta_S}$ for comparing strategies with a constant building step $h_i = \{0.06;\ 0.10\}$ mm, with a variable step at $\Delta\varphi_{NZ} = \{0;\ 5;\ 10;\ 15;\ 20\}$ and a given limit of $\Delta_{Limit} = \{0.06;\ 0.10\}$ mm.

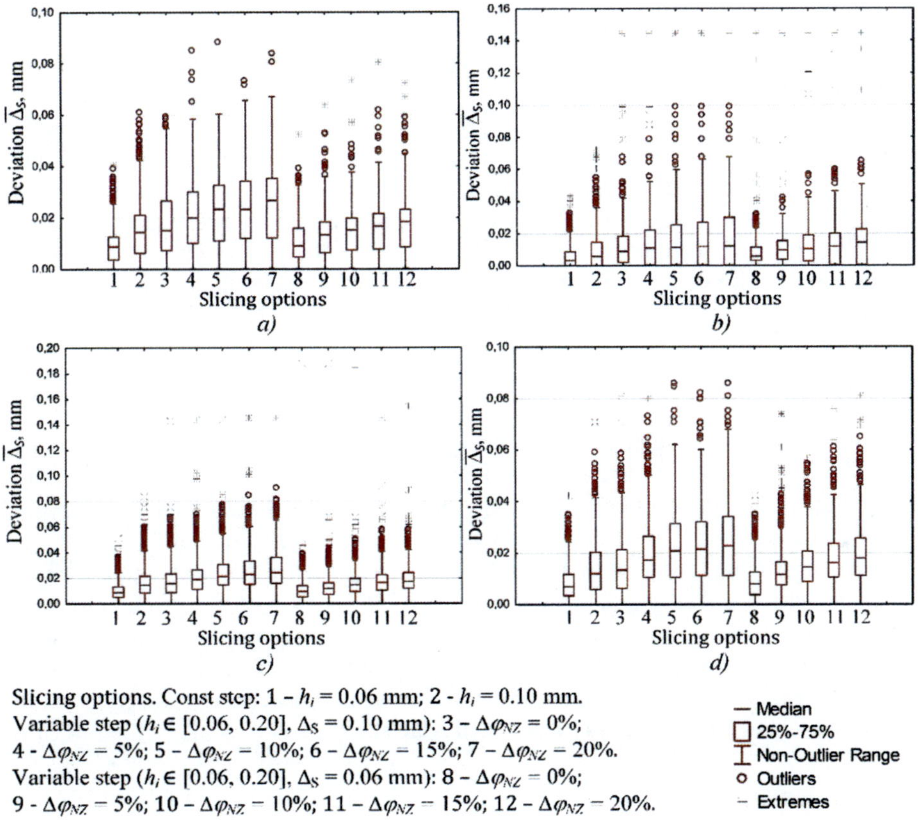

Fig. 3. Statistical layered analysis of arithmetic mean deviation from correct surface shape: a – shaft; b – auger; c – case; d – souvenir.

Comparative analysis of layered statistical characteristics (shown in Fig. 3) using the example of different variants of test 3D-models placement when slicing with variable step, no significant differences are observed, truncation value $\Delta\varphi_{NZ}$ does not significantly affect on distribution $\overline{\Delta_S}$. For most placement variants, the resulting distribution

$\overline{\Delta_S}$ when slicing with variable step approximately corresponds to slicing with constant build step $h_i = 0.06$–0.10 mm. This is especially typical with a limitation of $\Delta_{Limit} = 0.06$ mm, as the most typical for $h_i = 0.06$–0.10 mm.

According to research, the following conditions were formulated for the rational use of developed adaptive slicing in [4] for the 3D-models group:

- rational orientation of products according to the criterion of minimizing surface area with the greatest deviations from a correct surface shape;
- selection of truncation value $\Delta\varphi_{NZ} \in [0,\ 0.2]$, taking into account the permissible local increase in deviation from correct form;
- choice of the smaller value of distribution truncation $\Delta\varphi_{NZ}$ for complex products.

The revealed range of rational truncation of distribution of 3D-model adaptive slicing makes it possible to reduce build time up to approximately 38% compared to strategies with constant slice steps. The approach proposed in the article also makes it possible to assess the adaptive slicing effectiveness of the 3D models' group with sufficient reliability. Based on the increasing efficiency of the proposed algorithm for adaptive slicing, when setting rational orientation to products in the workspace, it can be assumed that its application for products group rationally placed on the platform, taking into account the developed algorithm features, will provide a more significant reduction in layers number.

5 Conclusions

Adaptive slicing of 3D-models group using statistical analysis of the distribution of angles between the Z-axis and surfaces normals that fall into the layer cross-section, taking into account their relative area, allows you to increase the productivity of layered building due to its rational combination with orientation of the product.

Adaptive slicing, taking into account a slight decrease in the distribution range of angles between the Z-axis and surface normals by $\Delta\varphi_{NZ} = 5$–20% with high requirements for the build accuracy for the considered group of complex products, allows reducing the build time by 33.8–37.6% depending on the chosen orientation.

Further research should identify the influence of product relative placement in the workspace on the effectiveness of developed adaptive slicing.

References

1. Zhang, Y., Bernard, A.: A KBE CAPP framework for qualified additive manufacturing. CIRP Ann. Manuf. Technol. **67**(1), 467–470 (2018)
2. Lin, J.Q., Wang, Y.B., Jing, X., Gu, Y.: Research on adaptive slicing algorithm of STL model in additive manufacturing. Mach. Des. Manuf. **2**(2), 51–53 (2018)
3. Gibson, I., Rosen, D., Stucker, B., Khorasani, M.: Additive manufacturing technologies. 3rd edn. Springer Nature Switzerland, Cham (Switzerland) (2021)
4. Garashchenko, Y., Zubkova, N.: Adaptive slicing in the additive manufacturing process using the statistical layered analysis. In: Ivanov, V., Trojanowska, J., Pavlenko, I., Zajac, J., Peraković, D. (eds.) DSMIE 2020. LNME, pp. 253–263. Springer, Cham (2020). https://doi.org/10.1007/978-3-030-50794-7_25

5. Wang, W., et al.: Saliency-preserving slicing optimization for effective 3D printing. Comp. Graph. Forum **34**(6), 148–160 (2015)
6. Kirschman, C.F., Jara-Almonte, C.C.: A parallel slicing algorithm for solid freeform fabrication processes. In: Marcus, H.L., et al. (eds.) Solid Freeform Fabrication Symposium 1992, pp. 26–33. University of Texas, Austin (1992)
7. Sabourin, E., Houser, S.A., Bohn, J.H.: Adaptive slicing using stepwise uniform refinement. Rapid Prototyp. J. **2**(4), 20–26 (1996)
8. Sikder, S., Barari, A., Kishawy, H.A.: Global adaptive slicing of NURBS based sculptured surface for minimum texture error in rapid prototyping. Rapid Prototyp. J. **21**(6), 649–661 (2015)
9. Zhou, M.Y., Xi, J.T., Yan, J.Q.: Adaptive direct slicing with non-uniform cusp heights for rapid prototyping. Int. J. Adv. Manuf. Technol. **23**(1–2), 20–27 (2004)
10. Jin, G.Q., Li, W.D., Gao, L.: An adaptive process planning approach of rapid prototyping and manufacturing. Robot. Comput. Integr. Manuf. **29**(1), 23–38 (2013)
11. Nadiyapara, H.H., Pande, S.: A review of variable slicing in fused deposition modeling. J. Inst. Eng. India Ser. C **98**(3), 387–393 (2016). https://doi.org/10.1007/s40032-016-0272-7
12. Li, Q., Xu, X.Y.: Self-adaptive slicing algorithm for 3D printing of FGM components. Mater. Res. Innov. **19**(sup5), S5-635-S5-641 (2015). https://doi.org/10.1179/1432891714Z.0000000001167
13. Pan, X.D., Chen, K., Zhang, Z.Y., Chen, D.F., Li, T.T.: Adaptive slicing algorithm based on STL model. Appl. Mech. Mater. **288**, 241–245 (2013). https://doi.org/10.4028/www.scientific.net/AMM.288.241
14. Pandey, P.M., Reddy, N.V., Dhande, S.G.: Real time adaptive slicing for fused deposition modelling. Int. J. Mach. Tools Manuf. **43**(1), 61–71 (2003)
15. Singhal, S.K., Jain, P.K., Pandey, P.M.: Adaptive slicing for SLS prototyping. Comput. Aided Des. Appl. **5**(1–4), 412–423 (2008). https://doi.org/10.3722/cadaps.2008.412-423
16. Sikder, S., Barari, A., Kishawy, H.A.: Control of NURBS-based surface error factor using a manufacturing cost optimization in rapid prototyping process. IFAC Proc. Volumes **46**(9), 1560–1565 (2013). https://doi.org/10.3182/20130619-3-RU-3018.00577
17. Gohari, H., Kishawy, H., Barari, A.: Adaptive variable layer thickness and perimetral offset planning for layer-based additive manufacturing processes. Int. J. Comput. Integr. Manuf. **34**(9), 964–974 (2021)
18. Mao, H.: Minimizing printing time and volumetric error by GPU-accelerated adaptive slicing. Manuf. Lett. **29**, 19–23 (2021). https://doi.org/10.1016/j.mfglet.2021.05.002
19. Siraskar, N., Paul, R., Anand, S.: Adaptive slicing in additive manufacturing process using a modified boundary octree data structure. J. Manuf. Sci. Eng. **137**(1), 011007 (2014). https://doi.org/10.1115/1.4028579
20. Hu, B., Jin, G., Sun, L.: A Novel adaptive slicing method for additive manufacturing. In: 2018 IEEE 22nd International Conference on Computer Supported Cooperative Work in Design ((CSCWD)), pp. 218—223 (2018). https://doi.org/10.1109/CSCWD.2018.8465247
21. Guoqiang, F., Jianzhong, F., Lin, Z., Shen, H., Jin, Y.: A polygons boolean operations-based adaptive slicing with sliced data for additive manufacturing. Proc. Inst. Mech. Eng. C J. Mech. Eng. Sci. **231**(15), 2783–2799 (2016). https://doi.org/10.1177/0954406216640576
22. Zhao, D., Guo, W.: Mixed-layer adaptive slicing for robotic Additive Manufacturing (AM) based on decomposing and regrouping. J. Intell. Manuf. **31**(4), 985–1002 (2019). https://doi.org/10.1007/s10845-019-01490-z
23. Chen, Q., Xu, J., Zhang, S.: Cylindricity and flatness optimization for mechanical parts in additive manufacturing based on tolerance adaptive slicing. Int. J. Adv. Manuf. Technol. **115**(11–12), 3839–3857 (2021). https://doi.org/10.1007/s00170-021-07271-4

24. Boschetto, A., Giordano, V., Veniali, F.: 3D roughness profile model in fused deposition modelling. Rapid Prototyp. J. **19**(4), 240–252 (2013). https://doi.org/10.1108/13552541311323254
25. Garashchenko, Y., Zubkova, N.: Evaluation of the forecasted efficiency of performance of rational orientation of the product in the workspace of additive installations. Cutting and Tools in Technological System **94**, 142–150 (2021). https://doi.org/10.20998/2078-7405.2021.94.16
26. Abdurajimov, L.: Adaptive slicing of product 3D-model in technology of rapid prototyping and manufacturing. Scientific notes of the Crimean Engineering and Pedagogical University. Technical Science 18, 15–20 (2009). [in Russian]

Optimal Conditions for Deformation of Stamping-Drawing Process from Aviation Materials

Anton Onopchenko, Oleksii Horbachov, Volodymyr Sorokin, Yuri Dudukalov, and Maksym Kurin(✉)

National Aerospace University "Kharkiv Aviation Institute", 17, Chkalova Street, Kharkiv 61070, Ukraine
m.kurin@khai.edu

Abstract. The main goal of theoretical investigations in the presented work is to identify optimal metal flow conditions. Most interested in practice is a determination of the main modes of the deformation process and their connection with external factors. In sheet metal stamping operations, plastic deformations provide a given character of formation change. Thus, there is a need to choose a theoretical calculation method corresponding to requirements and a particular case of stamping-drawing conditions. The most suitable methods for calculating stamping-drawing processes are based on a closed system of equations of continuum mechanics. With this approach, the deformable metal is considered as an ideal condition with averaged mechanical properties. Expressions are obtained that make it possible to calculate power parameters on the drawing rib. The fields of strain rate intensity are constructed for specific modes of sheet stamping-drawing. The constructed dependencies make it possible to visually assess the deformed state of material on the die corner of the tool die. Experimental research on stamping-drawing of four materials is presented. The experimental values of limit load are shown for all four materials.

Keywords: Sheet metal stamping-drawing · Deformation intensity · Process power parameters · Process innovation · Dislocation densities

1 Introduction

Corrosion-resistant, heat-resistant, and heat-resistant alloys are widely used in aircraft engine building. It has high strength combined with high toughness [1]. There are classified as difficult to machine materials [2]. Therefore, one of the main directions in developing modern technology for the production of aircraft engine parts is to reduce the volume of technological processes with the removal of metal layer [3] and its replacement by processes of the precise surface and metal forming of parts with subsequent minimal cutting. [4, 5]. The widespread use of parts manufactured by pressure treatment - stamping of thin-walled parts and blanks from a metal sheet one of the main directions in the development of mechanical engineering, which meets this requirement to the maximum extent possible [6].

V. Tonkonogyi et al. (Eds.): InterPartner 2022, LNME, pp. 109–118, 2023.
https://doi.org/10.1007/978-3-031-16651-8_11

Stamping-drawing is a standard punching operation in mechanical engineering [7] designed to produce hollow parts from flat and hollow workpieces of various shapes and sizes, such as tank bottoms, cylinder hemispheres, and aircraft engines nacelles, doors, hatches, various casings, caps, and other parts. The drawing process has a number of advantages [8, 9], such as high productivity, high accuracy of dimensions, and shape of parts, which makes it possible to obtain parts of complex geometry. These parts have sufficient strength and rigidity with a low specific weight compared to assembly units (welded and prefabricated). This one allows for the rational use of the source material.

The work's goal is to investigate technological factors' influence on the power parameters of the drawing process and the quality indicators of stamped parts.

Thus, researchers were devoted to increasing the efficiency of stamping-drawing of hollow parts from difficult-to-machine materials are relevant.

2 Literature Review

In sheet metal stamping operations, plastic deformations provide a given character of formation change. It usually occurs only in part of the workpiece – the deformation cell [10]. In the deformation zone, the field of stresses and strains are inhomogeneous. The stress in the deformation zone is a function of coordinates at each moment of deformation [11]. The value and distribution of stresses in the deformation zone depend on many factors associated with the dimensional characteristics of the workpiece and tool. There are contact conditions, temperature-rate conditions of deformation, etc. [12]. The degree of permissible deformation in sheet stamping operations is limited either by the destruction of the workpiece or by loss of stability, which leads to unacceptable distortion of the shape [13]. The stress field determines the deformation forces and sometimes a degree of permissible formation change in the deformation zone. Determining the distribution stresses and energy-power parameters in the deformation zone should be one of the main tasks when considering sheet-stamping operations [14]. It is necessary to develop mathematical models [15, 16] to predict the power characteristics of various types of pressure treatment, including sheet stamping, reflecting a relationship between functional characteristics of the process and technological parameters of the processing modes [17]. The model's correct construction is possible if a structural and logical scheme is developed and the methods and sequence of theoretical and experimental investigation studies are determined.

The fillet radius of the matrix has a significant effect [18] on such essential parameters of the drawing process as stress in the material, drawing force, formation of corrugations [19], thinning of the wall material, and limiting draw ratio and durability of matrix [20]. This necessitates further investigations of the plastic flow of metal on the waist edge of the tool die matrix and determine analytical dependencies that make it possible to predict the stress-strain state material on the waist rib and the drawing forces at the design stage. In this regard, it became necessary to develop a method for obtaining velocity fields of the sheet stamping process, which fully reflects the physics of the process. At the same time, it is quite simple and acceptable for engineering calculations. The most suitable methods for calculating plastic deformation processes are based on a closed system of equations of continuum mechanics. In this case, the deformed metal is considered an idealized continuous medium with average mechanical properties of actual metal.

3 Research Methodology

Calculation of the deformation cell requires the introduction of the continuity hypothesis. In this case, the main task of flow kinematics is to determine the particle velocity field. Since elastic-plastic deformation of sheet material inevitably entails the movement of various defects in the crystal structure, which leads to energy dissipation. It is an almost complete transformation into heat and significant forces of internal friction. The presence of significant forces of internal friction allows rotation of metal particles in the deformation process, and movement, in this case, cannot be considered potential.

As known, theoretical analysis of the majority of technological processes along with the conducted experiments allows us to determine the nature of dependence velocity of particles plastic deformable metal on coordinates. The velocity of particles in four-dimensional space can be represented through a velocity vector [21]:

$$\boldsymbol{V} = \upsilon_x \boldsymbol{i} + \upsilon_y \boldsymbol{j} + \upsilon_z \boldsymbol{k} + \upsilon_t \boldsymbol{n}. \tag{1}$$

The law constancy of a volume during the deformation is expressed by the continuity equation [17]:

$$\left.\begin{aligned} div\boldsymbol{V} &= 0 \\ rot\boldsymbol{V} &\neq 0 \end{aligned}\right\} \tag{2}$$

Using Eqs. (1) and (2), we can determine the functional dependence of speed on coordinates. Thus, the particles velocity field of material is determined, which makes it possible to calculate the strain rates and their intensity using the formulas:

$$\begin{aligned} \varepsilon_{q_1q_1} &= \frac{1}{H_1}\frac{\partial \boldsymbol{V}_{q_1}}{\partial q_1} + \frac{\boldsymbol{V}_{q_2}}{H_1H_2}\frac{\partial H_1}{\partial q_2} + \frac{\boldsymbol{V}_{q_3}}{H_1H_3}\frac{\partial H_1}{\partial q_3}, \\ \varepsilon_{q_1q_2} &= \frac{1}{H_2}\frac{\partial \boldsymbol{V}_{q_1}}{\partial q_2} + \frac{1}{H_1}\frac{\partial \boldsymbol{V}_{q_2}}{\partial q_1} - \frac{\boldsymbol{V}_{q_1}}{H_1H_2}\frac{\partial H_1}{\partial q_1} - \frac{\boldsymbol{V}_{q_2}}{H_1H_2}\frac{\partial H_2}{\partial q_1}, \end{aligned} \tag{3}$$

where q_1, q_2, and q_3 are orthogonal curvilinear coordinates.

In this case, the coupling equations hold:

$$\begin{gathered} x = x(q_1, q_2, q_3),\ y = y(q_1, q_2, q_3),\ z = z(q_1, q_2, q_3); \\ H_k = \sqrt{\sum_{i=1}^{3}\left(\frac{\partial x_i}{\partial q_k}\right)^2}; \\ \varepsilon_i = \frac{\sqrt{2}}{3}\sqrt{\left(\varepsilon_{q_1q_2} - \varepsilon_{q_2q_2}\right)^2 + \left(\varepsilon_{q_2q_2} - \varepsilon_{q_3q_3}\right)^2 + \left(\varepsilon_{q_3q_3} - \varepsilon_{q_1q_1}\right)^2 + \frac{3}{2}\left(\varepsilon_{q_1q_2}^2 + \varepsilon_{q_2q_3}^2 + \varepsilon_{q_3q_1}^2\right)^2}; \end{gathered} \tag{4}$$

where H_k – Lamé parameters.

Then, it is necessary to determine components of deformation to find the energy-power process parameters,

$$e_{11} = \int \varepsilon_{11} dt \quad e_{22} = \int \varepsilon_{22} dt$$

and deformation intensity,

$$e_i = \frac{\sqrt{2}}{3}\sqrt{(e_{11} - e_{22})^2 + (e_{22} - e_{33})^2 + (e_{33} - e_{11})^2 + \frac{3}{2}\left(e_{12}^2 + e_{23}^2 + e_{31}^2\right)}.$$

If the deformation is performed at a low rate, the rate function has a sufficiently small value compared to the energy dissipation function and can be neglected. In this case, the deformation work will be determined through the energy dissipation function:

$$A = \int_t \iiint_V EdVdt$$

Now let us determine the law of change in the velocity of material flow on the drawing rib during stamping-drawing. A die corner is a part of the torus that the system of parametric equations can describe:

$$\left.\begin{array}{l} x(t,\varphi) = (R + r\,cos(\omega t))cos(\omega t) \\ y(t,\varphi) = (R + r\,cos(\omega t))sin(\omega t) \\ z(t,\varphi) = \pm rsin(\omega t) \end{array}\right\}, \tag{5}$$

where r – the radius of the die corner; R – distance from the rotation axis of the torus to the axis of the generatrix; t – time of deformation; ω is constant (Fig. 1). The system of Eqs. (5) describes a toroidal surface in Cartesian coordinates. It is more convenient for us to go to the cylindrical system to search for components of the velocities.

After differentiation of the system (5) in time and transition to Euler coordinates, we obtain a system of equations describing the velocity field of metal displacements on the die corner.

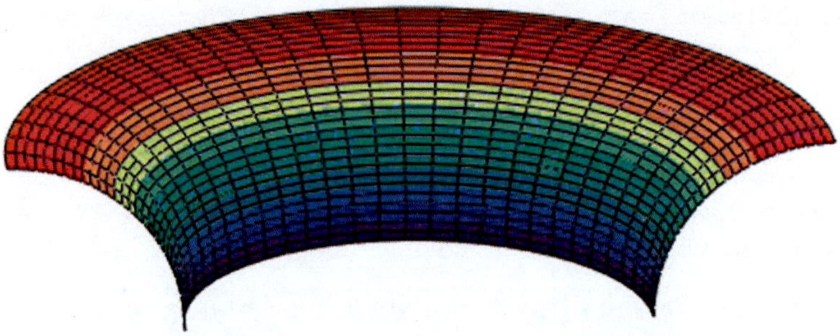

Fig. 1. General view of the toroidal surface of die corner.

Thus, the velocity field of the metal flow in the deformation zone has the following form:

$$\left.\begin{array}{l} V_\rho(\rho) = -\frac{V_0}{r}\sqrt{r^2 - (\rho - R)^2} \\ V_z(z) = -\frac{V_0}{r}\sqrt{r^2 - z^2} \\ V_\psi(\rho,\psi,z) = \left[\frac{V_0\left[r^2 - (R-\rho)(R-2\rho)\right]}{r\sqrt{r^2-(\rho-R)^2}} - \frac{V_0 z\rho}{r\sqrt{r^2-z^2}}\right]\psi \end{array}\right\} \tag{6}$$

Let us analyze the resulting system of Eqs. (6). System (6) has been obtained considering the initial conditions $V_\rho = -V_0$ at $\rho = R$ and $V_z = -V_0$ at $z = 0$. The velocity V_0

should be understood as metal flow velocity at the point where the flange passes into the die corner. As you can see, the radial V_ρ and vertical V_z components of the velocity depend on only one coordinate, while the transverse velocity V_ψ depends on all three. A negative value of the radial velocity component indicates that metal is moving towards the center.

Since the velocity components in the system (6) are parallel to the axes of the cylindrical system and mutually perpendicular, the velocity modulus can be calculated using the well-known formula:

$$V = \sqrt{V_\rho(\rho)^2 + V_\psi(\rho, \psi, z)^2 + V_z(z)^2}$$

The deformations and their intensities are calculated using the system (6) according to the above method. Thus, an integral picture of the plastic flow of the metal is being constructed on the die corner, which ultimately allows for calculating deformation work.

In our opinion, it is interesting to visualize the obtained equations under specific processing conditions graphically. Let us plot the dependencies of the radial (Fig. 2) and transverse (Fig. 3) components of the velocity with the following initial data: r = 3 mm, R = 15 mm, and V = 3 mm/s. These initial data correspond to a working part diameter matrix of 30 mm and the die corner of 3 mm.

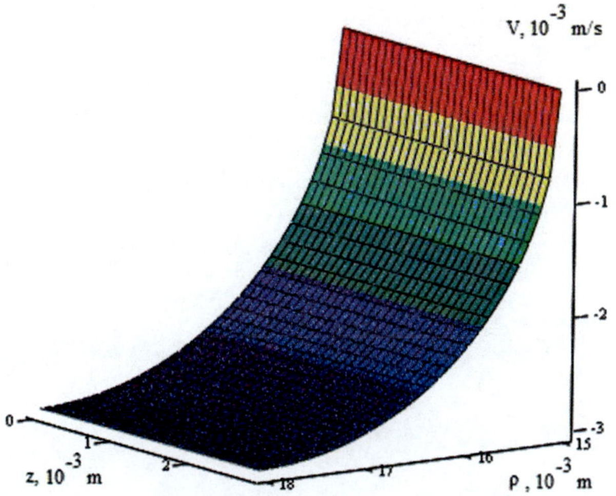

Fig. 2. The strain rate intensity on die corner ($\psi = 0$) radial component.

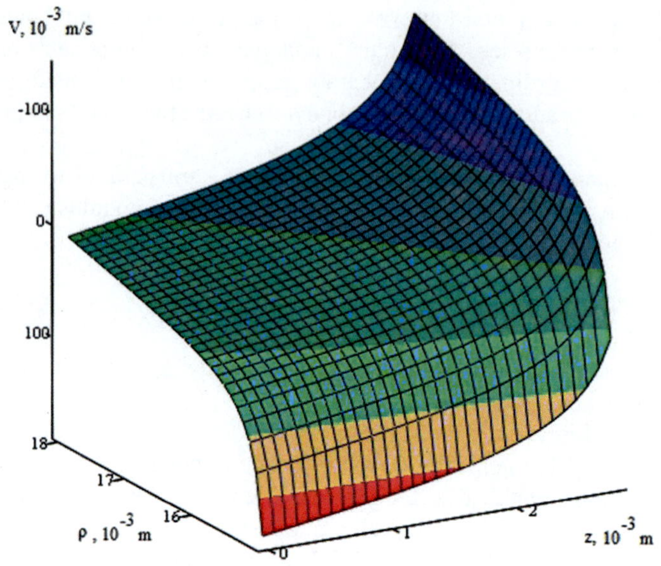

Fig. 3. The strain rate intensity on die corner ($\psi = \pi$) transversal component.

4 Results and Discussion

When choosing the investigation, we were guided by materials distribution in the aviation industry and general mechanical engineering. The drawing process was carried out in an adjustable tool die (Fig. 4) for experimental researchers. The breaking load was determined using a calibrated pressure gauge. Billets for experimental studies were carried out for three aluminum alloys AMg, AMc, D16T, and the titanium alloy OT4 (Fig. 5).

Fig. 4. Stamping-drawing die.

Fig. 5. Parts after breaking load (force).

The experimental values of limit load for all four materials are shown in Table 1.

Table 1. Drawing force dependence on workpiece material for diameter 63 mm.

№	Material grade	Maximum deformation force, H
1	AMg	14715
2	AMc	7553,7
3	D16T	21582
4	OT4	49344

The dependence breaking load of workpieces is shown in Fig. 6 for two standard materials sizes.

For aluminum alloys of medium strength and high ductility (AMg and AMc), the limiting force is more significant for the diameter of the workpiece 47 mm, while for the aluminum alloy of normal strength – D16T and titanium alloy OT4, an increase in the size of workpiece lead to an increase in the limiting pulling force, which is explained by high yield strength and strength of alloys D16T and OT4. In this case, the destruction of workpieces AMg and AMc occurs faster for diameters of 63 mm, and destruction occurs almost at the beginning of the deformation process at the formed bottom. Thus, the deformation process for ductile aluminum alloys is closer to the punching process. At the same time, stronger alloys have significantly greater deformations before fracture, which requires more effort.

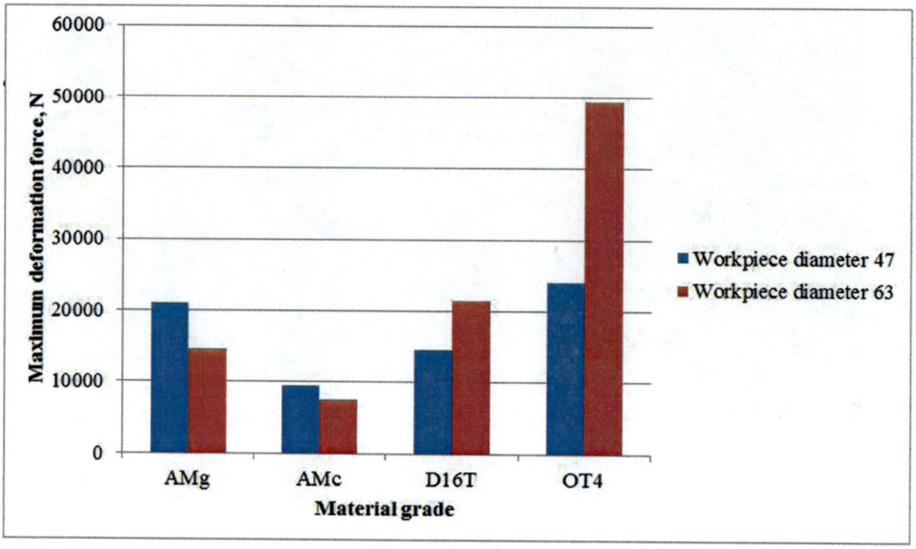

Fig. 6. Drawing force dependence on the diameter of the workpiece and material.

5 Conclusions

The necessity and relevance of the investigation have been established. The metal flow kinematics has been investigated in the constriction die corner stamping-drawing. The change law in the material flow velocity has been determined on the die corner during stamping-drawing. The method for calculating the energy-power parameters of the circular drawing process is proposed. It is based on a closed system of continuum mechanics equations. The fields of strain rate intensity are constructed for specific modes of sheet stamping-drawing. The constructed dependencies make it possible to visually assess the deformed state of material on the die corner of the tool die.

It has installed:

- breaking load for three aluminum AMg, AMc, and D16T and one titanium alloy OT4 for specific billet diameters;
- the maximum drawing force was recorded for the titanium alloy OT4 49344 N for a workpiece diameter of 63 mm, and the minimum for the AMg alloy – 4715 N;
- influence of billet diameter on the drawing force is diametrically opposite for ductile and strong alloys. So for the AMg and AMc alloys, the maximum forces were recorded with a workpiece diameter of 47 mm, and a minimum for 63 mm, and for D16T and OT4 – the maximum at 63 mm, and the minimum at 47 mm.
- based on the velocity field, we will obtain pictures of the intensities of the rates of deformations and their intensities. It is planned to obtain a general formula for calculating the deformation work, which will be presented as the sum of the deformation work of the flange, die corner, wall, and bottom.

References

1. Stachurski, W., Sawicki, J., Krupanek, K., Nadolny, K.: Numerical analysis of coolant flow in the grinding zone. I. J. Adv. Manuf. Technol. **104**(5–8), 1999–2012 (2019). https://doi.org/10.1007/s00170-019-03966-x
2. Grzesik, W.: Advanced Machining Processes of Metallic Materials: Theory, Modelling, and Applications, 2nd edn. Elsevier (2017)
3. Kombarov, V., Sorokin, V., Tsegelnyk, Y., Aksonov, Y., Fojtů, O.: Numerical control of machining parts from aluminum alloys with sticking minimization. Int. J. Mechatron. Appl. Mech. **1**(9), 209–216 (2021)
4. Bouchaâla, K., Ghanameh, M.F., Faqir, M., Mada, M., Essadiqi, E.: Numerical investigation of the effect of punch corner radius and die shoulder radius on the flange earrings for AA1050 and AA1100 aluminum alloys in cylindrical deep drawing process. Heliyon **7**(4), e06662 (2021). https://doi.org/10.1016/j.heliyon.2021.e06662
5. Vinod, L.H., Shivashankar, R.S.: Sheet metal forming processes – recent technological advances. Mater. Today Proc. **5**(1), 2564–2574 (2018). https://doi.org/10.1016/j.matpr.2017.11.040
6. Wang, K., et al.: High-efficiency forming processes for complex thin-walled titanium alloys components: State-of-the-art and perspectives. Int. J. Extrem. Manuf. **2**, 032001 (2020). https://doi.org/10.1088/2631-7990/ab949b
7. Deepak, V., Abhilash, O., Ravitej, Y.P., Veerachari, Abhinandan, L.: Design and development of progressive tool for mold tag. AIP Conf. Proc. **2316**, 030015 (2021). https://doi.org/10.1063/5.0038385
8. Bell, C., Corney, J., Zuelli, N., Savings, D.: A state of the art review of hydroforming technology. Int. J. Mater. Form. **13**(5), 789–828 (2019). https://doi.org/10.1007/s12289-019-01507-1
9. Trzepieciński, T.: Recent developments and trends in sheet metal forming department of materials forming and processing. Metals **10**(6), 779 (2020). https://doi.org/10.3390/met10060779
10. Kajikawa, S., Kuboki, T., Iizuka, T.: Flange compression using stepped punch for forming extremely deep cup with flange from aluminum alloy sheet. J. Mater. Process. Technol. **288**, 116835 (2021). https://doi.org/10.1016/j.jmatprotec.2020.116835
11. Luyen, T.T., Tong, V.C., Nguyen, D.T.: A simulation and experimental study on the deep drawing process of SPCC sheet using the graphical method. Alexandria Eng. J. **61**(3), 2472–2483 (2021). https://doi.org/10.1016/j.aej.2021.07.009
12. Prakash, V., Kumar, D.R.: Numerical simulation of warm deep drawing incorporating strain rate effect in sheet material properties. Mater. Today Proc. **18**(7), 2595–2602 (2019). https://doi.org/10.1016/j.matpr.2019.07.118
13. Basak, S., Panda, S.K., Lee, M.G.: Formability and fracture in deep drawing sheet metals: Extended studies for pre-strained anisotropic thin sheets. Int. J. Mech. Sci. **170**, 105346 (2020). https://doi.org/10.1016/j.ijmecsci.2019.105346
14. Singh, A., et al.: Prediction of earing defect and deep drawing behavior of commercially pure titanium sheets using CPB06 anisotropy yield theory. J. Manuf. Processes **33**, 256–267 (2018). https://doi.org/10.1016/j.jmapro.2018.05.003
15. Lal, R.K., Choubey, V.K., Dwivedi, J.P., Kumar, S.: Study of factors affecting springback in sheet metal forming and deep drawing process. Mater. Today Proc. **5**(2), 4353–4358 (2018). https://doi.org/10.1016/j.matpr.2017.12.002
16. Mahmoud, M., Bay, F., Muñoz, D.P.: An efficient multiphysics solid shell based finite element approach for modeling thin sheet metal forming processes. Finite Elem. Anal. Des. **198**, 103645 (2022). https://doi.org/10.1016/j.finel.2021.103645

17. Kurin, M.O.: Analysis the process of plastic deformation metal chip at non-free cutting. Metallofiz. Noveishie Tekhnol. **42**(3), 433–449 (2020). https://doi.org/10.15407/mfint.42.03.0433
18. Arhat, R., Puzyr, R., Shchetynin, V., Levchenko, R., Pedun, O.: The stress state of the workpiece at the radius of matrix rounding during drawing, considering the bending moment. In: Tonkonogyi, V., Ivanov, V., Trojanowska, J., Oborskyi, G., Pavlenko, I. (eds.) InterPartner 2021. LNME, pp. 479–488. Springer, Cham (2022). https://doi.org/10.1007/978-3-030-91327-4_47
19. Thor, M., Sause, M.G.R., Hinterhlzl, R.M.: Mechanisms of origin and classification of out-of-plane fiber waviness in composite materials – a review. J. Compos. Sci. **4**(3), 130 (2020). https://doi.org/10.3390/jcs4030130
20. Blala, H., Lang, L., Khan, S., Alexandrov, S.: Experimental and numerical investigation of fiber metal laminate forming behavior using a variable blank holder force. Prod. Eng. Res. Devel. **14**(4), 509–522 (2020). https://doi.org/10.1007/s11740-020-00974-9
21. Kurin, M.O.: Determination of the boundaries of plastic zone of metal deformation during the cutting. Prog. Phys. Met. **21**(2), 249–273 (2020). https://doi.org/10.15407/ufm.21.02.249

Manufacturing of the T-207 Prismatic Part Using Additive Manufacturing Technologies

Viktoriya Pasternak[1], Oleg Zabolotnyi[1](✉), Nataliia Zubovetska[1], Dagmar Cagáňová[2], and Ivan Pavlenko[3]

[1] Lutsk National Technical University, 75, Lvivska Street, Lutsk 43018, Ukraine
volynasi@gmail.com
[2] Comenius University in Bratislava, 10, Odbojárov Street, 820 05 Bratislava, Slovak Republic
[3] Sumy State University, 2, Rymskogo-Korsakova Street, Sumy 40007, Ukraine

Abstract. In this paper, the main problems that are associated with the mechanical processing of parts are investigated. The general view of the drawing of the part body type T-207 is presented, and the main methods of obtaining the workpiece are justified. Two basing methods were recorded along three planes (while forming a coordinate angle) and along the plane using two holes (processed according to the accuracy quality of H7 with landing). The chemical composition and basic mechanical properties of inhomogeneous materials of grey cast iron of the SCH 15 brand were studied. We found that material losses are reduced by 10–15% due to the correct method of manufacturing the part. We performed a correlation analysis of the average statistical data. The tool wear mechanism was presented using experimental results, where traces of abrasion were recorded. Some internal defects of the workpiece components were identified. We recommended minimal surface roughness and justified the maximum wear of tools. It was found that the surface roughness of samples and the main parameters obtained on their basis favorably affect the chips' size, shape, and brittleness. With the help of modern 3D modeling technologies of the part, the T-207 body can predict the main properties, such as high corrosion resistance, no rod defects, extreme accuracy, and general mechanical processing.

Keywords: Billet · Technological process · Basing · Sample surface · Roughness · Tool wear · 3D modelling · Sustainable manufacturing

1 Introduction

Any part's manufacturing technology consists of correctly choosing the most rational workpiece. The volume of its mechanical processing depends on the choice of the workpiece and the determination of its production method, shape, size of allowances, tolerances, accuracy, roughness, configuration, and surface quality. It should be noted that the choice of workpiece depends on the following main parameters: material, the shape of the part, its weight, dimensions, permissible working conditions of the part, as well as the scale of additive production. During mechanical processing, the material of

V. Tonkonogyi et al. (Eds.): InterPartner 2022, LNME, pp. 119–128, 2023.
https://doi.org/10.1007/978-3-031-16651-8_12

any part is usually determined by the designer based on the main performance characteristics of the heterogeneous material. It is also necessary to consider the fulfillment of all requirements for the accuracy and quality of the product, which are provided for in the technical conditions of the national and international standards. It should also be noted that the mechanical properties of inhomogeneous materials (grey cast iron SCH 15) differ significantly in their characteristics, which significantly affects the final parameters of the T-207 part. The main task in the manufacturing technology of workpieces is to maximize the approximation of their size and shape to the finished parts. Simultaneously, it should also be considered that in the process of additive pre-production, the part's material has already been determined, which largely determines the type of workpiece, the design of technological processes, machining, and the technology of its production in general [1]. Therefore, studying machine-building heterogeneous materials and their further mechanical processing with the intervention of new additive production technologies is one of the main research areas today. This, in turn, allows us to ensure a high level of development of mechanical engineering technology and improve the main indicators of an economic and industrial nature.

2 Literature Review

In work [2], the authors investigated the general effect of the magnetic field on the physical and chemical properties used in lubricating coolants under industrial conditions. In the first stage of the work, the authors magnetized coolants under stationary and standard conditions. In the second stage, the magnetization of the liquid that occurs in the middle of the materials was considered. After that, the difference between all the results obtained was analyzed to build diagrams. In [3], economic and technical aspects of the study of continuous and non-contact measurement of the workpiece diameter are presented. Certain production operations using the appropriate sensor are also considered. Works [4] include tribological behavior and surface roughness analysis, and crushed composites are treated with double heat treatment. However, the technical conditions according to standards are not fully considered. The analysis of works [5] suggests using a new patented approach in the production process. It should be noted that the disadvantage of this work is that the production process is described here only after abrasive treatment. The materials cover the optimization of the Taguchi method and the investigation of the effect of input parameters on surface roughness [6]. For the study, only the ranking method (DEAR) was used, which solves problems with a narrow range of applications. In work [7], an individual approach to manufacturing parts with structural elements of any purpose is proposed. The authors found that a certain thickness of the transition layer is formed in the interaction zone of inhomogeneous materials, which practically does not change with increasing exposure time. These studies made it possible to improve the quality indicators of parts with structural elements, in particular gears. Experimental papers [8] make appropriate measurements using a one-dimensional laser vibrometer, which was combined in parallel with a CNC milling machine. The purpose of this study is to establish comparative characteristics that occur between an ultrasonic and conventional machine. The authors of [9] proposed to study mainly batch planning in a workshop environment. At the same time, only energy-saving planning and processing completion time should be considered. The important point is that the priority of the product batch is set according to the upper and lower relationship that occurs in the product structure itself. As well as modeling low-cycle fatigue and imperfect external factors that affect the manufacture of parts during the machining process [10]. In the research work [11], the theoretical part is developed, and a probabilistic model is proposed for calculating surface roughness during magnetic abrasive treatment. However, the materials used have random dimensional characteristics and are randomly arranged on the surface of the tools. That is, the workpiece has an incorrect appearance and profile. A team of co-authors [12] proposed investigating the software's internal relationships between components. The calculation of individual segments of structurally heterogeneous materials was carried out considering all the features of the Smart-eye Software Product. It should be noted that the proposed algorithm for recognizing images of individual sections of the microstructure of component particles allows us to justify the surface and internal properties of samples. In scientific works [13, 14], composite and nanomaterials were modeled for additive manufacturing. The scientific results of [15, 16] represent design and processing parameters optimization. Roughness measurements on the surface of the processed parts were performed offline. Predicting roughness results was based on processing vibration signals recorded on a vertical and horizontal CNC machining center. The research theories [17, 18] justify the influence of temperature, force, tool wear, and the release of fine dust. These studies mainly aim to apply analytical and empirical models to assess cutting force indicators' reliability and theoretical prediction. It was also found that the speed of the spindle significantly affects the cutting temperature, because of which small particles are released, the specific cutting energy increases, and tool wear occurs. In [19], the roughness of the honing surface and some holes is optimized based on the GRA-RSM system. A big disadvantage of these experiments is that conventional internal grinding mainly carried out the surface treatment, which is prone to burning and increases a large number of cracks. Research works [20, 21] considered general methods for assessing the stability of technological processes and experimental diagnostics of the corresponding characteristics. We ensured the stability of product quality and accuracy indicators using the Fischer criterion and presented an overall assessment of mechanical processing in digital production conditions. However, the problems that arise during the mechanical processing of parts in the additive type of production are not fully covered. Since you need to pay more attention to the tool wear mechanism and internal workpiece components' internal defects that occur during processing. Because obtaining these results ensures minimal surface roughness of the samples, high corrosion resistance, high accuracy, and general mechanical processing.

3 Research Methodology

3.1 Material and Main Methods of Obtaining a Blank

This article aims to investigate abrasion traces and identify internal defects in the components of the workpiece (based on type housing T-207). Correlation analysis (quantitative analysis) of average mechanical processing data using 3D modeling was carried out. On the main surface roughness parameters studied, ensure the minimum surface roughness of the samples and determine the average tool wear value.

Body parts of the T-207 type are designed to accommodate assembly units and individual elements of parts. The T-207 housing ensures stable accuracy of the relative position of the part and mechanism, both in the static state and during the operation of any machine. Therefore, this type of part has sufficient rigidity. It should be noted that the part of the housing type has the main base surfaces. As a rule, they are presented in the form of planes with which they are connected to the frame and the rest of the housing if necessary. In Fig. 1. The drawing of the T-207 body part is presented.

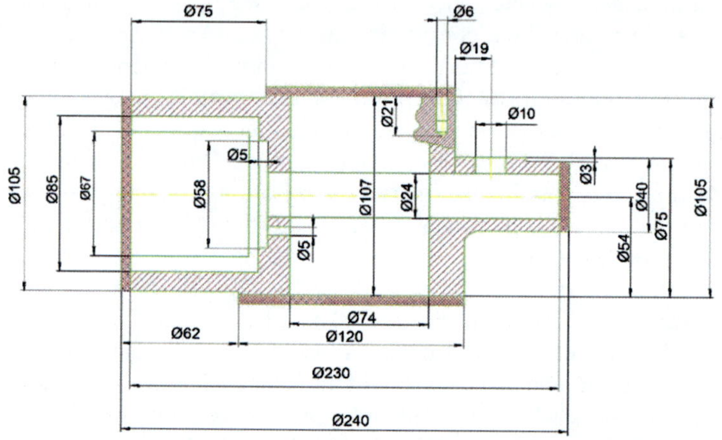

Fig. 1. General view 2D of the part drawing T-207 type housing.

It should also be noted that in this type of part, there are auxiliary base surfaces of holes and planes. In turn, the hole diameters were made with a tolerance field of H7 and R_a = 1.4–0.5 μm). The alignment tolerance of some holes was set within half the tolerance field for the diameter of the smallest hole of the housing part no more than 0.2–0.6 μm. At the same time, some permissible deviations and center-to-center distances were recorded, which lay within the acceptable norms according to the standard.

The blank of the T-207 body type part was made by casting into sand and clay molds. The advantage of this method is that it is the lowest cost of casting, and there is also the possibility of repeated use of research molds. Material – grey cast iron of the SCH 15 brand. This SCH 15 material is used only in those types of parts where high strength and corrosion resistance are required, as well as increased wear resistance. Mechanical processing of the body blank was determined by choice of bases and by the dimensional connections that function between the surfaces of the part. During the manufacturing process, two basing methods were recorded: 1) along three planes while forming a coordinate angle; 2) along the plane using two holes, which was processed according to the accuracy quality of H7 with a fit.

3.2 Correlation Analysis of Average Mechanical Processing Data

In order to improve the manufacturing technology and mechanical processing of any parts under additive manufacturing conditions, we investigated the chemical composition and mechanical properties of inhomogeneous SCH 15 materials. The results obtained are presented in Tables 1 and 2.

Table 1. Chemical composition of grey cast iron.

Carbon, %	Silicon, %	Manganese, %	Phosphorus, %	Sulphur, %
2,5…4,5	0,8…4,5	0,1…1,2	0,02…0,3	0,02…0,15

Table 2. Mechanical properties of SCH15.

Ultimate strength, MPa	Compression, MPa	Brinell hardness, HB
100…350	450…1400	143…289

Using the results obtained, we found out that material losses are reduced by 10–15% due to the correct method of manufacturing the part. At the same time, the complexity of mechanical processing is reduced by 1.5–2.0 times due to reduced allowances and high dimensional accuracy.

4 Results and Discussion

4.1 Integrity Analysis Based on SCH 15 Grey Cast Iron

The integrity of the surface of the coating of grey cast iron of the SCH 15 brand requires special attention since it should reduce the cost of the final product of use. It should be noted that the integrity of the surface of the already processed product depends on the wear of tools, processing parameters, and the cooling state. During this study, we found some limitations due to tool wear. With rapid tool wear, the surface roughness was reduced significantly. They also recorded that it is necessary to process the surface at lower feed speeds during turning. After reducing the cutting speed, we got much better surface roughness indicators. At the same time, it was necessary to control the increase in cutting speed so that this also did not lead to failure of tool wear. Figure 2 presents the tool wear mechanism of grey cast iron components SCH 15.

The study showed that the surface of the micro-grinder in both directions (horizontal and vertical) was smoother, which in general did not weaken the efficiency of the machining process at a qualitative level. In some places, the surface of components of inhomogeneous SCH 15 materials was torn due to the appearance of pores and poor etching of components (Fig. 3).

The results obtained showed that controlling the cutting force can lead to an increase in the surface roughness of inhomogeneous SCH 15 materials. In addition, variations in the SCH 15 material's microstructure allow favorable processing conditions for obtaining the desired result.

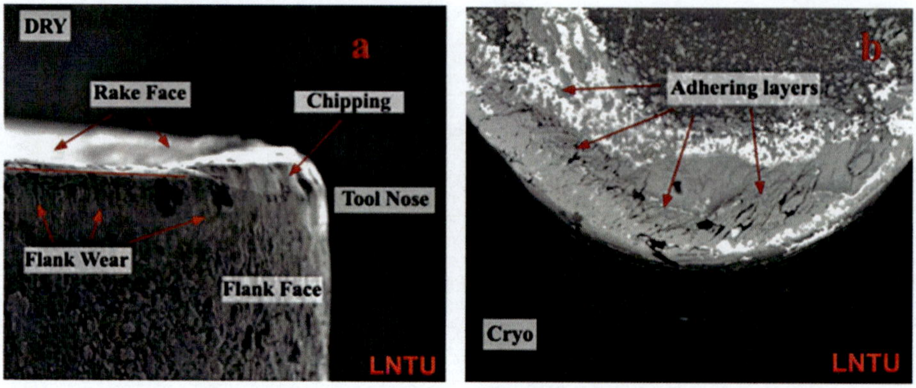

Fig. 2. Tool wear mechanism, where: a) traces of abrasion; b) internal defects of the workpiece components.

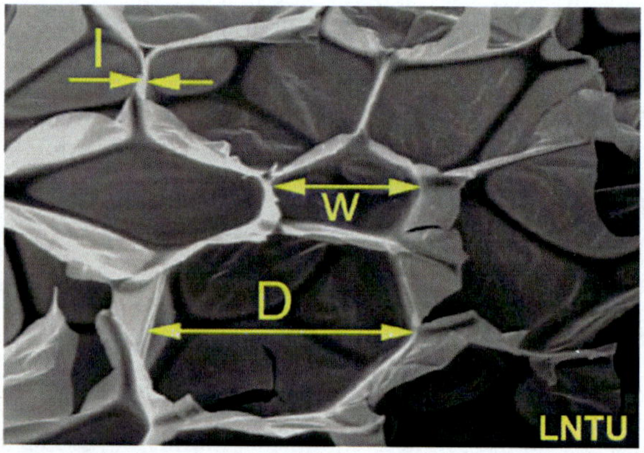

Fig. 3. Microstructure of inhomogeneous materials of grey cast iron SCH 15.

4.2 Basic Parameters of Surface Roughness of Samples Made of Inhomogeneous Materials

One of the essential points is the normalization of roughness parameters, which are calculated according to the indicators R_a, R_z, and KR_a. It should be noted that the roughness rationing of samples made of inhomogeneous materials was carried out according to GOST 9378–93. Table 3 shows the main parameters of surface roughness for samples SCH 15 recorded during the study.

It was found that the first criterion on the surface roughness of inhomogeneous materials SCH 15 depended on the choice of a turning plate, the boundaries of which ranged from $R_a = 1{,}08$ mm to $R_a = 2{,}02$ mm. The second criterion was the maximum lateral wear, the average value of which was: $VB = 0{,}148$ mm. Figure 4 presents practical results of roughness parameters of structurally inhomogeneous materials SCH 15.

Table 3. Basic parameters of surface roughness of midrange SCH 15 samples.

Plate type	Brand turning tools	F, (mm)	V_c, (m/min)	Feed rate, VB, (mm)	KB, (mm)	R_z, (μm)	R_a, (μm)	KR_a, (–)
CC1	M-08-100	0,22	101	0,5	0,122	9,34	1,82	1,82
	M-08-70		72	0,4	0,098	8,65	1,75	2,19
	M-08-50		50	0,6	0,145	12,2	1,99	2,43
CC2	U-08-100	0,2	100	0,4	0,058	7,64	1,87	2,19
	U-08-70		70,2	0,3	0,049	7,64	1,64	2,14
	U-08-50		51	0,4	0,072	7,99	1,75	2,01
CC3	U-03-70-2	0,21	69	0,7	0,033	13,82	2,02	1,42
	U-03-70-01	0,1	69	0,2	0,021	6,24	1,08	3,14

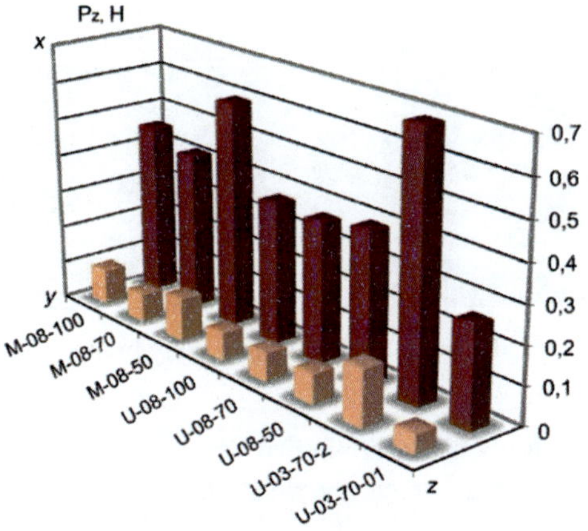

Fig. 4. Practical results of the main parameters of sample roughness.

It should be noted that according to the standard mentioned above, the error of deviation of the maximum wear lies in the range of up to 2 mm. This value allowed us to complete the test process since the results obtained fully correspond to the wear conditions according to the standard.

It should also be noted that the minimum surface roughness of samples and the main parameters obtained favorably affect the chip components' size, shape, and brittleness. This extends the service life of the cutting tool and allows you to determine the average cutting length, equal to LSC = ± 10 m.

4.3 3D Modelling of the T-207 Body Part Using CAD/CAM/CAE Technologies

Modern CAD/CAM/CAE technologies made it possible to simultaneously combine the workpiece of the part 2D modeling (Fig. 1) with the calculation of the design of the part itself in 3D modeling (Fig. 5). At the same time, creating a rational technological process for its production with a rapid transmission on CNC machines, as well as presenting it in 3D modeling. Quantitative results 3D modeling show an assessment of the following main parameters: friction force analysis, hole quality, residual stresses, burr size, specific cutting energy, tool breakage, torque, corrosive floor, fatigue fracture, hardness, working distance effect, topography, microhardness, mechanical properties, surface roughness, tool wear, cutting temperature, cutting force, microstructure, hour size, shape, and the morphology of chips of components made of grey cast iron grade SCH 15 is presented. Figure 5 presents 3D modeling of the T-207 body part.

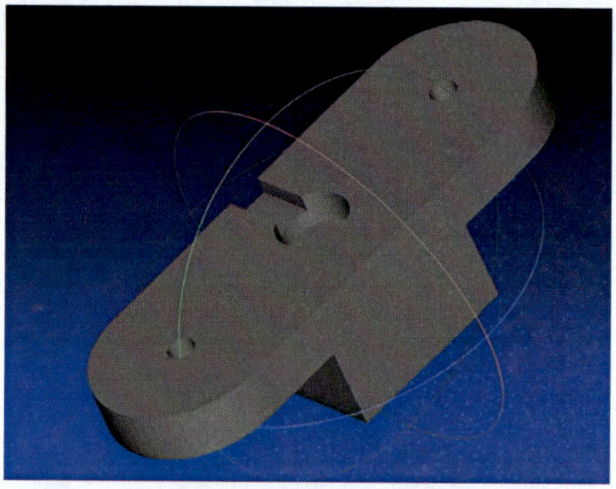

Fig. 5. 3D modeling of the part body brand T-207.

The simulated T-207 body has a number of advantages, in particular: high corrosion resistance, no rod defects, high accuracy, and mechanical processing in general. The housing made of grey cast iron of the SCH 15 brand is designed for fixing (fixing) the clips of any bearings and installing them in products and also occupies a significant segment in the range of foundry products, which are significant advantages for the manufacturing technology of additive (any) production. It should also be noted that structurally heterogeneous materials used in housing manufacturing technology are the most rational and economically justified approach from the point of view of additive manufacturing. It allows you to ensure the uniqueness of manufacturing products using standard sizes and technical characteristics, bringing production to a high-quality level.

5 Conclusions

After using the surface integrity analysis, the tool wear mechanism was presented, where traces of abrasion were recorded, and some internal defects of the workpiece components were identified.

Based on a number of studies, we made recommendations on the minimum surface roughness of samples and justified the maximum tool wear, the average value of which is VB = 0,148 mm. In addition, the average cutting length is LSC = ±10 m.

It was found that the surface roughness of the samples and the main parameters obtained on their basis favorably affect the size and shape of the chip components, as well as brittleness.

Modern CAD/CAM/CAE technologies have made it possible to predict high corrosion resistance, the absence of rod defects, extreme accuracy, and mechanical processing in general.

Acknowledgment. The research was partially supported by International Association for Technological Development and Innovations.

References

1. Ivanov, V.: Process-oriented approach to fixture design. In: Ivanov, V., et al. (eds.) DSMIE 2018. LNME, pp. 42–50. Springer, Cham (2019). https://doi.org/10.1007/978-3-319-93587-4_5
2. Odilov, E., Mardonov, U., Abdirakhmonov, K., Eshkulov, A., Rakhmatov, B.: Effect of magnetic field on the physical and chemical properties of flowing lubricating cooling liquids used in the manufacturing process. IIUM Eng. J. **22**(2), 327–338 (2021). https://doi.org/10.31436/iiumej.v22i2.1768
3. Krehel, R., Szentivanyi, P., Kocisko, M., Pollak, M.: Technical and economic description of the research on a measuring device for continuous measurement of the diameter and vibration of the workpiece during the machining process. Adv. Mater. Sci. Eng. **1**(2), 3373197 (2021). https://doi.org/10.1155/2021/3373197
4. Bovas Herbert Bejaxhin, A., Balamurugan, G., Sivagami, S., Ramkumar, K., Vijayan, V., Rajkumar, S.: Tribological behavior and analysis on surface roughness of CNC milled dual heat treated Al6061 composites. Adv. Mater. Sci. Eng. **1**(2), 3844194 (2021). https://doi.org/10.1155/2021/3844194
5. Lowe, A., Majumdar, K., Mavrokoridis, K., Philippou, B., Roberts, A., Touramanis, C.: A novel manufacturing process for glass THGEMs and first characterisation in an optical gaseous argon TPC. Appl. Sci. **11**, 9450 (2021). https://doi.org/10.3390/app11209450
6. Lam Khanh, N., Van Cuong, N.: Parameter selection to ensure multi-criteria optimization of the taguchi method combined with the data envelopment analysis-based ranking method when milling SCM440 steel. Eng. Technol. Appl. Sci. Res. **11**(5), 7551–7557 (2021). https://doi.org/10.48084/etasr.4315
7. Pasternak, V., Zabolotnyi, O., Ilchuk, N., Cagáňová, D., Hulchuk, Y.: Improvement of processes for obtaining titanium alloys for manufacturing parts with design elements. In: Tonkonogyi, V., Ivanov, V., Trojanowska, J., Oborskyi, G., Pavlenko, I. (eds.) InterPartner 2021. LNME, pp. 323–333. Springer, Cham (2022). https://doi.org/10.1007/978-3-030-91327-4_32
8. Hage, T., Fritzsche, M, Henkel, S., Bliedtner, J.: High-resolution measurement technology for the detection of complex process influences in machining operations. In: Proceedings of the EPJ Web of Conferences, vol. 255, p. 03003 (2021).https://doi.org/10.1051/epjconf/202125503003
9. Li, N., Feng, C.: Research on machining workshop batch scheduling incorporating the completion time and non-processing energy consumption considering product structure. Energies **14**, 6079 (2021). https://doi.org/10.3390/en14196079
10. Abarkan, I., Khamlichi, A, Shamass, R.: Numerical modeling of the low cycle fatigue: effect of manufacturing imperfections caused by machining process. In: Proceedings of the MATEC Web of Conference, vol. 349, p. 02011 (2021). https://doi.org/10.1051/matecconf/202134902011

11. Ikonnikov, A., Leonov, S.: Theoretical-probability model for calculating roofness in magnetic abrasive machining. In: Proceedings of the MATEC Web of Conferences, vol. 346, p. 01045 (2021).https://doi.org/10.1051/matecconf/202134601045
12. Zabolotnyi, O., Pasternak, V., Ilchuk, N., Huliieva, N., Cagáňová, D.: Powder technology and software tools for microstructure control of $AlCu_2$ samples. In: Ivanov, V., Trojanowska, J., Pavlenko, I., Zajac, J., Peraković, D. (eds.) DSMIE 2021. LNME, pp. 585–593. Springer, Cham (2021). https://doi.org/10.1007/978-3-030-77719-7_58
13. Berladir, K.V., Hovorun, T.P., Bilous, O.A., Baranova, S.V.: The modeling of the composition and properties of functional materials based on polytetrafluoroethylene. Funct. Mater. **25**(2), 342–347 (2018). https://doi.org/10.15407/fm25.02.342
14. Berladir, K.V., Hovorun, T.P., Sviderskiy, V.A., Rudenko, P.V., Vyshehorodtseva, M.E.: Nanostructural modification of polytetrafluoroethylene and its composition by energy influence. J. Nano-Electron. Phys. **8**(1), 01033-1-01033–5 (2016). https://doi.org/10.21272/jnep.8(1).01033
15. Abu-Mahfouz, I., Banerjee, A., Rahman, E.: Evolutionary optimization of machining parameters based on surface roughness in end milling of hot rolled steel. Materials **14**(19), 5494 (2021). https://doi.org/10.3390/ma14195494
16. Karpus, V.E., Ivanov, V.A.: Choice of the optimal configuration of modular reusable fixtures. Russ. Engin. Res. **32**, 213–219 (2012). https://doi.org/10.3103/S1068798X12030124
17. Elgnemi, T., Songmene, V., Kouam, J., Jun, M., Samuel, A.: Experimental investigation on dry routing of CFRP composite: Temperature, forces, tool wear, and fine dust emission. Materials **14**(19), 5697 (2021). https://doi.org/10.3390/ma14195697
18. Shvets, S.V., Machado, J.: Numerical model of cutting tool blade wear. J. Eng. Sci. **8**(2), A1–A5 (2021). https://doi.org/10.21272/jes.2021.8(2).a1
19. Tang, T., Liu, C., Wang, R.: Optimization of honing surface roughness of carburized holes based on GRA-RSM. Machines **9**, 291 (2021). https://doi.org/10.3390/machines9110291
20. Tchigirinsky, J., Chigirinskaya, N., Evtyunin, A.: Stability assessment methods of technological processes. In: Proceedings of the MATEC Web of Conferences, vol. 346, p. 03013 (2021). https://doi.org/10.1051/matecconf/202134603013
21. Ivanov, V., Dehtiarov, I., Denysenko, Y., Malovana, N., Martynova, N.: Experimental diagnostic research of fixture. Diagnostyka **19**(3), 3–9 (2018). https://doi.org/10.29354/diag/92293

Control of Thermomechanical Conditions for Working Surfaces of Products Made of Heterogeneous Materials at Finishing Operations

Maksym Kunitsyn(✉), Anatoly Usov, and Yuriy Zaychyk

Odessa Polytechnic National University, 1, Shevchenko Avenue, Odessa 65044, Ukraine
m.v.kunitsyn@op.edu.ua

Abstract. Various technological operations contribute to hereditary defects in the surface layer, e.g., non-metallic inclusions, flocks, air pores, microcracks of a shrinkage nature (smelting), deformation of crystalline grains, cracks of liquation origin (forging, broaching), carbide stitching, cement mesh, coarse-graininess, accumulation of carbides, chips, breaks, a grid of surface cracks, internal cracks, peeling (finishing operations), and cauterization. These defects, being stress concentrators, contribute to cracking during material processing and operation—especially significant losses in the national economy from marriage due to hereditary defects in finishing operations. The calculated dependences on the control of the thermomechanical state of the working surfaces of products made of materials of heterogeneous structure at finishing operations are established in the article. The control of the thermomechanical state of the processing conditions is determined, taking into account accumulated damages and heterogeneities of materials and alloys that are particularly prone to cracking during grinding. An algorithm has been developed for selecting technological conditions for processing materials with genetic heterogeneities that ensure maximum productivity while ensuring quality indicators.

Keywords: Temperature · Technological · Surface processing · Grinding · Layer · Control parameters · Industrial growth

1 Introduction

Establishing links between the essential operational properties of parts (e.g., wear resistance, fatigue and long-term strength, and contact stiffness) and technological parameters - the microrelief of the treated surface, microhardness, the presence of microcracks, chips, the depth of propagation of the hardened layer is one of the most critical tasks of mechanical engineering technology [1, 2].

The study of the influence of mechanical processing alone on the functional properties of products is insufficient since the types mentioned above of treatments (e.g., thermal, thermomechanical, and chemical-thermal) and especially the methods of obtaining

V. Tonkonogyi et al. (Eds.): InterPartner 2022, LNME, pp. 129–139, 2023.
https://doi.org/10.1007/978-3-031-16651-8_13

blanks make a significant contribution to changing the properties of the surface layer, which is subsequently subjected to machining [3, 4].

The complexity of the processes occurring in the metal of the surface layer subjected to mechanical processing and during the operation of these parts makes it necessary to consider the problem of technological heredity only at the final processing operation [5, 6].

The most common final processing method is grinding, which ensures manufacturing parts' high accuracy and high productivity [7]. However, the thermal stress of this type of processing affects the change in the thermophysical parameters of the materials being processed (strength limits, thermal conductivity) [8]. Therefore, the study of optimal temperature problems in the treatment area did not consider these factors.

Nevertheless, with the use of grinding, the appearance of cauterization, cracks, and tensile stresses in the surface layers of parts is associated, which significantly affects the reliability and durability of these parts during their operation [9, 10]. Thus, the purpose of this study is to study and control the thermomechanical state of the working surfaces of products made of a material of a heterogeneous structure during finishing operations, considering the above processing of products to eliminate cracking and burning on the treated surfaces.

Research and control of the thermomechanical state of the working surfaces of products made of a material of a heterogeneous structure at finishing operations, considering the types mentioned above of processing of products, to eliminate cracks and burns on the treated surfaces, constitute the essence of this work.

Achievement of this goal required the formulation and solution of the following main tasks:

1. Develop a mathematical model describing thermomechanical processes in the surface layer when grinding parts made of materials and alloys, taking into account their inhomogeneities that affect the formation of grinding defects and determining the control of technological parameters to eliminate these defects.
2. To develop an algorithm for selecting technological conditions for processing materials with genetic heterogeneities that ensure maximum productivity while ensuring the required quality indicators.

2 Literature Review

The problem of improving the quality of the surface layer of the polished products is currently being solved by the following methods [11, 12]: selection of grinding modes that are rational for a given material and the corresponding characteristics of the tool is carried out; grinding wheels and belts with an intermittent working surface are used; systems of automatic control of active cutting power are used; cooling lubricants are recommended, which significantly reduce the heat intensity of the grinding operation and thereby the likelihood of burns and cracks.

However, with the existing technology for producing parts from heterogeneous materials, including the appearance of composite materials, these methods do not wholly exclude defects arising in the surface layer [13]. On the contrary, it is facilitated by:

inevitable fluctuations in the allowance from errors of previous machining operations; microheterogeneity of the material itself, characterized by grain size, stacking faults, dislocations and structural transformations, warpage of parts during thermal and similar treatment, thermomechanical phenomena accompanying the grinding process and as a result of which burns, microcracks, structural transformations, and residual stresses appear on the treated surfaces [14].

The high thermal intensity of diamond abrasive processing processes leads to the fact that the thermal physics of these processes is often dominant in forming the qualitative characteristics of the processed surface [15]. The lack of control over the thermomechanical state of the working surfaces of articles made of heterogeneous materials at finishing operations does not allow avoiding the defects mentioned above on the treated surfaces [16].

3 Research Methodology

In this paper, we propose a model for describing the thermomechanical processes that form in the machined surfaces of materials of a heterogeneous structure at finishing operations, which allows optimizing the technological parameters to ensure the quality characteristics of the working surfaces of products. The model includes the heat conduction equation [17]:

$$\frac{\partial Q}{\partial t} = a\frac{\partial^2 Q}{\partial x^2} - b(y, t)\nu(t)\frac{\partial Q}{\partial y} \tag{1}$$

$$0 \le x \le S, 0 \le y \le L, 0 \le t \le T \tag{2}$$

The boundary conditions are as follows:

$$Q(x, 0, t) = q_r(x, t) \tag{3}$$

$$-\lambda\frac{\partial Q}{\partial x}\bigg|_{x=0} = a_1\big[U_1(y, t) - Q(0, y, t)\big] \tag{4}$$

$$\Lambda\frac{\partial Q}{\partial x}\bigg|_{x=S} = a_2\big[U_2(y, t) - Q(S, y, t)\big] \tag{5}$$

The initial conditions are as follows:

$$Q(x, y, 0) = q_0(x, y) \tag{6}$$

In formulas (1)–(6), $Q(x, y, t)$ is the temperature distribution function in the surface layers of the workpiece, which moves in the positive direction of the y-axis with the speed $\nu(t)$, which depends on the time t, $\nu(t) \ge 0$, $0 \le t \le T$. The thermomechanical state of the surface layer of the product is characterized by the temperature distribution function over the thickness of the material x, along with the length of contact of the tool with the machined surface $0 \le y \le L$ and in the contact time $0 \le t \le T$.

The thermophysical parameters of the product material are determined by setting the function: $b(y, t) = b > 0$, α, λ are the thermal diffusivity and thermal conductivity coefficients of the product material, and α is the heat transfer coefficient [18].

In this case, the processed surfaces of the products are in contact with the processed tool under the conditions:

$$Q(x, 0, t) = q_\Gamma(x, t), 0 \le x \le S, 0 \le t \le T$$

Moving through the processing zone, the surface layers of the products are heated. If the functions $b(y, t)$ and $\nu(t)$ are known, for each specific function $U(y, t)$, the following initial condition are considered:

$$Q(x, y, 0) = q_0(x, y)$$

It corresponds to a specific temperature distribution function over the product material $Q(x, y, t)$.

The task is to create such a control system for setting temperature regulators in the processing zone of the surface layers so that the deviation of the average temperature of the processed product leaving the processing zone is:

$$\overline{Q}(y, t) = \frac{1}{S}\int_0^S Q(x, y, t)dx, 0 \le y \le L, 0 \le t \le T \tag{7}$$

At $y = L$ from the proper temperature specified for the product's material, its structure did not change. For example, we require that the functionality of the product material:

$$I = \int_0^T \left|Q^*(t) - \overline{Q}(L, t)\right|^\gamma dt, \gamma \ge 1 \tag{8}$$

where $Q^*(t)$ is the preset temperature program for products leaving the processing zone, reaching its permissible value. Then it can get the minimum estimate of the deviation:

$$I = \max_{[0,T]}\left|Q^*(t) - \overline{Q}(L, t)\right| \tag{9}$$

Let us consider the process of heating the surface layer under the action of mechanical treatment, described by the following relations [18]:

$$c\rho\frac{\partial T}{\partial t} = \frac{\partial}{\partial x}\left(\lambda(T)\frac{\partial T}{\partial x}\right), x \in (0, l), t \in (0, \infty) \tag{10}$$

$$T(x, 0) = T^\circ = \text{const}, x \in [0, l] \tag{11}$$

$$\Lambda(T) = \frac{\partial T(l, t)}{\partial x} = \alpha[\nu(t) - T(l, t)], t \in \left[0, \bar{t}\right], 0 < \bar{t} < \infty \tag{12}$$

$$\frac{\partial T(0, t)}{\partial x} = -\frac{q}{\lambda(T)}, t \in \left[0, \bar{t}\right] \tag{13}$$

where T is the temperature (°C); t – time; c – heat capacity coefficient; ρ – density; λ – thermal conductivity coefficient; l – thickness of the surface layer; x – spatial coordinate; α – heat transfer coefficient; $\nu(t)$ is the control parameter, $\nu(t) \in V$, $V = \{\nu = \nu(t) : \nu(t) \in L_2[0, \bar{t}]\}$; q is the heat flux entering the surface layer.

In the interval of temperature variation $[T_1, T_2]$, the function $\lambda(t)$ is positive and, due to the thermophysical properties of the material, it has a bit derivative concerning T. In addition, we assume that within the framework of possible values of operating temperatures $T \in [T_1, T_2]$ the values of the function $\lambda(t)$ are determined by the expression:

$$0 < \beta_1 \le \lambda(T) \le \beta_2 \tag{14}$$

Under these conditions, the system of Eqs. (1)–(4) for each fixed $\nu(t) \in V$ has a generalized solution from the space $V_2^{1,0}(\Omega_i)$, where $\Omega_i = \{(x, t) : x \in (0, l), t \in (0, \bar{t})\}$.

The problem of thermoelasticity in a quasi-static formulation and under the assumption that α_Γ is the coefficient of linear expansion, E is the modulus of elasticity, does not depend on temperature, and is solved analytically [18, 19]. Analysis of thermal stresses shows that under the conditions of the problem under consideration, tensile stresses reach the highest values at depth and compressive ones – on the part's surface [20]. The thermal stress limits can be written as follows:

$$\frac{A_T E}{1-\psi}\left(-T(0,t) + \frac{1+3\Gamma}{l}\int_0^l T(\xi,t)d\xi - \frac{6\Gamma}{l^2}\int_0^l \xi T(\xi,t)d\xi\right) \le \sigma_1[T(0,t)] \tag{15}$$

$$\frac{A_T E}{1-\psi}\left(T(l,t) - \frac{1-3\Gamma}{l}\int_0^l T(\xi,t)d\xi - \frac{6\Gamma}{l^2}\int_0^l \xi T(\xi,t)d\xi\right) \le \sigma_2[T(l,t)] \tag{16}$$

where

$$\sigma_1[T(0,t)] = \begin{cases} \sigma_t[T(0,t)] - for \text{ fragile materials} \\ \sigma_{0.2}[T(0,t)] - for \text{ plastic materials} \end{cases}$$

$$\Sigma_2[T(l,t)] = \begin{cases} \sigma_c[T(l,t)] - for \text{ fragile materials} \\ \sigma_{0.2}[T(l,t)] - \text{ for plastic materials} \end{cases}$$

where ψ – Poisson's ratio; $\sigma_t, \sigma_c, \sigma_{0.2}$ – ultimate tensile, compressive, and yield strength, respectively.

The solution to the problem (1)–(4) will be the limit of solutions to problems (11)–(14) in the space $W_2^{1.0}(\Omega_i)$.

Since the function $\lambda(t)$ is positive, satisfies relation (5), and has a derivative bounded in T on the interval $[T_1, T_2]$, for any fixed control parameter $\nu(t) \in V$, the solutions T_{k+1} of the system of Eqs. (11)–(14) converge as $k \to \infty$ to the solution of system (1)–(4) is the average of type $W_2^{1.0}$.

Then the problem of optimal nonlinear heating of the surface layer of the product with restrictions on thermal stresses and the highest temperature is reduced by solving a system of linear ordinary differential equations:

$$\frac{dx}{d\tau} = A(\tau)x + B(\tau)u + D(\tau), \tau \in [0, T], x(0) = x_0 \ne 0_{RN} \tag{17}$$

with restrictions on phase variables and control parameters:

$$F_i(x, u, \tau) \leq 0, i = \overline{1, s} \tag{18}$$

where $x = x(\tau) = (x_1(\tau), \ldots, x_N(\tau))$ is an N-dimensional vector, $A(\tau)$, $B(\tau)$, $D(\tau)$ are the known dimension matrices $(N \times N)$, $(N \times 1)$, $(N \times N)$ with piecewise continuous coefficients; $u = u(\tau) \in U$ – control parameters.

4 Results and Discussion

The proposed approach to solving a nonlinear thermal conductivity problem with constraints was tested when checking for the adequacy of the control of the thermomechanical state of the working surfaces of products made of materials of heterogeneous structure at finishing operations. A plate of alloy *MAR-M200* with a thickness of $2l = 0.4$ m with an initial temperature of $T_0 = 200\,°C$ was processed in modes at which a temperature of $11000\,°C$ was formed in the processing zone. The maximum permissible temperature in the processing area should not exceed $7200\,°C$ for a minimum time, considering the restrictions on thermal stress and the temperature of the processed surface. The *MAR-M200* material is fragile. Therefore, the temperature in the treatment zone varied in the range [$7200\,°C$–$11000\,°C$]. The dependence of the ultimate strength on temperature was set in Table 1.

Table 1. The dependence of the ultimate strength on temperature.

Temperature, °C		20	720	1050	1100	1150
Tensile strength, MPa	Compression	1500	850	470	310	210
	Stretching	980	540	370	200	140

Furthermore, after the transition to dimensionless quantities, nonlinear relations approximated it using the least-squares method.

The dependence of the thermal conductivity coefficient on temperature was also set in Table 2.

Table 2. The dependence of the thermal conductivity coefficient on temperature.

Temperature, °C	20	200	500	600	700	800	900	1000
$\lambda(T)$, W / (m °C)	10.05	15.07	18.84	20.51	22.1	24.28	26.38	28.05

Furthermore, after the transition to dimensionless quantities, it was approximated by a linear function. It took six iterations in total to get the specified precision.

In Fig. 1 (a) shows the graphs of the dependences on the optimal control time, surface temperatures, and the primary material of the product after the sixth iteration.

The response time was 3.98 min; the optimal control has 135 switchings. Figure 1 (b) and Fig. 1 (c) show, respectively, the dependence of the compressive and tensile strengths and compressive and tensile thermal stresses from time to time at the optimal processing mode. As seen in Fig. 1 (b), the rate of temperature rise in the treatment zone is limited by tensile and compressive thermal stresses. Traditionally, only tensile thermal stresses and restrictions on the surface contact temperature were considered active.

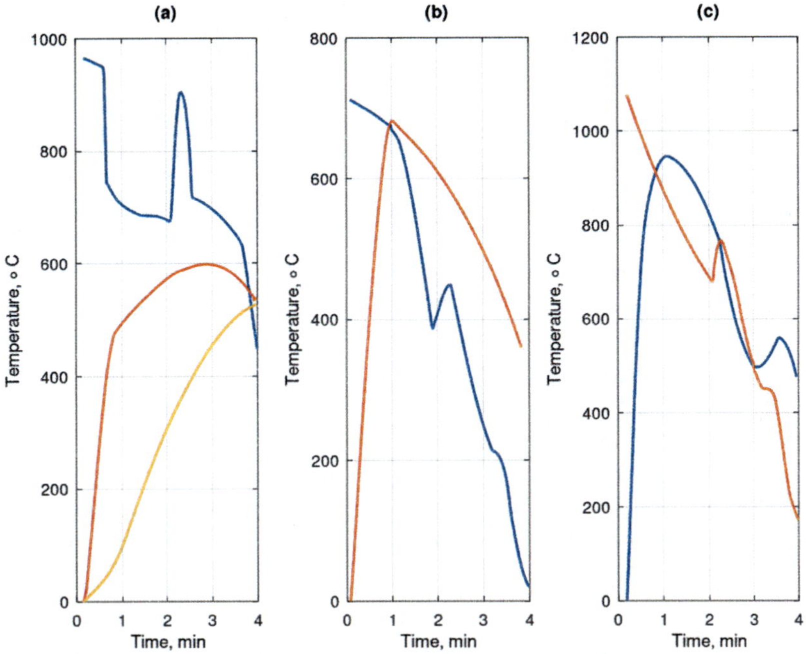

Fig. 1. (a) Graphs of time dependences of optimal control parameters (blue), surface temperatures (red), and center (yellow) after the sixth iteration; (b) Graphs of dependences of compressive strength (blue) and tensile strength (red) from the time at optimal heating mode; (c) Graphs of compressive (blue) and tensile (red) thermal stresses versus time at optimal heating mode.

The quality of the processed surfaces will be ensured if, using the technological control parameters, we select such processing modes, lubricating-cooling media, and tool characteristics such that the current values of grinding temperature $T(x, y, \tau)$, and heat flux $q(y, \tau)$, stresses $\sigma(M)$ And grinding forces P_y, P_z, and coefficient K_{1c} will not exceed their limiting values.

Implementation of limiting inequalities in terms of the values of the temperature itself and the depth of its propagation in the form [4]:

$$T(x, y, \tau) = \frac{C}{2\pi\lambda} \sum\nolimits_{k=0}^{n} H\left(\tau - \frac{kl}{\nu_{kp}}\right) H\left(\frac{L + kl}{\nu_{kp}}\right) \int\limits_{y_1}^{y_2} f\left(x, y, \tau, \tau'\right) d\tau' \leq [T]_M$$

$$T([h],0,\tau)=\frac{C}{2\pi\lambda}\sum_{k=0}^{n}H\left(\tau-\frac{kl}{\nu_{kp}}\right)H\left(\frac{L+kl}{\nu_{kp}}\right)\int\limits_{y_1}^{y_2}\psi\left(x,\tau,\tau'\right)d\tau'\leq[T]_{cp}$$

$$T_k(0,y,\tau)=\frac{C\nu_{kp}}{\pi\lambda\sqrt{\nu_g}}\int\limits_{a}^{\tau}\int\limits_{-e}^{e}\frac{\chi(\eta,t)e^{\frac{(y-\eta)^2}{4(\tau-t)}}}{2\sqrt{\pi(\tau-t)}}\times$$

$$\times\left\{\frac{1}{\sqrt{\pi(\tau-1)}}+\gamma e^{\gamma^2(\tau-t)}\left[1+\Phi\left(\gamma\sqrt{\tau-t}\right)\right]\right\}d\eta dt\leq[T]$$

$$T_k^{max}(L,0)=\frac{C\nu_{kp}\alpha}{\lambda l\nu_q^2}\sqrt{\frac{\alpha}{\pi}}\left[1-exp\left(-\frac{\nu_q\sqrt{Dt_{grind}}}{\alpha}\right)\right]\leq[T]$$

allows avoiding the formation of grinding burns and can serve as a basis for the design of grinding cycles by thermal criterion.

The processing of materials and alloys without grinding cracks can be ensured if the stresses that form in the zone of intense cooling are limited to the limiting values:

$$\Sigma_{\max}(x,\tau)=2G\frac{1+\nu}{1-\nu}\alpha_t T_k \text{erf}\left(\frac{x}{2\sqrt{\alpha\tau}}\right)\leq[\sigma_l]$$

In the case of the dominant influence of hereditary heterogeneity on the intensity of the formation of grinding cracks, it is necessary to use criteria, the structure of which includes deterministic relationships of technological parameters and the properties of the inhomogeneities themselves. As such, it can use the limitation of the stress intensity factor:

$$K=\frac{1}{\pi\sqrt{l}}\int\limits_{-e}^{e}\sqrt{\frac{l+t}{l-t}}\left\{\sigma_x,\sigma_y\right\}dt\leq K_{1c}$$

Alternatively, providing, using the technological control parameters, the limiting value of the heat flux at which the equilibrium of structural defects is maintained:

$$q^*=\frac{P_z\nu_{kp}\alpha_s}{\sqrt{Dt_{grind}}}\leq\frac{\sqrt{3}\lambda K_{1c}}{Hl\sqrt{\pi l}\sigma}$$

The conditions for defect-free grinding can be realized using information about the structure of the processed material. So, in the case of the general nature of structural imperfections of length $2l$, their regular location relative to the contact zone of the tool with the part, the equilibrium condition of the defect can be used as a criterion ratio in the form:

$$l_0\leq\frac{K_C^2}{x[GT_k(1+\nu)\alpha_t]^c} \tag{19}$$

In this formula, the technological part is connected with the value of the contact temperature T_k with the grinding conditions.

The above inequalities link the longitudinal characteristics of the temperature and force fields with the control and technological parameters. Furthermore, they define the range of combinations of these parameters that satisfy the obtained thermomechanical criteria. Simultaneously, the processed material's properties are considered, and the required product quality is guaranteed.

Based on the obtained criterion ratios, an algorithm was built to ensure the quality of the surface layer of parts during grinding, taking into account the maximum processing productivity (Fig. 2). The initial data of the control object (the technological process of grinding) are: physical and mechanical characteristics of the processed material; technical characteristics of processing equipment; machining modes: depth of cut, workpiece speed, transverse feed, the purpose of which is determined by the conditions for limiting the grinding temperature and heat flux, stresses, and grinding forces of the crack resistance coefficient, which will not exceed their limit values; characteristics of the selected tool (circle), affecting the heat stress of the processing process; the process of machining, described by Eqs. (1)–(6), the system of control relations (7)–(9), and thermoelastic stresses (15)–(16) formed in the treatment zone; the quality criteria of the machined surfaces of products are the fulfillment of the absence of structural changes in the machined material and the processing of materials and alloys without grinding cracks.

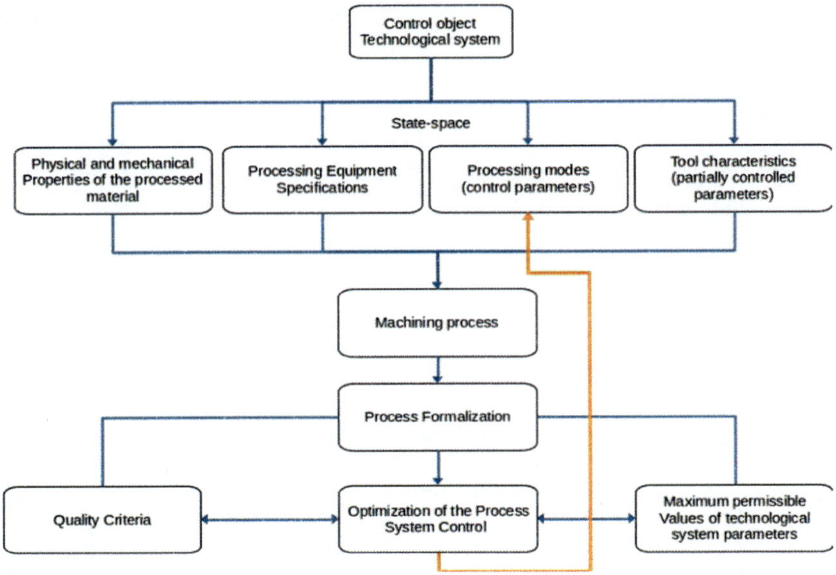

Fig. 2. Algorithm for ensuring machining quality with the technological system's optimal permissible parameters.

Using information about the structure of the material being machined, the prevailing nature of structural imperfections with the length of their regular location relative to the contact zone of the tool with the workpiece, it is possible to use the defect equilibrium condition in the form (19) as a criterion relation.

5 Conclusions

As a result, a scientific and technical problem was solved. It considers calculated dependencies to determine the influence of hereditary defects formed from previous operations on the crack resistance of the surface layer during grinding and the creation of control over the thermomechanical state of the working surfaces of products made of materials of heterogeneous structure at finishing operations. It also considers optimal technological processing conditions under the accumulation of damages and heterogeneity of materials and alloys, especially prone to cracking during grinding. It is of great technological importance for reducing scrap at finishing operations and improving the performance properties of various technological products of machines and mechanisms.

The scientific novelty of the presented research lies in the establishment of calculated dependencies to determine the influence of hereditary defects formed from previous operations on the crack resistance of the surface layer during grinding and the creation of optimal technological processing conditions, considering the accumulated damage and inhomogeneities of materials and alloys, especially prone to structural changes and cracking in the process of grinding:

1. The analysis of the thermomechanical state of the surface layer of products made of materials of the inhomogeneous structure during machining has been carried out;
2. Constructed the control of the thermomechanical state of the working surfaces of products, taking into account the nonlinearity of the heat tension in the processing zone;
3. Technological criteria have been developed for controlling the process of defect-free grinding, which are implemented based on the established functional relationships between the physical and mechanical properties of the processed materials and the main technological parameters.

References

1. Anderson, T.L.: Fracture Mechanics: Fundamentals and Applications, 4th edn. CRC Press, New York (2017)
2. Bofang, Z.: Thermal Stresses and Temperature Control of Mass Concrete. Elsevier Science (2013)
3. Barron, R.F., Barron, B.R.: Design for Thermal Stresses. Engineering Case Studies Online. Wiley (2011)
4. Vigak, V.M.: Control of thermal stresses and displacements in thermoelastic bodies. J. Sov. Math. **62**(1), 2506–2511 (1992)
5. Rapoport, E., Pleshivtseva, Y.: Optimal Control of Induction Heating Processes. CRC Press, New York (2006)
6. Shin, Y.C., Xu, C.: Intelligent Systems: Modeling, Optimization, and Control. CRC Press, New York (2017)
7. La Monaca, A., et al.: Surface integrity in metal machining. Part II: functional performance. Int. J. Mach. Tools Manuf. **164**, 103718 (2021). https://doi.org/10.1016/j.ijmachtools.2021.103718
8. Boley, B.A., Weiner, J.H.: Theory of Thermal Stresses. Dover Publications (2012)

9. Noda, N.: Thermal Stresses. CRC Press, New York (2018)
10. Hetnarski, R.B., Eslami, M.R.: Thermal Stresses – Advanced Theory and Applications. Springer, Netherlands (2008)
11. Barenblatt, G.I.: The mathematical theory of equilibrium cracks in brittle fracture. Adv. Appl. Mech. **7**, 55–129 (1962)
12. Appel, S., Wijker, J.: Simulation of Thermoelastic Behaviour of Spacecraft Structures: Fundamentals and Recommendations. Springer International Publishing, AG (2021). https://doi.org/10.1007/978-3-030-78999-2
13. Carrera, E., Fazzolari, F.A., Cinefra, M.: Thermal Stress Analysis of Composite Beams, Plates and Shells: Computational Modelling and Applications. Elsevier Science (2016)
14. Awrejcewicz, J., Krysko, V.A.: Elastic and Thermoelastic Problems in Nonlinear Dynamics of Structural Members: Applications of the Bubnov-Galerkin and Finite Difference Methods. Springer International Publishing (2020). https://doi.org/10.1007/978-3-030-37663-5
15. Nowacki, W.: Thermoelasticity. Elsevier Science (2013)
16. Eu, B.C.: Generalized Thermodynamics: The Thermodynamics of Irreversible Processes and Generalized Hydrodynamics. Springer, Netherlands (2006)
17. Dats, E., Mokrin, S., Murashkin, E.: Calculation of the residual stress field of the thin circular plate under unsteady thermal action. Key Eng. Mater. **685**, 37–41 (2016). https://doi.org/10.4028/www.scientific.net/KEM.685.37
18. Oborskiy, G.A., Daschenko, A.F., Usov, A.V., Dmitrishin, D.V.: System Modeling. Astroprint, Odessa (2013).(in Russian)
19. Pshenichnyui, B.N., Danilin, I.U.M., Danilin, J.M.: Numerical Methods in Extremal Problems. Mir Publishers (1978)
20. Malkin, S., Guo, C.: Grinding Technology: Theory and Application of Machining with Abrasives. Industrial Press (2008)

The Efficiency of Dynamic Vibration Dampers for Fine Finishing Boring

Alexandr Orgiyan[1], Vitalii Ivanov[2], Volodymyr Tonkonogyi[1], Anna Balaniuk[1](✉), and Vasyl Kolesnik[1]

[1] Odessa National Polytechnic University, 1, Shevchenko Avenue, Odessa 65044, Ukraine
annabalanyuk24@gmail.com
[2] Sumy State University, 2, Rymskogo-Korsakova Street, Sumy 40007, Ukraine

Abstract. Design and application of specialized structural devices that ensure a low vibration level for fine boring increase accuracy, performance, and reliability of machining equipment. The paper reviews the structures and comparative efficiencies of dynamic and impact oscillation dampers. Recommendations for optimization of some oscillation dampers tunning parameters for process operations with relative rotation of the tool and workpiece are provided based on the research conducted. The optimization criterion is assumed to be a reduction to the minimum of the peak amplitudes of the main system oscillations with mass in the entire frequency band characteristic of fine boring. The efficiency of oscillations damping is greatly influenced by changes in the parameters of the machining system with a dynamic oscillations dampener with viscous friction. In particular, the suppression of object oscillations using a vibration damper with a variable stiffness of its main mass fastening is studied. The results of studies of seven types of oscillation dampers for cantilever boring bars are presented. Comparison of the efficiency factors of boring bars oscillation dampers is assessed by the amplitude-frequency characteristics and the ultimate stiffness values of boring bars during boring.

Keywords: Damper efficiency · Dynamics · Shock · Cantilever · Boring bar · Amplitude · Frequency · Amplitude-frequency response characteristic · Oscillations · Process innovation · Industrial growth

1 Introduction

The efficiency of vibration damping can be evaluated by comparing the amplitudes A_0 and A_{dam} oscillations of the object before the damping device installation and with the installed damping device. The ratio of these amplitudes for the oscillation frequency ω is the damping efficiency factor $K_e(\omega) = A_0/A_{dam}$. Vibration damping is efficient if $K_e(\omega) > 1$. It is convenient to determine the $K_e(\omega)$ factor for a characteristic object point for analytical calculation or in the field experiment by comparing its corresponding amplitude-frequency response characteristics (AFRC) and, in the first turn, for resonance frequencies.

V. Tonkonogyi et al. (Eds.): InterPartner 2022, LNME, pp. 140–149, 2023.
https://doi.org/10.1007/978-3-031-16651-8_14

Application of vibration dampers is the most urgent for vibrations of a non-rigid, usually rotating cantilever tool of cylindrical shape with a length-diameter ratio exceeding 3.5–4.0. Vibration dampers are generally placed in a special cavity of a boring bar close to the place of a cutter mounting.

The work aims to study and compare the characteristics of different types of vibration dampers to determine the highest efficiency fine boring. The following study objectives are formulated: to determine the influence of the variability of the main system mass fastening stiffness on the damping efficiency; to identify design solutions and optimization of oscillation dampers parameters for fine boring.

2 Literature Review

Design solutions used to suppress vibration in metal-cutting machines and their model representations can be based on the following features.

The following is distinguished based on the design solution form. Devices and techniques change the elastic system parameters (structural elements with increased damping, weight, and/or stiffness; use of anti-vibration coatings). Thus, a damping dynamic cutting tool holder was designed [1]. Shock absorbers made of copper and brass are used to increase the damping capacity. In a study [2], the damping efficiency is increased by controlling the viscosity of the magnetorheological fluid. An approach for evaluating cutting forces and vibrations during machining was proposed in [3].

The structure also provides increased stiffness to the boring bar. In the research study [4], it was found that the chemical coating of the high-carbon steel (EN31) boring unit with nickel-phosphorus (Ni-P) ensures a significant increase in damping properties of the boring unit after heat treatment. In a research paper [5], it was established that using hybrid copper-zinc particles makes it possible to reduce vibrations during boring effectively. Other types of treatment were studied in [6, 7].

Devices attached to the elastic system (dampers and vibration dampers). For example, some publications describe the choice of damping material for boring bars to reduce cutter wear [8]. In the study [9], a shock damper is attached to the arm support bracket. This damper is installed in the boring bar's shank end and is made of phosphorous bronze.

According to the operating principle of the attached devices: oscillation absorbers (dampers) and other apparatuses [10, 11]. It should be noted that different materials are used for damping particle impacts: lead spheres, steel spheres, glass spheres, tungsten carbide pellets, lead and steel dust, and sand [12]. The effect on thermal treatment is analyzed in [13].

Dynamic oscillation dampers without damping and with damping are described. Shock oscillation dampers without damping and with damping. In the study [13], the designs of three cantilever boring bars, one of which was equipped with a shock damper with a spring-loaded shock mass, were studied experimentally. The shock damper provided an increase in metal removal rate by more than 100%. In the investigated boring process [14], tool vibrations lead to deterioration of the machined surface quality [15, 16], accelerated tool wear, noise generation, and reduction of the machine's service life due to the large length of the cutting tool holder's reach. The vibration resistance can be improved by increasing the system structural damping [17].

Composite action devices (shock-dynamic oscillation dampers with different characteristics of elastic mass bonding and damping). Carbide bars and rods with particle damping are widely used in metalworking, thus increasing the cantilever length when boring hardened steel [18].

Subject to energy dissipation characteristics at oscillations: dry friction; dampers that use changes in damping coefficients when using dry friction are implemented in [19]. Boring units include a substrate layer, damping layer, and a limiting layer, improving the boring process's dynamic characteristics.

Considering viscous friction [20] and internal friction [21], and combination of different friction types [10] are also the topical problem. According to the number of elements that make up the oscillation damper's mass, single-element, and multi-elemental (multi-mass) configurations have existed. In metalworking, both single-element and multi-elemental dampers are used [22]. It is shown that the amount of dissipated energy due to impact and friction is associated with the behavior of particle dampers. Parametric studies of oscillations were performed to evaluate the influence of such system parameters as masses ratio, recovery factor, container size, and excitation amplitude.

Boring bars are designed to suppress oscillations of different spatial forms: torsional oscillations; longitudinal or transversal oscillations [23]; torsional-transversal, and other complex oscillations.

Designed to suppress oscillations of different time forms: deterministic (harmonic, periodic pulse, polyharmonic) and random [22].

According to the possibility of tuning optimization during operation, it can be classified as unregulated (passive) [10] and self-regulated (active) and controllable vibration dampers (oscillation dampers with automatic change of parameters utilizing follow-up systems). Design models have been developed for predicting surface roughness, considering the cutting process's parameters and tool vibration parameters [24]. This paper deals chiefly with devices attached to a flexible system. Integrated processes for ensuring the reliability of the manufacturing system were analyzed in [25, 26].

3 Research Methodology

The experimental stand was designed to study the oscillations while boring by cantilever boring bars with dimensions: d = 25, 50, and 75 mm, l/d = 5…8 [27]. The testing stand was supplemented with a spectra analyzer to assess lateral harmonics levels. Fine boring tests were carried out on specimens made of steel 45 and cast iron SCh (gray cast-iron) 18. Cutting modes: $t = 0.05\ldots0.3$ mm, $s = 0.05$ mm/rev, v = 125 m/min – for cast-iron and v = 150 m/min – for steel. Boring bars d = 25 mm were installed on the boring head UAR 26, and d = 50 and 75 mm – on the boring head UAR 36. Parameters of dampers adjustment varied in the following range: diametrical clearance $2\Delta = 0.1\ldots1$ mm, disc pressure load N = 0…40 N.

Effect of changing the system's parameters with Dynamic Oscillation Damper with viscous friction on the oscillation damping efficiency.

Vibration dampers' parameters selection is difficult for real structures, especially in conditions of the object's variable parameters or oscillation spectrum. Let's introduce the following designations: M – the main system mass; C – stiffness factor of the spring that fastens together the mass M with the immovable base; m – the damper's added mass; c – stiffness factor for masses bonding m and M; b – viscous friction factor for masses bonding m and M; X and x – coordinates of masses M and m respectively; $P\sin\omega t$– constraining force, applied to the mass M (P – force amplitude, t– time); $\Omega_M = (C/M)^{1/2}$ and $\Omega_m = (c/m)^{1/2}$ – natural oscillation frequency of masses M and m respectively; $\mu = m/M$ – the damper's mass ratio; Δ – single-sided clearance between m and M; $\nu_M = \omega/\Omega_M$ and $\nu_m = \omega/\Omega_m$ – non-dimensional frequencies; $\chi_M = X/X_{CT}$– dimensionless amplitude of mass M; $\chi_m = x/X_{CT}$– dimensionless amplitude of mass m.

Let's consider the suppression of object oscillations using the damper whose stiffness changes relative to the average value C_0 according to a harmonic law with doubled frequency $2p$, corresponding, for instance, to the rotational frequency of the tool. The oscillations of the vibration damper are described by the equations (Fig. 1):

$$\begin{aligned} &Md^2X/dt^2 + CX + c(X - x) + b(dX/dt - dx/dt) = P\sin\omega t \\ &md^2x/dt^2 + c(x - X) + b(dx/dt - dX/dt) = 0 \end{aligned} \quad (1)$$

where $C = C_0 + \delta_C \sin 2pt$, a δ_C – amplitude of stiffness deviation from the average value C_0.

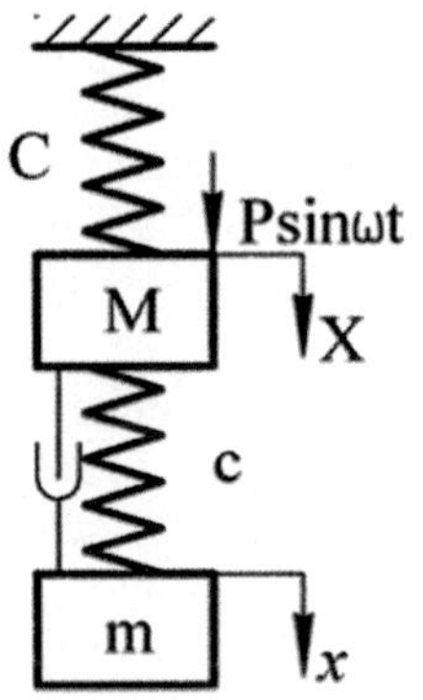

Fig. 1. Calculation model.

In the case of regular rotational frequencies characteristic of the boring bars p, which are by one order less than frequencies Ω_M (1), equations can be solved without considering the parametric effects. Therefore, we will limit our actions by finding amplitudes χ_M with stiffness C in the form of $C = KC_0$, where the factor K can assume values in compliance with congruence

$$K = 1 + (\delta_C/C_0)\sin 2pt \quad (2)$$

Then $\nu_M^2 = K^{-1}\nu_{M0}^2$, where $\nu_{M0}^2 = \omega^2 M/C_0$.

It allows to put down the solution of the system (3) as

$$\begin{aligned} \chi_M^2 = &\left[\left(1-\nu_m^2\right)^2 + 4\beta^2\nu_m^2\right]/\{[(1-\nu_m^2)(1-K^{-1}\nu_{M0}^2) - \\ &-\mu K^{-1}\nu_{M0}^2]^2 + 4\beta^2\nu_m^2\left[1-K^{-1}\nu_{M0}^2(1+\mu)\right]^2\}. \end{aligned} \quad (3)$$

The results of numerical calculations of the values $\chi_M(\nu_M)$ for the case of optimal adjustment and the damper mass value $\mu = 0.1$ are provided in Fig. 2. Data is provided for the damper's three positions relative to the stiffness axes: $K = 1.0$; $K = 1.1$; $K = 0.9$, which correspond to deviations from Co by 10% ($\delta = \pm 0.1$).

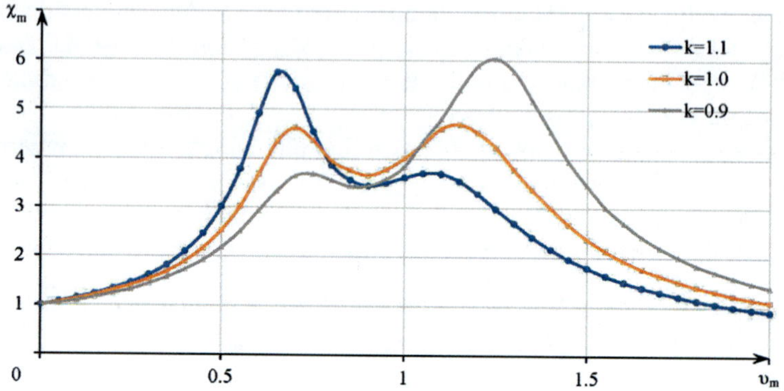

Fig. 2. Amplitude-frequency characteristics for the object stiffness value deviations.

Deviations of the main system stiffness from the mean value significantly affect the height of the graph peaks that correspond to oscillations close to the natural frequencies of the system; the amplitudes of the oscillations at the resonant frequency $\nu_{M0} = 1$ are not the lowest in any of the three cases of stiffness; the greatest damping efficiency corresponds to frequencies lower than $\nu_{M0} = 1$ (for the example considered close to $\nu_{M0} = 0.9$).

The dimensions of the added mass are limited by the tool design. Therefore, the damper mass m is usually assumed to be equal to 10–15% of the boring bar mass. When comparing it with the mass M of the elastic spindle-boring bar system reduced to the cutter, such a damper mass is characterized by the μ factor, whose values vary from 0.1 to 0.5 and more, depending on the dimensions of the boring bar. Calculations of the oscillation damping efficiency indicate that the increase in damping mass above $\mu = 0.1 - 0.15$ increases the damping efficiency to a small extent.

According to the operating principle, single-element dampers designs (Fig. 3a) can be referred to as impact dampers without attenuation or attenuation in case of clearance filling with liquid or gas. The clearance size, impact stiffness, and viscosity characteristics are optimized for these damper types. The axial elastic contracting of the mass (Fig. 3b) turns this damper into a shock-dynamic damper with the possibility of varying the spring stiffness. When the clearance is filled with rubber, polyvinylchloride paste, or other materials (Fig. 3c), an elastic-viscous bonding of the mass emerges, and the damper operating principle changes. The damper becomes a dynamic damper with interrelated elastic-viscous characteristics that cannot be separately optimized. The multi-elemental damper in Fig. 3d, unlike the one shown in Fig. 3c, has one more way to increase efficiency: by searching for the optimal number of n elements in the form of disks. The damper in Fig. 3e, similar to the Kennametal model, is shock-dynamic, and its efficiency depends on the clearance, stiffness, and the masses pressure load.

The shock-dynamic damper in Fig. 3f allows separate optimization of elastic-viscous characteristics of the individual elements' bondings. Elastic spacers 1 covered with a viscous substance are placed between discs (assembly A of Fig. 3), and annular grooves 2 are provided to retain this viscous substance. In the shock-dynamic damper (Fig. 3g), the masses are separated by rigid partitions, which exclude direct interaction between them.

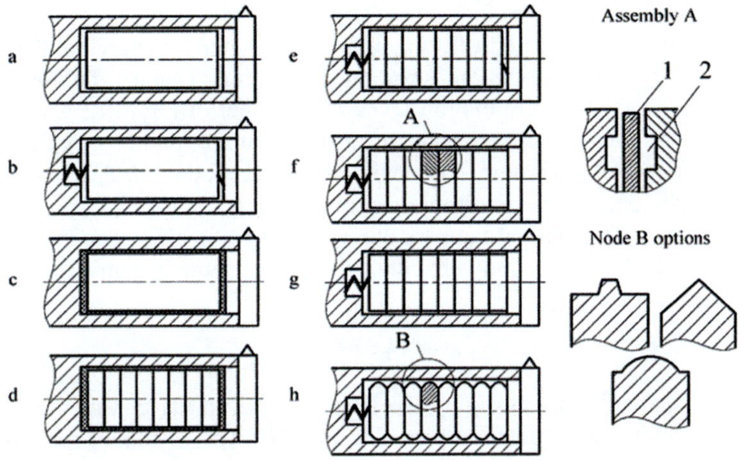

Fig. 3. Cantilever tools oscillation damper types: a – shock; b – shock-dynamic; c – dynamic with attenuation; d – multi-elemental dynamic with attenuation;e – multi-elemental shock-dynamic; f – multi-elemental shock-dynamic with optimization of elastic-viscous coupling between the elements; g – multi-elemental shock-dynamic with rigid partitions between the elements; h – shock- dynamic with optimization of shock characteristics.

The oscillation damping effect depends significantly on the shock pulse characteristic. To increase the shock efficiency, the shock pulse vector shall be aligned with the symmetry axis of the disk and its center of gravity. For this purpose, the shock surface of the disk is located in the middle section of the disk cylindrical surface and performed in the shape of an annular protrusion (Fig. 3h). The shape of the protrusion cross-sectional section that affects the contact area during impaction (options for the shape of protrusions are shown for B node in Fig. 3) is selected experimentally based on the efficiency of oscillations damping. As demonstrated by the experiments described below, a vibration damper with such disks is characterized by a wide range of optimization possibilities and high efficiency.

Shock-dynamic damping can be strengthened in dampers designs where the possibility of arbitrary flat movements of the disks is excluded, and the disks can move only in the direction of the cutting force.

4 Results and Discussion

The assessment results for the influence of the number n elements on the damping efficiency indicate the non-monotonicity of such dependence. The damper installed on different boring bars with the optimal number of $n_{opt} = 8$, reduces the amplitudes during cutting by 3–5 times compared to a single-element one.

When studying frequency characteristics, the efficiency is measured by the efficiency factor $K_e(A_{\max})$, which is not a function of frequency ω, but a function of the maximum amplitudes of the system oscillations without a damper $A_{0\max}$ and with a damper $A_{dam\max}$:

$$K_e(A_{\max}) = A_{0\max}/A_{dam\max} \tag{4}$$

In accordance with the increase in $K_e(A_{\max})$ factor during a joint varying of a diametral clearance 2Δ and N effort for the disc's compression, optimal values of these parameters are established (Fig. 4) for Oscillation Dynamic Damper of e type: $2\Delta_{opt} = 0.3$ mm and $N_{opt} = 6.5$ N. It can be observed that near the optimum (point Opt.), the efficiency changes slightly when the parameters deviate from the optimum. It creates favorable conditions for the operation of multi-elemental oscillation dampers.

For the study of oscillation dampers during cutting, the efficiency factor $K_e(C^{lim})$ was also used, which is a function of the limit stiffnesses of the system without a damper $C_0{}^{lim}$ and with damper $C_{dam}{}^{lim}$:

$$K_e\left(C^{\lim}\right) = C_0^{\lim}/C_{dam}^{\lim} \tag{5}$$

The stiffness was varied by changing the length of the boring bar, and the limit values were set following the method provided by [27]. The results of experiments on varying the boring clearance indicated a virtually complete coincidence of the optimum value of $2\Delta_{opt} \approx 0.3$ mm with the clearance determined as per Amplitude-Frequency Response Characteristic. The optimal values of the disk compression force N = 6.4 N were determined in the experiments.

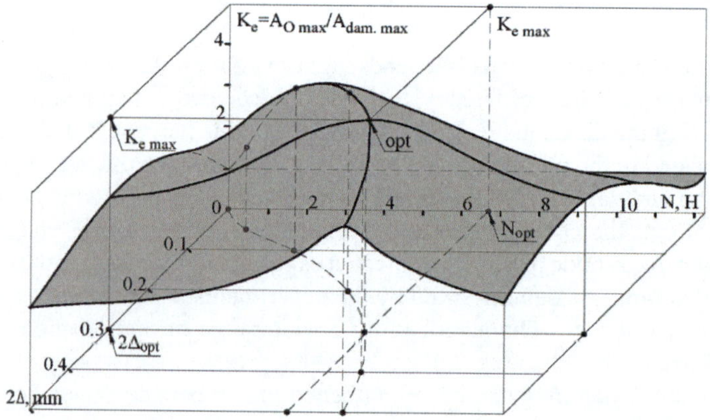

Fig. 4. Influence of 2Δ diametral clearance and pressure load N on the efficiency of *e*-type oscillation dynamic damper.

The results of experiments on varying the boring clearance indicated a virtually complete coincidence of the optimum value of $2\Delta_{opt} \approx 0.3$ mm with the clearance determined as per Amplitude-Frequency Response Characteristic.

Values graphs K_e – for efficiency of different oscillation types dampers (Fig. 5), built per amplitude-frequency characteristics $A_{0\,\max}/A_{dam\,\max}$ and limit boring stiffnesses $C_0^{\lim}/C_{dam}^{\lim}$, characterize the oscillation dynamic dampers (ODD) capacities from different perspectives.

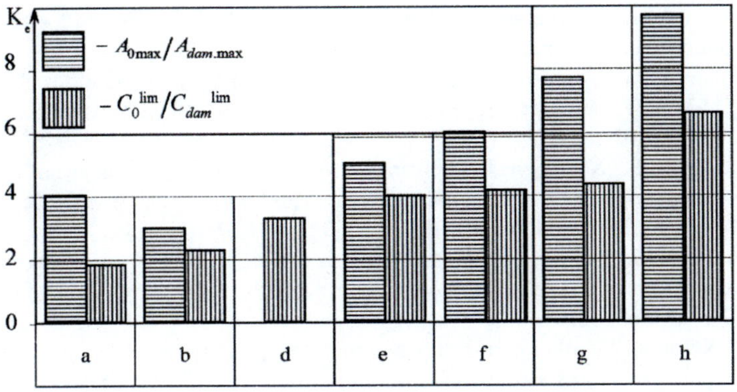

Fig. 5. Comparison of various types of dampers efficiency according to Fig. 3.

5 Conclusions

Both these relative characteristics indicate a significant advantage of multi-elemental dampers over single-element ones. Single-element (ODD) reduces the maximum oscillation amplitudes and reduces the ultimate stiffness of the boring bar by 3–4 times, while the multi-elemental ones – by 5–10 times. Type 3 damper is the most effective among the tested ODDs. Its efficiency at cutting is 3.5 times higher than that of the boring bars equipped with single-element ODD.

It should be noted that it is necessary to investigate the possibility of increasing the damper's efficiency via assessment of their parameter's optimization accuracy.

The settings of multi-elemental dampers with unequal characteristics of the elements and variable parameters of stiffness and damping of the main system links by the dampener will be studied in further research.

Acknowledgment. The scientific results have been partially obtained within the research project "Fulfillment of tasks of the perspective plan of development of a scientific direction "Technical sciences" Sumy State University" ordered by the Ministry of Education and Science of Ukraine (State Reg. No. 0121U112684) and project "Improvement of the Production Planning by Implementation of the Computer-Aided Fixture Design System" within the Joint Ukrainian-Slovak R&D Projects for the period of 2022–2023 funded by the Ministry of Education and Science of Ukraine (State Reg. No. 0122U002657). The research was partially supported by the Research and Educational Center for Industrial Engineering (Sumy State University) and International Association for Technological Development and Innovations.

References

1. Chockalingam, S., Ramabalan, S., Govindan, K.: Chatter control and stability analysis in cantilever boring bar using FEA methods. Mater. Today Proc. **33**, 2577–2580 (2020). https://doi.org/10.1016/j.matpr.2019.12.166
2. Biju, C.V., Shunmugam, M.S.: Performance of magnetorheological fluid based tunable frequency boring bar in chatter control. Meas. J. Int. Meas. Confederation **140**, 407–415 (2019). https://doi.org/10.1016/j.measurement.2019.03.073
3. Wojciechowski, S., Twardowski, P., Pelic, M.: Cutting forces and vibrations during ball end milling of inclined surfaces. Procedia CIRP **14**, 113–118 (2014). https://doi.org/10.1016/j.procir.2014.03.102
4. Chockalingam, S., Natarajan, U., Selvam, M., Cyril, A.G.: Investigation on machinability and damping properties of nickel–phosphorus coated boring bar. Arab. J. Sci. Eng. **41**(2), 669–676 (2015). https://doi.org/10.1007/s13369-015-1830-7
5. Chockalingam, S., Natarajan, U., George Cyril, A.: Damping investigation in boring bar using hybrid copper-zinc particles. JVC/J. Vib. Control **23**(13), 2128–2134 (2017). https://doi.org/10.1177/1077546315610946
6. Kostyk, K., et al.: Simulation of diffusion processes in chemical and thermal processing of machine parts. Processes **9**(4), 698 (2021). https://doi.org/10.3390/pr9040698
7. Zablotskyi, V., Moroz, S., Tkachuk, A., Prystupa, S., Zabolotnyi, O.: Influence of diamond smoothering treatment power parameters on microgeometry of working surfaces of conjugated parts. In: Tonkonogyi, V., et al. (eds.) InterPartner 2019. LNME, pp. 372–381. Springer, Cham (2020). https://doi.org/10.1007/978-3-030-40724-7_38
8. Ramesh, K., Alwarsamy, T., Jayabal, S.: Investigation of chatter stability in boring tool and tool wear prediction using neural network. Int. J. Mater. Prod. Technol. **46**(1), 47–70 (2013). https://doi.org/10.1504/IJMPT.2013.052789
9. Ramesh, K., Alwarsamy, T., Jayabal, S.: ANN prediction and RSM optimization of cutting process parameters in boring operations using impact dampers. J. Vibroengineering **14**(3), 1160–1175 (2012)
10. Rubio, L., Loya, J.A., Miguélez, M.H., Fernández-Sáez, J.: Optimization of passive vibration absorbers to reduce chatter in boring. Mech. Syst. Signal Process. **41**(1–2), 691–704 (2013). https://doi.org/10.1016/j.ymssp.2013.07.019
11. Merzliakov, I., Pavlenko, I., Chekh, O., Sharapov, S., Ivanov, V.: Mathematical modeling of operating process and technological features for designing the vortex type liquid-vapor jet apparatus. In: Ivanov, V., et al. (eds.) DSMIE 2019. LNME, pp. 613–622. Springer, Cham (2020). https://doi.org/10.1007/978-3-030-22365-6_61
12. Marhadi, K.S., Kinra, V.K.: Particle impact damping: Effect of mass ratio, material, and shape. J. Sound Vib. **283**(1–2), 433–448 (2005). https://doi.org/10.1016/j.jsv.2004.04.013
13. Thomas, M.D., Knight, W.A., Sadek, M.M.: Impact damper as a method of improving cantilever boring bars. J. Eng. Ind. **97**(3), 859–866 (1975). https://doi.org/10.1115/1.3438693
14. Lawrance, G., Sam Paul, P., Varadarajan, A.S., Paul Praveen, A., Ajay Vasanth, X.: Attenuation of vibration in boring tool using spring controlled impact damper. Int. J. Interact. Des. Manuf. (IJIDeM) **11**(4), 903–915 (2015). https://doi.org/10.1007/s12008-015-0292-1
15. Lishchenko, N.V., Larshin, V.P., Pitel, J.: Vibrational impact on milled surface irregularities. J. Eng. Sci. **7**(1), A8–A16 (2020). https://doi.org/10.21272/jes.2020.7(1).a2
16. Ropyak, Y.L., Vytvytskyi, V.S., Velychkovych, A.S., Pryhorovska, T.O., Shovkoplias, M.V.: Study on grinding mode effect on external conical thread quality. IOP Conf. Ser: Mater. Sci. Eng. **1018**(1), 012014 (2021). https://doi.org/10.1088/1757-899X/1018/1/012014

17. Sims, N.D., Amarasinghe, A., Ridgway, K.: Particle dampers for workpiece chatter mitigation. In: Proceedings of the ASME 2005 International Mechanical Engineering Congress and Exposition. Manufacturing Engineering and Materials Handling, Parts A and B, pp. 825–832. ASME, Orlando, Florida, USA (2005). https://doi.org/10.1115/IMECE2005-82687
18. Suyama, D.I., Diniz, A.E., Pederiva, R.: The use of carbide and particle-damped bars to increase tool overhang in the internal turning of hardened steel. Int. J. Adv. Manuf. Technol. **86**(5–8), 2083–2092 (2016). https://doi.org/10.1007/s00170-015-8328-z
19. Song, Q., Shi, J., Liu, Z., Wan, Y., Xia, F.: Boring bar with constrained layer damper for improving process stability. Int. J. Adv. Manuf. Technol. **83**(9–12), 1951–1966 (2015). https://doi.org/10.1007/s00170-015-7670-5
20. Siddhpura, M., Paurobally, R.: A review of chatter vibration research in turning. Int. J. Mach. Tools Manuf **61**, 27–47 (2012). https://doi.org/10.1016/j.ijmachtools.2012.05.007
21. Alammari, Y., Sanati, M., Freiheit, T., Park, S.S.: Investigation of boring bar dynamics for chatter suppression. Procedia Manuf. **1**, 768–778 (2015). https://doi.org/10.1016/j.promfg.2015.09.059
22. Lu, Z., Lu, X., Masri, S.F.: Studies of the performance of particle dampers under dynamic loads. J. Sound Vib. **329**(26), 5415–5433 (2010). https://doi.org/10.1016/j.jsv.2010.06.027
23. Pratt, J.R., Nayfeh, A.H.: Chatter control and stability analysis of a cantilever boring bar under regenerative cutting conditions. Philos. Trans. R. Soc. A Math. Phys. Eng. Sci. **359**(1781), 759–792 (2001). https://doi.org/10.1098/rsta.2000.0754
24. Hessainia, Z., Belbah, A., Yallese, M.A., Mabrouki, T., Rigal, J.-F.: On the prediction of surface roughness in the hard turning based on cutting parameters and tool vibrations. Meas. J. Int. Meas. Confederation **46**(5), 1671–1681 (2013). https://doi.org/10.1016/j.measurement.2012.12.016
25. Varela, M.L.R., Putnik, G.D., Manupati, V.K., Rajyalakshmi, G., Trojanowska, J., Machado, J.: Integrated process planning and scheduling in networked manufacturing systems for I4.0: A review and framework proposal. Wireless Netw. **27**(3), 1587–1599 (2019). https://doi.org/10.1007/s11276-019-02082-8
26. Krol, O., Sokolov, V.: Research of modified gear drive for multioperational machine with increased load capacity. Diagnostyka **21**(3), 87–93 (2020). https://doi.org/10.29354/diag/126026
27. Oborskyi, G., Orgiyan, A., Tonkonogyi, V., Aymen, A., Balaniuk, A.: Study of dynamic impacts at combined operations of the thin turning and boring. In: Tonkonogyi, V., et al. (eds.) InterPartner 2019. LNME, pp. 226–235. Springer, Cham (2020). https://doi.org/10.1007/978-3-030-40724-7_23

An Improved Model for Integrated Management Systems

Morteza Rajabzadeh[1](✉), Viliam Zaloga[2], Oleksandr Ivchenko[2], Andrii Chepizhnyi[3], and Dmytro Hladyshev[2]

[1] Mahallat Institute of Higher Education, km 1 of Khomein Road, 3781151958 Mahallat, Iran
rajabzadeh.m@gmail.com

[2] Sumy State University, 2, Rymskogo-Korsakova Street, Sumy 40007, Ukraine

[3] Sumy National Agrarian University, 160, Herasyma Kondratieva Street, Sumy 40000, Ukraine

Abstract. Most manufacturing and service organizations and enterprises have recently established integrated management systems (IMS) to increase their competitive advantages in world markets. This study aims to increase the efficiency of the IMS implementation by developing an improved conceptual model for IMS and recommendations to classify management systems requirements, using analytical and comparative review and analysis of different scientific approaches in the field of IMS, graphical methods of system analysis. As a result of the performed research, five classes of requirements have been developed: specific, analogical, identical, and individually specific requirements of international requirements and additionally standard requirements of IMS. According to the supplied classification, zone of integration and zone of integrated requirements have been supplied, allowing distinguishing structural elements that conform to the distinct requirements of worldwide requirements. Definitions of concepts "integrated management system" and "common requirements of integrated management system" and for that reason, graphical conceptual improved model for developing IMS has been given.

Keywords: Integrated management system · Classification of requirements · Conceptual model · Life-cycle assessment · Manufacturing innovation

1 Introduction

Currently, most organizations are involved with the necessity of modernization of their management systems [1]. It is caused, foremost, by the quickly changeable requirements of consumers to their products or services. More and more, the expansion of products (services) sale markets conduces to the necessity of simultaneous implementation of requirements in the different management systems by the organization, including, for example, quality management system, ecological management, safety, and hygiene. Most of these systems concentrate on providing a competitive advantage for an organization that conduces to realization by it, the conception of optimization of its management system by integrating different requirements of international standards. Currently, an

V. Tonkonogyi et al. (Eds.): InterPartner 2022, LNME, pp. 150–162, 2023.
https://doi.org/10.1007/978-3-031-16651-8_15

organization's competitiveness is often considered from that position that it must support, as a rule, only the optimum use of all economic resources: financial, material, technological, and human [2]. At the same time, the accumulated experience for today testifies that only effective financial management and investments in physical assets cannot guarantee the substantial competitive attractiveness of the company (e.g., organization and enterprise).

More substantial advantages of organizations providing its competitiveness are formed, along with the effective use of economic resources, and due to intangible assets (e.g., adaptive strategic management; efficiency of business processes; part of the capital embodied in the personnel's knowledge and qualification; ability to retain and involve new clients; high corporate culture). All this is accompanied by involving investments, and it stimulates an organizational change in their use, particularly investments in information technologies. To manage these factors effectively, unite them into flexible structures capable of reformatting, depending on the queries of the external dynamic environment, all of them are allowed only by the so-called IMS of company, enterprise, and organization. In the general case, the concept of the management structures (systems) integration, firstly, refers to the difficult process of association in one unit of some management systems [3], and, secondly, it is related to the optimum method of organization exists in the conditions of global competition [4]. However, in the recognized understanding: IMS – is part of the system of the general management of the enterprise, meeting the requirements of two or more international standards on the management systems and functioning as a single unit [5]. Therefore, this paper aims to increase the efficiency of the IMS implementation by developing an improved conceptual model for IMS and recommending the classification of international standards requirements on management systems.

2 Literature Review

Guceva [6] systematized knowledge about the essence and concepts of the different management systems integration in a concrete organization and put forward a postulate: regardless of organization structure, the quantity of subdivisions and their functional appointment, one of the significant mechanisms of "integration" is a process of concordance and combining effort of all, without an exception, personnel of organization subdivisions on achievement to its general-purpose, directed on products (services) competitive advantage providing of all of its enterprises. In the general case, it is possible to consider IMS by analogy with a living organism, the viability of which is determined by the correct functioning of its separate organs and systems and the organism as a whole.

Shaw [7] illustrated a concept of IMS, including three standards in the areas of quality, safety, and environment.

Nováková [8] characterized the possibility of creating IMS by identifying common elements and specific requirements according to professional references ISO 9001:2015 and ISO 14001:2015.

According to Poltronieri [9], an instrument that assesses the degree of maturity of IMS should be employed in organizations for improvements in integration [10, 11], including infrastructure development [12]. According to a literature review conducted in this study, no research that takes the concept of maturity models as a point of support has focused on the evaluation of such maturity level in IMS.

Mellado et al. [13] developed the recommendations for the International Organization for Standardization (ISO) about the development of the common standard, containing requirements for developed IMS, and the possibility of its use for their certification. In their opinion, this standard must be based on management aspects, such as policy, planning, implementation, analysis, etc.

Chew [14], under the concept of IMS, suggested considering a management system of enterprise, which organically combines all the management subsystems: strategic management, quality management, environmental management, innovative management, safety, and labor protection. For a more effective perception of the integration process, the author suggested additionally entering the concept "area of synergy" of the main systems, which characterizes common requirements of two international standards.

It is shown by Armiagova et al. [15] that when solving integration problems arising up at development whether IMS of any organization or country, consideration of the following aspects is essential: basic motive forces, resulting in the process of integration; depth of integration; degree (fullness) of the integration of the different systems.

In this research, the authors, as an example, compared European countries' integration with the former USSR republics.

British Standards Institution (BSI) [16] regulated the management system requirements as a general framework (PAS 99) for the integration of different standards, in particular, ISO 9001:2000, ISO 14001:2004, OHSAS 18001:1999, ISO 22000:2005, ISO/IEC 20000:2005 and ISO/IEC 27001:2005.

Himicheva [17] interprets the concept of IMS as the association of the systems requirements, which are developed on the process-oriented requirements of standards by additive or multiplicative method, forming corresponding models of the integrated management system of product quality (IMS PQ). She developed the structural diagram of the process-oriented multiplicative model of IMS PQ.

Mourougan [18], under the concept of IMS, understands the part of the general management system of an organization that meets the requirements of two or more standards on management systems, functioning as a single unit, and is directed to the satisfaction of interested parties. He offers one of the possible IMS organization models developed based on four standards on management systems: ISO 9001, ISO 14001, OHSAS 18001, and SA 8000. The major condition of such IMS is the compatibility of the presented standards in it, in the first place, conditioned by the presence of identical requirements (for example, continual improvement, implementation of obligatory and legislative requirements, effective and efficient management, conducting management review, and internal audits).

It is possible to find three factors that can affect the development of IMS [19]: enhancement of the enterprise's overall features; the creation of frameworks for implementation of recognized standards for management systems subject to independent reviews (ISO 9001, ISO 14001, OHSAS 18001, SA 8000); the creation of an integrated scheme for independent controls of IMS.

Kania [20] defined the management system integration as: "the connection of the processes, procedures, and practices of the working of applied at the organization with the aim of its politics implementation which can be more effective in achieving aims resulting from the politics than the approach through separate systems". Khanna [21] indicates that the adoption of standards of management systems is a key issue in the development of IMS. An IMS integrates all components of management standards into one coherent system to facilitate achieving its purpose and mission.

Amor [22] proposed a process-based approach for implementing an integrating quality, environment, and security management system based on three aspects: process-based approach, risk management, and a global monitoring system used as integrating factors to satisfy three essential levels of integration, namely, correspondence, coordination, and integration. Ivanov et al. [23, 24] proposed technology for complex parts machining in multiproduct manufacturing. Also, comprehensive approaches in ensuring technological efficiency considering features in designing energy-efficient technological equipment were proposed in [25, 26].

Gankevich [27] determined the general processes of ISO 9001, OHSAS 18001, and ISO 14001, based on which may develop the IMS, including control of documents, control of records, and internal audit.

In today's established management practice, it's possible to call an integrated, management system consisting of three components that implement a set of requirements of products and processes quality (standard ISO 9001), environmental aspects of activities (standard ISO 14001), and the occupational safety and health of personnel (standard OHSAS 18001) [28].

Rebelo [29] defined two independent concepts related to the process of management systems integration: integration is the arrangement, concordance, and association of structures and functions in the holistic system; IMS is the association of two and more interrelated and interactive management systems directed to various enterprise developments.

Herewith, the author noted that at enterprises, the integration of management systems could be used in different combinations of standards, presenting them schematically as "petals", and it means that the set of different "petals" are necessary for the successful functioning of enterprise is IMS.

Santos [30] proposed and presented a methodology to support the Portuguese organizations in developing and structuring the integration process of their individualized management systems and consequently minimize problems that are generators of inefficiencies, value destruction, and loss of competitiveness.

During the conducted analysis of the results, it was determined that though the considered papers have used close to the definition of the concept "IMS", yet, It has to be stated that practically in none of them there are no answers to complex questions, consisting of the following: "What is the purpose of management systems integration?"; "What is the motive force of integration processes of management systems?"; "What happens when integrating various management structures (systems) in the process of analyzing their requirements?"; "How to remove the contradictions and nonconformities (at their presence) among the requirements of different standards when developing IMS?".

The absence of answers to these questions entails a lack of a system for determining and identifying needed requirements. Also, it results in the irrational use of all types of its resources and does not allow the desired economic effect [31], from the implementation of either essential complex standards or developed IMS, even in cases when the actuality of their implementation already makes no doubts. Works that are related to the development and implementation of the management system are a costly measure for any organization [5].

Therefore, this paper aims to increase the efficiency of the IMS implementation by developing the conceptual improved [5] model for IMS and recommendations for the classification of international standards requirements on management structures (systems).

3 Research Methodology

Our assumed integrated management system model includes the following: classification of requirements on management systems; determination of zone of integration and zone of integrated requirements; definitions of concepts "IMS" and "common requirements of IMS"; development of the graphical conceptual improved model for IMS.

The research methodology of this paper is based on the use of:

- the analysis of the importance and necessity of IMS in the Sect. "1 Introduction";
- the analytical and comparative review of different scientific approaches in the field of IMS in the Sect. "2 Literature Review";
- the analysis of conformity between different management systems requirements in the Sect. "4.1 Classification of Management Systems Requirements" (Table 1);
- the graphical methods of system analysis before integration process in the Sect. "4.1 Classification of Management Systems Requirements" (Fig. 1);
- the set theory of system characteristics in the Sect. "4.1 Classification of Management Systems Requirements";
- the graphical methods of system analysis after realization of the integration process in the Sect. "4.2 Structure of the Integrated Management System" (Fig. 2).

4 Results and Discussion

4.1 Classification of Management Systems Requirements

During the conduction of works related to the creation and realization of IMS, it is expedient to enter the classification of corresponding categories of all international standard requirements to which the created management system must conform. As an example, fragmentary, we will consider the issue of conformity of requirements (Table 1), contained only in some (characteristics) clauses of two international standards, ISO 9001 and OHSAS 18001 (we will designate them accordingly by the letters A and B), regulating the requirements of the quality management system [32] and occupational health and safety management system (OH&S MS) [33].

It is possible conditionally to classify the brought requirements in Table 1 to the following categories, designating them in relevant letters of Latin font:

Table 1. Fragment of conformity between OHSAS 18001 and ISO 9001.

A – ISO 9001 [32]		B – OHSAS 18001 [33]	
Clause	Contents of clause	Clause	Contents of clause
4.2.3	Control of documents (see the detailed requirements of clause 4.2.3 in the standard ISO 9001: 2008)	4.4.5	Control of documents (see the detailed requirements of clause 4.4.5 in the standard OHSAS 18001: 2007)
5.3	Quality policy (see the detailed requirements of clause 5.3 in the standard ISO 9001: 2008)	4.2	OH&S policy (see the detailed requirements of clause 4.2 in the standard OHSAS 18001: 2007)
7.5.3	Identification and traceability (see the detailed requirements of clause 7.5.3 in the standard ISO 9001: 2008)		
		4.5.3.1	Incident investigation (see the detailed requirements of clause 4.5.3.1 in the standard OHSAS 18001: 2007)

1. identical requirements (category F) – for example, according to requirements given in 4.2.3.a-g and 4.4.5.a-g, they are the same requirements of considered international standards as well as in the subject and on the object of management;
2. analogical (similar, like, etc.) requirements (category E) – for example, according to requirements given in 5.3.c and 4.2.d, they are the requirements of considered international standards that regulate the same subject but are related to the various objects of management;
3. specific requirements (category G) – for example, according to requirements given in 4.2.a, they are the requirements of considered international standards. They are characterized by their specific requirements of both the subject and the object of management. They are also related to the clauses in the structure of which are already included the "identical" and/or "analogical" requirements;
4. individually-specific requirements (category I) –according to requirements given in 7.5.3 and 4.5.3.1. They are the individual requirements characterized by only one of the considered international standards. Also, they are not related to the clauses in the structure, already included in the"identical" and/or "analogical" requirements.

The expenses of different types of organization resources on IMS development and, consequently, its economic efficiency and prospect from the point of providing its competitiveness, to a great extent, is determined by interrelation among categories of requirements of those subsystems of which it will be created. We will consider the procedure of determination of these interrelations and algorithm of development of predictive mathematical dependence of the developed IMS on the example of some organizations, having the management system (we will designate it by letter A), including already accepted in it and the simultaneously current following subsystems (Fig. 1):

1. quality management system, which conforms to the requirements of international standard ISO 9001 [32] (letter B);
2. occupational health and safety management system, conformable to the requirements of international standard OHSAS 18001 [33] (letter C);
3. other management systems and activities which are not related to the implementation of the requirements of international standards ISO 9001 [32] and OHSAS 18001 [33] (letter D).

It is possible to give the mathematical relationship of the brought management system as the following system of equations for this organization:

$$\begin{cases} \mathrm{A} = \mathrm{B} \cup \mathrm{C} \cup \mathrm{D} \\ B = I(B) \cup G(B) \cup E(B) \cup F(B) \\ C = I(C) \cup G(C) \cup E(C) \cup F(C) \\ \left.\begin{matrix} \mathrm{F(B)} = \mathrm{F(C)} \\ E(B) \approx E(C) \end{matrix}\right. \longrightarrow F(B) \cup E(B) \cong F(C) \cup E(C) \end{cases} \quad (1)$$

$$S_{(OHSAS\ 18001 \longrightarrow ISO\ 9001)} = \frac{E(C) + F(C)}{B} \times 100\%, \quad (2)$$

$$S_{(ISO\ 9001 \longrightarrow OHSAS\ 18001)} = \frac{E(B) + F(B)}{C} \times 100\%, \quad (3)$$

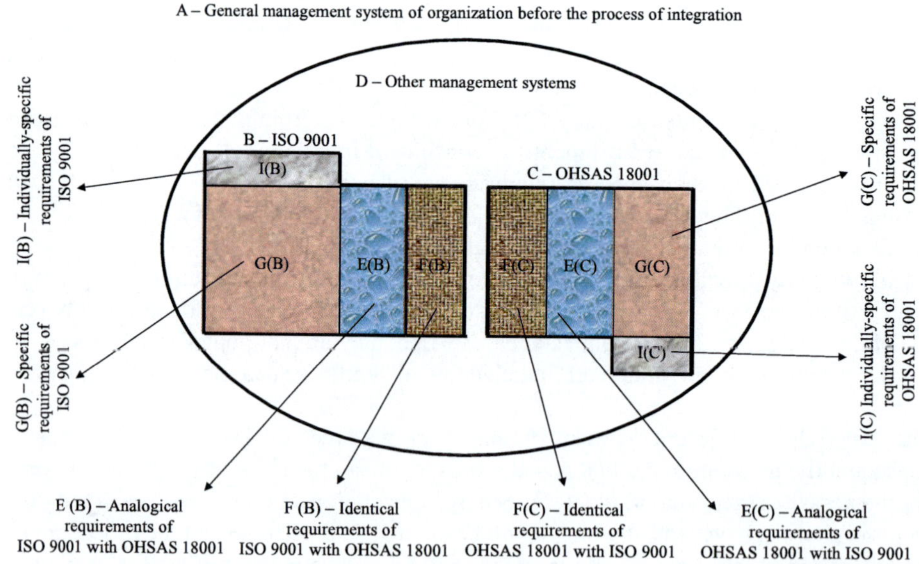

Fig. 1. Diagram of the management system of the organization and its subsystems.

where $S_{(OHSAS\ 18001\ \longrightarrow\ ISO\ 9001)}$ is the conformity degree of requirements of international standard OHSAS 18001 with requirements of ISO 9001; $S_{(ISO\ 9001\ \longrightarrow\ OHSAS\ 18001)}$ is the conformity degree of requirements of international standard ISO 9001 [32] with requirements of OHSAS 18001 [33].

4.2 Structure of the Integrated Management System

As well as any activity, the activity of an organization on the development and implementation of IMS must have a particular goal (motive force). In most cases, in many foreign enterprises, as a motive force, there was a desire to increase the effectiveness and efficiency of the general management of the organization [5] based on minimization of the use of all types of resources by the association of implementation of the conformable (analogical and identical) requirements of international standards on management systems. It gives us an occasion to approve a concept of conformity among the requirements of the different management systems based on their integration process.

At the same time, for the realization of this process, three base conditions are needed at least as follows:

1. minimum two subsystems (for example, two standards, two specifications, two sets of rules, or any combination of two of the indicated (or other) documents) on the management systems in accordance to which the IMS is developed;
2. the base of integration is the requirements of one of the standards (and/or specifications, set of rules, etc.) on the management systems according to which the IMS is developed;
3. the desire of top management and motivation of personnel of organization is in the necessity of these works.

A priori can consider that the first and third conditions are either consistently implemented practically or do not require the substantial attraction of enterprise resources for their realization. The level of enterprise resources attraction and, consequently, the level of expenses on the realization of the second condition to a great extent depends on the following factors:

1. The management system of an organization does not conform to the requirements of any international standards or any other documents on management systems. It has to develop and implement a system which would conform to the chosen (by a necessity) international standards and other documents. Such an approach is called "multiplicative";
2. the management system of an organization conforms to the requirements of one of the international standards or other documents on management systems, i.e., the organization wishes to "add" (to extend) its management system with the purpose of its conformity to the requirements of other international standards or documents. Such an approach is called "additive";
3. there are at least two management systems in an organization, each of which conforms to the requirement of that or other concrete international standards or documents, i.e., the purpose of the organization is an "association" of these systems to

the single integrated mechanism of achievement to purpose, in which, the systems making it accept the status of subsystems. Such an approach is called "synthesis".

Note. In principle, any state of enterprise's management system can be reduced into larger units given in the listed combination of factors.

In this paper as the primary state of the management system of organization, we will consider the third variant, i.e., two management systems operate in the organization in parallel (simultaneously), each of which conforms to the requirements of one of the international standards, for example, ISO 9001 [32] and OHSAS 18001 [33] accordingly (Fig. 1).

For conducting further theoretical research works in the integration of international standards requirements on management systems as a base of integration, we will use the requirements of international standard ISO 9001, considering the recommendations of the International Organization for Standardization (ISO), given in the introduction of this international standard in the clause "Compatibility with other management systems".

The graphic interpretation (conceptual model) of the results of activity on the integration of requirements of indicated international standards in the considered organization is presented in Fig. 2. It is expedient to define zone K(B, C) in Fig. 2 as a "zone of integration" – it is the complex of standard requirements of IMS, which is formed in the process of integration of the management structures (systems) as a result of the

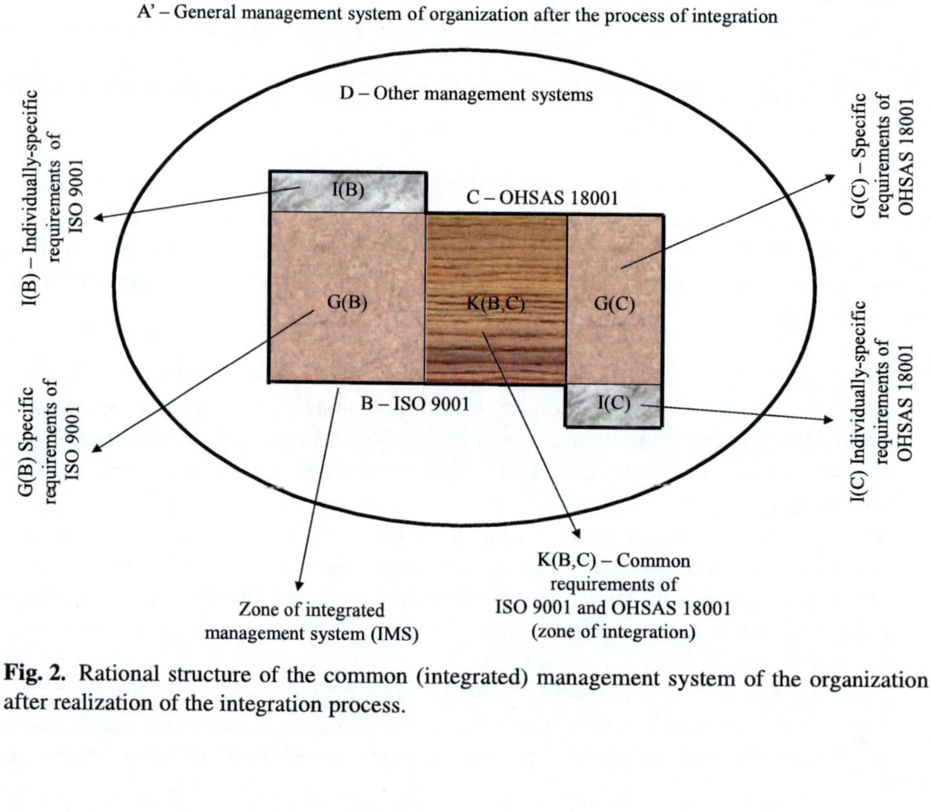

Fig. 2. Rational structure of the common (integrated) management system of the organization after realization of the integration process.

synthesis of identical and analogical requirements of the used subsystems (in this case, international standards B and C).

4.3 Discussion

Unfortunately, the single approach for the concept of IMS is absent, i.e., there is no single opinion about what should be understood under the word combination "integrated management system". Belkovcky and Khachaturov [34] outspeak an assumption that the tendency to integration, i.e., an association of the methods and establishment of the different sciences of their general regularity, is one of the features of sciences development at the present stage of social development in the way to "free society", to ideal society in which, all the cultural traditions get equal rights. Herewith common languages and methods are born. The science of management is no exception. The authors of this research indicate that modern management based on a system approach is engaged with the problems of all the processes complex integration, providing transformation of resources which disposes of humanity for the satisfaction of economic necessities of people and society. Thus, an integration method will be realized based on the situation approach considering the current synergistic interaction of all factors of the external and internal environment of the organization.

The significant differences between the proposed model of IMS in this paper and similar recommendations [9] are distinguishing the structural elements in the proposed model (Fig. 1 and 2), which conform to certain categories of international standards requirements, including identical, analogical, specific, individually specific requirements of international standards and common requirements of IMS [35], which can be efficient instruments for its documentation.

After comparing the obtained results with similar papers [8, 14, 17, 29], and [30], it can be said that, unlike other works that restricted the IMS in the zone of integration (partial requirements). It can also be proposed that the IMS includes all complete requirements of international standards based on which the IMS is formed (Fig. 2). The proposed model allows considering the specific and individually specific requirements of the standards and other normative documents forming the IMS [29], missing in other papers overall. Thus, the integration process of the management systems in the concrete organization is the development, realization, and maintenance of the IMS in the working state, which conforms to the requirements of international standards on given management systems [5].

According to the results of the conducted research works in the integration of management systems, the development of the concept of the integrated management system is offered – it is the management system in realization of international standards requirements in the different spheres of management by their integration based on a synthesis of the identical and analogical requirements of the given standards, characterized with the complex of common IMS requirements.

A concept of the common requirements of an integrated management system (the fifth category in the suggested classification of requirements – category K) can be formulated as follows: common requirements of the integrated management system (category K) – the requirements of the integrated management system, formed based on identical and analogical requirements of international standards and other documents.

The scientific novelties of this research are: the classification of international standards requirements on management systems into five categories of requirements as structural elements, including identical, analogical, specific, individually-specific requirements of international standards and common requirements of the integrated management system; the new conceptual model of IMS, in which concretely indicated the structural elements, conforming to the certain categories of international standards requirements.

5 Conclusions

In the present research, the various scientific works around the development and implementation of IMS were investigated. For example, the requirements of two international standards, ISO 9001 and OHSAS 18001, which regulate the requirements of quality management systems and occupational health and safety management systems, were considered. In this regard, the integration process of two management systems was analyzed. The following findings prove the scientific novelty of this research study.

Categorizing global requirements on management structures (systems) considering the possibility of their integration was offered. Five classes of requirements have been entered and found: identical, analogical, specific, individually specific requirements of global requirements, and common requirements of the IMS;

The zone of IMS was offered to break at least into two: zone of integration and zone of integrated requirements [35], which allowed different structural elements to comply with the unique requirements of worldwide requirements in keeping with their supplied classification.

The concepts of "integrated management system" and "common requirements of integrated management system" [35] had been given, which allowed the normalizing of terminology withinside the region of management systems integration.

A new conceptual (improved) model of IMS has been developed, which concretely indicates the structural elements, conforming to the certain categories of international standards requirements, and graphically shows the zone of IMS.

By comparing the results of this paper with other works, we conclude that our proposed model is more efficient than similar models.

Finally, it can be concluded that the employment of the above recommendations for IMS has a remarkable significant effect on the minimization of the use of all types of resources (e.g., financial, material, technological, and human) of organizations, companies, and enterprises. However, it is deserved that the research is expanded on documentation and audit of IMS as future studies.

Acknowledgment. The research was partially carried out within the R&D project "Fulfillment of tasks of the perspective plan of development of a scientific direction "Technical sciences" Sumy State University" (State Reg. No. 0121U112684).

References

1. Varela, M.L.R., Putnik, G.D., Manupati, V.K., Rajyalakshmi, G., Trojanowska, J., Machado, J.: Integrated process planning and scheduling in networked manufacturing systems for I4.0: A review and framework proposal. Wireless Netw. **27**(3), 1587–1599 (2019). https://doi.org/10.1007/s11276-019-02082-8
2. Petrenko, S.: Integrated management systems as the source of enterprise competitiveness increase. Rev. Gen. Manage. **12**, 66–78 (2010)
3. Kostiuchko, S., Polishchuk, M., Zabolotnyi, O., Tkachuk, A., Twarog, B.: The auxiliary parametric sensitivity method as a means of improving project management analysis and synthesis of executive elements. In: Miraz, M.H., Southall, G., Ali, M., Ware, A., Soomro, S. (eds.) iCETiC 2021. LNICSSITE, vol. 395, pp. 174–184. Springer, Cham (2021). https://doi.org/10.1007/978-3-030-90016-8_12
4. Hrdinova, G.: Integrated management systems as an inseparable part of corporate social responsibility strategy. https://inobr.misis.ru/3587hgc. Accessed 15 Sep 2012
5. Surkova, I.V.: The development of an integrated management system to ensure the quality stability and food safety. Foods and Raw Materials **3**(1), 111–119 (2015). https://doi.org/10.12737/11245
6. Guceva, T.V.: Integration as legal stage of development of the management systems. Manage. Russia and Abroad **5**, 1–5 (2003)
7. Shaw, O.: A to Z of integrated management systems. http://forum.iosh.co.uk/posts/m349654-Integration-of-18001-and-14001. Accessed 20 May 2012
8. Nováková, R.: Integration management system – new requirements of ISO 9001:2015 and ISO 14001:2015 standards. Prod. Eng. Arch. **13**(4), 35–40 (2016)
9. Poltronieri, C.F.: An instrument for the assessment of management systems integration. Gest. Prod., São Carlos **24**(4), 638–652 (2017)
10. Trojanowski, P.: Comparative analysis of the impact of road infrastructure development on road safety – a case study. Sci. J. Marit. Univ. Szczecin **63**(135), 23–28 (2020). https://doi.org/10.17402/436
11. Trojanowski, P., Trusz, A., Stupin, B.: Correlation between accidents on selected roads as fundamental for determining the safety level of road infrastructure. In: Ivanov, V., Trojanowska, J., Pavlenko, I., Rauch, E., Peraković, D. (eds.) DSMIE 2022. Lecture Notes in Mechanical Engineering, pp. 104–113. Springer International Publishing, Cham (2022). https://doi.org/10.1007/978-3-031-06025-0_11
12. Lasinska, N.: Hybrid management methodology for transport projects related to rolling stock. J. Eng. Sci. **8**(2), B7–B11 (2021). https://doi.org/10.21272/jes.2021.8(2).b2
13. Mellado, M.D., Jorgensen, T.H., Remmen, A: Integrated management systems. Technical Report. Aalborg University, Denmark (2004)
14. Chew, B.C.: The implementation of integrated management system in AUO SunPower Sdn. Bhd. Towards quality management system. Int. J. Bus. Technopreneurship **5**, 119–138 (2015)
15. Armiagova, A.A., Egorov, V.S., Pashkov, P.I.: Integrated management systems at small enterprises, Guidelines. http://www.klubok.net/Downloads-index-req-viewdownloaddetails-lid-338.html. Accessed 07 Aug 2012
16. PAS 99: Specification of common management system requirements as a framework for integration (2006)
17. Himicheva, A.I.: Formation of a system of regulatory requirements during the construction of an integrated standard. Bulletin of KNUTD **5**, 31–36 (2006)
18. Mourougan, S.: Auditing integrated management system for continuing suitability, sustainability and improvement. IOSR J. Bus. Manag. **17**, 1–14 (2015)

19. Bogetic, S.: Implementation of integrated management systems in the sector of small and medium enterprises. In: Proceedings of the 2nd International Quality Conference, pp. 1–8. Kragujevac (2008)
20. Kania, A.: Integration of management systems on the chosen example. J. Achievements Mater. Manuf. Eng. **35**, 204–210 (2009)
21. Khanna, H.: A survey on Indian experience on integrated management standards (IMS). Int. J. Qual. Res. **3**, 1–11 (2009)
22. Amor, N.B.: A new process-based approach for implementing an integrated management system: Quality, security, environment. In: Proceedings of the International Multi Conference of Engineers and Computer Scientists, p.p 1–6. Hong Kong (2009)
23. Ivanov, V., Dehtiarov, I., Pavlenko, I., Kosov, I., Kosov, M.: Technology for complex parts machining in multiproduct manufacturing. Manage. Prod. Eng. Rev. **10**(2), 25–36 (2019). https://doi.org/10.24425/mper.2019.129566
24. Ivanov, V., Pavlenko, I., Kuric, I., Kosov, M.: Mathematical modeling and numerical simulation of fixtures for fork-type parts manufacturing. In: Knapčíková, L., Balog, M. (eds.) Industry 4.0: Trends in Management of Intelligent Manufacturing Systems. EICC, pp. 133–142. Springer, Cham (2019). https://doi.org/10.1007/978-3-030-14011-3_12
25. Merzliakov, I., Pavlenko, I., Chekh, O., Sharapov, S., Ivanov, V.: mathematical modeling of operating process and technological features for designing the vortex type liquid-vapor jet apparatus. In: Ivanov, V., et al. (eds.) DSMIE 2019. LNME, pp. 613–622. Springer, Cham (2020). https://doi.org/10.1007/978-3-030-22365-6_61
26. Kostyk, K., et al.: Simulation of diffusion processes in chemical and thermal processing of machine parts. Processes **9**(4), 698 (2021). https://doi.org/10.3390/pr9040698
27. Gankevich, T.V.: Integration of international standards into the organization's management system. Das Management **4**, 54–56 (2010)
28. Dudnic, C.C.: Documentation of management systems. Integrated management system requirements. Das Management **5**, 66–69 (2010)
29. Rebelo, M.: A generic model for integration of quality, environment and safety management systems. TQM J. **26**, 143–159 (2014)
30. Santos, G.F.: Integration of standardized management systems: A dilemma? Systems **3**, 45–59 (2015)
31. Ivanov, V.: Process-oriented approach to fixture design. In: Ivanov, V., et al. (eds.) DSMIE 2018. LNME, pp. 42–50. Springer, Cham (2019). https://doi.org/10.1007/978-3-319-93587-4_5
32. ISO 9001: Quality management systems – Requirements (2008)
33. OHSAS 18001: Occupational health and safety management systems – Requirements (2007)
34. Belkovcky, A.N., Khachaturov, A.E.: Principles of integrated management, Guideline (2012)
35. Rajabzadeh, M.: Development of conceptual apparatus for integrated management systems. Part 2. Bull. Sumy State Univ. Ser. Econ. **2**, 54–63 (2013)

Machining Processes

Numerical Simulation of Grain Concentration Effect on Output Indicators of Diamond Grinding

Janos Kundrak[1], Vladimir Fedorovich[2], Ivan Pyzhov[2], Yevgeniy Ostroverkh[2], and Larisa Pupan[2](✉)

[1] Institute of Manufacturing Science, University of Miskolc, C/1 108 Miskolc-Egyetemváros, Miskolc 3515, Hungary

[2] National Technical University "Kharkiv Polytechnic Institute", 2, Kyrpychova Street, Kharkiv 61002, Ukraine

pupan2612@gmail.com

Abstract. The paper analyzes the influence of the diamond wheel grains concentration during grinding of difficult-to-machine materials (carbonado synthetic polycrystalline diamond CSPD, sitall AC-418, cemented carbide type DIN HG30) on the output indicators of the process. Using the SIMULIA Abaqus, SOLIDWORKS Simulation, and Ansys LS-DYNA software packages, the nature, magnitude, and location of equivalent stresses in the "diamond grain – bond" system were determined under thermal and force loading corresponding to the actual grinding process. Microlevel 3D models of the diamond-bearing layer of a wheel with different grain concentrations of 25%, 50%, and 100% were developed. The change in the localization area and the maximum stress with a change in the grain concentration are established. At a concentration of 25%, the maximum equivalent stresses are located in the volume of grains for all processed materials. An increase in concentration to 50% causes the appearance of equivalent stresses along the boundary of grain embedding in a bond, the value of which is 2 times higher than the stresses in the diamond grain. The highest stress level is observed at 100% grain concentration in the wheel for all types of processed materials. Based on the results of dynamic modeling, the dependences of the destruction volumes of diamond grains, bonds, and the processed material on the concentration of diamond in the grinding wheel and the type of the processed material were obtained. A method for predicting grinding productivity, diamonds' specific wear and specific consumption during the grinding of difficult-to-machine materials is proposed.

Keywords: Grinding productivity · Specific wear · 3D modeling · Equivalent stresses · Manufacturing innovation

V. Tonkonogyi et al. (Eds.): InterPartner 2022, LNME, pp. 165–175, 2023.
https://doi.org/10.1007/978-3-031-16651-8_16

1 Introduction

Increasing productivity and reducing processing costs are machine-building production's main tasks. Concerning grinding difficult-to-machine materials, one of the most important factors affecting these indicators is the concentration of diamond grains in the diamond-bearing layer of grinding wheels. The most promising method for studying the effect of grain concentration on the output grinding parameters is the simulation of the process by the finite element method, which makes it possible to determine the equivalent stresses in local areas of complex geometry under static and dynamic loading. This technique considers the tool's physical and mechanical properties, the processed materials, and the power and thermal factors of the actual grinding process.

The paper aims to analyze the effect of diamond grain concentration in a wheel on the localization and volumes of zones of controlled micro-destruction of diamond grains, bonds, and machined material to predict the productivity and specific consumption of the grinding wheel when processing different materials.

2 Literature Review

Since grinding is often the final stage in the process chain of finished products, there are particularly stringent requirements to improve the efficiency of this technology.

The works [1, 2] indicate the essential categories of parameters for increasing the productivity and efficiency of diamond grinding. These factors work together, and changing one or more categories changes the efficiency of the complete machining. The influence of equipment, processed materials, grinding wheel parameters (abrasive material, bond type, size, and concentration of abrasive grains), technological factors (wheel balancing, dressing frequency, coolant supply, cutting mode) on grinding productivity and efficiency is analyzed.

The subject of many studies is improving the efficiency of grinding with the diamond abrasive wheel as a method of precision finishing of difficult-to-machine materials. The process complexity and multi-parameter nature make modeling the most promising [3, 4].

In [5], the modeling of the efficiency of material removal during grinding is studied, considering the technological parameters (cutting speed and feed rate) and tool characteristics (grain size and concentration). Experimental data and statistical regression methods were used to analyze the process. The most significant effect of grain size on processing efficiency is shown. The proposed regression models can be used to predict the intensity of controlled material removal without experiments.

Features of 3D simulation of ultraprecise grinding of hard and brittle materials with diamond wheels are shown in [6, 7]. The main characteristics of grinding wheels are investigated, considering the probabilistic nature of the grain size and their distribution in the bond. The proposed methodology increases modeling accuracy and a possible expansion of the tool's scope.

To analyze deformations in the cutting zone, the finite element method (FEM) was used in the study [8]. The developed model of the grinding process was created to predict the deformation of the elements of the "abrasive tool – workpiece" system caused by cutting forces and temperature.

In [9, 10], a hybrid process of electric discharge diamond grinding (EDDG), which is a combination of electrical discharge machining (EDM) and grinding with a metal-bonded diamond wheel, is studied. Based on the simulation, the output indicators of the process (wheel wear rate, surface roughness of the processed material, and material removal rate) were obtained depending on various process factors such as pulse current, wheel rotation speed, size, and concentration of abrasive particles. Studies have shown the possibility of optimizing modeling input parameters (including the characteristics of abrasive grains) to ensure high process productivity and surface quality of the products made of difficult-to-machine materials.

The effect of changes in the individual grain's shape due to wear under the cutting forces is described in [11]. The results of model experiments using FEM agree well with the experiment.

The method of creating a model of the micro-grinding process based on the finite element method [12] makes it possible to study the process of change in the tangential and equivalent stresses in the machined material at various technological parameters of grinding. The technique minimizes simulation time, predicts, and optimizes output grinding parameters to ensure the required surface quality.

The authors of [13] note the complexity of grinding modeling due to the stochastic nature of the process. The uncertainty factor of grinding is mainly related to the uncertainty of grain morphology. The article proposes a hybrid FEM-ML approach for predicting the forces created by the action of a single grain and associated with a change in grain geometry during machining. The technique increases the accuracy of predicting the output grinding parameters.

Literature analysis showed that many studies of various aspects of grinding are performed by simulation. But there are practically no data on modeling the diamond grinding of difficult-to-machine materials to calculate the effect of the concentration of diamond grains of a wheel on the output indicators of the process.

3 Research Methodology

To conduct simulation experiments, models of the system "bond – grain – metal phase – pores – processed material (PM)" were developed (Fig. 1).

It is assumed that a fragment of a diamond-bearing layer in the form of a cube with a certain number of diamond grains, surrounded by a bond array, can be transferred to a diamond grinding wheel as a whole. Depending on the concentration of grains in the wheel, the number of grains in the accepted volume of the bond was 1, 2, and 4, corresponding to concentrations of 25%, 50%, and 100%, respectively. The principle of constructing a geometric 3D model of the diamond-bearing layer fragment and the model components is shown in Fig. 2.

The shape, dimensions, and properties of its elements, considered elastic solids, were considered when creating a model. The grain was designed with octahedral geometry as the most common shape of diamond crystals. The grain size varied depending on the grit size of the diamond wheel (50/40, 100/80, 125/100, 200/160). Local inclusions of the metal phase of composition $Ni_{39.6}Mn_{59.6}(Cr_3C_2)_{0.8}$ were created as arbitrarily oriented parallelepipeds.

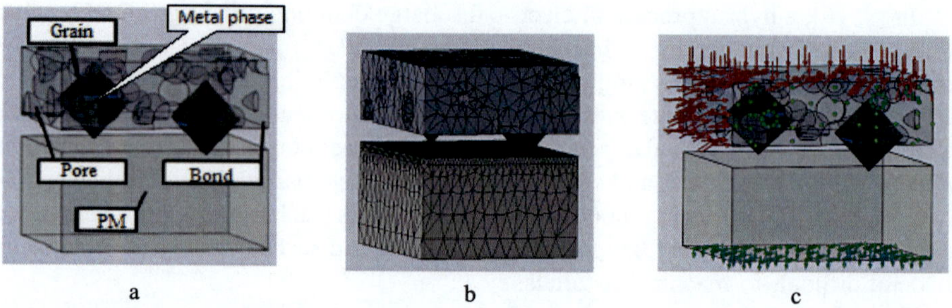

Fig. 1. 3D models (a), finite elements mesh (b) and loading diagram (c) in the study of grinding with a 50% grain concentration in the wheel.

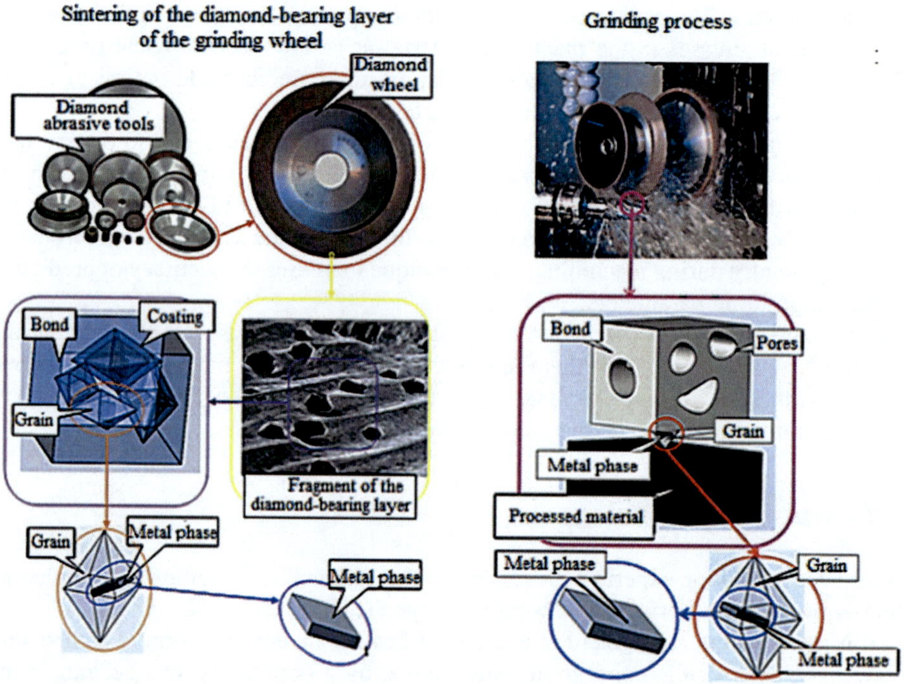

Fig. 2. Principle of designing 3D models of the diamond-bearing layer fragment of the wheel at the micro-level.

The bond was created as a prismatic fragment with dimensions of 1000 × 1000 × 500 μm, which corresponds to 100% grain concentration in the diamond-bearing layer. Diamond grains were randomly placed in the volume of the bond. Their number varied depending on the concentration of diamonds. To reproduce the natural structure of the grinding wheel, pores of arbitrary shape with dimensions of 80 … 100 μm were created in the bond.

The relative pore volume in the model varied according to the selected porosity values of 10%, 20%, 30%. The "processed material" element was modeled as a prismatic fragment with dimensions of 1000 × 1000 × 500 μm.

Finite element analysis was performed using SOLID eight-node elements. In ANSYS, the type of the finite element was selected for each system component. The generation of the finite element mesh and its selective refinement were specified. Hex Dominant and Tetrahedron elements were used for the metal phase mesh. Selective refinement of the finite element mesh was carried out in the places where diamond grains were embedded in a bond, where the grains contacted the workpiece, and where the metal phase was included. This approach made it possible to increase the accuracy of modeling the deformation of the model fragments with remote zones of edge effects.

The model was affected by a static uniaxial uniformly distributed load and temperature (Fig. 1), the values of which corresponded to the actual grinding process [14, 15]. The load values P_y and P_z corresponded to the normal pressure of 10 ... 40 MPa to simulate the pressing force of the wheel according to the technological parameters of diamond-abrasive machining. The uniform temperature load was 500 ... 800 °C according to the grinding conditions when using ceramic bonds. The feed motion and the wheel rotation were simulated by the longitudinal movement of the "bond – diamond grain" element along the "processed material" system element.

The calculation model used the physical and mechanical properties of the processed materials (carbonado synthetic polycrystalline diamond CSPD, sitall AC-418, cemented carbide type DIN HG30) [3, 16]. We also used the calculated information obtained by the authors on the temperature dependence of the properties of synthetic diamond (Fig. 3) and the dependence of the properties of the metal phase on the grinding temperature.

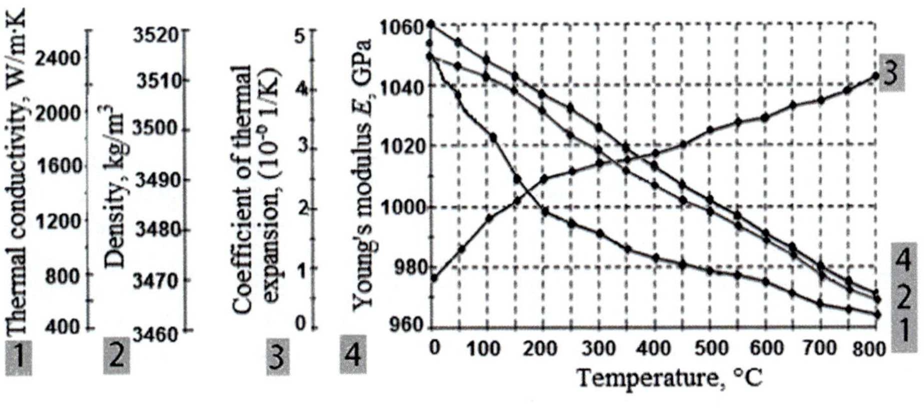

Fig. 3. Temperature dependences of the synthetic diamond properties.

The model for dynamic simulation of the grinding system using the LS-DYNA package is presented as objects oriented in space in a certain way and interacting with each other along the contact surfaces (processed material, bond, diamond grains and metal phase inclusions).

As a result of the contact interaction, stresses arise in the corresponding sections of the contact surfaces due to the deformation resistance of the grinding wheel's components. Elements of the system are destroyed when the ultimate strength is exceeded. Criteria and destruction schemes were set by the process model described in [3].

Since the dimension of the element is set by the mesh parameters before the calculation starts, the program can calculate the volume of destroyed elements of the system "wheel grain – bond – processed material" at each subsequent integration step. Elements in which the ultimate strength is exceeded are considered destroyed and excluded from the calculation. Thus, the volume of destruction of system elements per unit of time is determined. Figure 4 shows a diagram for calculating the destruction of diamond grains of the wheel per unit of time.

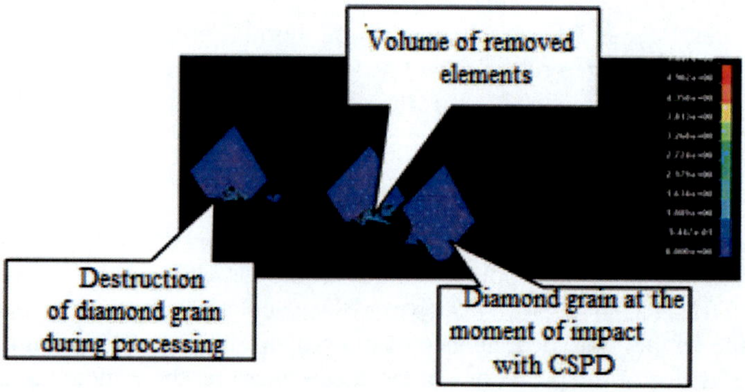

Fig. 4. Illustration of the destruction of the working diamond grain during grinding (the bond and the processed material are hidden for clarity).

4 Results and Discussion

Analysis of the SOLIDWORKS Simulation data indicates that the nature of localization and the level of stresses in the "grain – bond" contact zone significantly depend on the concentration of diamond grains in the wheel and the type of the processed material (its hardness).

For all studied materials (PM), when using wheels with a diamond grain concentration of 25%, the maximum equivalent stresses are observed in the volume of grains in the areas of metal phase distribution and within the boundaries of crystallite facets of diamond grains.

An increase in the diamonds concentration in the grinding wheel up to 50% extends the zone of maximum stresses along the boundary of the grains embedding in a bond (Fig. 5). The value of the equivalent stresses in the contact areas of grains with the bond is 2 times higher than the stress level in the diamond grain.

The largest areas with the maximum stress level are observed at 100% grain concentration in the wheel for all types of processed materials. The increase in the size of the regions of maximum equivalent stresses is explained by the superposition and overlapping of stress fields localized around diamond grains.

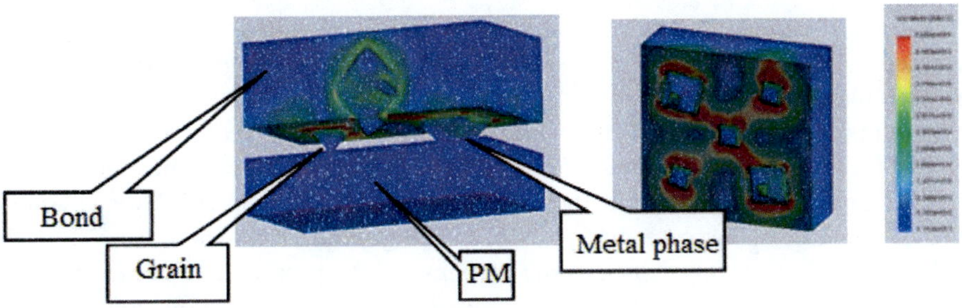

Fig. 5. Stress distribution diagrams for CSPD grinding (50% concentration of diamond grains in the wheel).

A comparative analysis of equivalent stresses during grinding CSPD, HG30, and AC-418 shows the highest level of equivalent stresses for the hardest material (carbonado synthetic polycrystalline diamond CSPD). In contrast to cemented carbide and sitall, a feature of grinding synthetic diamond polycrystal is the occurrence of maximum stresses in the bond of the wheel surrounding the grains, even at low grain concentrations in the wheel (25%). At 100% grain concentration, the equivalent stresses in the contact zone with the processed material reach the critical strength values of the diamond grains. This can play a positive role in the self-sharpening ability of diamond grains and, therefore, improve machining productivity.

To obtain mathematical dependences describing the relationship between the destruction of the structural elements of the grinding wheel and the output characteristics of diamond grinding, a series of model experiments were performed using LS-DYNA. All solid elements of the 3D system "bond – diamond grains – metal phase – processed superhard material", considered using SOLIDWORKS, were imported into LS-DYNA. The material properties of the system components, boundary conditions, and fracture models were specified to simulate the grinding process. The calculation model and boundary conditions are shown in Fig. 6.

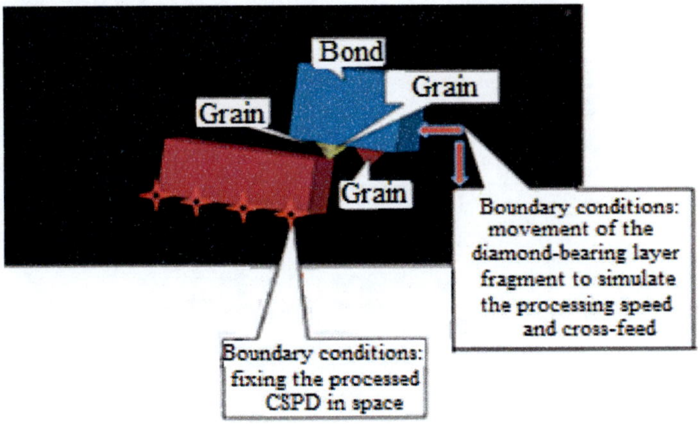

Fig. 6. Solid model exported to LS-DYNA.

Theoretical calculations of the volumes of destroyed grains, the wheel bond, and processed materials depending on the concentration of diamond grains and the type of processed materials are shown in Fig. 7.

For all processed materials, the minimum volume of destroyed diamond grains V_g is typical for the grain concentration of 25%. With an increase in the grain concentration from 50% to 100%, when grinding the hardest processed material, CSPD, the volume of grains destruction V_g increases by 7.4 times, when grinding HG30 and AC-418, by 4.5 and 4.0 times, respectively. Thus, as the hardness of the processed material increases, there is a tendency for the destruction volume of the grinding wheel diamond grains to increase with their concentration. A high concentration of grains in a wheel reduces the economic efficiency of the process.

The sum of the volumes of the destroyed wheel bond (V_b) and the destroyed grains (V_g) (Fig. 7, a, b) correlates well with the specific consumption of diamond grains, which is one of the most important grinding output indicators.

The analysis showed an increase in V_b with increasing grain concentration in the grinding wheel from 25% to 100% for all materials studied. The highest values of V_b are typical for CSPD during grinding, of which the bond is destroyed even at the minimum grain concentration of 25%. This is probably due to an increase in the load on individual grains and, as a result, the achievement of the critical stress in the bond zone localized around the diamond grains. At 100% grain concentration, the volume of the destroyed bond V_b during CSPD grinding exceeds the similar indicator for cemented carbide HG30

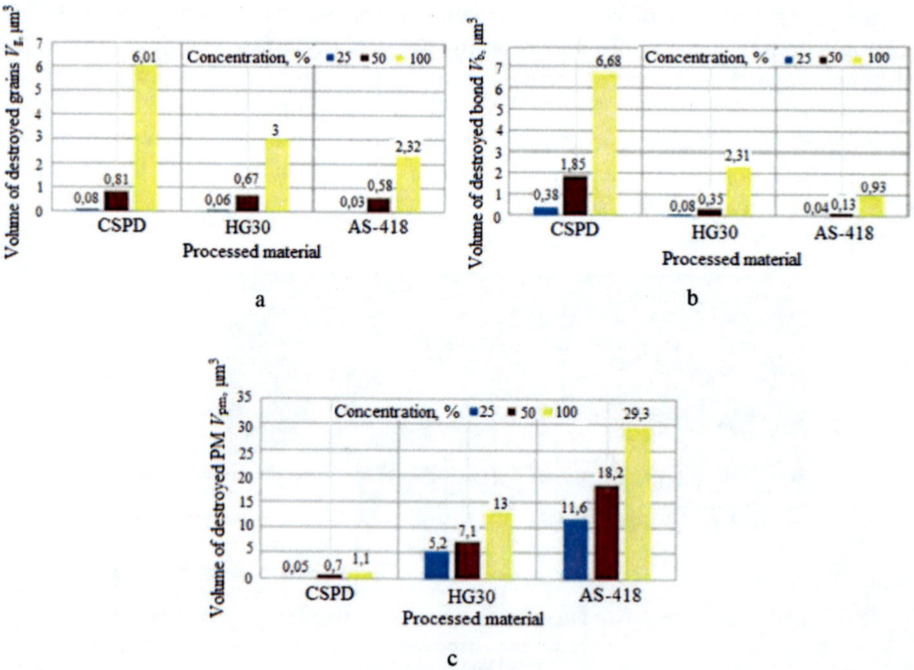

Fig. 7. Dependence of the volumes of destroyed diamond grains (a), bond (b), and PM (c) on the diamond concentration in the wheel and the type of PM.

by 2.9 times, compared to sitall AC-418 by 7.2 times. Increasing the concentration of grains leads to an increase in specific consumption of the wheel and an increase in the cost of machining.

The volume of the destroyed processed material V_{pm} (Fig. 7, c) can be used to estimate the grinding productivity. This indicator increases with the increasing concentration of diamond grains in the wheel for all processed materials. The increase in V_{pm} with increasing concentration from 25% to 100% is 22, 2.5, and 2.5 for CSPD, HG30, and AC-418, respectively. V_{pm} is a qualitative characteristic of machining productivity. It can be argued that the maximum productivity values are typical for the highest grain concentration in the grinding wheel (100%).

Based on the results of dynamic modeling, the specific wear of diamonds during grinding was determined, which characterizes the economic efficiency of the process. Specific diamond wear g_w is as follows:

$$g_w = V_g / V_{pm}, \tag{1}$$

where V_g, μm^3 is the volume of destroyed diamond grains at a certain time; V_{pm}, μm^3 is the volume of destroyed processed material during the same time interval.

The calculation showed that an increase in the grain concentration in a wheel leads to an increase in the specific wear of diamond grains for all processed materials. For example, changing the grain concentration from 25% to 100% during CSPD grinding causes a corresponding change in specific wear from 1.6 to 5.5, which correlates with the experimental data (1.1…8.3 $\mu m^3/\mu m^3$) [3].

5 Conclusions

Simulation of diamond grinding using SOLIDWORKS Simulation made it possible to estimate the stress state of the system "grain – bond – processed material". The stress distribution at the system elements' contacts and the maximum equivalent stresses are determined by the grain concentration in the wheel and the type of processed material.

At a grain concentration of 25%, the area of maximum stresses is concentrated mainly in the volume of diamond grains in the metal phase locations. An increase in the wheel's diamond grains' concentration of up to 50% shifts the maximum stress area to the boundary of the grains embedding in a bond, with their level increasing by 2 times compared to the grain volume. At 100% grain concentration, the equivalent stresses in the contact zone with the machined material reach the critical values of diamond grain strength. This tendency is characteristic of all the studied materials.

Based on the results of dynamic modeling of the diamond grinding process in LS-DYNA, the volumes of the destruction of the system elements (diamond grains V_g, wheel bond V_b, processed material V_{pm}), caused by the increase in equivalent stresses to a critical level, comparable to the ultimate strength of the corresponding material, were examined. Trends in V_g, V_b, and V_{pm} change depending on the type of processed material and the concentration of diamond grains in the wheel. A method for calculating the specific wear of grinding wheel diamonds g_w, i.e., one of the essential output indicators of the grinding process, is proposed.

The results can be used to predict the grinding productivity, specific grain wear, and specific wheel consumption to select the optimal concentration of diamond wheel grains when grinding difficult-to-machine materials (including superhard diamond materials) depending on the performance and efficiency requirements of the process.

The work results can be further used to develop an abrasive composite on a glass-ceramic bond for maximum use of diamond grain's potentially high cutting properties.

References

1. Konrad, W., Bleicher, F., Krajnik, P., Hoffmeister, H.-W., Brecher, C.: Recent developments in grinding machines. CIRP Ann. **66**(2), 779–802 (2017). https://doi.org/10.1016/j.cirp.2017.05.006
2. Zahedi, A., Khosravi, J., Azarhoushang, B.: Grinding efficiency and profile accuracy of diamond grinding wheels dressed with wire electrical discharge conditioning (WEDC). Int. J. Adv. Manuf. Technol. **117**(7–8), 2163–2171 (2021). https://doi.org/10.1007/s00170-021-07114-2
3. Mamalis, A.G., Grabchenko, A.I., Fedorovich, V.A., Grinko, S.A., Paulmier, D., Horvath, M.: Development of an expert system of diamond grinding of superhard polycrystalline materials considering grinding wheel. Int. J. Adv. Manuf. Technol. **17**(7), 498–507 (2001). https://doi.org/10.1007/s001700170150
4. Zhao, B., Zhang, S., Li, J.: Influence of surface functional parameters on friction behavior and elastic–plastic deformation of grinding surface in mixed lubrication state. Proc. Inst. Mech. Eng. J: J. Eng. Tribol. **233**(6), 870–883 (2018). https://doi.org/10.1177/1350650118806375
5. Pandiyan, V., Caesarendra, W., Glowacz, A., Tjahjowidodo, T.: Modelling of material removal in abrasive belt grinding process: a regression approach. Symmetry **12**(1), 99 (2020). https://doi.org/10.3390/sym12010099
6. Guo, B., Zhao, Q.: Ultra-precision machining of hard and brittle materials with coarse-grained grinding wheels. In: Zhang, J., Guo, B., Zhang, J. (eds.) Simulation and Experiments of Material-Oriented Ultra-Precision Machining. STME, pp. 201–236. Springer, Singapore (2019). https://doi.org/10.1007/978-981-13-3335-4_8
7. Yan, Y., Zhang, Z., Liu, J., Yan, H., Wang, X.: Study on the algorithm of three-dimensional surface residual material height of nano-ZrO_2 ceramics under ultra-precision grinding. Micromachines **12**, 1363 (2021). https://doi.org/10.3390/mi12111363
8. Schieber, C., Hettig, M., Zaeh, M.F., Heinzel, C.: 3D modeling and simulation of thermal effects during profile grinding. Prod. Eng. Res. Dev. **14**(5–6), 655–665 (2020). https://doi.org/10.1007/s11740-020-00983-8
9. Sharma, R., Gupta, A., Vates, U.K., Singh, G.K.: Electrical discharge diamond grinding (EDDG): a review. In: Prasad, A., Gupta, S.S., Tyagi, R.K. (eds.) Advances in Engineering Design. LNME, pp. 523–533. Springer, Singapore (2019). https://doi.org/10.1007/978-981-13-6469-3_49
10. Yunhai, J., Linxin, Z.: Research on electrical discharge grinding technics and tool's life of polycrystalline cubic boron nitride cutting tool. Proc. CIRP **68**, 637–642 (2018). https://doi.org/10.1016/j.procir.2017.12.146
11. Wöste, F., Siebrecht, T., Fast, M., Wiederkehr, P.: Geometric physically-based and numerical simulation of NC-grinding processes for the calculation of process forces. Proc. CIRP **86**, 133–138 (2019). https://doi.org/10.1016/j.procir.2020.01.022
12. Gu, Y., Zhu, W., Lin, J., Lu, M., Sun, J.: Investigation of silicon carbide ceramic polishing by simulation and experiment. Adv. Mech. Eng. **9**(11) (2017). https://doi.org/10.1177/1687814017729090

13. Lerra, F., Candido, A., Liverani, E., Fortunato, A.: Prediction of micro-scale forces in dry grinding process through a FEM-ML hybrid approach. Int. J. Precis. Eng. Manuf. **23**(1), 15–29 (2021). https://doi.org/10.1007/s12541-021-00601-2
14. Sałata, M.: The analysis of the influence of technological parameters on the grinding temperature in the single-pass grinding process of solid carbide end mill flutes. Adv. Sci. Technol. Res. J. **16**(1), 190–202 (2022). https://doi.org/10.12913/22998624/143483
15. Chen, S., Liu, X., Wan, L., Gao, P., Zhang, W., Hou, G.: Effect of V2O5 addition on the wettability of vitrified bond to diamond abrasive and grinding performance of diamond wheels. Diam. Relat. Mater. **102**, 107672 (2020). https://doi.org/10.1016/j.diamond.2019.107672
16. Vornberger, A., et al.: Influence of cemented carbide composition on cutting temperatures and corresponding hot hardnesses. Materials **13**(20), 4571 (2020). https://doi.org/10.3390/ma13204571

Wave Nature of the Abrasive Granules Action on the Surface of Parts During Vibration Processing

Andrii Mitsyk[1](✉), Vladimir Fedorovich[2], and Anatoliy Grabchenko[2]

[1] Volodymyr Dahl East Ukrainian National University, 59-a, Central Avenue, Severodonetsk 93400, Ukraine
an.mitsyk@gmail.com

[2] National Technical University "Kharkiv Polytechnic Institute", 2, Kyrpychova Street, Kharkiv 61002, Ukraine

Abstract. The article aims to simulate the propagation of acoustic waves in pseudo-gas of abrasive granules caused by deflector oscillations to increase the energy action on the processed parts. The design of the deflector was developed. Also, a diagram of its location in the vibrating machine reservoir was given. The movement of the abrasive granules and their influence on the deflector was established. The movement of abrasive granules during vibration processing is similar to the movement of atoms or molecules of a gas. The dynamics of the deflector and its effect on the abrasive granules mass were presented. A scheme for forming waves generated by the deflector tab was developed. It is necessary to use deflectors with a large number of tabs to improve the efficiency of vibration treatment. The wave motion in a pseudo-gas of abrasive granules caused by an oscillating deflector was determined. The wave action generated by the deflector on the surface of the processed parts was established. Also, the velocity field in the wave generated by the deflector was given. The pressure of acoustic radiation carried out by a plane wave and the wave action generated by the deflector on the part's surface were determined. The geometric parameters of the wave action on the part's surface were indicated. A diagram of the geometric parameters of the interaction of the front of a cylindrical wave with a flat plate of the working medium deflector was presented.

Keywords: Vibration treatment · Vibrating machine reservoir · Processed parts · Working medium deflector · Acoustic wave · Abrasive granules · Pressure · Velocities field · Process innovation · Industrial growth

1 Introduction

When designing technologies for vibration finishing and grinding process of parts in engineering practice, we are faced with the need to select the optimal technological and design parameters of the process, such as the amplitude, frequency, and motion trajectory of the vibrating machine reservoir, the number of simultaneously processed parts, the

V. Tonkonogyi et al. (Eds.): InterPartner 2022, LNME, pp. 176–189, 2023.
https://doi.org/10.1007/978-3-031-16651-8_17

geometric shape and nature of the movement of the reservoir, etc. All these parameters ensure the required result of vibration finishing and grinding processing. Until now, these parameters were chosen experimentally, which was associated with significant material costs and additional consumption of electricity, water, chemical reagents, as well as the machine operating time of vibrating machines.

From a scientific point of view, the problems mentioned above regarding technology design in the development and implementation of vibration treatment have been little studied. In addition, insufficient attention has been paid to them because of the lack of a comprehensive mathematical approach to the propagation of periodic oscillations in a medium of abrasive granules and the wave effect of these oscillations on the surfaces of parts.

In this regard, the purpose of this article is to develop the theoretical foundations of the process of vibration finishing and grinding processing by means of mathematical simulation of the physical effects that occur during the wave action of oscillations of abrasive granules on the surface of the processed parts. This goal seems timely and relevant for studying processes of working with a free abrasive medium on purpose to their further introduction to metalworking enterprises.

The results of mathematical simulation of processes occurring during vibration treatment of parts are presented in the studies [1]. The simulation is based on the representation of an abrasive granules mass in the form of a pseudo-gas subject to the action of periodic oscillations. The abrasive granules in the pseudo-gas play the role of atoms. The oscillatory action of the flat and cylindrical walls of the vibrating machine reservoir on the pseudo-gas of abrasive granules leads to the appearance of flows in it similar to wave motions. The effects arising from vibration treatment can also lead to the formation of shock waves in the pseudo-gas [1].

This work is devoted to modeling the propagation of acoustic waves in pseudo-gas of abrasive granules caused by deflector oscillations to increase the energy impact on the processed parts. Also, the calculation results for the interaction of acoustic waves with the flat surface of the processed part are obtained.

2 Literature Review

In the practice of finishing and grinding processing, various studies assess the effect of the main process parameters on the resulting metal removal and the achieved surface roughness of the processed parts. These parameters are the forces of mutual pressure and the speed of relative movement of the abrasive medium granules and parts placed in the reservoir of the vibrating machine. These pressures and displacements depend on the technological characteristics of the material of the abrasive medium granules. They also depend on the physical and mechanical properties of the material of the processed parts [2].

The article presents modeling of the system's operation concerning the oscillating trajectory of the reservoir and the working medium. It also evaluates the medium speed and analysis of the system operation under various operating conditions [3].

Also known are the studies of the processes of vibration treatment of metal parts manufactured using additive manufacturing technologies. As a rule, such parts have a high surface roughness, which reduces their performance properties. In these studies, the physical and technological characteristics of the abrasive medium used are given, and the dependence of the achieved surface roughness of parts on the machine time during metal removal and rounding of sharp edges are presented [4].

The practical value of research is in studying ways to improve the productivity of grinding crankshaft journals by intensifying cutting conditions and optimizing the structure of machining cycles [5].

The authors note that with the help of vibration processing, parts are finished with vibrationally liquefied granular media. In particular, this method is used for surface finishing of parts in a limited space, for example, in the cavity of a one-piece cast body [6].

The paper presents the results of experimental studies of vibration treatment operations in the production of cutting tools and deteriorated parts. Cemented carbides are widely used as cutting and tooling materials. It is indicated that vibration treatment operations increase the surface quality of hard alloys, depending on their duration [7].

The paper presents a model of the process of finishing and grinding for a vibrating machine to study the three-dimensional movement of the medium under vibrational action. This work describes a vibrating machine using the discrete element method (DEM), which calculates the normal and tangential contact forces between the granules of the working medium. The DEM method makes it possible to assume the nature of the dynamic movement of individual granules inside the vibrating machine reservoir. The influence of such motion parameters as contact stiffness, friction, and damping of the medium was established to determine the critical parameters of the vibration exciter [8].

The authors propose an approach that allows evaluating metal removal by simulating the vibratory finishing process based on the working medium's hydrodynamics, which can be used to predict the surface roughness of processed parts [9].

The studies indicate that vibration processing is widely used to manufacture various nomenclature parts. Nevertheless, it is difficult to predict the accuracy of the processed part's shape and its surface's roughness. In the studies, a model for obtaining metal removal is proposed. It is based on the static overlap of the medium granules trajectories, calculated by the discrete element method [10].

The article indicates that the vibration treatment requires intensification to reduce machine time. This is achieved by increasing the frequency and amplitude of oscillations of the vibrating machine reservoir. When controlling the surface roughness of parts, it was shown that the amplitude of oscillations has a decisive effect with an increase in the processing time when the required technological result is achieved [11].

3 Research Methodology

The design of the deflector and its location in the vibrating machine reservoir. During vibration treatment, a deflector is often used. It is located in the center of the cylindrical part of the vibrating machine reservoir (Fig. 1). Structurally, the deflector is depicted in the form of a single "tab", the center of which freely rotates on an axis rotating together with the walls of the reservoir with amplitude A and frequency ω. The edges of the deflector are cylindrical surfaces with a radius of r_{d0}. The distance from the center of the axis of rotation of the deflector "tab" to its distant end is R_d. . The distance from the center of the cylinder (that forms the edge of the deflector to a point on the part) surface is r_d.

The circles indicate the abrasive granules, the arrows near which indicate the randomness of their movement. Deflector "tab" thickness is b. Digit *1* indicates the processed parts.

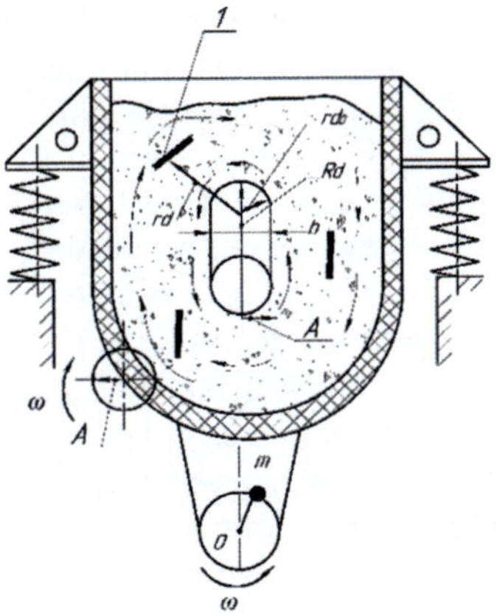

Fig. 1. The design of the deflector and its location in the vibrating machine reservoir.

We will consider the movement of abrasive granules during vibration treatment similar to the movement of atoms or molecules of a gas. In this case, the energy of motion of the granules are created by the oscillating walls of the vibrating machine reservoir and the surface of the deflector.

In a gas, transverse waves are impossible because shear deformations are not transmitted along the wave propagation. Therefore, we will consider only the movements of the deflector surface, which create longitudinal waves.

The movement of abrasive granules and their effect on the deflector. In the works [12] it is shown that as a result of the vibrating machine operation, the abrasive granules participate in a movement directed opposite to the rotation of the vibration exciter. The speed of this movement near a flat or cylindrical surface V_{b_r} is determined by the formula:

$$V_{b_r} = -\frac{A\omega}{2\pi}\left[cos\alpha^* - cos\alpha_0 - sin\alpha_0(\alpha_0 - \alpha^*)\right], \quad (1)$$

where α^*, α_0 – the pick-up and rebound angles of the granule. In this case, it is essential to know that the expression in parentheses in formula (1) cannot exceed unity.

Since the deflector can freely rotate around the axis (Fig. 1), it will oscillate with the axis and participate in a circular motion with the granule's mass. The angular velocity of this movement will be equal to $\Omega_d = \frac{V_{br}}{R}$. The maximum angular velocity resulting from the cyclical movement of the reservoir is equal to $\Omega_r = \frac{A\omega}{R}$. The following ratios determine the maximum accelerations that cause movements with these angular velocities:

$$a_d = \Omega_d^2 R = \frac{A^2\omega^2}{4\pi^2 R}; \; a_r = \Omega_r^2 R = \frac{A^2\omega^2}{R}. \tag{2}$$

A comparison of these values shows that the maximum acceleration created by the constant motion of the granules – a_d is almost 40 times less than the amplitude value of the acceleration created by the cyclic movement of the reservoir wall of the vibrating machine – a_r. Consequently, the influence of the acceleration a_d in the transition to rotating with an angular velocity of Ω_d, together with the deflector in the coordinate system can be neglected. That is, the deflector "tab" can be considered motionless and located in the inertial coordinate system, where Newton's first law is fulfilled.

Deflector Dynamics and Its Action on an Abrasive Granulesmass. The deflector axis is located on a rod that freely rotates synchronously with the walls of the vibrating machine reservoir with the same amplitude. From Fig. 1 can be seen that the axis of rotation of the deflector makes circular movements, the same as the movements of the cylindrical wall of the vibrating machine. Therefore, the axis of rotation of the deflector does not change its position relative to the cylinder's axis, coinciding with it. The impact of this zone of the deflector on wave processes is insignificant in the area where the processed parts are located due to the distance from this zone.

The end of the "tab" of the deflector makes a complex movement, consisting of longitudinal oscillations with amplitude A and oscillations centered on the end of the "tab" having the maximum angular deviation $\mathrm{arctg}\left(\frac{A}{R_d}\right) \approx \frac{A}{R_d}$. This equality is valid due to the smallness of the amplitude A in comparison with R_d. For the same reason, such oscillations can be neglected compared to oscillations with amplitude A.

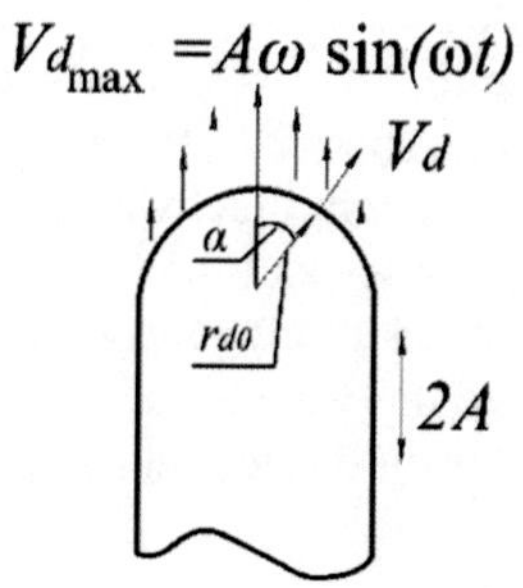

Fig. 2. Scheme of the formation of waves generated by the "tab" of the deflector.

Consequently, the solution to the problem of propagation of deflector oscillations in pseudo-gas from abrasive granules is reduced to finding the parameters of the wave formed by an "infinite" bar with a radiating surface in the form of a side wall of a cylinder with radius r_{d_0} (Fig. 2).

The radius of the rounded part of the deflector "tab" is r_{d_0}, A is the amplitude of the oscillational-rotational motion of the deflector (and the walls of the vibrating machine reservoir), $V_{d_{max}}$ is the speed of the oscillational motion of the Uvector r_{d_0}.

The arrows in the upper part (Fig. 2) conventionally show the direction of the deflector's waves emitted by the "tab". As noted above, transverse waves are not formed in the gas. Therefore, wave radiation will be caused only by surfaces orthogonal to the lateral cylindrical surface of the deflector, which are the oscillation velocities on the cylindrical surface $V_d = V\cos\alpha_{d_{max}}$.

Wave motion in pseudo-gas of abrasive granules caused by an oscillating deflector. To solve this problem, it is necessary to use the wave equation in polar coordinates:

$$\frac{1}{r_d}\cdot\frac{\partial}{\partial r_d}\left(r_d\frac{\partial U}{\partial r_d}\right)+\frac{\partial^2 U}{\partial^2\alpha}=\frac{1}{C^2}\cdot\frac{\partial^2 U}{\partial^2 t}, \tag{3}$$

where U is any characteristic of a sound wave [13].

Since the radiating surface moves according to the harmonic Helmholtz law, for the amplitude values of the function U expression (3) takes the form:

$$\frac{1}{r_d}\cdot\frac{\partial}{\partial r_d}\left(r_d\frac{\partial U}{\partial r_d}\right)+\frac{\partial^2 U}{\partial^2\alpha}+\frac{\omega^2}{C^2}U=0. \tag{4}$$

When tending to infinity, the following conditions must be met:

$$\lim_{r_d\to\infty} P=0;\ \lim_{r_d\to\infty} r_d\left(\frac{\partial P}{\partial r_d}-i\omega P\right)=0. \tag{5}$$

The solution of Eq. (4) for the pressure emitted by the cylinder surface, taking into account the conditions for pressure at infinity (5), will have the form [14]:

$$P=C_pH_1^{(1)}(kr_d)cos\alpha, \tag{6}$$

where C_p is a constant determined from the boundary condition, $H_1^{(1)}(kr_d)$ is the Hankel function of the first kind of the first order, which is a superposition of the Bessel and Neumann functions [15].

On the cylindrical surface of the "tab" of the deflector ($r_d = r_{d_0}$, Fig. 1), the following relationship should be satisfied for the pressure:

$$\frac{\partial P}{\partial r_d}=i\rho\omega V_d. \tag{7}$$

Here $V_d = V_{d_{max}}\cos\alpha$ is the velocity on the surface of the cylinder, perpendicular to it. Taking into account expressions (6) – (7) and the expression for V_d, we can derive the relation for the constant C_p:

$$C_p=\frac{i\rho\omega^2 A}{\left(\frac{\partial H_1^{(1)}}{\partial r_d}\right)}. \tag{8}$$

On the surface of the deflector, the length of the sound wave is:

$$\lambda = \frac{A\omega 2\pi}{\omega} = 2\pi A. \tag{9}$$

Since in the deflectors used for vibration treatment, there is $r_{d_0} \approx 20...30$ mm, then in the overwhelming majority of cases, the value is $kr_d \geq \frac{2\pi r_{d_0}}{\lambda} = \frac{r_{d_0}}{A} >> 1$. In this case, the Hankel function has the following asymptotic [14] and is represented by the expression:

$$H_1^{(1)}(kr_d) \approx \sqrt{\frac{2}{\pi kr_d}} e^{i\left(kr_d - \frac{3}{4}\pi\right)}. \tag{10}$$

Taking into account relations (8) and (10), the following expression can be written for the constant C_p:

$$C_p \approx -\frac{\rho\omega^2 A\sqrt{\pi kr_{d_0}}\left(\frac{i}{2r_{d_0}} - k\right)}{\sqrt{2}\left(k^2 + \frac{1}{4r_{d_0}^2}\right)} e^{-i\left(kr_{d_0} - \frac{3}{4}\pi\right)}. \tag{11}$$

Let us write down the final expression for the pressure, proceeding from relation (8), but without disclosing the constant C_p due to the cumbersomeness of the resulting expression.

$$P \approx C_p \sqrt{\frac{2}{\pi kr_d}} e^{i\left(kr_d - \frac{3}{4}\pi\right)} \cos\alpha. \tag{12}$$

From expression (12), the pressure generated by one "tab" of the deflector decreases slowly depending on the distance to it, like a square root.

The value expressed by relation (12) is the amplitude value of the pressure. The complete expression for the pressure in the wave created by one deflector "tab" in the pseudo-gas made of abrasive granules will be as follows:

$$P \approx C_p \sqrt{\frac{2}{\pi kr_d}} e^{i\left(kr_d - \frac{3}{4}\pi\right)} cos\alpha sin(\omega t - kr_d). \tag{13}$$

It should be noted that formula (13) is valid only in the range of angles $-\frac{\pi}{2} \leq \alpha \leq \frac{\pi}{2}$, since the flat walls of the deflector "tab", in our approximation, do not generate waves and for values of α outside the specified interval, the pressure will be equal to zero.

4 Results and Discussion

Wave action is generated by the deflector on the surface of the processed parts. Velocity field in a wave generated by a deflector. To determine the wave action generated by the deflector on the surface of the processed part, it is necessary to know the field of velocities created by this deflector. Let us find it from relation (7), which is valid for the amplitude value of pressure and velocity not only at $r_d = r_{d_0}$ but also in the case of any value of r_d:

$$V_r = \frac{-i}{\rho\omega} \cdot \frac{\partial P}{\partial r_d}. \tag{14}$$

For the tangential component, the following relation is valid [14]:

$$V_\tau = \frac{-i}{\rho\omega r_d} \cdot \frac{\partial P}{\partial \alpha}. \tag{15}$$

Let us write expressions (14) and (15), considering relations (11) and (12). To find the radial component of the velocity, we write the expression for the pressure:

$$P = \rho\omega^2 A \sqrt{\frac{r_{d_0}}{r_d}} \cdot \frac{\left(k - \frac{i}{2r_{d_0}}\right)}{\left(k^2 + \frac{1}{4r_{d_0}^2}\right)} e^{ik\left(r_d - r_{d_0}\right)} cos\alpha. \tag{16}$$

Using formula (14), we get:

$$V_r = i\omega A \sqrt{\frac{r_{d_0}}{r_d}} cos\alpha \frac{e^{\left(ik\left(r_d - r_{d_0}\right)\right)}}{\left(k^2 + \frac{1}{4r_{d_0}^2}\right)} \cdot \left(\frac{k}{2}\left(\frac{1}{r_d} - \frac{1}{r_{d_0}}\right) - i\left(k^2 + \frac{1}{4r_d r_{d_0}}\right)\right). \tag{17}$$

Further, by expanding the exponential and highlighting the real part of the expression (17), we obtain the formula for the dependence of the amplitude of the radial component of the wave velocity initiated by one "tab" of the deflector:

$$V_{r_{\text{real}}} = \omega A \sqrt{\frac{r_{d_0}}{r_d}} \times$$
$$\times cos\alpha \frac{\left[\left(\left(k^2 + \frac{1}{4r_d r_{d_0}}\right) cos\left(k\left(r_d - r_{d_0}\right)\right) - \frac{k}{2}\left(\frac{1}{r_d} - \frac{1}{r_{d_0}}\right) sin\left(k\left(r_d - r_{d_0}\right)\right)\right)\right]}{\left(k^2 + \frac{1}{4r_{d_0}^2}\right)}. \tag{18}$$

The real part of the tangential component of the velocity amplitude in the wave generated by the deflector is expressed by the following relationship:

$$V_{\tau_{\text{real}}} = \omega \frac{A}{r_d} \sqrt{\frac{r_{d_0}}{r_d}} \cdot \frac{\left[\frac{cos\left(k\left(r_d - r_{d_0}\right)\right)}{2r_{d_0}} - k sin\left(k\left(r_d - r_{d_0}\right)\right)\right]}{\left(k^2 + \frac{1}{4r_{d_0}^2}\right)} sin\alpha. \tag{19}$$

A plane wave carries out the pressure of acoustic radiation. The following relation determines the pressure of acoustic radiation generated by a plane wave:

$$P = \frac{I}{2C_1}\left[(1 + RR)\text{ctg}i - (1 - RR)\text{ctg}r\right] 2sin2i. \tag{20}$$

Here I is the intensity of the incident wave, C_1 is the speed of sound in a pseudo-gas of abrasive granules, i is the angle of incidence, r is the angle of refraction, RR is the reflection coefficient determined by the following relation [16]:

$$RR = \left(\frac{\rho_2 C_2 cosi - \rho_1 \sqrt{C_1^2 - C_2^2 sin^2 i}}{\rho_2 C_2 cosi + \rho_1 \sqrt{C_1^2 - C_2^2 sin^2 i}} \right)^2, \tag{21}$$

where C_2 – the speed of sound in the material of the part; ρ_1 – the density of the pseudo-gas of the abrasive granules; ρ_2 – the density of the material of the part.

At a higher, in comparison with the speed of sound in pseudo-gas and the speed of sound in the material of the part, and when a certain angle of incidence is exceeded, which is called critical, expression (21) becomes complex. This critical angle i_{cr} is determined by the following relationship:

$$i_{cr} = arcsin\left(\frac{C_1}{C_2}\right). \tag{22}$$

When the angle of incidence is equal to the critical angle of refraction r becomes equal to 90° $\left(r_{cr} = \frac{\pi}{2}\right)$. With a further increase in the angle of incidence, the acoustic wave no longer propagates from the first medium (pseudo-gas) to the second (part material). The wave is completely reflected from the surface of the part. In this case, the modulus of the reflection coefficient is equal to unity, and the angle of refraction r remains equal to $\frac{\pi}{2}$, and the reflected wave changes its phase with respect to incident one. The phase difference $\Delta\varphi$ is determined by the ratio [17]:

$$\Delta\varphi = -2\mathrm{arctg}\left[\frac{\sqrt{cos^2\theta - \left(\frac{C_1}{C_2}\right)^2}}{\frac{\rho_2}{\rho_1 sin\theta}}\right], \tag{23}$$

where instead of the angle of incidence, for convenience, the sliding angle is used – $\theta = \frac{\pi}{2} - i$.

The speed of sound in pseudo-gas of abrasive granules is approximately equal to $A\omega \approx 1$ m/s. And the density is $\rho_1 \approx 1000$ kg/m^3 [1]. The speed of sound in steel and bronze is ≈ 5000 m/s. The density of these metals is $\rho_2 \approx 7000$ kg/m^3. In our case, we have $i_{cr} = 2 \cdot 10^{-5}$ rad. That is, almost all angles of incidence of the acoustic wave on the surface of the part, which can be positioned with respect to the deflector's "tabs" as you like, will be supercritical. In this case, the modulus of the reflection coefficient – RR (21) is equal to one [18].

Calculations show that the phase difference, in this case, can vary from 2π to zero.

The pressure of acoustic radiation generated by a plane wave in the case of supercritical sliding angles is determined by expression (20), in which it should be taken into account that for any values of the sliding angle, the angle of refraction r remains equal to $\frac{\pi}{2}$, and the reflection coefficient takes the following form [18, 19]:

$$RR^* = Real\left(RRe^{i2\Delta\varphi}\right) = RRcos(2\Delta\varphi). \tag{24}$$

Considering the above, it is possible to write down the relationship for the pressure exerted by a plane wave on a flat surface:

$$P = \frac{I}{2C_1}(1 + RR^*)sin^2\theta. \tag{25}$$

In relation (25), I is the intensity of the incident wave, equal to $I = C_1\rho\frac{V^2}{2}$, where V is the maximum speed of oscillational motion in a pseudo-gas from abrasive granules [16].

Wave action is generated by the deflector on the surface of the part. In our case, the deflector's acoustic wave emitted by the "tab" is not plane. Expressions for the components of the velocities in the wave generated by the deflector (18), (19), and relations (23) – (25). They determine the pressure on the surface of the processed part and show that this wave decays depending on the distance r_d as a cylindrical one. In relations (18), (19), there is also a dependence on the angle α. In addition, the pressure depends on the sliding angle θ. Due to the proximity of the dimensions of the part and the distance r_d, the pressure at different points on the part surface will depend on different sliding angles (Fig. 3).

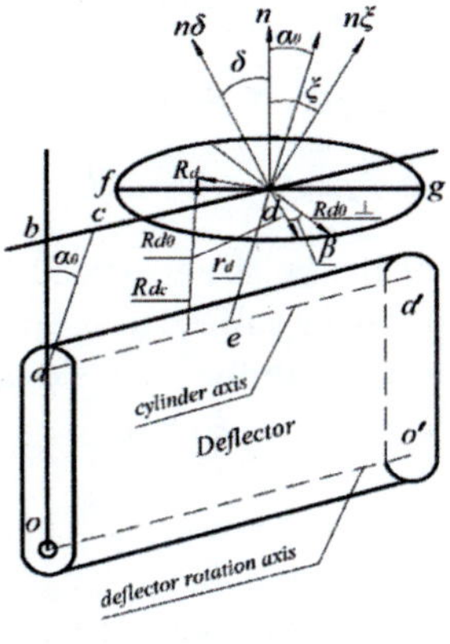

Fig. 3. Geometric parameters of the interaction of the front of a cylindrical wave with a flat plate of the working medium deflector.

θ – the angle of sliding of the acoustic wave incident on the center of the part, δ – the angle of incidence of the wave. The radius of the flat part is R_{d_0}. α is the angle between the plane formed on the axes a, a' and o, o', as well as by the plane in which the parallelogram *acde* lies (Fig. 2). The length of the deflector (a, a' or o, o') is much greater than the diameter of part $2R_{d_0}$, so the problem can be considered as flat, that is, it does not depend on the directions along the a, a' or o, o' axes. Since the parameters of the emitted wave do not change along the axis of rotation of the deflector and the axis of its cylindrical surface, the angular position of the part will be determined by the angles δ and ξ (Fig. 3).

Initially, we determine the dependence of the angle of sliding of the wave incident on the center of the part (point d) on the angles δ and ξ. Without loss of generality, we will assume that the unit vector n, perpendicular to the surface of the part (ort) are in a vertical position. In this case, in the center of the part $\xi_0 = 0$, $\delta_0 = 0$, $\alpha_0 = 0$. When turning through angle ξ, the position of the ort changes. The new ort position is shown with n_1 symbols. (Fig. 3) The value of the sliding angles – θ and the angle of incidence – δ also change. For any point on the plane of the part located at a distance R_d from its center, we can determine these values. Omitting simple but cumbersome expressions, we can write formulas for determining the new value of the sliding angle – θ_c depending on the angles δ and ξ:

$$\theta_c = \frac{\pi}{2} - \alpha_c;\ \alpha_c = \operatorname{arctg}\left(\sqrt{(\mathrm{tg}\xi_1)^2 + \left(\frac{\mathrm{tg}\ (\alpha - \delta)}{\cos\xi_1}\right)^2}\right);$$
$$\alpha = \operatorname{arctg}\left(tg(\alpha_0) - \frac{R_d \cos\beta}{r_d \cos\alpha_0}\right);\ \alpha_0 = \frac{\pi}{2} - \theta_0;\ \xi_1 = \arcsin(\sin\xi\cos\delta), \tag{26}$$

where α is the angle of incidence of the wave at the point under consideration for the vertical position of the vector n. Distances from axis aa' to any point on the surface of the part – Rd_c depending on δ and ξ: are determined by the following expressions:

$$RY = (r_d \cos\alpha_0 + R_d \cos\beta \sin\delta \cos\xi_1);$$
$$RX = (r_d \sin\alpha_0 - R_d \cos\beta \cos\delta);\ R_{d_c} = \sqrt{RX^2 + RY^2}. \tag{27}$$

Figure 4 shows the graphs of the pressure exerted by the wave generated by the deflector, depending on the radius on the surface of the part and the angle β (see Fig. 3).

The graphs in Fig. 4, a correspond to the values $\alpha = 0$, $\xi = 0$, $\delta = 0$, in 4, b – $\alpha = 2\frac{\pi}{3}$, $\xi = 0$, $\delta = 0$, in 4, c – $\alpha = 0$, $\xi = 0$, $\delta = \frac{2\pi}{3}$, in 4, d – $\alpha = 0$, $\xi = 0$, $\delta = \frac{\pi}{6}$.

For clarity of perception, the graphic charts are presented in two forms. Each left one depicts a surface formed by the magnitude of the pressure depending on the radius and angle. On each right one, there is a surface in coordinates R_d and β, on which the pressure value is qualitatively displayed in the form of color. Red represents the highest pressure, and purple the lowest. The color scale clearly shows the minima and maxima of the pressure value, which are "shaded" in the proper graphs.

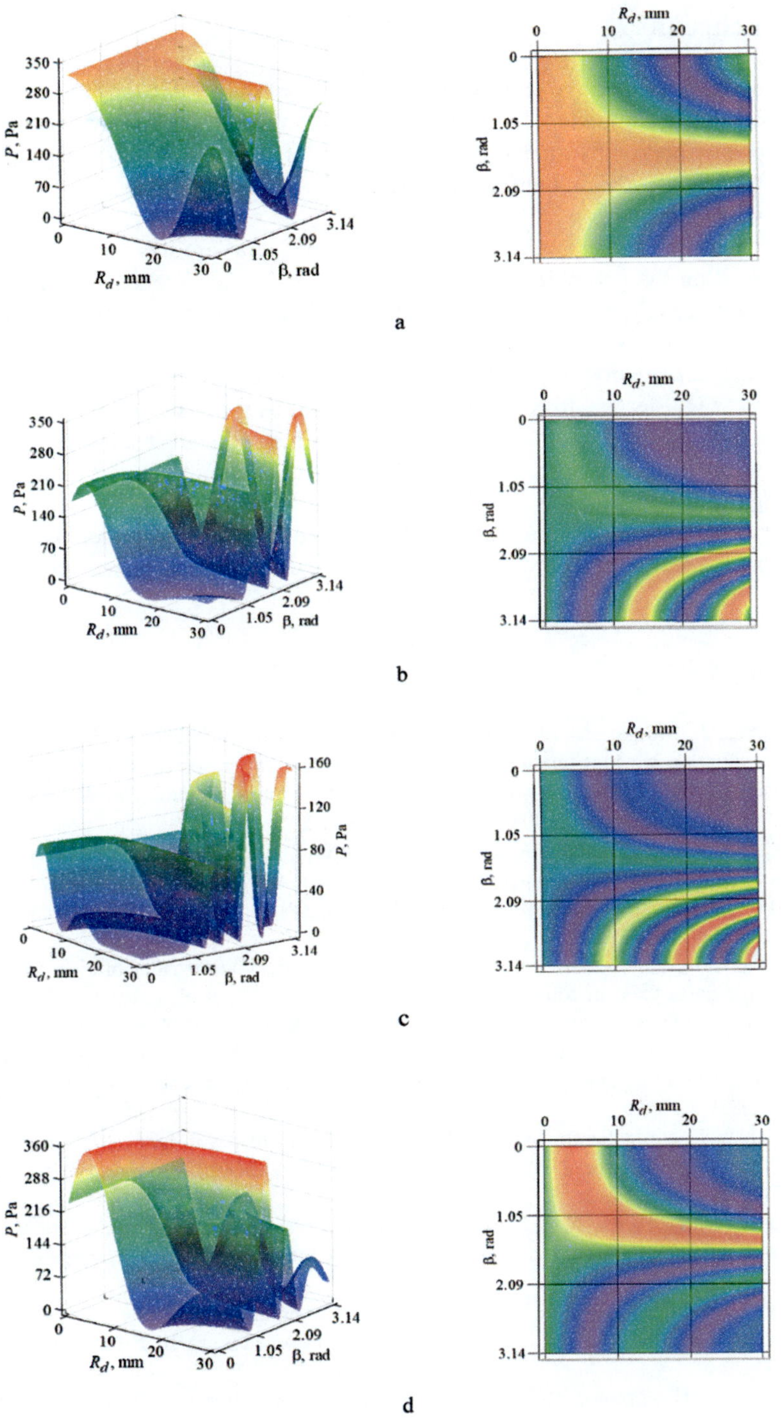

Fig. 4. Pressures on the surface of the part, exerted by the wave generated by the deflector, depending on the radius on the surface of the part and the angle β at different values of the angles α, ξ, δ.

5 Conclusions

The pressure exerted on the part by the wave action of the deflector decreases inversely with the distance from one "tab" of the deflector to the part. Therefore, to improve the efficiency of vibration treatment, it is necessary to use deflectors with a large number of "tabs". In this case, the amount of metal removal will be proportional to the number of "tabs" of the deflector.

Dependencies of the pressure initiated by the deflector oscillations on the radius R_d and angle β on the part surface are the result of wave interference in the medium of abrasive granules. The pressure on the part's surface changes abruptly, varying ten or more times.

Irregular distribution of pressure on the surface of the part from the wave action of the deflector will lead to irregular removal of metal over the surface by more than ten times, in the case of a fixed fixation of the part, and to spoilage. Therefore, when using a deflector, it is necessary to allow the part to rotate and move in the reservoir of the vibrating machine.

An abrupt change in pressure on the surface of the part, which is caused by interference, leads to the deformation of individual sections of planar parts with low rigidity.

References

1. Kundrák, J., Morgan, M., Mitsyk, A.V., Fedorovich, V.A.: The effect of the shock wave of the oscillating working medium in a vibrating machine's reservoir during a multi-energy finishing-grinding vibration processing. Int. J. Adv. Manuf. Technol. **106**(9–10), 4339–4353 (2020). https://doi.org/10.1007/s00170-019-04844-2
2. Lachenmaier, M., Dehmer, A., Trauth, D., Mattfeld, P., Klocke, F.: Influence of different input parameters on the contact conditions determing the surface integrity of workpieces in an unguided vibratory finishing process. Proc. CIRP **71**, 53–58 (2018). https://doi.org/10.1016/j.procir.2018.05.022
3. Borovets, V., Lanets, O., Korendiy, V., Dmyterko, P.: Volumetric vibration treatment of machine parts fixed in rotary devices. In: Tonkonogyi, V., et al. (eds.) InterPartner 2020. LNME, pp. 373–383. Springer, Cham (2021). https://doi.org/10.1007/978-3-030-68014-5_37
4. Atzeni, E., et al.: Performance assessment of a vibro-finishing technology for additively manufactured components. Proc. CIRP **88**, 427–432 (2020). https://doi.org/10.1016/j.procir.2020.05.074
5. Pavlenko, I., et al.: Parameter identification of cutting forces in crankshaft grinding using artificial neural networks. Materials **13**(23), 5357 (2020). https://doi.org/10.3390/ma13235357
6. Hao, Y., Yang, S., Li, X., Li, W., Wang, X.: Analysis of contact force characteristics of vibratory finishing within pipe-cavity. Granular Matt. **23**(2), 1–14 (2021). https://doi.org/10.1007/s10035-021-01089-3
7. Bergs, T., Müller, U., Barth, S., Ohlert, M.: Experimental analysis on vibratory finishing of cemented carbides. Manuf. Lett. **28**, 21–24 (2021). https://doi.org/10.1016/J.MFGLET.2021.02.004
8. Kang, Y.S., Hashimoto, F., Johnson, S.P., Rhodes, J.P.: Discrete element modeling of 3D media motion in vibratory finishing process. CIRP Ann. Manuf. Technol. **66**, 313–316 (2017). https://doi.org/10.1016/j.cirp.2017.04.092

9. Gaggar, S., Vasudevan, D., Kumar, P., Mitra, S.: Application of smoothed particle hydrodynamics for the simulation and analysis of vibratory finishing process. Int. J. Adv. Manuf. Technol. **108**(1–2), 183–190 (2020). https://doi.org/10.1007/s00170-020-05307-9
10. Makiuchi, Y., Hashimoto, F., Beaucamp, A.: Model of material removal in vibratory finishing, based on Preston's law and discrete element method. CIRP Ann. Manuf. Technol. **68**, 365–368 (2019). https://doi.org/10.1016/j.cirp.2019.04.082
11. Pandiyan, V., Castagne, S., Subbiah, S.: High frequency and amplitude effects in vibratory media finishing. Proc. Manuf. **5**, 546–557 (2016). https://doi.org/10.1016/j.promfg.2016.08.045
12. Mamalis, A.G., Grabchenko, A.I., Mitsyk, A.V., Fedorovich, V.A., Kundrak, J.: Mathematical simulation of motion of working medium at finishing–grinding treatment in the oscillating reservoir. Int. J. Adv. Manuf. Technol. **70**(1–4), 263–276 (2013). https://doi.org/10.1007/s00170-013-5257-6
13. Tikhonov, A.N., Samarskii, A.A.: Equations of Mathematical Physics. Pergamon Press Ltd., New York (1963)
14. Koshlyakov, N.S., Gliner, E.B., Smirnov, M.M.: Equations of Mathematical Physics in Partial Derivatives. Graduate school, Moscow (1970).[in Russian]
15. Bronshtein, I.N., Semendiaev, K.A.: Handbook of Mathematics for Engineers and Students of Technical Universities. Science, Moscow (1986).[in Russian]
16. Yavorsky, B.M., Detlaf, A.A., Lebedev, A.K.: Reference Book on Physics, 8th edn. ONIKS. World and education, Moscow (2006).[in Russian]
17. Isakovich, M.A.: General Acoustics. Science, Moscow (1973).[in Russian]
18. Efimov, A.P., Nikonov, A.V., Sapozhkov, M.A., Shorov, V.I.: Acoustics: Handbook. Radio and communication, Moscow (1989).[in Russian]
19. Sokolov, V.: Hydrodynamics of flow in a flat slot with boundary change of viscosity. In: Radionov, A.A., Gasiyarov, V.R. (eds.) ICIE 2021. LNME, pp. 1172–1181. Springer, Cham (2021). https://doi.org/10.1007/978-3-030-54817-9_136

Evaluation of a Decrease in Temperature Conditions upon Intermittent Grinding

Fedir Novikov[1], Andrii Hutorov[2](✉), Oleksii Yermolenko[1], Stanislav Dytynenko[1], and Yana Halahan[1]

[1] Simon Kuznets Kharkiv National University of Economics, 9-A, Nauky Avenue, Kharkiv 61166, Ukraine
[2] NSC "Institute of Agrarian Economics", 10, Heroyv Oborony Street, Kyiv 03127, Ukraine
gutorov.andrew@gmail.com

Abstract. The aim of the work is theoretical substantiation of the laws of cutting temperature formation during discontinuous grinding and conditions of its significant reduction to ensure high-productive and high-quality machine parts machining. It is established that with decreasing the length of the working shoulder of a discontinuous circle, the cutting temperature passes a minimum point, in which the lengths of the working shoulder and the hollow of the circle are equal. It is established that the lowest value of cutting temperature and the highest value of machining productivity is achieved at interrupted depth grinding with relatively low part speed and grinding width close to the circle height. Cutting temperature is lower, and machining productivity is higher than grinding with a full circle. We received the analytical dependence for determining the cutting temperature ratio at grinding by discontinuous and solid circles. It contains only one value: the number of contacts of working ledges of a discontinuous circle with a fixed section of a machined workpiece. The discrepancy between calculated and experimental values of the given cutting temperature ratio does not exceed 10%. It indicates the reliability of the obtained theoretical solution. The results of the research are recommended to be used in the development of high-performance technological processes of interrupted grinding of machine parts.

Keywords: Machining quality · Processed material · Working ledge of the intermittent circle

1 Introduction

It is well known that the use of intermittent grinding reduces the cutting temperature and improves the machining quality of machine parts compared to grinding with solid circles. The machining effect is achieved by eliminating the formation of burns, microcracks, and other temperature defects on the machined surfaces. It allows the method of intermittent grinding to be widely used in finishing grinding operations at aviation, instrumentation, and machine-building enterprises.

V. Tonkonogyi et al. (Eds.): InterPartner 2022, LNME, pp. 190–199, 2023.
https://doi.org/10.1007/978-3-031-16651-8_18

Meanwhile, when grinding products from hard-to-machine materials (ceramics, hard alloys, high-strength steels, welded materials, etc.), there is a need to remove more significant allowances with increased productivity. The application of intermittent grinding in these conditions becomes ineffective [1]. Therefore, based on international experience, designs of discontinuous grinding wheels and technologies of their application are constantly being improved to ensure increased productivity and exclude the formation of burns [2] and microstructural damage on machined surfaces [3]. In this regard, the segment wheel proposed in [4] deserves attention, which reduces the grinding temperature significantly below the critical limit and excludes the formation of burns on the machined surface (titanium alloy Ti-6Al-4V). It was also shown in [5] that by intermittent grinding of modern ceramics with a segment wheel of special design T-Tool, it is possible to reduce cutting force and temperature, to increase the material removal rate, especially when grinding SiC ceramics [6]. In the study [7], a new design scheme of a grinding wheel with a groove is proposed. In this case, only part of the groove on the wheel has slots, and the alternating grinding mode of groove and non-groove reduces the grinding temperature. At the same time, when creating new designs of discontinuous circles and technologies of their application, mainly the results of experimental studies are used, which allow obtaining only private solutions, valid for specific processing conditions and which do not give a general (generalizing) idea about the ways of reducing cutting temperature and increasing processing productivity. As is known, more general solutions can be obtained based on theoretical research. Therefore, the present work is devoted to establishing generalized theoretical solutions about the conditions of significantly reducing cutting temperature and increasing machining productivity during intermittent grinding. It will make it possible to find optimal solutions for further improvement of intermittent grinding technologies within a wide range of changes in the grinding mode parameters and characteristics of intermittent grinding wheels.

2 Literature Review

The issues of cutting temperature reduction during interrupted grinding are constantly paid much attention to in the scientific and technical literature. However, as practice shows, for example, in the operation of gear grinding, interrupted grinding reduces cutting temperature only in the range of 30–50% [8]. It is clearly insufficient for high-productive grinding, so interrupted grinding is mainly used in finish grinding operations [9]. Therefore, to reduce the cutting temperature in [10], practical recommendations for geometric characteristics selection of discontinuous profile of the working surfaces of abrasive wheels were developed. In the study [11], it was proposed to impregnate discontinuous wheels by contact method, and in [12], the effective composition of solid lubricant for impregnation of discontinuous wheels was proposed. It has been established that an important factor in temperature reduction in discontinuous grinding is the use of highly porous abrasive wheels [13], diamond wheels [14], and effective technological media [15]. To substantiate the conditions for reducing the grinding temperature in the research [16], a method of designing the optimal geometry of the discontinuous working surface of the wheel was developed. In the study [17], it was theoretically and experimentally established that the temperature at flat discontinuous grinding strongly depends on the meshing state of the wheel segments - the number of thermal pulses in the grinding area. In the research [18], the calculation of lengths of protrusions and troughs of the discontinuous wheel concerning thermal and dynamic factors is given. A generalized mathematical model of the thermal process in discontinuous grinding is developed in [19]. It was established that it is possible to reduce the grinding temperature by reducing the contact time of the working shoulder of the wheel with the workpiece, i.e., the length of a working shoulder of the wheel. The same result was obtained theoretically in [20, 21].

However, the above calculations do not consider the increase in the actual grinding depth because part of the machined material remains unremoved at the moment of interruption of the grinding process (at the moment of passing the wheel trough). It must be removed at the subsequent contact of the working shoulder of the wheel with the workpiece. Obviously, this should lead to an increase in cutting temperature. Therefore, reducing the length of the working shoulder of a wheel has an ambiguous effect on the cutting temperature during interrupted grinding. On the one hand, with a decrease in the length of the working shoulder of the wheel, the contact time of the working shoulder of the wheel with the workpiece decreases, which leads to a decrease in the cutting temperature.

On the other hand, the grinding depth increases, leading to an increase in the cutting temperature. Consequently, there must be an extremum (minimum) of the cutting temperature. Therefore, it is important to theoretically substantiate the conditions for the emergence of extremum (minimum) of cutting temperature in discontinuous grinding, taking into account the increase in the actual grinding depth and the extreme nature of cutting temperature change depending on the length of the working shoulder of the discontinuous wheel. Establishing conditions for significantly reducing cutting temperature and increasing machining productivity in discontinuous grinding is necessary. It will clarify the known theoretical solutions for choosing optimal conditions for interrupted grinding according to the temperature criterion.

3 Research Methodology

Calculation of the cutting temperature when grinding with a continuous and discontinuous wheel is made based on the calculation scheme of the parameters of planar grinding (Fig. 1). In it, the removed allowance is conventionally represented by a pack of infinitely thin adiabatic rods [8], which are cut by the grinding wheel with the speed V_R. It is assumed that all the heat released in the grinding process goes deep into the surface layer of the workpiece, i.e., along the adiabatic rods. In this case, the cutting temperature can be determined by the known simplified dependence [22]:

$$\theta = \frac{q \cdot l_2}{\lambda}, \tag{1}$$

where $q = \sigma \cdot V_R$ is the thermal flux density, W/m^2; $\sigma = N/Q$ – conventional cutting tension (machining energy intensity), N/m^2; N – grinding power, W; Q is the machining capacity, m/s^2; $l_2 = (2a \cdot \tau)^{0.5}$ – depth of heat penetration into the machined surface layer, m; $a = \lambda/(c \cdot \rho)$ – thermal conductivity coefficient of the processed material, m²/s; λ – heat transfer coefficient of the processed material, W/(m·deg); c – specific heat capacity of the processed material, J/(kg·deg); ρ – material density, kg/m³; τ – the time of cutting the adiabatic rod with the grinding wheel, s.

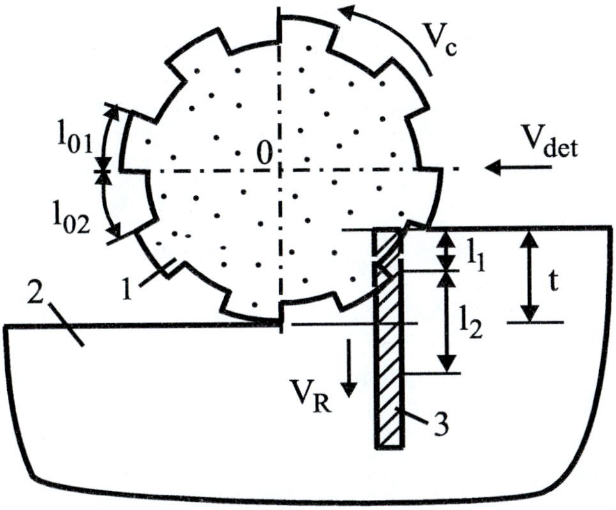

Fig. 1. Cutting temperature design for surface grinding, taking into account the cutting around adiabatic rods, the set of which represents a removable allowance: 1 - grinding wheel; 2 - processed material; 3 - adiabatic rod (l_1 is the length of the cut part of the adiabatic rod; l_2 is the depth of heat penetration into the surface layer of the workpiece; l_{01} is the length of the working protrusion of the intermittent circle; l_{02} is the length of the notch on the interrupted circle; V_c is the speed of the circle; V_{det} is the speed of the part; t – grinding depth).

The parameter $\sigma = N/Q$ is entered by experiment for the considered pair "grinding wheel – machined material" taking into account the experimentally established values of both N and Q parameters [22].

First, the cutting temperature when grinding with a continuous circle is determined. (see Fig. 1). The time τ of cutting an adiabatic rod by a grinding wheel is specified by two conditions: $\tau = l/V_{det} = t/V_R$, where $l = \sqrt{2t \cdot R_c}$ is the arc length between both wheel and workpiece, m; t – grinding depth, m; R_c is the circle radius, m; V_{det} – part speed, m/s.

Where:

$$V_R = \frac{t}{\tau} = \sqrt{\frac{t}{2R_c}} \cdot V_{det}. \tag{2}$$

After transformation, dependence (3) takes the following form:

$$\theta = \frac{\sigma}{\lambda} \cdot \sqrt{a \cdot Q_0 \cdot \sqrt{\frac{2t}{R_c}}}, \tag{3}$$

where $Q_0 = V_{det} \cdot t$ is the specific treatment capacity, m²/s.

From relation (3) it follows that it is possible to reduce the cutting temperature θ by reducing the parameters σ, Q_0, t and increasing R_c. In this case, the part speed V_{det} is determined by the dependence $V_{det} = Q_0/t$. Obviously, an effective scheme for intermittent grinding with a continuous wheel is multipass grinding with increased speed of the part [8, 19].

Under the condition of planar grinding with an intermittent circle, the cutting temperature θ is determined by dependence (1) taking into account $\tau = \tau_{01} = l_{01}/V_c$, where τ_{01} is the contact time of the intermittent circle working ledge with the material being machined, s; l_{01} is the working ledge of the intermittent circle, m; V_c – lap speed, m/s. The following condition determines the cutting speed of an adiabatic rod with an intermittent circle:

$$\Delta t = \tau_0 \cdot V_R = \tau_{01} \cdot V_{R_{01}}, \tag{4}$$

where Δt is a layer of material removed during the operation of one working ledge of an intermittent circle, m; $\tau_0 = \tau_{01} + \tau_{02}$ – is the total time of grinding zone by the working tooth (length l_{01}) and cutout (length l_{02}) of the intermittent circle, s; V_R is the cutting speed of the adiabatic rod with a continuous circle, m/s.

Based on condition (5), taking into account dependence (6), the following equation has been obtained:

$$V_{R_{01}} = \left(1 + \frac{l_{02}}{l_{01}}\right) \cdot \sqrt{\frac{t}{2R_c}} \cdot V_{det}. \tag{5}$$

Substituting dependence (5) and the expression $\tau = \tau_{01} = l_{01}/V_c$ in dependence (1), the cutting temperature in surface grinding with an intermittent circle is determined:

$$\theta = \frac{\sigma}{\lambda} \cdot V_{det} \cdot \left(\sqrt{l_{01}} + \frac{l_{02}}{\sqrt{l_{01}}}\right) \cdot \sqrt{\frac{a \cdot t}{R_c \cdot V_c}}. \tag{6}$$

Based on dependence (6), the working ledge length of the intermittent circle l_{01} has an ambiguous effect on the cutting temperature θ. . It is because by creating alternating working ledges and notches on the machined surface of the circle, the contact time of the intermittent circle (working ledge) with the workpiece is reduced [8]. Accordingly, the depth of heat penetration into the surface layer of the workpiece l_2 is reduced [23]. However, based on dependence (5), the rate of adiabatic rod cutting with an intermittent circle $V_{R_{01}}$ has been increased.

Thus, the parameters l_2 and $V_{R_{01}}$ with a change in l_{01} affect the cutting temperature θ contrarily. According to dependence (6), this predetermines the existence of an extreme cutting temperature value θ with a change in l_{01}. . To determine it, dependence (6) should be subjected to the necessary extremum condition: $\theta'_{l_{01}} = 0$. As a result, the extreme value of the parameter $l_{01} = l_{02}$ has been obtained.

The second derivative $\theta''_{l_{01}}$ at the extremum point of the function θ is negative. Therefore, the cutting temperature at the extremum point ($l_{01} = l_{02}$) takes the minimum value:

$$\theta_{min} = \frac{2\sigma}{\lambda} \cdot \sqrt{\frac{a \cdot Q_0 \cdot V_{det} \cdot l_{01}}{R_c \cdot V_c}}. \tag{7}$$

It is possible to reduce the minimum cutting temperature $\theta_{\min}$ by using deep-feed grinding with a relatively low part speed and a grinding width close to the circle's height [15]. The lower the speed of the workpiece $V_{\det}$, the higher the specific machining capacity Q_0 can be in interrupted grinding. The resulting analytical solution is fundamentally new, absent in scientific and technical literature. It is opposite to the analytical solution derived from dependence (3) obtained by surface grinding with a continuous circle.

According to this solution, the effective scheme of grinding with a continuous wheel is multipass grinding with increased speed of the part. Based on dependence (7), the main condition for effective application of discontinuous grinding in conditions of high-productive machining is the use of depth grinding with relatively low part speed and grinding width close to the wheel's height. It is the scientific novelty of the obtained theoretical solution.

To quantify the difference in cutting temperature when grinding with intermittent.

(θ_{int}) and continuous (θ_{sol}) circles, described by dependencies (8) and (3), let's consider their ratio:

$$\frac{\theta_{\text{int}}}{\theta_{sol}} = 2 \cdot \sqrt{\frac{V_{\det}}{V_c} \cdot \frac{l_{01}}{l}} = \sqrt{\frac{2}{n}}, \tag{8}$$

where $n = t/\Delta t$ is the number of contacts of the working ledges of the intermittent circle with the adiabatic rod until its complete cutting (see Fig. 1); $\Delta t = \tau_{01} \cdot V_{R_{01}} = \frac{l_{01}+l_{02}}{V_c} \cdot \sqrt{\frac{t}{2R_c}} \cdot V_{\det}$

Following dependence (8), it is possible to reduce the ratio $\theta_{\text{int}}/\theta_{sol}$ by reducing the parameters $V_{\det} = Q_0/t$ and l_{01}. This implies the application of a depth grinding scheme, since a significant reduction in l_{01} is limited by the strength of the working ledge of a discontinuous circle [8, 18, 19]. The increase in the value of n is due to the application of the effect of discontinuous grinding in generalized form.

4 Results and Discussion

It has been established by calculations (Table 1) that the parameter $A = \sqrt{l_{01}} + l_{02}/\sqrt{l_{01}}$ included in dependence (6), with an increase in l_{01} takes on an extreme (minimum) value under the condition $l_{01} = l_{02}$. Therefore, the grinding temperature takes a minimum value under the condition $l_{01} = l_{02}$.

Table 1. Calculated values of parameter A for $l_{02} = 10$ mm.

l_{01}, mm	0	5	10	15	20	25	30	35	40
A, $\text{mm}^{0,5}$	1	6.71	6.32	6.46	6.71	7.0	7.3	7.6	7.9

The parameter A slightly rises with increasing l_{01} at $l_{01} > l_{02}$. Therefore, the parameter l_{01} can be increased to reduce the broken circle's wear. At the same time, the grinding temperature increases slightly. In practical terms, the ratio l_{01}/l_{02}, as a rule, is set equal to 2–3 [8, 19]. According to Table 1, this increases parameter A and grinding temperature by only 6–15%.

Based on dependence (9) and Table 2, it is possible to significantly reduce the ratio $\theta_{int}/\theta_{sol}$ and cutting temperature in intermittent grinding, provided the V_{det} reduction and depth grinding application. The effect is achieved by increasing the n value determined from dependence (8).

Table 2. The calculated values of the ratio $\theta_{int}/\theta_{sol}$ and the value of n for the initial data: $V_c = 30$ m/s; $t = 0.1$ mm; $R_c = 100$ mm; $l_{01} = 20$ mm.

V_{det}, m/min	1	2	5	10	20	30
$\theta_{int}/\theta_{sol}$	0.1	0.14	0.22	0.315	0.446	0.546
n	200	100	40	20	10	7

From dependence (8) and Table 2, it has been concluded that at $n > 8$ it is possible to reduce the cutting temperature at deep intermittent grinding by 2–4 times or more compared to grinding with a continuous circle.

In [19], experimental data are given, according to which the condition $\theta_{int}/\theta_{sol} = 0.6$ is reached at a value of $n = 5$ ($l_{01} = l_{02} = 25$ mm; $V_{det} = 12$ m/min). It coincides with the calculated value $\theta_{int}/\theta_{sol}$ (see Table 3). The discrepancy between both calculated and experimental values is less than 10%, indicating the reliability of the obtained theoretical solution.

Table 3. Calculated ratio values $\theta_{int}/\theta_{sol}$.

n	2	4	5	6	8	10	20	40
$\theta_{int}/\theta_{sol}$	1	0.7	0.63	0.58	0.5	0.447	0.316	0.22

The theoretical research outcomes have been used on the surface grinding operation of C8 carbide plates with intermittent diamond circle 12A2 45° 150 × 42 × 10 × 5 × 32 AC6 160/125 4 B1–13 (number of working ledges – 26) under deep grinding conditions. It has been found that there were no burns, microcracks, and other temperature defects on the machined surfaces with increasing grinding depth up to 0.4 mm. These temperature defects appeared at small grinding depths of up to 0.02 mm (multipass grinding range) during grinding with a conventional diamond circle, i.e., at fairly low machining productivity. As a result, the intermittent diamond circle employment allowed a substantial increase in machining productivity while providing high machining quality.

Based on this work [24], the cutting temperature is particularly increased in comparison with multipass grinding in the conditions of deep grinding with an intermittent circle. It rather limits the use of deep grinding with a continuous circle in the operations of high-capacity grinding by the profile copying method instead of the conventional low-capacity grinding by the roughing method [23, 25]. Hence, the use of discontinuous grinding circles (and increasing the value of n) in this operation opens up the unique technological prospect for reducing the cutting temperature, enhancing the quality, and machining productivity.

Hence, the given work reveals the mechanism of a more substantial decrease of cutting temperature in intermittent grinding compared with established grinding with a continuous circle from a fundamentally unique analytical point of view. It makes it feasible to extend the technological prospects of intermittent circles' advantageous use in the operations of machine parts' and products' heavy-duty grinding employing heavy-duty materials.

5 Conclusions

In work, a fundamentally new theoretical solution to cutting temperature reduction in discontinuous grinding conditions is obtained. It is shown that reducing the length of the working ledge of the discontinuous wheel has an ambiguous effect on the cutting temperature. On the one hand, it leads to a reduction of contact time of working shoulder of a discontinuous disk with a workpiece and, accordingly, to a reduction of cutting temperature. On the other hand, it leads to an increase in the actual grinding depth, falling on the working ledge of the wheel, and, accordingly, to an increase in the cutting temperature. As a result, there is an extremum (minimum) of cutting temperature, at which the lengths of the working ledge and the trough of the discontinuous wheel are equal. It is established that the lowest value of cutting temperature and the highest value of machining productivity is achieved at discontinuous deep grinding with relatively low workpiece speed and grinding width close to the wheel's height. In this case, the cutting temperature is lower, and the machining productivity is higher than in grinding with a continuous wheel.

A rather simple analytical dependence for specifying the ratio of the cutting temperature when grinding with intermittent and continuous circles has been specified. It was revealed that it includes solely one value – the number of working ledges' contacts of an intermittent circle with a fixed workpiece cross-section. The more of them, the lower the given ratio of cutting temperatures and the higher the grinding efficiency with an intermittent circle, achieved in deep grinding conditions. It has been established that the discrepancy between estimated and surveyed values of this cutting temperature ratio does not surpass 10%. It is a reliable indicator of the theoretical explanation in the given work. The method of specifying the prerequisites for reducing the grinding temperature with intermittent circles has been facilitated thanks to its application. There is no need for intricate numerical estimations, being the basis of comprehended approaches. The research outcomes are advised for usage during the high-duty technological processes' growth of both machine parts' and products' intermittent grinding made of heavy-duty materials.

References

1. Phuong, N.T., Giang, N.T.P., Dong, N.T.: A research on the affect of technological parameters on cutting temperature when machining use segmented grinding wheel. Int. J. Electron. Commun. Comput. Eng. **8**(3), 208–212 (2017)
2. Ribeiro, F.S.F., et al.: Grinding assessment of workpieces with different interrupted geometries using aluminum oxide wheel with vitrified bond. The Int. J. Adv. Manuf. Technol. **108**(3), 931–941 (2020). https://doi.org/10.1007/s00170-020-05500-w
3. Rodriguez, R.L., et al.: Grinding process applied to workpieces with different geometries interrupted using CBN wheel. The Int. J. Adv. Manuf. Technol. **107**(3–4), 1265–1275 (2020). https://doi.org/10.1007/s00170-020-05122-2
4. Handa, D., Kumar, S., Babu, S., Surendran, T., Sooraj, V.S.: Simulation of Intermittent grinding for Ti-6Al-4V with segmented wheel. Mater. Today: Proc. **44**(1), 2537–2542 (2021). https://doi.org/10.1016/j.matpr.2020.12.626
5. Tawakoli, T., Azarhoushang, B.: Intermittent grinding of advanced ceramic with the T-tool grinding wheel. Adv. Mater. Res. **126–128**, 615–620 (2010). https://doi.org/10.4028/www.scientific.net/AMR.126-128.615
6. Tawakoli, T., Azarhoushang, B.: Theoretical and experimental investigation of intermittent grinding of SiC with a segmented grinding wheel. Int. J. Abras. Technol. **4**(1), 90–99 (2011). https://doi.org/10.1504/IJAT.2011.039005
7. Ding, N., Jiang, S., Duan, J., Liu, C., Cui, S.: Design of new slotted structured grinding wheel. J. Phys: Conf. Ser. **1635**, 012013 (2020). https://doi.org/10.1088/1742-6596/1635/1/012013
8. Yakimov, O.V., Usov, A.V., Slobodyanik, P.T.: Thermal Physics of Mechanical Processing. Odesa, Astroprint (2000)
9. Bogutsky, V., Novoselov, Y., Shron, L.: Calculating the profile of intermittent grinding wheel for the sharpening teeth of the broach. MATEC Web Conf. **224**, 01003 (2018). https://doi.org/10.1051/matecconf/201822401003
10. Yakimov, O., Bovnegra, L., Tonkonogyi, V., Vaysman, V., Strelbitsyi, V., Sinko, I.: Influence of the geometric characteristics of the discontinuous profile working surfaces of abrasive wheels for precision and temperature when grinding. Cutting Tools Technol. Syst. **94**, 115–125 (2021). https://doi.org/10.20998/2078-7405.2021.94.13
11. Tonkonogyi, V., Yakimov, A., Bovnegra, L., Sidelnykova, T., Dašić, P.: The use of intermittent wheels, impregnated by the contact method to reduce the thermal stress of the grinding process. IOP Conf. Ser.: Mater. Sci. Eng. **708**(1), 012034 (2019). https://doi.org/10.1088/1757-899X/708/1/012034
12. Tonkonogyi, V., Sidelnykova, T., Dašić, P., Yakimov, A., Bovnegra, L.: Improving the performance properties of abrasive tools at the stage of their operation. In: Karabegović, I. (ed.) NT 2019. LNNS, vol. 76, pp. 136–145. Springer, Cham (2020). https://doi.org/10.1007/978-3-030-18072-0_15
13. Kalashnikov, A.S., Morgunov, Y.A., Kalashnikov, P.A., Filippov, V.V.: Features of intermittent profile grinding cylindrical gears. Izvestiya MGTU MAMI **7**(1–2), 51–54 (2013). https://doi.org/10.17816/2074-0530-68008
14. Lavrinenko, V.I., Novikov, M.V.: Superhard Materials in Machining, vol. Bakul. Institute for Superhard Materials, Kyiv (2013).[in Ukrainian]
15. Larshin, V.P., Lishchenko, N.V., Pitel, J.: Intermittent grinding temperature modeling for grinding system state monitoring. Appl. Aspects Inform. Technol. **3**(2), 58–73 (2020). https://doi.org/10.15276/aait.02.2020.4
16. Li, H.N., Axinte, D.: On the inverse design of discontinuous abrasive surface to lower friction-induced temperature in grinding: an example of engineered abrasive tools. Int. J. Mach. Tools Manuf **132**, 50–63 (2018). https://doi.org/10.1016/j.ijmachtools.2018.04.006

17. Fang, C., Xu, X.: Analysis of temperature distributions in surface grinding with intermittent wheels. The Int. J. Adv. Manuf. Technol. **71**(1–4), 23–31 (2013). https://doi.org/10.1007/s00170-013-5472-1
18. Tonkonogiy, V., Yakimov, A., Bovnegra, L., Beznos, S., Dobrovolskiy, V.: Reduction of the heat factor in flat abrasive grinding. Tech. Sci. Technol. **4**, 16–26 (2017)
19. Sipaylov, V.A.: Thermal Processes During Grinding and Surface Quality Control. Mechanical engineering, Moscow (1978).[in Russian]
20. Sizyi, Iu.A., Stalinskii, D.V.: Dynamics and Thermal Physics of Grinding. State Enterprise UkrNTC "Energostal", Kharkiv (2016)
21. Oborskii, G.A., Dashchenko, A.F., Usov, A.V.: Systems Modeling. Odessa, Astroprint (2013)
22. Novikov, F.V., Yakimov, A.V.: Physical and Mathematical Theory of Materials Processing and Mechanical Engineering Technologies, vol. 2. Thermal Physics of Cutting Materials. Odessa, ONPU (2003)
23. Novikov, F., et al.: Determining the conditions for decreasing cutting force and temperature during machining. Eastern-Eur. J. Enterp. Technol. **6**(1), 41–50 (2019). https://doi.org/10.15587/1729-4061.2019.183882
24. Rowe, W.B., Jin, T.: Temperatures in high-efficiency deep grinding. Ann. CIRP **50**(1), 205–208 (2001). https://doi.org/10.1016/S0007-8506(07)62105-2
25. Zaborowski, T., Ochenduszko, R.: Grinding burns in the technological surface of the gear teeth of the cylindrical gears. Mechanik **10**, 135–139 (2017). https://doi.org/10.17814/mechanik.2017.10.135

Influence of Back Rake Angle of a Threading Cutter on the Drill-String Tool-Joint Pitch Diameter

Oleh Onysko(✉), Vitalii Panchuk, Volodymyr Kopei, Lolita Pituley, and Tetiana Lukan

Ivano-Frankivsk National Technical University of Oil and Gas, 15, Karpatska Street, Ivano-Frankivsk 76019, Ukraine
onysko.oleg@nung.edu.ua

Abstract. Most drill-string tool-joint require strong and reliable workpieces and require precision. Modern manufacturers of tools for turning thread make them with null rake angle, which significantly reduces their efficiency and makes it impossible to process the tool-joint made of high-strength steels. The proposal put forward by the authors is to apply the negative value of this angle. However, this idea needs a theoretical investigation of the accuracy of the resulting nut and pin connection. In this work, the developed algorithm for calculating the thread profile depending on the value of the back rake parameter of the threading cutter tool is used, and on its basis, the possible divergence from the tool-join pitch diameter is based on the possible divergence from the tool-join pitch predicted. If the pin is made with a cutter of an anterior angle of −5°, and the cut of the box is machined with a conventional cutter, the deviation of the pitch diameter of the NC23 connection can reach 16% of the tolerance.

Keywords: Process innovation · Industrial growth · Tapered thread · Accuracy · Lathe tool · Machining

1 Introduction

In drill–strings, the most important part is the special tapered threaded connection of a pin and a box called a tool-joint. The reliability of this joint depends on the strength of the box and the pin, as well as the accuracy of their cutting. The main process used for the manufacture of box and pin is turning, in which the workpiece rotates. A special cutting tool changes the position along the thread axis (Fig. 1). One of the most important tools geometric parameters is the back rake angle γ, which is determined at the nose of the cutting edge (Fig. 2). However, well-known world manufacturers of cutting tools produce thread cutters only with a null-value of that angle [1]. This is perhaps because research on the influence of cutter geometric parameters on the thread accuracy of the cut and its stability has not been widely publicized.

V. Tonkonogyi et al. (Eds.): InterPartner 2022, LNME, pp. 200–210, 2023.
https://doi.org/10.1007/978-3-031-16651-8_19

Thus, it is clear that there is a need for research related to the process of manufacturing tool-joint and the simultaneous increase of their operational parameters, namely the mechanical properties of drill pipes [2], drill bits [3, 4], and thread accuracy, which includes the accuracy of the pitch, profile and pitch diameter [5].

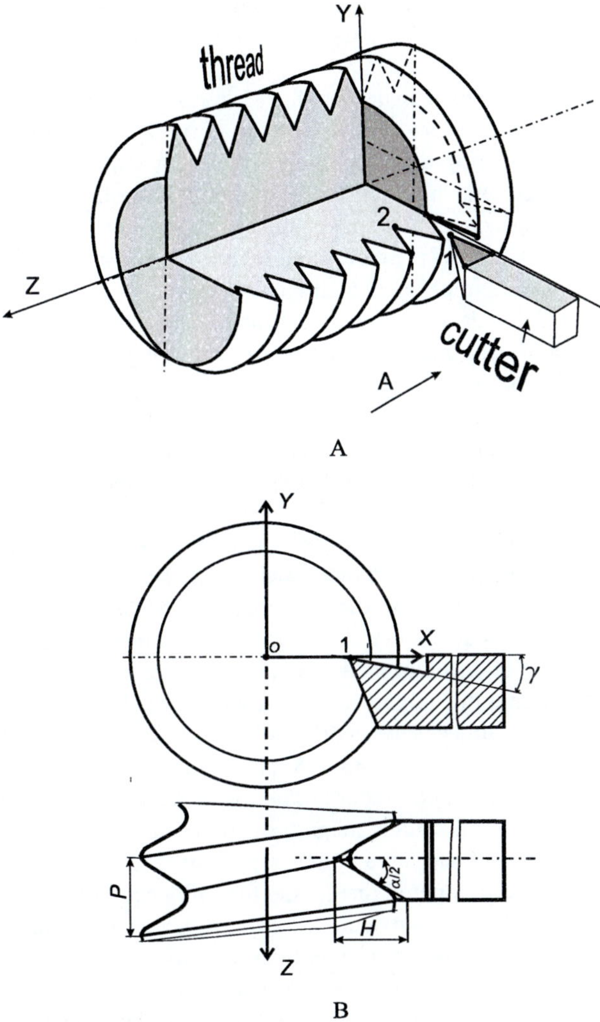

Fig. 1. Scheme of the threading by lathe tool cutter: A – General view, B – scheme of formation of the back rake angle corner (View A).

So, the research aim of this work is to investigate the influence of the main geometric parameters of lathe tool cutter-back rake angle on thread pitch accuracy and to obtain the theoretical prediction data of the deviation from pitch diameter nominal or its tolerance.

2 Literature Review

Some works demonstrate technological and design techniques to improve the mechanical properties of the thread. The influence of mechanical treatment on surface roughness is substantiated in the article [6]. However, it is provided by the process of grinding and not turning the thread. The influences of machining mistakes on the stress state of the tapered tool-joint are studied in [7]. But this article does not include the thread accuracy date.

Moreover, studying composite materials [8, 9] in terms of their machinability should also be extended. The detailed study [10] aims to carve a variable pitch, increasing the casing connection's bearing capacity in an oil and gas well. Improvements to the drill pipe's thread connections and their calculation are given in [11, 12]. In [13], the dependence of the casing connection's mechanical properties on the thread's variable pitch is shown. Article [14] shows the increased strength of the turned high-torque thread connection due to its design improvement at the face ends.

However, these articles do not refer to the manufacture of cut surfaces with tool cutting edge, which dominate in the production of drill string tool-joint. Therefore, an important study is an article [15] that shows the dependence of the chip formation process on the parameters of multi-cutter turning. Computer simulation to ensure the accuracy of milling cut is presented in [16]. No less important research is what is said in the article [17], the study of the machinability of high-strength stainless steel. Article [18] connects the effectiveness of experiments on manufacturing oil and gas pipe tool-joint assortments with the tool cutter rake angle. In [19], the direct dependences of the stability of the lathe cutter on the applied cutting force are shown. Article [20] analyses the deviations in the accuracy of the obtained thread due to the deformation process during machining. The study [21] shows the deformation results during turning depending on the tool cutter rake angle. The influence of different methods of turning a lathe cutter on the efficiency of threading in a high-strength stainless steel part was studied in [22]. The parameters of high-speed cutting on stainless steel workpieces are studied in [23, 24]. Modulated Tool Path (MTP) Machining for Threading Applications is cowered in [25]. But that application does not include the computer prediction of the thread accuracy. In [26], a study of the influence of the rake angle magnitude on the stress and temperature at the carbide insert cutting edge of the lathe cutter is presented. Theoretical studies of the process of turning a tool-joint tapered thread and the influence of the anterior angle on its angle of elevation are discussed in detail in [27]. Predictive algorithmic calculations of the accuracy of the keyway profile are presented in [28], including a great back rake value for the possibility of obtaining a thread made of stainless steel [29]. The actuality of the proposed methodology application is substantiated in [30, 31].

All these studies prove the importance of the turning process for manufacturing threads and cover the influence of machining parameters on high-strength threaded parts, including the pin and box of the tool-joint. Among the parameters, an important place is given to the back rake angle of the cutter. First of all, it is necessary to establish the theoretical algorithm of the functional influence of a back rake angle of a tool on deviations of the obtained thread pitch diameter.

3 Research Methodology

Studies [27] closely show the dependence of the accuracy of the tool-joint on the value of the rake angle of the cutter. However, the article considers only one parameter of the thread accuracy – the profile. According to the standard API 7, the thread's profile angle is 30°, and its tolerance is $\pm 40'$. Other parameters of the thread that regulate its accuracy are the pitch P and the pitch diameter d_{pitch}.

Other important parameters of the tool-joint tapered thread: The tool–joint tapered thread parameters are h - the working height of the thread profile, H - is theoretical height, b – theoretical root value, φ – screw angle. A bold line shows the profile of the thread, and the obtained straight sections of the profile are marked between the points D and G and between points E and F. The half profile angle - 30° is the angle between the straight side sections and the axis of symmetry of the threaded turn. The following equation describes these rectilinear sections:

$$z_1 = tg\frac{\alpha}{2}x \tag{1}$$

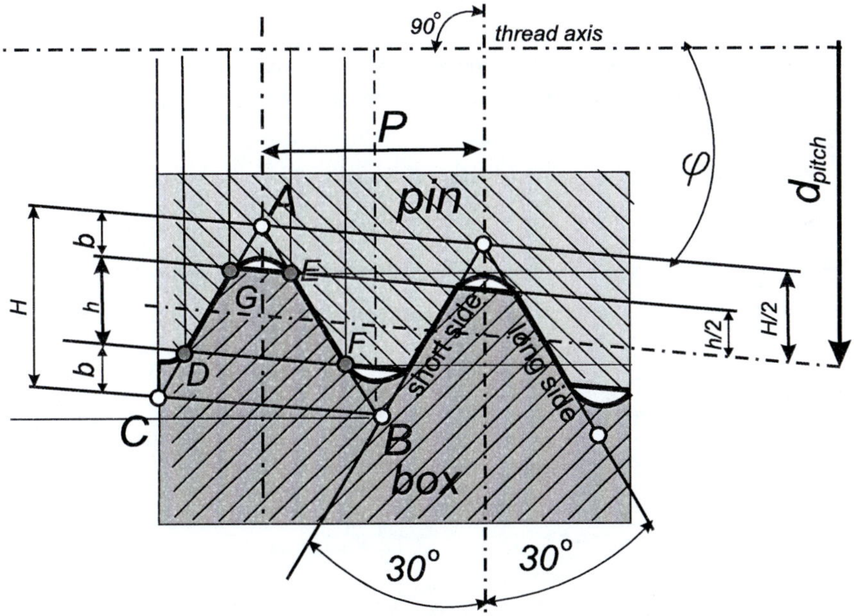

Fig. 2. The scheme of the tool-joint tapered thread according to the standard API 7.

However, during turning, there is a thread, the surface of which is different from the standard and is described as a convoluted surface [28]. The axial section will be curved rather than rectilinear [29].

The formula describing this cross-section can be explained based on using a cylindrical coordinate system, the axis of the thread Z. The YX plane will serve as the plane of the axial section of the thread (Fig. 3).

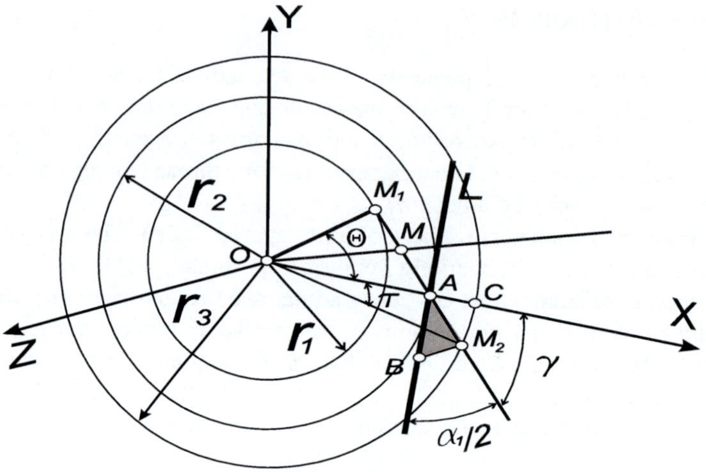

Fig. 3. Layout of the rake surface of the thread cutter in cylindrical coordinates. Thread parameters: r_1– the radius of the main cone; r_2 – the radius of the roots; r_3 – radius of crests; $\alpha_1/2$- half profile angle of the thread, *L*- straight section of cutting edge.

As a result of the application of the scheme, the equation of the axial section of a convoluted helical surface - theoretically corresponding to conditions of turning of thread is received [29]:

$$Z_2(x) = tg\left(\frac{\alpha_1}{2}\right)x\frac{\sin\tau}{\sin\gamma} - \frac{P}{2\pi}\tau \tag{2}$$

where:

$$\tau = \gamma - \arcsin\left(\frac{r_2\sin\gamma}{x}\right);$$

P – thread pitch; γ – back rake angle;

$$\frac{\alpha_1}{2} = arctg\left(\frac{P + \Delta_{\max} - \Delta_{\min}}{2H}\right), \tag{3}$$

where:

$$\Delta\frac{r_{min}^{2\sin^2\gamma}}{1+\cos\gamma_{max}};$$

$$\Delta_{\min} = \frac{r_{\min}^2\sin^2\gamma}{r_{\max} + \sqrt{r_{\max}^2 - (r_{\min}\sin\gamma)^2}},$$

where: $r_{min} = r_2$, $r_{max} = r_3$.

Since a negative back rake angle of −5° is often recommended for machining high-strength steel thread, this parameter is used in [28]. The calculation results for the thread NC23 for the long and short sides (Fig. 2) are presented in Table 1.

Table 1. Theoretical predicted results of half profile angle calculation of the Tapered Thread NC23.

Parameter	Turn 1	Turn last
$\alpha_{long\ side}$	29,90°	29,97°
$\alpha_{short\ side}$	30,09°	30,10°

4 Results and Discussions

4.1 Results of Calculations

As a result of the application of formulas and application of the visual algorithm developed by the authors, the coordinates of the profile points of the NC23 thread section are calculated according to formula (1) - X and Z_1 coordinates, according to formula (2), (3) X and Z_2 coordinates. The fragments of the visual algorithmic program are presented in Figs. 4, 5 are made with a step of 0.1 mm along the X axis. Since there are more than 50 rows, the fragments are built on the principle: *A*-first two rows; *B* – four middle rows and 2 last rows (C).

X	Z	Z1	Z2	Z-Z1	Z2-Z1
27,160	0,000	0,000	0,000	0,000	0,000
27,260	0,058	0,058	0,058	0,000	0,000

A

29,860	1,560	1,560	1,570	0,005	0,012
29,960	1,620	1,620	1,630	0,005	0,012
30,060	1,680	1,670	1,690	0,006	0,012
30,160	1,740	1,730	1,740	0,006	0,013

B

32,560	3,130	3,120	3,140	0,008	0,021
32,660	3,180	3,180	3,200	0,008	0,022

C

Fig. 4. Projected thread profile NC23 short side: A – 2 first calculated rows; B – middle row; C – last 2 rows.

According to the obtained forecast calculations, it is proposed to carry out the connection scheme provided that the box is made with a cutter with zero angle γ value and the pin according to the table. That is, on the long side, the half profile angle is less than the nominal value, and on the short side - more than the nominal value. For illustrating the explanation, the marked coordinates are used: for example, for point *A* (*XaZa*), for point *B* (*XbZb*), and for *C* - (*XcZc*). Other coordinate points on the *X* axis are common to the profile of the box thread and the pin. The *Z* coordinates are different for the box and the pin (Fig. 6).

X	Z	Z1	Z2	Z-Z1	Z2-Z1
27,160	0,000	0,000	0,000	0,000	0,000
27,260	0,058	0,058	0,058	0,000	0,000

A

30,060	1,670	1,670	1,670	-0,005	-0,003
30,160	1,730	1,730	1,730	-0,005	-0,003
30,260	1,780	1,790	1,790	-0,006	-0,003
30,360	1,840	1,850	1,840	-0,006	-0,003

B

32,860	3,280	3,290	3,290	-0,010	-0,006
32,960	3,340	3,350	3,340	-0,010	-0,006

C

Fig. 5. Projected thread profile NC23 long side: A- 2 first calculated rows, B- middle rows, C - last 2 rows.

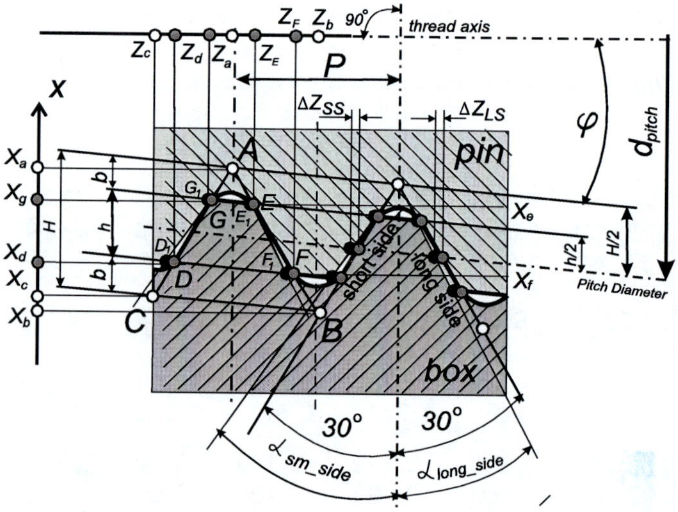

Fig. 6. – Projected thread profile NC23 short side: *A*- 2 first calculated rows, *B* - middle rows, *C* last 2 rows.

Black pinpoints, and gray box profile points are also marked on the middle diameter. The differences between them are marked on the diagram as Z_{ss} and Z_{Ls}. Their values correspond to the differences *Z2-Z1* (Fig. 4, 5) on fragments B (highlighted by frames). Thus, on the pitch diameter, the difference between the coordinates of the pin and the box at the short side of the profile of thread $Z_{ss} = 0{,}012$ mm, and at the long side – $Z_{Ls} = -0.003$ mm. It is due to the large difference $Z_{ss} = 0.012$ mm and different profiles of the threads of the box and the pin that they have a mismatch of pitch diameters by the value of Δd_{pitch} (Fig. 7, 8).

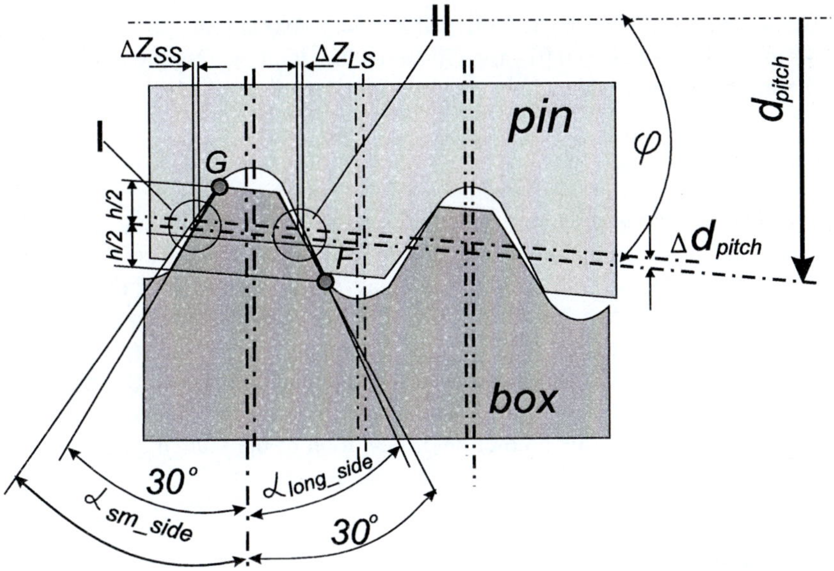

Fig. 7. The scheme of combination of threaded profiles of the box and the pin.

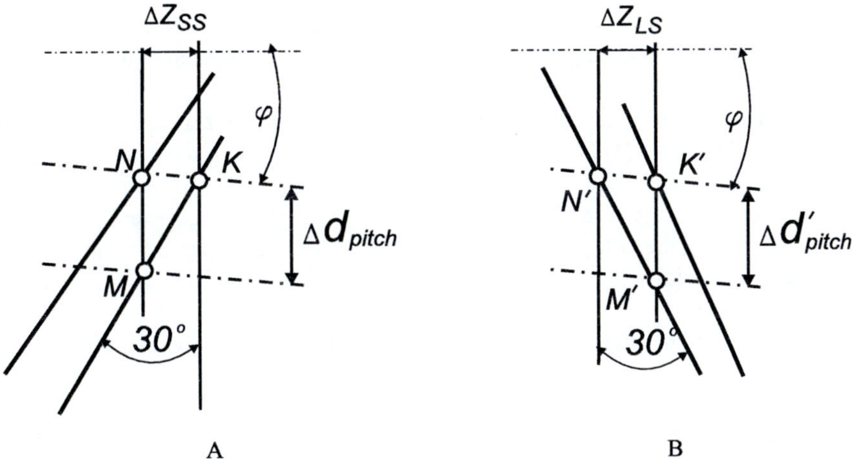

Fig. 8. Scheme for calculating the deviation of the pitch diameter of the connection.

Figure 8(*A* and *B*) are the enlarged fragments I and II of Fig. 7, respectively. Using them, and assuming that the figures *KNM* and *K'N'M* are right triangles, we can calculate:

$$\Delta_{dpitch} = 2 \cdot \Delta_{zss} \cdot ctg30^{\circ} \quad (4)$$

$$\Delta_{dpitch'} = 2 \cdot \Delta_{zLs} \cdot ctg30^{\circ} \quad (5)$$

After substitution Zss = 0.012 mm, and ZLs = − 0.003 mm, the values of Δdpitch = 0.04 mm, and Δ′dpitch = 0.01 mm. Given that the tolerance for the pitch diameter is ± 0.25 mm, the resulting deviations are: 16% and 4%, respectively.

4.2 Discussion

The nearest and too close prediction calculations give simile results but of the influence of the back rake angle on thread profile only in [28]. The predicted accuracy can reach the value from −0.03° to +0.1°, which is equivalent to 4,5–17% of the tolerance on the profile angle using lathe machining (±0,67°). So we have resumed the complicated influence of the back rake tool parameter of the lathe cutter on the tapered string-grid thread accuracy. Both results are possible if the cutter with a non-zero back rake angle is applied separately for the pin or box. In the case of using the same tool parameter for both parts of the tool-joint the prediction of deviation of the pitch diameter and profile angle is near zero.

5 Conclusions

The deviation of the pitch diameter of the tool-joint tapered thread of a pin is due to the cutter with a non-zero value of the back rake angle.

The essential factor that affects the change in the pitch diameter of the tapered thread is the tool cutter back rake angle because, with its increase, the pitch diameter deviation also increases.

The increase of the deviation of tapered thread pitch diameter functionally depends on the number (size or value) of tool-joint: with its decreasing, the pitch diameter deviation increases.

So, if it is necessary to produce the tool-joint for high-strength steel drill pipes, it does mean applying the lathe tool-cutters with back-rake angle up to ±5° and decreasing the accuracy. Especially for the minor size of drill pipe (e.g., NC23), the tolerance part will be up to 16%.

In future research, we plan to conduct detailed research on any drill-string tapered thread.

Acknowledgment. The authors are grateful to the Ministry of Science and Education of Ukraine for the grant to implement project D-2-22-P (State Reg. Number 0122U002082).

References

1. Threading. Tread turning tools. Sundwik Coromant, https://www.sandvik.coromant.com/en-us/products/pages/thread-turning-tools.aspx. Last accessed 20 Dec 2021
2. Shatskyi, I., Velychkovych, A., Vytvytskyi, I., Seniushkovych, M.: Analytical models of contact interaction of casing centralizers with well wall. Eng. Solid Mech. **7**(4), 355–366 (2019). https://doi.org/10.5267/j.esm.2019.6.002

3. Bazaluk, O., Velychkovych, A., Ropyak, L., Pashechko, M., Pryhorovska, T., Lozynskyi, V.: Influence of heavy weight drill pipe material and drill bit manufacturing errors on stress state of steel blades. Energies **14**(14), 4198 (2021). https://doi.org/10.3390/en14144198
4. Danil'chenko, Y.: Increasing the accuracy of rotation of high-speed spindle units on anti-friction bearings. Sov. Eng. Res. **7**(7), 61–63 (1987)
5. Ropyak, L.Y., Pryhorovska, T.O., Levchuk, K.H.: Analysis of materials and modern technologies for PDC drill bit manufacturing. Prog. Phys. Met. **21**(2), 274–301 (2020). https://doi.org/10.15407/ufm.21.02.274
6. Severinchik, N.A., Kopei, B.V.: Inhibitive protection of steel drill pipes from corrosion fatigue. Sov. Mater. Sci. **13**(3), 318–319 (1978)
7. Ya Ropyak, L., Vytvytskyi, V.S., Velychkovych, A.S., Pryhorovska, T.O., Shovkoplias, M.V.: Study on grinding mode effect on external conical thread quality. IOP Conference Series: Materials Science and Engineering **1018**(1), 012014 (2021). https://doi.org/10.1088/1757-899X/1018/1/012014
8. Pryhorovska, T., Ropyak, L.: Machining Error Influence on Stress State of Conical Thread Joint Details. In: Proceedings of the International Conference on Advanced Optoelectronics and Lasers, CAOL, pp. 493–497 (2019). DOI: https://doi.org/10.1109/CAOL46282.2019.9019544
9. Budnik, A.F., Rudenko, P.V., Berladir, KV., Budnik, O.A.: Structured nanoobjects of polytetrafluoroethylene composites. J. Nano-Electron. Phys. **7**(2), 02022 (2015)
10. Berladir, K.V., Sviderskiy, V.A.: Designing and examining polytetrafluoroethylene composites for tribotechnical purposes with activated ingredients. Eastern-European J. Enterp. Technol. **6**(84), 14–21 (2016). https://doi.org/10.15587/1729-4061.2016.85095
11. Dong, L., Wang, J., Zhu, X.: Design and mechanical behavior study of ultrahigh-torque variable pitch casing joint. Adv. Mech. Eng. **11**(1), 1–12 (2019). https://doi.org/10.1177/1687814018814084
12. Tutko, T., Dubei, O., Ropyak, L., Vytvytskyi, V.: Determination of Radial Displacement Coefficient for Designing of Thread Joint of Thin-Walled Shells. In: Ivanov, V., Trojanowska, J., Pavlenko, I., Zajac, J., Peraković, D. (eds.) DSMIE 2021. LNME, pp. 153–162. Springer, Cham (2021). https://doi.org/10.1007/978-3-030-77719-7_16
13. Dong, L., Wang, J., Zhu, X.: Design and mechanical behavior study of ultrahigh-torque variable pitch casing joint. Adv. Mech. Eng. **11**(1), 168781401881408 (2019). https://doi.org/10.1177/1687814018814084
14. Zhu, X., Zhang, Z.: Design of an ultra-high torque double shoulder drill-pipe tool joint for extended reach wells. Nat. Gas. Ind. **4**(5), 374–381 (2017). https://doi.org/10.1016/j.ngib.2017.10.002
15. Khoshdarregi, M.R., Altintas, Y.: Generalized modeling of chip geometry and cutting forces in multi-point thread turning. Int. J. Mach. Tools Manuf **98**, 21–32 (2015). https://doi.org/10.1016/j.ijmachtools.2015.08.005
16. Fromentin, G., Döbbeler, B., Lung, D.: Computerized Simulation of Interference in Thread Milling of Non-Symmetric Thread Profiles. Procedia CIRP **31**, 496–501 (2015). https://doi.org/10.1016/j.procir.2015.03.018
17. Sonawane, G.D., Sargade, V.G.: Machinability Study of Duplex Stainless Steel 2205 During Dry Turning. Int. J. Precis. Eng. Manuf. **21**(5), 969–981 (2020). https://doi.org/10.1007/s12541-019-00305-8
18. Wang, J., Mathew, M., Li, X., Huang, C., Zhu, H.: Experimental Study on Cutting Characteristics for Buttress Thread Turning of 13%Cr Stainless Steel. Key Eng. Mater. **443**, 262–267 (2010). https://doi.org/10.4028/www.scientific.net/KEM.443.262
19. Zhang, G., Guo, C.: Modeling Flank Wear Progression Based on Cutting Force and Energy Prediction in Turning Process. Procedia Manufacturing **5**, 536–545 (2016). https://doi.org/10.1016/j.promfg.2016.08.044

20. Koleva, S., Enchev, M., Szecsi, T.: Compensation of the deviations caused by mechanical deformations during machining of threads. Procedia Manufacturing **13**, 480–486 (2017). https://doi.org/10.1016/j.promfg.2017.09.066
21. Baizeau, T., Campocasso, S., Fromentin, G., Rossi, F., Poulachon, G.: Effect of rake angle on strain field during orthogonal cutting of hardened steel with c-BN tools. CIRP **31**, 166–171 (2015). https://doi.org/10.1016/j.procir.2015.03.089
22. Costa, C.E., Polli, M.L.: Effects of the infeed method on thread turning of AISI 304L stainless steel. J. Braz. Soc. Mech. Sci. Eng. **43**(5), 1–9 (2021). https://doi.org/10.1007/s40430-021-02978-7
23. González, L.H., Ahmed, Y.S., Rodríguez, R.P., Robledo, P.Z., Mata, M.G.: Selection of machining parameters using a correlative study of cutting tool wear in high-speed turning of AISI 1045 steel. J. Manuf. Mater. Process. **2**(4), 66 (2018). https://doi.org/10.3390/jmmp2040066
24. Ivanov, V., Pavlenko, I., Kuric, I., Kosov, M.: Mathematical modeling and numerical simulation of fixtures for fork-type parts manufacturing. In: Knapčíková, L., Balog, M. (eds.) Industry 4.0: Trends in Management of Intelligent Manufacturing Systems. EICC, pp. 133–142. Springer, Cham (2019). https://doi.org/10.1007/978-3-030-14011-3_12
25. Berglind, L., Ziegert, J.: Modulated tool path (MTP) machining for threading applications. Procedia Manuf. **1**, 546–555 (2015). https://doi.org/10.1016/j.promfg.2015.09.029
26. Yanda, H., Ghani, J.A., Haron, C.H.C.: Effect of rake angle on stress, strain and temperature on the edge of carbide cutting tool in orthogonal cutting using FEM simulation. ITB J. Eng. Sci. **42**(2), 179–194 (2010). https://doi.org/10.5614/itbj.eng.sci.2010.42.2.6
27. Onysko, O., Panchuk, V., Kopei, V., Havryliv, Y., Schuliar, I.: Investigation of the influence of the cutter-tool rake angle on the accuracy of the conical helix in the tapered thread machining. J. Phys.: Conf. Ser. **1781**(1), 012028 (2021). https://doi.org/10.1088/1742-6596/1781/1/012028
28. Panchuk, V., Onysko, O., Kotwica, K., Barz, C., Borushchak, L.: Prediction of accuracy of the tapered thread profile made by lathe. J. Eng. Sci. **8**(2), E1–E6 (2021). https://doi.org/10.21272/jes.2021.8(2).b1
29. Onysko, O., Medvid, I., Panchuk, V., Rodic, V., Barz, C.: Geometric modeling of lathe cutters for turning high-precision stainless steel tapered threads. In: Ivanov, V., Trojanowska, J., Pavlenko, I., Zajac, J., Peraković, D. (eds.) DSMIE 2021. LNME, pp. 472–480. Springer, Cham (2021). https://doi.org/10.1007/978-3-030-77719-7_47
30. Ivanov, V.: Process-oriented approach to fixture design. In: Ivanov, V., et al. (eds.) DSMIE 2018. LNME, pp. 42–50. Springer, Cham (2019). https://doi.org/10.1007/978-3-319-93587-4_5
31. Ivanov, V., Dehtiarov, I., Denysenko, Y., Malovana, N., Martynova, N.: Experimental diagnostic research of fixture. Diagnostyka **19**(3), 3–9 (2018). https://doi.org/10.29354/diag/92293

Features of Deformation Mechanics in the Deformation Zone During Deforming Broaching of Cast Iron Workpieces

Ihor Shepelenko(✉), Yakiv Nemyrovskyi, Sergii Mahopets, Oleksandr Lizunkov, and Ruslan Osin

Central Ukrainian National Technical University, 7, Universytetskyi Avenue, Kropyvnytskyi 25006, Ukraine
kntucpfzk@gmail.com

Abstract. In order to create a combined technology for processing cast iron products using deforming broaching, the processes occurring in the deformation zone have been investigated from the point of view of exhaustion of the plasticity resource. A technique has been developed for studying the stress-deformed state parameters of the surface layer during processing by deforming broaching of workpieces made of SCH20 cast iron. The presence of local plastic deformation zones in the contact zone under critical contact pressures has been proved. The appearance of a zone of local plastic deformation creates conditions for intensive depletion of the plasticity resource of the processed material. The regularities of changes in the plasticity resource in all zones of the deformation zone have been established. The deformation mechanics in the deformation zone have been studied, and the ways of optimizing the tool's design and the process of deforming broaching by the resource of plasticity used parameter have been established. The design of a deforming tool is proposed, which makes it possible to increase the plasticity resource used when processing cast iron products due to the presence of a curved section, the parameters of which correspond to the zone of local plastic deformation.

Keywords: Plasticity resource · Non-contact zone · Contact zone · Local zone · Process innovation

1 Introduction

The development of world mechanical engineering is inextricably linked with modern technologies for processing parts, which improve the quality of their working surfaces by achieving optimal operational properties [1]. An essential indicator of the quality of the surface layer is its ability to deform plastically, estimated by the resource of used plasticity [2]. This parameter is significant when processing the surface of the product made of low-plastic materials, for example, products made of graphite-containing cast irons [3].

V. Tonkonogyi et al. (Eds.): InterPartner 2022, LNME, pp. 211–221, 2023.
https://doi.org/10.1007/978-3-031-16651-8_20

The increasing requirements for the functional properties of working surfaces stimulate the development of technologies that combine the advantages of various processing methods [4]. This determines the relevance of complex combined technologies development that allows the creation of a working surface with predetermined operational properties [5].

Let us consider one of these technologies for machining holes in cast iron parts, which includes deforming broaching operations (DBR) [6] and finishing antifriction treatment (FANT) [7]. This technology should ensure the presence of an equilibrium roughness favorable concerning the microrelief wear resistance, hardening of the surface layer to a very significant depth, and increasing antifriction properties [8]. As a result of such processing, the running-in time of rubbing surfaces is minimized, abrasion in the friction pair is reduced, and the mechanism's durability increases [9].

To effectively use the joint technology of DBR and FANT, it is necessary to carefully study the issues of plastic deformation of the working surface of the material under study. Given the low ability of cast iron to plastically deform, the study of the plasticity resource of a low-plastic material and the search for ways to increase it is especially relevant.

2 Literature Review

The efficiency of using the process of deforming broaching in processing parts made of plastic materials has been proven [10]. Simultaneously, the use of this operation in processing low-plastic materials, for example, cast irons, is limited. The perspectives for processing cast iron workpieces by DBR are evidenced by the fact that, according to the data of work [11], in the contact zone of the deforming element with the processed surface, conditions arise that are close to the conditions of uniform compression. This, as indicated in work [12], allows the microdefects to heal and prevents the process of brittle fracture. The work [13] provides examples of the successful deformation of cast iron products when solving a specific production task.

The studies by the authors [14] made it possible to develop and implement a new computational and experimental technique during the deformation of composite samples from cast iron. This made it possible, under certain conditions, to perform deformations of cast-iron samples at significant negative values of the stiffness index of the stress state η. As a result, a section of the plasticity diagram was completed using the values range of the specified coefficient $-1 \leq \eta \leq -5$. Thus, for the first time, significant plastic deformation before failure was achieved – $e_{ult} > 80\%$.

It is known that in the contact zone with DBR, the stress state is close to all-sided compression. This allows us to assume that the surface of a cast iron workpiece can be successfully deformed plastically [15]. Research [16] of the stress-deformed state (SDS) in the deformation zone revealed the presence of plastic deformation local zones. It turned out that these zones are the places of damage accumulation. The zone of local plastic deformation behind the contact area has a particularly significant negative effect on the exhaustion of the plasticity resource. This is due to the presence of tensor of tensile stresses σ_r, σ_z and σ_φ in it, which leads to intensive exhaustion of the plasticity resource and the appearance of micro-fractures in the surface layer. In this regard, studies of the processes and phenomena occurring in the deformation zone seem very topical. Separate research also deserves the study of the plasticity of the treated surface.

The relevance of such studies made it possible to formulate the purpose of this work–the study of the processes occurring in the deformation area during the processing of holes made of low-ductility cast iron SCH20 by deforming broaching, as well as the establishment of the main patterns of changes in the plasticity resource in this case.

To achieve the goal of the work, the following tasks are formulated:

- to develop a methodology for studying SDS during deforming broaching of a cast-iron billet using the finite element method;
- establish the change in plasticity resources in all zones of the deformation zone;
- determine the possibilities of managing the residual plasticity resource.

3 Research Methodology

The development of a methodology for studying the stress-strain state of a cast-iron billet consisted in establishing the conditions for modeling the deforming broaching of a cast-iron billet. For this, the object of study was initially set - a bushing made of SCH20 cast iron (Fig. 1), as well as the modes of deforming broaching: the speed of movement of the deforming element $V = 0.5$ mm/s, the angle of the working cone of the deforming element $\alpha = 4°$, the nominal interference $a = 0.05$ mm.

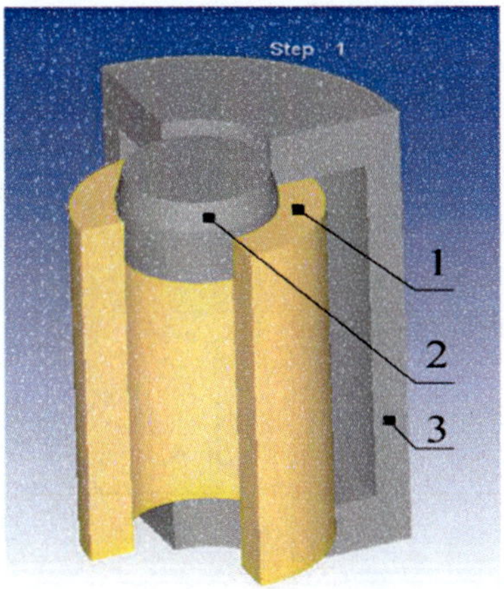

Fig. 1. Scheme of processing: 1 – cast iron billet; 2 – deforming element; 3 – base [17].

The properties of the studied material were set by the experimentally obtained compression curve and by the hardness HB = 1.7 GPa, Poisson's ratio $\mu = 0.27$, and Young's modulus $E = 1.6 \times 10^5$ MPa.

The study of the SDS of the surface layer was carried out at points P1, P11, and P2. Note that point P1 is located on the surface of the workpiece, and point P11 is at a depth of 0.01 mm. The distance between points P1 and P2 is 0.25 mm, which corresponds to the thickness of the coating at FANT. The process of simulating deforming pulling was carried out with a *Step*. The number of steps is 1–115.

4 Results and Discussion

Let us consider the processes occurring in the deformation zone and changes in its sections of the SDS parameters and factors depending on it: η (stiffness coefficient of the stress state), σ (hydrostatic pressure), e_0 (deformations intensity), $\Delta\psi$ (increment of deformation at each cycle), ψ (plasticity resource), σ_z (an axial component of the stress tensor).

Let us analyze the phenomena occurring in the deformation zone, starting with *Step 43*, considering the SDS changes and the parameters depending on it the depth of occurrence in the active layer. As is known [18], the deformation zone begins from the non-contact zone in front of the contact area. According to the modeling results (Fig. 2), the stress state coefficient for this zone is $\eta = -1.73$ and slightly increases to the value $\eta = -2$ at *Step 50* (the beginning point of the contact zone).

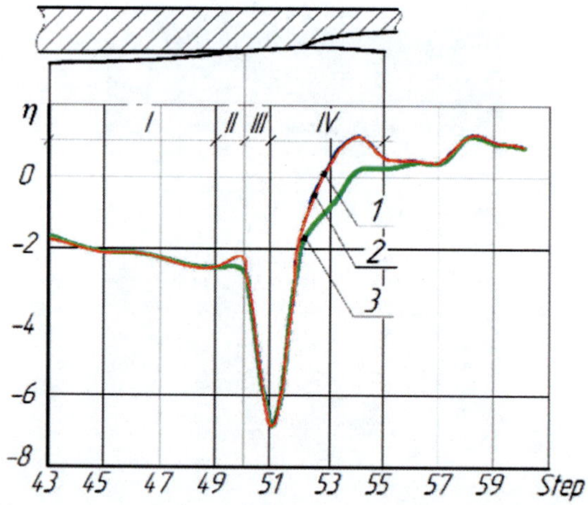

Fig. 2. Dependence of the stiffness coefficient of the stress state η on the depth of the occurrence of the point, relative to the treated surface when processing products from cast iron SCH20: I, IV – non-contact zone; II, III – contact zone; 1 – point 1; 2 – point 11; 3 – point 3.

These values correspond to the stress state – compression under plane deformation conditions and are characterized by the presence of hydrostatic pressure negative value in the range of $\sigma = -400$ MPa (Fig. 3), a relatively small value of accumulated deformation reaching the value $e_0 = 0.02$ (Fig. 4) and the value of axial compressive stresses $\sigma_z = -250 \div 650$ MPa, increasing towards the end of the zone (Fig. 5).

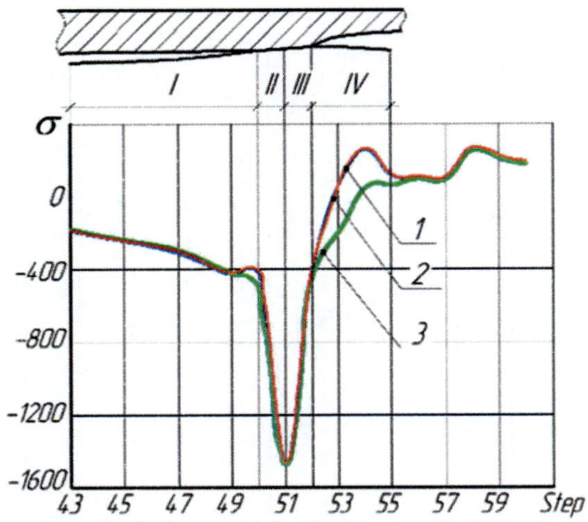

Fig. 3. Dependence of hydrostatic pressure σ on the depth of the occurrence of the point relative to the treated surface when processing products from cast iron SCH20: I, IV – non-contact zone; II, III – contact zone; 1 – point 1; 2 – point 11; 3 – point 3.

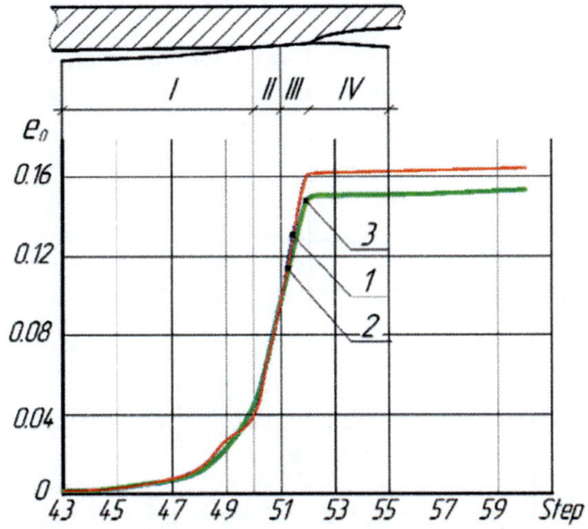

Fig. 4. Dependence of accumulated deformation e_0 on the depth of the occurrence of the point, relative to the treated surface when processing products from cast iron SCH20: I, IV – non-contact zone; II, III – contact zone; 1 – point 1; 2 – point 11; 3 – point 2.

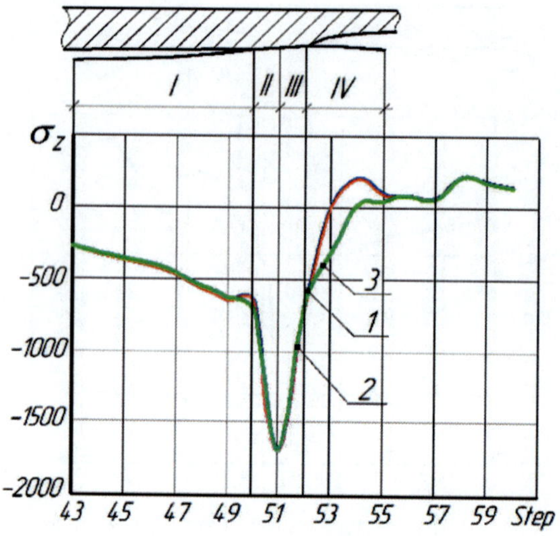

Fig. 5. Dependence of the stress tensor axial component σ_Z on the depth of the occurrence of the point, relative to the treated surface when processing products from cast iron SCH20: I, IV – non-contact zone; II, III – contact zone; 1 – point 1; 2 – point 11; 3 – point 2.

In the considered zone, some accumulation of damage occurs, which is reflected by the growth of the resource of plasticity used ψ to the value $\psi = 0.08$ (Fig. 6) and the increase in the parameter $\Delta\psi$ to the value $\Delta\psi = 0.02$ (Fig. 7).

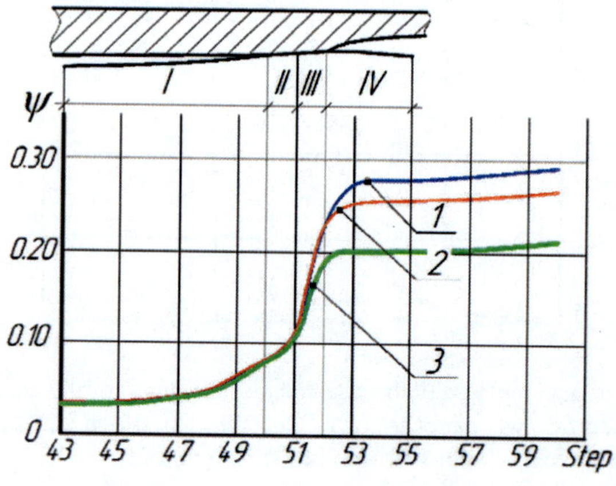

Fig. 6. Dependence of the resource of plasticity used ψ on the depth of the occurrence of the point, relative to the treated surface: I, IV – non-contact zone; II, III – contact zone; 1 – point 1; 2 – point 11; 3 – point 3.

It should be noted that the value of the considered parameters in this area practically does not depend on the depth of the occurrence of the studied point (Fig. 2–7).

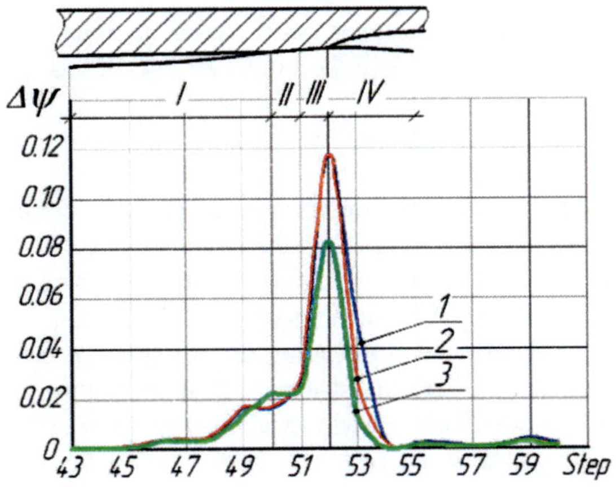

Fig. 7. Dependence of plasticity increments $\Delta\psi$ on the depth of the occurrence of the point, relative to the treated surface: I, IV – non-contact zone; II, III – contact zone; 1 – point 1; 2 – point 11; 3 – point 3.

A somewhat different situation is observed when the non-contact zone passes into contact zone II (*Step 50*). In this case, the negative values of hydrostatic pressure sharply increase from the value $\sigma \approx -500$ MPa (*Step 50*) to the value $\sigma \approx -1450$ MPa (*Step 51*) (Fig. 3).

In turn, the negative value of the stiffness coefficient of the stress state also increases to the value $\eta = -7$ (*Step 51*) (Fig. 2). This indicates the material transition in contact zone II into a state of powerful volumetric compression, which is also confirmed by an increased negative value of $\sigma_z = -1700$ MPa (*Step 51*) (Fig. 5).

All-sided volumetric compression determines the absence of microdefects growth due to their healing, which practically leads to the values invariability of the resource of plasticity used in this zone (Fig. 6, 7). At the same time, the accumulated deformation increases sharply to the value $e_0 = 0.1$ (*Step 51*) (Fig. 4). Features of deformation processes in contact zone II provide an opportunity for the intense shear deformation implementation in the near-surface layer due to friction forces. The data of work [19] indicate that even for low-plastic cast iron, such deformation can exceed 170%, and for plastic materials (Armco-iron), in the presence of 40 deformation cycles, it exceeds 400% [20].

Next, consider the deformation process in contact zone III (*Step 52*). In this case, there is a sharp decrease in the negative value of hydrostatic pressure from the value $\sigma \approx -1450$ MPa (*Step 51*) to the value $\sigma \approx -400$ MPa (*Step 52*) (Fig. 3). The stiffness coefficient of the stress state behaves similarly, decreasing its negative values to the value $\eta = -1$ (*Step 52*) (Fig. 2).

A change in the conditions of volumetric compression causes some exhaustion of the plasticity resource to values in the surface layer (points P1 and P11) (Fig. 7) $\Delta\psi = 0.12$. In this case, the influence of the coordinates occurrence depth of the studied layer is observed for the point P2 $\Delta\psi = 0.08$ (Fig. 7). Likewise change, the values of the resource of plasticity used, which for the points P1 and P11 increase to the value $\psi =$

0.18 (Fig. 6). In this case, the accumulated deformation continues to increase somewhat, reaching its maximum value $e_O = 0.16$ for points P1 and P11 and $e_O = 0.15$ for point P2 (*Step 52*) (Fig. 4).

Axial stresses in this zone decrease their negative value to $\sigma_z = -500$ MPa, and the downward trend remains (Fig. 5).

Consider deformation in the non-contact zone behind the contact area from *Step 52* to *Step 55*. In this case, a sharp decrease in the negative value of hydrostatic pressure σ occurs. Moreover, decreasing its negative value, the hydrostatic pressure becomes positive to the value $\sigma = 220$ MPa for points P1 and P11 and begins to influence the value of σ depth of the occurrence of the point, for example, for point P2 $\sigma = 100$ MPa (Fig. 3). This confirms the earlier conclusions [21] about the presence of local plastic deformation zone in this area. This leads to axial tensile stresses (Fig. 5), which sharply reduces the surface layer's plasticity resource.

The stiffness coefficient of the stress state for points P1 and P11 acquires the value $\eta = +1$, and for point P2, $\eta = +0.5$ (Fig. 2).

These changes naturally cause sharp exhaustion of the plasticity resource $\Delta\psi = 0.12$ for points P1 and P11, and point P2 $\Delta\psi = 0.08$ (Fig. 7). It also increases the resource of plasticity used $\psi = 0.28$–0.25 for points P1 and P11, respectively, and for point P2 $\psi = 0.2$ (Fig. 6).

These changes occur with an insignificant increase of the accumulated deformation $e_O \approx 0.16$. It is noticeable only in the area of plastic deformation local zone (Fig. 4).

The studies indicate that the most dangerous zone concerning the $\Delta\psi$ parameter is the surface layer (point P1). Point P11 is close to it in the distance and hence in the value of this parameter. Point P2 has a much smaller increment $\Delta\psi$. Subsequently, this parameter decreases with an increase in the depth of subsequent points occurrence.

It is also possible to draw a meaningful conclusion that at a layer depth equal to the thickness of the antifriction coating $\Delta \approx 0.005$ mm, that is, from point P1 to point P11, the values of e_O, $\Delta\psi$, and ψ change by no more than 8%, which makes it possible to consider the SDS in such a coating is practically uniform. This fact can be used for determining and analyzing the coating SDS, such as the thickness applied to the working surface of the hole.

The deformation analysis shows that the primary accumulation of damage and the plasticity resource exhaustion occurs in the zone of local plastic deformation, which occurs at the transition point of the contact area into the non-contact zone. One of the ways to eliminate this drawback is to develop a tool design that allows influencing the zone of local plastic deformation. This tool must have an additional curvilinear section to exclude the influence of the specified local zone on the parameters $\Delta\psi$ and ψ. In the proposed tool design (Fig. 8), the working cone of tool 2, which interacts with workpiece 1, is mated with an additional curvilinear Sect. 3, the generatrix parameters of which are calculated based on the modeling results (Fig. 2–7).

This approach makes it possible to reproduce the non-contact zone with a local area of plastic deformation on the tool. The presence of a curvilinear section, the parameters of which correspond to the zone of local plastic deformation, makes it possible to influence this zone, change the SDS in it and eliminate its influence on the plasticity resource parameters.

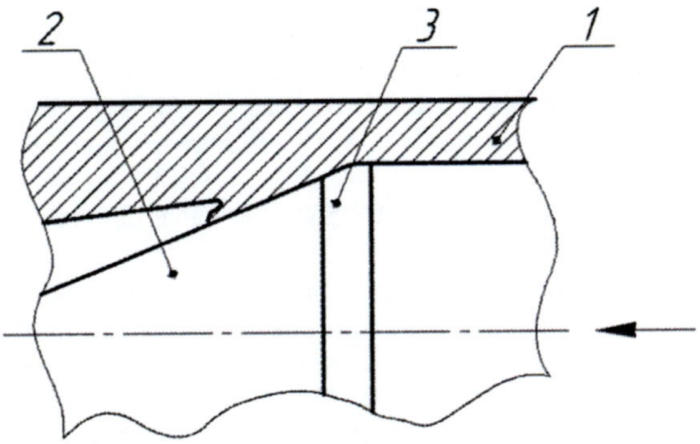

Fig. 8. Hole machining with a deforming element with a curvilinear section: 1 – processed part; 2 – tool working cone; 3 – curvilinear section.

5 Conclusions

The performed studies allowed us to conclude the following statements.

It has been proven that during deforming broaching in the deformation zone, the minor plasticity resource is observed on the surface of the machined hole. As the occurrence depth of the studied layer points increases, the resource of residual plasticity increases.

It was confirmed that in the contact zone, where the main deformation of the workpiece hole occurs, the stress state is close to all-side compression. This leads to the almost complete absence of plasticity resource exhaustion during plastic deformation, which makes it possible to carry out intense shear deformation in the near-surface layer due to friction forces.

Under conditions of critical contact pressures, a local zone of plastic deformation has been established, which creates the prerequisites for intensive depletion of the plasticity resource of the processed material. It is shown that in the contact zone, under critical contact pressures, the processed material flows out into the non-contact zone, creating conditions for forming a zone of local plastic deformation.

It was found that at a depth of the deformable layer, corresponding to points 1 and 2, the stress-deformed state is practically uniform, which can be used when studying the antifriction coating obtained by FANT.

A variant of plasticity resource control by improving the design of the deforming tool is proposed.

References

1. Hosseini, A., Kishawy, H.A., Moetakef-Imani, B.: Effects of broaching operations on the integrity of machined surface. Procedia CIRP **45**, 163–166 (2016). https://doi.org/10.1016/j.procir.2016.02.352
2. Kobets, A., et al.: Design of mated parts using polymeric materials with enhanced tribotechnical characteristics. Eastern-European J. Enterp. Technol. **5**(12 (107)), 49–57 (2020). https://doi.org/10.15587/1729-4061.2020.214547

3. Dobrotvorskiy, S., Balog, M., Basova, Y., Dobrovolska, L., Zinchenko, A.: Concept of the software for materials selection using .NET technologies. In: Tonkonogyi, V. et al. (eds.), Advanced Manufacturing Processes. InterPartner 2019. Lecture Notes in Mechanical Engineering, pp. 32–43. Springer, Cham (2020). https://doi.org/10.1007/978-3-030-40724-74
4. Smolentsev, E.V.: Design of Electric and Combined Methods of Processing. Moscow (2016). [in Russian]
5. Safonov, S.V., Smolentsev, V.P., Mandrykin, A.V.: The methodology of manufacturability testing in launching high-tech products of aerospace engineering into production. Mater. Today: Proc. **19**, 2526–2528 (2019). https://doi.org/10.1016/j.matpr.2019.08.176
6. Chernyavsky, O.V., Sivak, I.O., Lopatenko, S.G.: Determining the limit of process application of deforming broaching in the processing of cast iron. Improving the technical level of agricultural production and engineering, 82−85 (2016). [in Ukrainian]
7. Ragutkin, A.V., Sidorov, M.I., Stavrovskij, M.E.: Some aspects of antifriction coatings application efficiency by means of finishing nonabrasive antifriction treatment. J. Min. Inst. **236**, 239–244 (2019). https://doi.org/10.31897/PMI.2019.2.239
8. Kıvak, T.: Optimization of surface roughness and flank wear using the Taguchi method in milling of Hadfield steel with PVD and CVD coated inserts. Measurement **50**, 19–28 (2018)
9. Kirillov, O., Smolentsev, V., Yukhnevich, S.: The technology of combined processing of extruded materials. MATEC Web Conf. **224**, 1–6 (2018). https://doi.org/10.1051/matecconf/201822401004
10. Studenets', S.F., Eryomin, P.M., Chernyavs'kyi, O.V.: The influence of deformation conditions on the structure and hardening of cast iron surface layer in machining with combined carbide broaches. J. Superhard Mater. **37**(4), 282–288 (2015). https://doi.org/10.3103/S1063457615040085
11. Sheykin, S.Y., et al.: Improving the performance of hard-alloy deforming broaches using modified technological lubricants. J. Superhard Mater. **42**(4), 276–282 (2020). https://doi.org/10.3103/S1063457620040085
12. Ye Sheykin, S., et al.: On the contact interaction between hard-alloy deforming broaches and a workpiece during the shaping of grooves in the holes of tubular products. J. Superhard Mater. **43**(3), 222–230 (2021). https://doi.org/10.3103/S1063457621030096
13. Sheikin, S.E., Pogrelyuk, I.M., Sergach, D.A.: Modification of working surface of titanium components of friction units. J. Superhard Mater. **37**, 351–356 (2016). https://doi.org/10.3103/S1063457615050093
14. Shepelenko, I., Tsekhanov, Y., Nemyrovskyi, Y., Gutsul, V., Mahopets, S.: Compression mechanics of cylindrical samples with radial deformation limitation. In: Ivanov, V., Pavlenko, I., Liaposhchenko, O., Machado, J., Edl, M. (eds.) DSMIE 2021. LNME, pp. 53–62. Springer, Cham (2021). https://doi.org/10.1007/978-3-030-77823-1_6
15. Grushko, A.V.: Maps of materials in cold working by pressure. Vinnitsa (2015). [in Russian]
16. Kukhar, V.V., Grushko, A.V., Vishtak, I.V.: Shape indexes for dieless forming of elongated forgings with sharpened end by tensile drawing with rupture. SPP **284**, 408–415 (2018)
17. Shepelenko, I., Nemyrovskyi, Y., Chernovol, M., Kyrychenko, A., Vasylenko, I.: Deformation zone scheme clarification during deforming broaching. In: Ivanov, V., Trojanowska, J., Pavlenko, I., Rauch, E., Peraković, D. (eds.), Advances in Design, Simulation and Manufacturing V. DSMIE 2022. Lecture Notes in Mechanical Engineering, pp. 302−311. Springer, Cham (2022). https://doi.org/10.1007/978-3-031-06025-0_30
18. Sheykin, S.E., Pashchenko, E.A., Rostotsky, I.: Process lubricant for deforming drawing of pieces made of titanium. Metall. Mining **4**, 38–43 (2017)
19. Johnson, G.R.: A constitutive model and data for metal subjected to large strains, high strain and temperatures. In: Proc. of 7th Symposium on Ballistics, pp. 541–547. Hague (1983)

20. Ogorodnikov, V.A., Kiritsa, I.Yu., Muzyichuk, V.I.: Plasticity diagrams and peculiarities of their construction. Improving pressure treatment processes and equipment in metallurgy and mechanical engineering, pp. 251–255 (2016). [in Russian]
21. Rosenberg, O.A.: Technological mechanics of deforming broaching. Voronezh (2001). [in Russian]

Check for updates

Numerical Simulation of Cutting Forces in Face Milling

Heorhii Vyhovskyi[1], Mykola Plysak[1], Nataliia Balytska[1,2(✉)], Larysa Hlembotska[1], and Valentyn Otamanskyi[1]

[1] Zhytomyr Polytechnic State University, 103, Chudnivska Street, Zhytomyr 10005, Ukraine
tmkts_nno@ztu.edu.ua

[2] Technische Universität Dresden, 7a, Helmholtz Street, 01069 Dresden, Germany

Abstract. The work is devoted to the development of a finite element model of machining the workpiece flat surface from gray cast iron with one tooth face oblique milling cutter in the DEFORM-3D software. The model development methodology is described step by step. The preparation peculiarities of solid-state models of workpiece and cutting insert, finite-element mesh generation, as well as preprocessor setup are discussed. It is shown that to save calculation time, it is advisable to develop a 3D model of only a part of the workpiece with a cut from the path of the previous cutter insert, formed using the SolidWorks Motion module. 2 options of a finite element mesh with a different number of elements were considered. In addition, local refinement of the workpiece mesh in the area of interaction with the cutting insert was used. As a result of finite element simulation of the face milling process, the cutting forces in the feed range of 0.25…0.625 mm/tooth, with a cutting speed of 150 m/min and a cutting depth of 0.12 mm were calculated. The adequacy of the developed model is confirmed by comparison with the experimental results for the corresponding processing conditions. The finite element model of the face milling developed in DEFORM-3D will be used to further optimize the geometric parameters of the cutter to increase the efficiency of machining flat surfaces of grey cast iron parts.

Keywords: Finite elements · Mesh · 3D model · Cutting force · Feed

1 Introduction

Face mills of oblique cutting, equipped with superhard materials, have found wide application in the finishing of high hardness materials. One of the ways to improve such a tool is to determine the optimal geometric parameters of the inserts, which can be done by real experimental research, mathematical, physical modeling, or simulation of the cutting process. Simulation reduces research costs and achieves results with sufficient accuracy. In addition, a simulation allows the calculation of a number of the milling process characteristics (cutting forces, torques, stresses, temperatures, etc.).

Finite element analysis systems such as DEFORM, ABAQUS, LS-DYNA, ADVENT EDGE, ANSYS EXPLICIT DYNAMICS, etc. are increasingly used to study the cutting

V. Tonkonogyi et al. (Eds.): InterPartner 2022, LNME, pp. 222–231, 2023.
https://doi.org/10.1007/978-3-031-16651-8_21

processes. The Deform-3D software was chosen to model the face milling process. This system is characterized by the relative ease of use, user-friendly interface and allows getting a stable result.

In numerical modeling of machining processes, the main problem is the adequacy of the obtained results [1]. Different types of mechanical processing require different ways of setting the parameters of the finite element model, which affects the accuracy of the obtained results [2].

The research aim of this work was the development of a finite element model of oblique face milling of flat surfaces from gray cast iron with one tooth face oblique milling cutter. This model can be used to speed up the design process of the cutting tool, as well as to study the influence of insert geometric parameters and cutting conditions on the final machining efficiency characteristics.

2 Literature Review

Milling processes simulation is a handy tool that has been used successfully in research.

Johnson-Cook constitutive material model is the most common, yet simplest, model to describe the material behavior in machining. The main disadvantages of the Johnson-Cook model include its empirical basis and the lack of relationship between the strain rate and temperature in the plastic deformation process. The model also has shortcomings in the representation of the strengthening characteristics of all types of materials. Inaccuracies in determining the parameters included in the phenomenological equation can lead either to the impossibility of the task solving or to significant errors in determining the functional processing parameters (cutting forces, cutting temperatures, etc.) and surface quality parameters [1]. The material model has a significant effect on the results of the simulation and suitable selection of its parameters is essential for an accurate simulation of the cutting process.

Different types of metal cutting models are known. Each model corresponded to a given numerical formulation: Lagrangian, Arbitrary Eulerian-Lagrangian, and Couple Lagrangian-Eulerian [3, 4]. The issues of accuracy of the metal cutting models are discussed in [2]. It is shown that the accuracy of the models does not depend only on the constitutive model, but also how these metal cutting models deal with the material separation from the workpiece to form the chip.

Finite element (FE) modeling of milling workpieces from different types of materials in strictly orthogonal cutting condition is described in [5, 6]. The authors used the developed models to obtain cutting forces and stress fields, both in the tool and in the workpiece [5]. The comparison of the numerical results with the experiments when the titanium alloy Ti6Al4V machining is carried out in [6]. It shows that the level of the forces is mainly influenced by the material constitutive model, while the chip morphology is mostly impacted by the chip separation criterion.

In [7] the chip formation mechanism is investigated with the help of finite element simulation of the face milling process and the cutting forces are determined. The cutting forces and cutting temperatures when face milling of Ti-6Al-4V alloy were calculated by finite element analysis in [8]. In [9] ABAQUS software is using to simulate the 3D milling of GH4169 nickel base superalloy. The article [10] solves the problem of determining the power characteristics of face milling process by computer modelling in the DEFORM 3D environment. The influence of the rake face shapes of round inserts of face mills on the contact area of the chips with the insert, the conditions of the chip flow, the dynamic power loads both during the entry of inserts into the workpiece and after the cutting is investigated. The comparison of the simulated and real force data confirms this affirmation with an average deviation under 10%.

The temperature fields that occur in the tool and the workpiece during the cutting process determine the tool life and effect of the treated surface quality [11]. Therefore, such information is of great importance for the appointment of optimal milling modes. Nevertheless, temperature measurement during milling operations imposes a number of restraints to experimental methods, mostly related to the cutter rotational speed, variable chip thickness and intermittent action of the cutting edges [12]. The paper [13] presents a hybrid simulation system consisting of a geometric multiscale milling simulation and a finite element method kernel for solving problems of linear thermoelasticity.

Modelling allow to predict the resulting roughness [14], the surface topography generated in face milling operations [15], and various types of flatness deviation [16] for different flank wear values.

Consequently, modern studies of numerical simulations of milling processes focus mainly on the orthogonal cutting, most authors study end mills with positive cutting edge geometry. Therefore, the issue of numerical simulation of the oblique face milling process is an actual scientific and technical task and requires further research.

3 Research Methodology

The finite element model was developed in the DEFORM-3D software for the conditions of face oblique milling of grey cast iron workpiece with a single-insert tool equipped with superhard material (hexanite). The diameter of the mill is 250 mm. Processing conditions were adopted accordingly previous works of the authors to compare the experimentally obtained results on the cutting forces and assess the adequacy of the developed model.

The following settings were used for this calculation: Lagrangian incremental, number of simulation steps – 100, solver – conjugate gradient, iteration method – direct, environment temperature – constant, convection coefficient – constant, friction – shear, object type (workpiece – plastic, tool – rigid).

The development of the finite element model was carried out in the following sequence.

1. Development of a 3D model of the cutting insert (see Fig. 1): face is a flat surface, rake angle $\gamma = -10°$, cutting edge inclination $\lambda = -45°$, clearance angle in the direction of the cutting speed vector $\alpha = 12°$.

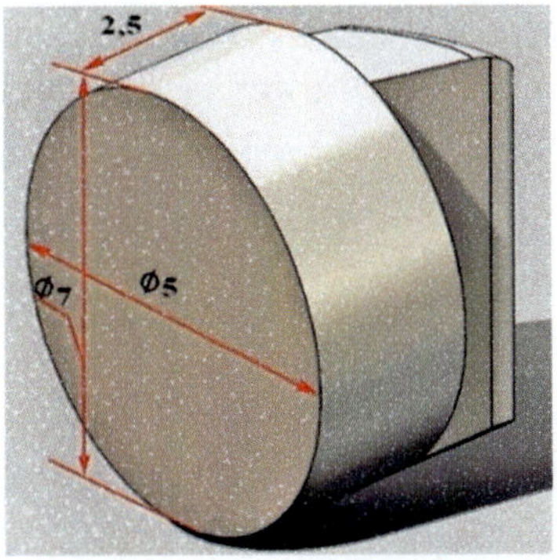

Fig. 1. Simplified 3D-model of the cutting insert

2. Setting the material properties of the cutting insert.
   Rigid – this condition defines the object as undeformed and retains the initial geometry assigned to it, and is used in most tasks to determine the cutting tool. At the same time, the calculation time is reduced compared to elastoplastic type.
3. Development of a workpiece 3D-model with a previous cut.
   Since the investigated face mill has a diameter of 250 mm, a contact angle between the insert and the workpiece will have a significant value. Simulation of all time of the cutting insert passage along the entire tool contact path will require significant calculation time and computer resources. Taking into account these factors, the simplified (clipped) workpiece 3D-model was developed with a cut from the path of the previous cutting insert (see Fig. 2).

   The workpiece 3D-model with a previously cut was developed in the Motion module of the SolidWorks system according to the method described in [17]. The face mill trajectory was realized using two motors: rotary (simulates the milling cutter rotation) and linear (simulates the workpiece feed).

   It should be noted that with different cutting modes, the workpiece will have different cut shape from the insert passage. To save the laboriousness of the model development process, one workpiece 3D-model was created and a cut was formed for the following conditions. The milling cutter speed, which was simulated by a rotating motor, was 160 rpm. The feed speed, which was simulated by a linear motor, was 1.05 mm/s, which corresponds to the feed rate of the workpiece 63 mm/min. This simplification does not cause large errors due to the small size of the cutting insert.

At the time of calculation, when the cutting insert coincided with the plane passing through the mill axis and with the workpiece symmetry plane in the feed direction, the cutting edge projection was created. The resulting sketches were edited by clipping unnecessary elements to form a cut cross section for the trajectory formation.

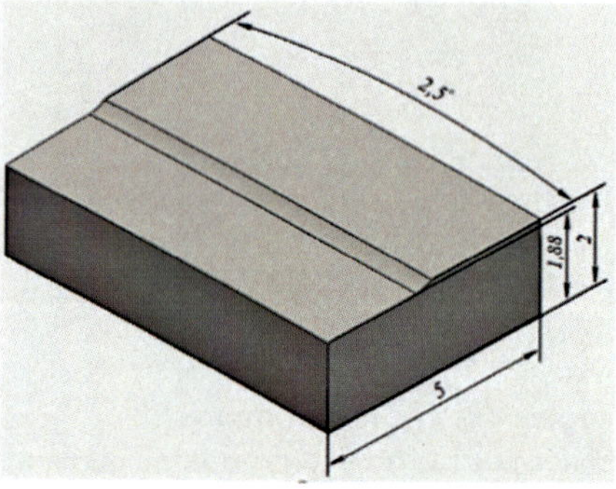

Fig. 2. 3D-model of a workpiece with a previous cut

4. Determining the behavior model of the machined material (Table 1).
   The results accuracy obtained in the face milling simulation depends on the correctness of the machined material characteristics. The behavior of the machined material is often described by a mathematical equation Johnson Cook when simulation machining processes. The equation takes into account the influence of stresses, strains, strain rates and temperatures on the material properties.
   The calculation took into account the mechanical behavior of the machined material, which is described by the Johnson-Cook equation:

$$\sigma = (A + B \cdot \varepsilon^{n}) \cdot \left(1 + Cln \cdot \left(\frac{\varepsilon^{\bullet}}{\varepsilon_0^{\bullet}}\right)\right) \cdot \left(\frac{\varepsilon^{\bullet}}{\varepsilon_0^{\bullet}}\right)^{\alpha} \cdot (D - E \cdot T^{*m})$$

where ε – is equivalent plastic deformation, $\varepsilon^{\bullet}$ and $\varepsilon_0^{\bullet}$ – is equivalent and reference rates of plastic deformation; *A, B, n, C* and *m* – the material constants of the basic Johnson-Cook equation, representing the yield strength, strain and strain rate, strain hardening and thermal softening coefficient. T_0, T_m – ambient temperature and melting point of the material, respectively: $T^{*m} = (T - T_0)/(T_m - T_0)$. The coefficients of the equation were taken according to [18].

Table 1. Johnson-Cook model parameters

Material	A, MPa	B, MPa	n	C	m	T*, °C	ρ, kg/m^3
Grey cast iron	573	380	0.17	0.034	0.12	1200	7200

5. Finite element mesh generation.
The DEFORM system offers two methods of mesh generation. "System Setup" - mesh settings performs by default, assigned in this parameters group during the mesh generation and its subsequent restructuring. The "User Defined" method allows for an independent selection of the surfaces to generate the mesh.

There are also two types of mesh density settings: "Relativ" - a constant number of elements in the object and the ratio of the maximum size of the element edge to the minimum ("Size ratio"); and "Absolute" - the ratio of the element size in the window to the element size in basic settings ("Size ratio").

The mesh of the cutting insert included 50000 finite elements (see Fig. 3, a). Two workpiece meshes contained 200000 and 400000 finite elements (see Fig. 3, b) and had a local refinement in the interaction area with the cutting insert.

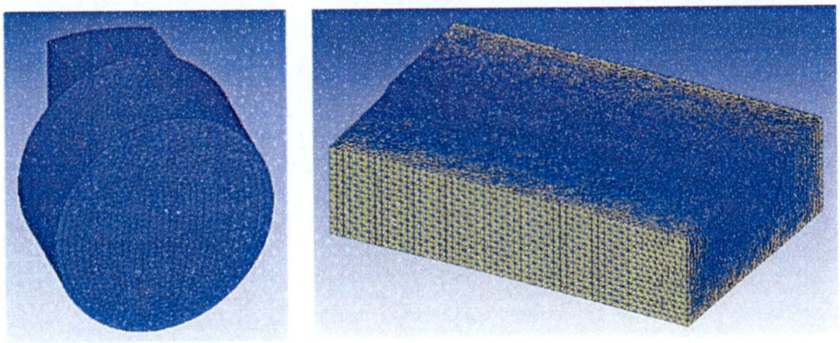

a b

Fig. 3. Finite element mesh: a – the cutting insert, b – the workpiece

To ensure the stability of the cutting process model, local reconstruction of the workpiece mesh with refinement in the area of the local shear surface was used to reduce the prediction error in the cutting zone.

To assess the influence elements mesh number of the workpiece on results accuracy were considered two options (see Fig. 4): with 200000 and 400000 number of elements in the interaction area between cutting insert and workpiece.

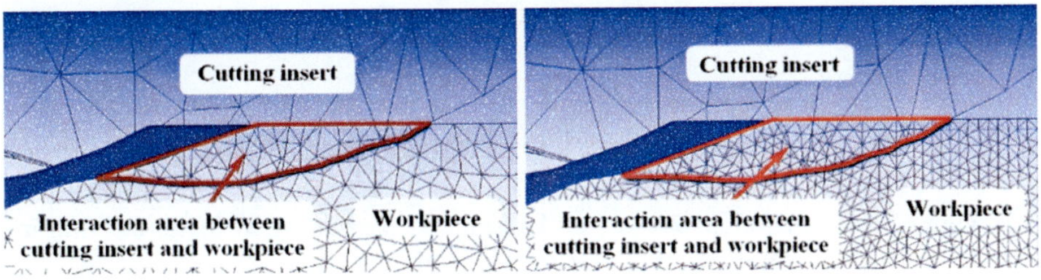

a b

Fig. 4. Workpiece finite element mesh: a – 200000 and b – 400000 number of elements in the interaction area between cutting insert and workpiece (feed per tooth 0,4 mm/tooth)

6. Contact conditions.
   When set up contact conditions were determined roles between objects: Master – the tool and Slave - the workpiece.
7. Determination of movement conditions.
   The cutting insert was set up on the relevant radial distance to the rotation axis (115 mm). The workpiece was fixed in all axes.

4 Results and Discussion

Simulation of the face milling process was performed for the aforecited cutting condition in range of feeds 0.25…0.625 mm/tooth, at depth of cut 0.12 mm and cutting speed 150 m/min. The calculated cutting forces acting on one cutter are presented in Fig. 5. To assess the influence of the element number of the workpiece mesh on results accuracy, two opinions with 200000 (Simulation 2) and 400000 (Simulation 1) elements number in the interaction area between cutting insert and workpiece were considered. Also, a local refinement of the workpiece mesh in the interaction area was applied.

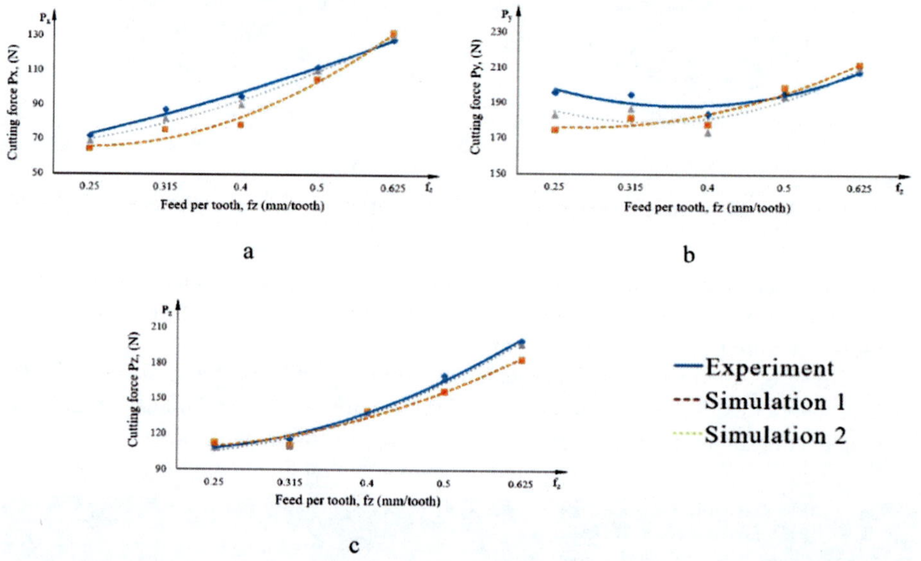

Fig. 5. The cutting force components in face milling of grey cast iron: a – the feed force *Px*, b – the horizontal force *Py* perpendicular to the feed, c – the vertical cutting force *Pz*

Figure 6 shows the relative errors of determining the cutting forces in face milling using DEFORM-3D software and experimental method.

As can be seen from the graphs, an increasing of workpiece finite elements number makes it possible to increase the calculating accuracy of the cutting force components by 2.8 times (*Px* component), 1.33 times (*Py* component), and 2.54 (*Pz* component).

The obtained results indicate that the developed model allows calculating cutting forces with an average error about the experimental values of 3.7% for the *Px* component,

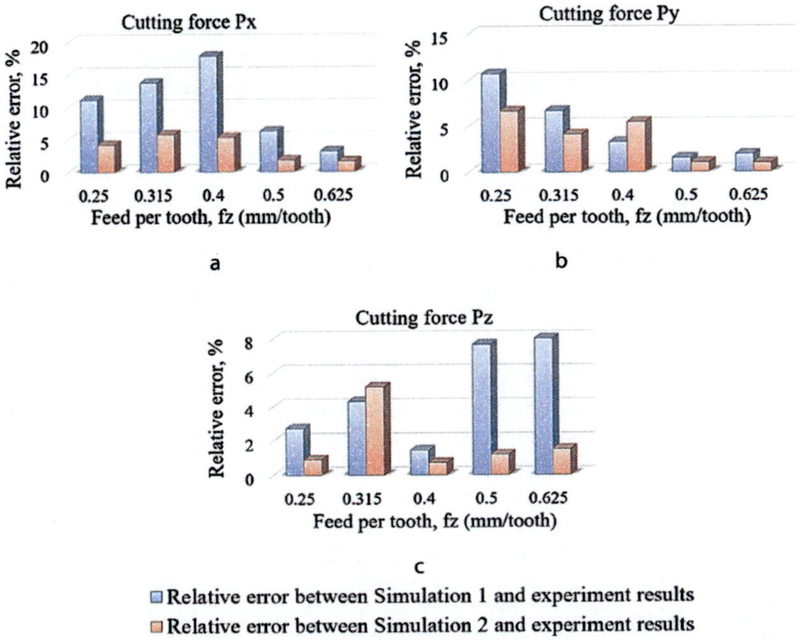

Fig. 6. The relative errors of determining the cutting forces in face milling using DEFORM-3D software and experimental method: a – the feed force *Px*, b – the horizontal force *Py* perpendicular to the feed, c – the vertical cutting force *Pz*

3.6% for the *Py* component, and 1.9% for the *Pz* component. In this way, the adequacy of the developed model of oblique face milling of grey cast iron can be considered as confirmed.

5 Conclusions

Using DEFORM-3D software a finite element model of machining the grey cast iron workpiece with oblique face single-tooth milling cutter with a diameter of 250 mm in the range of feeds 0.25…0.625 mm/tooth, at cutting speed of 150 m/min and depth of cut 0.12 mm was developed. The mechanical behavior of machined material is described by the Johnson-Cook equation, which parameters were determined on a literature review of relevant scientific works. The adequacy of the developed model is confirmed by compare with experimental results since the average relative error of modelling does not exceed 10.4%.

It is shown that for the saving calculation time it is effectually to develop a simplified workpiece 3D model with a previous cut using the SolidWorks Motion module.

The peculiarities of this work is the simulation of the oblique milling process with an insert with a radius cutting edge and the description of the method of developing of the workpiece 3D model with a preliminary cut, which allow to reduce calculation time and computer resources.

Therefore, the finite element model of the oblique face milling process developed in the DEFORM-3D software can be used for further cutter geometric parameters optimization to increase the efficiency of machining parts flat surfaces from grey cast iron.

References

1. Ducobu, F., Riviere, E., Filippi, E.: On the importance of the choice of the parameters of the Johnson-Cook constitutive model and their influence on the results of a Ti6Al4V orthogonal cutting model. Int. J. Mech. Sci. **122**, 143–155 (2017). https://doi.org/10.1016/j.ijmecsci.2017.01.004
2. Zhang, Y., Outeiro, J.C., Mabrouki, T.: On the selection of Johnson-Cook constitutive model parameters for Ti-6Al-4V using three types of numerical models of orthogonal cutting. Procedia CIRP **31**, 112–117 (2015). https://doi.org/10.1016/j.procir.2015.03.052
3. Vovk, A., Sölter, J., Karpuschewski, B.: Finite element simulations of the material loads and residual stresses in milling utilizing the CEL method. Procedia CIRP **87**, 539–544 (2020). https://doi.org/10.1016/j.procir.2020.03.005
4. Ducobu, F., Arrazola, P.-J., Rivière-Lorphèvre, E., Ortiz de Zarate, G., Madariaga, A., Filippi, E.: The CEL method as an alternative to the current modelling approaches for Ti6Al4V orthogonal cutting simulation. In: 16th CIRP Conference on Modelling of Machining Operations (16th CIRP CMMO) Procedia CIRP, vol. 58, pp. 245–250 (2017). https://doi.org/10.1016/j.procir.2017.03.188
5. Dobrotvorsky, S.S., Basova, E.V., Dobrovolskaya, L.G.: Computer design and simulation of technological processes of high-speed milling of hardened steels. Bull. Lviv Polytech. Natl. Univ. **822**, 1–6 (2015)
6. Ducobu, F., Rivière-Lorphèvre, E., Filippi, E.: Material constitutive model and chip separation criterion influence on the modeling of Ti6Al4V machining with experimental validation in strictly orthogonal cutting condition. Int. J. Mech. Sci. **107**, 136–149 (2016). https://doi.org/10.1016/j.ijmecsci.2016.01.008
7. Borysenko, D., Karpuschewski, B., Welzel, F., Kundrák, J., Felhő, C.: Influence of cutting ratio and tool macro geometry on process characteristics and workpiece conditions in face milling. CIRP J. Manuf. Sci. Technol. **24**, 1–5 (2019). https://doi.org/10.1016/j.cirpj.2018.12.003
8. Mehta, S., Singh, G., Saini, A., Singh, H.: Finite element analysis of face milling of Ti-6Al-4 V alloy considering cutting forces and cutting temperatures. Proceedings (2021). https://doi.org/10.1016/j.matpr.2021.10.061
9. Geng, G., Zhang, L., Xiao, M., Dong, X., Chen, K.: Finite element analysis and parameter optimization selection of high speed milling GH4169. In: Manufacturing Technology, vol. 20, no. 3, pp. 300–306 (2020)
10. Hlembotska, L., Balytska, N., Melnychuk, P., Melnyk, O.: Computer modelling power load of face mills with cylindrical rake face of inserts in machining difficult-to-cut materials. Sci. J. TNTU **93**(1), 70–80 (2019). https://doi.org/10.33108/visnyk_tntu2019.01.070
11. Benabid, F., Benmoussa, H., Arrouf, M.: A thermal modelling to predict and control the cutting temperature. The simulation of face-milling process. Procedia Eng. **74**, 37–42 (2018). https://doi.org/10.1016/j.proeng.2014.06.220
12. Lima, H.V., Campidelli, A.F.V., Maia, A.A.T., Abrão, A.M.: Temperature assessment when milling AISI D2 cold work die steel using tool-chip thermocouple, implanted thermocouple and finite element simulation. Appl. Therm. Eng. **143**, 532–541 (2018)

13. Schweinoch, M., Joliet, R., Kersting, P.: Predicting thermal loading in NC milling pro-cesses. Prod. Eng. Res. Devel. **9**, 179–186 (2015)
14. Tapoglou, N., Antoniadis, A.: 3-dimensional kinematics simulation of face milling. Meas.: J. Int. Meas. Confed. **45**(6), 1396–1405 (2012). https://doi.org/10.1016/j.measurement.2012.03.026
15. Arizmendi, M., Jiménez, A.: Modelling and analysis of surface topography generated in face milling operations. Int. J. Mech. Sci. **163**, 105061 (2019). https://doi.org/10.1016/j.ijmecsci.2019.105061
16. Pimenov, D.Y., Guzeev, V.I., Krolczyk, G., Mozammel Mia, S.: Modeling flatness deviation in face milling considering angular movement of the machine tool system components and tool flank wear. Precis. Eng. **54**, 327–337 (2018). https://doi.org/10.1016/j.precisioneng.2018.07.001
17. Hlembotska, L., Melnychuk, P., Balytska, N., Melnyk, O.: Modelling the loading of the nose-free cutting edges of face mill with a spiral-stepped arrangement of inserts. Eastern-Eur. J. Enterp. Technol. **1**(91), 46–54 (2018). https://doi.org/10.15587/1729-4061.2018.121712
18. Chernykh, D.M., Tkachenko, Y., Tsyganov, V.S.: Simulation of the machining process in order to optimize the operating parameters. Bull. Voronezh State Tech. Univ. **15**(1), 130–137 (2019)

Optimization of the Cutting Process Based on Thermophysical Characteristics

Serhii Zelynskyi, Gennadii Oborskyi, Volodymyr Tonkonogyi, and Maryna Holofieieva(✉)

Odessa Polytechnic National University, 1, Shevchenko Avenue, Odessa 65044, Ukraine
mgolofeyeva@gmail.com

Abstract. In the energy sector, aviation, and other branches of modern mechanical engineering, complex-profile parts are used to determine the entire product's technical and operational properties. In manufacturing such parts, modern structural materials are often used with high strength and heat resistance, which classifies them as difficult to machine. End mills allow the processing of such parts on CNC machines. At the same time, the processing time of one product can reach tens of hours. Despite the use of wear-resistant coatings, because of wear, the end mill has to be changed repeatedly, which reduces productivity and quality of manufacturing. In conditions of high-speed machining, it is essential to have an effective technique based on which you can quickly determine the optimal cutting speed. In the presented work, the optimal conditions are investigated for the temperature field distribution between the part's material and the tool in terms of tool wear. For this purpose, the dependences of specific heat capacity and thermal conductivity on temperature were analyzed. These dependencies have a pronounced extremal character with a maximum for the heat capacity and a minimum for the thermal conductivity. According to the authors, this explains the effect of the optimal cutting temperature, characterized by the minimum intensity of tool wear. The optimal cutting temperature and the corresponding optimal cutting speed were determined by the material being processed heating temperature, at which its specific heat capacity reaches its maximum value.

Keywords: Optimal cutting speed · Cutting temperature · Temperature field distribution · Sustainable manufacturing

1 Introduction

Scientific and technological progress in mechanical engineering requires using new materials with special properties. These materials have increased strength characteristics, high thermal stability, and corrosion resistance [1]. This applies to parts of the power industry and the aircraft industry (e.g., turbine blades, unicycles) that have a complex geometric shape, the processing of which, as a rule, occurs with end mills on CNC machines [2].

In manufacturing such highly loaded aircraft products, much attention is paid to the geometric parameters and parameters of the state of the surface layer, which determine

V. Tonkonogyi et al. (Eds.): InterPartner 2022, LNME, pp. 232–240, 2023.
https://doi.org/10.1007/978-3-031-16651-8_22

the quality of processing. Achieving high quality and productivity in modern conditions of digital products can only be ensured by ensuring optimal processing conditions, including cutting conditions [3].

The processing time of such parts on CNC milling machines can reach tens and hundreds of hours. At the same time, despite the presence of wear-resistant coatings, the wear of end mills can be so intense that the replacement or regrinding of the tool must be carried out several times during the processing of one part, which reduces the economic efficiency and quality of processing of such parts. Therefore, optimizing cutting conditions to achieve maximum tool life is relevant [2].

2 Literature Review

The contour milling is often the final form of surface shaping, where the specified geometric accuracy and surface roughness are provided, and residual stresses are formed that affect the technological deformations of parts. During end milling, the surface layer of parts is formed under the influence of force and temperature fields. Consequently, stable provision during milling of the parameters specified by the technical requirements for the operation, which are required for the quality of processing, especially for parts with a complex profile, such as blades, can be achieved by choosing rational cutting modes that affect the functional parameters of the process [4].

When processing such parts, end mills of small diameters with wear-resistant coatings are usually used, capable of operating in the operating temperature range of 973–1373 K, which corresponds to the temperature range for high-speed machining conditions [5].

As is known, of all cutting modes, the cutting speed has the most significant influence on the wear rate of the tool [6]. The dependence of the tool wear intensity on the cutting speed has a pronounced extreme character with a very narrow range of optimal cutting speeds. Determining the optimal cutting speed values is associated with labor-intensive tool life tests. The extreme character of such dependence has been known for a long time. However, an unambiguous explanation of this phenomenon based on the thermal physics of the cutting process has not yet existed [7].

The practical significance of studying this issue is essential for the following reasons: for parts with complex geometric surfaces, high requirements are imposed on the initial characteristics of the quality of the surface layer (roughness, depth and degree of hardening, level, and stability of residual stresses), which affect the performance of parts, fatigue strength, corrosion resistance. Processing on a CNC milling machine is often a finishing operation for such parts. There are close relationships between the wear parameters of the cutter and the quality indicators of the surface layer of parts [8, 9].

The problem of optimizing the cutting process during machining [10], the physics of cutting tool wear, considering thermophysical phenomena [11], is the subject of many works by scientists from many countries. The practical goal of these studies was to optimize cutting conditions, which provides the tool's lowest wear rate and, as a result, its maximum durability [12].

The prevailing direction of well-known works was associated with the need to test the tool for resistance by various methods [13, 14]. Despite the development of various, including accelerated methods for studying resistance, their common drawback is the performance of labor-intensive, expensive, and lengthy experimental studies [15].

One of the most progressive approaches to the practical optimization of cutting speed are methods that use the position of the constancy of the optimal cutting temperature, which is of great scientific and practical importance. As shown in [16], the optimal cutting temperature is a stable value when processing the same material and does not depend on the tool's and the workpiece's geometric parameters.

In the well-known works of scientists, the thermophysical nature of the existence of the optimal cutting temperature and the phenomena that occur when using it, providing the lowest wear rate of the tool, improving the quality of the surface layer, the occurrence of a "dip in plasticity" of the material being processed, and reducing the specific cutting energy, have not yet been explained [17].

3 Research Methodology

To explain the thermophysical phenomena during processing at the optimum temperature, an analysis was made for heat transfer between the part and the tool [18].

From the standpoint of thermal physics, it can be assumed that when machining at the optimum temperature [19], conditions are created that contribute to the best distribution of cutting heat between the workpiece and the tool, considering the effect of temperature on thermal resistance material characteristics.

As is known, according to the heat balance equation [20]:

$$Q = Q_c + Q_w + Q_t + Q_e \tag{1}$$

where Q – the total amount of heat during cutting; Q_c – heat escaping into chips; Q_w – heat absorbed by the workpiece; Q_t – heat absorbed by the tool; Q_e – heat leaving the environment.

The best will be the conditions under which $Q_w + Q_c$ will be maximum, and Q_t will be minimum. The amount of heat, and hence the cutting temperature, is determined by a set of parameters: the type of blade processing, the physical and mechanical characteristics of the processed and tool materials, tool geometry, cutting conditions, and others.

Obviously, for specific processing conditions, the heat balance and heat transfer conditions will be determined by the energy parameters of the cutting process and the thermophysical characteristics of the processed and tool materials. When analyzing various thermophysical characteristics of materials from temperature, the most significant interest in thermal field control is the heat capacity and thermal conductivity of the material being processed. This can be explained by the fact that the dependences of thermal conductivity and heat capacity on temperature are of a pronounced extreme nature, which was taken as a hypothesis to explain the existence of an optimal cutting temperature.

Based on the previous, the problem is reduced to finding the cutting temperature at which the optimal combination of specific heat capacity and thermal conductivity coefficient of the processed material is achieved.

Thermal conductivity characterizes heat transfer due to the energy interaction of microparticles (molecules, atoms, electrons). The thermal conductivity of a material, which is usually given by the coefficient of thermal conductivity λ (W $m^{-1}.K^{-1}$), characterizes the ability to conduct heat. The specific heat capacity C_p (J $kg^{-1}.K^{-1}$), characterizes the ability of the material to absorb the heat transferred to the body [21].

Optimal conditions are under which the best indicators of workpiece material machinability and tool wear resistance are simultaneously achieved [22]. These are conditions under which the maximum amount of heat will be locally concentrated in the cutting zone of the material being processed, contributing to its "softening" and decompression of crystalline bonds and the ability to transfer this heat into the tool material will be the worst. This condition, based on the physical meaning of the specific heat capacity C_p and the thermal conductivity coefficient λ, can be formulated as:

$$Optimum \Rightarrow \left\{ \frac{C_p = C_p \max}{\lambda = \lambda \min} \right\} \quad (2)$$

This condition, from the standpoint of the optimal combination of thermophysical characteristics of the processed material, can be written:

$$\left\{ \frac{C_p = C_{\max}}{\lambda = \lambda_{\min}} \right. \quad (3)$$

The experimentally obtained dependencies of the specific heat capacity and the thermal conductivity coefficient of pure iron on temperature were analyzed.

Figure 1 shows the temperature dependence of the specific heat capacity $C_p = f(\theta)$ of pure iron, constructed according to the data of [22], from which a maximum specific heat capacity at a certain temperature is clearly defined seen.

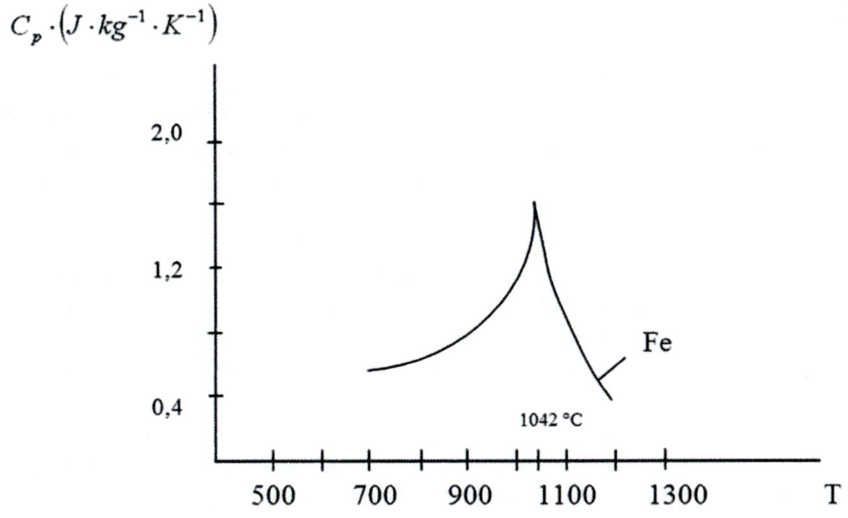

Fig. 1. Temperature dependence of the specific heat capacity of iron.

Since the basis of all steel is iron, the temperature dependencies on the heat capacity of the absolute majority of various grades of steel and alloys have a similar, extreme character [7].

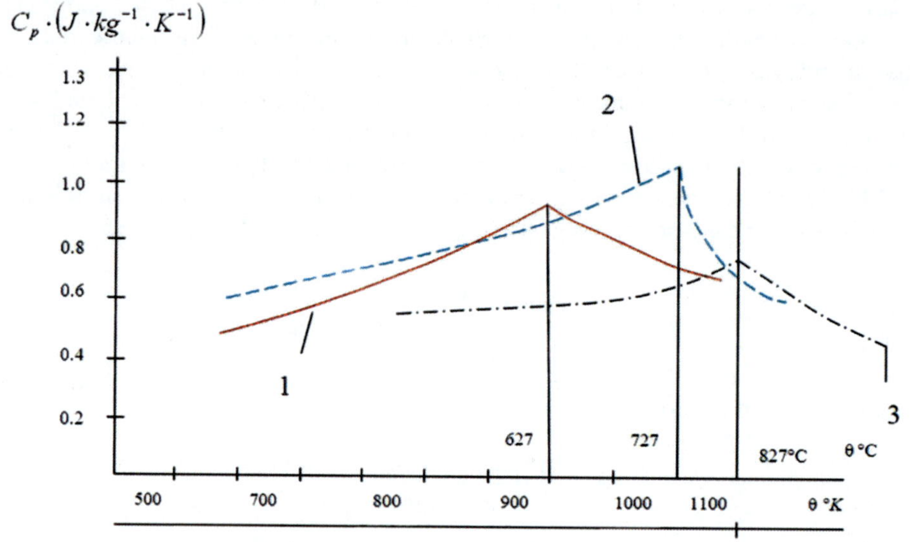

Fig. 2. Temperature dependence of the heat capacity of steels 1 – carbon steel; 2 – high-alloy steel AIS1446; 3 – low-alloy steel.

The temperature at which the maximum value of the specific heat capacity Cpmax is reached on the temperature dependence Cp = f (θ) corresponds to such a state of the processed material, which contributes to the absorption of heat to the greatest extent.

The minimum value of the thermal conductivity coefficient λmin on the temperature dependence λ = f (θ) corresponds to the best conditions for heat concentration from the cutting zone in the material being processed. For example, Fig. 3 shows the temperature dependence of the heat capacity Cp = f (θ) and thermal conductivity λ = f (θ) for steel 40 (carbon steel). The similar extreme nature of these dependencies is also characteristic of other steel grades. Moreover, for hard-to-cut materials, the extremeness of these dependencies is more pronounced.

However, the temperature at which $C_p = C_{max}$ and the temperature at which $\lambda = \lambda_{min}$ most often do not coincide. In connection with the above, the optimum cutting temperature should be in the range of these temperatures. Moreover, the smaller the value of this temperature range, the stronger the effect of processing with the optimum cutting temperature, characterized by a minimum intensity of tool wear, a «dip in plasticity» of the material being processed, and other phenomena.

As can be seen from the comparison of graphs of temperature dependences $C_p = f(\theta)$ and $\lambda = f(\theta)$ (Fig. 3), the dependence of specific heat on temperature has a more pronounced extremum than the dependence of thermal conductivity. This suggests that the heat capacity has a more significant influence on the effect of "ductility dip" than the coefficient of thermal conductivity by influencing the local cutting temperature.

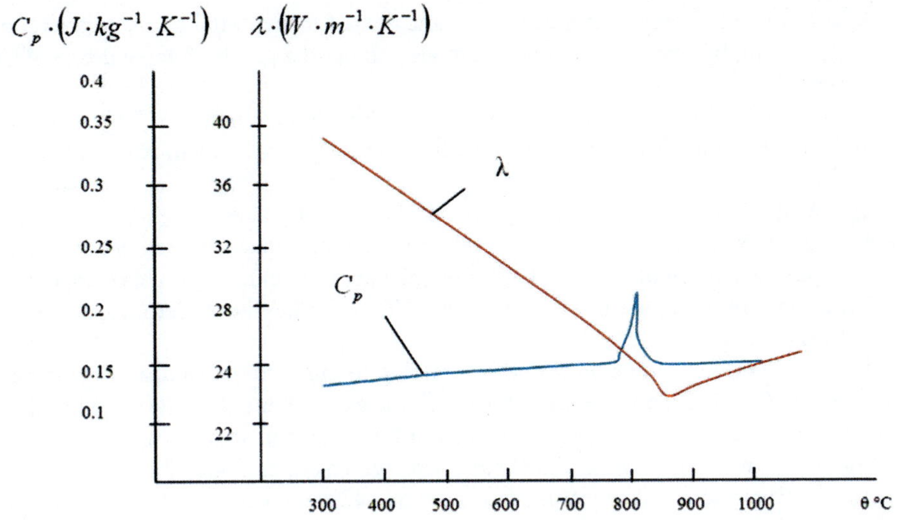

Fig. 3. Temperature dependence of heat capacity $C_p = f(\theta)$ and thermal conductivity $\lambda = f(\theta)$ for steel 40X.

The latter assumption is also confirmed by the well-known Cronenberg [20] dimensional ratio, which shows the relationship between the cutting temperature θ and the energy parameters of the cutting process, taking into account the thermophysical characteristics of the material being processed:

$$\theta = \frac{C_0 U_8 \cdot V^{0.44} \cdot A^{0.22}}{\lambda^{0.44} \cdot C_p^{0.56}} \tag{4}$$

where θ - cutting temperature; U8 – specific cutting energy; V – cutting speed; A – the cut area; λ – thermal conductivity of the processed material; Cp is the heat capacity of the processed material.

The more significant influence of the heat capacity of the machined material on the cutting temperature is confirmed by the greater value of the exponents at C_p than at λ. Based on the foregoing, we can conclude that the optimal cutting temperature can be determined by the heating temperature of the processed material at maximum heat capacity.

4 Results and Discussion

To implement the proposed method under production conditions, the optimal cutting speed can be determined by the heating temperature of the material being processed, at which the specific heat capacity is maximum.

The method is carried out as follows:

1. When cutting the material under study, the dependence of the cutting temperature on the cutting speed is plotted $T = f(V)$

2. To determine the dependence of the specific heat of the material on the temperature of its heating by one of the known methods and build a graph of dependence of Cp $= f(\theta)$.
3. The optimal cutting speed is determined according to the graph $T = f(V)$ according to the temperature at which the specific heat capacity takes on a maximum value.

This method of determining the optimal cutting speed makes it possible to exclude the cutting tool's labor-intensive and rather expensive resistance tests to determine the optimal speed when machining new high-strength materials and alloys. This is especially true when milling complex-profile parts on CNC machines in modern conditions of high-speed cutting.

The correlation of temperatures corresponding to the extreme values of the dependences $C_p = f(273\ K)$ and $\lambda = \varphi(273\ K)$ with the experimentally determined optimal cutting temperatures is shown in Fig. 4 when milling iron and steel.

Figure 4 shows that when turning "pure" iron with a cutter made of P20 carbide, the plasticity and hardness of the machined material on minimum values and internal stresses

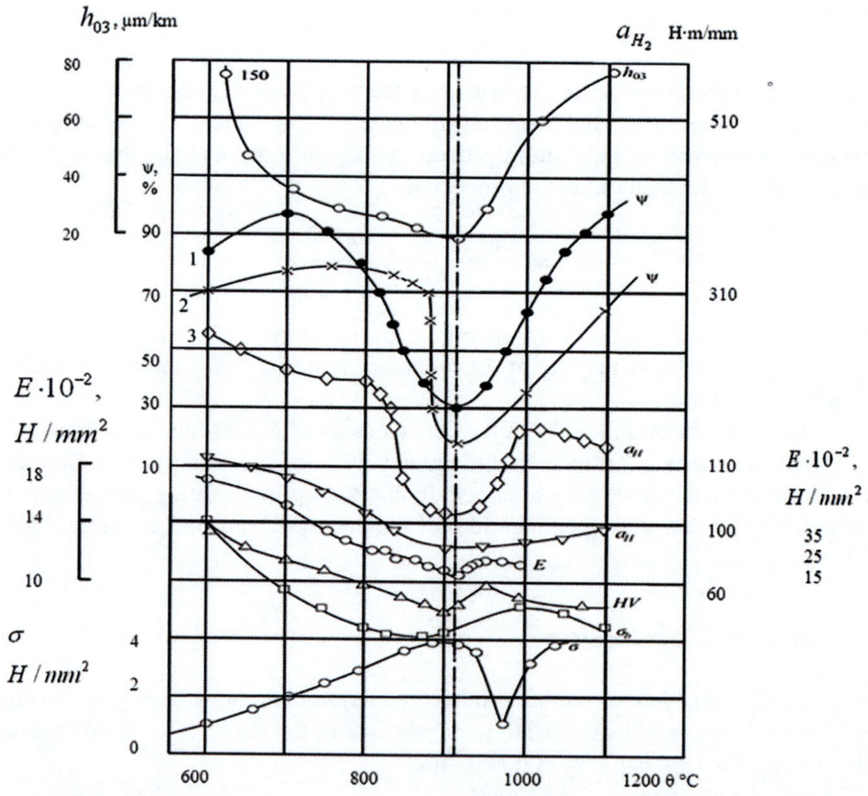

Fig. 4. The effect of temperature on the mechanical properties of technical iron and the wear rate of the cutter made of P20 carbide when turning a part made of steel. 1 - deformed iron; 2 - cast iron; 3 - pure iron (99.99%).

are maximum at the same temperature (1160 K). At the same temperature, the minimum intensity of cutter wear is achieved. As can be seen from the graph, the indicated temperature lies in the temperature range between the temperatures that correspond to $C_p = C_{pmax}$ and $\lambda = \lambda_{min}$. This confirms the conclusion that the temperature between C_p and λ creates a cumulative effect of thermophysical and mechanical conditions under which the tool wear rate is minimal.

5 Conclusions

The conditions for the optimal temperature field distribution between the tool and the workpiece during edge processing are formalized, taking into account the relationship between the thermophysical parameters of the material and temperature.

The extremal nature of the temperature dependencies of the specific heat capacity and thermal conductivity coefficient is shown with a maximum for the specific heat capacity and a minimum for the thermal conductivity coefficient, using the example of pure iron.

A method is proposed for determining the optimal cutting speed by the optimal cutting temperature, defined as the heating temperature of the processed material, at which its specific heat capacity has a maximum value.

The implementation of the proposed method makes it possible to determine the optimal cutting speed for high-speed machining of new hard-to-cut materials without the need for laborious durability tests, which makes it possible to reduce the complexity of determining the optimal cutting speed by 1.5–2.5 times, which will increase the cutting tool life in production conditions. 1.2–1.3 times, as well as the quality of the machined parts.

In the future, it is planned to search for optimal combinations of processed and tool materials in terms of the optimal combination of their thermophysical characteristics, as well as the creation of a database values the specific heat capacity of materials versus temperature, which will simplify the implementation of the method in production conditions.

Acknowledgment. The research was partially supported by International Association for Technological Development and Innovations.

References

1. Klimenko, S.A.: Superhard Materials. Obtaining and application: In 6 Volumes, vol. 5: Processing of Materials with a Blade Tool. Bakul ISM, Kyiv (2006)
2. Hochrainer, T., et al.: An integrated approach to the modeling of size-effects in machining with geometrically defined cutting edges. In: Proc. of 8th CIRP International Workshop on Modeling of Machining Operations, pp. 123–130 (2005)
3. Pasko, N.I.: Generalized Stochastic Model of Cutter Failures and Its Application. Publishing House of Tula, Tula (2016)
4. Wang, Z., Nakashima, S., Larson, M.: Energy efficient machining of titanium alloys by controlling cutting temperature and vibration. Procedia CIRP **17**, 523–528 (2014). https://doi.org/10.1016/j.procir.2014.01.134

5. Huang, P., et al.: Milling force vibration analysis in high-speed-milling titanium alloy using variable pitch angle mill. Int. J. Adv. Manuf. Technol. **58**, 153–160 (2012). https://doi.org/10.1007/s00170-011-3380-9
6. Fu, Z., Yang, W., Wang, X., Leopold, J.: An analytical force model for ball-end milling based on a predictive machining theory considering cutter runout. Int. J. Adv. Manuf. Technol. **84**(9–12), 2449–2460 (2015). https://doi.org/10.1007/s00170-015-7888-2
7. Shin, Y.C., Dandekar, C.: Mechanics and modeling of chip formation in machining of MMC. In: Davim, JPaulo (ed.) Machining of Metal Matrix Composites, pp. 1–49. Springer, London (2012). https://doi.org/10.1007/978-0-85729-938-3_1
8. Balogun, V.A., Edem, I.F., Mativenga, P.T.: Specific energy based characterization of tool specific energy based characterization of tool. Int. J. Sci. Eng. Res. **6**, 1674–1680 (2015)
9. Kostyk, K., et al.: Simulation of diffusion processes in chemical and thermal processing of machine parts. Processes **9**(4), 698 (2021). https://doi.org/10.3390/pr9040698
10. Shvets, S.V., Machado, J.: Numerical model of cutting tool blade wear. J. Eng. Sci. **8**(2), A1–A5 (2021). https://doi.org/10.21272/jes.2021.8(2).a1
11. Usov, A., Tonkonogyi, V., Dašic, P., Rybak, O.: Modelling of temperature field and stress-strain state of the workpiece with plasma coatings during surface grinding. Machines **7**(1), 20 (2019). https://doi.org/10.3390/machines7010020
12. Zhou, F., Wang, X., Hu, Y., Ling, L.: Modeling temperature of non-equidistant primary shear zone in metal cutting. Int. J. Therm. Sci. **73**, 38–45 (2013)
13. Qing, Z., Song, Z., Jianfeng, L.: Three dimensional finite element simulation of cutting forces and cutting temperature in hard milling of AISI H13 steel. Procedia Manuf. **10**, 37–47 (2017). https://doi.org/10.1016/j.promfg.2017.07.018
14. Ivanov, V., Pavlenko, I., Kuric, I., Kosov, M.: Mathematical modeling and numerical simulation of fixtures for fork-type parts manufacturing. In: Knapčíková, L., Balog, M. (eds.) Industry 4.0: Trends in Management of Intelligent Manufacturing Systems. EICC, pp. 133–142. Springer, Cham (2019). https://doi.org/10.1007/978-3-030-14011-3_12
15. Liu, W.W., Wang, D.F., Li, F., Chen, H., Wang, C.Z.: Research on milling parameters optimization based on surface residual stress for aviation stainless steel. Appl. Mech. Mater. **526**, 3–8 (2013). https://doi.org/10.4028/www.scientific.net/AMM.526.3
16. Makarov, A.D.: Optimization of Cutting Processes. Mechanical engineering, Moscow (1976)
17. Oborsky, G.O., Usov, A.V.: Influence of thermophysical phenomena on the dynamic stability of the cutting process. New Materials and Technologies in Metallurgy and Machine-Building. Sciences Magazine **14**(1), 36—41 (2014). https://doi.org/10.21062/ujep/x.2014/a/1213-2489/MT/14/1/36
18. Loveday, M.S.: Consideration of High Temperature Friction Measurement Uncertainty. CMMT(NH)070. http://resourse.npl.co.uk/cgi-bin/download (2000). Last Accessed 02 Jul 2022
19. Zhao, Z., Qian, N., Ding, W., Wang, Y., Fu, Y.: Profile grinding of DZ125 nickel-based superalloy: grinding heat, temperature field, and surface quality. J. Manuf. Process. **57**, 10–22 (2020)
20. Armarego, E.J.A., Brown, R.H.: The Machining of Metals. Prentice-Hall. Inc. (1969)
21. Yakimov, O.V., Usov, A.V., Slobodyanik, P.T., Iorgachev, D.V.: Thermophysics of Mechanical Processing. Astroprint, Odessa (2000)
22. Larikov, L.N., Yurchenko, Yu, F.: Thermal Properties of Metals and Alloys. Directory. Scientific Thought, Kyiv (2005)

Advanced Materials

Method for Evaluating the Resource of Diffusion Coatings Under the Fatigue Conditions

Natalia Artsibasheva, Tetiana Melenchuk(✉), Sergiy Chaban, Dmitriy Purich, and Oleksandr Kovra

Odessa Polytechnic National University, 1, Shevchenko Ave, Odessa 65044, Ukraine
tamelenchuk1508@gmail.com

Abstract. The operating conditions of many car parts require high wear resistance and fatigue strength of parts, which are achieved by various methods of surface hardening, and in particular, by chemical-thermal treatment. From the standpoint of preserving the geometry of complex-shaped parts and heredity in the structure from preliminary hardening treatments, the methods of low-temperature chemical-thermal treatment are becoming increasingly widely used. In this paper, nicotrated coatings are considered. The degree of heterogeneity was controlled by technological parameters of nicotration: temperature, saturation time, pressure, and composition of the saturation gas medium. By adjusting the technological parameters of nicotration for research, 3 types of layers of different geometry and degree of heterogeneity were selected. Analyzing the results on the energy content of the active diffusion zone of heterogeneous nicotrated layers, it was noted that the fractions of the energy content W(disl.) and W(facet) are largely determined by the structural state of the active surface layer. At the same time, not surface but deep diffusion layers differ in maximum energy consumption (minimum energy content). Based on the work performed, a calculation method for estimating the fatigue life of nicotrated layers is proposed, estimated by the tendency of the α-solid solution substructure to inhibit main cracks. It is shown that the deviation of the calculated and experimental fatigue resources is explained by the influence of dispersed particles in the second phases, which hinders the development of small cracks. Accordingly, the diffusion zones of the 2nd type, containing a high concentration of dispersed inclusions, have the maximum real fatigue life. It may be due to the influence of dispersed particles that restrains the propagation of small cracks. Since the concentration of these particles is higher for the 2nd type of layer, the experimental resource Ny for it will also be large.

Keywords: Nicotration · Heterogeneity · Cyclic loads · Energy assessment · Process innovation

1 Introduction

One of the most effective potential in terms of content and capabilities to improve car parts' reliability is methods for creating high-strength coatings on their working surfaces.

V. Tonkonogyi et al. (Eds.): InterPartner 2022, LNME, pp. 243–252, 2023.
https://doi.org/10.1007/978-3-031-16651-8_23

The scope of their use is determined by the external conditions of operation of vehicles, when fatigue, corrosion, erosion, and other types of impact are realized.

The most universal methods of applying coatings are chemical-thermal ones, moreover, from the point of view of maintaining the geometric dimensions of parts of complex shapes, as well as heredity in the structure obtained by preliminary hardening treatments [1]. Low-temperature saturation occupies the prevailing volume of use in the industry. Therefore, in this work, nicotrated coatings are considered. Because parts of motor vehicles during operation most often lose their performance due to fatigue and wear [2], then in the work tests were carried out for bending fatigue of nicotrated coatings with various degrees of heterogeneity of diffusion zones.

The damage of diffusion zones of nicotrated coatings under fatigue loading is predetermined by the crack propagation processes through the matrix α-solid solution at the stages of small main defects [3]. The resource R_W of a structurally heterogeneous nicotrated coating in the model proposed in this work is understood as some incubation period during the active layer's structure transformation, corresponding to the moment of formation of the main crack [4].

This work aims to develop a method for estimating the resource of diffusion coatings under conditions of cyclic loads. Thanks to these studies, it is possible to develop recommendations for optimizing the modes of nicotration of structural steels, which makes it possible to increase the reliability of car parts.

2 Literature Review

The damage kinetics of wear-resistant coatings is determined by the implementation of crack formation processes, which are based on the two-stage loosening effects, which consist of the initiation and propagation of cracks. All models of crack initiation assume, as the first stage of surface damage, the manifestation of microplasticity, which is predetermined by the physical and mechanical properties of the material. Griffiths [5] developed a fracture criterion for a solid body with a defect in the form of a crack. Griffith's model, despite being derived from the law of conservation of energy, is a force criterion. It does not take into account the energy dissipation during crack development, as a result of which it gives an overestimated value of the strength or critical length of the crack, and also excludes the initial stage of the appearance of a defect, does not consider destruction as the appearance and development of one or more free surfaces. Irwin and Keys [5] proposed to estimate the resistance of a material to crack propagation using the energy released during crack propagation.

The authors of [6] developed a brittle fracture model in which the Griffith model's limitations are excluded, and the Griffith formula is a special case. At the initial destruction stage, the formation of pores in the form of spheres is energetically favorable. The sphere, with a minimum surface, allows you to unload the maximum volume. The critical pore size is directly proportional to the surface energy and inversely proportional to the specific volume energy of elastic deformation.

It is shown that the relationship between the force, deformation and energy characteristics of fracture is linear in the elastic region.

The criterion-determining value is the value of the specific volumetric energy required to form or develop the existing fracture surface (crack length). It takes into account that not all of the energy expended in forming a new free surface goes to the formation of the surface [7].

Increasing the resistance to crack development can be implemented by creating materials with high plasticity indices, as well as materials with a structure that can inhibit the formation and development of the main crack, for example, due to its branching or stopping in front of the reinforcing elements of the material structure [8]. The crack resistance of such materials should be studied based on studies of energy and deformation fracture criteria because these criteria, in comparison with the force criterion, better represent the process of crack development when significant plastic deformations are realized at its tip [9]. The energy criterion combines the action of force and deformation. The deformation criterion is easily associated with the process of material damage [10] and monitored experimentally.

Any high-strength coating material is heterogeneous in its structure and contains macrodefects and elements of substructural imperfections. The type, geometry, topography, and density of defects are inherited from thermal, mechanical, and chemical-technological technologies [11]. In such a coating, the main mechanisms of surface microcrack initiation can be superimposed by the activating effects of stress concentration from macrodefects in the structure, as well as the weakening of the material as a result of the physical and chemical impact of the external environment [12]. In the first case, when calculating the unblocking stresses of critical substructural accumulations for the formation of a nucleus cracks, the internal stresses component should be considered [13]. In the second – a decrease in the number of defects in a critical cluster due to a decrease in surface energy during the adsorption of surface-active atoms or [14], as in the case of the presence of internal stresses, leveling of the critical unblocking stress due to a decrease in the cohesive strength of the crystal [15].

The crack formation and propagation processes in the general case require additional energy consumption, for example, friction for microplastic deformation.

The movement of substructural defects determines the kinetic processes of surface damage at the initial stages. Which are "locked" on the initial macrobarriers and gradually accumulate as the impact impulses increase. When a critical concentration of defects is created at the stoppers, a directed nucleating crack is formed, and its subsequent development occurs. Such structural loosening processes are activated by internal stresses from microdefects, tensile in the crack development planes, and the presence of initial active substructural defects in the material structure. Processes of crack branching. Their intersection forms a characteristic cellular loosening structure of the quasi-defective surface layer [16].

A decrease in the level of damage can be achieved either by increasing the stress of the formation of a critical nucleus or by slowing down a developing supercritical crack.

The following physical and structural changes in the coating material can be positive for cracking inhibition: achieving the optimal degree of macrodefects heterogeneity in the active layer, applying coatings with high surface energy and elastic modulus of the material, and choosing as base crystalline materials with a high number of slip systems and their strong blocking [17].

It is necessary to consider two factors: the possible deceleration of the crack and a change in its trajectory. For braking, it is necessary to create by technological methods extended stable fields of compressive stresses, structural barriers of a macro- and microscopic nature, provide for the braking of cracks with the help of other misoriented cracks, etc. [18].

A promising way is a structural way of blocking propagating cracks through the formation of an inhomogeneous structure over the coating cross-section by unique technological methods, as well as the creation of coherent microinclusions due to the development of effects in the material under mechanical, operational impact. All these methods consider the dependence of the energy intensity of the coating on the rate of crack propagation.

3 Research Methodology

As a heterogeneous system, nicotrated layers on high-quality steel 38CrNi3MoVA (analog 34NiCrMoV14-5 DIN) with varying degrees of heterogeneity were taken. The degree of heterogeneity is determined by the following parameters: average particle diameter, total particle volume, and particle density. The degree of heterogeneity was controlled by technological parameters of nicotration: temperature, saturation time, pressure, and composition of the saturation gas medium. By adjusting the technological parameters of nicotration, 3 types of layers of different geometry and degree of heterogeneity were selected for research [5].

The structural factor was studied using the standard method of quantitative metallography [7] and electron microscopy. The distribution of structural heterogeneity characteristics was preliminary studied: the sizes of phase inclusions, their total volume, and density over the depth of diffusion zones. It was found that statistically, the most probable size of inclusions for all types of layers is 0.1 μm. At the same time, the 2nd and 3rd types of diffusion zones are characterized by a very large number of particles 0.2–0.4 μm in size, which are topographically located along the boundaries of the α-phase. Inside the matrix grains in these layers are dispersed inclusions with sizes of 0.02–0.05 μm with distances between them of 0.01–0.02 μm. The volume fractions of these phases over the active layer's depth correlate with the microhardness distribution, according to which the main geometric characteristics of the diffusion layers were determined.

The nucleation and growth of cracks were studied using microscopic and electron scopic methods. Fatigue tests were carried out on the original installation, which makes it possible to obtain the stress state of a flat cantilever bend of various levels. In the course of tests, the elongation of fatigue cracks was studied using microscopy depending on the number of cycles.

4 Results and Discussion

The resource RW of a structure-heterogeneous nicotrated coating in the model proposed in this work means a certain incubation period during the active layer's structure transformation. It corresponds to the moment of formation of the main crack. At the same time, it is natural that the RW parameter includes the implementation of the stage of damage both at the level of substructural effects accompanying crack formation and directly related to small cracks [7, 10].

At the moment of the formation of the main crack, as shown by the studies in this work, the system's energy consumption drops sharply and approaches the specific surface energy F_S. Taking into account the quasi-static nature of the problem of crack propagation in a heterogeneous α-solid solution, its fatigue life from an energy standpoint can be estimated according to the expression:

$$R_W = (F_S - W)/F_S \quad (1)$$

where W is some energy content of the system.

At the same time, a low level of the initial energy of the W_0 content should be attributed to weakly heterogeneous diffusion zones, where a low level of effective distortions of the α-iron matrix phase is maintained. On the contrary, large values of the parameter W_0 characterize the energy content of highly heterogeneous systems in which the volumes of the α-solid solution are localized by particles of the second phases [13, 15].

In the proposed method, the parameter W, taking into account the energy additivity of substructural preparatory and fracture finishing processes, is estimated taking into account the dislocation $W_{(disl.)}$ and boundary $W_{(facet)}$ components:

$$W = W_{(disl.)} + W_{(facet)} \quad (2)$$

In the process of low-cycle fatigue, the energy content W will continuously increase. When it reaches the limiting level corresponding to the F_S value, the local volume is considered to be destroyed. The maximum system resource meets the condition:

$$R_w = 1 \text{ and is determined by the condition } W = 0 \quad (3)$$

The calculation of the energy component $W_{(disl.)}$ was carried out for the specific material volume of the structure in 1 cm 3 under the assumption that the lengths of dynamic, active dislocation loops weakly fixed by atomic segregations are commensurate with the average size of blocks D. Then we get:

$$W_{(disl.)} = ((2Gb\text{^}2)/4 \lg D/b - σ D) \rho \pi D b \quad (4)$$

where G - is Young's modulus; b - is the dislocation mismatch vector.

Local microdistortions σ are approximated in the model by the values of microdistortions Δa/a.

The values of the parameters characterizing the microdistortions Δa/a, the block size D, and the dislocation density ρ were calculated according to the corresponding X-ray diffraction patterns. The results of calculating the dislocation fraction of the energy content W(disl.) The depth of the diffusion zones is shown in Fig. 1.

The calculation of the fraction of the energy $W_{(facet)}$ contained in low-angle surface defects was carried out according to the dependence:

$$W_{(facet)} = F_{(S)} S(1 - \mu)/\mu \quad (5)$$

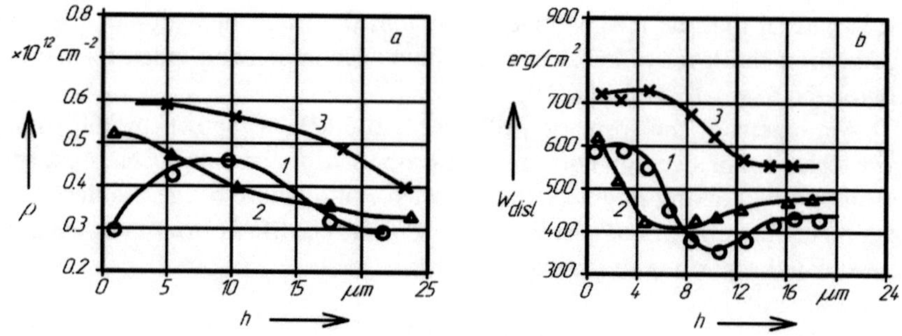

Fig. 1. Change in the dislocation density / a / and the dislocation component of stored energy / b / by the depth of the layer: a -1 type of layer; b-2 layer type; in-3 layer type.

It considered the specific length of the fragmentary boundaries S in the calculated volume of 1 cm^3 and a given Poisson's ratio μ. The determination of the parameter S was carried out taking into account its inverse proportionality to the cube of the average block size D. The specific surface energy of the "hidden" boundary was equated to the lattice energy of the base of the α-phase, equal to $F_S = 1950$ erg/cm^2. It calculated the component W(facet) shown in Fig. 2.

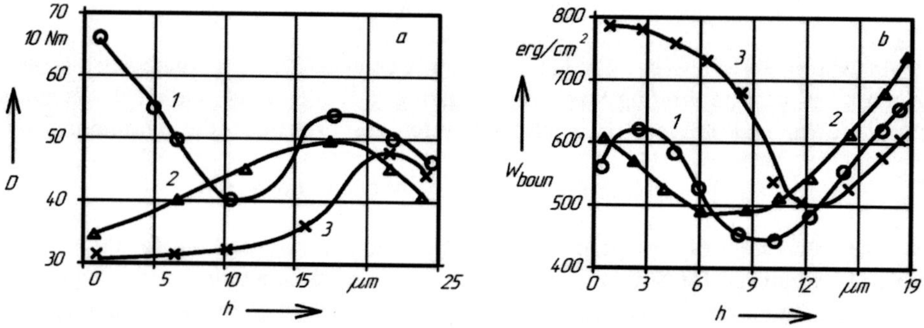

Fig. 2. Changing the size of the block D / a / and the marginal component of stored energy / b / for the depth of the layer: a - 1 type of layer; 6 - 2 type of layer; in - 3 type of a layer.

Analyzing the results on the energy content of the active diffusion zone of nicotrated layers of different heterogeneity, it was noted that the fractions of the energy content W(disl.) and W(facet). They are primarily determined by the structural state of the active surface layer. There is a clear difference between the two energy components with the dominant role of the boundary W for weakly heterogeneous structures. The energy content by defect types is leveled with an increase in the degree of heterogeneity, and the contribution of W(facet) and $W_{(disl.)}$ becomes almost the same.

The nature of the distribution of the dislocation component of the energy content $W_{(disl.)}$ (see Fig. 1, b) over the depth of active diffusion zones is practically the same for all three types of structures under consideration. In this case, the maximum energy intensity (minimum energy content) differs not on the surface but in the deep diffusion layers. The most heterogeneous layer of type 3 has up to 2.5 times higher energy content. On the other hand, the energy characteristics of the α - phase for the 1st and 2nd types of layers are very similar. As an analysis, it should also be noted that the distributions of the parameter $W_{(disl.)}$ And the total volume of particles of the second phases V over depth. Moreover, low levels of the V-factor on the surface for the 1st (V ~ 25%) and 2nd (V ~ 30%) types of layers correspond to lower values of the energy content in comparison with the 3rd type of layer (V ~ 42%).

The distribution of the boundary component of the energy content $W_{(facet)}$ (Fig. 2, b) according to the shape of the curves is similar to the considered dislocation one with the same quantitative energy anomalies for the third type of layer. The near-surface layers of active diffusion zones at a depth of 10–15 μm have the highest fatigue resistance. At the same time, the layers of the 1st and 2nd types have the maximum and similar energy intensity on the surface. In this case, it is quite natural. That the larger, according to the size distribution of inclusions of the second phases in-depth, particles for the 2nd (d_{av} = 0.2 μm) and 1st (d_{av} = 0.1 μm) types of nicotrized layers on the surface compared to the 3rd (d_{av} = 0.03 μm) correspond to a shorter length of phase boundaries S and provide lower levels of energy content.

Figures 3 and 4 show the depth distributions of 3 types of diffusion zones, respectively, for the total energy content W and fatigue life R_W.

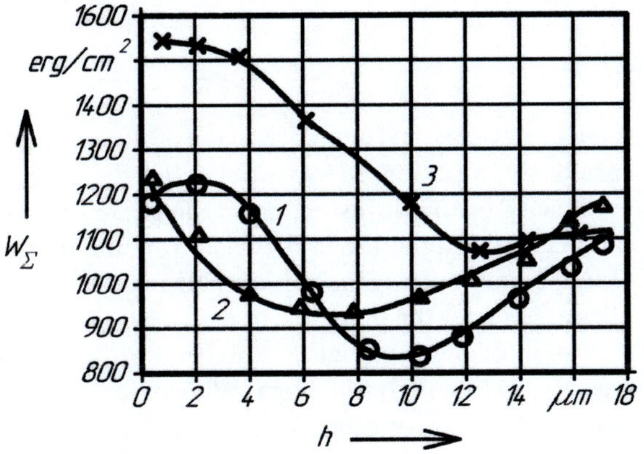

Fig. 3. Change of energy saving of various nicotrated layers, on the depth of a diffusion zone: 1-1 type of a layer; 2-2 layer type; 3-3 layer type.

A comparison of the calculated energy parameter W for surface (up to ~10 μm) layers shows that the energy state of the α matrix for low (type 1) and medium (type 2) heterogeneous structures is approximately the same, and the latter type has even up to 30% preference for energy intensity. At the same time, in deeper layers of the active diffusion zone (more than 10 μm), the energy W - characteristics for the layers are leveled, and this coincides with the averaging of V and d_{av} factors of structure heterogeneity at these depths.

The fatigue life of the α-solid solution on the surface is maximum for diffusion zones of the 1st and 2nd types (respectively 0.38 and 0.35) and decreases to 0.21 for the zone

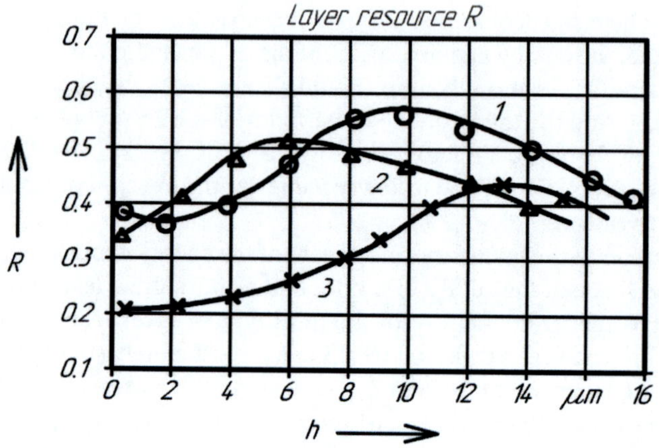

Fig. 4. Changes in the fatigue resource of picketed layers according to the depth of the diffusion zone: 1-1 type of layer; 2-2 layer type; 3-3 layer type.

of the 3rd type. The maximum resource value (respectively for 1, 2, and 3 layers is 0.55, 0.50, and 0.43) is achieved at depths of 5–15 μm. The calculated resources are leveled for layers more remote from the surface (depths more than 15 μm) [10].

5 Conclusions

Taking into account the purely classical approach to the destruction of α - solid solution, which does not consider the blocking effects from dispersed particles of the second phase, the resource capacity of the diffusion matrix decreases as saturation duration. Interesting is the fact that real fatigue life is obtained from the results of fatigue tests, maximum for the 2nd medium-heterogeneous layer. It may be due to restraining the propagation of small cracks under the influence of dispersed particles [1, 17].

Since the concentration of these particles is higher for the 2nd type of layer, the experimental resource N_y for it will also be large. With an increase in external stresses under high-cycle fatigue, the positive effect of dispersed precipitates on the damage kinetics correspondingly decreases [17], and the patterns in the change in the calculated R_W and practical resources converge.

Thus, based on the work carried out, a calculation method for estimating the fatigue life of nicotrated layers is proposed, calculated from the tendency of the substructure α - solid solution to inhibition of main cracks. It is determined that the resource of diffusion zones of the 1st and 2nd layers is approximately the same and is 0.40... 0.50 compared to the value of 0.20... 0.30 for the 3rd layer. The maximum design resource is achieved at a depth of 5...15 μm. It is shown that the deviation of the calculated and experimental fatigue resources is explained by the influence of dispersed particles in the second phases, which hinders the development of small cracks. Accordingly, the diffusion zones of the 2nd type, containing a high concentration of dispersed inclusions, have the maximum real fatigue life.

References

1. Troshchenko, V.T., Khamaza, L.A.: Stages of fatigue failure of metals and alloys. Supplement National Academy of Sciences of Ukraine **10**, 56–63 (2018). https://doi.org/10.15407/dopovidi2018.10.056
2. Goryk, A., Koval'chuk, S., Brykun, O., Chernyak, R.: Viscoelastic resistance of the surface layer of steel products to shock attack of a spherical pellet. Key Eng. Mater. **864**, 217–227 (2020). https://doi.org/10.4028/www.scientific.net/KEM.864.217
3. Ghasemi, M.H., Ghasemi, B., Semnani, R.M.: Wear performance of DLC coating on plasma nitrided Astaloy Mo. Diam. Relat. Mater. **93**, 8–15 (2019). https://doi.org/10.1016/j.diamond.2019.01.016
4. Kaminsky, A.A., Kurchakov, E.E.: On the pre-fracture zone at the tip of a crack of normal separation in a nonlinearly elastic orthotropic material. Appl. Mech. **55**(1), 26–43 (2019)
5. Shchelokova, M.A., Slobodyan, S.B., Dyrda, V.I.: Fractal approach to solid fracture mechanics. Geotechnical mechanics: Interdisciplinary. Coll. of Science Works **138**, 227–259 (2018)
6. Zuev, L.B., Barannikova, S.A., Danilov, V.I., Gorbatenko, V.V.: Plasticity: from crystal lattice to macroscopic phenomena. Prog. Phys. Met. **22**(1), 3–57 (2021). https://doi.org/10.15407/ufm.22.01.003
7. Roa, J., Sapezanskaia, I., Fargas, G., Kouitat, R., Redjaïmia, A., Mateo, A.: Influence of testing mode on the fatigue behavior of <111> austenitic grain at the nanometric length scale for TRIP steels. Mater. Sci. Eng. **A**(713), 287 (2018). https://doi.org/10.1016/j.msea.2017.12.047
8. Kindrachuk, M., Dukhota, O., Kharchenko, V., Stebeletska, N., Glovin, A.: Combined method of increase wear resistance details of tribo-mechanical systems. Problems of Friction and Wear **2**(95), 46–57 (2022). https://doi.org/10.18372/0370-2197.2(95).16556
9. Toboła, D., Kania, B.: Phase composition and stress state in the surface layers of burnished and gas nitrided Sverker 21 and Vanadis 6 tool steels. Surf. Coat. Technol. **353**, 105–115 (2018). https://doi.org/10.1016/j.surfcoat.2018.08.055
10. Artsybasheva, N.M., Melenchuk, T.M., Kovra, O.V., Berdiyev, B.: Study of the influence of the heterogeneity of welds on the resource of trailer holding systems. Collection of scientific papers ODATRYA **2**(13), 21–26 (2018)
11. Sharifahmadian, O., Mahboub, F.: A comparative study of microstructural and tribological properties of N-DLC/DLC double layer and single layer coatings deposited by DC-pulsed PACVD process. Ceram. Int. **45**, 7736–7742 (2019). https://doi.org/10.1016/j.ceramint.2019.01.076
12. Benedetti, M., Fontanari, V., Barozzi, M., Gabellone, D.: Low and high cycle fatigue properties of TRIP ultra high strength bainitic steel. Fatigue Fract. Eng. Mater. Struct. **40**(9), 1459 (2017). https://doi.org/10.1111/ffe.12589
13. Molodkin, V.B., et al.: The physical nature and new capabilities of use of effects of asymmetry of azimuthal dependence of total integrated intensity of dynamical diffraction for diagnostics of crystals with the disturbed surface layer and defects. Usp. Fiz. Met. **18**(2), 177–204 (2017). https://doi.org/10.15407/ufm.18.02.177
14. Corona-Gomez, J., Shiri, S., Mohammadtaheri, M., Yang, Q.: Adhesion enhancement of DLC on CoCrMo alloy by diamond and nitrogen incorporation for wear resistant applications. Surf. and Coat. Technol. **332**, 120–127 (2017). https://doi.org/10.1016/j.surfcoat.2017.10.050
15. Kharchenko, V.O., et al.: Multiscale modelling of self-organization of non-equilibrium point defects in irradiated α-zirconium. Usp. Fiz. Met. **18**(4), 295 (2017). https://doi.org/10.15407/ufm.18.04.295

16. Crawford O'Brien, E.C.H., Yeddu, H.K.: Multi-length scale modeling of carburization, martensitic microstructure evolution and fatigue properties of steel gearss. J. Mater. Sci. Technol. **49**, 157–165 (2020). https://doi.org/10.1016/j.jmst.2019.10.044
17. Artsibasheva, N.M., Grigorova, T.M., Churenova, D.I.: Investigation of the regularities of wear of heterogeneous structures of high-strength coatings. In: XXVIII International Scientific and Practical Conference: "Development of Science in the XXI Century" Part 1, pp. 91–97. Knowledge, Kharkiv (2017)
18. Mordyuk, B.N., Mikosyanchik, O.O., Mnatsakanov, R.G.: Structure-phase state and wear of Ni–Cr–B–Si–C coating on steel 1045 under friction conditions with the shear load component. Metallofiz. Noveishie Tekhnol. **42**(2), 175–195 (2020). https://doi.org/10.15407/mfint.42.02.0175

Effect of Ti-Zr Ligature on Microstructure and Mechanical Properties of Automotive Silumin

Kristina Berladir[1](✉), Tetiana Hovorun[1], Frantisek Botko[2], Oleksandr Gusak[1], and Yuliia Denysenko[1]

[1] Sumy State University, 2, Rymskogo-Korsakova Street, Sumy 40007, Ukraine
kr.berladir@pmtkm.sumdu.edu.ua

[2] Technical University of Kosice, 1, Bayerova Street, 08001 Presov, Slovak Republic

Abstract. Complex processing increases the mechanical and technological properties of secondary aluminum alloys, including refining and modification processes. Therefore, managing the structure of castings and the properties of silumin in the volume crystallization of castings is one of the most critical tasks in traditional casting processes. It can be achieved by introducing special additives, modifiers, and ligatures, which are widely used in the production and casting of aluminum alloys and are introduced into the charge or directly into the melt. Improving the quality of secondary silumin to the primary level is possible using Ti-Zr ligature. The effect of Ti-Zr ligature on the structure and mechanical properties of secondary aluminum alloy in the cast and the heat-treated state has been studied in this work. A linear dependence of the influence of the Ti-Zr ligature concentration on the hardness of silumin has been established. Heat treatment of the secondary aluminum alloy increased the hardness by 1.4–1.8 times and tensile strength by 1.3–1.5 times. The introduction of 0.1 wt. % Ti-Zr increases the hardness of alloy by 1.4 and 1.1 times in the cast and heat-treated states, respectively. The maximum value of tensile strength is observed with the introduction of 0.05 wt. % Ti-Zr. Adding Ti-Zr ligatures does not complicate the technology of alloys, has low cost, and can be used in foundry production.

Keywords: Secondary aluminum alloy · Ti-Zr · Energy efficiency · Cast state · Heat treatment · Microstructure · Process innovation

1 Introduction

Aluminum alloys are widely used due to their value for engineering the complexity of mechanical, physical, and corrosion properties [1]. Domestic and international experience in aluminum alloy metallurgy [2, 3] shows that obtaining high-quality alloys at minimal costs for its production is crucial in today's market economy.

V. Tonkonogyi et al. (Eds.): InterPartner 2022, LNME, pp. 253–263, 2023.
https://doi.org/10.1007/978-3-031-16651-8_24

The main factors determining the technical level are materials and structures that can facilitate the car, reduce fuel consumption, and increase economic impact and environmental safety [4]. Among the range of structural materials, they take second place after iron-based alloys because they have a low density, high strength, and corrosion resistance.

Developing efficient technologies for producing aluminum alloys modified with dispersed alloys is a task of modern metallurgical and foundry technologies [5], as well as for energy engineering [6, 7] and general machine building [8, 9]. In this case, the main direction in the search for new alloys is the versatility of their composition and the possibility of obtaining high-performance properties in the conditions of use of different casting technologies and directly in the cast or the heat-treated state [10].

Improving the mechanical properties of aluminum alloy castings is an essential theoretical and practical task for foundry professionals. According to the trends in aluminum processing in recent years, the production of secondary aluminum alloys from waste and scrap accounts for almost half of the total volume [11]. It determined the purpose and relevance of this paper, namely, the study of microstructures and mechanical properties of secondary modified silumin with Ti-Zr ligature.

2 Literature Review

Various methods are used to improve the quality of industrial products, from coatings [12, 13], non-ferrous alloys [14, 15], and high-entropy alloys [16] to composite materials [17, 18].

Complex treatment improves secondary aluminum alloys' mechanical and technological properties, including refining [19] and modification processes [20]. Modification and ligatures are used in producing and casting aluminum alloys introduced into the charge or directly into the melt [21]. Models of machine parts' surfaces were realized in [22, 23]. The impact of treatment and coating parameters on the working surfaces of conjugated parts were studied in [24, 25].

Kuz'min M. et al. [26] showed the prospects of producing Al-Si alloys by adding amorphous silica with the argon flow in the aluminum melt at high temperatures. A practical method for obtaining eutectic and super-eutectic Al-Si alloys by induction melting a mixture containing silica is presented. It increases the efficiency of the existing technological process of obtaining silumin by partially eliminating the energy-intensive stage of production of metallurgical silicon.

Gilev I. et al. [27] proposed the small Ti and Zr additions for restraining the recrystallization processes of Al–Cu alloys. After deformation and heat treatment, they investigated the influence of Ti and Zr additions on the macrostructure and mechanical properties of the Al–4%Cu alloy.

Lipinski T. [28] found the dependence of different ranges of Ti addition and melt number of the hypo-eutectic Al-9% Si Mg alloy on its microstructure and mechanical properties. Modifying such aluminum alloy firstly improved the alloy's properties, but with an increased number of melts, the microstructure and mechanical properties decreased.

Cherepanov A. et al. [29] proposed to classify modifiers into two fundamentally different groups. The first group includes positive additives in small concentrations up to 1% wt., such elements as Ti, Zr, B, Sb. The second group includes sodium, potassium, and their salts.

Romanchenko et al. [30] proposed an approach for studying the hardness of coatings. Mamzurina O. et al. [31] investigated the influence of Zr on the phase, chemical composition, and mechanical properties of aluminum-based alloys. Zr additives increase the hardness of the starting alloys due to the deposition of complex phases during the homogenization process.

In our previous paper [32], we analyzed the influence of Ti-Al ligature on increasing the properties of silumin AK5M2. It was shown that using non-deficient and cheap Al-Ti5 ligature in combination with heat treatment can significantly increase the properties of cast aluminum alloy products. So, we continued to study the effect of ligatures of other compositions, particularly Ti-Zr, on the microstructure and mechanical properties of silumin AK5M2.

3 Research Methodology

This work used an aluminum alloy grade AK5M2 as the matrix. The alloy was smelted in industrial production (LLC "RELIT", Sumy, Ukraine). Its chemical composition was determined by spectral analysis using reference control samples and chemical analysis (Table 1).

Table 1. The chemical composition of the alloy AK5M2 (wt. %)

The chemical composition of the base elements, wt. %						Impurities ≤ wt. %	
Al	Si	Cu	Mg	Mn	Ti	Fe	Zn
Basis	4,0–6,0	1,5–3,5	0,2–0,8	0,2–0,8	0,05–0,2	1,3	1,5

The iron content in the alloy corresponds to its concentration in the most common industrial silumin (DSTU 2839-94), produced from scrap and non-ferrous metal waste (DSTU 3211-95). The leading indicators of the material are as follows: short-term strength – 118–206 MPa; relative elongation – 0.5–2.0%; Brinell hardness – 65–75 MPa (DSTU 2839-94).

The modification was performed by the introduction of the ligature Ti-Zr (Table 2) в depending on its concentration (0.025; 0.05; 0.075; 0.100%).

Table 2. The chemical composition of Ti-Zr ligature (wt. %)

Element	Zr	Fe	Si	V	Ti
Amount	45	0,5	0,25	0,45	Remainder

The ligature Ti-Zr is used to increase some aluminum alloys' recrystallization temperature. Zirconium is added to aluminum alloys to prevent grain growth with increasing temperatures and in weld areas, to improve the weldability of the alloy, to reduce susceptibility to corrosion under stress, and to reduce the sensitivity to the cooling rate during quenching.

The material's melting was carried out in an electric resistance furnace under a layer of salt flux. Additives were introduced into the melt at a 720–740 °C temperature using a foundry bell. The melt was thoroughly mixed to ensure uniformity and complete absorption of the alloying elements. After modification, the silumin was kept in a furnace at 720 °C for 10 min, after which it was poured into a 50 mm diameter mold.

After that, the following heat treatment was performed, the graph of which is shown in Fig. 1.

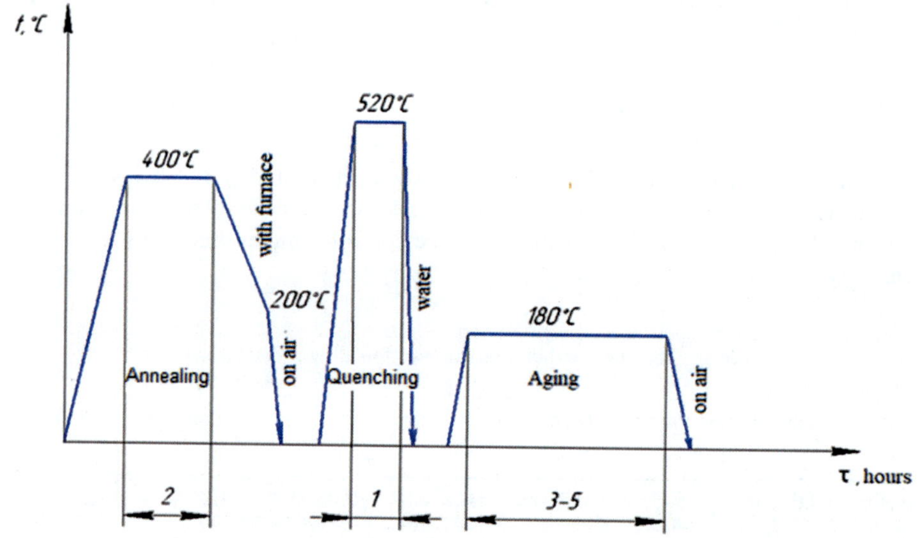

Fig. 1. The graph of heat treatment of test materials.

The studies were conducted on secondary silumin AK5M2 in the initial and modified state; in cast and heat-treated conditions used to manufacture medium-loaded details in the automobile industry.

The study of test specimens' properties included determining tensile strength and hardness as primary data on the material. The tensile strength of test materials was determined on samples: a diameter of 10 mm and a working length of 50 mm. The hardness of the samples was measured by the Brinell method.

The study of the microstructure of the test materials was performed using a MIM-7 metallographic microscope at 200-fold magnification and a scanning electron microscope of high-resolution Tescan-VEGA 3 on etched samples with Keller's reagent.

4 Results and Discussion

Silumin AK5M2 belongs to pre-eutectic aluminum alloys and contains a solid solution of Si in Al and eutectic with Si like the needles and plate inclusions (Fig. 2). Such microstructure reduces the strength and ductility of alloys.

Fig. 2. Microstructure of silumin AK5M2: a) – the cast state, b) – the heat-treated state.

It was found that alloying the alloy up to 0.05% Ti-Zr does not affect the macro-grain size; with its increase to 0.1%, the grain is ground twice (Fig. 3). The additional alloying of Ti-Zr enhances modifying the cast structure of silumin. At a total concentration of Ti and Zr appear excess aluminides of complex composition Al3(Zr, Ti). It was found that these ligatures positively affect the size of the dendritic cells and eutectic silicon. Despite the increased total content of Ti and Zr (0.1%) in the casting structure, there are no excess aluminides.

Table 3. The grain score of silumin AK5M2

Grain score of silumin	
Before heat treatment	After heat treatment
0.025 wt. % Ti-Zr	
3–4	4–5
0.050 wt. % Ti-Zr	
4–5	5–6
0.075 wt. % Ti-Zr	
4–5	5–6
0.100 wt. % Ti-Zr	
5–6	6–7

Since both the Ti and Zr are present in the alloy, in which the melting point is higher than the melting point of Al, they become centers of crystallization around which the grain of an α-solid solution of Si in aluminum is formed. The influence of the Ti-Zr on the inclusions form in the cast state and after heat treatment according to the proposed model (Fig. 1) is shown in Fig. 4.

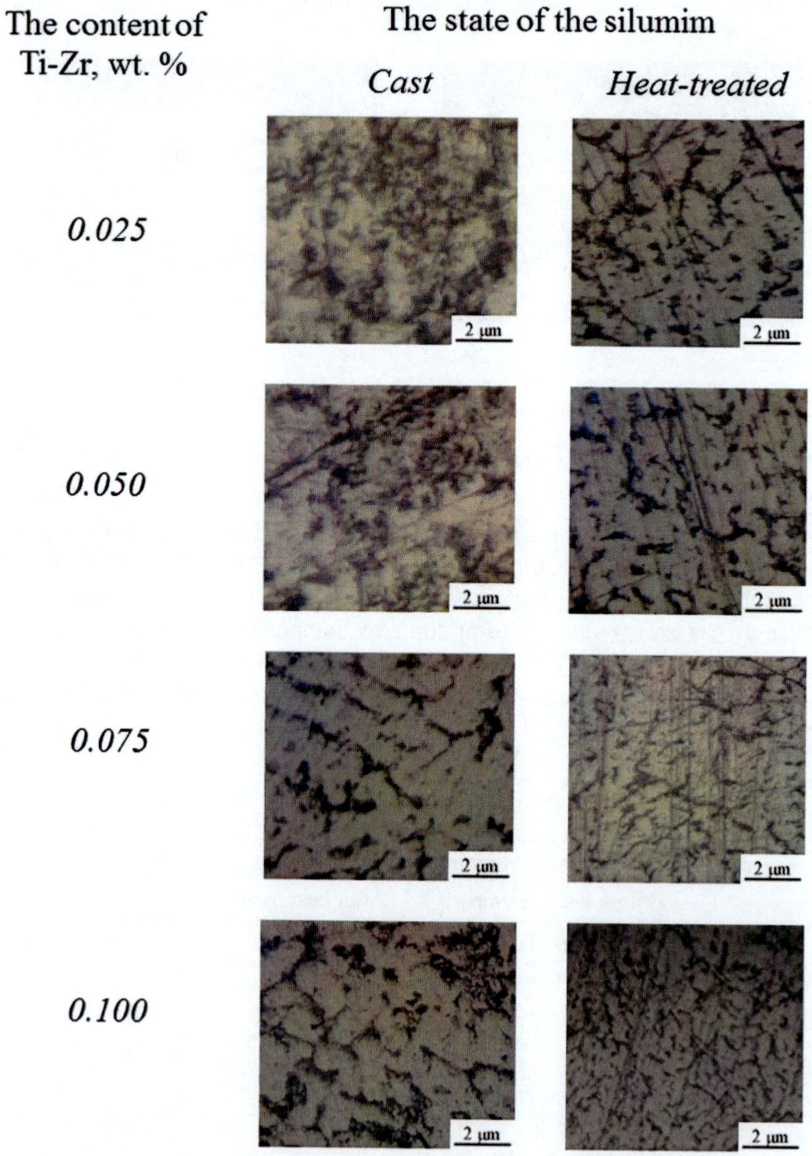

Fig. 3. Microstructure of aluminum alloy AK5M2 with Ti-Zr ligature.

The grain score of silumin depending of Ti-Zr ligature concentration and heat treatment is presented in Table 3.

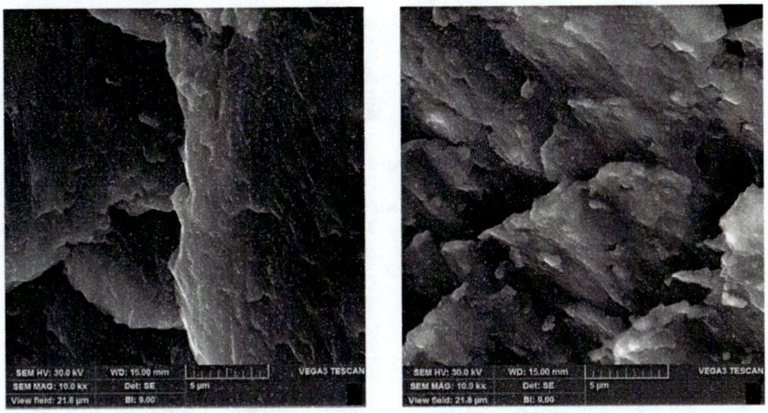

a) b)

Fig. 4. Microstructures of silumin with 0.10 wt. % Ti-Zr obtained by SEM: a) in the cast state; b) in the heat-treated state.

As a result of the studies, the dependences of Ti-Zr ligature concentration on the hardness and tensile strength of silumin in the cast and heat-treated state were determined (Fig. 5).

After analyzing the results obtained, we can note the following. A linear dependence of the influence of the ligature's concentration on the silumin's hardness was revealed. An increase in the concentration of the ligature leads to an increase in the hardness of the silumin in both the cast and the heat-treated state. The heat treatment of the samples contributes to the increase in the hardness by 1.4–1.8 times and the tensile strength by 1.3–1.5 times. Introduction the ligature of 0.1 wt. % Ti-Zr increases the hardness of silumin by 1.4 and 1.1 times in the cast and heat-treated conditions, respectively. The maximum tensile strength value is observed with the introduction of Ti-Zr ligatures of 0.05 wt. %.

Based on the obtained results, it is established that the introduction of modifying additives in the secondary aluminum alloys contributes to the increase of the alloys' physical, technical, and operational properties. The hardness of the investigated alloys increases by (18–24) %, the tensile strength – by (12–16) %, which is consistent with [32].

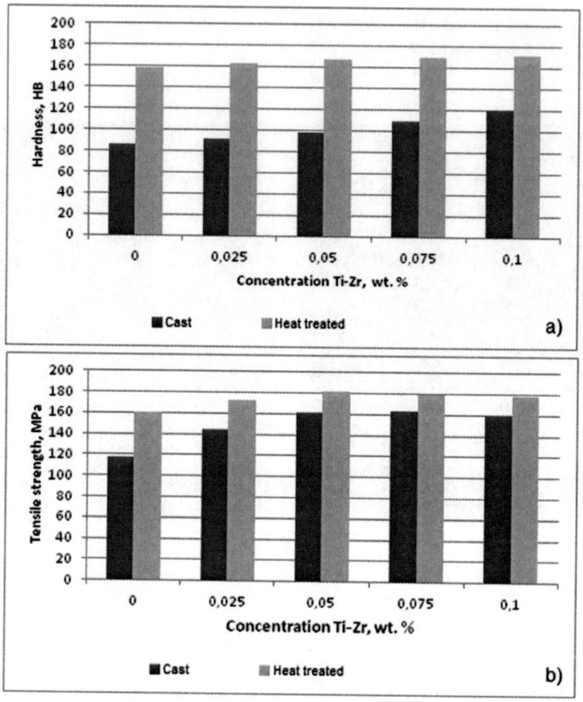

Fig. 5. The influence of Ti-Zr ligature on the hardness (a) and tensile strength (b) of silumin.

5 Conclusions

Based on the obtained results, it is established that introducing active modifying additives in secondary aluminum alloys improves the physical, technical and operational properties of alloys.

Improving the quality of secondary silumin to the primary level is possible using Ti-Zr ligature. The analysis of the obtained data allowed to formulate the following conclusions: a linear dependence of the effect of the concentration of Ti-Zr ligature on the hardness of silumin; increasing the concentration of the ligature leads to an increase in the hardness of silumin both in the cast and in the heat-treated state; heat treatment of materials leads to an increase in tensile strength by 1.3–1.5 times, the maximum value of which is observed with the introduction of a ligature of 0.05 wt. %; the addition of the Ti-Zr ligature modifier in the cast and heat-treated material increased the hardness by (18–24)%, tensile strength by (12–16)% by changing the shape and dispersion of the formed intermetallic phases, increasing the density and reducing porosity.

The obtained research results testify to the high efficiency of the proposed Ti-Zr ligature in creating cast aluminum alloys, which does not complicate the technology of alloy production, has low cost, and allows to recommend industrial applications.

Acknowledgments. The research was partially carried out within the project "Fulfillment of tasks of the perspective plan of development of a scientific direction "Technical sciences" Sumy State University" funded by the Ministry of Education and Science of Ukraine (State reg. no. 0121U112684)". The research was partially supported by the Research and Educational Center for Industrial Engineering (Sumy State University) and International Association for Technological Development and Innovations.

References

1. Varshney, D., Kumar, K.: Application and use of different aluminium alloys with respect to workability, strength and welding parameter optimization. Ain Shams Engineering Journal **12**(1), 1143–1152 (2021). https://doi.org/10.1016/j.asej.2020.05.013
2. Grimzin, I., Ponomarenko, O., Marynenko, D., Yevtushenko, N., Berlizeva, T.: The Technological Process of Obtaining Sand-Plaster Molds for Complex Thin-Walled Aluminum Castings. In: Tonkonogyi, V., et al. (eds.) InterPartner 2019. LNME, pp. 405–414. Springer, Cham (2020). https://doi.org/10.1007/978-3-030-40724-7_41
3. Vlasiy, O., Mazurenko, V., Ropyak, L., Rogal, O.: Improving the aluminum drill pipes stability by optimizing the shape of protector thickening. Eastern-European Journal of Enterprise Technologies **1**(7–85), 25–31 (2017). https://doi.org/10.15587/1729-4061.2017.65718
4. Joost, W.J.: Reducing vehicle weight and improving U.S. energy efficiency using integrated computational materials engineering. JOM **64**, 1032–1038 (2012). https://doi.org/10.1007/s11837-012-0424-z
5. Mavhungu, S.T., Akinlabi, E.T., Onitiri, M.A., Varachia, F.M.: Aluminum matrix composites for industrial use: advances and trends. Procedia Manufacturing **7**, 178–182 (2017). https://doi.org/10.1016/j.promfg.2016.12.045
6. Pavlenko, I., Liaposhchenko, A., Ochowiak, M., Demyanenko, M.: Solving the stationary hydroaeroelasticity problem for dynamic deflection elements of separation devices. Vibrations in Physical Systems **29**, 2018026 (2018)
7. Andrenko, P., Rogovyi, A., Hrechka, I., Khovanskyi, S., Svynarenko, M.: The Influence of the Gas Content in the Working Fluid on Parameters of the the Hydraulic Motor's Axial Piston. In: Ivanov, V., Pavlenko, I., Liaposhchenko, O., Machado, J., Edl, M. (eds.) DSMIE 2021. LNME, pp. 97–106. Springer, Cham (2021). https://doi.org/10.1007/978-3-030-77823-1_10
8. Panchenko, A., et al.: Development of the universal model of mechatronic system with a hydraulic drive. Eastern-European Journal of Enterprise Technologies **4**(7–94), 51–60 (2018). https://doi.org/10.15587/1729-4061.2018.139577
9. Ivanov, V., Pavlenko, I., Kuric, I., Kosov, M.: Mathematical Modeling and Numerical Simulation of Fixtures for Fork-Type Parts Manufacturing. In: Knapčíková, L., Balog, M. (eds.) Industry 4.0: Trends in Management of Intelligent Manufacturing Systems. EICC, pp. 133–142. Springer, Cham (2019). https://doi.org/10.1007/978-3-030-14011-3_12
10. Ponomarenko, O., Grimzin, I., Yevtushenko, N., Lysenko, T., Marynenko, D.: Advanced Technologies of Manufacturing Readily Removable Cores for Obtaining High-Quality Castings. In: Ivanov, V., Trojanowska, J., Pavlenko, I., Zajac, J., Peraković, D. (eds.) DSMIE 2021. LNME, pp. 565–574. Springer, Cham (2021). https://doi.org/10.1007/978-3-030-77719-7_56
11. Capuzzi, S., Timelli, G.: Preparation and melting of scrap in aluminum recycling: a review. Metals **8**, 249 (2018). https://doi.org/10.3390/met8040249
12. Imbirovych, N., Povstyanoy, O., Zaleta, O., Shymchuk, S., Priadko, O.: The Influence of Synthesis Modes on Operational Properties of Oxide Ceramic Coatings on Aluminum Alloys. In: Ivanov, V., Trojanowska, J., Pavlenko, I., Zajac, J., Peraković, D. (eds.) DSMIE 2021. LNME, pp. 536–545. Springer, Cham (2021). https://doi.org/10.1007/978-3-030-77719-7_53

13. Loskutova, T., et al.: Composition, Structure, and Properties of Ti, Al, Cr, N, C Multilayer coatings on AISI W1–7 alloyed tool steel. Coatings **12**, 616 (2022). https://doi.org/10.3390/coatings12050616
14. Santhosh, M.S., Sasikumar, R.: Influences of aluminium / E-glass volume fraction on flexural and impact behaviour of GLARE hybrid composites. J. Eng. Sci. **6**(1), C6–C10 (2019). https://doi.org/10.21272/jes.2019.6(1).c2
15. Zabolotnyi, O., Pasternak, V., Ilchuk, N., Huliieva, N., Cagáňová, D.: Powder Technology and Software Tools for Microstructure Control of AlCu2 Samples. In: Ivanov, V., Trojanowska, J., Pavlenko, I., Zajac, J., Peraković, D. (eds.) DSMIE 2021. LNME, pp. 585–593. Springer, Cham (2021). https://doi.org/10.1007/978-3-030-77719-7_58
16. Demchenko, M.V., Gaponova, O.P., Myslyvchenko, O.M., Antoszewski, B., Bychenko, M.M.: Microstructure and properties of AlCrFeCoNiCux high-entropy alloys. J. Eng. Sci. **5**(1), C11–C15 (2018). https://doi.org/10.21272/jes.2018.5(1).c3
17. Udovytska, Y.A., et al.: Effect of epoxy composite coatings on radiation stability and magnetic sensitivity of n-Si single crystals. Surf. Engin. Appl. Electrochem. **57**, 222–227 (2021). https://doi.org/10.3103/S1068375521020125
18. Budnik, A.F., Rudenko, P.V., Berladir, KV., Budnik, O.A.: Structured nanoobjects of polytetrafluoroethylene composites. J. Nano- Electron. Phys. **7**(2), 02022 (2015)
19. Peter, I., Varga, B.: Some considerations on the structure refinement in al-based alloys. In: Sivasankaran, S. (ed.) Aluminium Alloys - Recent Trends in Processing, Characterization, Mechanical Behavior and Applications. IntechOpen, London (2017). https://doi.org/10.5772/intechopen.71450
20. Shlyaptseva, A.D., Petrov, I.A., Ryakhovsky, A.P., Medvedeva, E.V., Tcherdyntsev, V.V.: Complex structure modification and improvement of properties of aluminium casting alloys with various silicon content. Metals **11**, 1946 (2021). https://doi.org/10.3390/met11121946
21. Savchenkov, S., Kosov, Y., Bazhin, V., Krylov, K., Kawalla, R.: Microstructural master alloys features of aluminum-erbium system. Curr. Comput.-Aided Drug Des. **11**, 1353 (2021). https://doi.org/10.3390/cryst11111353
22. Kostyk, K., et al.: Simulation of diffusion processes in chemical and thermal processing of machine parts. Processes **9**(4), 698 (2021). https://doi.org/10.3390/pr9040698
23. Kusyi, Y., Onysko, O., Kuk, A., Solohub, B., Kopei, V.: Development of the technique for designing rational routes of the functional surfaces processing of products. In: Karabegović, I., Kovačević, A., Mandžuka, S. (eds.) New Technologies, Development and Application V. NT 2022. Lecture Notes in Networks and Systems, vol 472, pp. 135–143. Springer, Cham (2022). https://doi.org/10.1007/978-3-031-05230-9_16
24. Zablotskyi, V., Moroz, S., Tkachuk, A., Prystupa, S., Zabolotnyi, O.: Influence of Diamond Smoothering Treatment Power Parameters on Microgeometry of Working Surfaces of Conjugated Parts. In: Tonkonogyi, V., et al. (eds.) InterPartner 2019. LNME, pp. 372–381. Springer, Cham (2020). https://doi.org/10.1007/978-3-030-40724-7_38
25. Bychkov, A.S., Kondratiev, A.V.: Criterion-based assessment of performance improvement for aircraft structural parts with thermal spray coatings. J. Superhard Mater. **41**(1), 53–59 (2019). https://doi.org/10.3103/S1063457619010088
26. Kuz'min, M.P., et al.: Obtaining of Al–Si foundry alloys using amorphous microsilica – crystalline silicon production waste. Journal of Alloys and Compounds **806**, 806–813 (2019). https://doi.org/10.1016/j.jallcom.2019.07.312
27. Gilev, I.O., Kotenkov, P.V., Popova, E.A., Shubin, A.B.: Effect of Ti and Zr additions on the structure and mechanical properties of the Al–4 % Cu alloy. In: AIP Conference Proceedings 2466, 060006 (2022). https://doi.org/10.1063/5.0088878

28. Lipiński, T.: Influence of Ti and melt number on microstructure and mechanical properties of Al-Si alloy on agriculture machine parts. In: Proceedings 17th International Scientific Conference "Engineering for Rural Development", pp. 1431–1436. Latvia University of Life Sciences and Technologies, Jelgava, Latvia (2018). https://doi.org/10.22616/ERDev2018.17.N242
29. Cherepanov, A.I., et al.: The effect of fast crystallization ligature modification on the microstructure of aluminum alloys J. Phys.: Conf. Ser. **1515**, 052043 (2020). https://doi.org/10.1088/1742-6596/1515/5/052043
30. Romanchenko, O., Lohunov, O., Kharlamov, Y., Sokolov, V., Krol, O.: Microindentation Hardness Testing of D – gun Sprayed Coatings. In: Cioboată, D.D. (ed.) ICoRSE 2021. LNNS, vol. 305, pp. 91–99. Springer, Cham (2022). https://doi.org/10.1007/978-3-030-83368-8_9
31. Mamzurina, O.I., et al.: Effect of Zr on microstructure and mechanical properties of the Al–Cu–Yb and Al–Cu–Gd Alloys. Metals **12**, 479 (2022). https://doi.org/10.3390/met12030479
32. Berladir, K., Hovorun, T., Gusak, O., Reshetniak, Y., Khudaybergenov, D.: Influence of Modifiers-Ligatures on the Properties of Cast Aluminum Alloy AK5M2 for the Automotive Industry. In: Ivanov, V., Trojanowska, J., Pavlenko, I., Zajac, J., Peraković, D. (eds.) DSMIE 2020. LNME, pp. 473–482. Springer, Cham (2020). https://doi.org/10.1007/978-3-030-50794-7_46

Check for updates

Synthesis of Copper Nanoparticles on Graphite Using Transient Glow-to-Arc Discharge Plasma

Andrii Breus, Sergey Abashin, Ivan Lukashov, Oleksii Serdiuk, and Oleg Baranov(✉)

National Aerospace University, 17, Chkalova Street, Kharkiv 61070, Ukraine
O.Baranov@khai.edu

Abstract. A transient glow-to-arc mode developed in a setup for ignition of the glow discharge plasma was employed to synthesize copper nanoparticles with 2 to 5 nm diameters on surfaces of complex 1D and 2D graphite nanostructures grown in craters formed during the arc generation. In the setup, a cathode and an anode were manufactured of graphite and copper, respectively. Graphite samples were arranged on the cathode. When igniting the plasma glow between the electrodes, the anode was considered as a source of a copper atomic flux, while the cathode served as a source of carbon atoms generated in abundant yield during the arc stage of the plasma discharge. Extreme conditions obtained on the surface of the samples subject to the arcing allowed the formation of complex carbon nanostructures densely covered by copper nanoparticles. TEM images obtained during a microscopic stage of the research revealed the sequential growth of the nanostructures, when 2D carbon nanostructures formed in the craters were covered by the copper nanoparticles that, in turn, served as the seeds for the formation of 1D carbon nanostructures on a body of the parent 2D carbon nanostructure. At that, the copper nanoparticles were covered by a few atomic layers of graphite during the growth of the complex carbon nanostructure.

Keywords: Copper nanoparticles · Carbon nanostructures · Plasma glow · Arc discharge · Process innovation

1 Introduction

Nowadays, metal nanoparticles are widely recognized in science and industry as a flexible instrument to change various properties of matter by changing the size, shape, and structure. Many applications have been proposed during the last years of the explorations, and copper nanoclusters attracted attention at that [1]. Barani et al. [2] investigated the thermal behavior of hybrid structures with graphene and copper nanoparticles and found that the conductivity shows a sharp increase at the graphene concentration of about 15% when the concentration of Cu is close to 40%. The effect was explained by the presence of Cu particles between the graphene nanosheets, resulting in the generation of the thermally conductive circuits. A review focused on the antibacterial reaction of Cu nanoparticles was presented by Al-Hakkani [3]. At the same time, the toxic behavior of copper oxide

V. Tonkonogyi et al. (Eds.): InterPartner 2022, LNME, pp. 264–273, 2023.
https://doi.org/10.1007/978-3-031-16651-8_25

nanoparticles on several plants resulted in preventing germination of seed, reduction in photosynthesis and respiration rate was studied by Rajput et al. [4], while Ghoto et al. reported advancement in the application of copper nanoparticles for colorimetric detection of dithiocarbamate pesticides [5]. Yang et al. produced a highly conductive and foldable electrode using controlled deposition of Cu nanoparticles [6]. Simultaneously, Popok et al. explored the ability of the nanoparticles to the plasmon resonance widely applied in optics, electronics, catalysis, and sensing [7]. Mechanical application of copper nanoparticles was studied by Borda et al., who investigated the tribological behavior of lubricants with additive containing the nanoparticles [8], and the results show that Cu nanoaggregates are quite helpful as the anti-wear substance in mineral oil with the reduction of wear up to 64%. The structures prepared by placing Cu particles between the graphene layers were successfully applied as a non-enzymatic sensor in an investigation conducted by Soganci et al. [9], while Qing et al. reported progress in developing a biosensor based on DNA-templated Cu nanoparticles [9]. Simultaneously, because Cu is a necessary element for metabolism in animals and plants, Ameh and Sayes, in their review [10], also concluded that the antibacterial potential of Cu particles allows them to be used in biomedicine and drug delivery in cancer therapy. Yet, the accumulation and toxicity of the particles should be considered at that.

2 Literature Review

A lot of methods of synthesis of Cu nanoparticles were reported, such as engineering on the carbon nanotubes with the purpose of their further antibacterial applications [11], biosynthesis by use of Shewanella Loihica platform [12], and synthesis on bentonite, which also supports the green protocols [13], a fungal-based synthesis for anticancer, antidiabetic and antibacterial activities [14], or phytosynthesis by dissolving CuSO4 in the corresponding plant extract [15].

However, most engineering methods are based on other approaches [16, 17]. One of the reasons to develop new methods is the instinctive oxidative power occurring at the synthesis of Cu under ambient conditions [18], which results in the application of the techniques with controlled oxygen supply, like the plasma-enhanced technologies that engage magnetron [19], arc [20], radiofrequency [21], or microwave discharges [22]. Liquid solutions are also engaged – thus, Phan et al. investigated the in-situ production of Cu particles on cotton fabrics by implementing spark discharge at room temperature via solution plasma [23] and discovered that the obtained copper nanoparticles encapsulated by the graphene shell were more stable in acid. In addition, the thickness of the graphene shell affected the stability. El-Khatib et al. [24] developed the arc discharge method in deionized water by applying stabilized alternating current between the rods; the nanoparticles were collected on the water's glass tube. A similar technique was applied by Wongrat et al. [25] for the purpose. Magnetron sputtering was employed by Kousal et al. [26], who observed by X-ray scattering the condensation of copper at a few mm gap from the target. The nanoparticles were captured preferentially within a region limited by the plasma torus where the nanoparticles reached the value of 90 nm while the nanoparticles of 10–20 nm escaped the region and formed a flux of the nanoparticles. Flexible generation of Cu particles from the scrap in arc discharge was proposed by Tharchanaa et al. [27].

Here, both electrodes were produced of graphite, and copper scrap was placed inside the anode. Under the plasma action, copper was evaporated, and the vapors were cooled down at the collisions with background gas, thus forming the copper nanoparticles that were found later on the walls of the reactor. A combined DC/RF setup was used by Orazbayev et al. [28] to grow Cu nanoparticles with a diameter of 50–500 nm. In the setup, the copper electrode was simultaneously biased by RF and DC power, while another electrode was grounded. It was assumed that the additional DC voltage enhances the sputter yield. The particles were accumulated by a silicon plate placed under the grounded electrode. Theoretical modeling of a DC thermal plasma torch operating at atmospheric pressure confirmed that this approach is also a perspective for synthesizing Cu particles from the solid phase [29]. This paper reports a method of formation of copper nanoparticles with sizes as small as 2 to 5 nm obtained on surfaces of 2D carbon nanostructures grown using a DC argon plasma discharge.

3 Research Methodology

A plasma chamber to ignite glow discharge was engaged to conduct the experiments on the growth of copper nanoparticles. Graphite samples (diam. 8 mm, the thickness of 5 mm) were put on a cathode made of graphite (diam. 35 mm, the thickness of 4 mm) and exposed to the action of plasma. An anode was made of copper (diameter of 15 mm, the height of 4 mm). The electrodes were mounted in the chamber (diameter of 300 mm, the height of 350 mm) connected to an argon line, and a pressure of 120 Pa was maintained. A schematic of the setup is shown in Fig. 1.

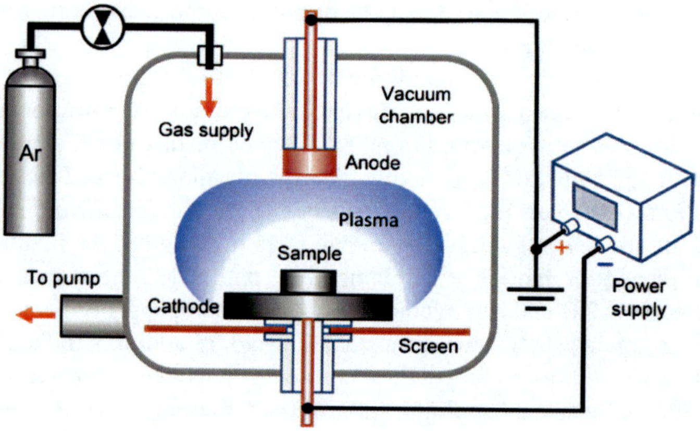

Fig. 1. A schematic of the setup to grow copper nanoparticles on a graphite sample in the glow discharge environment.

The glow discharge was maintained at the voltage between the electrodes of 720 V and a discharge current of 0.18 A. In this mode, the samples were heated to red with a temperature of about 750 °C. During the whole ion treatment cycle, unipolar arcs appeared on the sample's surface, which caused the glow transition to the arc discharge.

After the plasma treatment, the samples remained in the chamber for 30 min, then transferred for examination using scanning electron microscopy (SEM) to study the results of the plasma processing on the sample.

4 Results and Discussion

As was mentioned, throughout the experiment, a significant number of breakdowns were observed, i.e., the transition from the glow to the arc discharge. As a result, almost all nanostructures were found in the craters, formed at the action of arcs on the sample's surface. SEM images of the sample surface in a crater are shown in Fig. 2.

The general view of the nanostructures in the depth of the arc discharge craters is shown in Fig. 2, a, and they appeared to be a bunch of one-dimensional nanostructures. Nanostructures closer to the surface unaffected by the arc have a different appearance (Fig. 2, b), and they have relatively large copper particles on their tips. Clusters of nanostructures can also be found on the surface near the crater (Fig. 2, c), but this was quite a rare observation. An enlarged view of the tip of the nanostructures shown in Fig. 2, a, exhibits their length of about 500 nm and a diameter of 40 nm (Fig. 2, d).

The complex nature of the nanostructures obtained in argon plasma led to the study using transmission electron microscopy (TEM) by use of the Jeol JEM-2100 microscope (Fig. 3 and Fig. 4). A general view of the TEM image of the nanostructure from Fig. 2, a is shown in Fig. 3, a. It was found that only the upper part of the nanostructure is an array of 1D nanostructures with a density of 400 μm^{-2} (Fig. 3, b).

The enlarged view of 1D nanostructures allowed determining their diameters – about 30 nm (Fig. 3, c), as well as to detect of a significant number of copper nanoparticles with a diameter of about 5 nm, which are covered with 4–5 atomic layers of carbon (Fig. 3, d). Although most copper particles have been removed from the tips of the carbon nanostructures after the treatment required to use TEM instruments, 350 nm copper particles still can be found on the tips of the 1D carbon nanostructures, also covered by the dense array of copper nanoparticles of 2–5 nm in diameter (Fig. 3, e). The electron diffraction measurements confirmed the chemical composition of the nanoparticles (Fig. 3, f).

Additionally, the lower part of the nanostructure, which is adjacent to the graphite sample, i.e., the "roots" of the nanostructure (Fig. 4, a), was studied. An enlarged view (Fig. 4, b) of the "roots" reveals that the nanostructures in Fig. 3, a are composed of folded sheets (2D) of graphite nanostructures, at least in the lower part, which is a significant result for the development of theoretical models of growth of such nanostructures.

The inverse view in Fig. 4, c shows the nanostructure covered with a dense array of copper nanoparticles, which is also a significant result, both from a practical point of view (synthesis of catalytic nanoparticles for energy applications) and from a theoretical point of view – obviously the carbon nanoparticles grow under such a nanoparticle and form the branches on the carbon nanostructure. The enlarged view also confirms the existence of 2–5 nm copper particles on the entire surface of the carbon nanostructure, even near the tip (Fig. 4, d). Figure 4, e shows carbon nanostructures with copper particles of 20 nm at the tips, and the enlarged view of the tip of the carbon nanostructure confirms the presence of an array of copper particles with a size of 2 nm (Fig. 4, f).

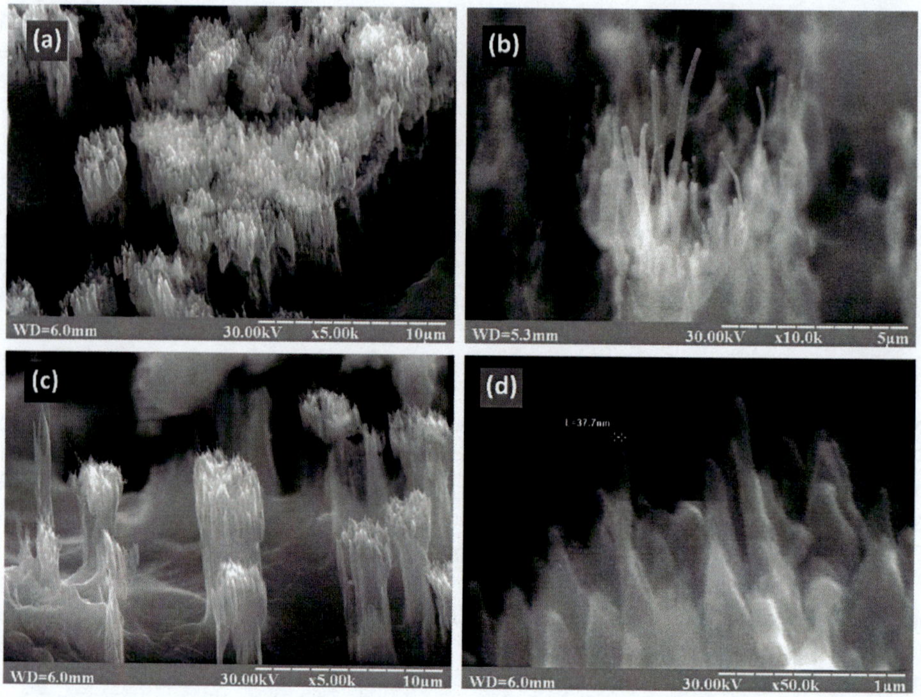

Fig. 2. SEM image of a graphite sample processed for 30 min in an argon plasma (120 Pa, 720 V, and 0.18 A): a – general view of nanostructures in the depth of the craters of the arc discharge; b – magnified view of the nanostructures closer to the surface unaffected by the arc revealing their complex structure; c – accumulation of nanostructures on the surface near the crater; d – enlarged view of the tips of the nanostructures shown in Fig. 2, an exhibiting their dimensions with a length of 500 nm and a diameter of 38 nm.

The latter result means a continuous synthesis of copper nanoparticles throughout the whole procedure of the carbon nanostructure growth. Another significant result of the TEM study of the carbon nanostructures is the determination of the morphology of the nanostructures – they are not hollow like carbon nanotubes. Thus, their growth can be described by the same mechanism proposed for forming copper oxide nanowires [18]. The only difference is that the active growth zone, also located at the tip of the nanostructure, is not a defective surface of this tip but a copper particle attached to it.

The unambiguous association of the carbon nanostructures with Cu particles may greatly simplify the solution of the problem of the origin of nucleation of carbon nanostructures, which is caused by Cu particles, and their density defines the density of carbon nanostructures on the graphite sheets. Moreover, a copper nanoparticle on the side surface of the carbon nanostructure is a source for the growth of a next 1D carbon nanostructure on the body of the "parent" carbon nanostructure.

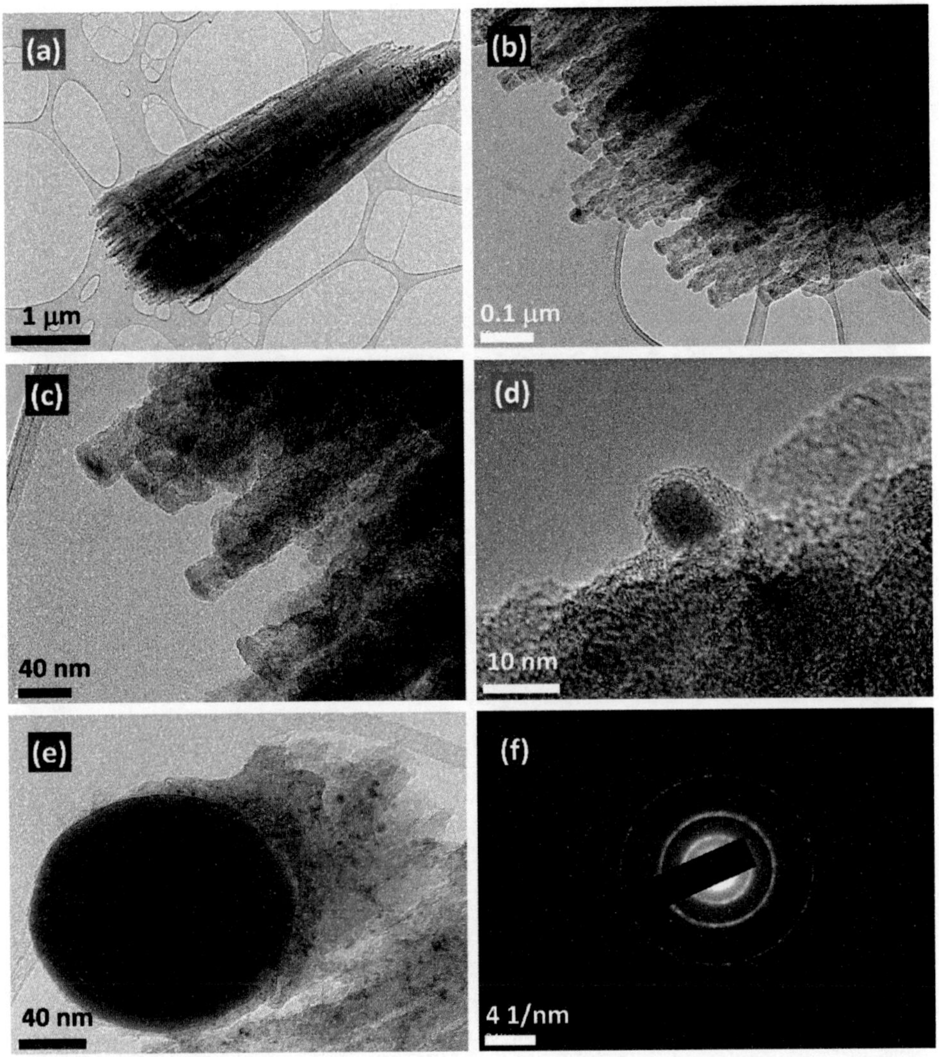

Fig. 3. TEM image of the nanostructure from a sample treated for 30 min in an argon plasma (120 Pa, 720 V, and 0.18 A): a – general view; b – the tip of the nanostructure, which detects an array of 1D nanostructures with a density of 400 μm^{-2}; c – enlarged view of 1D nanostructures with diameters of about 30 nm; d – copper nanoparticle with a diameter of 5 nm that is covered with 4–5 atomic layers of carbon; e – a copper particle with a diameter of 350 nm at the tip of the 1D carbon nanostructure that is covered with a dense array of copper nanoparticles with a diameter of 2–5 nm; f – picture of the electron diffraction that reveals the presence of the copper nanoparticles.

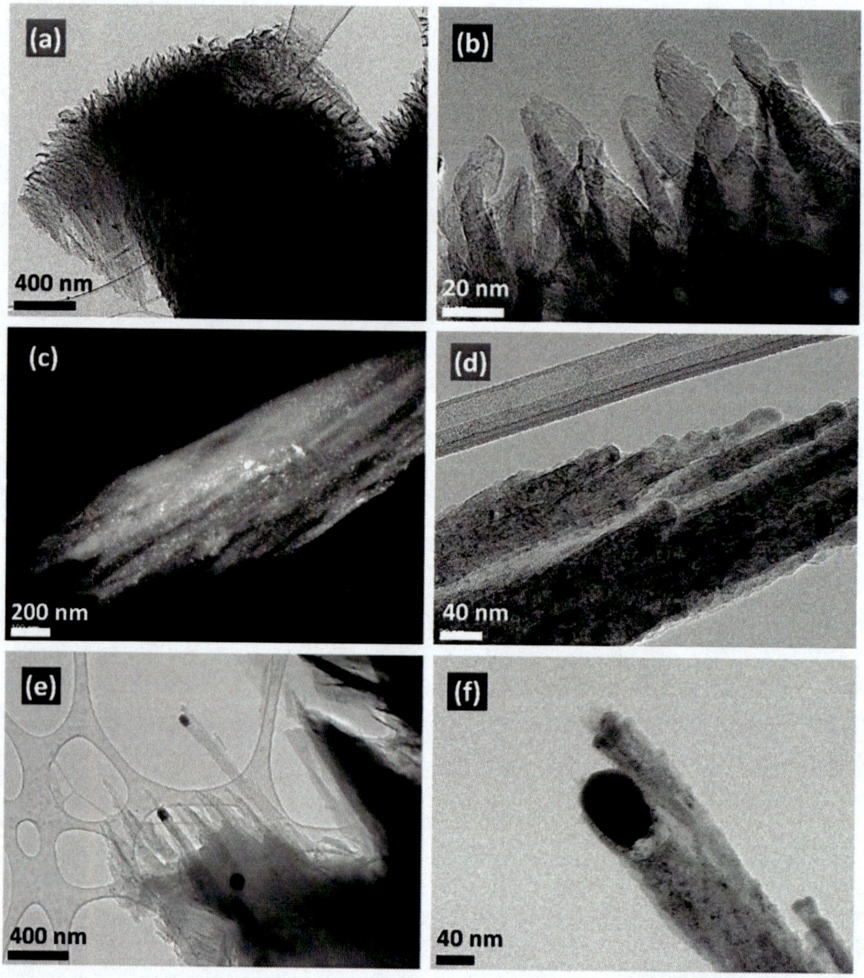

Fig. 4. TEM image of the nanostructure from a sample treated for 30 min in an argon plasma (120 Pa, 720 V, and 0.18 A): a – the lower part of the nanostructure ("roots"); b – enlarged view of the "roots", which reveals that the nanostructures in Fig. 3, a, are composed of the folded sheets of 2D nanostructures; c – inverse view showing the nanostructure covered with a dense array of the copper nanoparticles; d – enlarged view showing the presence of copper nanoparticles with a size of 2–5 nm on the entire surface of the carbon nanostructure; e – carbon nanostructures with the copper particles with dimensions of about 20 nm at the tips; f – enlarged view of the tip of the carbon nanostructure with an array of copper particles of 2 nm.

The proposed method allows obtaining the copper particles protected from oxidation because a few carbon layers shell them. The method can be considered as environmental-friendly [11–13] while it is more cost-effective since it does not require the application of potentially hazardous chemistry [15], additional media like fluids [23–25] complicating the setup structure, and relies on rather simple plasma-generating technique thus avoiding the implementation of expensive plasma generators [21, 22, 28], or use of magnetic fields, which can affect the dispersion of the nanoparticles in the production volume [19, 26, 30]. Moreover, the obtained results provide additional insight into the mechanisms of the formation of copper nanoparticles [29, 31, 32].

5 Conclusions

The experiment observed the formation of dense arrays of copper nanoparticles with a diameter of 2 to 5 nm on a surface of carbon nanostructures where the transient glow-to-arc mode on plasma discharge was observed. In the setup designed for the ignition of the glow plasma, graphite samples were arranged on a graphite cathode, while an anode was made of copper and served as a source of copper vapors generated from ion sputtering. During the transition of the glow discharge to arc, extreme conditions were obtained in the places on the cathode where the arc was initiated. Superposition of the fluxes of copper and carbon atoms in these regions ensured the generation of the graphite nanostructures where copper particles served as catalysts. At that, the particles deposited on the surface of the graphite nanostructures play a role in the nucleation centers for the formation of other graphite nanostructures, thus making possible the growth of the complex composite graphite-copper nanostructures obtained in the experiment. The obtained tree-like carbon nanostructures, which are covered with a dense array of copper nanoparticles, can be a raw material for further production of the catalyst from the copper nanoparticles.

Acknowledgment. The research was partly sponsored by the NATO Science for Peace and Security Programme under grant id. G5814 project NOOSE. A. Breus, S. Abashin, and O. Baranov acknowledge the support from the project funded by the National Research Foundation of Ukraine under grant agreement No. 2020.02/0119.

References

1. Din, M.I., Rehan, R.: Synthesis, Characterization, and applications of copper nanoparticles. Anal. Lett. **50**(1), 50–62 (2017)
2. Barani, Z., et al.: Thermal properties of the binary-filler hybrid composites with graphene and copper nanoparticles. Adv. Func. Mater. **30**, 1904008 (2020)
3. Al-Hakkani, M.F.: Biogenic copper nanoparticles and their applications: a review. SN Applied Sciences **2**, 505 (2020)
4. Rajput, V.D., et al.: Effects of copper nanoparticles (CuO NPs) on crop plants: a mini review. BioNanoScience **8**(1), 36–42 (2017). https://doi.org/10.1007/s12668-017-0466-3
5. Ghoto, S.A., Khuhawar, M.Y., Jahangir, T.M., Mangi, J. ul D.: Applications of copper nanoparticles for colorimetric detection of dithiocarbamate pesticides. Journal of Nanostructure in Chemistry **9**, 77–93 (2019)
6. Yang, Y., Huang, Q., Payne, G.F., Sun, R., Wang, X.: A highly conductive, pliable and foldable Cu/cellulose paper electrode enabled by controlled deposition of copper nanoparticles. Nanoscale **11**, 725 (2019)
7. Popok, V.N., et al.: Gas-aggregated copper nanoparticles with long-term plasmon resonance stability. Plasmonics **16**(2), 333–340 (2020). https://doi.org/10.1007/s11468-020-01287-4

8. Borda, F.L.G., de Oliveira, S.J.R., Lazaro, L.M.S.M., Leiroz, A.J.K.: Experimental investigation of the tribological behavior of lubricants with additive containing copper nanoparticles. Tribol. Int. **117**, 52–58 (2018)
9. Soganci, T., Ayranci, R., Unlu, G., Acet, M., Ak, M.: Designing sandwich type single-layer graphene decorated by copper nanoparticles for enhanced sensing properties. J. Phys. D Appl. Phys. **53**, 255105 (2020)
10. Ameh, T., Sayes, C.M.: The potential exposure and hazards of copper nanoparticles: a review. Environ. Toxicol. Pharmacol. **71**, 103220 (2019)
11. Seo, Y., et al.: Engineering copper nanoparticles synthesized on the surface of carbon nanotubes for anti-microbial and anti-biofilm applications. Nanoscale **10**, 15529–15544 (2018)
12. Lv, Q., et al.: Biosynthesis of copper nanoparticles using shewanella loihica PV-4 with antibacterial activity: novel approach and mechanisms investigation. J. Hazard. Mater. **347**, 141–149 (2018)
13. Issaabadi, Z., Nasrollahzadeh, M., Sajadi, S.M.: Green synthesis of the copper nanoparticles supported on bentonite and investigation of its catalytic activity. J. Clean. Prod. **142**, 3584–3591 (2017)
14. Noor, S., et al.: A fungal based synthesis method for copper nanoparticles with the determination of anticancer, antidiabetic and antibacterial activities. J. Microbiol. Methods **174**, 105966 (2020)
15. Vijayakumar, G., et al.: Phytosynthesis of copper nanoparticles using extracts of spices and their antibacterial properties. Processes **9**, 1341 (2021)
16. Hasanin, M., Al Abboud, M.A., Alawlaqi, M.M., Abdelghany, T.M., Hashem, A.H.: Ecofriendly synthesis of biosynthesized copper nanoparticles with starch-based nanocomposite: antimicrobial, antioxidant, and anticancer activities. Biological Trace Element Research **200**, 2099–2112 (2022). https://doi.org/10.1007/s12011-021-02812-0
17. Fernández-Arias, M., et al.: Copper nanoparticles obtained by laser ablation in liquids as bactericidal agent for dental applications. Appl. Surf. Sci. **507**, 145032 (2020)
18. Baranov, O., Košiček, M., Filipič, G., Cvelbar, U.: A deterministic approach to the thermal synthesis and growth of 1D metal oxide nanostructures. Appl. Surf. Sci. **566**, 150619 (2021)
19. Baranov, O., Romanov, M., Fang, J., Cvelbar, U., Ostrikov, K.: Control of ion density distribution by magnetic traps for plasma electrons. J. Appl. Phys. **112**(7), 073302 (2012)
20. Baranov, O., Fang, J., Rider, A., Kumar, S., Ostrikov, K.: Effect of ion current density on the properties of vacuum arc-deposited TiN coatings. IEEE Trans. Plasma Sci. **41**(12), 3640–3644 (2013)
21. Nakysbekov, Z., Buranbayev, M., Aitzhanov, M.B., Suyundykova, G.S., Gabdullin, M.T.: Synthesis of copper nanoparticles by cathode sputtering in radio-frequency plasma. Journal of nano- and electronic physics **10**(3), 03010 (2018)
22. Sreeju, N., Rufus, A., Philip, D.: Microwave-assisted rapid synthesis of copper nanoparticles with exceptional stability and their multifaceted applications. J. Mol. Liq. **221**, 1008–1021 (2016)
23. Phan, P.Q., et al.: In situ synthesis of copper nanoparticles encapsulated by nitrogen-doped graphene at room temperature via solution plasma. RSC Adv. **10**, 36627–36635 (2020)
24. El-Khatib, A.M., Doma, A.S., Abo-Zaid, G.A., Badawi, M.S., Mohamed, M.M., Mohamed, A.S.: Antibacterial activity of some nanoparticles prepared by double arc discharge method. Nano-Structures & Nano-Objects **23**, 100473 (2020)
25. Wongrat, E., Wongkrajang, S., Chuejetton, A., Bhoomanee, C., Choopun, S.: Rapid synthesis of Au, Ag and Cu nanoparticles by DC arc-discharge for efficiency enhancement in polymer solar cells. Mater. Res. Innovations **23**(2), 66–72 (2019)
26. Kousal, J., et al.: Magnetron-sputtered copper nanoparticles: lost in gas aggregation and found by in situ X-ray scattering. Nanoscale **10**, 18275–18281 (2018)

27. Tharchanaa, S.B., Priyanka, K., Preethi, K., Shanmugavelayutham, G.: Facile synthesis of Cu and CuO nanoparticles from copper scrap using plasma arc discharge method and evaluation of antibacterial activity. Mater. Technol. **36**(2), 97–104 (2021)
28. Orazbayev, S.A., Utegenov, A.U., Zhunisbekov, A.T., Slamyiya, M., Dosbolayev, M.K., Ramazanov, T.S.: Synthesis of carbon and copper nanoparticles in radio frequency plasma with additional electrostatic field. Contrib. Plasma Phys. **58**, 961–966 (2018)
29. Boselli, M., Gherardi, M., Colombo, V.: 3D modelling of the synthesis of copper nanoparticles by means of a DC transferred arc twin torch plasma system. J. Phys. D Appl. Phys. **52**, 444001 (2019)
30. Shyrokyi, Y., Kostyuk, G.: Investigation of the influence of crystallization energy on the size of nanostructures during copper ion-plasma treatment. In: Nechyporuk, M., et al. (eds.) ICTM 2021. LNNS, pp. 57–66. Springer, Cham (2022). https://doi.org/10.1007/978-3-030-94259-5_6
31. Shyrokyi, Y., Kostyuk, G.: Erosion processes on copper electrodes applied to growth of nanostructures in plasma. In: Ivanov, V., Trojanowska, J., Pavlenko, I., Rauch, E., Peraković, D. (eds.) Advances in Design, Simulation and Manufacturing V. DSMIE 2022. Lecture Notes in Mechanical Engineering, pp. 494–503. Springer, Cham (2022). https://doi.org/10.1007/978-3-031-06025-0_49
32. Breus, A., Abashin, S., Serdiuk, O., Baranov, O.: Linking dynamics of growth of copper oxide nanostructures in air. In: Nechyporuk, M., et al. (eds.) ICTM 2021. LNNS, pp. 555–564. Springer, Cham (2022). https://doi.org/10.1007/978-3-030-94259-5_47

Fatigue Strength of Steel Samples After Friction Treatment

Volodymyr Gurey[1(✉)], Ihor Hurey[1,2], Tetyana Hurey[1], and Weronika Wojtowicz[2]

[1] Lviv Polytechnic National University, 12, Bandera Street, Lviv 79013, Ukraine
Volodymyr.I.Hurei@lpnu.ua

[2] Rzeszow University of Technology, 12, Powstancow Warszawy al., 35-959 Rzeszow, Poland

Abstract. The work aims to study the influence of the tools' shape of the working surface used during friction treatment of samples' cylindrical surfaces on the formation of the strengthened layers and their effect on fatigue strength in the study on air and corrosion medium. The strengthened white layers with a nanocrystalline structure are formed in the samples' surface layers during friction treatment. The research results of the influence of the frictional treatment on samples' working surfaces made of Steel C45, Steel 41Cr4, and Steel CT80 (quench-hardening and low-temperature tempering) on fatigue strength during tests on pure bending with rotation in the air and corrosive media (3% aqueous NaCl solution) are presented. The thickness of the strengthened layer after treatment was from 120–130 μm to 180–190 μm. The microhardness of the strengthened layer is 1.7–1.9 times higher than the base metal. Residual compressive stresses are formed in the surface layers of the strengthened samples, which decrease with the depth of the layer. The fatigue strength of Steel C45 after friction treatment increased 1.4–1.7 times, Steel 41Cr4 – 1.3–1.5 times, and Steel CT80 – 1.4–1.6 times was shown by experiments. The fatigue strength has increased sharply and ranges from 3.4 to 4.2 times on Steel C45 and 6.7 times on Steel CT80, as shown in studies using the corrosive media. The obtained strengthened layers with a nanocrystalline structure can be used to increase the durability of parts with working cylindrical surfaces that work with cyclic loads in corrosive mediums.

Keywords: Industrial growth · Corrosive media · White layer · Nanocrystalline structure

1 Introduction

A complete characteristic of products is their quality, i.e., the set of characteristics of products that determine the ability to satisfy human needs, production, and so on. Modern equipment operates at high speeds and loads, under cyclic loads, and often in the corrosive medium.

The products' operational characteristics are determined by the condition of the working surfaces and the surface layer of the metal. The destruction of parts begins from their surface, and there is an accumulation of various defects. In the process of cyclic

V. Tonkonogyi et al. (Eds.): InterPartner 2022, LNME, pp. 274–283, 2023.
https://doi.org/10.1007/978-3-031-16651-8_26

loading of parts in their surface layers accumulates, the internal defects of the metal structure, which are the stress concentrators. The formation and propagation of micro-cracks in the direction to the center takes place, leading to a decrease in the durability and failure of parts [1].

Reliability of parts' work is defined by the quality of the working surfaces and metal's surface layer and is characterized by geometrical parameters and physical and mechanical properties. The geometric parameters of the working surfaces are determined by the surface profile's roughness, waviness, and load-bearing capacity and depend on the quality of surface treatment. The properties of the surface layer metal depend on its chemical and phase composition, structure, stress state, etc. The correspondence of the characteristics of the working surfaces' quality and the surface layer with the performance properties show that the working surfaces and surface layers must have high hardness, fine structure, residual compressive stresses, and others. The specified properties of the surface layers can be obtained by methods that reduce the grain size from the macro to the nano level during the treatment process [2].

2 Literature Review

During operation, the cyclic loads act into the rotating machines' parts, which are much smaller than the value of the strength limit of materials [3, 4], and there are fatigue loads [5, 6]. Fatigue failure is one of the most dangerous types of destruction of machine parts in operation [4, 7]. Cracks, during cyclic loading, are usually generated from the surface or in the machine parts' surface layers [8, 9].

Studies have shown [10, 11] that the value of fatigue strength is significantly influenced by the condition of both machine parts' working surfaces and the metal's surface layer [12, 13]. In this case, the parameters of surface stereometry (roughness, waviness, technological stress concentrators, etc.) and the characteristics of the surface layer (hardness, structure, texture, grain size, residual stresses, thickness of the reinforced (strengthened) layer, etc.) are decisive [14, 15].

During fatigue failure of parts, the processes that take place in the surface layer are considered as the interaction between the microstructure, deformations, and mechanical state of the metal parts [16]. Slowing down the development of damaged parts that are subject to cyclic loads is an important task and should be addressed at the stage of the design and manufacture of machine parts [17, 18].

Using various technological methods of strengthening the working surfaces, it is possible to improve the quality of surface layers and increase the durability and reliability of machine parts in operation [18–22]. Such technological processes contribute to the formation of high-quality strengthened layers that retard the formation and propagation of fatigue cracks.

To ensure the appropriate reliability of products during their operation, especially during fatigue, it is necessary to study the stress state of the surface layers [23]. The stress state of the metal, which is formed in the surface layers of the parts during processing, significantly affects the performance properties of the parts [24].

Surface strengthening of machine parts' working surfaces using various technological methods increases the durability of parts that work in conditions of fatigue failure during cyclic loads, especially under corrosive medium [9, 10]. Surface plastic deformation methods are used as surface strengthening methods [2, 23, 24]. Also, to improve the performance properties of machine parts, it is necessary to use technological methods of surface treatment, during which strengthened surface layers with the nanocrystalline structure are formed on the treated surfaces of machine parts [1, 2, 23, 24]. Such methods are technological methods of processing using highly concentrated energy sources [25]. These include laser [26], plasma [27], electron beam [28], cutting [29, 30], friction treatment [31, 32]. When using these methods of surface strengthening, the metal surface layer of the machined parts' surfaces is modified [1, 2, 33]. The formation of strengthened surface layers is due to heating and subsequent cooling at high metal speeds [34]. During the friction treatment, the shear deformation of the metal's surface layers of the treated surfaces is additionally intensive. The surface layers change the structure, physico-mechanical and electrochemical characteristics of the metal, increase their hardness, decrease the grain size, change the chemical and phase composition of the metal, the magnitude, and sign of residual stresses, etc., which significantly affect the performance of machine parts during operation [35, 36].

A highly concentrated energy flow is formed in the tool's contact area with the treated surface during its high-speed friction at the friction treatment process. Strengthened layers with a nanocrystalline structure – white layers are formed in the surface layers of the machined parts. White layers have increased hardness and viscosity, dislocation density, and residual austenite, crushed grain, and residual compressive stresses are formed compared to the base metal [35, 36].

The work aims to study the influence of the tools' shape of the working surface used during friction treatment of samples' cylindrical surfaces on the formation of the strengthened layers and their effect on fatigue strength in the study on air and corrosion medium.

3 Research Methodology

The most common type of deformation in cyclic loading of parts is bending with rotation. The surface layer of metal receives tension and compression of the same magnitude for one revolution of the part under these load conditions. Therefore, tests for fatigue strength, which mimic this type of deformation, are carried out in the laboratory.

Studies of fatigue strength were performed according to the method [3] on cylindrical samples at a pure bending with rotation and with a load frequency of 50 Hz both in air and in a corrosive medium (3% aqueous NaCl solution) (Fig. 1). The test base for samples on air was 20 million cycles and in a corrosive medium – 50 million. In the quench-hardened and low-temperature tempered state, samples for the study of fatigue and corrosion-fatigue strength were made of Steel C45, Steel 41Cr4 and Steel CT80. The diameter of the working part of the samples was 10 mm, and the length was 66 mm (Fig. 2).

The operation of strengthening (hardening) of the samples' working parts by friction treatment was performed on a special device mounted on a lathe. The special device with an autonomous drive for rotating the tool was mounted on the carriage instead of the tool post [34]. The tool with a smooth working surface and the tool with multidirectional inclined grooves on its work surface were used. As a technological medium for friction treatment, mineral oil with active additives containing polymers was used. During the process of friction treatment by the tool with multidirectional inclined grooves on its work surface, the technological medium was fed directly into the contact area of the sample with the tool through the holes made in the tool's body. Holes for feeding (delivering) the technological medium go out into the grooves. The technological medium was fed into the samples' treatment area by flowing when using the tool with a smooth working part.

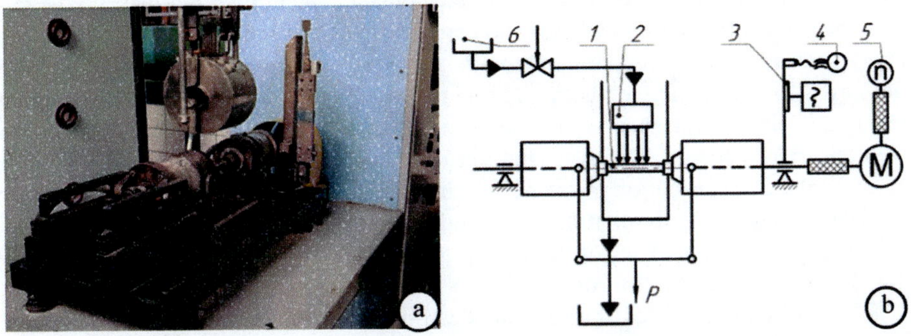

Fig. 1. View (a) and scheme (b) of the device for the study of the fatigue strength of cylindrical samples in air and in a corrosive medium: 1 – sample; 2 – medium supply collector; 3 – strain gauge and registration device; 4 – support to establish the initial value of the deflection; 5 – cycle counter; 6 – tank with corrosion medium.

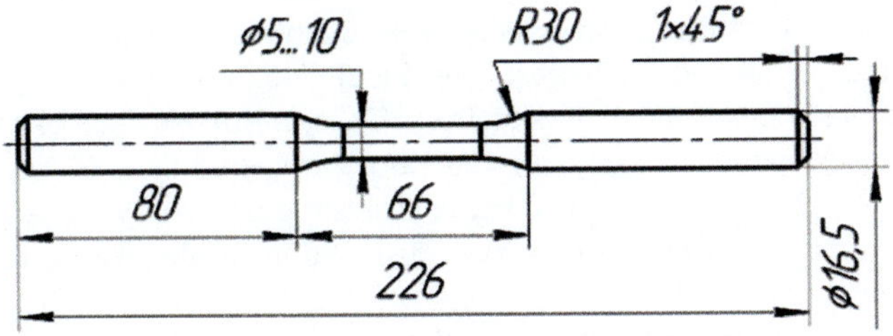

Fig. 2. Sample for fatigue strength study.

Also, for comparison, non-reinforced (without strengthening) samples were used, and an aluminum-oxide abrasive wheel ground their working part.

4 Results and Discussion

Metallographic studies showed that the thickness of the strengthened layer (white layer) obtained on Steel C45 after friction treatment by using the tool with a smooth working surface was 120–130 μm, and by using the tool with multidirectional inclined grooves on its working surface – 180–190 μm. The thickness of the strengthened layer on Steel

41Cr4 was 170–190 μm and 230–240 μm, and on Steel CT80 – 240–250 μm and 300–320 μm, respectively. The microhardness of the strengthened layer increased on Steel C45 to 8.1–9.1 GPa against 4.7 GPa of the main structure, on Steel 41Cr4 – increased to 9.1–9.7 GPa against 5.3 GPa, and on Steel CT80 – increased to 10.3–11.2 GPa against 6.1 GPa, respectively.

X-ray analysis showed that the grain size of the strengthened layer near the treated surface is 20–40 nm, with a smooth transition in depth to the structure of the base material.

During friction treatment by using the mineral oil with active additives containing polymers as the technological medium, in the contact zone of the tool-disc and the part under the action of high temperatures and pressures, thermal and mechanical destruction of polymers takes place. It creates macroradicals with their subsequent depolymerization, releasing active hydrogen, carbon, and other elements that make up the technological medium. Their adsorption with the subsequent mass transfer in superficial layers on juvenile surfaces occurs. The hydrogen released during this process facilitates the processes of plastic deformation of the surface layers, which increases the thickness of the strengthened layer, its microhardness, and other characteristics.

X-ray diffraction analysis of the strengthened layers (white layers) showed that the white layer has a high carbon content and on the studied steels near the surface was 1.3–1.8%. As the depth increases, the amount of carbon decreases to its original value. Due to the white layer's increased carbon content, the residual austenite amount also increases. Thus, in the white layer obtained on Steel CT80, the amount of residual austenite near the surface reached almost 40%, which decreased with depth to the initial value (about 5%).

The density of dislocations characterizes the stress state of the metal. With increasing dislocation density, the mechanical characteristics of the metal increase significantly. Studies have shown that the highest density of dislocations is observed in the white layer obtained on Steel CT80 and the lowest – on Steel C45. With the strengthened layer's depth, dislocations' density decreases sharply. Thus, on the surface of the white layer obtained on quench-hardened and low-temperature tempered Steel CT80, the density of dislocations was $2–2.6 \cdot 10^{11}$ cm^{-2}, and in the base structure – only $0.15–0.2 \cdot 10^{11}$ cm^{-2}. The same pattern of changes in the density of dislocations is observed in the white layers obtained on other studied steels.

Using the tool with multidirectional inclined grooves on its work surface does not significantly affect the amount of residual austenite and the density of dislocations in the surface layer of the metal.

Friction treatment is an effective technological method to increase the wear resistance of steel [29, 32], also effectively increasing its fatigue strength in the air and corrosive media. The fatigue strength of Steel C45 after friction treatment increased by 1.4–1.7 times, Steel 41Cr4 – 1.3–1.5 times, and Steel CT80 – 1.4–1.6 times have been shown by experiments (Fig. 3).

The fatigue strength increased sharply and ranged from 3.4 to 4.2 times on Steel C45 and to 6.7 times on Steel CT80 in a corrosive medium (3% aqueous NaCl solution) were studied. It should be noted that the frictional treatment by using the tool with multidirectional inclined grooves on its work surface increases the fatigue strength much

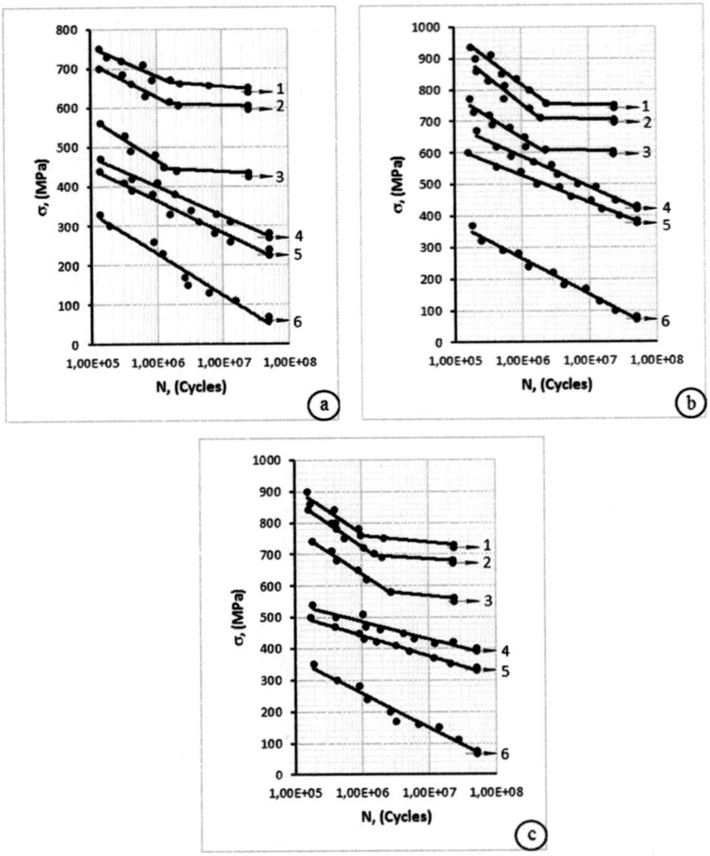

Fig. 3. Fatigue failure curves of samples made of Steel C45 (a), Steel 41Cr4 (b), and Steel CT80 (c) in the quench-hardened and low-temperature tempered state after friction treatment on air (1, 2, 3) and corrosive (3% NaCl solution) medium 4, 5, 6): 1, 4 – original samples; 2, 5 – friction treatment, tool with the multidirectional inclined grooves on its work surface; 3, 6 – friction treatment, tool with the smooth working surface.

more than the frictional treatment by using the same processing regimes and the tool with a smooth working surface.

Fatigue strength of steels in the air research depends mainly on the surface roughness of the samples, the magnitude and sign of residual stresses, and the physical and mechanical properties of the metal. In a corrosive medium, surface roughness is of secondary importance, the main importance being the electrochemical properties of the surface layers of the product. Also of great importance are the residual stresses, much less – the strength of the metal. The high strength of the metal obtained after quench-hardening and low-temperature tempering often impairs the fatigue and corrosion fatigue strength. The white layer mainly has residual compressive stresses and more favorable electrochemical properties, increased viscosity, and resistance to the formation and spread of cracks compared to martensite of conventional quench-hardening.

After the operation of frictional treatment near the samples' surface formed, residual compressive stresses were shown by studies. With depth, the magnitude of residual stresses decreases, and they become tensile. The zone of change the sign of residual stresses is at a depth that significantly exceeds the thickness of the white layer and passes through the metal base structure. When using the frictional treatment with a tool with multidirectional inclined grooves on its work surface, the depth and magnitude of residual stresses are greater than when using the tool with a smooth working surface (Fig. 4).

During the aluminum-oxide abrasive grinding of quench-hardened steels, the partial tempering of the metal's surface layer may take place, and the residual tensile stresses are formed near the treated surface, and electrode potentials of the ground surface change, which is the main reason for lowering their fatigue and corrosion-fatigue strength.

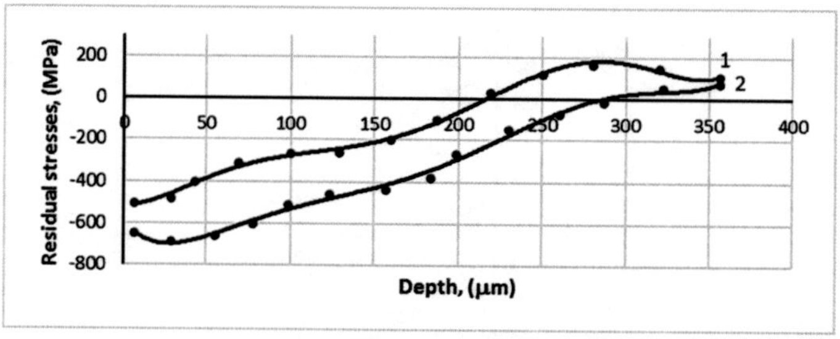

Fig. 4. Residual stresses of Steel 41Cr4 samples (quench-hardening and low-temperature tempering) after friction treatment using the tool with a smooth working surface (1) and with multidirectional inclined grooves on the working surface (2).

Analysis of fractures of strengthened samples by friction treatment after fatigue failure showed that cracks arise at the boundary of the change in the sign of residual stresses and their transition to tensile. In studies in corrosive medium, cracks arise mainly from the sample's surface. When quench-hardened and low-temperature tempered steels are strengthened, there is a zone under the white layer with reduced hardness, less than the martensite of the core. The zone with low hardness is formed because the metal is heated to lower temperatures to form a white layer during strengthening but sufficient to release the pre-hardened structure. This zone is a buffer layer where plastic shifts occur under cyclic loading. These shifts reduce load peaks and reduce the possibility of cracking and peeling of the white layer.

The whole complex of physical and mechanical properties of white layers (increased microhardness, viscosity, residual austenite, dislocation density, residual compressive stresses, etc.) leads to blocking the formation and propagation of microcracks in the surface layers of machine parts. Due to the greater homogeneity of the white layer's structure and its better electrochemical characteristics, the corrosion processes of the metal surface are slowed down in a corrosive medium.

In operation, machine parts can work at both low and high temperatures. To assess the stability of the strengthened layers during operation, studies were conducted after additional heat treatment, which consisted of artificial ageing – heating to 150 °C and holding for 2 h, and cooling in liquid nitrogen for 2 h. Experiments have shown that artificial aging and cold treatment slightly increase the fatigue and corrosion-fatigue strength destruction of strengthened samples. In all studied steels, this increase was approximately 6–8%.

Studies have shown that frictional treatment significantly increases the fatigue strength of the studied steels in both air and corrosive medium. When using the tool with multidirectional inclined grooves on its working surface, the fatigue and corrosion fatigue strength during frictional treatment is greater than when strengthening using the tool with a smooth working surface. The physical and mechanical properties of the obtained strengthened layer play a significant role.

5 Conclusions

Increasing the thickness and microhardness of the obtained strengthened white layer with a nanocrystalline structure is possible using the tool with multidirectional grooves on its working surface.

In the formed strengthened layers with a nanocrystalline structure, the tool with multidirectional grooves on the working surface forms residual compressive stresses greater in magnitude ($\sigma \approx -700$ MPa) and depth of occurrence ($\approx$300 μm).

Experiments showed that the fatigue strength of Steel C45 after friction treatment increased by 1.4–1.7 times, Steel 41Cr4 – by 1.3–1.5 times, and Steel CT80 – by 1.4–1.6 times. During studies in a corrosive medium (3% NaCl aqueous solution), the fatigue strength increased sharply and ranged from 3.4–4.2 times on steel 45 to 6.7 times on Steel CT80. It should be noted that friction treatment with the tool with multidirectional grooves on its working surface increases the fatigue strength more than strengthening in the same processing modes with the tool with a smooth working surface.

Strengthened layers with a nanocrystalline structure can be used to increase the durability of parts with working cylindrical surfaces that work with cyclic loads both in air and in the corrosive medium.

References

1. Dwivedi, D.K.: Surface Engineering. Enhancing Life of Tribological Components. Springer, New Delhi (2018). https://doi.org/10.1007/978-81-322-3779-2
2. Quintino, L.: Overview of coating technologies. In: Miranda, R. (ed.) Surface Modification by Solid State Processing, pp. 1–24. Elsevier, Cambridge (2014). https://doi.org/10.1533/9780857094698.1
3. Troshchenko, V.T.: Deformation and destruction of metals under multicycle loading. Scientific Thought, Kyiv (1981). (in Russian)
4. Pronikov, A.S.: Reliability of machines. Mechanical Engineering, Moscow (1978). (in Russian)
5. Pineau, A., McDowell, D.L., Busso, E.P., Antolovich, S.D.: Failure of metals II: fatigue. Acta Materialia **107**, 484–507 (2016). https://doi.org/10.1016/j.actamat.2015.05.050
6. Santecchia, E., et al.: A review on fatigue life prediction methods for metals. Adv. Mater. Sci. Eng., 9573524 (2016). https://doi.org/10.1155/2016/9573524

7. Guo, Y.B., Warren, A.W., Hashimoto, F.: The basic relationships between residual stress, white layer, and fatigue life of hard turned and ground surfaces in rolling contact. CIRP J. Manuf. Sci. Technol. **2**, 129–134 (2010)
8. Reshetov, D.N.: Workability and reliability of machine parts. Higher School, Moscow (1974). (in Russian)
9. Ivanova, V.S., Terentyev, V.F.: The nature of metal fatigue. Metallurgy, Moscow (1975). (in Russian)
10. PerezMora, R., PalinLuc, T., Bathias, C., Paris, P.C.: Very high cycle fatigue of a high strength steel under sea water corrosion: a strong corrosion and mechanical damage coupling. Int. J. Fatigue **74**, 156–165 (2015). https://doi.org/10.1016/j.ijfatigue.2015.01.004
11. Sayid, A.A., El-Kashif, E., Adly, M.A., Morsy, M.A., Abdelkawy, A.: Fatigue behavior of surfaced C45 steel. J. Eng. Appl. Sci. **68**(1), 1–11 (2021). https://doi.org/10.1186/s44147-021-00016-w
12. Pedersen, M.M.: Introduction to metal fatigue. Aarhus University (2018)
13. Leitner, M., Simunek, D., Maierhofer, J., Gänser, H.P., Pippan, R.: Retardation of fatigue crack growth in rotating bending specimens with semi-elliptical cracks. Metals **9**, 156 (2019)
14. Cicero, S., Álvarez, J.A.: Fracture, fatigue, and structural integrity of metallic materials. Metals **9**, 913 (2019). https://doi.org/10.3390/met9080913
15. Troshchenko, V.G.: Fatigue of metals at inhomogeneous stress state. Message 1. Methods of stress assessment and research results. Probl. Strengths **2**, 5−24 (2010)
16. Withers, P.J.: Residual stress and its role in failure. Rep. Prog. Phys. **70**, 2211–2264 (2007)
17. Matsevity, V.M., Kazak, I.B., Vakulenko, K.V.: To the physical mechanism of destruction of metals. Probl. Mech. Eng. **5–6**(11), 87–96 (2008)
18. Larikov, L.N.: Healing of defects in metals. Scientific Thought, Kyiv (1980). (in Russian)
19. Ropyak, L.Y., Pryhorovska, T.O., Levchuk, K.H.: Analysis of materials and modern technologies for PDC drill bit manufacturing. Prog. Phys. Metals **21**(2), 274–301 (2020). https://doi.org/10.15407/ufm.21.02.274
20. Prysyazhnyuk, P., et al.: Development of the composite material and coatings based on niobium carbide. East. Eur. J. Enterp. Technol. **6**(12–96), 43–49 (2018). https://doi.org/10.15587/1729-4061.2018.150807
21. Shatskyi, I.P., Perepichka, V.V., Ropyak, L.Y.: On the influence of facing on strength of solids with surface defects. Metallofizika i Noveishie Tekhnologii **42**(1), 69–76 (2020). https://doi.org/10.15407/mfint.42.01.0069
22. Kalchenko, V.V., Yeroshenko, A.M., Boyko, S.V., Ignatenko, P.L.: Development and research of thermoplastic methods for hardening details. Naukovyi Visnyk Natsionalnoho Hirnychoho Universytetu **2**, 53–60 (2020)
23. Yushchenko, K.A., Borysov, Yu.S., Kuznetsov, V.D., Korzh, V, M.: Surface engineering. Scientific Tthought, Kyiv (2007). (in Russian)
24. Suslov, A.G., Bezyazychny, V.F., Panfilov, Yu.V., et al.: Surface engineering of details. Mashynostroenie, Moscow (2008). (in Russian)
25. Inoue, A., Hashimoto, K.: Amorphous and Nanocrystalline Materials. Preparation, Properties, and Applications. Springer, Heidelberg (2001). https://doi.org/10.1007/978-3-662-04426-1
26. Bartkowska, A.: Production and properties of FeB-Fe2B-Fe3(B, C) surface layers formed on tool steel using combination of diffusion and laser processing. Coatings **10**(11) (2020). https://doi.org/10.3390/coatings10111130
27. Korzhyk, V., Tyurin, Yu., Kolisnichenko, O.: Theory and practice of plasma-detonation technology of surface hardening metal products. PC Technology Center, Kharkiv (2021)
28. Lukaszkowicz, K.: Review of nanocomposite thin films and coatings deposited by PVD and CVD technology. In: Rahman, M. (ed.) Nanomaterials, pp. 145−163 (2011). https://doi.org/10.5772/25799

29. Fangyuan, Z., Chunzheng, D., Wei, S., Ju, K.: Effects of cutting conditions on the microstructure and residual stress of white and dark layers in cutting hardened steel. J. Mater. Process. Tech. **266**, 599–611 (2019)
30. Zhang, F., Duan, C., Wang, M., Sun, W.: White and dark layer formation mechanism in hard cutting of AISI52100 steel. J. Manuf. Process. **32**, 878–887 (2018). https://doi.org/10.1016/j.jmapro.2018.04.011
31. Vilaca, P.: Friction surfacing. In: Miranda, R. (ed.) Surface Modification by Solid State Processing, pp. 25–72. Elsevier, Cambridge (2014). https://doi.org/10.1533/9780857094698.25
32. Gurey, V., Hurey, I.: The effect of the hardened nanocrystalline surface layer on durability of guideways. In: Tonkonogyi, V., et al. (eds.) InterPartner 2019. LNME, pp. 63–72. Springer, Cham (2020). https://doi.org/10.1007/978-3-030-40724-7_7
33. Shaw, L.L.: The surface deformation and mechanical behavior in nanostructured alloys. In: Whang, S.H. (ed.) Nanostructured Metals and Alloys. Processing, Microstructure, Mechanical Properties and Applications, pp. 481–506. Woodhead Publishing, Cambridge (2011)
34. Gurey, V., Shynkarenko, H., Kuzio, I.: Mathematical model of the thermoelasticity of the surface layer of parts during discontinuous friction treatment. In: Ivanov, V., Pavlenko, I., Liaposhchenko, O., Machado, J., Edl, M. (eds.) DSMIE 2021. LNME, pp. 12–22. Springer, Cham (2021). https://doi.org/10.1007/978-3-030-77823-1_2
35. Kyryliv, V.I., Gurey, V.I., Maksymiv, O.V., Hurey, I.V., Kulyk, Yu.O.: Influence of the deformation mode on the force conditions of formation of the surface nanostructure of 40KH steel. Mater. Sci. **3**(57), 422–427 (2021). https://doi.org/10.1007/s11003-021-00556-9
36. Hurey, I., Hurey, T., Lanets, O., Dmyterko, P.: The durability of the nanocrystalline hardened layer during the fretting wear. In: Ivanov, V., Pavlenko, I., Liaposhchenko, O., Machado, J., Edl, M. (eds.) DSMIE 2021. LNME, pp. 23–32. Springer, Cham (2021). https://doi.org/10.1007/978-3-030-77823-1_3

New Complex Treatment to Ensure the Operational Properties of the Surface Layers of Machine Parts

Kateryna Kostyk[1](✉), Xinlei Chen[1], Viktoriia Kostyk[2], Oleg Akimov[1], and Yurii Shyrokyi[3]

[1] National Technical University "Kharkiv Polytechnic Institute", 2, Kyrpychova Street, Kharkiv 61002, Ukraine
Kateryna.Kostyk@khpi.edu.ua
[2] Donbas State Engineering Academy, 72, Akademichna Street, Kramatorsk 84313, Ukraine
[3] National Aerospace University named by N.Ye. Zhukovsky "KhAI", 17, Chkalov Street, Kharkiv 61070, Ukraine

Abstract. To obtain high-performance indicators of machine parts, the development of diffusion saturation of the surface with various atomic elements for the formation of a complex surface layer structure is currently relevant. A complex treatment was proposed to form a composite structure on the surface of steel parts, consisting of sequential cementation, nitrocementation, and boriding. The material of the study is 38Cr2MoAl steel. Charcoal with activators was used for cementation. Nitrocementation was carried out in a carbamide medium with activators. A mixture of amorphous boron with potassium tetrafluoroborate, boron nitride, and dolomite was used for boriding. It is established that the temperature of the complex treatment has the most direct effect on the change in the depth of the hardened steel layer. The depth of the boride layer increases with increasing temperature. This paper established the dependencies of the depths of the diffusion layers of steel on the temperature after complex processing. It was found that the greatest microhardness in boride is FeB (22 GPa), with the lowest microhardness in Fe_2B (18 GPa).

Keywords: Process innovation · Hardening complex treatment · Cementation · Nitrocementation · Boriding · Composite structure · Surface hardening · Diffusion layer

1 Introduction

It is relevant to improve the quality of machine parts and tools, which is manifested in increasing their service life. The solution to this problem can be to improve the surface's quality and properties by developing new methods of hardening the metal surface. Methods of thermal surface hardening are being developed with an emphasis on the hardening of metallic materials by warm shot blasting, warm laser impact shot blasting, and thermal ultrasonic surface hardening [1]. Combined thermodynamic effects in comparison

V. Tonkonogyi et al. (Eds.): InterPartner 2022, LNME, pp. 284–293, 2023.
https://doi.org/10.1007/978-3-031-16651-8_27

with conventional surface hardening methods aim to improve the mechanical properties of products by increasing the surface hardness, wear resistance, the thickness of layers, and fatigue durability [2]. However, they do not allow obtaining high indicators of surface hardness with high plasticity of the core of the part in combination with high operational properties of the surface layers. Thus, the demand for high-strength steels is increasing daily in all industries. In turn, this caused the need to improve various indicators of the surface layers. Modifying surface layers is the most versatile way to improve properties such as wear resistance and surface hardness [3]. Methods of surface modification, in which protective coatings are formed on steel surfaces, usually consist of carbides, nitrides [4], or carbonitrides [5]. However, they also do not allow obtaining high-performance indicators of metal products' performance properties [6]. Therefore, further developments of diffusive surface saturation with various atomic elements are currently relevant due to the complex effect on the surface layers of machine parts and tools [7].

2 Literature Review

In industry, the most common method of hardening the surface of parts is a thermochemical treatment to obtain the functional properties of products [8]. One of the popular methods of thermochemical treatment of steels and alloys is to obtain hardened surface layers by gas nitriding with carbonation and gas carbonitrization [9]. In order to change the microstructure and increase wear resistance after gas nitriding, laser modification is used [10] with and without remelting [11]. To change the tribological characteristics of medium-carbon steels, anodic plasma electrolytic nitrocementation in a carbamide electrolyte is used [12]. At the same time, it is also possible to increase the corrosion resistance of low-carbon steels [13] due to significant changes in its structure [14] during electrolyte-plasma saturation of the anode with nitrogen and carbon [15]. This method is accompanied by physicochemical features of diffusion saturation of structural steels [16].

Boriding in solid [17], liquid [18], and gaseous media [19] is an effective method of hardening various products and significantly improves not only the mechanical properties of steels but also the chemical ones. However, the heterogeneity of mechanical properties at the interface of the phases FeB-Fe_2B can cause brittle fractures during shear [20]. The reduction of the brittleness of the boron layer as a whole is achieved in the case of the use of concentrated energy sources, such as laser boriding [21], plasma arc welding [22], plasma electrolytic boriding [23], electric spark boron deposition [24].

The best way to increase the wear resistance of metal products is boriding [25], corrosion properties [26] of steel parts, and high-energy alloys [27] due to the formation of solid borides. A decrease in the coefficient of friction and, accordingly, less wear was shown by boron alloys that used nanoscale powders [28]. In the process of borating tool steel using a nanobore, an increase in the thickness of the boride layer [29] with a high hardness of the formed borides was observed. An increase in the temperature and duration of boriding leads to an increase in wear resistance. Liquid boriding of high-strength cast iron with spherical graphite is also effective in forming a boride layer and phases [30]. Evaluation of fatigue damage as a result of cyclic spherical contact of nitrided and borated steel gave the following data: the nitride layer without chips withstands applied mechanical loads, and the boride layer with a greater thickness has large cracks with moderate cracking [31]. It was also found that when using a dispersed powder mixture during nitriding of steel, a low micrograin is formed in the surface layer, which makes it possible to double the wear resistance of steels and, consequently, significantly increase the reliability and durability of products [32].

One of the effective ways to reduce the fragility of diffusion layers is the formation of layers with a composite structure. In addition to borides, it is proposed to form several additional phases, which are located in the layer arbitrarily or in an orderly manner. Chemically obtained nickel-boron coatings have wear resistance [33], comparable to a hard chrome coating [34] due to its various physicochemical properties. The presence of niobium with chromium and boron contributes to the formation of corrosion-resistant coatings [35]. Chromium-nickel nanocomposite coatings with carbon are used to obtain the necessary mechanical, electrochemical, and tribocorrosion properties [36]. However, interestingly, as the carbon component was added and, accordingly, increased, the adhesive strength in the resulting coatings gradually deteriorated. A comparison of various thermochemical coatings obtained after such treatments as boriding, titanization [37], and borotitanization [38] showed that the parts made of tool steel are resistant to wear and corrosion.

A two-stage thermochemical treatment with borotitan is proposed for a nickel-based superalloy, which improves wear resistance by almost ten times compared to an untreated alloy [39].

Recently, more and more attention has been paid to the use of inexpensive atomic saturation elements (carbon, nitrogen, and boron) and processes for obtaining hardened diffusion layers due to complex chemical and thermal treatment: boronitrotization [40], boron carburization [41], saturation with nitrogen, boron, and carbon [42].

3 Research Methodology

The material of the study is 38Cr2MoAl steel. The steel class is heat-resistant and relaxation-resistant. Nitrided parts made of this steel are widely used in industry, such as gears, rollers, bushings, etc., operating at temperatures up to 450°. Chemical composition of steel: 0.35–0.42% C, 1.35–1.65% Cr, 0.15–0.25% Mo, 0.7–1.1% Al, 0.2–0.45% Si, 0.3–0.6% Mn, up to 0.3% Ni, up to 0.3% Cu, up to 0.025% S, up to 0.025% P, ~95% Fe.

The hardening complex treatment consisted of cementation, nitrocementation, and boriding carried out sequentially. The modes are shown in Table 1.

Charcoal with activators was used for cementation. Nitrocementation was carried out in a carbamide medium with activators. A mixture of amorphous boron with potassium tetrafluoroborate, boron nitride, and dolomite was used for boriding.

For all treatments, heating was carried out in a chamber furnace of various durations at different temperatures according to the modes indicated in Table 1.

The final treatment was quenching from the boriding temperature with cooling in oil and temperature treatment at 200 °C with air cooling for partial relief of internal stresses.

Table 1. Modes of complex processing of alloy steel 38Cr2MoAl.

Mode number	Cementation	Nitrocementation	Boriding
1	Temperature 800 °C with a duration of 2 h	Temperature 550 °C with a duration of 5 h	Temperature 800 °C with a duration of 2 h
2	Temperature 850 °C with a duration of 2 h		Temperature 850 °C with a duration of 2 h
3	Temperature 900 °C with a duration of 2 h		Temperature 900 °C with a duration of 2 h
4	Temperature 950 °C with a duration of 2 h		Temperature 950 °C with a duration of 2 h

4 Results and Discussion

After complex treatment according to the modes indicated in Table 1, a composite structure of the reinforced layer is formed, which consists of successively arranged carbide, carbonitride, and boride layers (Fig. 1). Behind the boride layer is a layer of carbonitrides, which was formed after low-temperature nitrocementation. Compounds $Fe_2(N, C)$, $Fe_3(N, C)$, and $Fe_4(N, C)$ were formed in this transition zone. With low nitrogen saturation, light discharge in the form of particles or a grid can be constructed on the basis of cementite $Fe_3(N, C)$. The formation of cementite-based carbonitrides is facilitated not only by the high activity of the saturating medium but also by an increase in the process temperature. Nitrogen, which diffuses into steel along with carbon, significantly affects the degree of saturation of the surface layer with carbon and the depth of carbon diffusion. In the transition carbonitride zone (Fig. 1), the release of Fe_4N-phase nitride with a thickness of approximately 1–2 μ was observed. During low-temperature nitrocementation of 38Cr2MoAl steel, ε-phases with alloying elements are formed.

As you can see (Fig. 1), behind the ε-phase is a layer of carbonitrides, the depth of the layer of which varies depending on the temperature. Below the carbonitride layer is a carbide layer consisting of Fe_3C carbide and alloying element carbides.

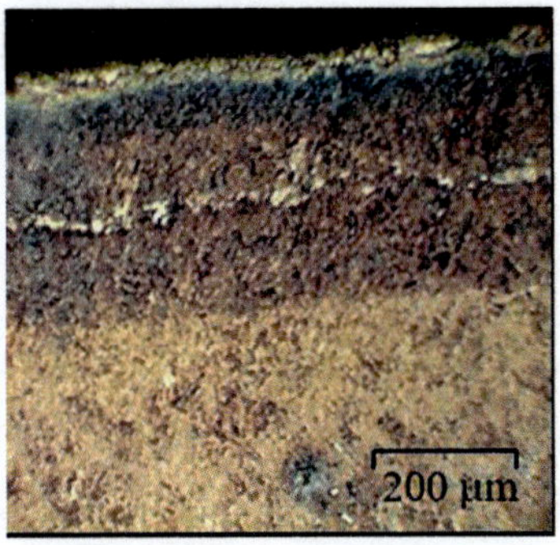

Fig. 1. Microstructure of the surface diffusion layer of structural alloy steel 38Cr2MoAl after complex processing according to mode No. 1.

Significant grinding of carbonitride and boride particles is observed with increasing processing temperature (Fig. 2). An increase in temperature to 950 °C leads to the formation of a larger boride layer and a transition zone, which is of great importance for reducing the brittleness of the solid surface layer of borides. Under the boride layer is a transition zone with a large number of dispersed borides, nitrides, carbides, and carbonitrides of iron and alloying elements.

Under the boride layer, when processed according to modes No. 3 and 4, there is a larger number of dispersed nitro- and carbon-boride particles, which appear to be small white dots in Fig. 2.

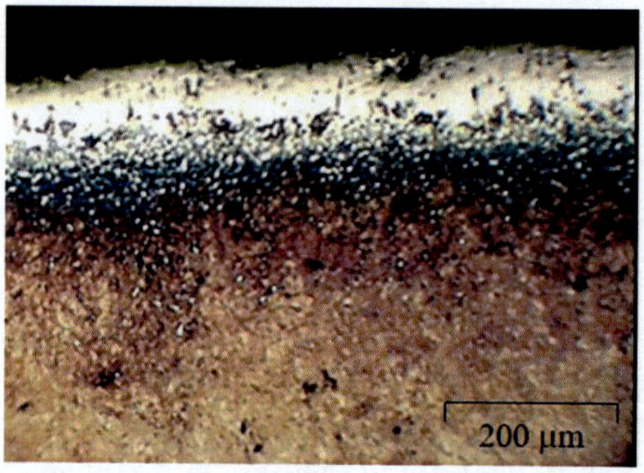

Fig. 2. Microstructures of the surface diffusion layer of steel 38Cr2MoAl after complex processing according to mode No. 3.

It was found that with the developed complex treatment, the temperature of diffusion saturation directly affects the formation of the depth of the diffusion layer in steel (Fig. 3).

The dependence of the depth of diffusion layers on the temperature in normalized form during various developed complex strengthening chemical and thermal treatments is shown in formulas (1–4), which were obtained by approximating experimentally obtained data with a second-degree polynomial.

The depth of the diffusion layer of borides during the final strengthening treatment, which consisted in atomic saturation of the surface layers with boron:

$$h = 0.0002t^2 + 0.078t - 186.5. \quad (1)$$

Depth of the reinforced diffusion layer obtained by low-temperature nitrocementation during complex processing:

$$h = 0.0025t^2 - 2.925t + 823.75. \quad (2)$$

Depth of the diffusion layer, which was obtained due to the technological mode of carburization:

$$h = 0.042t^2 - 61.06t + 22260. \quad (3)$$

The total depth of the composite diffusion layer was obtained after complex processing to improve the functional properties of products, which consisted of sequential saturation with atomic elements:

$$h = 0.04t^2 - 59.6t + 22275. \tag{4}$$

where h is the thickness of borides in normalized form; t is the boriding temperature in normalized form.

The dependence of the diffusion layers depth for 38Cr2MoAl steel on the temperature after complex processing is shown in Fig. 3.

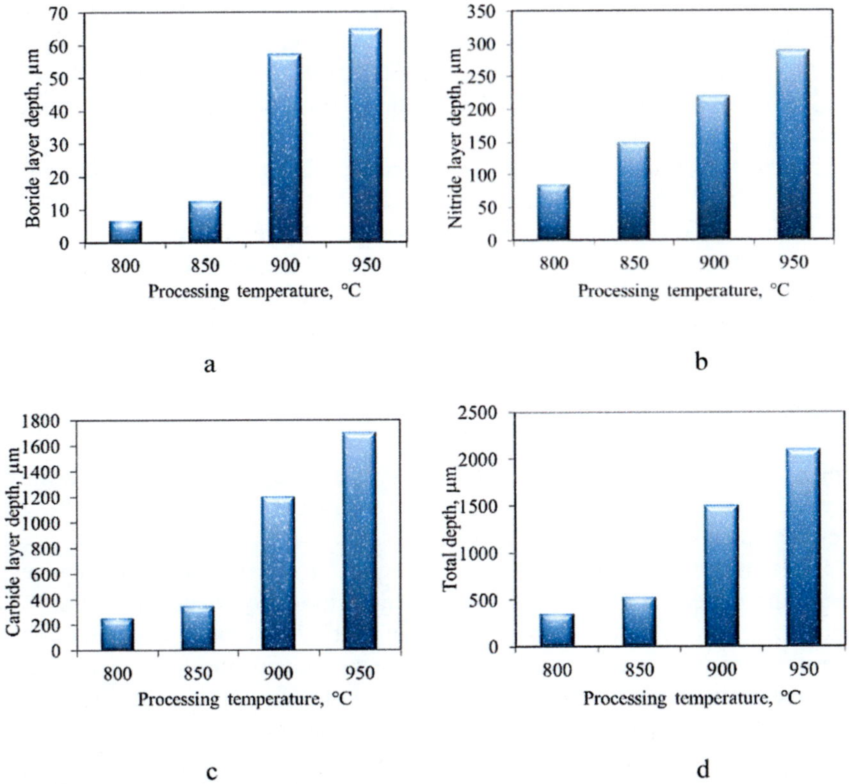

Fig. 3. Dependence of the depth of diffusion layers of 38Cr2MoAl steel after each chemical and thermal treatment and the total depth on temperature.

Since the depth of diffusion layers and transition zones differs during complex processing under different modes, microhardness distribution from the surface to the core also changes after each mode.

Boride has the highest microhardness in *FeB* (22 GPa), while Fe_2B has the lowest microhardness (18 GPa). The depth of the boride layer increases with increasing temperature. The depth of the boride layer at a temperature of 800 °C is 7 μ, at a temperature of 850 °C is 15 μ, and at temperatures of 900 and 950 °C is 55 and 65 μ, respectively (Fig. 3, *a*). A similar change in layer depths occurs with carbonite and carbide layers.

At a temperature of 800 °C, the depth of the carbonitride layer is 85 μ; at a temperature of 850 °C is 140 μ, and at temperatures of 900 and 950 °C, the depth of the carbonitride layer is 220 and 300 μ, respectively (Fig. 3, *b*). The depth of the carbide layer at a temperature of 800 °C is 250 μ, at a processing temperature of 850 °C is 350 μ, at surface hardening temperatures of 900 and 950 °C, the depth of the carbide layer is 1200 and 1700 μ (Fig. 3, *c*).

As it was found, the microhardness on all surfaces is equally high, but the depth of the transition layer is significantly different, which greatly affects the properties of steel. During complex treatment according to mode No 1, which consisted of carburization at a temperature of 800 °C for 2 h, nitrocementation at a temperature of 550 °C for 5 h, followed by boriding at a temperature of 800 °C for 2 h, and according to mode No 2 (Table 1), which consisted of cementation at 850 °C for 2 h, nitrocementation at 550 °C for 5 h, followed by boriding at 850 °C for 2 h, the transition zone is small, which can lead to chipping of the diffusion layer.

In complex treatment under mode No 3 (carburization at 900 °C for 2 h, nitrocementation at 550 °C for 5 h, followed by boriding at 900 °C for 2 h) and mode No 4 (carburization at 950 °C for 2 h, nitrocementation at 550 °C for 5 h, followed by boriding at 950 °C for 2 h), the transition zone is much larger. It improves the mechanical properties of steel – strength, hardness, endurance limit, yield strength, and impact strength. Anti-corrosion resistance and scuff resistance are also increased.

5 Conclusions

Complex treatments were developed, which consisted in consistently strengthening the surface layers of steels with atomic elements to ensure the operational properties of the final product at a high level.

Complex treatments of 38Cr2MoAl steel (sequential carburization, nitrocementation and boriding under various modes) create compositionality of the structure to form a boride layer with a transition zone containing nitrides, carbonitrides, and carbides. The resulting transition zone makes it possible to increase the functional properties of machine parts and tools by reducing the micro-fragility of the diffusion layer. A rational mode of complex processing was chosen, namely according to mode No 4, which includes cementation at a temperature of 950 °C for two hours, nitrocementation at a temperature of 550 °C for five hours, boriding at a temperature of 950 °C for two hours to obtain the highest surface hardness of twenty-two GPa with a maximum total diffusion layer of 2.1 mm.

Mathematical models of the formation of diffusion layers and changes in the boron diffusion coefficient after complex saturation with atomic elements are obtained, which make it possible to predict the necessary properties of the surface layers of alloy steel from the temperature of complex treatment.

Acknowledgment. The authors would like to acknowledge the financing of the National Research Foundation of Ukraine under grant agreement No. 2020.02/0119.

References

1. Liu, J., Ye, C., Dong, Y.: Recent development of thermally assisted surface hardening techniques: a review. Adv. Ind. Manuf. Eng. **2**, 100006 (2021). https://doi.org/10.1016/j.aime.2020.100006
2. Kusmanov, S.A., Silkin, S.A., Smirnov, A.A., Belkin, P.N.: Possibilities of increasing wear resistance of steel surface by plasma electrolytic treatment. Wear **386**, 239–246 (2017). https://doi.org/10.1016/j.wear.2016.12.053
3. Kusmanov, S.A., Dyakov, I.G., Kusmanova, Y.V., Belkin, P.N.: Surface modification of low-carbon steels by plasma electrolytic nitrocarburising. Plasma Chem. Plasma Process. **36**(5), 1271–1286 (2016). https://doi.org/10.1007/s11090-016-9724-3
4. Kusmanov, S.A., Smirnov, A.A., Silkin, S.A., Belkin, P.N.: Modification of low-alloy steel surface by plasma electrolytic nitriding. J. Mater. Eng. Perform. **25**(7), 2576–2582 (2016). https://doi.org/10.1007/s11665-016-2134-0
5. Belkin, P.N., Yerokhin, A., Kusmanov, S.A.: Plasma electrolytic saturation of steels with nitrogen and carbon. Surf. Coat. Technol. **307**, 1194–1218 (2016). https://doi.org/10.1016/j.surfcoat.2016.06.027
6. Ivanov, V., Pavlenko, I., Kuric, I., Kosov, M.: Mathematical modeling and numerical simulation of fixtures for fork-type parts manufacturing. In: Knapčíková, L., Balog, M. (eds.) Industry 4.0: Trends in Management of Intelligent Manufacturing Systems. EICC, pp. 133–142. Springer, Cham (2019). https://doi.org/10.1007/978-3-030-14011-3_12
7. Kostyk, K., et al.: Simulation of diffusion processes in chemical and thermal processing of machine parts. Processes **9**(4), 698 (2021). https://doi.org/10.3390/pr9040698
8. Pashechko, M., Józwik, J., Dziedzic, K., Karolus, M., Usydus, I.: Surface hardening of HS6-5-2 quick-cutting steel in the course of chemical thermal treatment. Mater. Sci. **52**(6), 834–840 (2017). https://doi.org/10.1007/s11003-017-0028-4
9. Panfil, D., Kulka, M., Wach, P., Michalski, J., Przestacki, D.: Nanomechanical properties of iron nitrides produced on 42CrMo4 steel by controlled gas nitriding and laser heat treatment. J. Alloy. Compd. **706**, 63–75 (2017). https://doi.org/10.1016/j.jallcom.2017.02.220
10. Alaa, F.I.I., Akimov, O., Kostyk, K.: Development of a combined technology for hardening the surface layer of steel 38Cr2MoAl. East Eur. J. Adv. Technol. **2**(11), 56–62 (2017). https://doi.org/10.15587/1729-4061.2017.100014
11. Kulka, M., Panfil, D., Michalski, J., Wach, P.: The effects of laser surface modification on the microstructure and properties of gas-nitrided 42CrMo4 steel. Opt. Laser Technol. **82**, 203–219 (2016). https://doi.org/10.1016/j.optlastec.2016.02.021
12. Mukhacheva, T.L., Belkin, P.N., Dyakov, I.G., Kusmanov, S.A.: Wear mechanism of medium carbon steel after its plasma electrolytic nitrocarburising. Wear **462**, 203516 (2020). https://doi.org/10.1016/j.wear.2020.203516
13. Kusmanov, S.A., Grishina, E.P., Belkin, P.N., Kusmanova, Y.V., Kudryakova, N.O.: Raising the corrosion resistance of low-carbon steels by electrolytic-plasma saturation with nitrogen and carbon. Met. Sci. Heat Treat. **59**(1–2), 117–123 (2017). https://doi.org/10.1007/s11041-017-0114-0
14. Kostyk, K.O., Kostyk, V.O., Kovalev, V.D.: Strengthening the surface layer of tools with state-of-the-art technologies. Prog. Phys. Met. **22**(1), 78–102 (2021)
15. Belkin, P.N., Kusmanov, S.A.: Plasma electrolytic nitriding of steels. J. Surf. Invest. **11**(4), 767–789 (2017). https://doi.org/10.1134/S1027451017020045
16. Kusmanov, S.A., Kusmanova, Y.V., Smirnov, A.A., Belkin, P.N.: Modification of steel surface by plasma electrolytic saturation with nitrogen and carbon. Mater. Chem. Phys. **175**, 164–171 (2016). https://doi.org/10.1016/j.matchemphys.2016.03.011

17. Campos-Silva, I., et al.: Evolution of boride layers during a diffusion annealing process. Surf. Coat. Technol. **309**, 155–163 (2017). https://doi.org/10.1016/j.surfcoat.2016.11.054
18. Kusmanov, S.A., Tambovskiy, I.V., Sevostyanova, V.S., Savushkina, S.V., Belkin, P.N.: Anode plasma electrolytic boriding of medium carbon steel. Surf. Coat. Technol. **291**, 334–341 (2016). https://doi.org/10.1016/j.surfcoat.2016.02.062
19. Delai, O., Xia, C., Shiqiang, L.: Growth kinetics of the FeB/Fe2B boride layer on the surface of 4Cr5MoSiV1 steel: experiments and modelling. J. Mater. Res. Technol. **11**, 1272–1280 (2021). https://doi.org/10.1016/j.jmrt.2021.01.109
20. Hernández-Ramírez, E.J., Guevara-Morales, A., Figueroa-López, U., Campos-Silva, I.: Wear resistance of diffusion annealed borided AISI 1018 steel. Mater. Lett. **277**, 128297 (2020). https://doi.org/10.1016/j.matlet.2020.128297
21. Kulka, M., Mikolajczak, D., Makuch, N., Dziarski, P., Miklaszewski, A.: Wear resistance improvement of austenitic 316L steel by laser alloying with boron. Surf. Coat. Technol. **291**, 292–313 (2016). https://doi.org/10.1016/j.surfcoat.2016.02.058
22. Sigolo, E., Soyama, J., Zepon, G., Kiminami, C.S., Botta, W.J., Bolfarini, C.: Wear resistant coatings of boron-modified stainless steels deposited by plasma transferred arc. Surf. Coat. Technol. **302**, 255–264 (2016). https://doi.org/10.1016/j.surfcoat.2016.06.023
23. Belkin, P.N., Kusmanov, S.A.: Plasma electrolytic boriding of steels and titanium alloys. Surf. Eng. Appl. Electrochem. **55**(1), 1–30 (2019). https://doi.org/10.3103/S106837551901006X
24. Wei, X., et al.: Facile preparation of nanocrystalline Fe2B coating by direct electro-spark deposition of coarse-grained Fe2B electrode material. J. Alloy. Compd. **717**, 31–40 (2017). https://doi.org/10.1016/j.jallcom.2017.05.081
25. Kondul, B., Cetin, M.H.: Increasing the wear resistance of railway switches with boron coating and analysis of tribological performance by ANOVA method. Wear **488**, 204132 (2022). https://doi.org/10.1016/j.wear.2021.204132
26. Sezgin, C.T., Hayat, F.: The effects of boriding process on tribological properties and corrosive behavior of a novel high manganese steel. J. Mater. Process. Technol., 117421 (2022). https://doi.org/10.1016/j.jmatprotec.2021.117421
27. Karakaş, M.S., et al.: Microstructure, some mechanical properties and tribocorrosion wear behavior of boronized $Al_{0.07}Co_{1.26}Cr_{1.80}Fe_{1.42}Mn_{1.35}Ni_{1.10}$ high entropy alloy. J. Alloy. Compd. **886**, 161222 (2021). https://doi.org/10.1016/j.jallcom.2021.161222
28. Günen, A.: Tribocorrosion behavior of boronized $Co_{1.19}Cr_{1.86}Fe_{1.30}Mn_{1.39}Ni_{1.05}Al_{0.17}B_{0.04}$ high entropy alloy. Surf. Coat. Technol., 127426 (2021). https://doi.org/10.1016/j.surfcoat.2021.127426
29. Erdoğan, A.: Investigation of high temperature dry sliding behavior of borided H13 hot work tool steel with nanoboron powder. Surf. Coat. Technol. **357**, 886–895 (2019). https://doi.org/10.1016/j.surfcoat.2018.10.066
30. Kul, M., Oskay, K.O., Temizkan, A., Karaca, B., Kumruoğlu, L.C., Topçu, B.: Effect of boronizing composition on boride layer of boronized GGG-60 ductile cast iron. Vacuum **126**, 80–83 (2016). https://doi.org/10.1016/j.vacuum.2016.01.021
31. Escobar-Hernández, J., Rodríguez-Castro, G.A., Arzate-Vázquez, I., Meneses-Amador, A., Morón, R.C.: Fatigue damage assessment from cyclic spherical contact of borided and nitrided H13 steel. Mater. Lett. **285**, 129118 (2021). https://doi.org/10.1016/j.matlet.2020.129118
32. Kostyk, K., Kostyk, V., Akimov, O., Kamchatna-Stepanova, K., Shyrokyi, Y.: Ensuring the high strength characteristics of the surface layers of steel products. In: Tonkonogyi, V., Ivanov, V., Trojanowska, J., Oborskyi, G., Pavlenko, I. (eds.) InterPartner 2021. LNME, pp. 292–301. Springer, Cham (2022). https://doi.org/10.1007/978-3-030-91327-4_29
33. Vitry, V., Hastir, J., Mégret, A., Yazdani, S., Yunacti, M., Bonin, L.: Recent advances in electroless nickel boron coatings. Surf. Coat. Technol. **429**, 127937 (2021). https://doi.org/10.1016/j.surfcoat.2021.127937

34. Lentz, J., Röttger, A., Großwendt, F., Theisen, W.: Enhancement of hardness, modulus and fracture toughness of the tetragonal (Fe, Cr) 2B and orthorhombic (Cr, Fe) 2B phases with addition of Cr. Mater. Des. **156**, 113–124 (2018). https://doi.org/10.1016/j.matdes.2018.06.040
35. Santana, D.A., et al.: Wear-resistant boride reinforced steel coatings produced by non-vacuum electron beam cladding. Surf. Coat. Technol. **386**, 125466 (2020). https://doi.org/10.1016/j.surfcoat.2020.125466
36. Xu, X., Sun, J., Xu, Z., Li, Z., Su, F.: Microstructure, electrochemical and tribocorrosion behaviors of CrCN nanocomposite coating with various carbon content. Surf. Coat. Technol. **411**, 126997 (2021). https://doi.org/10.1016/j.surfcoat.2021.126997
37. Hovorun, T., et al.: Improvement of the physical and mechanical properties of the cutting tool by applying wear-resistant coatings based on Ti, Al, Si, and N. J. Eng. Sci. **8**(2), C13–C23 (2021). https://doi.org/10.21272/jes.2021.8(2).c3
38. Karakaş, M.S.: Tribocorrosion behavior of surface-modified AISI D2 steel. Surf. Coat. Technol. **394**, 125884 (2020). https://doi.org/10.1016/j.surfcoat.2020.125884
39. Günen, A., et al.: Effect of borotitanizing on microstructure and wear behavior of Inconel 625. Surf. Coat. Technol. **311**, 374–382 (2017). https://doi.org/10.1016/j.surfcoat.2016.12.097
40. Gómez-Vargas, O.A., Solis-Romero, J., Figueroa-López, U., Ortiz-Domínguez, M., Oseguera-Peña, J., Neville, A.: Boro-nitriding coating on pure iron by powder-pack boriding and nitriding processes. Mater. Lett. **176**, 261–264 (2016). https://doi.org/10.1016/j.matlet.2016.04.135
41. Kusmanov, S.A., Tambovskii, I.V., Naumov, A.R., D'yakov, I.G., Kusmanova, I.A., Belkin, P.N.: Anodic electrolytic-plasma borocarburizing of low-carbon steel. Prot. Met. Phys. Chem. Surf. **53**(3), 488–494 (2017). https://doi.org/10.1134/S2070205117030121
42. Kusmanov, S.A., Silkin, S.A., Belkin, P.N.: Effect of plasma-electrolytic polishing on the corrosion resistance of structural steels after their anodic saturation with nitrogen, boron, and carbon. Russ. J. Electrochem. **56**, 356–364 (2020). https://doi.org/10.1134/S1023193520040084

Functional Evaluation of Surface Texture in Laser Selective Melted Inconel 718 Alloy Parts Processed by Shot Peening

Dmytro Lesyk[1,3,4](✉), Vitaliy Dzhemelinskyi[1], Silvia Martinez[2], Dariusz Grzesiak[4], and Bohdan Mordyuk[1,3]

[1] National Technical University of Ukraine "Igor Sikorsky Kyiv Polytechnic Institute", 37, Prosp. Peremohy, Kyiv 03056, Ukraine
lesyk_d@ukr.net

[2] University of the Basque Country, 48013 Bilbao, Spain

[3] G.V. Kurdyumov Institute for Metal Physics of the NAS of Ukraine, 36, Academician Vernadsky Boulevard, Kyiv 03142, Ukraine

[4] West Pomeranian University of Technology, 17, al. Piastów, 70310 Szczecin, Poland

Abstract. The test plane Inconel 718 alloy specimens were fabricated using a selective laser melting (SLM) technique. The SLM-manufactured plane parts are characterized by excessive surface waviness and surface roughness. The post-printing surface modification, such as shot peening (SP), was applied to eliminate the surface defects and increase the surface hardness in the Inconel 718 alloy parts, producing a near-surface compressive residual stress. The peening intensity magnitudes ranged from 6 to 10 C Almen, providing a full surface coverage. The surface texture, roughness, and waviness parameters, as well as the functionality-related surface parameters for the SLM-fabricated and SP-treated lateral surfaces, were analyzed and compared. The effect of the peening pressure on the macro-hardness/microhardness and hardening intensity of the SLM-built parts was also examined. The results showed that a further increase in the peening pressure (0.6 MPa) resulted in higher area roughness parameters, increasing a significant maximum pit height Sp parameter. Compared to the SLM-fabricated part, the optimized SP treatment induced an increase in the hardening intensity by ~70%, providing a hardening depth of ~300 μm. The formed wavy/chaotic surface texture without surface defects may be appreciable to enhance the corrosion and wear behavior in the SLM-fabricated final parts.

Keywords: Process innovation · Laser powder bed fusion · Shot peening · Roughness · Waviness · Skewness · Kurtosis · Hardening intensity

1 Introduction

The laser powder bed fusion technique is an advanced additive manufacturing technology that contributes to forming complexly-shaped metal components from a digital three-dimensional (3D) design. Despite the vast potential of the laser powder bed

V. Tonkonogyi et al. (Eds.): InterPartner 2022, LNME, pp. 294–305, 2023.
https://doi.org/10.1007/978-3-031-16651-8_28

fusion technology, post-printing surface modification is mostly required to eliminate the surface defects and increase the mechanical properties in the 3D-printed metal parts, including nickel-based alloys. The corrosion-resistant nickel-based alloys are applied for manufacturing the safe parts, which work in extreme environments subjected to high mechanical loads. The mechanical surface treatment is usually applied to enhance the functionality-related properties and surface integrity of metallic parts manufactured by laser 3D printing techniques. Therefore, improving surface properties in the 3D-printed superalloys is a crucial issue in the additive manufacturing industry.

2 Literature Review

Inconel (IN) alloy products manufactured by powder bed fusion 3D printing technologies contain good strength and high-temperature performance, making superalloys an attractive choice for critical applications. Today, the complexly shaped and lightweight nickel superalloy end-use parts printed by the laser powder bed fusion or selective laser melting (SLM) [1] and electron beam melting (EBM) [2] techniques are promising for the aerospace, aviation, automotive, and nuclear industries. At the same time, the powder bed fusion technologies reduce material usage and have lower environmental impacts in comparison with traditional manufacturing methods.

During the SLM process, thin layers of fine metallic powder are melted along the build direction by a scanned high-power laser beam in an atmosphere of inert gases, forming solid metal components [3]. Currently, several alloys are successfully printed by the SLM process [4]. In particular, the SLM method is used to manufacture small-sized and/or highly complexly shaped metal products in the aerospace [5] and medical [6] industries.

It should also be noted that there are challenges in fabricating fully dense components using some powders. It is well-known that an inhomogeneous dendritic structure [7] and a rough surface [8] are formed in the SLM-built metal parts. Moreover, excessive residual porosity occurs in the sub-surface layer [9].

On the one hand, some defects in the 3D-printed parts can be reduced by the SLM process settings (such as scanning strategy [10], powder variation [4], etc.). Nevertheless, the mechanical or chemical surface post-treatments combined with the hot isostatic pressing and/or heat treatments are frequently required to improve surface properties in the final metal parts printed by SLM. Recently, various mechanical processes for surface treatment, such as ultrasonic peening or finishing [11], sandblasting [12] or shot peening (SP) [13], water jet shot peening [14], water jet cavitation peening [15], laser shock peening [16], laser polishing [17], barrel/vibratory finishing [18], were used for surface modification of the SLM-fabricated metal components.

The above-mentioned surface modification post-treatments for additive manufacturing can reduce the sub-surface porosity [19] and cracking [20], changing the surface texture coupled with functionality-related surface roughness characteristics [21]. The formed surface texture/relief, waviness, and structure strongly influence the wear/corrosion resistance and endurance limit of the SLM-printed metallic end-use parts. In particular, Balbaa et al. [22] reported that compared to the SLM-fabricated sample, the fatigue life of the IN 625 and IN 718 nickel-based alloy samples was respectively increased by 110% and 105% after severe SP due to the compressive residual stresses formation and surface roughness reduction. As a result, surface texture characterization is indispensable in evaluating the functionality-related properties of mechanical post-processed surfaces [23]. The evaluation of the functionality-related performance with surface roughness parameters of the SP-processed superalloy parts manufactured by the SLM technique is virtually absent.

This study aimed to study the surface morphology/texture, roughness, waviness, macrohardness, microhardness, and hardening intensity of the SLM-printed IN 718 alloy parts peened by severe shot peening. Particular attention was paid to comparing the functionality-related surface texture and roughness parameters with different SP parameters.

3 Research Methodology

The plane Inconel 718 superalloy parts with base dimensions of about 3.5 mm x 30 mm x 40 mm were printed by SLM with a build angle of 90° using a powder with a particle size of 15–45 μm. The nominal chemical composition of the IN 718 powder (weight %) is as follows: 17–21% Cr, 18–20% Fe, 4.75–5.5% Nb, 2.8–3.3% Mo, 0.65–1.2% Ti, 0.2–0.8% Al, <0.35% Si, <0.35% Mn, <0.3% Cu, <0.015% P, <0.015% S, 0.02–0.08% C, and balance Ni.

The Renishaw AM400 laser system with a 400 W fiber laser was applied to fabricate the test parts [24]. The IN 718 alloy parts were fabricated with a powder layer thickness of 60 μm, a laser power of 200 W, a scanning speed of 700 mm/s, a laser spot size of 70 μm, and a distance between scanned lines of 90 μm. At the same time, a 5 mm width stripes scanning strategy was implemented during SLM [25].

The severe shot peening (SP) was carried out using industrial equipment [13]. The SP process was carried out at a treatment duration of 60 s, with peening pressure of 0.4 MPa (SP1 regime), 0.5 MPa (SP2 regime), and 0.6 MPa (SP3 regime).

The surface morphology of the SLM-printed and SP-peened IN 718 alloy samples was evaluated and compared using an SEM-106I scanning electron microscope (SEM) and Taylor Hobson Form Talysurf 120 profilometer. The area roughness parameters (Sa, Sp, Sv, Sq, and Sz) were evaluated according to ISO 25178. The surface roughness (Ra and Rt) and waviness (Wa, Wp, Wv, Wq, Wz, and Wt) profile parameters were also analyzed according to ISO 4287 standard [26]. The functionality-related surface roughness/waviness parameters (Rpk, Rk, Rvk, Rsk/Ssk/Wsk, and Rku/Wku) were estimated to predict the wear [27] and corrosion [28] behavior after the SP post-processing.

The surface macrohardness (HV_{10}) and near-surface microhardness ($HV_{0.025}$) were measured using a Dia-Testor 2Rc and Leica VMHT tester, respectively. The intensity under surface hardening ($I_{hard.}$) was evaluated by the macrohardness magnitudes of the SLM-printed and SP-peened specimens by the equation:

$$I_{hard.} = \frac{I_{HV}^{peen.} - I_{HV}^{print.}}{I_{HV}^{print.}}, \tag{1}$$

In all cases, a total of 6–8 measurements were performed, and the averaged magnitudes were presented.

4 Results and Discussion

4.1 Surface Morphology

The surface morphology/texture coupled with two-dimensional profiles of the SLM-printed and SP-treated side surfaces is presented in Figs. 1 and 2. The severe plastic strain induced by the applied SP treatment promoted the irregular texture/microrelief formation on the peened surface.

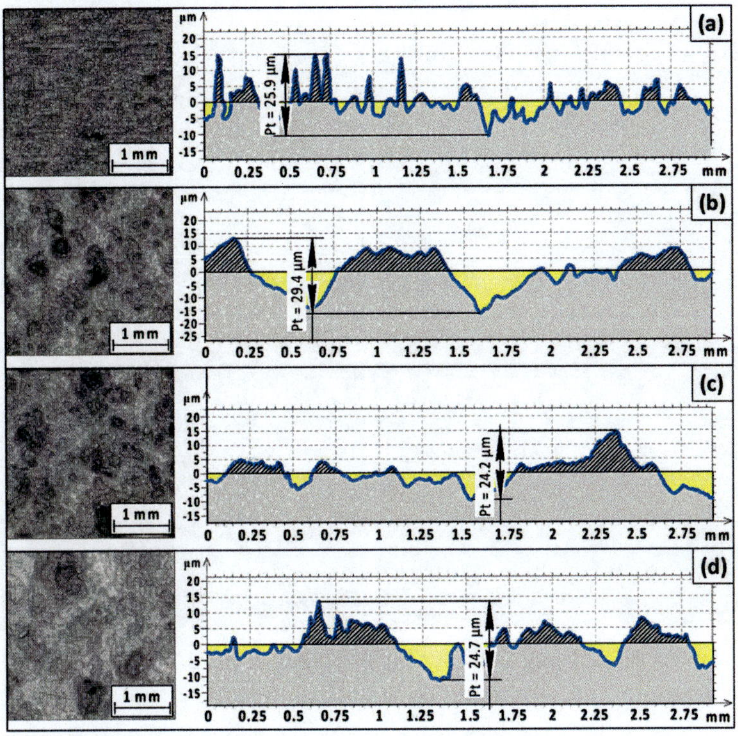

Fig. 1. Surface texture and profiles in the SLM-fabricated IN 718 alloy parts in the initial state (a) and SP1 (b), SP2 (c), and SP3 (d) post-processed samples.

Compared to the selective laser melted nickel-based alloy parts (*Sa* = ~4.5 µm, *Sz* = 71.3 µm), the average arithmetic mean height *Sa* and the maximum height *Sz* parameters of the surface texture are lower after the SP1 (*Sa* = 3.86 µm, *Sz* = 38.4 µm) and SP2 (*Sa* = 4.2 µm, *Sz* = 39.1 µm) regimes (Fig. 3). However, a further increase in the peening pressure resulted in higher *Sa* and *Sz* area roughness parameters, increasing a significant maximum pit height *Sp* value. It should also be noted that the values of the valleys and peaks height of the surface texture are the most homogeneous after the SP2 treatment (Fig. 1 and Fig. 3).

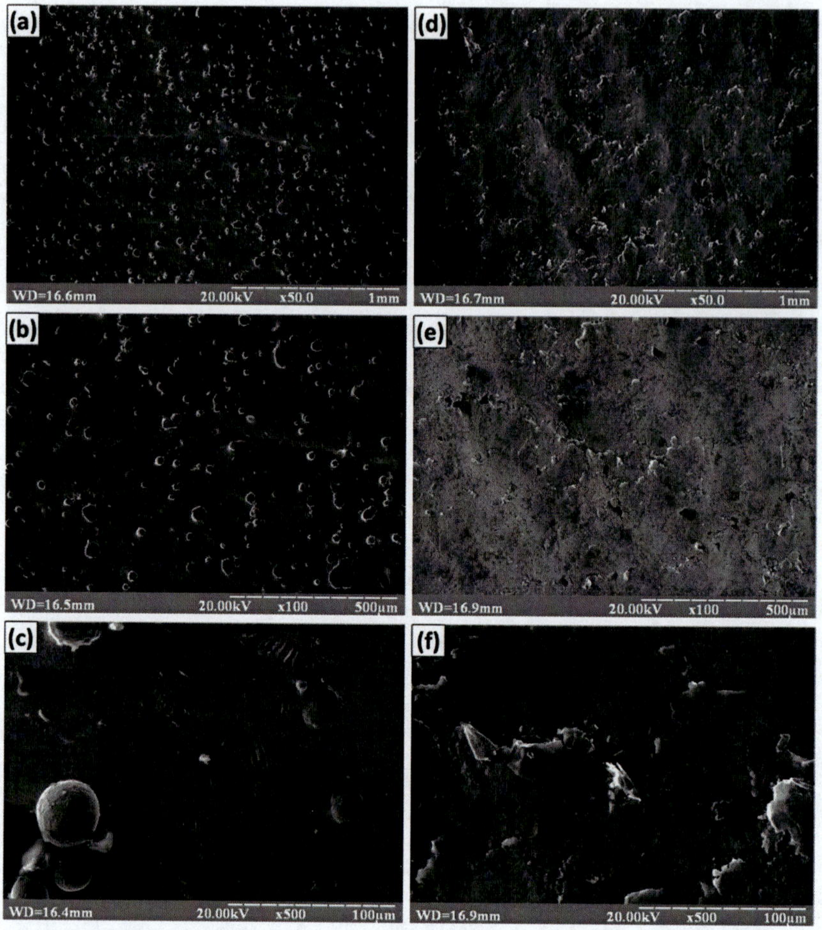

Fig. 2. Surface morphology of the SLM-fabricated (a–c) and SP2-treated (d–f) samples.

Figure 2 confirms that the cold working process removed the defects from the surface. This observation agrees with the conclusions of the number of works [11, 14, 29], where the partially melted or spattering powder particles were demolished and eliminated by the severe shot peening. Moreover, the shot peening process reduces the residual porosity by 15–30% in the near-surface layers [19].

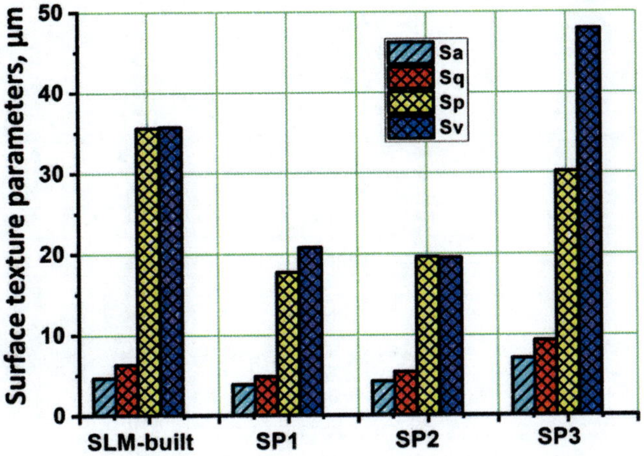

Fig. 3. Surface texture parameters of the SLM-fabricated and SP-treated IN 718 alloy samples.

4.2 Surface Waviness and Bearing Ratio Parameters

The SLM-manufactured side surfaces are characterized by excessive surface waviness (Table 1) and surface roughness (*Ra* = ~4.0 µm, *Rt* = ~32.0 µm) (Fig. 4a) [24].

The average surface waviness parameters are summarized in Table 1. The shot peening led to an increase in surface waviness parameters regardless of the used SP regimes. Increasing the peening pressure value caused an increase in the surface waviness parameters. The SP3 process induced a double increase in the *Wa* parameter compared to the SLM-fabricated part (Table 1).

The SP post-processing analyzed the Abbott-Firestone/bearing ratio curves of the SLM-fabricated superalloy parts to obtain more detailed surface roughness data (Fig. 4 and Fig. 5). The bearing ratio parameters are given in Fig. 5a.

Table 1. Surface waviness parameters.

Regime number	Wa (µm)	Wq (µm)	Wp (µm)	Wv (µm)	Wz (µm)	Wt (µm)
SLM-built	2.075	2.40	2.420	2.652	5.072	7.705
SP1	2.418	2.721	3.448	3.210	6.658	8.837
SP2	2.497	2.931	3.570	3.194	6.761	12.151
SP3	4.390	5.108	4.665	6.812	11.347	18.072

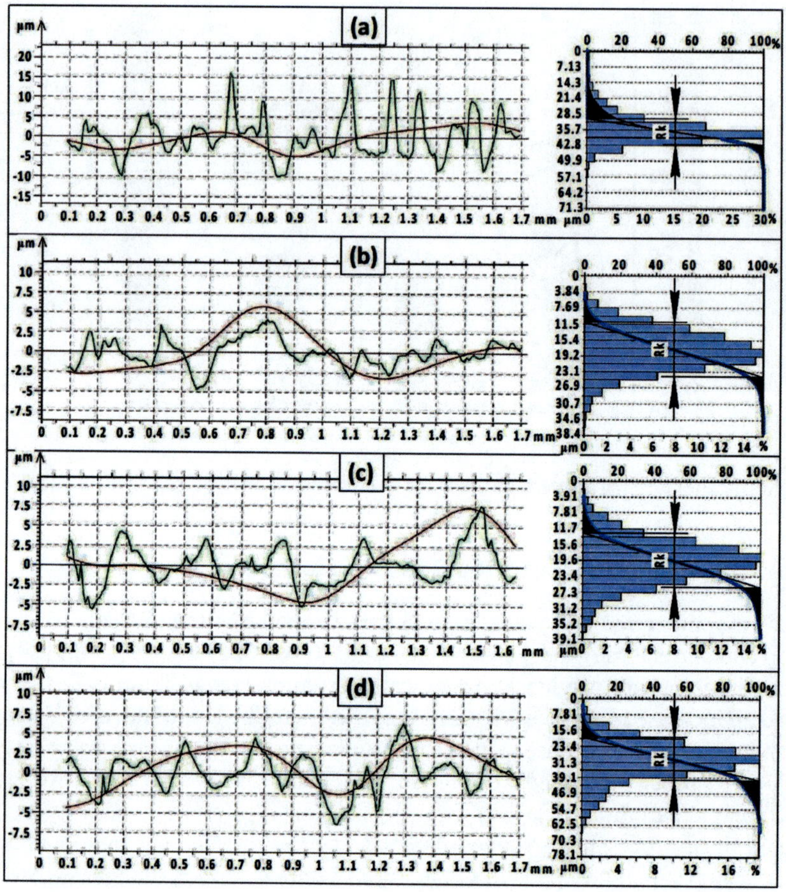

Fig. 4. Surface roughness/waviness profiles and bearing area curves in the SLM-fabricated IN 718 alloy parts in the initial state (a) and SP1 (b), SP2 (c), and SP3 (d) post-processed samples.

The lowest magnitudes of *Rpk*, *Rk*, and *Rvk* bearing ratio parameters were obtained for the SP1-treated samples. The SP1 regime is expected to be more efficient in promoting a quicker formation of the equilibrium roughness because the profile running-in stage becomes shorter (Fig. 5). Further increase in the peening pressure led to higher *Rk* parameter values, increasing the *Rpk* parameter values. Moreover, the SP2-treated surface texture is evidently more wear-resistant than the SLM-fabricated samples (Fig. 5b).

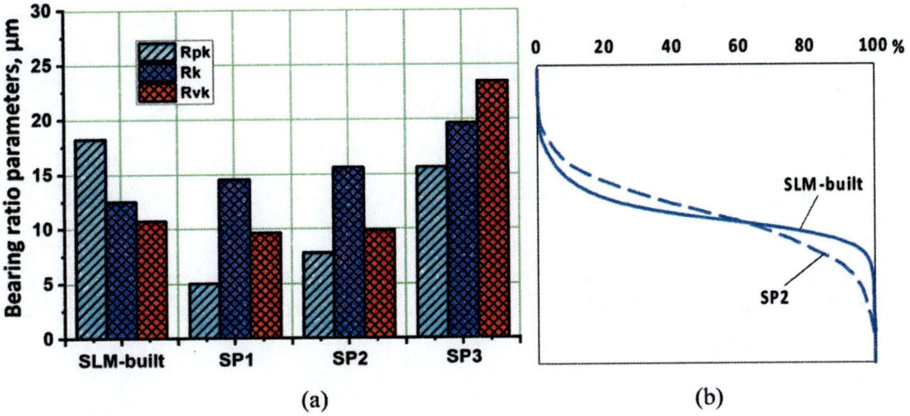

Fig. 5. Bearing ratio parameters (a) and material ratio curves (b) of the SLM-fabricated and SP-treated IN 718 alloy samples.

4.3 Skewness and Kurtosis Assessment

Analyzing the data presented in Table 2, it may be stated that the SP-treated surface is characterized by a slight predominance of rounded peaks (negative *Rsk*/*Ssk* parameter values). In contrast, the unprocessed SLM-fabricated surface contains comparatively more sharp peaks and valleys (positive *Rsk*/*Ssk* parameter values).

Additionally, the skewness magnitudes are lower than zero, indicating the surface texture after the SP treatment includes a large number of deep valleys and a small number of peaks [30]. Increasing the SP peening intensity caused an increase in the deep valley values (Fig. 4, Tables 1 and 2).

Table 2. Skewness and kurtosis parameters of roughness and waviness profiles.

Regime name	Rsk (µm)	Ssk (µm)	Rku (µm)	Wsk (µm)	Wku (µm)
SLM-built	1.099	1.18	4.596	0.157	1.72
SP1	−0.053	−0.143	2.560	−0.231	2.481
SP2	−0.094	−0.152	2.635	−0.311	2.610
SP3	−0.113	−0.485	2.670	−0.649	2.723

The average magnitudes of the kurtosis parameter, which describes the probability of flattening the profile for the SLM-built side surfaces ($Rku = 4.596$ µm, $Sku = 5.43$ µm > 3), indicates that the side surface is characterized by occasional sharp peaks and valleys (Fig. 4). As a result, the SLM-built parts may suffer from the uniform and pitting corrosion [20, 31, 32]. Contrary, the values of *Rku* parameters are within the range 2.56–2.67 µm < 3 after the shot peening. The kurtosis values after the SP process are correlated well with earlier works (*Rku* parameters very close to 3). The *Wku* values also showed a similar trend (Table 2).

According to the obtained result (*Rsk* < 0 and *Rku* < 3), the micropitting (pitting corrosion) may be avoided in the SML-built surface textures formed by the SP post-processing. At the same time, the SP-formed chaotic surface texture should also be evaluated and compared using a non-contact surface metrology technique.

4.4 Hardness and Hardening Intensity

The sub-surface microhardness was analyzed at a depth of ~25 μm from the top surface. The cross-section microhardness values were respectively achieved by ~560 $HV_{0.025}$, ~650 $HV_{0.025}$, and ~630 $HV_{0.025}$ after SP1, SP2, and SP3, providing the hardening depth of 220–330 μm (Fig. 6).

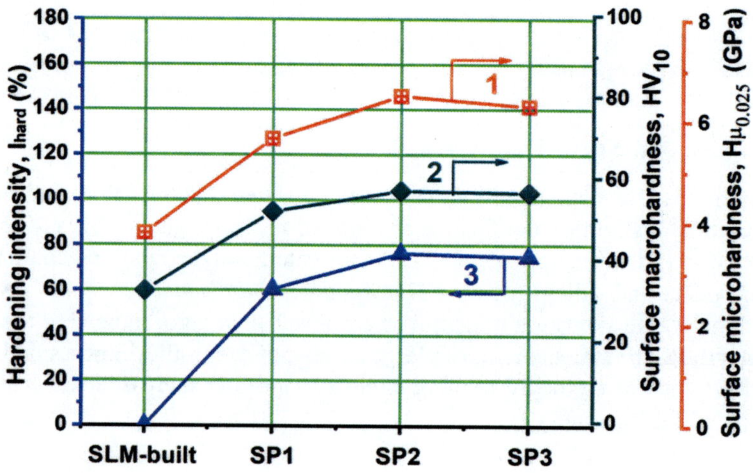

Fig. 6. Surface hardness and hardening intensity of the SLM-fabricated and SP-treated IN 718 alloy samples.

The hardening depth is ~300 μm after the SP2 treatment. The SP treatment of the used peening pressure demonstrates a saturation of the strain-induced hardening of the surface at the levels of ~56 HV_{10}/~650 $HV_{0.025}$ (Fig. 6). The surface macrohardness (marked by red color) correlates well with a cross-section microhardness (marked by dark cyan color).

The hardening intensity (marked by blue color) was increased by ~75% after the SP2 treatment compared to the SLM-built surface (~32 HV_{10}). It is attributed to grain size refinement and dislocation density increase [19, 22, 29]. Based on the results of hardness and hardening intensity (Fig. 6), it can be concluded that the SP2 regime is recommended for the surface treatment of the SLM-printed superalloy parts.

The results showed that the SP treatment induced the specific texture on the surface and enhanced the surface hardness could be beneficial for the wear/corrosion resistance and prolonged fatigue life of the SLM-printed IN 718 alloy components.

5 Conclusions

The shot peening surface treatment was applied to enhance the nickel superalloy parts printed by the SLM process. The effects of the SP surface mechanical treatment on the surface texture, roughness, waviness, functional parameters, and hardening intensity were studied. The results of this work allow drawing the following conclusions.

The SLM-fabricated flat parts are characterized by a rough surface (*Ra*/*Sa* is ~4.0 μm and *Wa* is ~2.0 μm), containing high peaks and deep valleys.

Compared to the untreated SLM-built IN 718 alloy parts, the average arithmetic mean height *Sa* parameter of the surface texture is lower after the SP1 (Sa = ~3.9 μm) and SP2 (Sa = ~4.2 μm) regimes and higher after SP3 (Sa = ~7.1 μm). The optimized SP treatment formed the wavy surface texture in the SLM-fabricated parts with a large number of deep valleys and a small number of peaks. Unlike the SLM-fabricated surface texture, the formed surface texture without surface defects can avoid pitting corrosion and enhance the wear-resistant of the IN 718 alloy.

The surface macrohardness values were respectively achieved by 51.7 HV_{10}, 56.8 HV_{10}, and 56.3 HV_{10} after SP1, SP2, and SP3. The SP-treated surface showed 70–80% higher hardening intensity than the SLM-fabricated surface, providing a hardening depth of about 300 μm under optimized conditions.

The study of the corrosion behavior of the SLM-fabricated and SP-treated IN 718 alloy is planned for further research.

Acknowledgments. This study was financially supported by the DAAD program (Grant Number 91695844). Partial support by the Ulam NAWA program (Grant Number BPN/ULM/2021/1/00153) is also acknowledged.

References

1. Yap, C.Y., et al.: Review of selective laser melting: materials and applications. Appl. Phys. Rev. **2**, 041101 (2015). https://doi.org/10.1063/1.4935926
2. Wang, X., et al.: Review on powder-bed laser additive manufacturing of Inconel 718 parts. Proc. IMechE Part B J. Eng. Manuf. **231**, 1890–1903 (2016). https://doi.org/10.1177/0954405415619883
3. Spears, T.G., Gold, S.A.: In-process sensing in selective laser melting (SLM) additive manufacturing. Integr. Mater. Manuf. Innov. **5**, 16–40 (2016). https://doi.org/10.1186/s40192-016-0045-4
4. Vock, S., Klöden, B., Kirchner, A., Weißgärber, T., Kieback, B.: Powders for powder bed fusion: a review. Prog. Addit. Manuf. **4**, 383–397 (2019). https://doi.org/10.1007/s40964-019-00078-6
5. Seabra, M., et al.: Selective laser melting (SLM) and topology optimization for lighter aerospace components. Procedia Struct. Integr. **1**, 289–296 (2016). https://doi.org/10.1016/j.prostr.2016.02.039
6. La Fé Perdomo, I., et al.: Selective laser melting: lessons from medical devices industry and other applications. Rapid Prototyp. J. **27**, 1801–1830 (2021). https://doi.org/10.1108/RPJ-07-2020-0151
7. Wang, X., et al.: Microstructure and yield strength of SLM-fabricated CM247LC Ni-Superalloy. Acta Mater. **128**, 87–95 (2017). https://doi.org/10.1016/j.actamat.2017.02.007

8. Tian, Y., et al.: Influences of processing parameters on surface roughness of Hastelloy X produced by selective laser melting. Addit. Manuf. J. **13**, 103–112 (2017). https://doi.org/10.1016/j.addma.2016.10.010
9. Karimi, P., Raza, T., Andersson, J., Svensson, L.-E.: Influence of laser exposure time and point distance on 75-μm-thick layer of selective laser melted Alloy 718. Int. J. Adv. Manuf. Technol. **94**(5–8), 2199–2207 (2017). https://doi.org/10.1007/s00170-017-1019-1
10. Serrano-Munoz, I., et al.: The residual stress in as-built Laser Powder Bed Fusion IN718 alloy as a consequence of the scanning strategy induced microstructure. Sci. Rep. **10**, 14645 (2020). https://doi.org/10.1038/s41598-020-71112-9
11. Gale, J., et al.: Application of ultrasonic peening during DMLS production of 316L stainless steel and its effect on material behavior. Rapid Prototyp. **23**, 1185–1194 (2017). https://doi.org/10.1108/RPJ-09-2016-0140
12. Löber, L., et al.: Comparison of different post processing technologies for SLM generated 316l steel parts. Rapid Prototyp. **19**, 173–179 (2013). https://doi.org/10.1108/13552541311312166
13. Lesyk, D., Martinez, S., Pedash, O., Dzhemelinskyi, V., Mordyuk, B.: Combined thermo-mechanical techniques for post-processing of the SLM-printed Ni-Cr-Fe alloy parts. In: Ivanov, V., Trojanowska, J., Pavlenko, I., Zajac, J., Peraković, D. (eds.) DSMIE 2020. LNME, pp. 295–304. Springer, Cham (2020). https://doi.org/10.1007/978-3-030-50794-7_29
14. Soyama, H., Okura, Y.: The use of various peening methods to improve the fatigue strength of titanium alloy Ti6Al4V manufactured by electron beam melting. Mater. Sci. **5**, 1000–1015 (2018). https://doi.org/10.3934/matersci.2018.5.1000
15. Sasaki, H., et al.: Cavitation erosion resistance of the titanium alloy Ti–6Al–4V manufactured through additive manufacturing with various peening methods. Wear **462–463**, 203518 (2020). https://doi.org/10.1016/j.wear.2020.203518
16. Lan, L., et al.: Effects of laser shock peening on microstructure and properties of Ti–6Al–4V titanium alloy fabricated via selective laser melting. Materials **13**, 3261 (2020). https://doi.org/10.3390/ma13153261
17. Boban, J., et al.: Polishing of additive manufactured metallic components: retrospect on existing methods and future prospects. Int. J. Adv. Manuf. Technol. **121**, 83–125 (2022). https://doi.org/10.1007/s00170-022-09382-y
18. Khorasani, M., et al.: On the role of wet abrasive centrifugal barrel finishing on surface enhancement and material removal rate of LPBF stainless steel 316L. J. Manuf. Process. **59**, 523–534 (2020). https://doi.org/10.1016/j.jmapro.2020.09.058
19. Damon, J., et al.: Process dependent porosity and the influence of shot peening on porosity morphology regarding selective laser melted AlSi10Mg parts. Addit. Manuf. **20**, 77–89 (2018). https://doi.org/10.1016/j.addma.2018.01.001
20. Volpato, G., et al.: A comprehensive literature review on laser powder bed fusion of Inconel superalloys. Addit. Manuf. **55**, 102871 (2022). https://doi.org/10.1016/j.addma.2022.102871
21. Boschetto, A., et al.: Post-processing of complex SLM parts by barrel finishing. Appl. Sci. **10**, 1382 (2020). https://doi.org/10.3390/app10041382
22. Balbaa, M., et al.: Improvement of fatigue performance of laser powder bed fusion fabricated IN625 and IN718 superalloys via shot peening. J. Mater. Process. Technol. **304**, 117571 (2022). https://doi.org/10.1016/j.jmatprotec.2022.117571
23. Zeng, Q., et al.: Correlating and evaluating the functionality-related properties with surface texture parameters and specific characteristics of machined components. Int. J. Mech. Sci. **149**, 62–72 (2018). https://doi.org/10.1016/j.ijmecsci.2018.09.044
24. Lesyk, D., Martinez, S., Mordyuk, B., Dzhemelinskyi, V., Lamikiz, A.: Surface finishing of complexly shaped parts fabricated by selective laser melting. In: Tonkonogyi, V., et al. (eds.) InterPartner 2019. LNME, pp. 186–195. Springer, Cham (2020). https://doi.org/10.1007/978-3-030-40724-7_19

25. Lesyk, D., Martinez, S., Dzhemelinkyi, V., Lamikiz, A.: Additive manufacturing of the super-alloy turbine blades by selective laser melting: surface quality, microstructure and porosity. In: Karabegović, I. (ed.) NT 2020. LNNS, vol. 128, pp. 267–275. Springer, Cham (2020). https://doi.org/10.1007/978-3-030-46817-0_30
26. Astaraee, A.H., et al.: Evaluating the homogeneity of surface features induced by impact-based surface treatments. Materials **14**, 3476 (2021). https://doi.org/10.3390/ma14133476
27. Peyrac, C., et al.: Shot peening for surface topography optimization to avoid micro pitting. In: Proceedings of Materials of International Conference on Shot Peening – ICSP-13, 2017049 (2017)
28. Ba, E.C.T., et al.: Investigation of the effects of skewness R_{sk} and kurtosis R_{ku} on tribological behavior in a pin-on-disc test of surfaces machined by conventional milling and turning processes. Mater. Res. **24**, e20200435 (2021). https://doi.org/10.1590/1980-5373-MR-2020-0435
29. Maamoun, A.H., et al.: Influence of shot peening on AlSi10Mg parts fabricated by additive manufacturing. Manuf. Mater. Process. **2**, 40–56 (2018). https://doi.org/10.3390/jmmp2030040
30. Kozior, T., et al.: Waviness of freeform surface characterizations from austenitic stainless steel (316L) manufactured by 3D printing-selective laser melting (SLM) technology. Materials **13**, 4372 (2020). https://doi.org/10.3390/ma13194372
31. Tang, Y., et al.: Corrosion behavior of a selective laser melted Inconel 718 alloy in a 3.5 wt.% NaCl solution. J. Mater. Eng. Perform. **30**, 5506–5514 (2021). https://doi.org/10.1007/s11665-021-05909-8
32. Luo, S., et al.: Microstructural evolution and corrosion behaviors of Inconel 718 alloy produced by selective laser melting following different heat treatments. Addit. Manuf. **30**, 100875 (2019). https://doi.org/10.1016/j.addma.2019.100875

Investigation of the Corrosion Resistance of Porous Permeable Materials with Protective Coatings

Oleksandr Povstyanoy[1], Natalia Imbirovych[1](✉), Volodymyr Posuvailo[2], Oleg Zabolotnyi[1], and Tatyana Artyukh[3]

[1] Lutsk National Technical University, 75, Lvivska Street, Lutsk 43018, Ukraine
n.imbirovych@lntu.edu.ua

[2] Physico-Mechanical Institute H.V. Karpenka NAS Ukraine, 5, Naukova Street, Lviv 79060, Ukraine

[3] Kyiv National Taras Shevchenko University, 90A, Vasilkivska Street, Kyiv 03022, Ukraine

Abstract. It is known that porous permeable materials are subject to corrosion during filtration and cleaning of aggressive media, which leads to failure. That is why the urgent task is to increase their corrosion resistance. This article is devoted to studying the corrosion resistance of PPM with combined protective coatings applied to their surface in the form of table salt and hydrochloric acid. The corrosion resistance of PPM with combined protective coatings applied in solutions of sodium chloride and hydrochloric acid has been studied. Combined protective coatings were formed by electric arc spraying and aluminum alloy methods on the surface of PPM and subsequent plasma electrolytic treatment. As a result, a combined protective coating is formed on the surface of the PPM. Based on potentiodynamic mode, the corrosion potentials in a solution of table salt move in the positive direction when a combined coating is applied to the surface of the PPM, which indicates a decrease in the corrosion activity of the surface. The corrosion currents of the combined protective coating for all the studied systems are reduced by 3 orders of magnitude, and an increase in the ratio of the cathode to anode current density reduces the rate of corrosion of the coating by another five times.

Keywords: Corrosion · Combined coating · Manufacturing innovation · Corrosion current · Corrosion potential · Corrosive environment

1 Introduction

The current stage of development of materials science and powder metallurgy is accompanied by the emergence of new ideas for creating and improving the technology for obtaining porous penetrating materials (PPM), which, first of all, are aimed at achieving high-performance characteristics and reducing their cost.

Today, recycling productions are relevant, as the problem of greening has gained large-scale importance. Therefore, we have developed a technology for manufacturing

V. Tonkonogyi et al. (Eds.): InterPartner 2022, LNME, pp. 306–316, 2023.
https://doi.org/10.1007/978-3-031-16651-8_29

filters from PPM, which are waste products from the production of bearings. Therefore, our technology belongs to recyclable. The materials from which the filters are made according to our technology are steel 1.3505, characterized by high hardness, strength, wear resistance, and insufficient resistance to corrosion [1]. Therefore, searching for methods of increasing the corrosion resistance of such materials is an essential and relevant decision today. In order to solve this problem, it is necessary to apply a protective coating, which will significantly increase the corrosion resistance of PPM in the conditions of filtration and cleaning of aggressive media [2].

Corrosion of PPM causes significant damage when used for cleaning aggressive media. This damage is determined not so much by the cost of PPM but rather by the cost of corrosion protection, repair work, and losses due to the temporary termination of the normal functioning of filter equipment. The relevance of PPM corrosion protection issues does not decrease from year to year but rather increases.

In order to ensure timely and reliable protection of PPM from corrosion, it is necessary to pay more attention to researching the risk of corrosion and protection against PPM and improve the quality and reliability of protective coatings applied to them.

Currently, issues related to the process of interaction during joint movement are being considered. The creation and selection of functional materials for parts manufacturing [3] are based on solving interrelated problems based on the study of friction mechanics and physicochemical phenomena occurring on the surface.

This work aimed to investigate the corrosion resistance of pressed samples from PPM made from 1.3505 powder with a protective coating.

2 Literature Review

The contact of solid interaction during pressing of samples with PPM is observed only in certain zones, the size and density of which depend on the value of the applied load and the stress-deformed state of the contacts. These contacts depend on the micro-roughness's geometric shape and the surface layer's mechanical properties.

To protect the pressed samples from PPM against corrosion, it was decided to develop combined electrometallization coatings [4, 5]. For the application of corrosion-resistant protective electrometallization coating on structural parts, the method of vacuum-arc spraying and plasma-electrolyte oxidation (PEO) is widely used, which is one of the most modern and promising methods for obtaining protective layers on the surface of metals and alloys with a complex of important characteristics. The properties of the obtained coatings are determined by the electrolyte's composition and the PEO process's modes.

The combination of coatings makes it possible to increase strength [6], hardness [7], wear and corrosion resistance [8, 9], ductility and contact resistance [10], dynamic loads and mechanical vibrations [11]. Thus, due to the ceramics' chemical and phase composition and structure, oxide coatings obtained on valve metals by plasmoelectrolyte oxidation have high functional properties. The versatility of oxide-ceramic coatings contributes to their wide application in many industries, and the range of processed parts is constantly increasing. However, the possibilities of the method have not yet been fully explored. Work continues to improve the process of oxide-ceramic synthesis itself; new electrolytes are being developed, and new power sources are being created. However, a significant disadvantage of the PEO process is the limitation of the metals that can be coated.

Expanding the scope of application of the plasmoelectrolyte oxidation method for other metals and alloys is possible using a complex technology of gas-thermal spraying of wires or aluminum-based powders, followed by plasmoelectrolyte oxidation. In addition, the use of such a comprehensive technology allows you to restore worn parts and extend the life of the equipment.

Plasma electrolyte oxidation (PEO) [12], known as ANOF ("Anodishen Oxidation unter Funkenentladung") in Germany [13, 14], ASD ("Anodic Spark Deposition") in the USA, Europe, and China [15, 16], MAO ("Microarray Oxidation") in Eastern Europe, and PEO ("Plasma Electrolytic Oxidation") in the USA. Switzerland [17] has been developing a new intensive metal anode processing technology in the last twenty years. The process is carried out under surface spark discharges due to electro-and plasmochemical reactions. It is used both for cleaning metal surfaces and applying various coatings, the properties of which are determined by the composition of the electrolyte and the electrolysis mode. Being essentially a technology similar to traditional anodizing, PEO does not require either deep cooling or Vi-valence chromium compounds, which makes this method very environmentally attractive [18].

3 Research Methodology

In this paper, samples made from steel powder BBS15+graphite powder were studied. The powder was mixed in a mixer for four hours, after which the steel powder was pressed into the mold for one minute (Fig. 1).

Fig. 1. Compressed samples of the composition BBS15+graphite.

Then the blanks were placed in an induction furnace and sintered at a temperature of 1200 C. Surface protection of samples from corrosion was carried out in two stages: an electrometallization coating was applied to the sample, then synthesis with PEO method was carried out.

The electrometallization coating was sprayed using a vacuum-arc method. The essence of this method of coating is that a jet of compressed gas (usually air) [19] is sprayed into small particles that fall at high speed in the molten or plastic state on the prepared surface, deformed during impact, welded or stick to the surface, forming a coating. The coating is sprayed using the electro-arc method [20].

To implement the process of metal oxide synthesis, an installation was designed and manufactured, which consists of a power part, control and measurement console, a diode-thyristor converter with a control circuit, and working baths.

The PEO process occurs in an electrolytic plasma of spark (microarray) discharge in an electrolyte based on distilled water. Oxygen formed as a result of dissociation of water partially reacts with the metal and forms an oxide on its surface. The coating formation process occurs in an electric field, an energy source for plasmochemical reactions at the working electrode. According to the voltage increase in the process, three zones can be distinguished: I - Faraday, II - spark discharges, III - micro-arc discharges (Fig. 2).

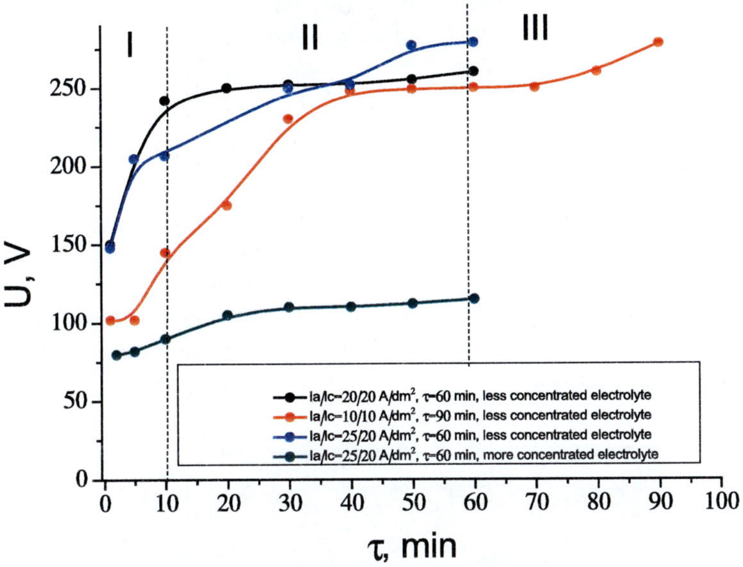

Fig. 2. Kinetics of the anode process: I-Faraday part, II-sparking part, III-microarray discharge part.

In the Faraday part, electrochemical processes are known as anodizing take place. Oxygen involved in the synthesis of the oxide film is formed as a result of electrolysis. The breakdown is carried out due to the injection of electrons from the band gap into the conduction band under the action of an electric field.

This process can also be stimulated by electrons released due to electrochemical reactions at the anode. In the breakdown channel, the temperature increases strongly, the electrolyte dissociates and ionizes, and a plasma clot is formed, in which plasmochemical reactions of oxide synthesis are realized.

Figure 2 shows the change in voltage over time during the synthesis process at the anode. According to the conducted studies, it is possible to establish the stages of the synthesis process and the influence of synthesis modes on it. The study of the anode curves shows that a higher electrolyte concentration reduces the synthesis voltage but stabilizes the process, resulting in the synthesis occurring uniformly.

After the proposed surface treatment, a sample was obtained, the general view of which is presented in Fig. 3.

Fig. 3. PPM sample with combined coating.

The corrosion properties of alloys and their coatings were studied according to polarization curves taken in potentiodynamic mode on potentiostats PY-50 and IPC-PRO+PK IBM PC according to a standard three-electrode scheme with a chlorinated comparison electrode and an auxiliary platinum (Fig. 4). On the electrode, a working area was selected, the area of which was $S = 1 \cdot 10^{-4}$ m^2, the remaining surfaces were insulated with epoxy varnish.

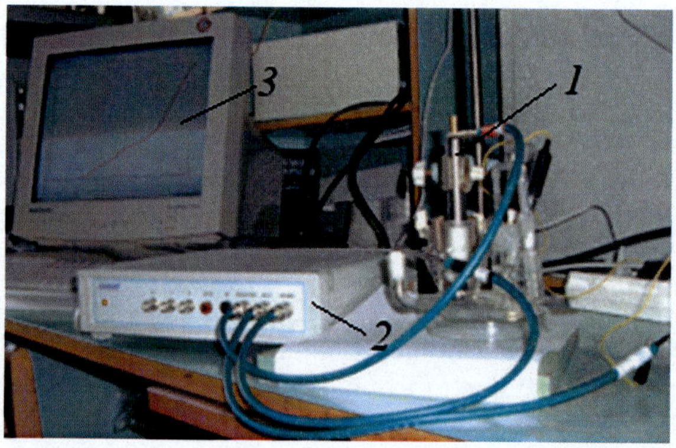

Fig. 4. General view of installation IPC-PRO+PK IBM PC: 1 - a cell with a chlorinated reference electrode and an auxiliary platinum one, 2 - potentiostat, 3 - polarization curve.

By building the polarization curves, the surface of the alloys was activated up to −2 B by the cathode. After cathodic polarization, continuous potentiodynamic polarization was immediately performed at a rate of a potential change of 5 MV/s. The corrosive media were 10% NaCl and 10% HCl. Corrosion currents and potentials were determined to characterize the corrosion resistance. Polarization curves were recorded after the stationary potential was established. Corrosion currents and potentials were determined graphically. In addition, the samples were examined using a MIM-10 microscope with a computer surface fix.

4 Results and Discussion

During the operation of anodized alloys in a corrosive environment, narrow channels with increased ionic conductivity are formed in some regions of the cover in the PPM. Such areas become foci of local corrosion damage to the PPM [21]. Active anions (for example, chlorine and water) in these places penetrate through the anode film and interact with the alloy.

Plasmoelectrolytic coatings are inert. However, if there are pores in them that are channels of spark discharges, corrosion destruction of the metal or alloy is also possible. All this reduces the functional properties of the coating and reduces the service life of the alloy. In this regard, the corrosion resistance of plasma-electrochemical oxidoceramic coatings (OCC) on zirconium and titanium alloys in different corrosive media was studied, and their corrosion currents were determined used to judge the rate of corrosion in PPM.

The rate of electrode reactions that cause the corrosion process can be judged from electrochemical studies. In this regard, we studied the electrochemical behavior of PPM in the initial state and with an oxide-ceramic coating in acidic media that are corrosive to PPM.

Of the solutions of table salt and hydrochloric acid used in corrosion studies, the medium containing sodium ions is more aggressive for PMT. After the formation of OCC, corrosion processes are significantly inhibited. In a hydrochloric acid environment, the corrosion current of the coating decreases by order of magnitude compared to the matrix y. In a more aggressive environment of sodium chloride, the corrosion current decreases by two orders of magnitude, and the corrosion potential in the salt solution is shifted in the positive direction, indicating a decrease in the corrosion activity of the surface.

It is known that high corrosion resistance is achieved by the formation of a thin oxide film. However, in severe operating conditions and in the presence of particularly aggressive media (used in work), these films, due to their small thickness and low hardness, do not provide high anti-corrosion properties.

The corrosion currents of the I_{cor} (based on which the material's corrosion is judged by the corrosion rate) are reduced for all the studied systems. Moreover, the higher effect of creating OCC is manifested in more aggressive environments.

Analysis of the obtained polarization curves shows that the formation of OCC affects the electrochemical parameters of PPM corrosion, i.e., the values of corrosion currents decrease. This change is due to less dissolution of the alloy components.

Polarization curves can evaluate the corrosion resistance of coatings. The corrosion potential of corrosion-resistant PPM under the action of a 10% HCl solution has a negative value of −370 mV (Fig. 5), and the corrosion potential is 80 mA/m^2. First of all, the reduction of corrosion currents means the corrosion resistance observed in the two investigated aggressive systems.

During the formation of the OCC PPM, the corrosion current decreases by 1…2 orders of magnitude depending on the composition of the electrolyte in which the coating was synthesized. Thus, the synthesized low-concentration electrolyte containing alkali and Liquid Glass has the lowest corrosion resistance of those synthesized in more concentrated electrolytes and has a value of $1{,}2 \cdot 10^{-3}$ A/m^2. With a further increase in the electrolyte concentration, the corrosion currents of i_{cor}OCC decrease Fig. 5).

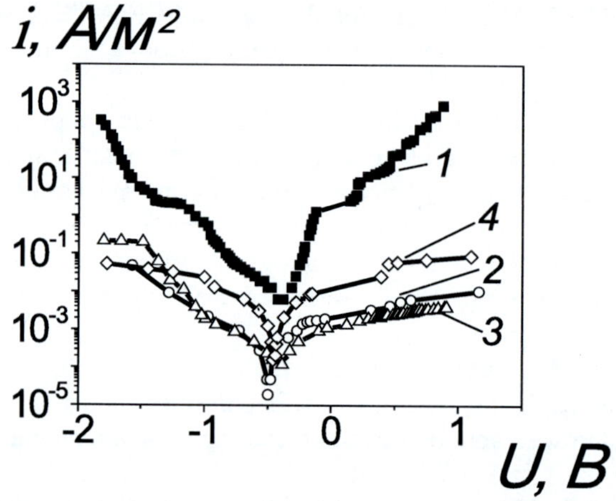

Fig. 5. Polarization curves of PM (1) and its OCC (2, 3) in solution 10%HCl (see Table 1).

Table 1. Potential and current of corrosion of the PMT and coating under various oxidation modes.

№	Corrosive environment	Electrolyte composition, g/l				I_a/I_c A/dm^2	τ, min	U_{cor}, B	i_{cor}, A/m^2
		KOH	p.c.	CrO_3	H_2O_2				
1	10%NaCl	–	–	–	–	–	–	0,48	0,82
2		10	15	–	–	20/20	20	0,41	$9{,}09 \cdot 10^{-4}$
3		10	15	–	–	20/30	40	0,24	$1{,}63 \cdot 10^{-4}$
4		10	15	0,1	–	20/30	40	0,26	$6{,}20 \cdot 10^{-4}$
1	10%HCl	–	–	–	–	–	–	0,37	$8{,}07 \cdot 10^{-2}$
2		10	15	–	10	20/20	30	0,41	$0{,}21 \cdot 10^{-4}$
3		10	15	–	–	20/20	30	0,50	$1{,}39 \cdot 10^{-4}$
4		3	2	–	–	20/20	20	0,45	$12{,}07 \cdot 10^{-4}$

The coating obtained in an electrolyte of the composition 10 g/l KOH + 15 g/l liquid glass (l.g.) has a corrosion current of 1,39 10^{-4} A/m^2, and when introduced into the electrolyte of 10 g/l, H_2O_2 reduces to a value of 1,21 · 10^{-4} A/m^2.

This result is obviously due increasing thickness of the coating on PPM [22].

In the more aggressive environment of table salt, the values of the current and corrosion potential of unprotected PPM are 0,82 A/m^2 and −0,48 B. Since the coating obtained in a low-concentration electrolyte did not perform well in the less aggressive environment of hydrochloric acid, it was impractical to study such a coating in further experiments.

So, we studied the covers, the modes of which are given in Table 1. According to the data analysis, three orders of magnitude reduce the corrosion currents of the OCC for all the studied systems.

The coating synthesized in an electrolyte of 10 g/l KOH + 15 g/l l.g has a corrosion current value of 9,09 · 10^{-4} A/m^2. However, increasing the ratio of the current density of the cathode to the anode reduced the rate of corrosion of the coating by another five times.

Adding 0,1 g/l CrO_3 to such an electrolyte did not significantly increase the corrosion current value (Fig. 6, curve 4).

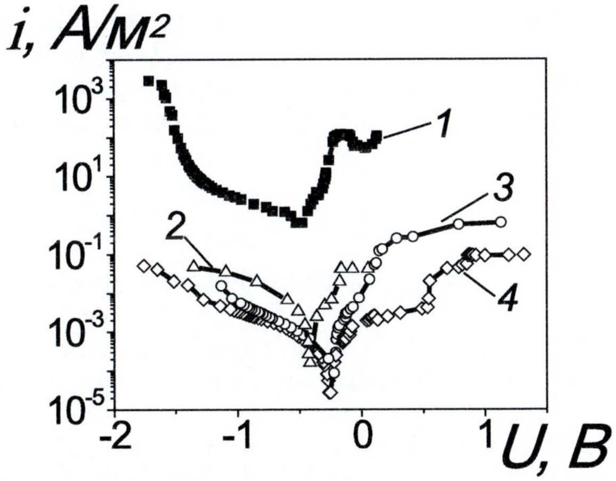

Fig. 6. Polarization curves of PPM (1) and its oxide-ceramic coating (2, 3, 4) in solution 10%NaCl (see Table 1).

Corrosion potentials in a solution of table salt move in a positive direction when all OCC are synthesized, which also indicates a decrease in the corrosion activity of the surface.

Figure 7 shows the PPM cover after corrosion in aggressive environments under anode polarization. PMT undergoes co-destruction in aggressive environments. However, the cover, even from the appearance assessment, does not show corrosive damage.

The figure shows a photo of OCC after corrosion in a highly aggressive 10% NaCl environment. In a solution of hydrochloric acid, the general appearance of the cover does not change, and its properties can be studied in more detail using polarization curves.

Fig. 7. General view of the oxide-ceramic coating on PPM after corrosion in media: a – 10%NaCl; b – 10%HCl.

Consequently, coatings synthesized in higher concentrations of electrolytes have better corrosion resistance [23, 24]. Introducing hydrogen peroxide into the electrolyte increases the corrosion resistance, and chromium oxide significantly reduces it. The most significant corrosion resistance can be achieved by synthesizing a coating in an electrolyte of 10 g/l KOH + 15 g/l liquid glass at high current densities.

5 Conclusions

Based on the analysis of the data, it was found that three orders of magnitude reduce the corrosion currents of the OCC for all studied systems. The coating synthesized in an electric roll of 10 g/l KOH + 15 g/l l.g has a corrosion current value of $9.09 \cdot 10^{-4}$ A/m^2. However, an increase in the cathode-to-anode current density ratio reduced the corrosion rate of the coating by another five times.

Adding 0.1 g/l CrO_3 to such an electrolyte does not significantly increase the value of the corrosion current. Corrosion potentials in a solution of table salt move in a positive direction when all OCC are synthesized. This fact also indicates a decrease in the corrosion activity of the surface.

Consequently, coatings for PMM that are synthesized in higher concentrations of electrolytes have better corrosion resistance, i.e., the anticorrosive effect of coatings increases with increasing aggressiveness of the corrosive medium. The introduction of hydrogen peroxide in electrolytes increases the corrosion resistance, and chromium oxide significantly reduces it.

It was experimentally established that reducing the electrolyte concentration from 10 g/l KOH and 15 g/l l.g and 10 g/l H_2O_2 to 10 g/l of alkali and 15 g/l of liquid glass reduces the corrosion current of the coating by 5 times.

It is also found that at higher ratios of anode and cathode currents, the corrosion currents of the coatings are reduced by 6 times, and the most significant corrosion resistance can be achieved by synthesizing the coating in an electrolyte of 10 g/l KOH + 15 g/l of liquid glass at higher current densities.

References

1. Snizhko, L.O.: The nature of anodic gas at plasma electrolytic oxidation. Prot. Met. Phys. Chem. Surf. **50**(6), 705–708 (2014). https://doi.org/10.1134/S2070205114060215
2. Shatskyi, I.P., Ropyak, L.Y., Makoviichuk, M.V.: Strength optimization of a two-layer coating for the particular local loading conditions. Strength Mater. **48**(5), 726–730 (2016). https://doi.org/10.1007/s11223-016-9817-5
3. Ivanov, V., Vashchenko, S., Rong, Y.: Information support of the computer-aided fixture design system. In: CEUR Workshop Proceedings, vol. 1614, pp. 73–86 (2016)
4. Sakhnenko, N.D., Ved', M.V., Androshchuk, D.S., et al.: Formation of coatings of mixed aluminum and manganese oxides on the AL25 alloy. Surf. Eng. Appl. Electrochem. **52**, 145–151 (2016). https://doi.org/10.3103/S1068375516020113
5. Hovorun, T.P., et al.: Physical-mechanical properties and structural-phase state of nanostructured wear-resistant coatings based on nitrides of refractory metals Ti and Zr. Funct. Mater. **26**(3), 548–555 (2019). https://doi.org/10.15407/fm26.03.548
6. Ivanov, V., Dehtiarov, I., Denysenko, Y., Malovana, N., Martynova, N.: Experimental diagnostic research of fixture. Diagnostyka **19**(3), 3–9 (2018). https://doi.org/10.29354/diag/92293
7. Shvets, S.V., Machado, J.: Numerical model of cutting tool blade wear. J. Eng. Sci. **8**(2), A1–A5 (2021). https://doi.org/10.21272/jes.2021.8(2).a1
8. Kostyk, K., et al.: Simulation of diffusion processes in chemical and thermal processing of machine parts. Processes **9**(4), 698 (2021). https://doi.org/10.3390/pr9040698
9. Hovorun, T.P., Berladir, K.V., Bilous, O.A., Lyubich, O.I., Vorobiov, S.I.: Development of alloy resistant in conditions of abrasive wear. Funct. Mater. **28**(1), 170–177 (2021). https://doi.org/10.15407/fm28.01.170
10. Ivanov, V., et al.: Numerical simulation of the system "fixture–workpiece" for lever machining. Int. J. Adv. Manuf. Technol. **91**(1–4), 79–90 (2016). https://doi.org/10.1007/s00170-016-9701-2
11. Ivanov, V., Pavlenko, I., Kuric, I., Kosov, M.: Mathematical modeling and numerical simulation of fixtures for fork-type parts manufacturing. In: Knapčíková, L., Balog, M. (eds.) Industry 4.0: Trends in Management of Intelligent Manufacturing Systems. EICC, pp. 133–142. Springer, Cham (2019). https://doi.org/10.1007/978-3-030-14011-3_12
12. Falcade, T., et al.: Electrodeposition of diamond-like carbon films on titanium alloy using organic liquids: corrosion and wear resistance. Appl. Surf. Sci. **263**, 18–24 (2012). https://doi.org/10.1016/j.apsusc.2012.08.052
13. Kurz, P., Krysmann, W., Schneider, H.G.: Application fields of ANOF layers and composites. Cryst. Res. Technol. **21**(12), 1603–1609 (1986). https://doi.org/10.1002/crat.2170211224
14. Kaluđerović, M.R., Schreckenbach, J.P., Graf, H.-L.: Titanium dental implant surfaces obtained by anodic spark deposition – from the past to the future. Mater. Sci. Eng., C **69**, 1429–1441 (2016). https://doi.org/10.1016/j.msec.2016.07.068
15. Curto, B.D., Diamanti, M.V., Pria, P.D., Sbaiz, F., Cigada, A.: Anodic Spark Deposition treatments to increase reliability of Ti6Al4V modular prostheses. J. Appl. Biomater. Biomech. **7**(3), 153–159 (2009)

16. Panov, V.A., Vasilyak, L.M., Vetchinin, S.P., et al.: Spark channel propagation in a microbubble liquid. Plasma Phys. Rep. **42**, 1074–1077 (2016). https://doi.org/10.1134/S1063780X16110064
17. Sikdar, S., Menezes, P.V., Maccione, R., Jacob, T., Menezes, P.L.: Plasma electrolytic oxidation (PEO) process – processing, properties, and applications. Nanomaterials **11**(6), 1375 (2021). https://doi.org/10.3390/nano11061375
18. Mohedano, M., Mingo, B.: Special issue: plasma electrolytic oxidation (PEO) coatings. Coatings **11**(1), 111 (2021). https://doi.org/10.3390/coatings11010111
19. Merzliakov, I., Pavlenko, I., Chekh, O., Sharapov, S., Ivanov, V.: Mathematical modeling of operating process and technological features for designing the vortex type liquid-vapor jet apparatus. In: Ivanov, V., et al. (eds.) DSMIE 2019. LNME, pp. 613–622. Springer, Cham (2020). https://doi.org/10.1007/978-3-030-22365-6_61
20. Kuroda, S., Kawakita, J., Watanabe, M., Katanoda, H.: Warm spraying—a novel coating process based on high-velocity impact of solid particles. Sci. Technol. Adv. Mater. **9**(3), 033002 (2008)
21. Voitovych, A.A., Pokhmurs'ka, H.V., Student, M.M., Student, O.Z.: Microstructure and abrasive-wear resistance of the vibration-deposited metal of core wires of the basic Fe–Cr–B system. Mater. Sci. **52**(3), 365–370 (2016). https://doi.org/10.1007/s11003-016-9965-6
22. Povstyanoy, O., Zabolotnyi, O., Rud, V., Kuzmov, A., Herasymchuk, H.: Modeling of processes for creation new porous permeable materials with adjustable properties. In: Ivanov, V., et al. (eds.) DSMIE 2019. LNME, pp. 456–465. Springer, Cham (2020). https://doi.org/10.1007/978-3-030-22365-6_46
23. Tkachuk, V., Bozhydarnik, T., Rechun, O., Karavayev, T., Merezhko, N.: Assessment of the quality of alternative fuels for gasoline engines. In: Ivanov, V., et al. (eds.) DSMIE 2019. LNME, pp. 871–881. Springer, Cham (2020). https://doi.org/10.1007/978-3-030-22365-6_87
24. Povstianoi, O.Y., et al.: Optimization of the properties of multilayer porous permeable materials. Mater. Sci. **56**(4), 530–535 (2021). https://doi.org/10.1007/s11003-021-00460-2

Quality Assurance

Influence of Drill Microgeometry on the Quality of the Machined Fiberglass Surface

Yuriy Adamenko[1](✉), Yuriy Besarabets[1], Serhii Maidaniuk[1], Oleksandr Plivak[1], and Dmytro Adamenko[2]

[1] National Technical University of Ukraine "Igor Sikorsky Kyiv Polytechnic Institute", 37, Peremohy Avenue, Kyiv 03056, Ukraine
adamenko.yuriy@lll.kpi.ua
[2] WAGO GmbH, 32423 Minden, Germany

Abstract. Composite Fiber Reinforced Plastics (CFRP) with glass fibers are widely used as they combine high toughness and specific strength, good resistance to corrosion, fatigue, temperature, and moisture. Drilling composite materials is one of the most common and complicated operations of composite machining. The complexity of drilling consists of the cutting tool's low durability and the formation of defects on the part surface. Tool wear is accompanied by changes in the microgeometry of the cutting edge, i.e., the surface between the blade's front and back surfaces. The microgeometry of the cutting edge and, consequently, the contact conditions of the tool with the material constantly change in the process of cutting, which affects the technological forces and machining quality. Experimental studies of fiberglass drilling with an HSS spiral drill bit were performed in this work. The influence of cutting edge microgeometry on the machining quality parameters has been established, and criteria for their evaluation have been selected. The information obtained as a result of the study will be used to analyze drill geometry and cutting modes on the microgeometry of the cutting edge, cutting forces, and processing quality, as well as to optimize drills and solve problems arising in composite drilling.

Keywords: Drill geometry · Composites · Tool wear · Delamination · Sustainable manufacturing

1 Introduction

Due to low-cost fiberglass plastics being widely used in construction, shipbuilding, automobile construction, and production of household items, their share in the structure of plastics application is constantly growing. The technology of manufacturing parts from composite materials makes it possible to approximate their shape to the shape of finished products. Still, it is impossible to avoid machining in most cases. The most widespread machining operations are cutting, milling of ledges and hatches, and drilling holes for bolt and rivet connections.

V. Tonkonogyi et al. (Eds.): InterPartner 2022, LNME, pp. 319–329, 2023.
https://doi.org/10.1007/978-3-031-16651-8_30

The high abrasive capacity of reinforcing fibers and low thermal conductivity of the matrix leads to intensive mechanical wear of the drill bit, increased cutting forces, and, consequently, deterioration of the quality parameters of the machined hole.

Work to improve tool durability is mainly carried out in three directions. The first is the use of more wear-resistant tool materials, for example, hard alloys, diamonds, and various coatings, the second is the improvement of the design of cutting tools, optimization of geometric parameters of the cutting part, preparation of cutting edges and the like and the third is the improvement of tool operation technology - the choice of cutting modes, use of overlays on the part, use of cryogenic processing, vibrations, etc.

Because of the significant differences in CFRP properties, there is no standard methodology for selecting machining parameters, so there is a need to specify tool geometry and cutting modes for each combination of cutting and tooling materials. The correct choice of machining parameters increases the tool durability and quality of the machined surface.

To improve the efficiency of hole machining, it is necessary to establish how the geometric parameters of a cutting tool, elements of cutting modes affect the wear of the cutting part, cutting forces, and the actual quality parameters of the part.

The cutting edge's progressive blunting occurs during glass reinforcing fibers' drilling. The intensity of wear on the rake and flank faces of the tool during machining is not the same due to differences in the physical processes that occur on these surfaces. It leads to a constant change in the shape and dimensions of the cutting edge elements.

Several research carried out by Ramireza et al. [1] and Henerichs [2] have shown a strong correlation between the rounding of the cutting edge resulting from wear, technological forces, and machining defects. The condition of the cutting edge, its sharpness or blunting, significantly affects the interaction between the tool and material in the cutting zone and all process parameters - chip formation, temperature, cutting forces, and, as a consequence, the quality of the machined part surface.

In practice, the most common way to describe blunting of a cutting edge is to control only one parameter - the wear on the back surface. Maegawa et al. [3] established that cutting edge wear during drilling is accompanied by an increase in the rounding of the cutting edge and a continuous decrease of the rake angle. However, one parameter alone does not provide information about the characteristic profile of the worn edge. For some time, the radius of rounding of the cutting edge of a worn tool was considered a sufficient criterion for blunting. However, it was found that this radius tends to stabilize at a specific value and does not increase with a further increase of the cutting path [4].

The purpose of this work is to experimentally investigate the influence of microgeometry of the cutting edge of a high-speed steel twist drill bit on the quality of the surface of the hole in the fiberglass workpiece - the size of the diameter size, delamination, and roughness. It will allow establishing criteria for evaluating drill microgeometry and quality parameters of the machined surface, which is reasonable to apply for research on improving the efficiency of fiberglass drilling.

2 Literature Review

Numerous studies confirm that for successful cutting of composite fibers and to avoid damage to the part, the cutting edge of the tool must be sharp, as proven by Henerichs et al. [5], Hocheng et al. [6], and Sauer et al. [7]. A blunted cutting edge leads to increased friction and associated heat generation, increased cutting forces, and deterioration of machining quality.

There are standards for determining the criterion of tool wear, but they do not present parameters that characterize the microgeometry of the edge, that is, the shape of the transition line from the Rake face to the flank face. Therefore, various criteria for evaluating the cutting edge's shape and methods of measurement can be found in the literature.

To estimate the cutting edge profile, four parameters are introduced in [8]: $\Delta r, \phi, S_\gamma, S_\alpha$. The parameter Δr is defined as the minimum distance from the top of a perfect sharp blade to the actual tool profile. The angle ϕ describes the displacement of the cut point to the rake or to the flank face relative to the bisector of the wedge angle. Parameters S_γ and S_α define distances from the ideal vertex to the intersection of the curve with rake and flank surfaces. Geometry asymmetry is determined by the K-factor $K = S_\gamma / S_\alpha$ and the average roundness of the cutting edge $\overline{S} = (S_\gamma + S_\alpha)/2$.

Raj and Karunamoorthy [9] proposed five parameters that characterize the microgeometry of the cutting edge profile: peak flatting, cutting edge flatting, flank rounding depth, flank rounding width, and flank wear.

In [10], a set of simplified parameters was developed to quantitatively describe the cutting edge microgeometry in the following structure: «straight line – ellipse – straight line». The chipping region is located along the front surface and is characterized by an anterior angle γ^* close in value to the initial angle γ. The pressing region has a rounded shape and is characterized by the semi-axes of the ellipse l_γ and l_α. The workpiece material in this zone is not sheared off but pressed into the workpiece. The tool's flank face contacts the workpiece in the spring back region. This zone is characterized by the current clearance angle (*) value and the elastic recovery height. In [11], the parameters proposed in [10] are used to study the microgeometry of tools when machining carbon plastics.

It was also proposed to use the area of a flat figure located between the original profile and the worn one at a specific moment of machining [12] to describe the microgeometry. The authors of [13] believe that the evaluation of wear geometry should be described by a simple one-parameter geometric model, which allows to estimate the change in the shape of the cutting edge and predict tool durability. As the main criterion for measuring tool wear, it is suggested to use its weight loss in the cutting process.

After machining holes in a polymer composite, it is necessary to evaluate the quality of machining of the hole. The most commonly used parameters are diameter accuracy, material delamination, and machined surface roughness. The value of delamination is the area of material destruction around the hole at the entrance of the drill bit into the workpiece and at the hole's exit. Criteria are needed to quickly and reliably assess the amount of material delamination and decide the workpiece's suitability. The spread of the damage can be evaluated using nondestructive testing methods. Some examples are using a toolmaker's microscope, ultrasound techniques, acoustic emission, enhanced radiography, C-Scan, or Computerized Tomography [14]. As a result, an image is obtained, from which a quantitative assessment of the cleavage is performed, using coefficients that consider the diameter size of the cleavage, its shape, area, and other parameters.

In many modern studies, particularly in Devitte et al. [15], He et al. [16], and Singaravel et al. [17], the parameter F_d proposed in [1] is used for quick quantitative assessment of delamination. The parameter F_d is called the stratification coefficient and is determined by the formula:

$$F_d = \frac{D_{max}}{D_0} \tag{1}$$

where D_{max} is the maximum delaminated diameter, and D is the nominal hole diameter. Given delamination, a factor is simple and convenient, but it does not always adequately reflect the degree of damage because it does not consider the areas of damage. Therefore, in some works, for example [18] two-dimensional (2D) delamination factor F_{da}, which is defined as the ratio of the delamination area around the hole to the nominal area of the drill [19] is used.

$$F_{da} = F_d + \left(\frac{A_d}{A_{max} - A_0}\right)(F_d^2 - F_d) \tag{2}$$

where F_d – is the delamination factor based on Eq. (1); A_d – is the damaged area around the bore; A_{max} – the area of the circle corresponding to the maximum diameter of the delamination of the hole D_{max} and A_0 – is the nominal area of a drill.

The first term in (2) represents the contribution of the crack size, and the second term represents the contribution of the damaged region around the hole.

A parameter called the equivalent delamination coefficient was developed [20]. It is calculated as the ratio of the equivalent delamination diameter D_e and nominal hole diameter D_0. A complex parameter for assessing the quality of a hole was developed that includes five parameters [10]. Two parameters account for cracking, delamination, and delamination in the vicinity of the drilled hole, and another three parameters describe the scale and number of uncut fibers inside the hole.

The choice of quality assessment parameters depends on various factors - the shape and size of the damage, the time and allowable cost to perform the assessment, the task at hand, etc.

3 Research Methodology

The use of HSS tools for fiberglass processing usually is undesirable because of intensive wear in conditions of high abrasive impact of reinforcing fibers. However, in the present work, HSS drills have been chosen due to significantly lower prices than drills from hard alloys, and they are widely used in conditions of single and small-scale production of civilian products. A fiberglass plastic consisting of polyester resin and fiberglass filaments with a diameter of 0.016 mm was adopted as the workpiece to be machined. The workpiece was a square cross-section profile with dimensions 50 × 50 mm and a thickness of 5 mm. The outer layers of the blank were made from glass fabric. In the middle part, the arrangement of the bundles had a chaotic or random filling, which can be conventionally considered isotropic. The tests were performed on a vertical drilling machine at the following modes: cutting speed 17.8 m/min, feed 0.1 mm/rev.

An HSS twist drill with ground chip grooves was selected as a cutting tool. The drill was sharpened using a boron nitride grinding wheel on two planes. The geometric parameters of the cutting part of the drill bit were chosen as follows: drill diameter 8 mm, helix angle 28°, point angle 120°, clearance angle 20°, chisel edge angle 55°.

The cutting edge profile was measured in sections perpendicular to the main cutting edge (Fig. 1). The step between the sections along the cutting edge was selected as 0.5 mm. The edge measurements were performed by the contact method using a surface roughness measuring instrument. The angle of the cone of the stylus was 90°, and the radius of the tip was 10 μm. During the measurement, the drill was placed in the fixture so that the main cutting edge occupied a horizontal position. Measurements were performed along the drill bit surface. Since the wedge angle is less than 90, it is impossible to measure the profile in one setup without changing the position of the tool. Therefore, at the initial stage, edge measurements were performed in two stages in the same section - separately on the flank face side and on the rake face side. Preliminary experiments showed that the change in the edge profile occurs mainly on the rear surface side, so further measurements were performed in this zone.

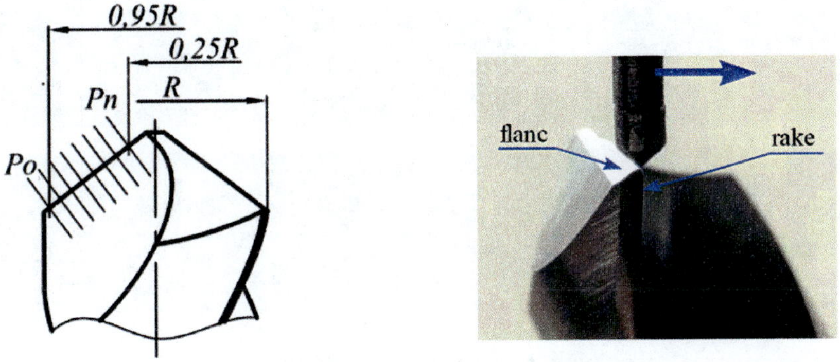

Fig. 1. Cutting edge profile measurement scheme.

The diameter of the hole was measured by the two-point method using a clock-type indicator with the division value of 0.01 mm in four sections (Fig. 2).

The roughness of the machined surface was measured in each hole in four sections using a MarSurf PS1 profilometer. The arithmetical mean deviation of the assessed profile *Ra* was chosen to assess the roughness of the hole surface. To evaluate the peel-up at the entrance and push-out parameters F_d and F_{da} were calculated by formulas (1) and (2).

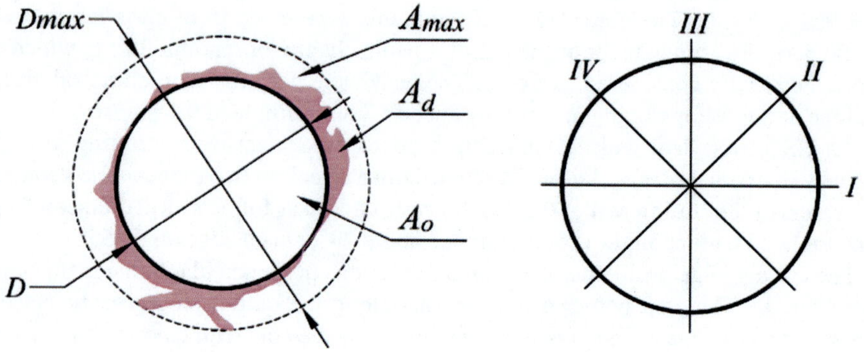

Fig. 2. Hole diameter and layering measurement scheme.

4 Results and Discussion

During the experiment, 80 holes were processed. The parameters were measured through a certain number of machined holes. The view of the holes at the entrance and the exit from the hole is shown in Fig. 3.

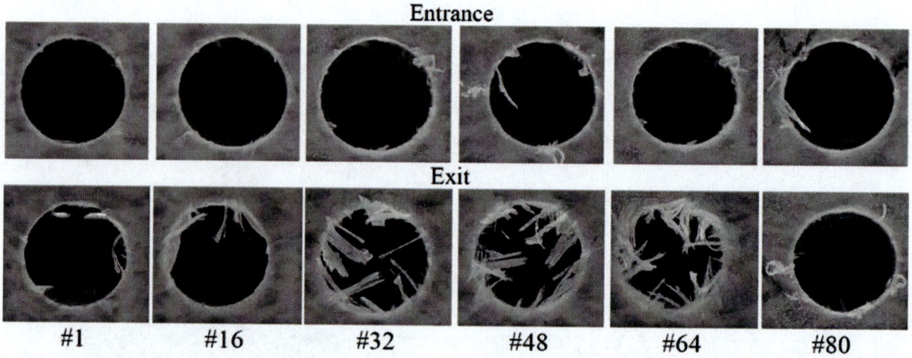

Fig. 3. State of machined holes.

After processing the images, the values of the parameters F_d and F_{da} were obtained (Fig. 4). As can be derived from the figures, the delamination at the tool inlet into the workpiece remains almost constant throughout the entire series of experiments, and the delamination at the outlet gradually increases, which is associated with the wear of the cutting edges of the drill bit. Therefore, the parameter F_{da}, calculated from the area of stratification is more objective compared to the parameter F_d.

Graphs of changes in the nominal actual diameter of the hole and roughness parameter Ra are shown in Fig. 5. Drill bit wear has little effect on the diameter of the machined hole: the average diameter of the first hole is 8.01 mm, and that of the last hole is 8.02 mm. It is noteworthy that the diameters of the same hole, measured in different directions, can differ significantly from each other and be larger or smaller than the nominal drill

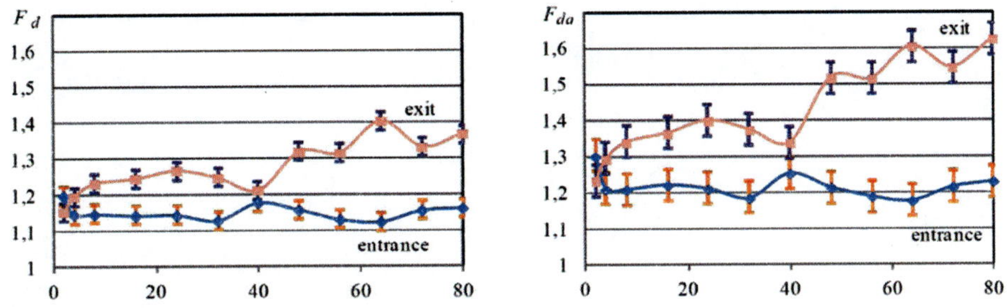

Fig. 4. Dependence of delamination parameters F_d and F_{da} on the number of machined holes.

diameter. It is due to the machined material's elastic properties and the reinforcement pattern.

The roughness parameter *Ra* practically did not depend on the degree of drill bit blunting and was 3…5 μm (Fig. 5).

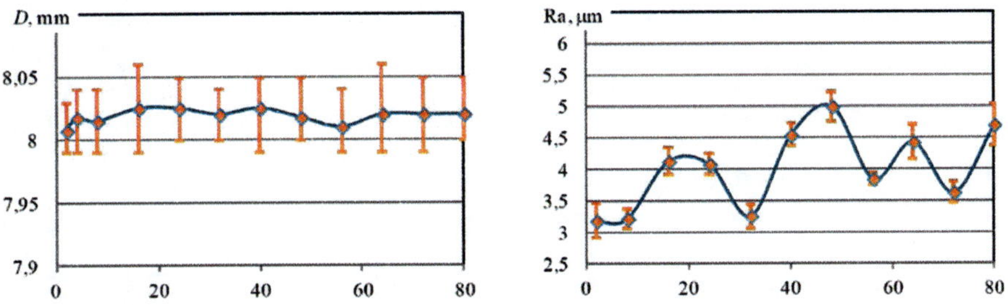

Fig. 5. Dependence of hole diameter and roughness on the number of drilled holes.

The change in the quality of the machined holes is the result of wear on the cutting edge of the drill bit. The top line represents the profile resulting from the sharpening of the cutting part (Fig. 9). Below are the profiles that are formed at different points in time as the tool wears. The edge has a distinctly asymmetrical profile. Near the front surface, an area has formed a circle of radius can describe that r_n. Further toward the rear surface, the profile has a rectilinear section resulting from the contact of the rear surface with the elastically restored machined surface. The cutting edge's shape indicates an intense abrasion along the rear surface caused by exposure to glass fibers.

To describe the microgeometry of the cutting edge, three parameters were chosen that reflect the features of the wear area to a greater extent: cutting-edge radius *CER*, flank wear *FW* and the value of total wear on the back surface *l* (Fig. 6).

The graph of change in microgeometric parameters of the cutting edge is shown in Fig. 7. The radius of rounding of the cutting edge r_n of the original profile immediately after sharpening is 5…8 μm. Then, there is a period of intensive wear. The radius value stabilizes at the level of 15…25 μm and does not change till the complete blunting of a drill. It agrees with the results of [21]. That is, the edge radius, which occurs naturally due to wear, will be determined by several factors, such as the grade of the tool material,

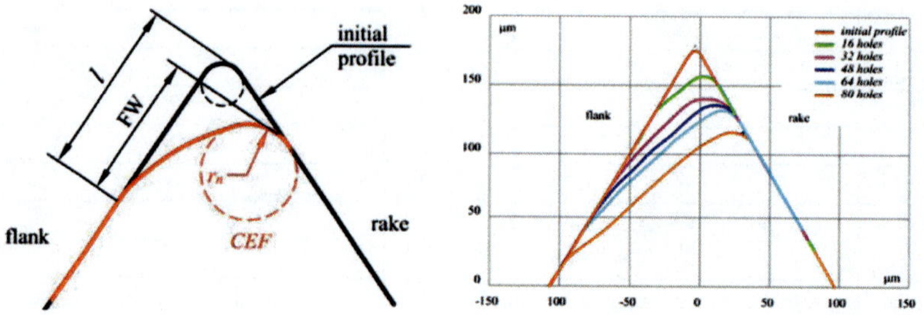

Fig. 6. Changing the cutting edge profile during drilling and parameters of the cutting edge profile of the drill bit.

fiber diameter, beam size, fiber adhesion to the matrix material, workpiece pliability, etc. If the diameter is larger or smaller than this radius, the worn edge will assume its natural wear radius. Thus, the rounding radius of the cutting edge has almost no effect on the change in the quality parameters of the machined surface.

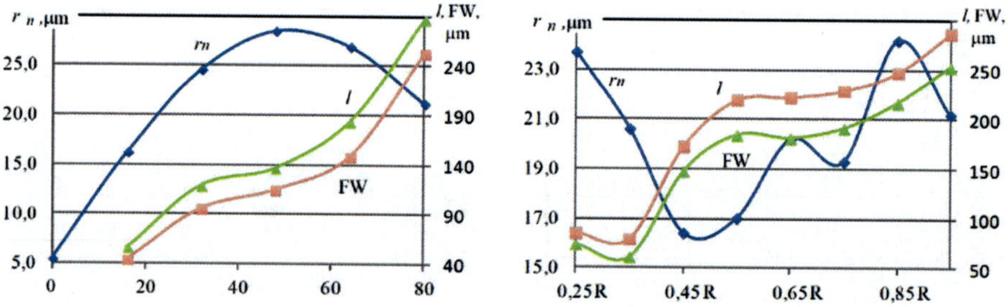

Fig. 7. Dependence of edge microgeometry on the number of drilled holes in the cross-section at radius 0.95R and change of profile parameters along the cutting edge after 80 holes.

The cutting edge wear on the back surface (parameter l_{fw}) continuously increases with the number of machined holes and significantly affects the delamination of the hole at the inlet and outlet of the workpiece. The parameter l_{fw} characterizes the length of the wear area at a specific moment of time and does not reflect the character of change of the wear value from the initial state, as l_{fw} is counted from the top of the edge, which is constantly displaced into the tool body due to wear. Therefore for the description of complete wear, the parameter l was introduced, which is counted from a constant base. When determining the parameter from the drawing, it is convenient to use the tip of the cutting edge in the initial position immediately after sharpening the drill. In the case of measuring the parameter on a tool microscope, it is enough to draw a baseline parallel to the cutting edge on the back surface and enter the measurements from it. The presented parameter correlates well with material delamination during drilling.

A specific dependence of the cutting edge microgeometry on the radius of the considered cutting edge point was established. So the values of the wear parameters l and FW constantly increase as the drill radius increases and reach the highest values at the outer diameter of the drill. During the cutting process, a zone of compressed material is formed in front of the cutting edge. After the fibers are cut, there is a sharp elastic recovery of the material, resulting in contact between the machined surface and the back surface of the drill bit. When drilling carbon fiber reinforced plastic with carbide drills, a similar dependence of wear on the rear surface on the radius of cutting edge points was obtained in [1].

The cutting conditions in the central zone and at the outer diameter of the drill differ not only because of the cutting speed but also because of the change in the front and rear angles along the cutting edge. Creating favorable cutting conditions is a key to improving tool life.

It is noteworthy that the curvature radius r_n along the cutting edge changes insignificantly and varies within 10...22 μm. In this case, it is difficult to establish the dependence of r_n on the rotation radius of the considered point of the cutting edge. Obviously, the radius of rounding of the cutting edge depends less on the cutting speed and more on the conditions of blade self-sharpening.

5 Conclusions

As a result of the work performed, several conclusions can be made. The wear of a high-speed steel drill bit blade occurs predominantly on the posterior surface, indicating a high abrasive capacity of the glass fibers of the material. For the selected pair of tool and work material, the cutting edge profile of the partially worn drill bit cannot be described by a single circle.

During the wear process, at least two areas are formed on the cutting edge profile: one in the form of an arc of a circle in contact with the layer to be cut and the other in the form of a wear chamfer on the back surface. The rounding radius in the initial period increases from the initial value of 5...8 μm to 15...25 μm and then stabilizes at this natural level until the drill bit blunts. The rounding radius along the cutting edge from the outside diameter toward the rotation axis changes insignificantly, despite the significant change in cutting speed at the points in question.

The diameter of the holes from the beginning of hole machining and to the blunting of the drill bit changed insignificantly. The influence of many factors causes the difference in diameter values in different measurement planes - the direction of fiberglass reinforcement, elastic recovery of the machined surface, and cutting edge wear.

The amount of peel-out at exit is greater than the amount of peel-up at entry. The amount of material delamination at the drill exit depends mainly on the back surface's wear. No significant effect of the cutting edge rounding radius on delamination was found.

To determine the amount of wear on the back surface, it is advisable to take a fixed base on the back surface rather than the tip of the edge as a reference point because the position of the tip of the edge is constantly changing during drill bit wear, which distorts the measurement result.

Further research is needed to study the effect of tool material properties, machined material, geometric parameters of the blade, modes and processing conditions on the cutting edge microgeometry, cutting forces, and the machined surface quality of the machined surface polymer composites.

References

1. Ramireza, C., Poulachona, G., Rossia, F., M'Saoubib, R.: Tool wear monitoring and hole surface quality during CFRP drilling. Procedia CIRP **13**, 163–168 (2014)
2. Henerichs, M.: Drilling of CFRP with special cutting edge micro-geometry. Swiss Federal Institute of Technology Zurich (ETH), Zurich (2015). (in Deutsch)
3. Maegawa, S., Morikawa, Y., Hayakawa, S., Itoigawa, F., Nakamura, T.: Mechanism for changes in cutting forces for down-milling of unidirectional carbon fiber reinforced polymer laminates: modeling and experimentation. Int. J. Mach. Tools Manuf. **100**, 7–13 (2016)
4. Sheikh-Ahmad, J.Y.: Machining of Polymer Composites. Springer, Boston (2009). https://doi.org/10.1007/978-0-387-68619-6
5. Henerichs, M., Voß, R., Kuster, F., Wegener, K.: Machining of carbon fiber reinforced plastics: influence of tool geometry and fiber orientation on the machining forces. CIRP J. Manuf. Sci. Technol. **9**, 136–145 (2014)
6. Hocheng, H., Tsao, C.: Comprehensive analysis of delamination in drilling of composite materials with various drill bits. J. Mater. Process. Technol. **140**, 335–339 (2003)
7. Sauer, K., Witt, M., Putz, M.: Influence of cutting edge radius on process forces in orthogonal machining of carbon fibre reinforced plastics (CFRP). Procedia CIRP **85**, 218–223 (2019)
8. Denkena, B., Friemuth, T., Fedorenko, S., Groppe, M.: Money is made at the cutting edge. Werkzeuge, 24–26 (2002). (in Deutsch)
9. Raj, S., Karunamoorthy, L.: A new and comprehensive characterisation of tool wear in CFRP drilling using micro-geometry and topography studies on the cutting edge. J. Manuf. Process. **32**, 839–856 (2018)
10. Voss, R.: Fundamentals of Carbon Fibre Reinforced Polymer (CFRP) machining. Swiss Federal Institute of Technology Zurich (ETH), Zurich (2017)
11. Seeholzer, L., Voss, R., Grossenbacher, F., Kuster, F., Wegener, K.: Fundamental analysis of the cutting edge micro-geometry in orthogonal machining of unidirectional Carbon Fibre Reinforced Plastics (CFRP). Procedia CIRP **77**, 379–382 (2018)
12. Montoya, M., Calamaz, M., Gehin, D., Girot, F.: Evaluation of the performance of coated and uncoated carbide tools in drilling thick CFRP/aluminium alloy stacks. Int. J. Adv. Manuf. Technol. **68**(9–12), 2111–2120 (2013). https://doi.org/10.1007/s00170-013-4817-0
13. Havin, G., Zhiwen, H.: Modeling of the instrumental microgeometry in the process of its wearing at cutting of composite materials. Cut. Tools Technol. Syst. **92**, 208–224 (2018)
14. Durão, L., Tavares, J., Gonçalves, D.: Damage evaluation of drilled carbon/epoxy laminates based on area assessment methods. Compos. Struct. **96**, 576–583 (2013)
15. Devitte, C., Souza, G., Souza, A., Tita, V.: Optimization for drilling process of metalcomposite aeronautical structures. Sci. Eng. Compos. Mater. **28**, 264–275 (2021)
16. He, Y., Zhou, X., Zou, P., Liu, K.: Study of complex helical drill with a "S" type chisel edge tip orient to carbon fiber reinforced plastics. Adv. Mater. Sci. Eng. **2021**, 5175515 (2021). https://doi.org/10.1155/2021/5175515
17. Singaravel, B., Radhika, M., Mohammed Asif, M., Reddy, K.P.: Analysis of hole quality errors in drilling of GFRP composite. IOP Conf. Ser. Mater. Sci. Eng. **1057**, 012069 (2021)
18. Wei, Y., An, Q., Ming, W., Chen, M.: Effect of drilling parameters and tool geometry on drilling performance in drilling carbon fiber–reinforced plastic/titanium alloy stacks. Adv. Mech. Eng. **8**(9), 1–16 (2016)
19. Davim, J., Rubio, J., Abrão, A.: A novel approach based on digital image analysis to evaluate the delamination factor after drilling composite laminates. Compos. Sci. Technol. **67**(9), 1939–1945 (2007)

20. Tsao, C., Kuo, K., Hsu, I.: Evaluation of a novel approach to a delamination factor after drilling composite laminates using a core-saw drill. Int. J. Adv. Manuf. Technol. **59**(5–8), 617–622 (2012)
21. Madhavan, V., Lipczynski, G., Lane, B., Whitenton, E.: Fiber orientation angle effects in machining of unidirectional CFRP laminated composites. J. Manuf. Process. **20**(2), 431–442 (2015)

A General Approach for Tolerance Control in Quality Assessment for Technology Quality Analysis

Oleksandr Kupriyanov[1](✉), Roman Trishch[2], Dimitar Dichev[3], and Kateryna Kupriianova[4]

[1] Ukrainian Engineering Pedagogics Academy, 16, Universytetska St., Kharkiv 61003, Ukraine
a_kupriyanov@uipa.edu.ua

[2] National Aerospace University «Kharkiv Aviation Institute», 17, Chkalov St., Kharkiv 61070, Ukraine

[3] Technical University of Gabrovo, 4, H. Dimitar St., Gabrovo 5300, Bulgaria

[4] Kharkiv National University of Radio Electronics, 14, Nauky Ave., Kharkiv 61166, Ukraine

Abstract. The paper proposes an approach to the analysis of production technology based on the proximity of the obtained accuracy to the desired parameters. Such an assessment of technology in terms of quality will allow for a reasonable choice of the best technology and assign technological parameters which can be applied in the conditions of Industry 4.0. The approach builds on the control system proposed earlier by the authors based on the use of the desirability function and the postulation that there is the best parameter from a constructive point of view, at which the best quality is achieved. The projection of the dimension distribution density function onto the desirability function gives the desired function called the dimension desirability distribution density function. Analytical equations for constructing this function have been obtained both for the general case and for the cases of linear and parabolic desirability functions. In the work, the dimension desirability distribution density function has been illustrated with a specific example for different values of the shape index of the desirability function and different technological processes from the point of view of quality. It has been shown that, although the desirability function can be constructed for any shape parameters, the shape factor for the dimension desirability distribution density function equal to 1 is the best, and outside the range 0.3–1.2, the desirability distribution density function does not allow analyzing the production technology in terms of suitability.

Keywords: Tolerancing · Actual dimension · Quality system · Desirability function · Continuous quality improvement · Industrial innovation

V. Tonkonogyi et al. (Eds.): InterPartner 2022, LNME, pp. 330–339, 2023.
https://doi.org/10.1007/978-3-031-16651-8_31

1 Introduction

The traditional tolerance system, which has become widespread in mechanical engineering, greatly simplifies production organization. Cooperation in production and the search for spare parts in the repair of products are greatly facilitated. The tolerance system is a compromise between structural designers and technologists. At the same time, it is desirable for the first of them to reduce the range of permissible values to the minimum values, and for the second, it is fraught with a significant increase in the production process cost.

The traditional tolerance system is not conducive to improving production at the workshop level since, for the production link, it is enough to adhere to the specified accuracy simply. Therefore, the concept of continuous quality improvement, especially effective at the workshop level, is not used.

The authors have proposed a control system for continuous quality improvement [1]. Its essence lies in the fact that the dimension suitability receives not discrete – good or bad – values, but continuous, which allows for a more flexible assessment of the proximity of the actual dimension to the desired one. The proposed system allows assessing each dimension separately and the part and assembly unit as a whole in terms of quality.

It is valuable to estimate the production process from the point of view of the proximity of dimensional accuracy to the desired design accuracy. Such an assessment would make it possible to compare various production technology options to choose the best one and theoretically substantiate the choice of production parameters in terms of obtaining the maximum share of products of the best quality.

2 Literature Review

The dimension distribution density function $f(x)$ of the parts is relatively well understood. In mechanical engineering, dimensions can have a normal distribution, equal probability, isosceles triangle, eccentricity, and composition. It is crucial for technologists to assess the production technology and to draw conclusions about its sufficient or insufficient accuracy in terms of grouping the actual dimensions relative to the optimal one.

In order to compare different technologies, numerical data should be converted to a dimensionless value. Jr. Harrington [2] was one of the first to propose using the desirability function for assessing production processes, thus offering a generalized indicator of quality. Later, G. Derringer and R. Suich proposed a mathematical description of the desirability function for the case when the variables have both lower and upper constraints, and there is also an optimal value [3].

Subsequently, the authors proposed other mathematical descriptions of the desirability function. These mathematical models had both certain advantages and disadvantages over the above. The model proposed in [4], along with its advantages, has lost the ability to consider the form of the desirability function and does not consider negative values.

Desirability functions are used in a wide variety of industries. They are one of the foundations of quality management in an enterprise and are designed to summarize various quality indicators or bring them to a standard scale [5].

The statistical approach to quality has become widespread. It allows you to significantly reduce the cost of production since parts are manufactured with wider tolerances, and rejection takes place at the stage of control of the entire product. With this approach, it is not guaranteed that all products will receive guaranteed quality, but only most of them [6]. This approach is especially effective when there are reliable methods for controlling the quality of the product as a whole.

For many decades, the quality system has been the backbone of the quality system [7]. Recently, however, quality has been interpreted more broadly, not only as compliance with the requirements of regulatory documents but also as compliance with consumer expectations, and even includes production safety [8].

The Japanese quality system makes extensive use of the quality loss function. Its meaning is that the technological characteristics of the product, which are considered the best in terms of quality, are set. The further the actual characteristics are from the desired ones, the greater the loss will be [9]. Such a system contributes to the improvement of quality, but consumers do not have a clear idea of the quality of the resulting product.

In the case of dimensional chains, the accuracy of the master link can be obtained at the processing stage and assembly stage. Computer selection of parts into sets allows obtaining sufficient resulting accuracy when using inaccurate components [10].

The inertial tolerance system further develops selective picking, which is popular in high-precision production [11]. Its essence lies in the absence of fixed tolerance limits, and the advantage is considerable flexibility in organizing the picking process.

The paper [12] proposes a methodology to improve investment efficiency by jointly optimizing product design, planning, and manufacturing for a product family. Allocation of tolerances as a sub-task of product design concerning process planning is solved in conjunction with long-term production planning.

In [13], an algorithm to optimize the distribution of tolerances in a dimensional chain based on various relevant cost criteria is proposed. An example from the aviation industry has illustrated the problem.

Different process options can significantly affect the quality and cost of a product. Developers prefer to use tight tolerances to ensure quality. Manufacturers prefer wide tolerances to reduce cost. To search for compromise solutions, it is proposed to use the objective function [14].

In contribution [15], attention is paid to optimizing the cost of tolerances during machining if the dimensional chains are interconnected. In this case, changing one tolerance affects several dimensional chains at once.

When assigning tolerances, we should permanently evaluate the percentage of rejects. The paper [16] compares various methods of scrap evaluation for tolerance optimization and provides recommendations to researchers and engineers.

Digital technologies allow applying virtual models of products, called digital twins. Such models allow the design stage to get an idea about the production and use of the product, which helps more correctly choose the technological modes of production [17]. The work [18] analyzes three approaches to tolerance analysis, namely tolerance stacks, vector loops, and tolerance analysis, based on a slight bias compared with classical tolerance analysis.

In the early stages of technology development, decision-making is based on rough predictions. The article [19] presents a data mining approach for knowledge-based agile manufacturing design. Knowledge of previous successful projects is used, which is extrapolated to the design project.

In Industry 4.0, control processes are predominantly automated [20]. It uses machine learning, a comprehensive account of the impact of all technological parameters on quality [21].

The desirability function *K(x)* can be represented as [1]:

$$K(x) = \begin{cases} 1 - \left(\frac{x-ko}{ei-ko}\right)^{\frac{1}{\beta}}, & x \leq ko, \\ 1 - \left(\frac{x-ko}{es-ko}\right)^{\frac{1}{\beta}}, & x \geq ko. \end{cases} \quad (1)$$

where *ko* – the desired value of the function; *ei, es* – the boundaries of the allowable function values, which may coincide with the boundaries of the tolerance field; β is the shape index when deviating from the desired dimension.

3 Research Methodology

Suppose we project the dimension distribution density function f(x) onto the desirability function K(x). In that case, we get a function that characterizes the probability distribution density of obtaining a dimension with a particular suitability value (Fig. 1). We will call this function the desirability density distribution function *g(y)*. Each of the combinations of the desirability function and the dimension distribution density function gives a different kind of dimension desirability density distribution function.

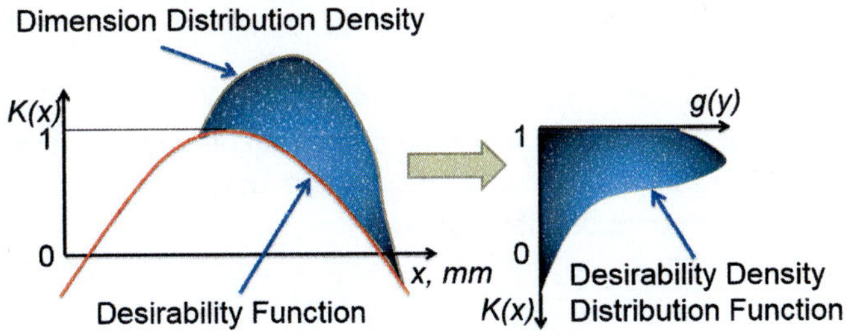

Fig. 1. Dimension desirability density distribution function.

The desirability density distribution function *g(y)* characterizes the production technology of parts in terms of quality. On the graph of this function, parts with dimensions close to the desired one are located in the area of arguments close to the unity element. Positive values of the arguments indicate the tolerability of such dimensions, and negative values – about intolerability.

This function can be built using statistical modeling. In order to obtain the distribution *g(y)* for each value on the dimension distribution curve (Fig. 2, point a), its suitability value (Fig. 2, point b) is found, multiplied by the probability of such p values, and the value of the function $g(y)$ is obtained (Fig. 2, point c). As the value of the maximum desirability distribution is removed from 1, and its density decreases, the quality of technology decreases, and negative values indicate a defective part.

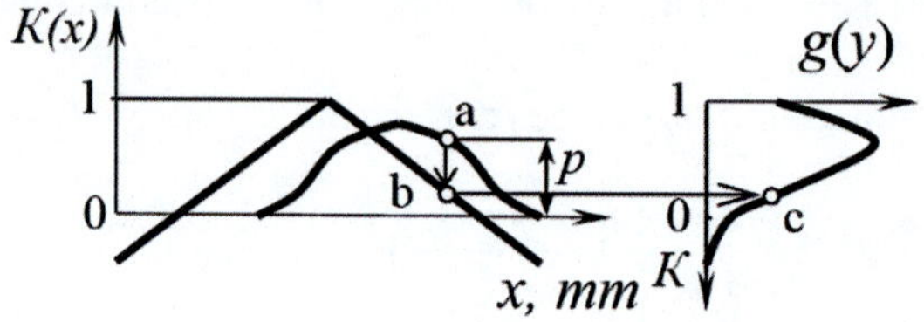

Fig. 2. Sequence of plotting the desirability density distribution function.

In practice, obtaining the desirability density distribution function in an analytical form is a topical problem. We will obtain the dimension desirability density distribution function for the case when the dimension distribution density obeys the normal law, with the arithmetic mean of the dimensions a and the standard deviation σ:

$$f(x) = \frac{1}{\sigma\sqrt{2\pi}} e^{-\frac{(x-a)^2}{2\sigma^2}} \tag{2}$$

The dimension desirability density distribution function for the case when dimensioning distribution density is normal (2), and the desirability function (1) has the form:

$$g(y) = \frac{1}{\sigma\sqrt{2\pi}} \left\{ \left|\beta\,(ko - ei)\,(1-y)^{\beta-1}\right| e^{-\frac{(ko-(ko-ei)\,(1-y)^{\beta}-a)^2}{2\sigma^2}} + \right.$$
$$\left. + \left|\beta\,(ko - es)\,(1-y)^{\beta-1}\right| e^{-\frac{(ko-(ko-es)\,(1-y)^{\beta}-a)^2}{2\sigma^2}} \right\}, \tag{3}$$

with $(-\infty < y \leq 1)$.

It is remarkable that for the case when the dimension distribution density is normal (2) and the desirability function is linear ($\beta = 1$), the desirability density distribution function also obeys the normal distribution law:

$$g(y) = \frac{1}{\sigma_G\sqrt{2\pi}} e^{-\frac{(y-a_G)^2}{2\sigma_G^2}} \tag{4}$$

where

$$a_G = \begin{cases} (a - ei)/(ko - ei), & x \leq ko, \\ (a - es)/(ko - es), & x \geq ko. \end{cases} \tag{5}$$

$$\sigma_G = \begin{cases} \sigma/(ko - ei), & x \leq ko, \\ \sigma/(ko - es), & x \geq ko. \end{cases} \tag{6}$$

Thus, the arithmetic mean value of the dimensions a_G of the desirability density distribution function is equal to the desirability value at point a. The root-mean-square deviation of dimensions σ_G of the desirability density distribution function differs from σ only by a multiplier (see (6)).

The dimension suitability distribution density function for the case when the dimension distribution density function is normal, and the desirability function is a parabola has the form:

$$g(y) = \frac{(es - ko)}{2\sigma\sqrt{2\pi(1-y)}}\left(e^{-\frac{(ko-(es-ko)\sqrt{1-y}-a)^2}{2\sigma^2}} + e^{-\frac{(ko+(es-ko)\sqrt{1-y}-a)^2}{2\sigma^2}}\right), \quad (7)$$

with $(-\infty < y \leq 1)$.

Formula (7) was derived under the assumption that $ei > c$ and $es < b$.

4 Results and Discussion

Let us use the example of the influence of the shape index of the dimension desirability β on the dimension desirability function from [1]. The tolerance Ø90F8$\left(^{+0,09}_{+0,036}\right)$ is selected as an example. We have, respectively, $ei = 90.036$, $es = 90.09$. Let the desired dimension be slightly offset from the tolerance field's middle and equal to $ko = 90.055$ mm. The graphs of the dimension desirability function are shown in Fig. 3 for the cases $\beta_1 = 2$, $\beta_2 = 1$, $\beta_3 = 0.5$, and $\beta_4 = 0.05$.

We plot the density function graphs for the dimension desirability distribution for the same four values of the rate of decrease in the dimension desirability: $\beta_1 = 2$, $\beta_2 = 1$, $\beta_3 = 0.5$, and $\beta_4 = 0.05$. The dimension distribution is assumed to be normal with the parameters a and σ. We plot the desirability values on the abscissa and the density of the dimension desirability distribution – on the ordinate axis. The simulation was done in Maple.

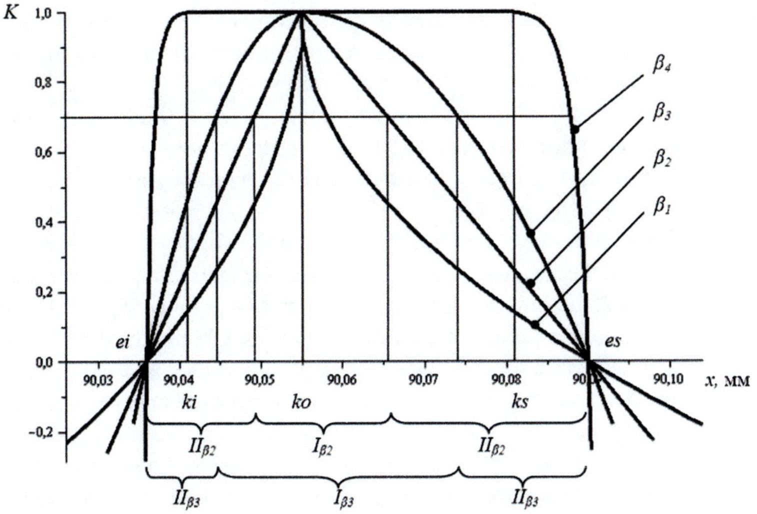

Fig. 3. Desirability function for various parameters of the β shape.

The simulation was carried out for four cases of input data. Example Fig. 4a) illustrates the technological case when 6σ is equal to the dimension tolerance T. Example Fig. 4b) illustrates the case with a technological margin of accuracy when $6\sigma << T$. Example Fig. 4c) illustrates a significant shift of the root-mean-square dimension relative to the desired one, Fig. 4d) – insufficient manufacturing accuracy, $6\sigma >> T$.

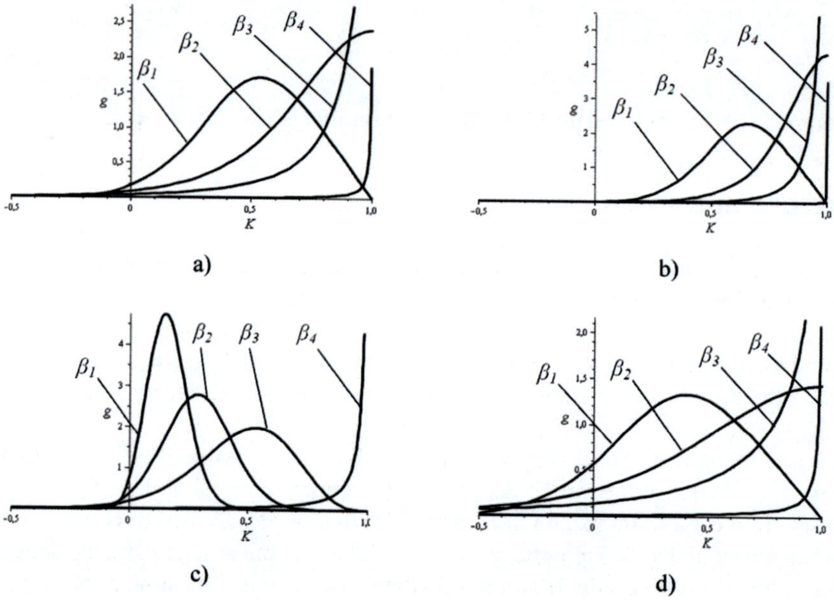

Fig. 4. The dimension suitability distribution density graphs.

Example Fig. 4a) illustrates the graphs of the dimension desirability distribution density for the classical case. When the center of the normal distribution curve $a =$ 90.055 mm coincides with the desired dimension, the dimension spread within 6σ is equal to the tolerance field, while the standard deviation $\sigma = 0.009$ mm. In this case, the number of parts with dimensions close to the optimum is maximum and decreases with distance from the center of the distribution curve, reaching almost zero values at the boundaries of the tolerance field.

Analysis of the graphs shows that not all values of the shape exponent of the dimension desirability function give a physically substantiated desirability density distribution function. So, for $\beta_1 = 2$, the function g has a maximum at $K(x) = 0.55$, which is logically wrong. At $\beta_4 = 0.05$, the maximum of the function g is real, but in the direction opposite to the maximum, it decreases too quickly, reaching zero values at $K(x) = 0.75$, which does not correspond to the actual dimension value distribution. For $\beta_2 = 1$ and $\beta_3 = 0.5$, the graphs of the function g are logically real, they have a maximum $K(x) = 1$ and reach zero values at $K(x) \approx 0$. There are practically no parts outside the positive suitability.

Example Fig. 4b) illustrates the desirability density distribution function for the case of manufacturing with a significant technological margin of accuracy. There are initial data: the center of the dimension distribution curve coincides with the desired dimension, and the root-mean-square deviation is reduced to $\sigma = 0.005$ mm. Thus, the resulting dimensions are grouped around the desired one. But there are no parts with dimensions close to the boundaries of the tolerance field. The remarks about the values of β_1 and β_4 are also valid for these inputs.

Examples Fig. 4c) and 4d) are constructed for the initial data characterizing the technology, in which there are parts with negative suitability, i.e., defective ones.

In the example in Fig. 4c), the center of the dimension distribution curve is displaced relative to the desired one, and it is equal to $a = 90.08$ mm, with the standard root-mean-square deviation $\sigma = 0.005$ mm. Such parameters of the distribution of the dimensions of parts are typical for the case of incorrect machine settings, while the production accuracy is sufficient. Analysis of the graphs shows that for $\beta_4 = 0.05$, the graph of the function g did not change its maximum, which is logically incorrect. For $\beta_2 = 1$ and $\beta_3 = 0.5$, the graphs of the function g are correct, and they have maxima in the area $K(x) \approx 0.5$. Most parts have a suitability value of 0–0.8. There are practically no parts with the desired dimensions, and, simultaneously, an insignificant number of the parts have negative suitability values, i.e., they are defective.

In the example in Fig. 4d), the center of the dimension distribution curve coincides with the desired one, but the scattering of dimensions is too large. The standard root-mean-square deviation $\sigma = 0.015$ mm. Such parameters of the dimension distribution of parts are typical for the case of insufficient production accuracy. Based on the statistical meaning, the suitability distribution density function should have a maximum in the range of values $g = 1$, and a significant number of the parts should have negative suitability. Functions with the shape exponent $\beta_2 = 1$ and $\beta_3 = 0.5$ are in good agreement with these provisions. The functions constructed for $\beta_1 = 2$ and $\beta_4 = 0.05$ do not allow assessing the technology in terms of quality.

5 Conclusions

The paper proposes a concept and tests the mathematical apparatus for constructing the desirability density distribution function, which characterizes the production process from the point of view of the proximity of the obtained characteristics to the desired ones. The scientific novelty of this approach is that it makes it possible to quantitatively analyze the production process as a whole and further manage it within the framework of Industry 4.0. The desirability density distribution function can be used for both dimensional accuracy and other quality parameters: shape error, rough edges, chipping, and hardness.

The paper shows that, although the desirability function $K(x)$ can be constructed for any values of the shape index with a deviation from the desired size β; in order to analyze the quality of production technology, using the desirability density distribution function, the β values should be limited to the range 0.3–1.2. When β is used outside this range, the desirability density distribution function does not allow determining the quality of the technological process.

Based on practical convenience, it should be recognized that the optimal value $\beta = 1$ since, in this case, the desirability function is linear, which significantly simplifies the mathematical apparatus of both its construction and the desirability density distribution function derived from it.

The estimation of the actual dimensions proposed in this model based on the degree of approximation of the link dimensions to the nominal characteristic can also be used in the method of calculating dimensional chains with a guaranteed value of the suitability of the closing link, which can be the basis for further research.

References

1. Kupriyanov, O., Trishch, R., Dichev, D., Bondarenko, T.: Mathematic model of the general approach to tolerance control in quality assessment. In: Tonkonogyi, V., Ivanov, V., Trojanowska, J., Oborskyi, G., Pavlenko, I. (eds.) InterPartner 2021. LNME, pp. 415–423. Springer, Cham (2022). https://doi.org/10.1007/978-3-030-91327-4_41
2. Harrington, J.: The desirability function. Ind. Qual. Control **21**(10), 494–498 (1965)
3. Derringer, G.C., Suich, D.: Simultaneous optimization of several response variables. J. Qual. Technol. **12**(4), 214–219 (1980)
4. Kim, K.-J., Lin, D.: Simultaneous optimization of mechanical properties of steel by maximizing exponential desirability functions. Appl. Statist. **48**(3), 311–325 (2000)
5. Ngoi, B.K.A., Seow, M.S.: Tolerance control for dimensional and geometrical specifications. Int. J. Adv. Manuf. Technol. **11**, 34–42 (1996). https://doi.org/10.1007/BF01177182
6. Leblond, L., Pillet, M.: Conformity and statistical tolerancing. Int. J. Metrol. Qual. Eng. **9**, 1 (2018)
7. Wagner, T., Trautmann, H.: Integration of preferences in hypervolume-based multiobjective evolutionary algorithms by means of desirability functions. IEEE Trans. Evol. Comput. **14**(5), 688–701 (2010)
8. Trisch, R., Gorbenko, E., Dotsenko, N., Kim, N., Kiporenko, A.: Development of qualimetric approaches to the processes of quality management system at enterprises according to international standards of the ISO 9000 series. Eastern-Eur. J. Enterprise Technol. **4**(3–82), 18–24 (2016). https://doi.org/10.15587/1729-4061.2016.75503
9. Babu, J.R., Asha, A.: Minimising assembly loss for a complex assembly using Taguchi's concept in selective assembly. Int. J. Prod. Qual. Manag. **15**(3), 335–356 (2015)
10. Kupriyanov, O., Lamnauer, N.: Improvement of the assembling technology for precision joints using the dimensional information. In: Tonkonogyi, V., et al. (eds.) InterPartner 2020. LNME, pp. 52–60. Springer, Cham (2021). https://doi.org/10.1007/978-3-030-68014-5_6
11. Pillet, M., Maire, J.L.: Inertial tolerancing and acceptance sampling. Int. J. Eng. Innovative Technol. **2**(10), 265–271 (2013)
12. Tsutsumi, D., et al.: Towards joint optimization of product design, process planning and production planning in multi-product assembly. CIRP Ann. **67**(1), 441–446 (2018). https://doi.org/10.1016/j.cirp.2018.04.036
13. Diet, A., Couellan, N., Gendre, X., Martina, J., Navarro, J.-P.: A statistical approach for tolerancing from design stage to measurements analysis. Procedia CIRP **92**, 33–38 (2020). https://doi.org/10.1016/j.procir.2020.05.171
14. Etienne, A., et al.: Cost engineering for variation management during the product and process development. Int. J. Interact. Des. Manuf. (IJIDeM) **11**(2), 289–300 (2016). https://doi.org/10.1007/s12008-016-0318-3
15. Heling, B., Aschenbrenner, A., Walter, M., Wartzack, S.: On connected tolerances in statistical tolerance-cost-optimization of assemblies with interrelated dimension chains. Procedia CIRP **43**, 262–267 (2016). https://doi.org/10.1016/j.procir.2016.02.031
16. Hallmann, M., Schleich, B., Heling, B., Aschenbrenner, A., Wartzack, S.: Comparison of different methods for scrap rate estimation in sampling-based tolerance-cost-optimization. Procedia CIRP **75**, 51–56 (2018). https://doi.org/10.1016/j.procir.2018.01.005

17. Schleich, B., Anwer, N., Mathieu, L., Wartzack, S.: Shaping the digital twin for design and production engineering. CIRP Ann. **66**(1), 141–144 (2017). https://doi.org/10.1016/j.cirp.2017.04.040
18. Schleich, B., Wartzack, S.: A quantitative comparison of tolerance analysis approaches for rigid mechanical assemblies. Procedia CIRP **43**, 172–177 (2016). https://doi.org/10.1016/j.procir.2016.02.013
19. Kretschmer, R., Pfouga, A., Rulhoff, S., Stjepandic, J.: Knowledge-based design for assembly in agile manufacturing by using data mining methods. Adv. Eng. Inform. **33**, 285–299 (2017). https://doi.org/10.1016/j.aei.2016.12.006
20. Muller, T., Pillet, M., Maire, J.L., Pillet, D.: Why and how to move from SPC (Statistical Process Control) to APC (Automated Process Control). In: Aifaoui, N., et al. (eds.) CMSM 2019. LNME, pp. 33–40. Springer, Cham (2020). https://doi.org/10.1007/978-3-030-27146-6_4
21. Singer, G., Cohen, Y.: A framework for smart control using machine-learning modeling for processes with closed-loop control in Industry 4.0. Eng. Appl. Artif. Intell. **102**, 104236 (2021). https://doi.org/10.1016/j.engappai.2021.104236

Application of Advanced Packaging Technology

Alona Kysylevska[1](✉), Konstantin Babov[1], Tatiana Bezverkhniuk[1], and Ihor Prokopovych[2]

[1] State Institute «Ukrainian Scientific-Research Institute of Medical Rehabilitation and Balneology of the Ministry of Health of Ukraine», 6, Lermontovsky Lane, Odessa 65014, Ukraine
kisilevskaya07@gmail.com

[2] Odessa Polytechnic National University, 1, Shevchenko Ave., Odessa 65044, Ukraine

Abstract. The article substantiates the technology of industrial packaging of natural healing hydrogen bicarbonate sodium waters of medium and high mineralization containing iron. The technological solution is based on adding a solution of ascorbic acid by injecting natural water directly into the pipe at the well during its extraction using a unique installation. Ascorbic acid, forming a complex compound with iron, stabilizes the chemical composition of water. The results of physical and chemical studies proved the quality of packaged healing water when stored for 16 months. Recommendations have been developed for improving the production technology of natural healing water, boric bicarbonate sodium water of medium and high mineralization, which contains iron. Results of the work provide a reason to apply this technology of stabilization of sodium bicarbonate natural healing waters of medium and high mineralization, which contain iron, with ascorbic acid to preserve the quality of packaged water in comparison with the primary natural source and to use them in non-resort conditions for the treatment and prevention of diseases.

Keywords: Water quality · Natural healing water · Natural mineral water · Ascorbic acid · Iron · Stabilization technology

1 Introduction

Nowadays, the culture of natural mineral waters (MW) consumption and natural healing waters (HW) has grown, which are an alternative to drinking treatment with natural healing waters at the resort.

Requirements for MW and their packaging process at the European level are presented in Directive 2009/54/EC, according to which MW is a food product. Any designation of MW properties related to preventing and treating human diseases is prohibited. But it is allowed to put on the labeling notations related to the content of components and the effects of exposure to the body based on the results of clinical analyzes of MW. The directive prohibits the processing of MW, except for the extraction or addition of carbon dioxide and the separation of unstable elements: iron, manganese, sulfur, arsenic – oxidation with ozone-enriched air and fluorides [1].

V. Tonkonogyi et al. (Eds.): InterPartner 2022, LNME, pp. 340–349, 2023.
https://doi.org/10.1007/978-3-031-16651-8_32

Packaged HW are considered medicinal products according to Directive 2001/83/EC. They are being sold in the pharmacy chain. Medical indications and contraindications should be developed for these waters. MW and HW are used to prevent and treat many diseases [2].

During packaging MW and HW, it is essential to preserve their medicinal properties due to the gas and ionic composition containing trace elements. Equally important is the packaging technology to ensure that the composition of the packaged waters matches the natural water from the well (source). Only if the medicinal properties of MW are preserved during their production can one count on obtaining products of the appropriate quality. In particular, this applies to MW and HW containing biologically active components compounds and dissolved gases [3].

Specific biologically active components and compounds of MW, in particular iron and dissolved gases, such as carbon dioxide (CO_2), are complex components in terms of the technology for the production of packaged MW, which can affect their content – significantly reduced.

MW's most common iron states are ferrous and trivalent iron (Fe^{2+} and Fe^{3+}). When packaging MW, a brown precipitate of iron hydroxide ($Fe(OH)_3$) may form, making a non-marketable appearance for the product. The iron in the MW is absorbed much better than that in the drug, so it is essential to keep it in the MW in the divalent form [4].

De-ironing of natural waters by air can lead to a number of consequences that will worsen their quality:

- increasing water pH (over 7) is necessary to precipitate iron compounds. For naturally gassed MW and HW, this oxidation can only occur through the extraction of CO2, and this can lead to the formation and precipitation of carbonates;
- removal of insoluble $Fe(OH)_3$ particles by filtration can lead to a change in the number of water microbiota and post-bacterial proliferation;
- de-ironing by air can cause denitrification of water with the formation of nitrites during water storage, temporary stoppage of the filter, etc. [1].

The MW iron removal technology with ozone-enriched air is more applicable for drinking water [5]. However, concerning MW and HW it has several disadvantages:

- a series of HW and MW contain bromides, which can be oxidized to bromates due to ozone and exceed the norm established by Directive 2003/40/EC [6];
- ozone can affect the quantitative and qualitative composition of MW and HW microbiota and, in the presence of organic substances in them, cause the formation of by-products of their oxidation [1].

To prevent the formation of $Fe(OH)_3$ in bottles, some manufacturers subject MW and HW to aeration by prolonged settling in open containers, which leads to oxidation of Fe^{2+} and the formation of $Fe(OH)_3$ precipitate even before MW enters the bottles. In this case, MW and HW are degassed, and the chemical equilibrium is shifted towards the formation of calcium carbonate. At the same time, $Fe(OH)_3$ adsorbs CO_2, and carbon dioxide content in water decreases. Therefore, the use of this technology is erroneous and significantly affects the chemical composition of MW and HW. More complicated but correct is the stabilization of Fe^{2+} compounds, which is achieved by regulating the pH of MW and HW, adding antioxidants or complexing agents, and, more often, food acids. The most rational method of stabilizing iron is with the help of ascorbic acid (AA). This method is practiced in producing packaged ferrous (>10 mg/l) HW but is not used to create packaged sodium bicarbonate with lower iron content.

CO_2 in this HWs is a natural stabilizer of their chemical composition. A decrease in its content during the packaging process leads to the fact that there is a violation of the carbonate balance and the need for additional stabilization of ferrous iron.

The work aims to develop a technology for stabilizing the chemical composition of carbonate bicarbonate sodium HW containing iron with ascorbic acid to preserve their quality for their packaging.

2 Literature Review

Bicarbonate sodium MW medium and highly mineralized (5…10 g/l) are used to prevent and treat gastrointestinal tract diseases [7]. The internal use of bicarbonate MW, depending on the method of administration, reduces or stimulates the secretion of hydrochloric acid, affects the motility of the stomach and biliary tract [8], and reduces cholesterol levels [9]. The group of moderately mineralized bicarbonate sodium MW in Ukraine is represented mainly by boron (>35 mg/l) and carbonic (>0.5 g/l) waters. Almost all of these waters contain iron.

The leading supplier of iron compounds in MW is the processes of chemical weathering and dissolution of rocks. Iron reacts with MW compounds, forming a complex of compounds in the water in a dissolved, colloidal, and suspended state.

Glandular MW is indicated for iron deficiency [8], in pregnant women, especially in treating anemia [10]. The Italian Geothermal Alliance recommends using MW with an iron content >1 mg/l for anemia and iron deficiency [11].

According to the Order of the Ministry of Health of 02.06.2003 No. 243, the balneological norm of iron in MW in Ukraine is 10 mg/l. According to the "Quality Criteria of the European Association of Spas" –20 mg/l. In Poland (according to the Law on Geological and Mining Law) and Bulgaria (according to the Decree of the Minister of Health of August 3, 1987 No. 14) – >10 mg/l. For MWs, according to Directive 2009/54/EC, an iron content >1 mg/l entitles the MW to be labeled with this information; deironing of MW with air with ozone is possible, but the addition of anything other than CO_2 is prohibited.

In Eastern Europe, there was a technology for producing ferrous MW with the addition of food acids (citric or ascorbic). AA was usually added to MW. AA is a powerful water-soluble antioxidant that performs the biological functions of a reducing agent and a coenzyme in some metabolic processes. At the same time, AA can act as a chelator of metal ions [12]. The effect of AA, which enhances iron absorption, is more pronounced than that of other organic acids due to its ability to reduce Fe^{3+} to Fe^{2+}. Based on the limited data available, other organic acids may only be effective at acid to Fe molar ratios exceeding 100 [13].

There are separate studies on adding food acids to stabilize iron ions, but predominantly in drinking water. At the same time, iron was also added at 10 mg/l. Thus, authors from Brazil [14] and [15] have shown in clinical studies that drinking water enriched with iron stabilized by AA effectively prevents childhood anemia. Scientists from India [16] experimentally showed that the optimal concentration of iron (5 mg/l) and AA solution could improve anemic conditions. The authors [17] demonstrated that AA has an effect by increasing iron absorption. AA in the intestine enhances the absorption of iron and selenium from food [18].

It is necessary to work out the optimal technology for stabilizing the chemical composition to produce packaged MW with iron content. In the presence of AA, iron entirely remains in the divalent form, while citric and tartaric acids partially convert Fe^{2+} compounds into soluble Fe^{3+} compounds.

There were few attempts to produce such MWs. These were mainly glandular (> 10 mg/l) MW. For the production of ferruginous carbonate (16 mg/l) moderately mineralized (9.2…9.4 g/l) chloride-bicarbonate magnesium-sodium water of the Malkinskoye deposit, the authors [19] experimentally determined the optimal concentration of AA in the amount of 70…100 mg/l. The authors of [20] stabilized the composition of MW ferrous bicarbonate magnesium-calcium ferruginous (45…60 mg/l) Martial water by adding a mixture of 0.65 g of citric acid and 0.35 g of AA per 1 L of MW and treating the vials with hot steam at a pressure of 8 atm. However, this technology has not been developed to produce packaged carbonated sodium hydrogen carbonate HW with lower iron content.

3 Research Methodology

The study consists of 7 stages (Fig. 1):

- I – selection of the object of research. The object of the study was chosen to be the carbonated boron highly mineralized hydrogen carbonate sodium HW "Polyana Kvasova" (well No. 7-rz, the village of Polyana, Transcarpathian region), which contains iron and is a known HW in Ukraine and abroad;
- II. At the stage of developing the general principles of the technology for stabilizing the chemical composition of carbon dioxide sodium bicarbonate HW with AA, which contains iron, a stabilizing acid – AA – was selected, and its required concentration was determined by the concentration of iron in HW;
- III. At this stage, the installation of the supply of AA stabilizing solution to the HW pipe at the well was developed, taking into account that the supply of AA solution should be continuous and taking into account the percentage of the gas factor;
- IV – practical implementation of the technology of stabilization of the chemical composition of HW of hydrocarbon sodium HW with iron using AA. At this stage, they received HW, stabilized AA;
- V – quality research (determination of physical and chemical parameters) of HW with the addition of AA at all critical points of the HW production process (directly at the factory). Quality indicators (physico-chemical) of stabilized HW were studied at all critical points of the technological scheme of production (from the well, after installation of AA stabilization, tank with HW, and finished products). The content of CO_2,

hydrogen carbonate ions (by titration), iron ions (photometrically), hydrogen index (pH), and oxidation-reduction potential (Eh) (potentiometrically) were determined;
- VI – stationary laboratory studies (in the testing laboratory (Odesa) of physical and chemical parameters (extended list) of packaged HW;
- VII – a study of the stability of physico-chemical indicators of HW with AA packaged in 1.5 l PET bottles and 0.5 l glass bottles during 16 months of storage.

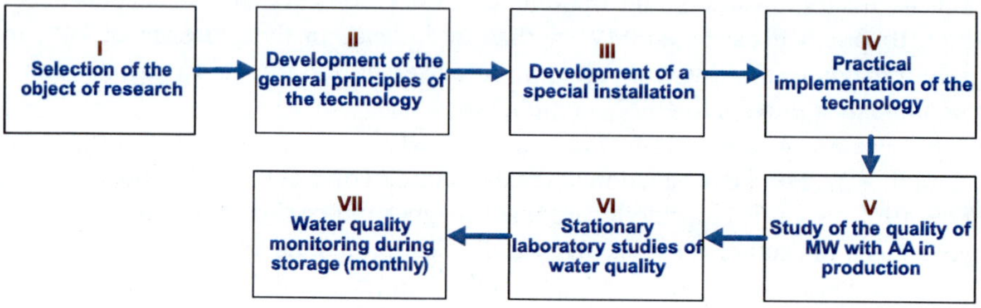

Fig. 1. Stages of research.

The formula determined the concentration of iron ions (mg/l):

$$X = \frac{V_1 n \cdot 28 \cdot 1000}{V_2}, \quad (1)$$

where V_1 – is the volume of a solution of complex III spent on the titration of iron ions, ml; V_2 – a volume of water, ml; n is the normality of the solution of complex III; 28 is the gram equivalent of Fe.

4 Results and Discussion

According to the developed technology, HW is fed from the well through a pipe through a unique installation with AA solution, which is injected directly into the water, taking into account the percentage of the gas factor. Installation (Fig. 2) includes a mass flow meter, a pH meter, and a dosing device. The concentration of the AA solution is calculated from the iron content and its consumption. After the dosing unit, MW is fed into a sealed collection tank (it is possible to use tank trucks with a system for maintaining excess pressure 0.02…0.05 MPa, which provides CO_2). Filling of the collector or tank trucks is carried out gradually until the air is completely expelled, which is checked by the cloudiness of the $Ca(OH)_2$ solution.

According to the results of physical and chemical studies, the packaged MW "Polyana Kvasova" is classified according to its composition as boric (orthoboric acid content – 228 mg/l), highly mineralized (10.37 g/l) bicarbonate sodium water. According to the calculation, 26.40 mg/l of AA was added in MW. The results of monitoring the physicochemical composition of MA are presented in Table 1.

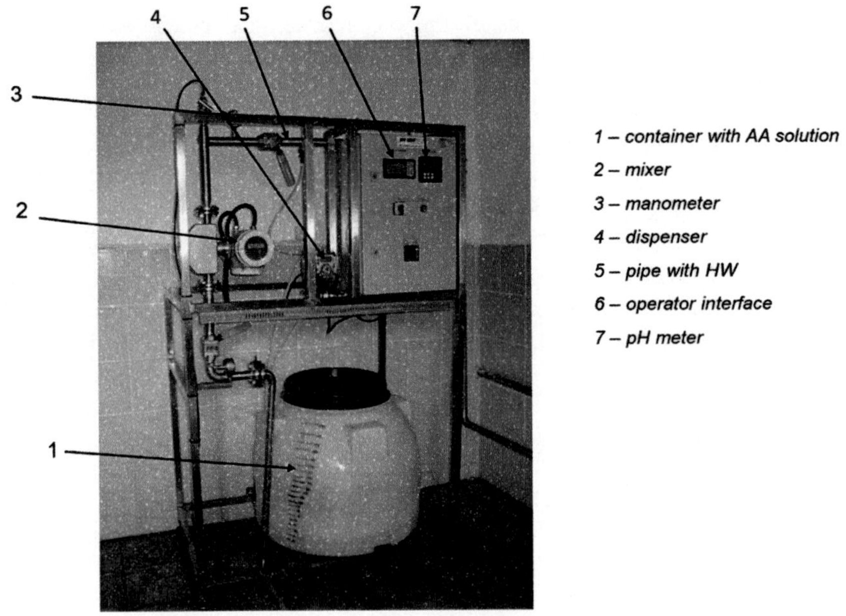

Fig. 2. A view of the HW stabilization plant with AA.

Table 1. Results of monitoring the quality of MW "Polyana Kvasova" during its production.

Indicator	Content, mg/l							
	Before stabilization with AA solution		After stabilization with AA solution		Storage tank		Final product **	
	PET	Glass bottles	PET	Glass bottles	PET	Glass bottles	PET	Glass bottles
CO_2 *	3061.20	2176.8	2458.80	1983.80	1588.90	1641.38	4040.00	4224.70
HCO_3^-, *	6588.0	6820.0	6588.0	6893.0	6466.0	6893.0	6466.0	6893.0**
Fe^{2+}*(Fe^{3+})	0.90	0.85	0.90	0.90	0.90	0.90	0.90	0.90*(0.70+0.10)
pH	6.55***	6.55***	6.65***	6.65***	–	–	6.25	6.30
Eh, mV	+290***	+290***	+250***	+250***	–	–	+330	+330

Note: * – research at the plant; ** – water storage – 12 days (stationary research); *** – water storage – 2 days.

As can be seen from Table 1, the physicochemical properties of MW stabilized by AA are basically preserved. The content of CO_2 in the tank is lower (lost during transportation); still, in the finished product, it increases due to the addition of CO_2, which acts as an additional mandatory preservative, to the bottles during packaging. As a result, the pH in the finished product decreases. The content of bicarbonate ions fluctuates somewhat during MW pumping. Stabilizing the chemical composition is the stable content of Fe^{2+}; however, Fe^{3+} appeared in the glass container. After AA stabilization, Eh decreases.

Next, we studied the dynamics of the physicochemical characteristics of MW "Polyana Kvasova" (highly carbonated) in PET and glass bottles during 16 months of storage (every 3 months, and the content of CO_2 and iron – monthly).

PET bottles. During storage, the pH of the water changed slightly – 6.25…6.70, Eh fluctuated within +290… +370 mV. The concentration of HCO_3 ions decreased (from 6.893 g/l to 6.527 g/l), the content of CO_2 and orthoboric acid decreased from 0.48% to mass to 0.29% to mass and 228 mg/l to 159 mg/l.

Glass bottles. The water pH changed slightly – 6.25…6.70, Eh fluctuated within +290… +370 mV. The concentration of HCO_3 ions slightly decreased (from 6.893 g/l to 6.405 g/l), while the content of CO_2 and orthoboric acid decreased from 0.567% to mass to 0.429% to mass and from 207 mg/l to 1. The content of nitrite ions was <0.008 mg/l; the content of nitrate ions – <0.34 mg/l in both types of containers. The content of ammonium ions in PET bottles was varied (from 3.18 mg/l to 6.89 mg/l), and in glass bottles (from 7.90 mg/l to 8.14 mg/l). Its content in MW is not standardized.

It was interesting to study the dynamics of preserving the ferrous form of iron during 16 months of storage (Fig. 3 and Fig. 4).

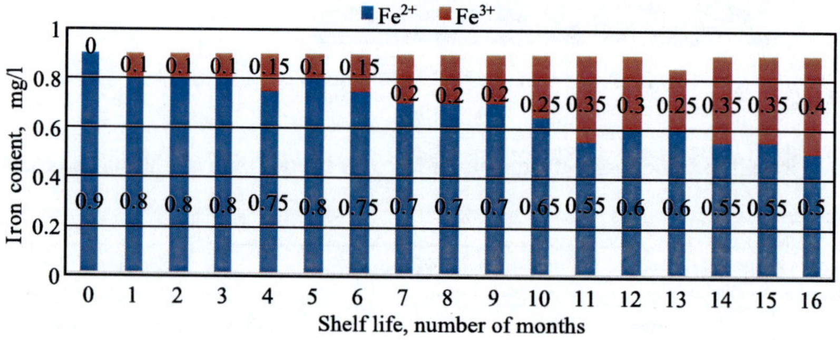

Fig. 3. Dynamics of iron content in MW "Polyana Kvasova" (highly carbonated) in PET bottles.

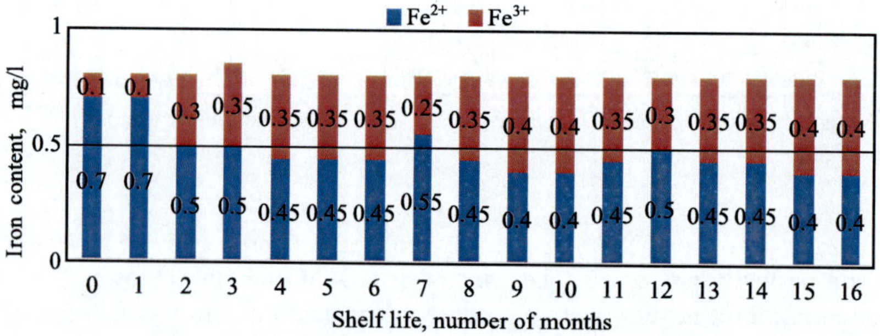

Fig. 4. Dynamics of iron content in MW "Polyana Kvasova" (highly carbonated) in glass bottles.

From Fig. 3 and 4, it can be seen that ferrous iron was preserved in the glass bottle in the same amount as in the PET bottle – 0.4 mg/l. It should be noted that the trivalent form began to appear at the beginning of storage. But a significant amount of iron remained stabilized with ascorbic acid until the end of storage.

It is known that the PET material allows ultraviolet rays and oxygen to enter the bottle, and CO_2 to the outside, which reduces the quality of the water and reduces the shelf life. This phenomenon is because the high-molecular structure of polyethylene terephthalate is not an obstacle for small gas molecules compared to polymer chains. Analyzing the data in Fig. 5 and 6, we observe more CO_2 loss in the PET bottle (29.58%) compared to the glass bottle (24.34%), confirming the disadvantages mentioned above of using the PET material. Therefore, this fact should be considered when packing, and CO_2 should be added to more than 0.48 wt.%. The concentrations of other components (ferrous iron, bicarbonates, and orthoboric acid) decreased almost equally in both PET and glass bottles. The most significant losses in both types of containers occurred in Fe^{2+} content.

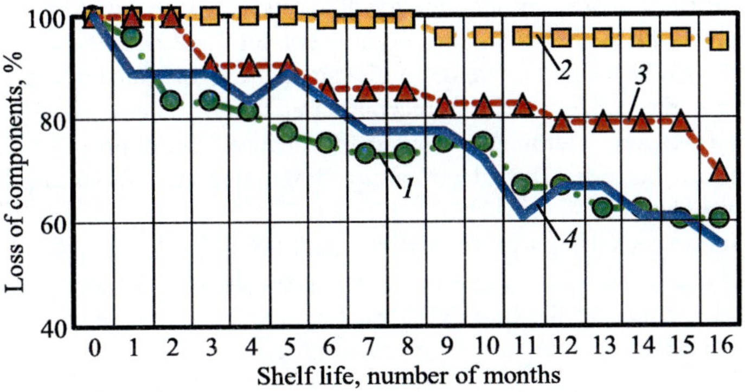

Fig. 5. Loss of basic and biologically active components and compounds of MW "Polyana Kvasova", packed in a PET bottle: *1* – CO_2; *2* – HCO_3^-; *3* – H_3BO_3; *4* – Fe^{2+}.

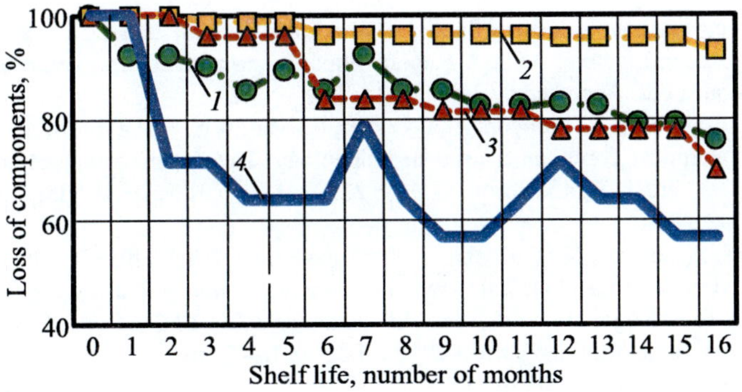

Fig. 6. Loss of basic and biologically active components and compounds of MW "Polyana Kvasova", packed in a glass bottle: *1* – CO_2; *2* – HCO_3^-; *3* – H_3BO_3; *4* – Fe^{2+}

Taking into account the results of the work for industrial packaging of any medium and highly mineralized sodium bicarbonate natural healing waters that contain iron, it is possible to recommend the use of the technology of stabilizing their chemical composition with ascorbic acid. To minimize the inhibition of the stabilization process due to the presence of a significant concentration of hydrocarbons, it is recommended to use a concentration of ascorbic acid that is 2 times greater than that calculated by the iron content. Natural healing water stabilized with ascorbic acid is recommended to be packaged in glass bottles to reduce carbon dioxide loss over time.

5 Conclusions

Applying packaging technology of natural healing highly mineralized boron bicarbonate sodium water, which contains iron, using ascorbic acid leads to stabilization of its chemical composition. The results of physico-chemical studies proved the preservation of water quality during storage for 16 months.

To optimize the packaging technology of the tested water, it is recommended to increase the amount of the stabilizing component for PET bottles—ascorbic acid to 50 mg/l (there will be a slight decrease in pH) since the significant content of hydrocarbons in the water neutralizes it and inhibits the stabilization process. It is worth switching to producing packaged water in glass containers because it better preserves carbon dioxide and chemical balance. The added carbon dioxide content should be increased to $>0.50\%$ by weight.

The proposed technology of stabilizing medium and highly mineralized sodium hydrogen carbonate natural healing waters, which contain iron with ascorbic acid, allows for maximally preserving the quality of packaged water compared with the primary natural source and to use them in non-resort conditions for the treatment and prevention of diseases.

References

1. Agence Française pour la Sécurité Sanitaire des Aliments. Guidelines for the safety assessment of natural mineral waters. AFSSA Report. https://www.anses.fr/fr/system/files/EAUX-Ra-EauxMinerales.pdf. Accessed 02 July 2022
2. Morer, C., Roques, C.-F., Françon, A., Forestier, R., Maraver, F.: The role of mineral elements and other chemical compounds used in balneology: Data from double-blind randomized clinical trials. Int. J. Biometeorol. **61**(12), 2159–2173 (2017). https://doi.org/10.1007/s00484-017-1421-2
3. Kysylevska, A., Babov, K., Gushcha, S., Prokopovich, I., Nasibullin, B.: Using the specific molarity indicator of the chemical parameters of mineral waters in assessing their biological effects. In: Tonkonogyi, V., et al. (eds.) InterPartner 2020. LNME, pp. 823–832. Springer, Cham (2021). https://doi.org/10.1007/978-3-030-68014-5_80
4. Dutra-de-Oliveira, J.E., Marchini, J.S., Lamounier, J., Almeida, C.A.: Iron-fortified drinking water studies for the prevention of children's anemia in developing countries. Anemia **2011**, 815194 (2011). https://doi.org/10.1155/2011/815194
5. Martinelli, M., Giovannangeli, F., Rotunno, S., Trombetta, C.M., Montomoli, E.: Water and air ozone treatment as an alternative sanitizing technology. J. Prev. Med. Hyg. **58**(1), E48–E52 (2017). https://doi.org/10.15167/2421-4248/jpmh2017.58.1.757

6. Al-Ansi, S.A., Othman, A.A., Al-Tufail, M.A.: Bromate pollutant in ozonated bottled Zamzam water from Saudi Arabia determined by LC/ICP-MS. J. Environ. Sci. Health Part A **46**(13), 1529–1532 (2011)
7. Petraccia, L., et al.: Water, mineral waters and health. Clin. Nutr. **25**(3), 377–385 (2006)
8. Dumitrescu, M., et al.: Benefits of crenotherapy in digestive tract pathology (Review). Exp. Ther. Med. **23**(2), 122 (2022). https://doi.org/10.3892/etm.2021.11045
9. Quattrini, S., Pampaloni, B., Brandi, M.L.: Natural mineral waters: Chemical characteristics and health effects. Clin. Cases Miner. Bone Metab. **13**(3), 173–180 (2016)
10. Albertini, M., Dacha, M., Teodori, L., Conti, M.: Drinking mineral waters: Biochemical effects and health implications – the state-of-the-art. Int. J. Environ. Health **1**(1), 153–169 (2007)
11. Unione Geotermica Italiana. Geotermia, notizario dell' Unione Geotermica Italiana. VIII. https://www.unionegeotermica.it/notiziari/UgiNotiziario26.pdf. Accessed 24 Jan 2022
12. Timoshnikov, V.A., Kobzeva, T.V., Polyakov, N.E., Kontoghiorghes, G.J.: Redox interactions of vitamin C and iron: Inhibition of the pro-oxidant activity by deferiprone. Int. J. Mol. Sci. **21**(11), 3967 (2022). https://doi.org/10.3390/ijms21113967
13. Teucher, B., Olivares, M., Cori, H.: Enhancers of iron absorption: Ascorbic acid and other organic acids. Int. J. Vitam. Nutr. Res. **74**, 403–419 (2004). https://doi.org/10.1024/0300-9831.74.6.403
14. De Almeida, C.A., et al.: Effect of fortification of drinking water with iron plus ascorbic acid or with ascorbic acid alone on hemoglobin values and anthropometric indicators in preschool children in day-care centers in Southeast Brazil. Food Nutr. Bull. **26**(3), 259–265 (2005). https://doi.org/10.1177/156482650502600302
15. De Almeida, C.A.N., De Mello, E.D., Ramos, A.P.R., João, C.A., João, C.R., Dutra-de-Oliveira, J.E.: Assessment of drinking water fortification with iron plus ascorbic acid or ascorbic acid alone in daycare centers as a strategy to control iron-deficiency anemia and iron deficiency: A randomized blind clinical study. J. Trop. Pediatr. **60**(1), 40–46 (2014). https://doi.org/10.1093/tropej/fmt071
16. Chaturvedi, R., et al.: Iron-rich drinking water and ascorbic acid supplementation improved hemolytic anemia in experimental Wistar rats. Int. J. Food Sci. Nutr. **65**(7), 856–861 (2014). https://doi.org/10.3109/09637486.2014.918589
17. Krisnanda, R.: Vitamin C Helps in the absorption of iron in iron deficiency anemia. J. Penelitian Perawat Profesional **2**(3), 279–286 (2020). https://doi.org/10.37287/jppp.v2i3.137
18. Solomons, N.W., Viteri, F.E.: Biological interaction of ascorbic acid and mineral nutrients. In: Seib, P.A., Tolbert, B.M. (eds.) Ascorbic Acid: Chemistry, Metabolism, and Uses, pp. 551–569. American Chemical Society, Washington, D. C. (1982). https://doi.org/10.1021/ba-1982-0200.ch023
19. Leonova, V.G., Remneva, G.A., Murashko, E.N.: New mineral water for industrial bottling. Beer and Drinks **1**, 44–45 (2005). (in Russian)
20. Sheremetyev, O.V.: Method for the industrial manufacture of mineral healing-table hydrocarbonate magnesium-calcium iron martial water. Patent of Russia No 2623155C1 (2017)

The Advanced Technology of Economic Efficiency Evaluation

Natalia Lishchenko[1], Vasily Larshin[2(✉)], Artem Mochuliak[3], and Victor Marchuk[4]

[1] Technical University of Kosice, 1, Bayerova Street, 08001 Prešov, Slovakia
[2] Odessa Polytechnic National University, 1, Shevchenko Avenue, Odessa 65044, Ukraine
vasilylarshin@gmail.com
[3] Technical University of Kosice, 1/9, Letná Street, 04200 Košice, Slovakia
[4] Lutsk National Technical University, 75, Lvivska Street, Lutsk 43018, Ukraine

Abstract. No quotient allows going directly from technical savings to saving time or money. It requires a special procedure, depending on the specific case and conditions for making the cost-benefit evaluation decision. One of such possible procedures is developed in this paper. The procedure is based on analyzing the boundary between a useful and useless technical solution. The lack of a technical-and-economic model encourages irresponsibility and reckless use of valuable resources. A case study for new sol-gel corundum 3SG46Hs12Vs grinding wheel instead of conventional white electro-corundum 25AF46L6V one has shown the former's efficiency despite the lower price of the latter. This case study given in the paper can be an example of the effectiveness of evaluating other technical solutions in science, engineering, and technology. The relationship between categories of technical efficiency (potential ratio of physical output to physical input, productivity (product quantity per unit of time), and effectiveness (real ratio of physical output to physical input) is investigated. A new technical and economic quotient has been formulated, which shows what kind of productivity gain a new technical solution in science, engineering, and technology should bring so that at least a minimal economic benefit from this technical solution appears.

Keywords: Economic growth · Investment decisions · Economic performance · Initial investment · Payback period · Rate-of-return · Profit rate · Profitability level · Cost-benefit evaluation · Gear grinding · Technical efficiency · Grinding wheels

1 Introduction

One of the difficulties when seeking to advance the technological aspects of equipment or a process is to overcome the so-called economic constraints. For example, wheel and labour costs per part must be cut while maintaining or improving target quality (part accuracy, its surface finish and integrity) to accomplish economic viability in grinding. The economic approach to assessing production activities' viability (feasibility) has many advantages. It allows converting individual technological advantages of a process

V. Tonkonogyi et al. (Eds.): InterPartner 2022, LNME, pp. 350–359, 2023.
https://doi.org/10.1007/978-3-031-16651-8_33

into comparable (and uniform for all) units of time and cost. Any process must be judged using the overall costs accounting approach and not by initial costs alone, which may be high. For example, the high initial costs can lead to lower overall costs. The general problem is now obtaining a cost model sensitive to the effects of process structure and parameters.

Today, investments in new machinery can be expensive and risky. Many formulas embodying various interpretations of return-on-investment have been used in engineering for a great many years. No one method offers the final answer to all decisions.

2 Literature Review

An indication of which methods are now being used in industry to justify capital purchases was revealed in a study conducted by the University of Michigan's Institute of Science and Technology [1]. The study included a survey showing the most important investment decision methods (Fig. 1).

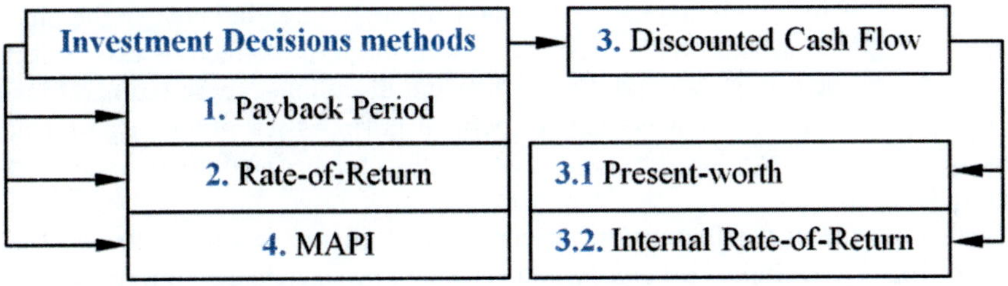

Fig. 1. Investment decisions methods review [1].

These methods, however, do not recognize the economic value of time. Besides, no one method mentioned in Fig. 1 offers the final answer to all decisions, and there is no rule for choosing the desired cost model.

The payback period method is widely used in evaluating scientific and production activities. Economic analysis includes assessing costs and benefits, the ratio between which determines the payback period [2].

In the 1970s, it was noticed that the cost of large-scale manufacturing systems was rising at a high rate, but many projects failed or resulted in defective products. This situation was called the software crisis, in which the cost of eliminating defects was the most important [3].

An innovative approach to evaluating investments from an economic point of view takes into account the costs and benefits in terms of security [4]. It allows the concepts of security and risk to be translated into monetary terms by integrating known indicators and measures from economic theory. They are net present value, internal rate of return, and payback period.

Engineers must always consider the capital and operating costs associated with each technical decision they make. For experienced engineers, this becomes pretty much natural, but beginners will need to acquire such experience [5].

Modeling energy sustainability is one of the criteria for making investment decisions since this sustainability is associated with economic efficiency [6]. For example, for the financial evaluation of renewable energy sources, four categories of indicators are distinguished [7]: (1) indicators of net present value, internal rate of return, and payback period; (2) an indicator of the equalized cost of electricity; (3) return on investment indicator (ROI); and (4) other metrics. In addition, the economic model of discounted cash flows is used [8].

The methods discussed earlier [1] are further discussed using examples [9, 10], as well as the MAPI method finds a new application [11]. However, in all of the above cases, i.e., in the range of sources [5–11], there are no engineering methods of economic justification of technological solutions.

For decades, the so-called single-task construction robots have become a topic for economic research. The analysis of the relationship between the costs and benefits of using these robots can be used to assess the net economic benefits compared to conventional construction methods. It contributes to the widespread use of these robots [12] and, in addition, is taken into account in the economic analysis of projects [13].

In the economic literature, many terms are used to express efficiency concepts. For example, "technical efficiency" is estimated through the ratio between the physical volume of production and physical costs. In turn, the term "effectiveness" is usually associated with the result obtained. But in different disciplines, these terms (efficiency and effectiveness) are not always clearly defined [14]. As for program performance indicators, they usually reflect changes in the results obtained and the objectives of the program [14, 15].

Evaluating the cost-effectiveness of commercial energy is similar to cost-benefit analysis [16, 17]. Technical efficiency is determined by technical solutions [18, 19] or innovations [20, 21], e.g., using the impregnated grinding wheels [18] or by special grinding wheels, as it will be shown in the "results" section.

Larger firms use more sophisticated economic methods. At the same time, a high level of economic analysis contributes to accelerating corporate growth and increasing the profitability of these firms [22, 23]. However, these are higher organizational tasks compared to engineering techniques. An example of such a technique can be establishing a link between improving the operational properties of an abrasive tool at the stage of its operation and increasing the efficiency of the grinding process [24].

The literature review has shown that the inefficient activity in science, technology, and production is an analog of Sisyphean labor, that is, useless, meaningless, and unable to complete activities. The mathematical models available in the literature for assessing the economic efficiency of technical solutions do not allow converting technical resource savings into time savings. In turn, converting time savings into production (technological) costs is not difficult. For example, according to a well-known method through the cost of a machine-hour. For this reason, the paper aims to develop a cost model and an appropriate practical algorithm for converting technical resource savings into time savings and creating a practical methodology for evaluating the effectiveness of research and production activities based on the development of a quantitative criterion for this converting.

3 Research Methodology

The annual economic effect E_a is calculated by the following equation [25]

$$E_a = [(C_1 - C_2) - E_n(K_2 - K_1)]A_2, \tag{1}$$

where C_1, C_2 – unit cost of production for the basic and new production options, UAH; E_n is the normative coefficient of efficiency of capital investments. For example, with a payback period of five years, it will be $E_n = 0.2$; K_2, K_1 are the specific capital investments (in production assets) per unit of output for the basic and new production options, UAH; A_2 is the volume of products produced according to the new option, that is, using new technology, mechanization, automation of production (or something of this sort) in the billing year, in natural units (e.g., in pieces of parts).

Equation (1) is a universal statement of the problem, but the transformations performed below allow obtaining a new cost model for engineering applications.

Let's accept the following designations:

$$A_2 = A_n,\ C_1 = C_b,\ C_2 = C_n,\ K_1 = K_b/A_n,\ K_2 = K_n/A_n,$$

where K_b and K_n are the total capital costs, not given per unit of output. Then we have

$$E_a = \left[(C_b - C_n) - \frac{E_n}{A_n}(K_n - K_b)\right]A_n. \tag{2}$$

For $E_a = 0$ we can write

$$\left[(C_b - C_n) - \frac{E_n}{A_n}(K_n - K_b)\right] = 0.$$

The minimum required increase in theoretical productivity at which the economic efficiency according to formula (2) is equal to zero can be determined from the condition

$$(C_b - C_n) = \frac{E_n}{A_n}(K_n - K_b). \tag{3}$$

In turn

$$C_b - C_n = C_{m-h} \cdot (T_b - T_n), \tag{4}$$

where C_{m-h} is the cost of a machine-hour, UAH per hour; T_b and T_n the technical norm of time for the basic and new options, respectively, hours.

Of all the listed components, the most objective, i.e., which has the property of being calculated, is the main (technological) time, which can be determined by calculation for the basic T_b and new T_n options.

For example, for a profile gear grinding operation on a Höfler RAPID 1250 machine, these time components for the base (T_b) and new (T_n) options in seconds can be found using the following formulas, which follow from the structure of the gear grinding cycle on this machine

$$T_b = \left(\frac{B + l_1 + l_2}{V_{fb}}\frac{z_v}{t_{vb}} + T_{IND}\right)z + kT_D, \tag{5}$$

$$\mathrm{T}_n = \left(\frac{\mathrm{B} + l_1 + l_2}{V_{fn}} \frac{z_v}{t_{vn}} + T_{IND} \right) z + kT_D, \tag{6}$$

where B is the width of gear to be ground; l_1 and l_2 are the overtravels at the start and end of grinding, respectively, mm; V_{fb}, V_{fn} are the axial feed according to the basic and new options, mm/s; z_v is the nominal grinding stock allowance, mm; t_{vb}, t_{vn} are the grinding depth in vertical terms according to the basic and new options, mm; T_{IND} is the time for dividing the gear, which includes the retraction of the grinding wheel, the discrete rotation of the gear, and the entry of the grinding wheel into the gap between the gear teeth, s; z, k are the number of gear teeth and the number of grinding wheel' dressings, respectively; T_D is the time of one dressing of a grinding wheel, s.

Formulas (5) and (6) can be represented in the following form

$$T_b = \left(\frac{\mathrm{B} + l_1 + l_2}{Q'_{wb}} z_v + T_{IND} \right) z + kT_D, \tag{7}$$

$$T_n = \left(\frac{\mathrm{B} + l_1 + l_2}{Q'_{wn}} z_v + T_{IND} \right) z + kT_D, \tag{8}$$

where $Q'_w = t_v \cdot V_f$ is the specific material removal rate, mm^3/(s·mm).

Substituting formulas (7) and (8) into Eq. (4), we obtain

$$\mathrm{C}_b - \mathrm{C}_n = \mathrm{C}_{m-h} \cdot \ (\mathrm{T}_b - \mathrm{T}_n) = \mathrm{C}_{m-h}$$
$$\times \left[\left(\left(\frac{\mathrm{B} + l_1 + l_2}{Q'_{wb}} z_v + T_{IND} \right) z + kT_D \right) - \left(\left(\frac{\mathrm{B} + l_1 + l_2}{Q'_{wn}} z_v + T_{IND} \right) z + kT_D \right) \right]. \tag{9}$$

After mathematical transformations of Eq. (9), we get.

$$\mathrm{C}_b - \mathrm{C}_n = \mathrm{C}_{m-h} \cdot z(\mathrm{B} + l_1 + l_2) z_v \left(\frac{1}{Q'_{wb}} - \frac{1}{Q'_{wn}} \right). \tag{10}$$

Substituting Eq. (10) into Eq. (3), we will have.

$$\mathrm{C}_{m-h} \cdot z(\mathrm{B} + l_1 + l_2) z_v \left(\frac{1}{Q'_{wb}} - \frac{1}{Q'_{wn}} \right) = \frac{\mathrm{E}_n}{\mathrm{A}_n} (\mathrm{K}_n - \mathrm{K}_b). \tag{11}$$

Solving Eq. (11) concerning the parameter Q'_{wn}, we obtain.

$$Q'_{wn} = \frac{Q'_{wb} \cdot \mathrm{C}_{m-h} \cdot z(\mathrm{B} + l_1 + l_2) z_v}{\mathrm{C}_{m-h} \cdot z(\mathrm{B} + l_1 + l_2) z_v - Q'_{wb} \cdot \frac{\mathrm{E}_n}{\mathrm{A}_n} (\mathrm{K}_n - \mathrm{K}_b)}. \tag{12}$$

Let's accept $l_1 = l_2 = l_0$, then $l_1 + l_2 = 2l_0$ and therefore, Eq. (12) can be rewritten as

$$\frac{1}{Q'_{wn}} = \frac{1}{Q'_{wb}} - \frac{\mathrm{E}_n (\mathrm{K}_n - \mathrm{K}_b)}{\mathrm{A}_n \cdot z[\mathrm{B} \cdot z_v + 2l_0 \cdot z_v] \mathrm{C}_{m-h}}, \tag{13}$$

or

$$Q'_{wn} = \frac{1}{\frac{1}{Q'_{wb}} - \frac{E_n(K_n - K_b)}{A_n \cdot z[B \cdot z_v + 2l_0 \cdot z_v]C_{m-h} \cdot}}. \tag{14}$$

The values $\frac{1}{Q'_{wb}}$ and $\frac{1}{Q'_{wn}}$ characterize the complexity of the grinding operation according to the basic and new options. The values Q'_{wb} and Q'_{wn} characterize the productivity of the grinding stock allowance removal.

Thus, from formula (13) follows that the new labor input should be less than the basic labor input by the value of

$$\frac{E_n}{A_n} \frac{K_n - K_b}{z\left[V'_{w\max} + 2l_0 \cdot z_v\right]C_{m-h}} = \delta, \tag{15}$$

where $V'_{w\max} = B \cdot z_v$ is the specific material removal which characterizes the grinding wheel wear, mm^3/mm.

It can be seen that the value δ is determined by the difference in costs for the new and basic options of gear grinding wheels on the Höfler RAPID 1250.

4 Results and Discussion

Initial data are listed below for evaluating the effectiveness of gear grinding technology with new abrasive grinding wheels – white electro-corundum 25AF46L6V and sol-gel corundum 3SG46Hs12Vs – both having sizes of 400 × 40 × 127 mm.

1. The cost of a machine-hour is $C_{m-h} = 0.25$ UAH per second, i.e., $C_{m-h} = 900$ UAH per hour.
2. The number of gear teeth $z = 40$.
3. Gear width $B = 24$ mm.
4. The overtravels at the start and end of grinding: $l_1 = l_2 = l_0 = 7.87$ mm.
5. Grinding stock allowance in normal terms: $z_n = 0.77$ mm. The same in vertical terms: $z_v = 1.73 \cdot z_n = 1.73 \cdot 0.77$ mm $= 1.3347$ mm.
6. The specific material removal rate at the grinding depth $t_v = 0.026$ mm is $Q'_{wb} = t_v \cdot V_f = 0.026 \cdot \frac{5000}{60} = 2.17$ mm^3/(s ·mm).
7. Normative coefficient of efficiency of capital investments, $E_n = 1$, i.e., the cost of purchasing and operating grinding wheels are paid off in one year.
8. Tool costs for the basic and new options $K_b = 27 \cdot P_{qwb}$ and $K_n = 27 \cdot P_{qwn}$, respectively. The cost of grinding wheels: for basic wheel $P_{qwb} = 400$ UAH (electro-corundum), for new one $P_{qwn} = 10{,}000$ UAH (sol-gel corundum).

The one-part machining time on the Höfler RAPID 1250 machine is 1.5 h. On a CNC machine Höfler RAPID 1250, 15 parts (gears) are ground in 3 shifts. The number of working days in a year is 249. Therefore, 15 parts (i.e., gears) multiplied by 249 days gives 3735 parts that will be ground for a year.

To determine the specific capital investments (in production assets) per unit of production for the basic and new options, we take the same number of dressings of grinding wheels and the same number of wheels per year.

Since the diameter of the wheel decreases by 0.1 mm at each edit, the number of dressings of the grinding wheel during the processing time of one part according to the basic and new options will be 14 for three stages of the same operation i.e., 6 + 6 + 2.

In this case, the diameter of the grinding wheel will accordingly decrease by 1.4 mm. Considering that the diameter of the grinding wheel decreases from 400 mm (new wheel) to 204 mm (critically worn wheel), then the grinding wheel life expressed by the number of gears machined for the base and new options, respectively, will be (400 – 204)/1,4 = 140 parts.

Let's find the main time for the basic and new options using formulas (7) and (8), respectively, with additional data: $T_D = 15$ s; $k = 14$; $T_{IND} = 2$ s. We get $T_b = 21.115$ min (or 1267 s) and $T_n = 16.488$ min (or 989.307 s).

Substitute the indicated data into Eq. (15) we get

$$Q_w' = \left(\frac{1}{2.17} - \frac{1}{3735}\frac{27 \cdot 10000 - 27 \cdot 400}{40[24 \cdot 1.3347 + 2 \cdot 7.8653 \cdot 1.3347]0.25}\right)^{-1} = 3.027\ \text{mm}^3/(\text{s} \cdot \text{mm}).$$

The technological efficiency coefficient $\mu = Q_{wn}'/Q_{wb}'$ that shows the increase in the parameter Q_{wn}' in comparison with Q_{wb}' will be

$$\mu = \frac{Q_{wn}'}{Q_{wb}'} = \frac{3.027}{2.17} = 1.4.$$

Thus, an increase in the parameter Q_w' by 1.4 times (or by 42%) is a condition for the minimum necessary increase in the productivity of grinding stock allowance removal with the new sol-gel corundum grinding wheel.

If the coefficient $\mu = 1$ (Fig. 2), when $K_n = K_b$,i.e., if the cost of a new grinding wheel (new option) is equal to the cost of a wheel accepted at the factory (base option), then the minimum possible increase in productivity is zero. The coefficient $\mu < 1$ when $K_n < K_b$, , i.e., the cost of a new grinding wheel (new option) is less than the cost of a factory-accepted wheel (basic option). In this case, there is no requirement to increase performance.

Checking the correctness of the calculations is performed according to the formula $E_a = C_{m-h} \cdot \Delta T - E_n(K_n - K_b) = 0$, the expected annual economic effect, which was previously taken as zero, should be just so. We get 0.25·277.9 – 1·(72.289 – 2.892) = 0.078, i.e., practically zero, which was required to be proved.

Since we save on one part $\Delta T = T_b - T_n = 21.115 - 16.488 = 4.627$ min (1267 – 989.307 = 277.59 s), then we save 4.627 x 3735 = 17281.8 min on the annual production program, i.e., 288 h.

It is possible to calculate the number of parts (gears) that can be manufactured according to the new version used. It will be (when the one part machining time on the Höfler RAPID 1250 machine is 1.5 h): 288/1.5 = 192 parts.

And so, basic option – 3735 parts; new option – 3735 + 192 = 3927 parts.

Thus, the grinding performance has increased by (3927 – 3735)/3735 = 0.051 or by 5.1%.

The established influence of the price P_{gwn} of the grinding wheel on the coefficient μ of its technological efficiency allows putting an unambiguous relationship between the price P_{gwn} of a grinding wheel and its potential performance characteristics $\mu = Q'_{wn}/Q'_{wb}$ (Fig. 2a). For example, we can see in Fig. 2a the following data:

$P_{gw\ n}$, UAH	400	5000	10000	15000	20000
μ	1.0	1.2	1.4	1.8	2.8

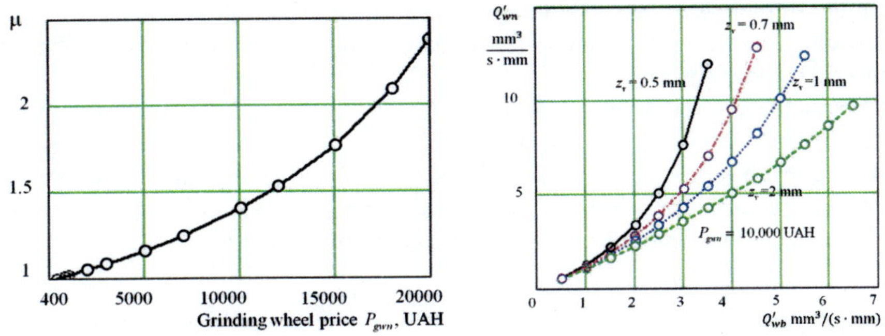

a b

Fig. 2. Dependencies $\mu(P_{wn})$ (*a*) and $Q'_{wn}(Q'_{wb}, z_v)$ at $P_{gw\ n} = 10\ 000$ UAH and $0.5 \leq z_v \leq 2.0$ mm (*b*).

It can be seen that the higher the price $P_{gw\ n}$ of the new version grinding wheel, the greater the performance of defect-free gear grinding it must provide.

No less interesting is the dependence of Q'_{wn} on Q'_{wb} at $P_{gw\ n} = 10\ 000$ UAH and $0.5 \leq z_v \leq 2.0$ mm. For example, we can see in Fig. 2, *b* the following data:

At $Q'_{wn} = 5$ mm^3/(s·mm)

z_v, mm	0.5	0.7	1.0	2.0
Q'_{wb}, mm^3/(s·mm)	2.5	2.9	3.4	4.0

At $Q'_{wn} = 10$ mm^3/(s·mm).

z_v, mm	0.5	0.7	1.0	2.0
Q'_{wb}, mm^3/(s·mm)	3.4	4.1	5.0	6.7

It can be seen that the larger the grinding stock allowance z_v, the smaller the difference between the value of Q'_{wn} and value of Q'_{wb}, but it always Q'_{wn} is greater than Q'_{wb}.

5 Conclusions

There is no some quotient that allows going directly from technical savings to saving time or money. It requires a special procedure, depending on the specific case and conditions for making the cost-benefit evaluation decision. The paper develops one of such possible procedures. The procedure is based on analyzing the boundary between a useful and useless technical solution.

The lack of a technical-and-economic model encourages irresponsibility and reckless use of valuable resources. A case study for new sol-gel corundum 3SG46Hs12Vs grinding wheel instead of conventional white electro-corundum 25AF46L6V one has shown the former's efficiency despite the lower price of the latter. The case study can be an example of the effectiveness of evaluating other technical solutions in science, engineering, and technology.

The relationship between categories of technical efficiency, productivity (product quantity per unit of time), and effectiveness is investigated. A new technical and economic quotient has been formulated, which shows what kind of productivity gain a new technical solution in science, engineering, and technology should bring so that at least a minimal economic benefit from this technical solution appears.

References

1. Steffy, W.: Investment Decisions. Automation, USA (1974)
2. De Silva, C.W.: Mechatronics: A Foundation Course, 1st edn. CRC Press, Boca Raton (2010)
3. Bishop, R.H. (ed.): Mechatronics: An Introduction. CRC Press, Boca Raton (2017). https://doi.org/10.1201/9781420037241
4. Reniers, G., Talarico, L.; Paltrinieri, N.: Dynamic risk analysis in the chemical and petroleum industry. Evolution and interaction with parallel disciplines in the perspective of industrial application. Chapter 16. In: Cost-Benefit Analysis of Safety Measures. Butterworth Heinemann, pp. 195–205 (2016). https://doi.org/10.1016/B978-0-12-803765-2.00016-0
5. Moran, S.: An Applied Guide to Process and Plant Design, 1st edn. Butter Worth-Heinemann, United Kingdom (2015)
6. Dincer, I.: Chapter 6 – Sustainability modeling. In: Energy Sustainability, pp. 119–164 (2020). https://doi.org/10.1016/B978-0-12-819556-7.00006-1
7. Delapedra-Silva, V., Ferreira, P., Cunha, J., Kimura, H.: Methods for financial assessment of renewable energy projects: A review. Processes **10**(2), 184 (2022). https://doi.org/10.3390/pr10020184
8. Mauboussin, M.J., Callahan, D.: Everything is a DCF model. Morgan Stanley Investment Management Journal, 1–12 (2021)
9. Panigrahi, A., Vachhani, K., Sisodia, M.: Application of discounted cash flow model valuation: The case of exide industries. Journal of Management Research and Analysis, 170–179 (2021)
10. Top 6 methods used for equipment replacement studies. https://www.businessmanagementideas.com/industries/top-6-methods-used-for-equipment-replacement-studies/7302. Accessed 03 July 2022
11. Abbott, R.A., Ring, E.A.: The MAPI method—Its effects on productivity: An alternative is needed. J. Manuf. Syst. **2**(1), 15–30 (1983). https://doi.org/10.1016/S0278-6125(83)80005-3
12. Hu, R., et al.: A simple framework for the cost-benefit analys is of single-task construction robots based on a case study of a cable-driven facade installation robot. Buildings **11**(1), 8 (2021). https://doi.org/10.3390/buildings11010008

13. Cost-benefit analysis for development: A practical guide. Asian Development Bank, Philippines (2013). https://www.adb.org/documents/cost-benefit-analysis-development-practical-guide. Accessed 03 July 2022
14. On efficiency and effectiveness: some definitions. Media and Publications, Melbourne (2013). https://www.pc.gov.au/research/supporting/efficiency-effectiveness. Accessed 03 July 2022
15. Making choices in health: WHO guide to cost-effectiveness analysis. World Health Organization, Geneva (2003). https://apps.who.int/iris/handle/10665/42699. Accessed 03 July 2022
16. Hart, P.R., Liu, B: Methodology for evaluating cost-effectiveness of commercial energy code changes. USA (2015). https://doi.org/10.2172/1213016
17. SDC how-to-note financial and economic analysis of projects with a focus on cost benefit analysis (CBA) and cost effectiveness analysis (CEA). Quality Assurance (2015). https://www.shareweb.ch. Accessed 03 July 2022
18. Gierszewska, E., Nadolny, K.: Economicanalysis and risk assessment in the process of implementing innovative products on the example of impregnated grinding wheels. J. Mech. Energy Eng. **3**(43), 289–294 (2019). https://doi.org/10.30464/jmee.2019.3.4.289
19. Parmacli, D.: Algorithm of calculating the annual economic effect from implementing different types of innovation in a business organization. Olsztyn Econ. J. **7**(2), 239–248 (2012). https://doi.org/10.31648/oej.3426
20. Working Party of National Experts on Science and Technology Indicators. The Impact of R&D Investment on Economic Performance: A Review of the Econometric Evidence. Organization for Economic Co-operation and Development, Paris (2015). Accessed 03 July 2022
21. Herbst, A.F.: Capital Asset Investment: Strategy. Tactics and Tools. Jonh Wiley & Sons, LTD, USA (2003)
22. Andrews, G.S., Butler, F.: Criteria for major investment decisions. Investment Anal. J. **15**, 31–37 (1986). https://doi.org/10.1080/10293523.1986.11082246
23. Haka, S.F.: A review of the literature on capital budgeting and investment appraisal: 697–728 (2006). https://doi.org/10.1016/S1751-3243(06)02010-4
24. Tonkonogyi, V., Sidelnykova, T., Dašić, P., Yakimov, A., Bovnegra, L.: Improving the performance properties of abrasive tools at the stage of their operation. In: Karabegović, I. (ed.) NT 2019. LNNS, vol. 76, pp. 136–145. Springer, Cham (2020). https://doi.org/10.1007/978-3-030-18072-0_15
25. Chukin, S.A., et al.: Calculations of the economic efficiency of organizational and technical measures in mechanical engineering. Moscow (1986). (in Russian)

Modernization of the Internal Audit Process Using a Risk-Based Approach at an Industrial Enterprise

Liudmyla Perperi[1](✉), Gennadii Oborskyi[1], Ganna Goloborodko[1], Vladimir Gugnin[1], and Oleg Prokopovych[2]

[1] Odessa Polytechnic National University, 1, Shevchenko Ave., Odessa 65044, Ukraine
perperi.l.m@op.edu.ua
[2] Odeskabel PJSC, 144, Mykolaivs'ka doroga, Odessa 65102, Ukraine

Abstract. This study aims to determine a strategy for modernizing the integrated management system internal audit process using a risk-based approach to ensure the audit effectiveness and efficiency through the use of the PDCA cycle and quality tools. The researched object is the integrated management system internal audit process. To achieve this goal, the existing methods on conducting an internal audit of integrated systems at various companies were investigated, and their disadvantages and advantages have been identified. The internal audit process development essence, features, concept, and theoretical aspects are determined, taking into account the analysis of the real state-of-art with regulatory support to the integrated system at an industrial enterprise. We recommend a sequence of actions for an audit program planning process with risk and opportunities assessment and its implementation in the audit process management scheme. A causal diagram of internal audit functioning and effectiveness is built to determine how much some specific factors affect the internal audit effectiveness compared to other factors, and the Pareto diagram is used.

Keywords: Integrated management system · Sustainable manufacturing · Risk · Ishikawa diagram · Pareto diagram

1 Introduction

Currently, one of the most important tasks for enterprises with integrated quality management systems (IMS) is implementing and maintaining these systems that ensure the industrial production of stable quality. The competitiveness of IMS should consider the enterprise specifics and ensure the product development and production costs minimizing, provided that the quality requirements are met. The integrated system is subject to a periodic internal audit as one of the tools for monitoring, analyzing, and improving the entire system's performance. Internal audit is an integral and important element of Management Control [1, 2].

V. Tonkonogyi et al. (Eds.): InterPartner 2022, LNME, pp. 360–369, 2023.
https://doi.org/10.1007/978-3-031-16651-8_34

To ensure compliance with modern requirements and achieve sustainable development goals, enterprises integrate their management systems in various combinations: QMS (Quality Management System, ISO 9001), EMS (Environmental Management System, ISO 14001), and OHSAS (Occupational Health and Safety Assessment Series, ISO 18001) [3], OH&S (Occupational Health and Safety Management System, ISO 45001). Today, many enterprises worldwide are certified according to the above standards and conduct internal audits regularly. However, the regulatory framework for various activity fields is constantly changing. Management systems based on management theories and methodologies continuously improve through innovative solutions. The internal audit process should add value to the organization, improve the external image, and significantly contribute to stakeholder relations.

Nowadays, leading specialists in the field of quality elaborated various methods for internal audits of enterprises' quality management systems, but of high relevance is considering possible ways to improve the integrated management systems internal audit process by taking into account the specifics of enterprises' activity areas and their organizational structures.

2 Literature Review

Different schemes for integrating management systems apply different approaches to quality in different areas of production. For example, the authors note that the implementation of vide variety of integrated systems combined with the methodologies of Total Quality Management (TQM) [4], Lean Six Sigma lean production (LSS) [5], Total Production Maintenance (TPM), Toyota Production System (TPS) and through the use of quality tools such as the Shewhart-Deming Plan-Do-Check-Act cycle (PDCA), allowed to reduce the defective production in various industries thus rendered them constantly competitive and profitable [6].

If we consider the use of IMS from a system point of view, the work [7] analyzed the feasibility of implementing these systems, and the integration methods classification has been determined by the multi-level management aspect, namely on the system and operational levels. It allowed identifying the advantages and disadvantages of various methods, as obviously now more in common practice at enterprises is using QMS according to ISO 9001 as an Integration Center.

The authors [3, 8] researched the introduction of internal audit procedures for integrated management systems ISO 9001, ISO 14001, and ISO 18001. It was emphasized that the internal audit management ISM introduction targeted to meet the requirements of internal and external stakeholders, thus identifying the advantages of audit conducting, namely, achieving key performance indicators and eliminating losses and complexity, referring to both the auditors' insufficient expertise and organizational structures' complexity.

The work of the authors [9] is aimed at defining general models of integration of several MS. In the article, a thorough analysis of 16 models was carried out, and it was determined that the tasks and purposes are common. The difference lies in the methodology of approaches to integration. The use of a graphical presentation of the PDCA cycle is proposed. According to the authors, the expected facilitation of the integration process is the similarity between the PDCA cycle and the requirements of the updated standards, which will lead to a reduction in the costs of implementing IMS.

Analyzing the above, we can conclude that various methods of the PDCA cycle integrated into management were implemented with positive results and achieved cost and defect percentage reduction, as well as contributed to improving the quality of processes and industrial production concerning environmental aspects. It should also be noted that the management systems integration is a methodological challenge to companies, and this process reasons refer to different goals in achieving key performance indicators.

3 Research Methodology

In the quality field, any system is subject to constant inspections, control, checks and quality audits [10]. The internal audit is particularly interesting for enterprises that should group audits of products, services, and processes. This scientific and practical research aims to determine the strategy for modernizing the internal audit process, whose urgency is due to the publication of ISO 45001:2018 international standard replacing OHSAS 18001:2007 and the upcoming management system certification for compliance with ISO 45001 requirements. OH&S is directed to implement and maintain an occupational health and safety management system. There is a need to disclose the internal audit process's theoretical basis, considering the ISO 19011:2018 international standard requirements. This standard introduces a new term "combined audit", conducted in one organization, where two or more management systems are checked simultaneously and used to assess ISM compliance. It is because the release of ISO 19011:2018 third version was preceded by a change in the main standards of management systems, including ISO 9001, ISO 14001, and ISO 45001. The standard's third version, such as the previous one, is also based on the PDCA cycle. In the standard's new version, valid in Ukraine since January 1, 2021, in addition to changes in terminology, the following changes took place: applying a risk-based approach to audit principles; the sections on audit program managing and conducting an audit are more detailed; added subparagraph 5.3, which deals with the risks and opportunities assessment in the audit program development, removed is Appendix A on auditors' special knowledge and skills.

The internal audit process concept relies on describing key provisions that provide a general understanding and comprehension of its implementation specificity. For the development and implementation of the audit procedure, participation in the organization of the planning process, verification of the results of the audit, documentation, and monitoring was assigned. Therefore, that concept's main provisions adopted are: applying the PDCA cycle as an iterative process that brings the internal audit program management closer to the desired result according to the set goals each time the cycle ends; defining a process management model; applying the Ishikawa diagram to establish the risk-based planning for conducting an efficient and effective audit; using the Pareto diagram to determine the severity of factors that may affect low audit efficiency.

4 Results and Discussion

To determine the approach to improving the audit process, it is necessary to analyze Sect. 5 of the standard ISO 19011 related to audit program management (AP). A new clause 5.3, "identifying and assessing risks and opportunities for the audit program", has been introduced. Clause 5.4 has also changed both in structure and essence. Attention should also be paid to clause 6.3.2, which states that when developing an audit plan, the audit team head should consider: the team composition and its overall competence; appropriate sampling methods; opportunities to improve the audit activities' effectiveness and efficiency; probabilities of failure to achieve audit goals resulting from an erroneous planning stage. Also has also been changed Sect. 7, dealing with the criteria for evaluating auditors, which should be qualitative (knowledge, special training, personal qualities) and quantitative (work experience, number of audits conducted). Summing up, it should be noted that the changes introduced in the new version mainly concern two elements: an approach based on risk assessment and the reference to documented information, which in particular can be processed by systems using information and communication technologies. The company constantly undergoes internal audits following the approved plan, which states that the main task is to monitor the functioning of the management system in all activity areas. However, it should be noted that the company's activities are aimed at continuous improvement, including in the organization of its processes, making changes to the internal audit process a high priority.

Let's consider some processes from the viewpoint of performing a certain technological sequence of actions (at the operational level). Such a formal approach rarely reveals real activities complexity at the analysis stage, for example, when it comes to end-to-end processes covering the activities of some company's several divisions different in their functions and subordination [11]. The basic model's internal document regulating process activities contains the following components: inputs, outputs, management actions, and resource support [12, 13]. The purpose of building a model for managing the internal audit process is to provide the necessary description of components that make up the modeled process and determine the nature of these components' relationship. For the implementation of the process, one should compile a combined audit program that is focused on the ISO 9001, ISO 14001, and ISO 45001 standards and determine the sequence of audit program management process operations based on the PDCA cycle (Fig. 1). At the "do plan" stage, in the first step (clause 5.2), the audit client must ensure that the objectives of the audit program are set to determine the directions of planning and conducting the audit, and must ensure the effective implementation of the audit program.

The goals of AP should correspond to the Strategic Development Plan, Policy, and the enterprise's integrated management system goals. In the second step (clause 5.3), it is necessary to determine which existing risks and opportunities may be associated with the audit program and may affect the achievement of its goals. In the third step (clause 5.4), the audit program manager must: set the audit scope following the established goals; identify and present to the audit client the risks and opportunities considered in the development of the audit program; ensure the selection of audit groups, defining their responsibilities, authority and powers, and leadership support, as necessary; provide relevant documented information (including AP records).

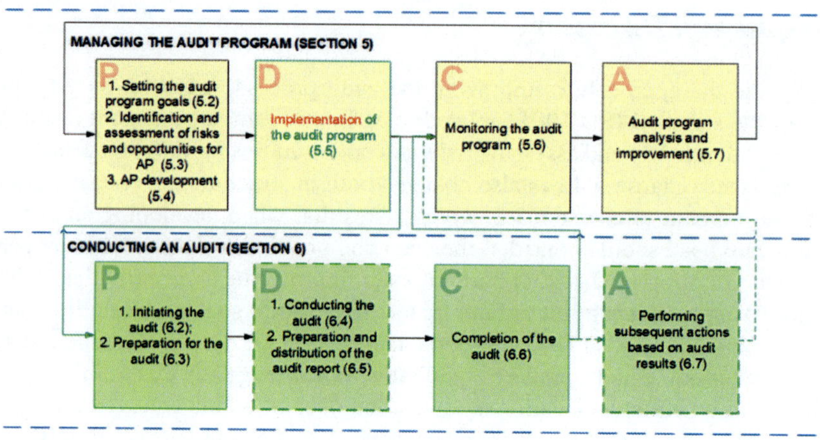

Fig. 1. Diagram of the audit program management process (Compiled by the authors based on ISO 19011).

At the "do execute" AP stage, it is necessary to perform operational planning and all activities coordinated within the program, inform interested parties which audit program segments relate to their concerns, including related risks and opportunities, and periodically notify them of the AP implementation using established channels of external and internal communications, clause 5.5).

The "do exit" from the AP execution stage lies at the beginning of the implementation of the audit stage carried out according to the PDCA cycle. In the planning stage first step, the responsibility for conducting the audit should be on the designated head of the audit team until the audit is completed.

To start an audit, one must follow such action sequence: establish contact with the audited entity and ensure that the audit objectives can be achieved (clause 6.2). In the second step, when preparing for the audit, we must analyze the relevant documented ISM data to collect information for a better understanding of the functions within the processes, operations, and preparation of audit activities (with jobs distribution among the audit team members) and relevant auditor's work documents (for example, using quality tools such as checklists, Turtle diagram, etc.). The audit team head should adopt a risk-based approach at the planning stage to ensure an effective schedule and coordination of the audit and effectively achieve the audit goal (clause 6.3).

At the "do perform" stage, the first step is related to audit activities conducted in a certain sequence: conducting an introductory meeting, data exchange during the audit, audit data availability and accessibility, analysing documented information, collecting and verifying information, forming audit disclosures, preparing audit conclusions (clause 6.4). The second step is devoted to preparing for sending out the audit report. The report is sent to the relevant stakeholders following the audit program/plan (clause 6.5).

The audit is completed at the "do check" stage (clause 6.6). The audit is completed after all planned audit activities have been completed or under another agreement with the audit client (for example, an unforeseen situation may arise, never allowing the audit to be completed as planned). At the action stage, the following actions are done based on the audit results: corrections, corrective actions, and opportunities for improvement, with verification of completion and these actions effectiveness assessment (clause 6.7). In Fig. 1, this stage is shown with a dotted line because verification can also be part of a subsequent audit. At the stage of AP management verification, it is monitored, and, if necessary, changes (clause 5.6) are made based on the audit results. At the action stage, the AP is analyzed and improved (clause 5.7), and conclusions based on those results are used as input data for the program's continuous improvement process. Implementing modern practices for evaluating the AP management process components: planning, conducting, and monitoring will help improve the internal audit process as a whole.

At the same time, the internal audit process, concerning the requirements of modern approaches to ISM, should take into account all factors affecting the internal audit process's effectiveness and efficiency.

The format of the approach defined by the concept was revised by identifying such "main factors" that can affect the internal audit process effectiveness (using the Ishikawa causal diagram Quality Tool, Fig. 2): quality control, personnel, audit process, methods (documentation). The main factors categories are determined based on the analysis of compliance with ISM standards requirements and internal knowledge of work processes in the enterprise. The method used in the Ishikawa diagram provides visualization of causal relationships between the analyzed object and factors affecting it. It offers a systematic approach to determining the causes of problems' probability [14, 15].

In this paper, the Ishikawa diagram was used to identify possible risks influencing the quality of internal audit effectiveness and efficiency and the probability of getting an incorrect audit report, which may affect the organization's activities as a whole and lead to negative consequences.

Further, using the expert ranking method, the factors (causes) were determined, $n = 12$), which have the most significant impact on the object of research, namely on the internal audit process effectiveness degree that can lead to unsatisfactory results. A commission of four experts was established ($m = 4$) who had theoretical knowledge and practical experience in this field.

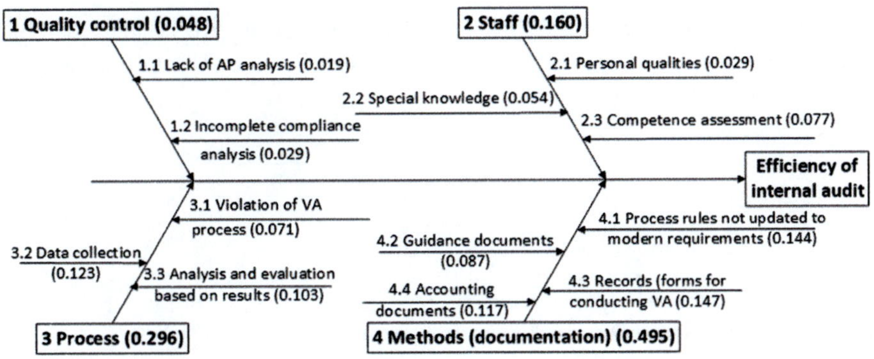

Fig. 2. Causal diagram of the effectiveness of the internal audit process.

Expert opinions were collected through a questionnaire survey. These experts do assess the severity of the factor by assigning them a rank number. If the expert recognizes several factors as equivalent, they are assigned the same rank number. Based on the questionnaire survey data, a summary rank matrix is compiled. The expert ranking matrix is presented in Table 1.

Table 1. The matrix of expert ranking.

Causes of an event	Experts				Sum of ranks	$d_i - d_a$	$(d_i - d_a)^2$
	1	2	3	4			
1.1	1	3	2	3	9	−17	289
1.2	2	2	1	1	6	−20	400
2.1	3	1	3	2	9	−17	289
2.2	4	4	5	4	17	−9	81
2.3	5	7	6	6	24	−2	4
3.1	7	6	4	5	22	−4	16
3.2	11	10	9.5	8	38.5	12.5	156.25
3.3	9	8	8	7	32	6	36
4.1	10	12	12	11	45	19	361
4.2	6	5	7	9	27	1	1
4.3	12	11	11	12	46	20	400
4.4	8	9	9.5	10	36.5	10.5	110.25
$\sum$	78	78	78	78	312		2143.5

Note: d_i is the sum of assessments determined by each of the experts on the i-th indicator; d_a is the average value of the sum of estimates d_i, determined for each indicator

To estimate the average degree of expert opinions consistency, the Kendall coefficient of concordance W was applied for the case when there exist related ranks (the same rank values in the estimates of one expert):

$$W = \frac{S}{\frac{1}{12} \cdot m^2 \left(n^3 - n\right) - m \cdot \sum T_i},$$

$$T_i = \frac{1}{12} \cdot \sum \left(t_l^3 - t_l\right).$$

where T_i – the number of types of elements repeated in estimates of the i-th expert, t_l - number of elements in l link for i-th expert. $T_3 = [(2^3–2)]/12 = 0.5$, $\sum T_i = 0.5$.

$$W = \frac{2143,5}{\frac{1}{12} \cdot 4^2 \left(12^3 - 12\right) - 4 \cdot 0.5} = 0.94.$$

$W = 0.94$ indicates a high degree of expert opinions consistency.

The Pearson matching criterion determines the estimation of the significance of the concordance coefficient:

$$\chi^2 = \frac{S}{\frac{1}{12} \cdot mn(n+1) + \frac{1}{n-1} \cdot \sum T_i},$$

$$\chi^2 = \frac{2143.5}{\frac{1}{12} \cdot 4 \cdot 12(12+1) + \frac{1}{12-1} \cdot 0.5} = 41.26.$$

The calculated criterion χ^2 is compared with a tabular value at a given significance level $\alpha = 0.05$. Considering that the calculated value χ^2 is equal to 41.26, which is more than the tabulated value (19.67514), then $W = 0.94$, which value is not random, the results obtained can be used in further studies.

The indicator of factors severity is calculated based on the obtained rank sums (Table 1). The survey matrix transformed into a rank matrix is shown in Table 2.

Table 2. Rank matrix.

Parameter name	Causes of an event											
	4.3	4.1	3.2	4.4	3.3	4.2	2.3	3.1	2.2	2.1	1.1	1.2
Sum of ranks	46	45	38.5	36.5	32	27	24	22	17	9	9	6
Weight	14.7	14.4	12.3	11.7	10.3	8.7	7.7	7.1	5.4	2.9	2.9	1.9
Accumulated weight	14.7	29.2	41.5	53.2	63.5	72.1	79.1	86.9	92.3	95.2	98.1	100

Pareto analysis is illustrated by the Pareto diagram (Fig. 3). The factors of quality problems are plotted along the abscissa axis in descending order, and the accumulated weight is plotted along the ordinate axis in quantitative terms.

As can be seen from the diagram, according to experts, reasons 4.1–4.4, 3.2–3.3, and 2.3 have the most significant impact on the internal audit process effectiveness. It was noted that the use of the Turtle Chart as a tool contributed to a clearer understanding of the components of the process, which were grouped into categories: inputs, personnel, equipment and facilities, instructions and procedures, key indicators (metrics), outputs, and their functioning in the system.

The questions used to assess the components of the process were compiled based on the best practices of leading companies in the industry. They contained recommendations with minimum requirements to close the content of the questions and a list of possible objective evidence leading to audit evidence.

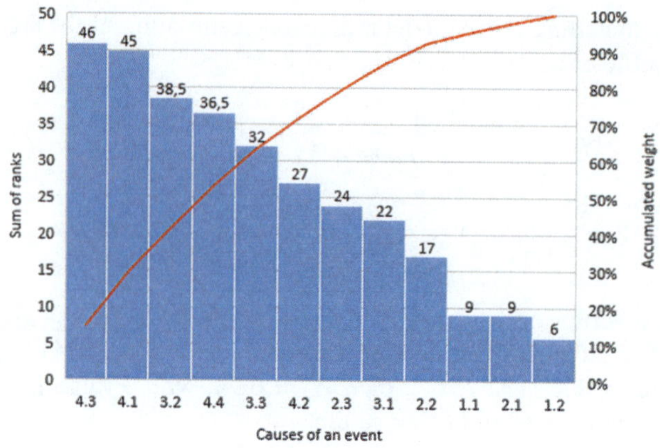

Fig. 3. Pareto diagram.

5 Conclusions

The essence, features, and concept of internal audit have been defined. A model for managing the internal audit process is proposed, and these model components are defined. The scheme of management process in the audit program for compliance with the international standard requirements is considered. Existing problems were analyzed to improve the "Internal Audit" process's effectiveness. The reasons affecting the "Internal Audit" process's effectiveness were identified using the Ishikawa causal diagram. A Pareto diagram is built, rendering it possible to identify the most significant factors of processes that affect the internal audit process.

It is revealed that the processes representation is most effective when their description principles are applied to all processes and using the same consistent terminology and visualization. Conducting an internal audit of processes using the PDCA process approach allows you to clearly understand the sequence of actions at each cycle stage and what exactly should be reflected in the documented information. It significantly improves clarity and minimizes discrepancies in relevant information by users and stakeholders. To implement this, the audit team is invited to make changes to the reporting forms, which will facilitate the analysis of the ISM to identify actual information about the real state of the enterprise.

References

1. Starzyńska, B., Bryke, M.: Compilation of the best practices for auditing the sustainable development of organizations. In: Tonkonogyi, V., et al. (eds.) InterPartner 2020. LNME, pp. 833–842. Springer, Cham (2021). https://doi.org/10.1007/978-3-030-68014-5_81
2. Nikityuk, V., Karamavrova, T., Lebedynets, V.: The self-inspections (internal audits) process as a part of the pharmaceutical quality system: formation of a risk-based approach to internal audits planning. J. Pharm. Pharmacol. **7**, 385–397 (2019). https://doi.org/10.17265/2328-2150/2019.07.004

3. Domingues, J.P.T., Sampaio, P., Arezes, P.M.: Analysis of integrated management systems from various perspectives. Total Qual. Manag. Bus. Excell. **26**(12), 1311–1334 (2015). https://doi.org/10.1080/14783363.2014.931064
4. Hailu, H., Mengstu, S., Hailu, T.: An integrated continuous improvement model of TPM, TPS and TQM for boosting the profitability of manufacturing industries: An innovative model & guideline. Manag. Sci. Lett. **8**, 33–50 (2018). https://doi.org/10.5267/J.MSL.2017.11.002
5. Sreedharan, V., Rajasekar, S., Kannan, S., Arunprasad, P., Trehan, R.: Defect reduction in an electrical parts manufacturer: A case study. TQM J. **30**(6), 650–678 (2018). https://doi.org/10.1108/TQM-03-2018-0031
6. Ahmed, S., Shahadat, M.: Theory driven real time empirical investigation on joint implementation of PDCA and 5S for performance improvement in plastic moulding industry. Aust. J. Basic Appl. Sci. **3**, 3825–3835 (2009)
7. Luo, H., Li, G., Li, C.: Research on integration method of integrated management system. Open Autom. Control Syst. J. **7**, 1802–1807 (2015). https://doi.org/10.2174/1874444301507011802
8. Hassan, N.A., Mohamad Zailani, S.H., Hasan, H.A.: Integrated internal audit in management system: A comparative study of manufacturing firms in Malaysia. TQM J. **32**(1), 110–126 (2020). https://doi.org/10.1108/TQM-03-2019-0077
9. Pauliková, A., Lestyánszka-Škurková, K., Kopilcáková, L., Zhelyazkova-Stoyanova, A., Kirechev, D.: Innovative approaches to model visualization for integrated management systems. Sustainability **13**, 8812 (2021). https://doi.org/10.3390/su13168812
10. Starzyńska, B., Bryke, M., Diakun, J.: Human lean green method–a new approach toward auditing manufacturing & service companies. Sustainability **13**(19), 10789 (2021). https://doi.org/10.3390/su131910789
11. Perperi, L., Oborskyi, H., Goloborodko, G.: Multi-level model of process management. Bull. NTU "KhPI" Series: New Solutions Mod. Technol. **32**(1254), 74–79 (2017). https://doi.org/10.20998/2413-4295.2017.32.12
12. Yu Kataev, M., Bulysheva, L.A., Xu, L.D., Loseva, N.V.: Modeling business processes of the social insurance fund in information system runa WFE. IOP Conf. Ser. Mater. Sci. Eng. **142**, 012135 (2016). https://doi.org/10.1088/1757-899X/142/1/012135
13. Dumas, M., La Rosa, M., Mendling, J., Reijers, H.A.: Fundamentals of Business Process Management, 2nd edn. Springer, Berlin, Heidelberg (2018). https://doi.org/10.1007/978-3-662-56509-4
14. Ilie, G., Ciocoiu, C.N.: Application of fishbone diagram to determine the risk of an event with multiple causes. Manage. Res. Pract. **2**(1), 1–20 (2010)
15. Jen, R.: Visualizing risk management–a new approach. In: Proceedings of the PMI Global Congress 2010. Project Management Institute, America, Washington, DC (2010)

Improvement of Operational Parameters for High-Precision Tribosystems

Alexander Stelmakh[1,2], Ruslan Kostunik[2], Sergii Shymchuk[3], Natalia Zaichuk[3], and Anatolii Tkachuk[3](✉)

[1] Beijing Institute of Technology, 5, Zhongguancun Street, Haidian Qu, Beijing, China
[2] National Aviation University, 1, Liubomyra Huzara Avenue, Kiev 03058, Ukraine
[3] Lutsk National Technical University, 75, Lvivska Street, Lutsk 43018, Ukraine
a.tkachuk@lntu.edu.ua

Abstract. In producing plunger pairs of high-pressure fuel pumps, product quality management has a problem of improving their durability and reliability. To improve the performance of products, it is advisable to pay special attention to the tribological characteristics of the friction surfaces that are part of such products, the conditions of their mutual contact, and lubrication. Using a scientific and experimental approach, friction pairs simulating the plunger pairs of high-pressure fuel pumps and aerial gases and diesel fuels used as working lubricants are investigated. The studies take into account the dual nature of friction. The obtained results do not contradict elastohydrodynamic and adhesion-deformation theories. A set of research equipment has been developed, demonstrating the occurrence of secondary currents in the contact zones and the effect of tribocavitation. In the process of friction, microbubbles dissolve in the diffuse contact region. It has been established that with certain types of microgeometric relief, contact load, and a temperature, it is possible to set precision surfaces, which may explain some types of failures of high-pressure fuel pumps and create conditions for expanding the temperature range of these products.

Keywords: Industrial growth · Contact pressure · Friction pair · Lubricating medium · Operational properties · Precision tribosurfaces

1 Introduction

Modern production tends to increase product durability, performance, and other operational and technological parameters [1]. Simultaneously, a change in the serial production may occur, and conditions for its rapid changeover to new types of products may be created, but the problem of controlling the accuracy and quality of production remains relevant [2].

Solutions to this and similar problems are inextricably linked to an increase in the production culture, a decrease in energy consumption, and the level of CO_2 emissions both at the production stage and during the operation of precision products [3]. An essential role in improving the performance of products is played by the methodological

V. Tonkonogyi et al. (Eds.): InterPartner 2022, LNME, pp. 370–379, 2023.
https://doi.org/10.1007/978-3-031-16651-8_35

approach and the use of a laboratory and methodological base, which, together with modern software and control systems, makes it possible to obtain and analyze the main tribotechnical characteristics of the studied friction pairs in real-time with the necessary accuracy and reliability to improve their operational performance characteristics [4, 5].

Simultaneously, the study of tribocontact parameters, contact pressures, and lubrication regimes are priority tasks, the solution of which is devoted to a number of theoretical and practical works in modern tribology [6, 7].

2 Literature Review

It is well known that 70...80% of wear of parts of friction units of ground and aircraft engines occurs at the time of their launch [8, 9]. From the standpoint of the adhesive-hydrodynamic model of friction and wear, this is explained by the fact that all friction nodes (from the moment of their movement to the exit to normal operation) pass areas of low speeds, if the limit friction mode is implemented [10, 11]. Under such conditions, the adhesive interaction and wear of the friction surfaces occur. Given the dual nature of friction, the mechanism of contact interaction in relative contact of friction surfaces should take into account physical models of surface phenomena [12, 13], microrelief of friction surfaces, depending on the type of mechanical and strengthening treatment [14, 15] and properties of lubricants and media. When developing precision tribosystems, new theories and approaches to friction and lubrication should be developed, models should be created and experimental studies of contact interaction should be conducted [16, 17]. Equally important is the position of increasing the reliability and durability of machines through friction modifiers, ensuring the operation of tribosystems, as well as the use of traditional and alternative fuels and lubricants and their impact on tribotechnical performance [18, 19].

There is a large amount of theoretical and experimental work in surface engineering, contact interaction of precision friction pairs, contact hydrodynamics, tribology of boundary lubrication [20]. However, almost no practical work combines the results of modern tribological achievements and technologies for the manufacture and assembly of high-precision products [21] to ensure elevated quality and management [22].

3 Research Methodology

The object of the study was plunger pairs, which are part of high-pressure fuel pumps (HPFP). Usually, the finishing operations for cleaning precision surfaces should be grinding operations with diamond pastes to the required clearance, which does not always provide the required surface accuracy and the ability to control the clearance [23, 24]. Therefore, a number of studies were conducted on precision polished tribo-surfaces in the mode of limiting friction on the developed laboratory research and testing equipment. Aviation gasoline TC-1 and diesel fuels of various supply series were used as lubricants [25].

The studied surfaces were treated with diamond pastes, by stepwise grinding with pastes of different grain sizes, to the required roughness (Fig. 1), which was monitored on a laser differential phase microscope profilometer (LDPMP) [26].

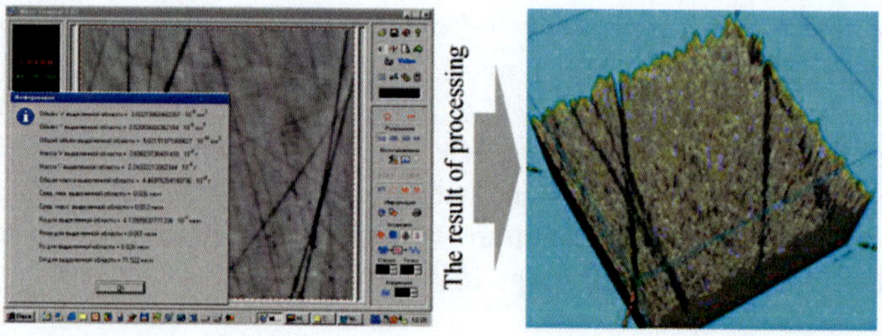

a) b)

Fig. 1. General view of the studied surface a) and microrelief b) of the studied surfaces on LDPMP.

High-pressure fuel pumps operate in different climatic zones at different temperatures, so an important scientific and practical task is to study the dynamics of boundary layers [27] under lubrication with different lubricants and ensure their trouble-free operation in a wide range of temperatures [28, 29].

Under operating conditions, it is stated that PHFP must operate in the temperature range from +40 °C to –30 °C. Very often, at temperatures below –20 °C, such pumps already fail due to the properties and quality of fuel the quality of the surfaces of the plunger pairs, and the gaps between them [30].

To study the dynamics of the boundary layers, a model friction pair was used: 100Cr6 steel - quartz glass. This allowed us to visually observe and investigate the occurrence of secondary flow in the contact zone (Fig. 2, a) and the effect of tribocavitation (Fig. 2, b) in the aviation medium at the speed of the model shaft (counter sample) ω [26, 31].

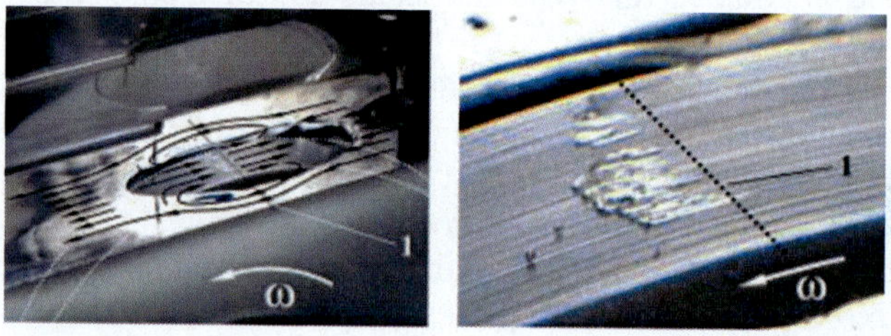

a) b)

Fig. 2. Visual studies of the model friction pair 100Cr6 steel - quartz glass in the environment of aviation gas: a) the occurrence of secondary flow in the contact zone; b) the effect of tribocavitation.

In addition, phase transitions were studied on the model friction pair at the rotational shaft frequency ω [32], particularly, all stages of nucleation, formation, and growth of gas-air mixture in dynamic equilibrium, tribocavitation process (as a partial case of cavitation), as well as partial dissolution of microbubbles in the diffuse (expanding) contact area [4, 25], Fig. 3.

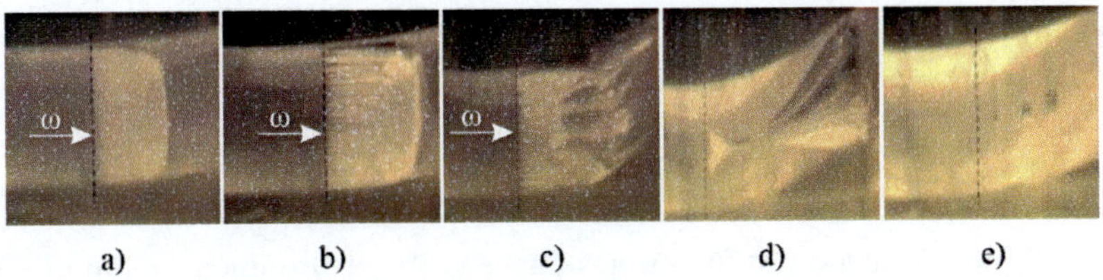

Fig. 3. Stages of formation of gas cavities at rarefaction of boundary layers in the diffuse region of tribocontact: a) nucleation at the initial moment of friction; b) growth and dynamic equilibrium with increasing speed; c) the emergence of cavitation processes; d, e) partial dissolution of gas cavities after cessation of friction.

In the framework of elastohydrodynamic friction theory, the calculated contact stresses that occur during surface compression are identified with the pressures in the boundary layers during friction [33].

That is, the pressure in the lubricating layer of the elastohydrodynamic contact is always higher than atmospheric pressure. Proponents of the adhesion-deformation approach hold the same opinion, where dynamic processes in the boundary layers are neglected [5, 34]. However, it is essential to investigate and experimentally measure the actual local pressures in the lubricating layers of the tribocontact during friction in the dynamics. The measurement results were recorded using an original computer program (Fig. 4).

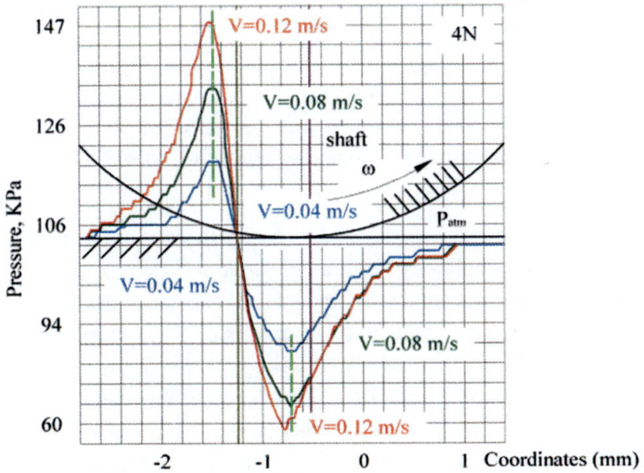

Fig. 4. The curve of the distribution of contact pressures of a series of experiments at a sliding speed of 0.251 m/s and different axial loads.

The model friction pair uses a receiving device as a slit with rounded edges to measure the pressure in the contact area. Local pressures in the contact zone were determined by scanning along the entire length of the contact (Fig. 4, b). The scan length, speed, and test coordinate are set in the program and controlled accordingly. After analyzing the results shown in Fig. 4, there is a natural decrease in pressure in the confuser contact area and an increase in the degree of rarefaction in the diffuser area [26, 35].

4 Results and Discussion

The study of the formation of the air-gas phase in the environment of diesel fuel was performed by modeling. Thus, a layer of diesel fuel (painted dark with 80W–90 oil) was applied to the polished surfaces made of 100Cr6 steel and quartz glass. When compressing these surfaces, the studied medium's extrusion into the studied surfaces' microrelief was observed, and when the load was removed, gas cavities in the boundary layer were formed (Fig. 5).

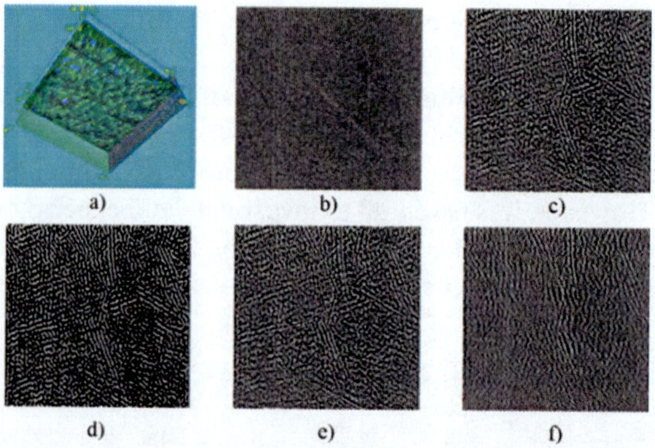

Fig. 5. Formation of planes with boundary layers of opaque lubricating medium between polished surfaces 100Cr6 - quartz: a) the general appearance of the polished surface; b) compressed surfaces with a force of 1.0 MPa; c) compressed surfaces with a force of 5.0 MPa; d) compressed surfaces with a force of 10.0 MPa; e) the surface 10 min after unloading; f) surfaces 24 h after unloading.

The studies were performed on LDPMP. After surface compression to 1.0 MPa, the contact of the surface roughness vertices of steel and quartz samples was observed (Fig. 5, b), and the contact area was about 40% of the contour contact area. When the load increases to 5.0 MPa, common planes with remnants of the opaque lubricating medium are formed between the elastically deformed vertices (Fig. 5, c). Further increase in the load to 10.0 MPa led to some expansion of the area of the contacting vertices and a decrease in the area of the cavities filled with lubricating medium (Fig. 5, d). When visually assessing the total contact area, in this case), it was increased to 60%. After the load was removed, there was a slow increase in the volume of lubricating media (Fig. 5 e, f). It should be noted that after removing the load of 10 MPa, even after 24 h, the adhesion forces between the surfaces were so strong that it was possible to separate them only by shifting them. This study allows us to explain the contact interaction of high-pressure fuel pumps' precision surfaces and make new assumptions about improving their performance.

Research on friction and wear of selected lubricants was performed on a laboratory friction device ASK-01M with linear contact (Fig. 6, a), the design features of which allow providing constant instantaneous contact stresses. Ensuring constant contact stresses is because all the studied parameters of tribocontact: the point of friction, load axis, the axis of rotation of the model shaft, and the axis around which the fixed flat sample oscillates, intersect at the center of masses of the counterbody (Fig. 6, b). The friction device controls the radial deviations of the counter-sample. The tests were carried out by mutual contact of surfaces of 100Cr6 steel in the environments of aviation gases and diesel fuels of different delivery series, considering the secondary structures (Fig. 6).

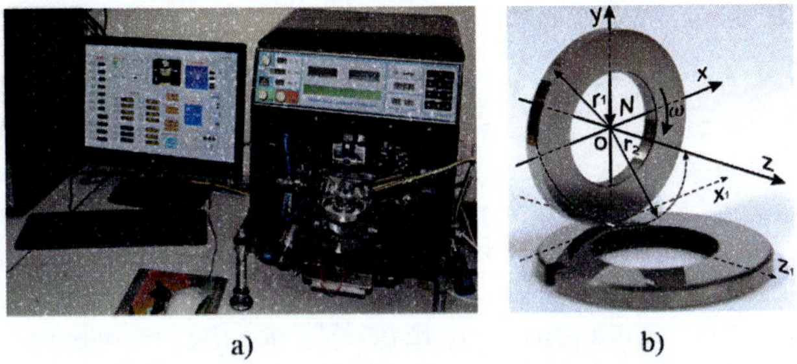

Fig. 6. General view of the friction machine with linear contact ASK-01M a) and a model sliding tribosystem that provides constant instantaneous contact stress b).

At the initial stages - 500 m of friction path (Fig. 7 a, b), secondary structures were developed (the so-called running-in period). In Fig. 7 c), quasi-stable secondary structures are obtained, which were obtained after running in (3000 m of friction). The arrows show the direction of friction.

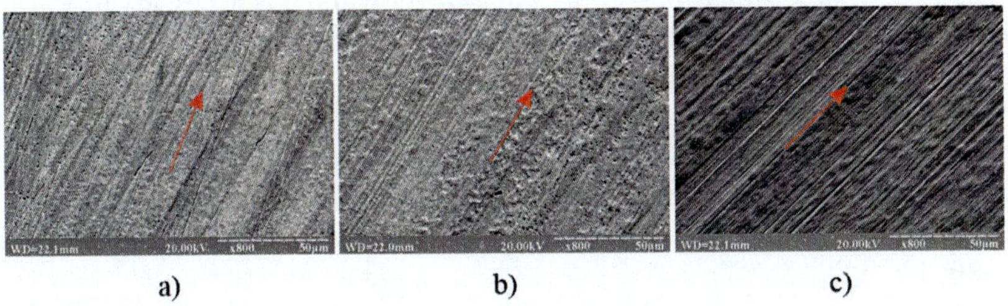

Fig. 7. General view of friction surfaces (magnification × 800) with developed secondary structures: a, b) in the initial stages; c) quasi-stable after the running-in process.

The test results of lubricants (aviation and diesel fuels) selected for research are shown in Fig. 8. The amount of wear was measured on an LDPMP. The results are presented as the dependences of wear I, μm on the friction path L, m of polished surfaces of 100Cr6 steel in selected test media.

As shown in Fig. 8, environment No. 5 has the best antiwear properties. The wear of the surfaces, when used for lubrication of the samples of the studied environments No. 2, No. 3, and No. 4 differs insignificantly (within 0.2 μm). However, the intensity

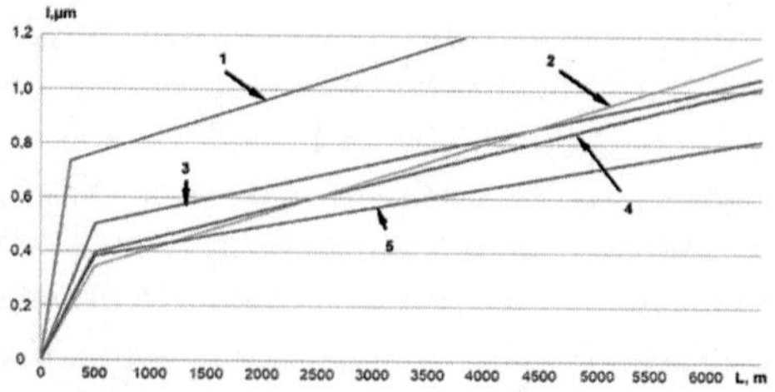

Fig. 8. The results of studies of antiwear properties of selected lubricants: 1, 2, 4 - diesel fuels from different manufacturers and different series of deliveries (selected at random); 3–5 - aviation gasoline TC-1 of different series of delivery (selected at random).

of wear in environment No. 2 is relatively high, compared to the intensity of wear on friction surfaces in environments No. 3 and 4.

Therefore, media No. 2–5 can be recommended as working media for use in operational conditions to lubricate high-precision tribo-units with small clearances (for example, a high-pressure fuel pump). When using sample No. 1, it is advisable to consider the machine's operating conditions in more detail since the anti-wear properties of this environment are low, compared to others studied in our work.

5 Conclusions

The recommended option to increase the efficiency, durability, and other operational parameters of complex high-precision products with precision tribosystems, in particular plunger pairs of high-pressure fuel pumps, is the use of a comprehensive experimental scientific and methodological approach. Experimental studies show that to increase the wear resistance of precision tribo-surfaces of plunger pairs, it is necessary to pay special attention to their microrelief (including its spatial configuration) and create appropriate technologies for their manufacture and control. One of the options for technological quality management of functional surfaces of plunger pairs is the development of high-tech operations and introducing additional finishing and strengthening operations. This will create microgemetric parameters, the configuration of the surface layer, which will not have the phenomenon of setting at the running stage in surfaces and temperatures below –20 °C.

According to the results of these studies, the wear resistance of friction surfaces and the range of operating temperatures depends on the properties of lubricating media. Particularly, when developing the technology for machining the parts of the plunger pair, the following statements should be taken into account:

1. Phase transformations in the boundary layers of lubricants – because they depend on the stage of nucleation of the gas-air mixture;
2. Dynamic phenomena in the boundary layers of lubricating media affect the state of dynamic equilibrium in the contact zone and the initialization of the process of tribocavitation. The conducted studies show that with a change in only the speed of

rotation of the movable couplings within 0.12…0.04 m/s, the maximum contact pressures change by 30 kPa (20%), which affects the wear resistance and the possibility of adhesion of the friction surfaces;
3. Backflows and heat flow in friction dynamics - leading to the formation of zones with insufficient oil film thickness and, as a result, oil starvation, especially taking into account that the contact contour area can vary by 20% or more depending on the contact pressures;
4. Viscosity-temperature and antiwear characteristics directly affect the service life of the friction assembly. Considering that the wear of surfaces in the same type of lubricating media (Fig. 8) differs by almost 50%, to ensure the necessary durability of specific high-precision products, it is advisable to recommend using only certain lubricating media of proven manufacturers.

The conducted research is a prerequisite for developing scientifically based recommendations for improving the design features of products, choosing methods for processing friction surfaces, and creating an engineering methodology for assigning processing modes. This approach will allow to expand the regulatory limits of high-pressure fuel pumps and predict the resource of their trouble-free operation.

Acknowledgment. The research was partially supported by International Association for Technological Development and Innovations.

References

1. Ovdiienko, O., Hryhorak, M., Marchuk, V., Bugayko, D.: An assessment of the aviation industry's impact on air pollution from its emissions: Worldwide and the Ukraine. Environ. Socio-Econ. Stud. **9**(2), 1 (2021). https://doi.org/10.2478/environ-2021-0006
2. Ivanov, V., Dehtiarov, I., Pavlenko, I., Kosov, I., Kosov, M.: Technology for complex parts machining in multiproduct manufacturing. Manage. Prod. Eng. Rev. **10**(2), 25–36 (2019). https://doi.org/10.24425/mper.2019.129566
3. Holmberg, K., Erdemir, A.: The impact of tribology on energy use and CO2 emission globally and in combustion engine and electric cars. Tribol. Int. **135**, 389–396 (2019). https://doi.org/10.1016/j.triboint.2019.03.024
4. Senin, P.V., Ionov, P.A., Stolyarov, A.V., et al.: Device for tribotechnical tests of friction pairs. J. Frict. Wear **41**, 141–145 (2020). https://doi.org/10.3103/S1068366620020130
5. Darovskoy, G.V., Krotov, V.N., Polyakov, V.N., et al.: Modeling the hydrodynamic friction mode on Amsler type friction testing machines. J. Frict. Wear **40**, 223–228 (2019). https://doi.org/10.3103/S1068366619030036
6. Marchuk, V.Y., Kindrachuk, M.V., Mirnenko, V.I., Bashta, O.V., Fedorchuk, S.V.: Physical interpretations of internal magnetic field influence on processes in tribocontact of textured dimple surfaces. J. Nano- Electron. Phys. **11**(5), 05013 (2019). https://doi.org/10.21272/jnep.11(5).05013
7. Lyashenko, I.A., Popov, V.L.: Transition between modes of adhesion and sliding friction in contacts of axially symmetric bodies. J. Frict. Wear **40**, 39–45 (2019). https://doi.org/10.3103/S1068366619010124

8. Kindrachuk, M., Dukhota, O., Tisov, O., Kharchenko, V., Naumenko, N.: Improving the wear resistance of heavyduty elements in tribomechanical systems by a combined laser-thermochemical processing method. Eastern-Eur. J. Enterp. Technol. **3**(12(111)), 6–13 (2021). https://doi.org/10.15587/1729-4061.2021.231595
9. Mnatsakanov, R.G., Mikosianchyk, O.A., Yakobchuk, O.E., et al.: Lubricating properties of boundary films in tribosystems under critical operation conditions. J. Mach. Manuf. Reliab. **50**, 229–235 (2021). https://doi.org/10.3103/S1052618821030110
10. Dmitrichenko, N.F., Milanenko, A.A., Savchuk, A.N., et al.: Improving the efficiency of lubricants by introducing friction modifiers for tracked vehicles under stationary conditions of friction. J. Frict. Wear **37**, 441–447 (2016). https://doi.org/10.3103/S1068366616050044
11. Kozdrach, R.: The innovative research methodology of tribological and rheological properties of lubricating grease. Tribol. Ind. **43**(1), 117–130 (2021). https://doi.org/10.24874/ti.941.08.20.11
12. Myshkin, N.K., Goryacheva, I.G., Grigoriev, A.Y., et al.: Contact interaction in precision tribosystems. J. Frict. Wear **41**, 191–197 (2020). https://doi.org/10.3103/S1068366620030113
13. Tsukanov, I.Y., Shcherbakova, O.O., Mezrin, A.M., et al.: The tribological characteristics and microgeometry of antifriction alloy surfaces in the running-in period. J. Frict. Wear **41**, 12–17 (2020). https://doi.org/10.3103/S106836662001016X
14. Zabolotnyi, O., Bozhko, T., Machado, J., Yarmoliuk, S., Zaleta, O.: Influence of the cutting temperature on the surface layer quality when grinding sintered porous materials. In: Tonkonogyi, V., Ivanov, V., Trojanowska, J., Oborskyi, G., Pavlenko, I. (eds.) InterPartner 2021. LNME, pp. 455–465. Springer, Cham (2022). https://doi.org/10.1007/978-3-030-91327-4_45
15. Zabolotnyi, O., Bozhko, T., Halchuk, T., Zaleta, O., Cagáňová, D.: Investigation of the surface layer hardness when grinding sintered porous workpieces. In: Ivanov, V., Trojanowska, J., Pavlenko, I., Rauch, E., Peraković, D. (eds.) DSMIE 2022. Lecture Notes in Mechanical Engineering, pp. 355–364. Springer International Publishing, Cham (2022). https://doi.org/10.1007/978-3-031-06025-0_35
16. Juostas, A., Jotautienė, E., Greco, C.: Experimental evaluation of the tribotechnical properties of engine oils for combine harvesters. J. Frict. Wear **42**, 11–16 (2021). https://doi.org/10.3103/S1068366621010128
17. Pavlenko, I.: Static and dynamic analysis of the closing rotor balancing device of the multistage centrifugal pump. Appl. Mech. Mater. **630**, 248–254 (2014). https://doi.org/10.4028/www.scientific.net/AMM.630.248
18. Groetsch, D., Motzet, R., Voelkel, K., Pflaum, H., Stahl, K.: Analysis of oil distribution and reduction of axial force due to oil supply in a multi-plate clutch. Tribol. Ind. **44**(2), 268–282 (2022). https://doi.org/10.24874/ti.1168.08.21.11
19. Shvets, S.V., Machado, J.: Numerical model of cutting tool blade wear. J. Eng. Sci. **8**(2), A1–A5 (2021). https://doi.org/10.21272/jes.2021.8(2).a1
20. Dmitrichenko, N.F., Milanenko, A.A., Hluhonets, A.A., et al.: A technique for forecasting the durability of rolling bearings and the optimum choice of lubricants under flood-lubrication and oil-starvation conditions. J. Frict. Wear **38**, 126–131 (2017). https://doi.org/10.3103/S1068366617020076
21. Kostyk, K., et al.: Simulation of diffusion processes in chemical and thermal processing of machine parts. Processes **9**(4), 698 (2021). https://doi.org/10.3390/pr9040698
22. Varela, M.L.R., Putnik, G.D., Manupati, V.K., Rajyalakshmi, G., Trojanowska, J., Machado, J.: Integrated process planning and scheduling in networked manufacturing systems for I4.0: A review and framework proposal. Wireless Netw. **27**(3), 1587–1599 (2019). https://doi.org/10.1007/s11276-019-02082-8

23. Slobodyan, B.S., Malanchuk, N.I., Martynyak, R.M., Lyashenko, B.A., Marchuk, V.E.: Local sliding of elastic bodies in the presence of gas in the intercontact gap. Mater. Sci. **50**(2), 261–268 (2014). https://doi.org/10.1007/s11003-014-9716-5
24. Lyashenko, I.A., Metlov, L.S., Khomenko, A.V., et al.: Nonequilibrium kinetics of phase transitions in the boundary friction mode. J. Frict. Wear **33**, 244–252 (2012). https://doi.org/10.3103/S106836661204006X
25. Stwl'makh, A.U., Kostyunik, R.E., LBadir, K.K.: Desorption-adhesion mechanism of wear under boundary lubrication. J. Frict. Wear **35**(1), 16–24 (2014). https://doi.org/10.3103/S1068366614010097
26. Stelmakh, A., Kostunik, R., Radzievskyi, V., Shymchuk, S., Zaichuk, N.: An increase in tribocharacteristics for highly loaded friction units of modern equipment. In: Ivanov, V., Trojanowska, J., Pavlenko, I., Rauch, E., Peraković, D. (eds.) DSMIE 2022. Lecture Notes in Mechanical Engineering, pp. 504–518. Springer International Publishing, Cham (2022). https://doi.org/10.1007/978-3-031-06025-0_50
27. Merzliakov, I., Pavlenko, I., Chekh, O., Sharapov, S., Ivanov, V.: Mathematical modeling of operating process and technological features for designing the vortex type liquid-vapor jet apparatus. In: Ivanov, V., et al. (eds.) DSMIE 2019. LNME, pp. 613–622. Springer, Cham (2020). https://doi.org/10.1007/978-3-030-22365-6_61
28. Berladir, K., Hovorun, T., Gusak, O.: Strengthening of the NKV type centrifugal pump's shaft by chemical-thermocycling treatment. In: Ivanov, V., Trojanowska, J., Pavlenko, I., Zajac, J., Peraković, D. (eds.) DSMIE 2021. LNME, pp. 525–535. Springer, Cham (2021). https://doi.org/10.1007/978-3-030-77719-7_52
29. Barandich, K.S., Vysloukh, S.P., Antonyuk, V.S.: Ensuring cyclic durability of parts during finishing turning with a tool made of cubic boron nitride tools. Superhard Materials **3**, 67–78 (2018)
30. Dmytrychenko, N., Khrutba, V., Savchuk, A., Hlukhonets, A.: Using mathematical, experimental and statistical modeling to predict the lubricant layer thickness in tribosystems. In: Palagin, A., Anisimov, A., Morozov, A., Shkarlet, S. (eds.) MODS 2019. AISC, vol. 1019, pp. 39–49. Springer, Cham (2020). https://doi.org/10.1007/978-3-030-25741-5_5
31. Marchuk, V., Kindrachuk, M., Krysak, Y., Tisov, O., Dukhota, O., Gradiskiy, Y.: The mathematical model of motion trajectory of wear particle between textured surfaces. Tribol. Ind. **43**(2), 241–246 (2021). https://doi.org/10.24874/ti.1001.11.20.03
32. Mordyuk, B.N., Mikosyanchik, O.O., Mnatsakanov, R.G.: Structure-phase state and wear of Ni–Cr–B–Si–C coating on steel 1045 under friction conditions with the shear load component. Metallofiz. Noveishie Tekhnol. **42**(2), 175–195 (2020). https://doi.org/10.15407/mfint.42.02.0175
33. Kindrachuk, M., Volchenko, D., Balitskii, A., et al.: Wear resistance of spark ignition engine piston rings in hydrogen-containing environments. Energies **14**, 4801 (2021). https://doi.org/10.3390/en14164801
34. Zaichuk, N., Shymchuk, S., Tkachuk, A., Feshchuk, Y., Szczot, J.: Structure and properties of surface bandage shelves for the gas turbine engine's blades. In: Ivanov, V., Trojanowska, J., Pavlenko, I., Zajac, J., Peraković, D. (eds.) DSMIE 2021. LNME, pp. 602–612. Springer, Cham (2021). https://doi.org/10.1007/978-3-030-77719-7_60
35. Litvinenko, S.V., Ilchenko, L.M., Kolenov, S.O., Smirnov, E.M., Molochko, P.V., Skryshevsky, V.A.: Laser scanning for sensing and study the operation of semiconductor devices. In: Proceedings of the 2008 4th International Conference on Advanced Optoelectronics and Lasers, pp. 441–443. IEEE (2008). https://doi.org/10.1109/caol.2008.4671918

The Taxonomy Approach for Engineering Students' Outcomes Assessment

Olena Titova[1](✉), Petro Luzan[2], Qudrat Q. Davlatzoda[3], Iryna Mosia[2], and Maryna Kabysh[2]

[1] Dmytro Motornyi, Tavria State Agrotechnological University, 18, B. Khmelnytskyi Avenue, Melitopol 72312, Ukraine
olena.titova@tsatu.edu.ua

[2] Institute of Vocational Education of the National Academy of Pedagogical Sciences of Ukraine, 98a, Vito-Lithuanian Lane, Kyiv 03045, Ukraine

[3] Tajik Technical University named after academician M. Osimi, 10, academicians Rajabov's Avenue, Dushanbe 734042, Republic of Tajikistan

Abstract. The paper presents the analysis of B. Bloom's taxonomy approach to describe learning objectives in connection with program learning outcomes and the assessment methodology and tools for analysis of engineering students' cognitive skills of different levels. The research aimed to develop some recommendations for designing testing items that could be realized for multilevel assessment of the results of engineering students' competency-based professional training. That was performed considering the evidence, which reflected the theory and practice of B. Bloom's taxonomy use for learning outcomes, learning activities, and assessment strategy description for engineering education. The verbal means were analyzed and used to develop the recommendations, which could be applied to formulate the learning outcomes and tasks at different cognitive levels. The application of the taxonomy approach required the development of test items provided for six cognitive categories. A set of multiple choice and open-ended tasks was developed to perform a multilevel assessment of the results of the specific course mastering by agricultural engineering students. An experiment was performed to study how the formulation and the test item type influence the assessment results. It showed the effectiveness of differential ability of the approach when taxonomy is used to assess students' learning outcomes.

Keywords: Learning outcomes · Test · Engineering students · Sustainability education · Assessment · Evaluation · Validity · Test items · Education policy

1 Introduction

Benjamin S. Bloom [1] introduced the original term taxonomy into education in 1956. It was an attempt to build a classification of educational objectives in the cognitive domain for teachers, administrators, and research workers who dealt with curricular and assessment issues. The six categories of educational objectives were hierarchically

V. Tonkonogyi et al. (Eds.): InterPartner 2022, LNME, pp. 380–390, 2023.
https://doi.org/10.1007/978-3-031-16651-8_36

arranged in a framework that meant that achievement of the next (more complex) skill or ability is based on the achievements of the skills of the lower level.

The taxonomy has become an understandable language for learning goals description. That facilitated educators' communication and cooperation through defining subject matter and grade levels. Nowadays, the taxonomy is used to determine broad educational goals stated in standards for a particular course or curriculum in the specific meaning. In the aspect of assessment, the taxonomy becomes a framework for connecting educational objectives, learning activities, and test items. It also reveals any limitations in the breadth and depth of a particular educational course or curriculum [2].

To assess students' learning outcomes, the original taxonomy proposed using test items (mostly of multiple choice) provided for each cognitive category. In that case, a common framework, where standard received vocabulary is applied, could help classify students' learning outcomes, indicate what an item is being measured, and promote the exchange of testing approaches and test items.

2 Literature Review

Continuous improvement of the higher engineering education system requires the search for new approaches to improving the quality of professional training. One of them is related to the development of methods, tools, and technologies for learning outcomes assessment. The scientists concluded that the acquisition of knowledge [3] and further control of student's achievements [4] should be carried out through a system of educational tasks developed based on certain didactic principles [5]. The leading prescription for designing educational testing tasks is the principle of their consistent complication [6]. In this regard, a level system approach to describing student's achievements was proposed [1, 2]. It allowed to group learning outcomes by levels of learning activities which characterize student achievements as abilities to differentiate, recognize; unconsciously reproduce; consciously reproduce; act in a familiar situation; act in an unfamiliar situation [7], highlighting the possibility of developing engineering students' cognitive activities [8], to improve the effectiveness of the assessment [9] especially when the process is automated [10].

The revision of the original taxonomy [2] was a two-dimensional framework that included Knowledge and Cognitive Processes. The latter resembles the six categories with the change in their names to their verb forms. The categories were also hierarchically arranged but not as rigidly as in the original taxonomy. The publications on the point analysis showed that the framework could be successfully used to classify objectives, activities, and assessments and provide a clear, concise, visual representation of a particular course or its units.

There are findings on the learning outcomes-based methodology development and testing [11]. The approach could be used to produce a summative assessment at any educational level. The European Qualifications Framework (EQF) can be used to define the learning outcomes. The assessment questions are recommended to design according to Bloom's taxonomy. It is possible to trace the tendency among teachers to increase the amount of higher-level test items and decline lower-level questions in course assessments [12]. That could foster cognitive skills development and become a precondition of an effective application of Bloom's taxonomy of program learning outcomes in the higher-level cognitive skills assessment process. For this research, some technics for examining the validity of assessment tools to evaluate student's learning and attainment results were considered [13].

As B. Bloom's original taxonomy combines specific learning outcomes (knowledge, understanding, application) with mental operations (analysis, synthesis, evaluation) and employs the goals-results approach, the scientists consider it to be quite effective in designing the categories of learning (and control) tasks in the context of systemic development of the student's personality [14]: considering the effectiveness of an 'outcome-based teaching model' [15] and engineering students' learning expectations [16].

Considering the findings analyzed, the following research gaps have been revealed. There is a lack of specific methods and approaches for designing testing items that could provide a multilevel check of engineering students learning outcomes at different cognitive levels. Those required methods and approaches made up the scientific novelty of the current research. It was solved by applying the content of the Hydraulic Drive engineering area. Therefore, the purpose of the current study is:

- to apply the approach of B. Bloom's taxonomy to meet the requirements of the principle of consistent complication of engineering student's educational and cognitive activities;
- to develop recommendations to the verbal means, which could be used to formulate the learning outcomes and test items for six cognitive levels to assess the results of competency-based training of agricultural engineering students;
- to make a comparative analysis of multiple choice and open-ended tasks differential ability while they are being used to assess agricultural engineering students' learning outcomes.

3 Research Methodology

To achieve the research aim, a set of theoretical and empirical methods were applied. To clarify the state of the research problem and determine the points for the current research, the analysis of scientific sources, the study of qualification standards, and educational and professional programs were performed. The comparison enabled to study of scientific approaches to problem-solving. Analysis and synthesis were used to develop test items to assess the results of agricultural engineering students' competency-based training considering each cognitive level according to B. Bloom's taxonomy. Empirical methods included observation and testing in determining the validity of tools and making adjustments to the assessment methodology. The pedagogical experiment method allowed the study of the differential ability of the developed tools for assessing students' learning outcomes and confirming the adequacy (identity) of the test items. The introduction of cognitive levels enabled to develop a set of test items that considered the principle of consistent complication of educational and cognitive activities during the professional competency-based training of agricultural engineering students.

The research methodology was also based on the principles of formative assessment, which effectively provided feedback, and students master the ability to plan, analyze and regulate their own learning activities, identify causes of difficulties, and possible ways to overcome them.

The advantage of the proposed technique was in its ability to be successfully used for designing tasks in the disciplines of engineering fundamentals (descriptive geometry and engineering graphics, materials science, physics, theoretical foundations of electrical engineering, etc.) as well as professional and practical disciplines (machinery and equipment design, maintenance and repair, automation, etc.).

In the current research, we employed educational theory and practice data to define the term 'engineering students' learning outcomes' as a set of knowledge, skills, abilities, and other competencies acquired by a person while learning a certain program. The knowledge, skills, and abilities identified, quantified, and measured were also considered [17]. We followed the approach for program learning outcomes formulation, which takes into account the internationally recognized reference positions and recommendations of international quality assurance agencies for the students from different fields [10, 12], as well as clear criteria for agile [18] and constant [19] evaluation of students' learning results.

The recommendations to the learning outcomes formulation were developed employing the results of the scientific research on the problem, dealing with the curriculum integration issues [20], design and assessment of learning courses [21], requirements to assessment experts [22], and the most effective forms for test items [23].

Some peculiar examples of the learning outcomes were retrieved from the recent scientific publications in a specific engineering area (remembering the terms [24, 25], understanding the operational processes [26, 27] as well as practical application [28]) and formulated according to the B. Bloom's taxonomy.

4 Results and Discussion

When designing a program, the teacher should consider that learning outcomes characterize what the student must be able to demonstrate after the successful completion of the program (knowledge, skills, ability to perform, etc.). As a full set of program learning outcomes expresses the characteristics of the program, they should be:

- specific to provide a sufficient level of detail, written in plain language;
- objective and formulated neutrally, avoiding excessive ambition and subjectivity;
- achievable and realistic in terms of time and resources needed to achieve them;
- useful that means that they should be perceived as meeting the level of higher education and the social requirements/expectations;
- relevant to meet the qualification requirements;
- have the nature of standards to determine the standard requirements to be met by learning outcomes [22, 23].

The next stage of the research tasks solution was devoted to verbal means. The language of formulation of program learning outcomes is especially important.

According to the scientists' findings, the wording should contain five main elements:

1. active verbal form (for example, a graduate demonstrates an ability, can demonstrate knowledge, demonstrates a skill);
2. indication of the type of learning outcome (knowledge, skills, other competencies);
3. the thematic area of learning outcomes (special or general, subject area, special skills, etc.);
4. the expected standard or level that is planned to be achieved as a result of training;
5. the scale and context of the learning outcome.

Based on the theoretical and practical experience, we could state that learning outcomes should also reflect the appropriate type of learning activity. That means that it should reflect whether the learning outcome relates to the acquisition of knowledge, understanding, mechanical skills, or professional status. The name of the learning outcome also contains information about the expected level of studies, such as breadth, depth, and complexity.

According to the original Bloom's taxonomy [1] and the revised framework [2], the cognitive sphere was considered a hierarchy of six successive levels of complexity: 1. Knowledge / Remember. 2. Comprehension / Understand. 3. Application / Apply. 4. Analysis / Analyze. 5. Evaluation / Evaluate. 6. Synthesis / Create. Therefore, the test items should be formulated to reflect a definite cognitive level, which comes after the previous one and determines the student's current level according to the complexity of learning outcomes (Fig. 1).

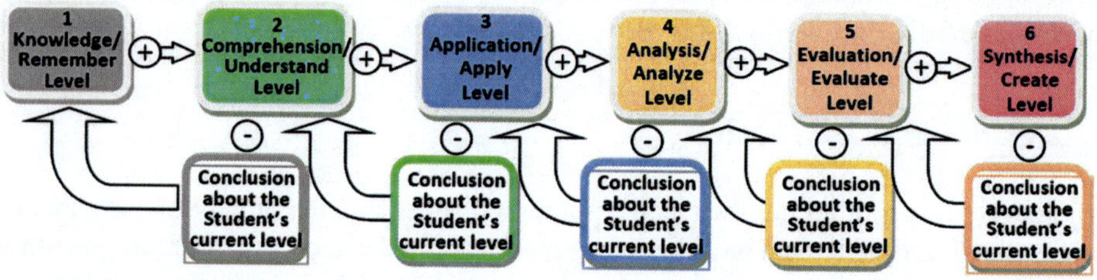

Fig. 1. The algorithm for application of the levels of complexity of learning outcomes while designing a test.

The first group of learning outcomes is marked as the Knowledge / Remember level. That includes student's ability to memorize or reproduce facts (terms, specific facts, methods and procedures, basic concepts, rules, and principles, etc.) without the need to understand them. The examples of learning outcomes determining could be used to formulate the testing items. At this cognitive level, it is recommended to use definite verbs: recognize the symbols of hydraulic devices (for example, the task could be formulated as Which letters in Fig. 2 label the pump, hydraulic motor, hydraulic cylinder, safety valve, distributor, throttle, filter, heat exchanger in the diagrams?); name the main elements of hydraulic drives of mechatronic systems of agricultural machinery; streamline operations of hydraulic devices inclusion in diagrams of a hydraulic drive of a mechatronic system; name the device that controls the flow of working fluid; describe the general structure of hydraulic devices; name advantages of application of the hydraulic drive of a mechatronic system in comparison with the mechanical drive, etc.

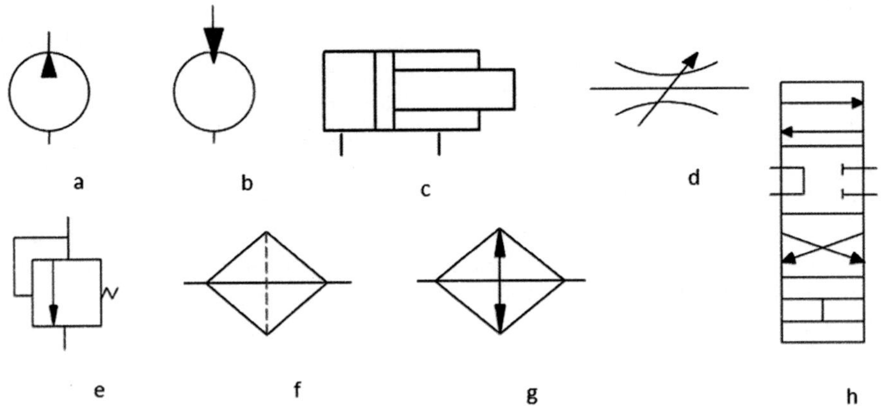

Fig. 2. Symbols of hydraulic devices used in diagrams.

The second cognitive level is Comprehension / Understand. The level states the student's ability to understand and interpret what is learned. It means the ability to explain facts, rules, and principles; turn verbal material into, for example, mathematical expressions; predict future consequences based on the acquired knowledge. The following verbs are used to formulate learning outcomes at this level: classify hydraulic drives of mechatronic systems and find out the conditions of their application; describe the working processes occurring in the elements of the hydraulic drive; explain the principles of operation of hydraulic drive elements; explain the difference between the work processes that take place in the pump and the hydraulic motor; explain the principle of operation of a hydraulic drive of a mechatronic system; establish a connection between hydraulic devices and the working fluid during the operation of the mechatronic system with a hydraulic drive, etc.

The next hierarchical level is Application / Apply, which considers the ability to use the studied material in new situations, for example, to apply ideas and concepts to solve specific problems. The learning outcomes could be determined at this level as the following: build the basic hydraulic scheme of a hydraulic drive of a mechatronic system; choose the diagram of the hydraulic drive of the mechatronic system of agricultural machinery depending on the conditions of application; find the relationship between the power of a hydraulic drive and the nominal pressure of the working fluid; choose the necessary standard size of hydraulic machines depending on the diagram of the hydraulic drive; select and calculate hydraulic machines with the most suitable functional parameters for the definite hydraulic drive diagram; find the dependence of the efficiency on the power of the hydraulic drive; develop and construct the basic hydraulic diagram of the hydraulic drive of the mechatronic system taking into account the selected hydraulic machines and hydraulic devices, etc.

At the Analysis / Analyze level, a student must demonstrate an ability to break down information into components, understand their relationships and organizational structure, see mistakes and flaws in the logic of reasoning, the difference between facts and consequences, and assess the relevance of data. The examples could include: analyzing the use of a hydraulic drive for active working tools and running systems of agricultural machinery; evaluating the selected sizes of hydraulic machines; compare the working processes that take place in hydraulic pumps and hydraulic motors in the operating conditions of the hydraulic drive; study the initial characteristics of hydraulic machines depending on their design; evaluate the modes of operation of a hydraulic drive depending on the regulations and operating conditions, etc.

In the revised taxonomy [2], the Evaluation / Evaluate level was put before the last one as a necessary precondition for the abilities of the highest, creative cognitive level. It includes assessing the material's importance for a specific purpose. The learning outcomes can be formed as follows: assess how the traction and energy performance of a tractor with a mechanical transmission changes depending on the agricultural background on which the agricultural work is performed; draw a conclusion about the expediency of using a tractor of this traction class on one or another agricultural background; assess how the traction and energy performance of a tractor with a hydrovolume transmission changes depending on the agricultural background on which the agricultural work is performed; make a conclusion about the expediency of using a tractor of this traction class with a hydrovolume transmission on one or another agricultural background; compare the traction and energy performance of the tractor with mechanical and hydrovolume transmissions when working on different agricultural background, on which the agricultural work is performed; recommend the optimal type of agricultural background and type of a transmission, for which it is advisable to use a tractor of this traction class with the best values of traction efficiency.

The highest cognitive level is Synthesis / Create. At this level, students demonstrate the ability to combine parts together to obtain a whole with a new system property. The following verbs are used to formulate learning outcomes related to synthesis: identify and formulate problems directly associated with the introduction of a hydraulic drive for active working tools and running systems of mobile agricultural machinery; generalize the reasons for insufficient use of hydraulic drives in mechatronic systems of agricultural machinery; classify active working tools of agricultural machinery according to the need of improvement through the introduction of a hydraulic drive; offer hydraulic devices that require development and design for the hydraulic drive for mechatronic systems of agricultural machinery; summarize the advantages and disadvantages of using hydraulic drives of active working bodies and running systems of agricultural machinery, etc.

The results were discussed considering the findings on the problem, which showed high results after the application of the new assessment methodology based on the student's outcomes [11]. Our findings coincided with conclusions about the comprehension level test items [12] and the application of a cognitive model based on B. Bloom's taxonomy to provide the consequent complication of the engineering test items [29]. The obtained approach could also become a preliminary one in the chain of creation of adaptive tests for automated testing [30].

The current research also provides the results of the analysis on the question of how learning outcomes assessment depends on the type of the test items. A set of test items for learning information of different levels was developed for that purpose. It included multiple choice test items and items of open-ended format. The pedagogical experiment was attended by the 4th-year agricultural engineering students from two groups, A and B, that made up 50 students from Dmytro Motornyi Tavria State Agrotechnological University. The respondents were offered a set of 18 open-ended and 18 multiple choice test items, developed according to the levels of B. Bloom's taxonomy. For the sake of purity of the experiment, students in group A first performed multiple choice test items and then open-ended ones; students in group B worked with multiple choice test items after performing open-ended tasks.

The students' learning outcomes assessment results according to 6 levels were processed using a statistics application. Statistical analysis of pedagogical experiment data showed that with a probability of 95%, it could be stated that the distributions of students in the A group and B group were consistent with each other. The fact was justified by applying the Mann Whitney Wilcoxon (Mann Whitney U) test. For all possible cases, its empirical value was less than a critical one. In addition, the results of students' learning outcomes assessment were compared with each other (Table 1).

Table 1. The number of correctly performed multiple choice test items (MC) and open-ended tasks (OE), differentiated according to B. Bloom's taxonomy.

No	Test items level	Amount of correct answers			
		A (MC)	A (OE)	B (MC)	B (OE)
1	Knowledge / Remember	68	71	66	69
2	Comprehension / Understand	48	53	47	53
3	Application / Apply	37	39	36	40
4	Analysis / Analyze	23	25	22	24
5	Evaluation / Evaluate	15	17	14	16
6	Synthesis / Create	8	10	9	12

In particular, the experiment results showed that among 150 simplest multiple choice test items of the Knowledge level, respondents solved 89.3% correctly. At the same time, the correct results for open-ended tasks made up 93.3%. In turn, provided 150 multiple choice test items for the Synthesis level gave 11.3% of correct answers. The share of correctly performed similarly in content open-ended tasks was 14.6%.

5 Conclusions

Considering the levels of learning material in the context of B. Bloom's taxonomy, it was possible to streamline the process of acquiring knowledge from simple to complex, from reproductive learning to productive and creative. That could provide proper testing procedures and tools for assessing the quality of engineering students professional training. The analysis of recent scientific publications revealed the verbal means for learning outcomes formulation and the formulation of the test items for one specific area of agricultural engineering students professional training. Thus, the proposed approach to assessing the results of competence-oriented training of engineering students enabled to differentiate clearly the levels of educational material mastery. The comparative analysis of multiple choice and open-ended tasks implementation showed that the obtained results were compared with a probability of 95%.

Prospects for further scientific research could be associated with developing a methodological system for assessing the quality of engineering students professional training.

References

1. Bloom, B.S., Engelhart, M.D., Furst, E.J., Hill, W.H., Krathwohl, D.R.: Taxonomy of educational objectives: The classification of educational goals. Handbook I: cognitive domain. David McKay Company (1956)
2. Anderson, L.W., Krathwohl, D.: A Taxonomy for Learning, Teaching, and Assessing: a Revision of Bloom's Taxonomy of Educational Objectives. Longman (2001)
3. Hernandez-de-Menendez, M., Morales-Menendez, R.: Technological innovations and practices in engineering education: a review. Int. J. Interactive Design and Manufacturing (IJIDeM) **13**(2), 713–728 (2019). https://doi.org/10.1007/s12008-019-00550-1
4. Felder, R.M., Brent, R.: Designing and teaching courses to satisfy the ABET engineering criteria. J. Eng. Educ. **92**, 7–25 (2003). https://doi.org/10.1002/j.2168-9830.2003.tb00734.x
5. Nagayev, V., Danchenko, I., Mitiashkina, T., Kyrepin, V.: Administrative fundamentals of ecological competence forming in agricultural engineering students under conditions of their professional training. In: Tonkonogyi, V., Ivanov, V., Trojanowska, J., Oborskyi, G., Pavlenko, I. (eds.) InterPartner 2021. LNME, pp. 697–706. Springer, Cham (2022). https://doi.org/10.1007/978-3-030-91327-4_67
6. Sharunova, A., Wang, Y., Kowalski, M., Qureshi, A.J.: Applying Bloom's taxonomy in transdisciplinary engineering design education. Int. J. Technol. Des. Educ. **32**(2), 987–999 (2020). https://doi.org/10.1007/s10798-020-09621-x
7. Sharunova, A., Butt, M., Qureshi, A.J.: Transdisciplinary design education for engineering undergraduates: mapping of bloom's taxonomy cognitive domain across design stages. Procedia CIRP **70**, 313–318 (2018). https://doi.org/10.1016/j.procir.2018.02.042
8. Granello, D.H.: Encouraging the cognitive development of supervisees: using Bloom's taxonomy in supervision. Couns. Educ. Superv. **40**(1), 31–46 (2000)
9. Firman, F., Baedhowi, B., Murtini, W.: The effectiveness of the scientific approach to improve student learning outcomes. Int. J. Active Learning **3**(2), 86–91 (2018)
10. Azuma, M., Coallier F., Garbajosa J.: How to apply the bloom taxonomy to software engineering. Software Technology and Engineering Practice 2003. In: Eleventh Annual International Workshop on Software Technology and Engineering Practice, pp. 117–122 (2003). https://doi.org/10.1109/STEP.2003.13
11. Violante, M.G., Moos, S., Vezzetti, E.: A methodology for supporting the design of a learning outcomes-based formative assessment: the engineering drawing case study. Eur. J. Eng. Educ. **45**(2), 305–327 (2020). https://doi.org/10.1080/03043797.2019.1622653
12. Verenna, A.-M.A., Noble, K.A., Pearson, H.E., Miller, S.M.: Role of comprehension on performance at higher levels of Bloom's taxonomy: Findings from assessments of healthcare professional students. Anat. Sci. Educ. **11**(5), 433–444 (2018)
13. El-Hassan, H., Hamouda, M., El-Maaddawy, T., Maraqa, M.: Curriculum-based exit exam for assessment of student learning. Eur. J. Eng. Educ. **46**(6), 849–873 (2021). https://doi.org/10.1080/03043797.2021.1920892

14. Radmehr, F., Drake, M.: Revised Bloom's taxonomy and integral calculus: unpacking the knowledge dimension. Int. J. Math. Educ. Sci. Technol. **48**(8), 1206–1224 (2017). https://doi.org/10.1080/0020739X.2017.1321796
15. Wang, L.H.: Construction of college english blended teaching model: an outcome-based education approach. Open Access Library Journal **8**, 1–16 (2021). https://doi.org/10.4236/oalib.1107652
16. Goel, S., Sharda, N.: What do engineers want? Examining engineering education through Bloom's taxonomy. In: Proc. of the 15th Annual Conference for the Australasian Association for Engineering Education, ED524509, pp. 1–14. Toowoomba, Queensland, Australia (2004)
17. Ferris, T.L.J., Aziz, S.M.: A Psychomotor Skills Extension to Bloom's Taxonomy of Education Objectives for Engineering Education. Exploring Innovation in Education and Research, CEER-2005, pp. 1–5 (2005)
18. Castillo-Barrera, F.E., Amador-García, M., Pérez-González, H., Martínez-Pérez, F.E.: Agile Evaluation of the Complexity of User Stories Using the Bloom's Taxonomy. In: 2017 International Conference on Computational Science and Computational Intelligence (CSCI), pp. 1047–1050 (2017). https://doi.org/10.1109/CSCI.2017.182
19. Bhargav, H., Akalwadi, G., Pujari N.V.: Application of blooms taxonomy in day-to-day examinations. In: 2016 IEEE 6th International Conference on Advanced Computing (IACC), pp. 825–829 (2016). https://doi.org/10.1109/IACC.2016.157
20. Ketchman, K., Dancz, C.L.A., Burke, R.D., Parrish, K., Landis, A.E., Bilec, M.M.: Sustainable engineering cognitive outcomes: examining different approaches for curriculum integration. J. Professional Issues in Eng. Education Practice **143**(3), 04017002 (2017). https://doi.org/10.1061/(ASCE)EI.1943-5541.0000324
21. Britto, R., Usman, M.: Bloom's taxonomy in software engineering education: A systematic mapping study. In: 2015 IEEE Frontiers in Education Conference (FIE), pp. 1–8 (2015). https://doi.org/10.1109/FIE.2015.7344084
22. Näsström, G.: Interpretation of standards with Bloom's revised taxonomy: a comparison of teachers and assessment experts. Int. J. Research Method Educ. **32**(1), 39–51 (2009). https://doi.org/10.1080/17437270902749262
23. Qamar, S.Z., Kamanathan, A., Al-Rawahi, N.Z.: Teaching product design in line with Bloom's taxonomy and abet student outcomes. In: 2016 IEEE Global Engineering Education Conference (EDUCON), pp. 1017–1022 (2016). https://doi.org/10.1109/EDUCON.2016.7474677
24. Panchenko, A., Voloshina, A., Panchenko, I., Pashchenko, V., Zasiadko, A.: Influence of the shape of windows on the throughput of the planetary hydraulic motor's distribution system. In: Ivanov, V., Pavlenko, I., Liaposhchenko, O., Machado, J., Edl, M. (eds.) DSMIE 2021. LNME, pp. 146–155. Springer, Cham (2021). https://doi.org/10.1007/978-3-030-77823-1_15
25. Voloshina, A., Panchenko, A., Titova, O., Pashchenko V., Zasiadko A.: Experimental studies of a throughput of the distribution systems of planetary hydraulic motors. IOP Conf. Ser.: Mater. Sci. Eng. 1021, 012054 (2021). https://doi.org/10.1088/1757-899X/1021/1/012054
26. Panchenko, A., Voloshina, A., Luzan, P., Panchenko I., Volkov S.: Kinematics of motion of rotors of an orbital hydraulic machine. IOP Conf. Ser.: Mater. Sci. Eng. 1021, 012045 (2021). https://doi.org/10.1088/1757-899X/1021/1/012045
27. Voloshina, A., Panchenko, A., Titova, O., Panchenko I.: Changes in the dynamics of the output characteristics of mechatronic systems with planetary hydraulic motors. J. Phys.: Conf. Ser. 1741, 012045 (2021). https://doi.org/10.1088/1742-6596/1741/1/012045
28. Panchenko, A., Voloshina, A., Boltianska, N., Pashchenko, V., Volkov, S.: Manufacturing error of the toothed profile of rotors for an orbital hydraulic motor. In: Tonkonogyi, V., Ivanov, V., Trojanowska, J., Oborskyi, G., Pavlenko, I. (eds.) InterPartner 2021. LNME, pp. 22–32. Springer, Cham (2022). https://doi.org/10.1007/978-3-030-91327-4_3

29. Gierl, M.J., Lai, H., Turner, S.R.: Using automatic item generation to create multiple-choice test items. Med. Educ. **46**, 757–765 (2012). https://doi.org/10.1111/j.1365-2923.2012.04289.x
30. Sorrel, M.A., Barrada, J.R., de la Torre, J., Abad, F.J.: Adapting cognitive diagnosis computerized adaptive testing item selection rules to traditional item response theory. PLoS ONE **15**(1), e0227196 (2020). https://doi.org/10.1371/journal.pone.0227196

A Universal Quality Control System on Machine-Building Enterprises

Nadezhda Yefimenko[1], Morteza Rajabzadeh[2](✉), Viliam Zaloga[3], Denys Fesenko[3], and Olga Ryasnaya[4]

[1] Bohdan Khmelnytsky National University of Cherkasy, 81, Shevchenko Blvd., Cherkasy 18031, Ukraine

[2] Mahallat Institute of Higher Education, 1 km of Khomein Road, 3781151958 Mahallat, Iran
rajabzadeh.m@gmail.com

[3] Sumy State University, 2, Rymskogo-Korsakova St., Sumy 40007, Ukraine

[4] Sumy National Agrarian University, 160, Herasyma Kondratieva St., Sumy 40000, Ukraine

Abstract. The machine-building is the main branch of the national economy. Its development is the determinative factor for increasing the product quality of many other branches of the national economy. Product quality is the essential criterion of a machine-building enterprise activity, which determines its success indicators in the market. This paper aims to study the design issue of a quality control system that enables machine-building enterprises to manufacture high-quality market-competitive products. Previous scientific investigation regarding statistical product quality control methods frequently proved their inadequacy and inefficiency in all their life-cycle stages. As a result, this research suggests the basic principles for creating a universal integrated product quality control system, using the theory of order statistics and the application of asymptotic and symmetric distributions of extreme values of random variables. While ensuring reassigned (required) dimensional accuracy, this system reduces series production, allows a large amount of information from a small number of control samples, and saves time and expenses on manufacturing machine-building products.

Keywords: Batch production · Statistical methods · Industrial growth · Sustainable manufacturing

1 Introduction

The machine-building is the main branch of the national economy. Its development is the determinative factor for increasing product quality in many other branches of the national economy, such as the oil and gas processing industries. Simultaneously, modern machine-building can be characterized by decreased series production, expanded products, and increased manufacturing accuracy through high technologies that use high-precision equipment and, often, nanometric measurement instruments. The management system of a modern machine-building enterprise should be built in such a way as to ensure the fulfillment of the primary operational function – the development and production of

V. Tonkonogyi et al. (Eds.): InterPartner 2022, LNME, pp. 391–399, 2023.
https://doi.org/10.1007/978-3-031-16651-8_37

scientifically advanced products (works, services) necessary for the customer promptly, with a given level of quality and an agreed price [1].

Product quality is an essential criterion of a machine-building enterprise activity, which determines its success indicators in the market [2]. The survival tendencies in a competitive market dictate the product quality level increase, where the enterprise sets the pace of technological production development, promotes innovation, accelerates the process of improving production efficiency, and an appropriate policy of enterprise resource consumption is implemented [3]. In modern market relations, the competition between enterprises is growing at the level of product quality [2].

When new machines (products) are manufactured, or previously produced ones are improved, when their quality system is being planned, it is necessary to use the machines (products) quality indicators [2]. A standard quality indicator is a quantitative expression of one or more product properties related to the product manufacturing, operation, and consumption [4]. To perform product quality control, the indicator information on all stages of the product life-cycle, especially at the manufacturing phase, is essential [5].

There are two types of product quality indicators: individual and complex. Individual quality indicators are related only to one product's properties [2]. Let's consider these indicators concerning machine-building products. So, the internal combustion engine is characterized by the following indicators: power, speed, specific fuel consumption per power unit, etc. [2]. Individual quality indicators are used in the engineer's products when studying the characteristics of a new part or modernization of a previously mastered one, as well as during technical and technological control of their production [2]. Automation of individual control processes is carried out without developing a general concept of the entire control system, creating duplicate unsynchronized data sources, and not excluding duplicate functions of structural units [1].

So, complex indicators are created for a complete assessment of the machine (product) quality only by their most important properties [6], which are used in the technical and economic planning of machine-building production [7] to assess the quality dynamics for a certain time [8].

On the other hand, it is necessary to note that today the most frequently used methods for collecting and processing quality data, also the methodological and normative provision (standard) of manufacturing technologies, as a rule, are oriented toward mass production. That is, they are intended for relatively vast masses of statistical data. These current methods reduce their efficiency in the present-day, typically small-scale, and individual production because they require more material and time costs. When paired with reduced series production and increased accuracy, efficient quality control is impossible without advanced and economic data collection methods.

Therefore, an objectively applied scientific issue emerges, which consists of creating a quality control system of machine-building products that can process both small-scale quality data and a set of quality indicators together. This issue has more importance for the national economy since its solution allows for making high-quality market-competitive products with minimum expenses. This article aims to develop a universal integrated quality control system that enables machine-building enterprises to make a high-quality market-competitive product under conditions of reduced series production and increased dimensional accuracy.

2 Literature Review

The current state of the scientific and technical issue of quality control of the machine-building product is a logical follow-up of the works below, which dedicated their scientific investigations to quality issues by applying statistical methods.

Dwiartono [9] proposed a method of quality control using statistical quality control to monitor the product production process to comply with the standards set by the company, which is engaged in the two-wheeler automotive spare parts component industry in Indonesia. This method used "check sheet tools", "histograms", "pareto diagrams", and "fishbone charts". The data obtained using control chart analysis show that the production of product type 04G22 at a company is still under control.

According to research works [10, 11], the control chart is the essential element of quality control. The control chart is the fundamental element of quality to monitor each stage of the production process and achieve the purpo.

Madanhire [12] studied the statistical process control (SPC) tools in the Zimbabwean manufacturing systems with the broad aim of upgrading them to improve on quality and cost-effectiveness, prevent problems, control their production processes, and ultimately increase profits and customer satisfaction, suggesting to use of SPC properly.

Zan [13], to realize intelligent SPC, proposed a pattern recognition method based on a multilayer bidirectional extended short-term memory network (Bi-LSTM) for quality control, which uses feature learning to realize end-to-end histogram pattern recognition (HPR) and control chart pattern recognition (CCPR).

Ivanov et al. [14] substantiated an experimental diagnostics approach for ensuring the quality of manufactured products.

Hrvačić [15], in his study, with the aim of quality improvement for the manufacturing process of a defense product manufacturing company, proposed the SPC, based on the tools such as $\overline{X}$R chart, $\overline{X}$S chart and the p-chart. The results showed that the process is mainly capable but not stable.

Oberoi [16] indicates that process variation needs to be reduced to improve quality and productivity. Their research discusses the need for SPC, the types, the procedure of plotting control charts, and process capability confirmation by SPC as a quality control tool.

According to Godina [17], to improve the quality of the technological process using SPC in the automotive industry, a brief review of concepts related to the methodology aims to demonstrate all the advantages associated with SPC use as a method for improving quality and reducing waste was made.

Wiecha [18] presented the case study of implementation of the example computer-aided quality assurance (CAQ) system, which allows for meeting the requirements of IATF 16949:2016 and the VDA 6.1 standards in the field of SPC. This system is aimed at inspecting the quality of products, using tools like goods inward inspection, SPC, management of measuring equipment, suppliers rating, and attribute charts [19].

Shehata [20] applied SPC tools to an actual case study of a cement bags factory to fully analyze the process and provide recommendations and suggestions for reducing the the percentage of defective bags. The results presented that small and medium companies can use SPC tools to solve such kinds of quality issues of their products and services.

Hajej [21] proposed an integrated model of production policy, maintenance strategy, and quality control, considering the impact of the production variation and control chart indicators on the optimum maintenance plan. The advantage of this approach is to consider the best combination between the production, maintenance, and quality policies and the information given by the control chart based on established analytical relationships.

Raheem [22], in his study, evaluated the process of production of Champion Breweries Plc., located at Aka Offot, Uyo, Akwa Ibom State, Nigeria, so he used the standard SPC charts for variable characteristics, $\overline{X}$, R charts and for attributes, P-chart.

Barakat [23] suggested a new method for constructing an efficient point predictor for future order statistics when the sample size is a random variable. The suggested point predictor is based on some characterization properties of the distributions of order statistics.

The native and foreign authors described scientific provisions on sampling control for quality assessment and mechanical processing control. Also, the examples of sampling control applications to optimize the technological processes have been reviewed in their works. At the same time, those papers lack recommendations for working with relatively small samples, typical for most machine-building enterprises (producing pumps, compressors, etc.) within the oil and gas industry. In other words, there are no recommendations that allow determining consistent patterns that are bound to rational usage of statistical data. One of the reasons for this situation is the absence of efficient methods of determining the distribution law of actual values for quality indicators, which are generally random variables, and the lack of approaches to derive an efficient and unbiased assessment of quality indicators. Therefore, this article aims to develop a universal integrated quality control system (considering all aspects of the product's lifecycle stages). This system enables machine-building enterprises to make a high-quality market-competitive product under conditions of reduced series production and increased dimensional accuracy.

3 Research Methodology

Statistical methods of quality assessment and control, which have been developed by native and foreign researchers so far, can be applied predominantly to the vast amount of data regarding quality indicators and are generally mass-production-oriented. It is caused primarily by applying the methods based on grouping the values, using the indicator assessment established on the distribution law (as a rule – normal distribution) of quality indicators as random variables. The analysis of modern standards' scientific and technological level on quality control methods [24] shows that they apply solely to the incredibly stable and sustained technological processes. These standards operate the large size of sampling data under conditions of the normal distribution law of quality indicators that scattering of dimensions after applying crude processing methods conforms to it.

The quality of machine-building products, including the oil and gas industry and instrument-making, is characterized by a range of indicators. The most crucial is dimensional accuracy: size accuracy, form accuracy, and accuracy of surfaces' relative position. Under the influence of random factors, while processing a batch of half-manufactured parts, the dimensions of which in most cases are random variables, the real surface size of each processed part (including due to technological heredity) also become random variables. At the same time, it is known that creating the quality control system for a product in the presence of random components of phenomena and processes requires statistical methods to be used [25].

If the statistical methods are used properly, it is possible to control the product quality in almost all its manufacturing life-cycle stages as a unified (integrated) technological system using efficient operational data on each phase. Efficient operational data comprises the size and quality of an appropriate control sample. When we talk about operational data, we imply data concerning phenomena under control, collected while measuring a random sample. The larger the sample, can be collected more data regarding a particular product quality indicator. Using standard statistical processing methods, the small-scale sample obtained from operational data is insufficient to control the product quality. This article suggests developing a method that would allow a relatively large amount of necessary data to be obtained without expanding the control sample size.

Since there is a lower limit for an indicator' unbiased assessment dispersion of the distribution law of a random variable, we could write:

$$D(x) \geq \frac{1}{I(x)} \tag{1}$$

where $D(x)$– is an indicators' unbiased assessment dispersion of the distribution law of a random variable;

$I(x)$ – is a data quantity regarding an indicator x, , which comprises the conservative data quantity ($I_c(x)$) and operational data ($I_o(x)$).

The inequation (1) shows that the decrease of dispersion $D(x)$ requires an increase in the data quantity. The analysis of fundamental provisions of statistics has shown that the solution for the issue of obtaining an immense amount of data from the smaller size control samples can be found in the theory of order statistics.

This is true if the $\boldsymbol{x}_1, \ldots, \boldsymbol{x}_n$– a statistical sample size ($\boldsymbol{n}$) from the general population with the distribution $\boldsymbol{F}(\boldsymbol{x})$ has an entropy other than zero, then to obtain $\boldsymbol{F}(\boldsymbol{x})$ with zero entropy, the sample order $\boldsymbol{x}_1 \leq \ldots \leq \boldsymbol{x}_n$, which is called ordinal statistical function will be sufficient. So, the suggested variable ordering $\boldsymbol{x}$ gives additional data regarding the process without changing (expanding) the sample size. Therefore, order statistics theory may be the basis for creating a quality control methodology using a small quantity of statistical data and developing a corresponding referential-normative provision [23]. In this regard, Montgomery [26] proposed a detailed presentation of the modern statistical methods for quality control and improvement.

4 Results and Discussion

The quality indicators analysis for an optimal decision regarding operating the enterprise's current quality system under certain conditions is challenged. This is because every machine-building product manufacturing life-cycle stage has its quality indicators. They are generally established by normative documents of different levels and periods of an issue (especially if various units of measurement are involved). A unified, integrated quality assessment system for products and technological processes using a composite indicator should be created to successfully solve the problem described above. It could be developed if an artificial metric based on pointwise and interval values is created for all quality indicators of a product and its processes. To solve this issue, we could apply the asymptotic distribution of extreme values of random variables. The limit distribution for the highest sample value ($-\infty \prec x \prec +\infty$) is as follows:

$$L_h(x) = exp(-exp(-x)) \tag{2}$$

For the symmetric distribution, which describes the dispersion of quality indicators that are generally random variables within machine-building manufacturing, the extreme values are distributed mutually symmetrically, i.e., the distribution of the least sample value is as follows:

$$L_l(x) = 1 - exp(-exp(x)) \tag{3}$$

Since in distribution (2), the maximum value of the measured variable is used, a lower quality estimate will be obtained by formula (3). Thus, applying the distribution of the least value (3) is suggested to obtain the upper-quality estimate. Therefore, the quality indicator interval ($L_i(x)$) can be obtained:

$$L_i(x) = \left[exp(-exp(-x))\right] - \left[1 - exp(-exp(x))\right] \tag{4}$$

As a pointwise assessment, the average value of the interval estimation ($\boldsymbol{L_a(x)}$) can be used:

$$L_a(x) = (L_h(x) + L_l(x))/2 \tag{5}$$

The choice of assessment type depends on the product's level of importance. The composite value of the quality indicator of the entire product (L) consists of a set of indicators. When working with pointwise assessments, it is suggested to find the value by applying the average geometric estimation formula that comprises the quality indicators (L_{aj}):

$$L = \sqrt[m]{\prod\nolimits_{j=1}^{m} L_{aj}} \tag{6}$$

At the same time, when working with interval assessments, it is suggested to find the value by applying the interval geometric estimation formula that comprises the quality indicators (L_{ij}):

$$L = \sqrt[m]{\prod\nolimits_{j=1}^{m} L_{ij}} \tag{7}$$

These expressions (6, 7) show that if at least one of the estimates comprising the quality indicators (L_j) equals zero, the quality of L-th composite indicator will equal zero. However, $L = 1$ only when all $L_j = 1$.

Table 1. Universal integrated quality control system for products and processes.

Assessment type	Level of importance	Quality indicators	
		Individual	Composite
Pointwise	High	$L_h(x) = exp(-exp(-x))$	$\frac{1}{L} = \frac{1}{m}\sum_{j=1}^{m}\frac{1}{L_{hj}}$
	Average	$L_a(x) = \left[exp(-exp(-x))\right] + \left[1 - exp(-exp(x))\right]/2$	$L = \sqrt[m]{\prod_{j=1}^{m} L_{aj}}$
	Low	$L_l(x) = 1 - exp(-exp(x))$	$L = \frac{1}{m}\sum_{j=1}^{m} L_{lj}$
Interval	General	$L_i(x) = \left[exp(-exp(-x))\right] - \left[1 - exp(-exp(x))\right]$	$L = \sqrt[m]{\prod_{j=1}^{m} L_{ij}}$

Therefore, the suggested method allows a unified, universal integrated quality control system for all products or processes (Table 1) to be developed, which applies to all product life-cycle stages.

For instance, the suggested method (system) can be applied to enterprises' mechanical processing in the following way. They take the measurements of individual parts quality indicators (e.g., dimensional accuracy, roughness of surfaces, and runout values). In order of products manufacturing, they draw a pointwise (for appropriate importance levels) and interval (composite) temporal changes diagram for quality indicators, then, based on the given diagram, the decision to reject, accept, or correct (if possible) the product (process) is made.

5 Conclusions

The obtained results enable the following conclusions to be made. The study of native and foreign research works containing scientific provisions of sampling data control for the quality assessment of mechanical processing and the review of the experience of its application with the aim of technological processes optimization has shown nearly a lack of information with the help of which the patterns regarding the rational use of statistical information could be determined.

Through the theory of order statistics using an asymptotic and symmetrical distribution of extreme values of random variables, the method of developing a universal integrated product quality control system was created, which allows the series production to be reduced and the dimensional accuracy to be increased based on the immense amount of information from smaller control samples size to save time and costs on the machine-building enterprises.

Additionally, the creation of such a system enables the synthesis of sufficient necessary normative documentation in statistical methods of quality control; to determine the priority of the document development, and ensures the interrelation of their requirements. Thus, a system of normative documents ensures the efficiency of the product quality management within the machine-building field, under a small quantity of statistical information, for example, in devices and machines manufacturing for the oil and gas industry. This system was built according to the module-adaptive principle [27], in which the quality control is provided by operating a unified, universal integrated assessment system of products and processes on all their life-cycle stages.

The paper proved that the system of relevant standards must be developed according to the module-adaptive principle and based on the system and process approaches. The normative document should be a holistic, orderly, and structured object that regulates the requirements, norms, and rules of quality control and management with minimum material, time, and labor costs.

Acknowledgment. The research was partially carried out within the R&D project "Fulfillment of tasks of the perspective plan of development of a scientific direction "Technical sciences" Sumy State University" (State Reg. No. 0121U112684).

References

1. Emanakov, I.V.: Application of standardization and lean production methods for the development of management systems at machine-building enterprises. In: Proceedings of the International Scientific and Practical Conference Strategy of Development of Regional Ecosystems "Education-Science-Industry", pp. 123–129. Atlantis Press International B.V., Paris (2021)
2. Konyuhov, V.Y., et al.: Machine-building enterprise performance and quality improvement tools. IOP Conf. Series: Materials, Science and Engineering 1064, 012021 (2021). https://doi.org/10.1088/1757-899X/1064/1/012021
3. Bondaruk, A.M.: Automated quality management systems in technological processes. Monograph. Ufa (2017)
4. Khokhlova, G.I.: Problems and factors affecting the development of innovation activities in Russia. Manage. Economic Syst. **3**(33), 10–17 (2018)
5. Ivanov, V., Vashchenko, S., Rong, Y.: Information support of the computer-aided fixture design system. CEUR Workshop Proceedings **1614**, 73–86 (2016)
6. Zaloga, V., Yashyna, T., Dynnyk, O.: Analysis of the theories for assessment of the quality management product efficiency. J. Eng. Sciences **5**(2), B1–B6 (2018). https://doi.org/10.21272/jes.2018.5(2).b1
7. Skrynkovskyy, R.M., Yuzevych, V.M., Kataev, A.V., Pawlowski, G., Protsiuk, T.B.: Analysis of the methodology of constructing a production function using quality criteria. J. Eng. Sciences **6**(1), B1–B5 (2019). https://doi.org/10.21272/jes.2019.6(1).b1
8. Shishkin, I.F.: Metrology, Standardization and Quality Management, 1st edn. Standards, Moscow (2018)
9. Dwiartono, A.I.: Application of statistical quality control for product 04G22 on PT. Maruichi Indonesia. Solid State Technol. **63**(4), 4966–4976 (2020)
10. Li, J.: Application of statistical process control in engineering quality management. IOP Conf. Ser.: Earth Environ. Sci. **831**, 012073 (2021). https://doi.org/10.1088/1755-1315/831/1/012073
11. Ivanov, V.: Process-oriented approach to fixture design. In: Ivanov, V., et al. (eds.) DSMIE 2018. LNME, pp. 42–50. Springer, Cham (2019). https://doi.org/10.1007/978-3-319-93587-4_5
12. Madanhire, I.: Application of statistical process control in manufacturing industry in a developing country. In: 13th Global Conference on Sustainable Manufacturing, pp. 580–583. Elsevier B.V., Amsterdam (2016). https://doi.org/10.1016/j.procir.2016.01.137

13. Zan, T., Liu, Z., Su, Z., Wang, M., Gao, X., Chen, D.: Statistical process control with intelligence based on the deep learning model. Appl. Sci. **10**, 308 (2020). https://doi.org/10.3390/app10010308
14. Ivanov, V., Dehtiarov, I., Denysenko, Y., Malovana, N., Martynova, N.: Experimental diagnostic research of fixture. Diagnostyka **19**(3), 3–9 (2018). https://doi.org/10.29354/diag/92293
15. Hrvačić, I.: Design and implementation of SPC systems in defense industry manufacturing process. Periodicals Eng. Natural Sciences **6**(1), 27–40 (2018). https://doi.org/10.21533/pen.v6i1.163
16. Oberoi, H.S.: SPC (Statistical Process Control): a quality control technique for confirmation to ability of process. Int. Res. J. Eng. Technol. **3**(6), 666–672 (2016)
17. Godina, R.: Quality improvement with statistical process control in the automotive industry. Int. J. Industrial Eng. Manage. **7**(1), 1–8 (2016)
18. Wiecha, L.: Statistical process control and CAQ systems as a tools assuring quality in the automotive industry. Multidisciplinary Aspects of Production Eng. **2**(1), 336–344 (2019). https://doi.org/10.2478/mape-2019-0033
19. Kent, R.: Quality Management in Plastics Processing, 1st edn. Elsevier Ltd, Oxford (2016)
20. Shehata, A.S.: Implementation of statistical process control techniques to reduce the defective ratio: a case study. In: Proceedings of the International Conference on Industrial Engineering and Operations Management, pp. 666–675. IEOM Society International, Washington DC (2018)
21. Hajej, Z., Nyoungue, A.C., Abubakar, A.S., Mohamed Ali, K.: An integrated model of production, maintenance, and quality control with statistical process control chart of a supply chain. Appl. Sci. **11**, 4192 (2021). https://doi.org/10.3390/app11094192
22. Raheem, M.A.: Application of statistical process control in a production process. Science J. Applied Mathematics Statistics **4**(1), 1–11 (2016). https://doi.org/10.11648/j.sjams.20160401.11
23. Barakat, H.: Predicting future order statistics with random sample size. AIMS Mathematics **6**(5), 5133–5147 (2021). https://doi.org/10.3934/math.2021304
24. Bezyazychny, V.F.: The Basics of Machine-Building Technology, 3rd edn. Innovative engineering, Moscow (2020)
25. Yefimenko, N.A.: The method of expert assessment of product quality control at machine-building enterprises. Bulletin of the Cherkasy State Technol. University **4**, 162–170 (2020). https://doi.org/10.24025/2306-4412.4.2020.218208
26. Montgomery, D.C.: Introduction to Statistical Quality Control, 8th edn. Wiley, New York (2019)
27. Karpus, V.E., Ivanov, V.A.: Choice of the optimal configuration of modular reusable fixtures. Russ. Eng. Res. **32**(3), 213–219 (2012). https://doi.org/10.3103/S1068798X12030124

Mechanical Engineering

Vibration Damping of Lifting Mechanisms

Andrii Boiko(✉), Elena Naidenko, and Yalin Wang

Odessa Polytechnic National University, 1, Shevchenko Avenue, Odessa 65044, Ukraine
a.o.boyko@op.edu.ua

Abstract. The article considers the need to implement dynamic modes - acceleration and deceleration by lifting mechanisms, with given restrictions on speed, acceleration, and its derivative. The optimal diagram was presented according to the criterion of the maximum speed for the lifting mechanism. Also, the optimal movement diagram was presented under restrictions on the loss number of the lifting mechanism components. A generalized structural diagram has been synthesized, which makes it possible to analyze nonlinear oscillatory lift systems, as objects of regulation, with arbitrary parameters of the lifting mechanisms and arbitrary laws of their control. Recommendations are proposed for improving the quality of the dynamic modes of lifting mechanisms due to vibration damping. The solution of minimizing the vertical vibrations of the elements of lifting mechanisms arising in dynamic modes by forming the required mechanical characteristics of the engine with specific rigidity parameters is considered. Dependencies of the damping coefficient of the equivalent vibrational link on the rigidity of the mechanical characteristics, the recommended values of the rigidity of the mechanical characteristics on the values of the mass ratio, and the maximum value of the damping coefficient on the mass ratio are given. It is shown that the choice of the required rigidity of the mechanical characteristics of the electric motor of the lifting mechanism increases the damping coefficient and reduces the oscillation in transient modes.

Keywords: Lift mechanism · Movement diagram · Optimal control · Elastic coupling · Vibration damping · Mechanical rigidity · Infrastructure investment

1 Introduction

The main feature of the lifting mechanisms is the need to implement dynamic modes - acceleration and deceleration of the cab, with the specified limits of speed, acceleration, and its derivative. The provision of optimal comfortable conditions for transportation is achieved due to the formation by the lifting mechanism of the required speed change diagrams in the continuation of the movement [1]. Such diagrams of movement are characterized as optimal, and, traditionally, under optimality is understood the condition for the implementation of the minimum time of movement of the lifting mechanism, taking into account the fulfillment of the mandatory restrictions on the magnitude of the speed v (ω), acceleration a (ε), and jerk r (ρ) [2].

The influence of elastic links of rope gears manifested during the operation of elevator hoisting mechanisms guarantees that the actual motion diagram of the cabin differs

V. Tonkonogyi et al. (Eds.): InterPartner 2022, LNME, pp. 403–413, 2023.
https://doi.org/10.1007/978-3-031-16651-8_38

from the required optimal diagram. A difference in motion parameters from the target negatively affects the quality of elevator control, increase dynamic loads on the lifting mechanisms, contribute to the accumulation of fatigue stress in the kinematic chains, lead to premature failures and unplanned equipment downtime, and increases the cost of repairs and operation.

The present study's research aims to synthesize analytical relations and graphical dependences, characterizing the influence of elastic connections on parameters of elevator lifting mechanisms motion and further formation of optimal motion diagrams in terms of speed, considering the elasticity of connecting ropes.

2 Literature Review

The generally accepted optimal diagram in terms of the speed criterion for the lifting mechanism is shown in Fig. 1, a. The known laws of change in linear speed, acceleration, and jerk, when the elevator car moves according to the generally accepted optimal diagram, can be written

$$v = k_1 \frac{r_n}{2} t^2 + k_2 a_n t + k_3 \frac{a_n^2}{2r_n} + k_4 v_n, \tag{1}$$

$$a = k_3 r_n t + k_2 a_n, \tag{2}$$

$$r = k_1 r_n, \tag{3}$$

where r_n, a_n, v_n – are the nominal linear values of the cabin's jerk, acceleration and speed, m/s^3, m/s^2, m/s; $k_1 \ldots k_4$ – values of functions for each section of the motion diagram.

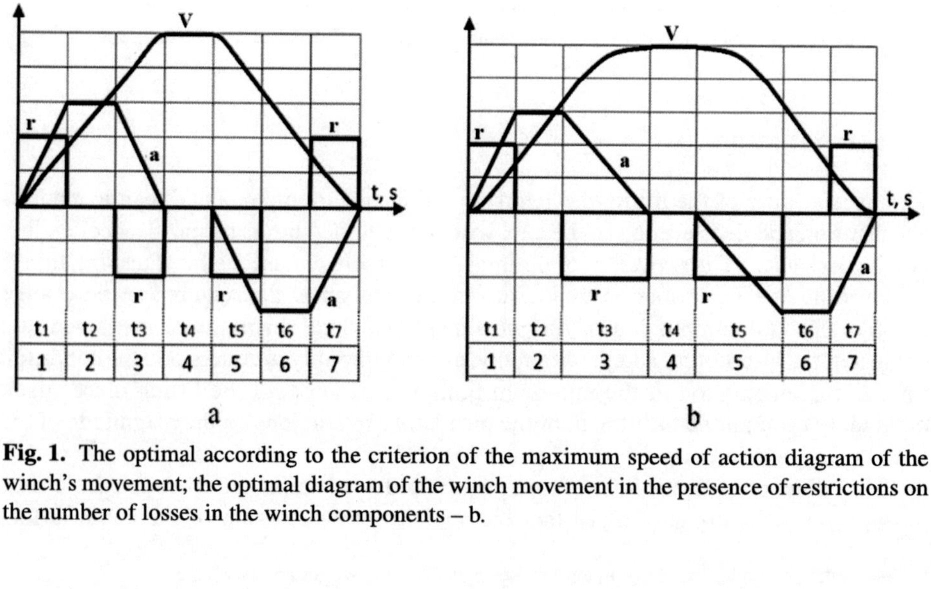

Fig. 1. The optimal according to the criterion of the maximum speed of action diagram of the winch's movement; the optimal diagram of the winch movement in the presence of restrictions on the number of losses in the winch components – b.

It should be noted that the generally accepted typical motion diagrams are also optimal in terms of speed if there are only local restrictions on the magnitude of speed, acceleration, and jerk. If along with these constraints, there are constraints of the integral type (for example, on the magnitude of losses in the winch), the optimal motion diagrams become somewhat different.

A distinctive feature of the laws of optimal control of the movement of lifting mechanisms with restrictions on power losses in the presence of sections. The acceleration changes according to a linear law, but the jerk is less than the maximum. It is determined by the conditions for heating the elements of the lifting mechanism. Typical motion diagrams for this case are shown in Fig. 1, b. A comparison of the diagrams in Fig. 1a and Fig. 1b shows that the presence of restrictions on the magnitude of losses does not significantly affect the nature of the generally accepted optimal speed diagram of the car's motion. Simultaneously, it leads to a more complex control. There are also other recommendations for the formation of optimal motion diagrams. Such diagrams, despite not being optimal according to the generally accepted maximum performance criterion, are of particular interest and deserve attention.

Analysis of publications shows that the problem of developing optimal control of electromechanical systems is of interest to many researchers. In work [3], optimal control is realized using the synthesis method of an optimal regulator. The problem of optimal control by the criterion of the speed of response and simultaneous damping of load oscillations by the mechanism of translational motion with a suspended load is considered in work [4], and the turning mechanism is considered in the article [5]. The vibration damping control [6] in the publication [7] is proposed using the states of the suspended load, in [8] joint swinging, in [9] - using the active change in the length of the suspension string. In publication [10], the suspended load is modeled as a double pendulum. The synthesis of lifting mechanisms and control systems for the drive of an asynchronous motor of lifting mechanisms is presented in [11]. The vibration analysis is considered in [12], and recommendations for reducing vibration in lifting mechanisms are given in publications [13, 14]. In [15], a method for suppressing and controlling horizontal vibrations of a high-speed elevator is proposed. The paper deals with the issues related to bridge-type cranes [16]. The operation was considered in the vertical plane [17, 18]. Vertical loads loaded the mechanism in paper [19], and the horizontal inertial loads were considered in [20], which considered the computational scheme.

3 Research Methodology

3.1 Synthesis of Design Schemes of Lifting Mechanisms

To study the properties and capabilities of the lifting mechanism, its actual kinematic diagram is replaced by the corresponding calculated one. The following features can be used depending on the kinematic diagram of a particular machine and the type of tasks being solved. A design diagram based on the representation of the lifting mechanism in the form of a rigid system with reduced moments of inertia and moments of resistance; design scheme based on the presentation of the lifting mechanism in the form of an elastic two-, three- or n-mass system; calculation scheme based on the representation of the mechanical part in the form of a system with distributed parameters, etc.

When applying the scheme in the form of a rigid system, its motion is described by the equation

$$(m_1 + m_2 + m_3)\frac{dV}{dt} = F - (m_1 - m_3)g - F_{c1} - F_{c3}, \quad (4)$$

where $F = \frac{M}{R}$ – force developed by the engine of the mechanism on the rim of the traction sheave, N; M – the moment developed by the engine, Nm; R – traction sheave radius, m; F_{c1}, F_{c3} – frictional forces in the guides of the cab and counterweight, N; g – acceleration of gravity, m/s^2; m_1, m_2, m_3 – reduced to the motor shaft the masses of the cabin with a load, the engine with a gearbox and a traction sheave, the counterweight, respectively, kg.

When determining the masses m_1, m_2 и m_3 it is obligatory to take into account the distributed mass of the ropes. To implement the optimal laws of speed and acceleration change, the developed force must change according to the law

$$F = k_1(m_1 + m_2 + m_3)r_n t + k_2(m_1 + m_2 + m_3)a_n + (m_1 - m_3)g \quad (5)$$

where r_n, a_n, v_n – nominal linear values of jerk, acceleration, and car speed, m/s^3, m/s^2, m/s; $k_1 - k_4$ – logical switching functions.

Based on Eqs. (4) and (5), a synthesized generalized structural scheme, which allows performing analysis of nonlinear oscillating elevator systems as control objects, with arbitrary parameters of lifting mechanisms and arbitrary laws of their control, is shown in Fig. 2.

Taking into account the inertia of the rope branches acting as transmitting links, the transfer functions $H_1(p)$ and $H_3(p)$ take the form.

$$H_1(p) = \frac{m_1}{m_2}\left(\frac{T_{k12}^2 p^2 + T_{12}p + 1}{T_{01}^2 p^2 + T_{12}p + 1}\right), \quad (6)$$

$$H_3(p) = \frac{m_3}{m_2}\left(\frac{T_{k23}^2 p^2 + T_{23}p + 1}{T_{03}^2 p^2 + T_{23}p + 1}\right), \quad (7)$$

where $T_{k12} = \frac{m_{k12}}{\sigma C_{12}}$, $T_{k23} = \frac{m_{k23}}{\sigma C_{23}}$ – time constants characterizing the inertial properties of the transmitting links, s; m_{k12}, m_{k23} – the weight of rope branches, kg.

$$H_2(p) = \frac{m_1}{C_{12}}\left(\frac{p^2}{T_{01}^2 p^2 + T_{12}p + 1}\right), \quad (8)$$

The use of a synthesized structural diagram allows:

- to determine the influence of the elastic connections of the system on the movement of the lifting mechanisms of passenger elevators;
- taking into account the elastic properties of the mechanical transmission, calculate the actual diagrams of the movement of passenger elevators, determining the maximum values of speed, acceleration, and jerk, with arbitrary control laws;

- to identify the influence of the system parameters (including control and disturbing influences) on its oscillation and to solve the problem of minimizing the oscillatory processes of the elements of lift hoisting mechanisms.

After considering the parameters of the actual lifting mechanism, the effect of the fourth mass can be neglected in the analytical analysis.

The aforementioned mechanical systems are lumped-parameter systems. An accurate analysis of distributed-parameter systems is complicated because partial differential equations describe them, so in most cases, a distributed-parameter system is reduced to an equivalent lumped system. In this case, the distributed mass of the ropes is taken into account when determining the masses m_1, m_2 и m_3. This approach gives good results even in the study of mine hoisting machines in which the rope length is considerable.

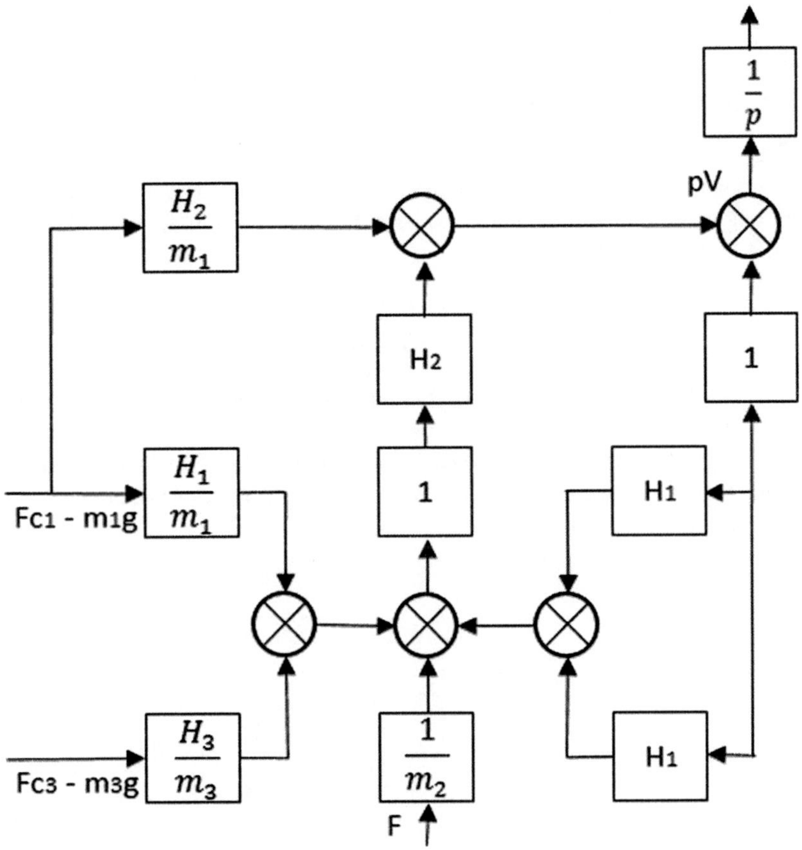

Fig. 2. Design structural diagram of an elastic three-mass lift system.

3.2 Influence of Elastic Connections on Optimal Motion Diagrams

When analyzing a mechanical system with an elastic connection, it is necessary to take into account dissipative forces, i.e., forces under the action of which its total mechanical energy decreases (that is, dissipates) on the system, passing into other, non-mechanical forms of energy, for example, heat. In principle, the action of dissipative forces in a mechanical system can be considered in various ways. However, it should be noted that none of the proposed methods for considering the effect of dissipation fully consider the complex processes of damping rope vibrations. At the same time, the method of accounting based on the use of the vibration damping coefficient, or the time constant, equivalent to this coefficient, is most widely used in the study of lifting mechanical systems with elastic connections (T_{12}).

4 Results and Discussion

4.1 Formation of the Recommended Rigidity of the Mechanical Characteristics of the Electric Motor of the Mechanism

The problem of minimizing the vertical vibrations of the elements of lifting mechanisms arising in dynamic modes can be solved by forming the required mechanical characteristics of the engine with specific stiffness parameters. Thus, the transfer function describing the internal feedback between the speed of the first mass of a two-mass system ω_1 and the static load moment *M*.

$$H_{\omega}(p) = \beta, \tag{9}$$

where β – rigidity of the mechanical characteristics of the winch motor, Nms ($\beta =$ const, if the mechanical characteristic at the working area is considered linear).

In Fig. 3, for example, two typical mechanical characteristics are shown - linear with $\beta =$ const (1) and “excavator” (2), relevant for use in lifting systems, consisting of two sections with different stiffness parameters.

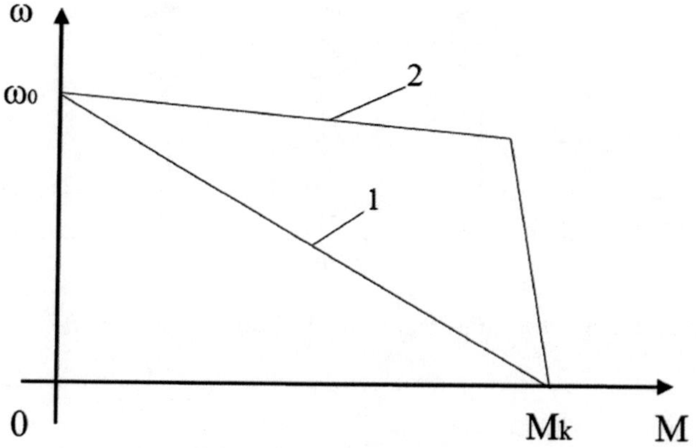

Fig. 3. Typical mechanical characteristics.

If dissipation is not considered ($b = \text{const}$), but the rigidity of the mechanical characteristic is finite ($0 < \beta < \infty$), the equivalent damping coefficient $\xi > 0$ indicates the implementation of vibration damping of the mechanical system. It is also evident that vibration damping is absent in cases of absolutely soft or rigid mechanical characteristics ($\beta = \infty$ and $\beta = 0$). Consequently, there is a certain rigidity of the mechanical characteristic β, at which the vibration damping is most significant, so when the damping coefficient of the corresponding oscillatory element ξ takes the maximum value ξ_{max}. From the characteristic equation of a two-mass equivalent mechanical system, the equalities can be obtained.

$$\left.\begin{array}{c} \frac{J_1 J_2}{\beta C_{12}} = \frac{k}{\Omega_0^3} \\ \frac{J_1 J_2}{\beta C_{12}} = \frac{k}{\Omega_0^3} \\ \frac{J_2}{C_{12}} = \frac{1}{\Omega_0^2}(2k\xi + 1) \\ \frac{J_1+J_2}{\beta} = \frac{1}{\Omega_o}(k + 2\xi) \end{array}\right\}, \tag{10}$$

where Ω_o- vibrating link frequency, 1/s; k - the ratio of the time constants of the inertial and oscillatory components, p.u.

A general solution of Eqs. (10) makes it possible to relate the actual parameters of the lifting mechanism: J_1, J_2, β, C_{12} with fictitious values Ωo, ξ, and k, which characterize the quality indicators of its work. The expression can find the damping factor

$$\xi = 0,5\left(\sqrt{\frac{(k^2+1)^2}{4k^2} + \frac{J_2}{J_1}} - \frac{k^2+1}{2k}\right) \tag{11}$$

To find the value of k, at which the damping coefficient ξ takes the boundary value, it is necessary to solve the following equation:

$$\frac{\partial \xi}{\partial k} = 0. \tag{12}$$

Provided that $k \neq 0$, its transformation leads to the form.

$$\frac{k^4 - 1}{2k\sqrt{\frac{(k^2+1)^2}{4k^2} + \frac{J_2}{J_1}}} = k^2 - 1. \tag{13}$$

The solution to the equation is obvious - $k = 1$. The maximum value of the damping factor can be found

$$\xi_{max} = 0,5\left(\sqrt{\gamma} - 1\right), \tag{14}$$

where the mass ratio is $\gamma_{12} = \frac{J_1+J_2}{J_1}$, p.u.

From (14), it follows that if $\gamma \geq 9$, the maximum damping coefficient $\xi_{max} \geq 1$, and the mechanical system ceases to be oscillatory. It is possible to ensure a decrease in vertical vibrations (ξ_{max}) by realizing the required rigidity of the mechanical characteristics of the winch motor using a controlled converter. In this case, the value of the required stiffness can be calculated, taking into account the substitution ξ_{max} from (14) and $k = 1$.

$$\beta = J_1 \cdot \Omega_2 \cdot \gamma^{\frac{3}{4}} \tag{15}$$

where $\Omega_2 = \sqrt{\frac{C_{12}}{J_2}}$ - vibration frequency of a single-mass mechanical part.

4.2 Influence of the Mechanical Characteristics Rigidity on the Vibrations Damping in Dynamic Modes

Figure 4, 5, 6 show the dependences obtained at various values of γ using a physical model, which confirms the above mathematical relations.

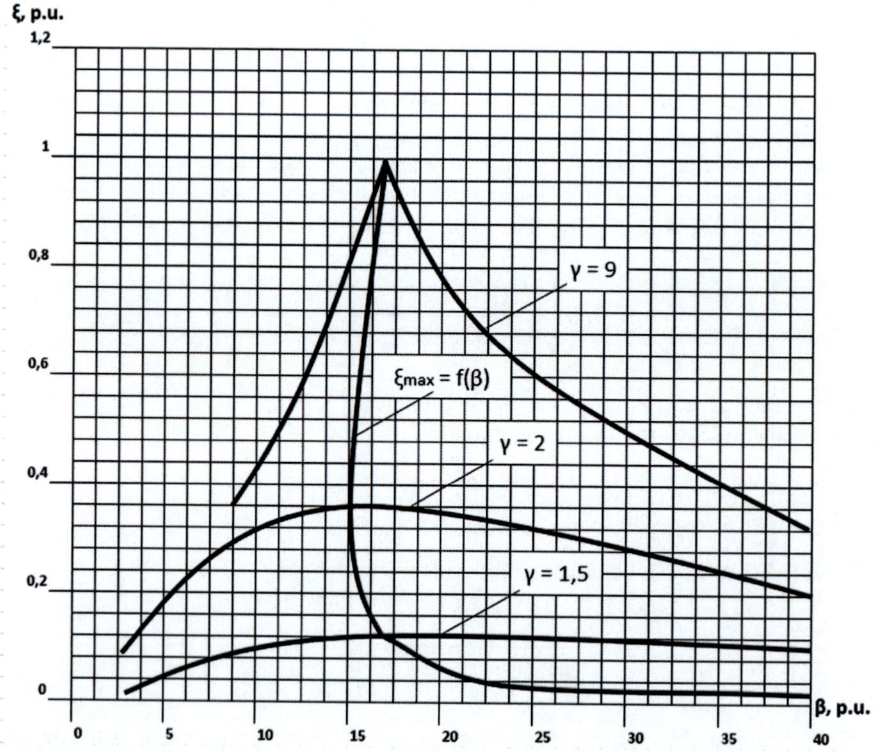

Fig. 4. Dependences of the equivalent vibrational link's damping coefficient from the mechanical characteristics' rigidity.

It should be noted that a complete absence of oscillations characterizes the boundary case corresponding to $\gamma = 1$ and $\xi = 0$, but it has no physical meaning since it indicates the absence in the equivalent circuit of a two-mass system of the second mass ($J_2 = 0$). The study's results (Fig. 4) confirm that the choice of the rigidity of the mechanical characteristics of the winch motor increases the damping coefficient of the mechanical lift

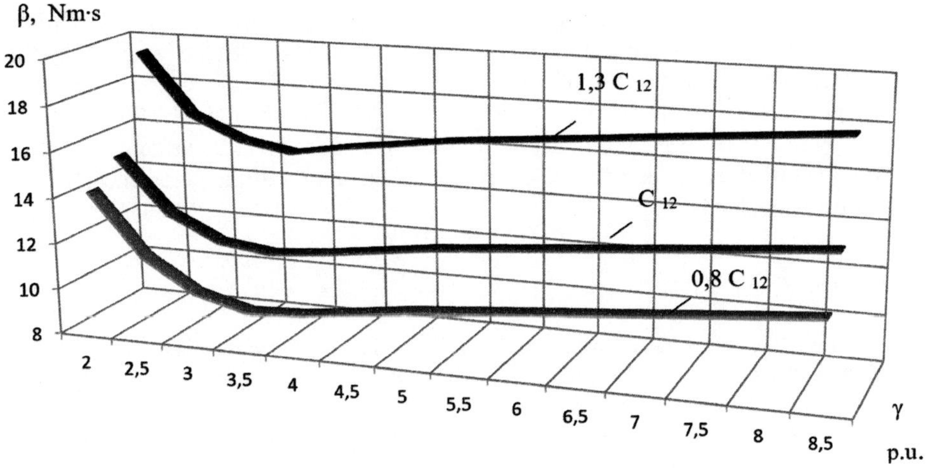

Fig. 5. Dependencies of the recommended values of the rigidity of mechanical characteristics on the value of the mass ratio.

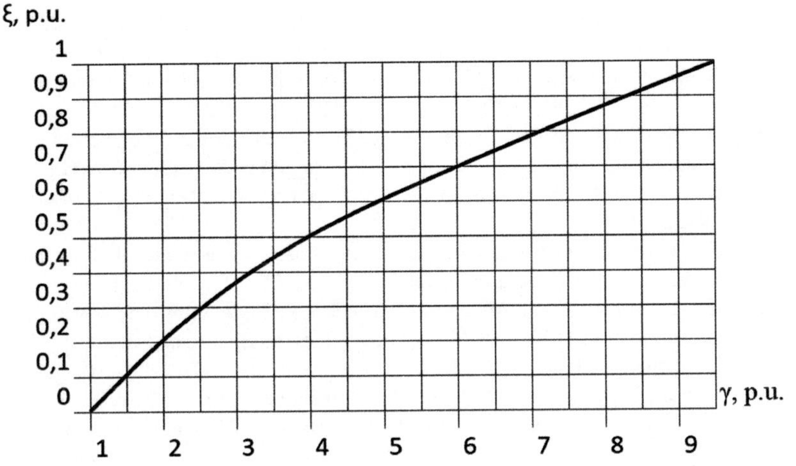

Fig. 6. Dependencies of the maximum damping coefficient value on the mass ratio.

system and reduces its oscillation in transient modes. The improvement in the quality of the transient process is directly proportional to the value of the mass ratio of the equivalent two-mass system - γ, and for the studied passenger elevators, it is 15–43%.

5 Conclusions

For lifting mechanisms, at the stage of their calculation, it is required to select rational parameters for dynamic vibration damping and minimization of the vibration level of the lift equipment.

Recommendations are offered for improving the quality of dynamic modes of lifting mechanisms due to vibration damping. It is shown that the choice of the required rigidity of the mechanical characteristics of the electric motor of the lifting mechanism increases the damping coefficient and reduces the oscillation in transient modes by 35–72%.

Further research aims to introduce corrective negative feedback in the control system, studying its effect on the damping coefficient and oscillation of the mechanical system of the lifting mechanism.

References

1. Zudilova, T.V., Ivanov, S.E., Ivanova, L.N.: The automation of electromechanical lift for disabled people with control from a mobile device. In: 2017 Computing Conference, p. 668–674. IEEE (2017). DOI: https://doi.org/10.1109/SAI.2017.8252167
2. Shrivastava, N., Pande, A., Lele, J., Kampassi, K.: Embedded Control System for Self Adjusting Scissor Lift. In: 2018 Fourth International Conference on Computing Communication Control and Automation (ICCUBEA), p. 1–5. IEEE (2018). DOI: https://doi.org/10.1109/ICCUBEA.2018.8697800
3. Pyatibratov, G., Danshina, A., Altunyan, L.: Optimal Force Compensating Control of Robotic Lifting Mechanisms. In: 2019 International Russian Automation Conference (RusAutoCon), p. 1–5. IEEE (2019). DOI: https://doi.org/10.1109/RUSAUTOCON.2019.8867811
4. Bai, W.W., Ren, H.: Horizontal positioning and anti-swinging control tower crane using adaptive sliding mode control. In: 2018 Chinese Control And Decision Conference (CCDC), p. 4013–4018. IEEE (2018). DOI: https://doi.org/10.1109/CCDC.2018.8407820
5. Naidenko, E., Bondar, O., Boiko, A., Fomin, O., Turmanidze, R.: Control Optimization of the Swing Mechanism. In: Tonkonogyi, V., Ivanov, V., Trojanowska, J., Oborskyi, G., Pavlenko, I. (eds.) InterPartner 2021. LNME, pp. 13–21. Springer, Cham (2022). https://doi.org/10.1007/978-3-030-91327-4_2
6. Yoshikawa, M., Iwatani, A., Ishikawa, J.: Damping control of suspended load for truck cranes in consideration of control input dimension. In: Duy, V., Dao, T., Zelinka, I., Kim, S., Phuong, T. (eds) AETA 2017 - Recent Advances in Electrical Engineering and Related Sciences: Theory and Application. AETA 2017. Lecture Notes in Electrical Engineering, vol. 465, p. 436–446. Springer, Cham (2017). DOI: https://doi.org/10.1007/978-3-319-69814-4_42
7. Kodani, N., Oushi, S., Takahashi, R., Hirata, H.: Transport control of a jib crane with a rotating cargo. Transactions of the Institute of Electrical Engineers of Japan. Series C 136, 821–831 (2016)
8. Donner, P., Buss, M.: Cooperative Swinging of Complex Pendulum-Like Objects: Experimental Evaluation. IEEE Trans. Rob. **32**(3), 744–753 (2016)
9. Anderle, M., Michiels, W., Čelikovský, S., Vyhlídal, T.: Damping a pendulum's swing by string length adjustment – design and comparison of various control methods. In: 2019 American Control Conference (ACC), p. 4399–4405. IEEE (2019). DOI: https://doi.org/10.23919/ACC.2019.8814293
10. Watanabe, K., Yoshikawa, M., Ishikawa, J.: Damping Control of Suspended Load for Truck Cranes in Consideration of Second Bending Mode Oscillation. In: IECON 2018 - 44th Annual Conference of the IEEE Industrial Electronics Society, p. 4561–4568. IEEE (2018). DOI: https://doi.org/10.1109/IECON.2018.8591232
11. Chernenko, M.Y., Kucher E.S., Kamysheva, E.Y.: High-Speed Passenger Lift Model Development. In: 2018 International Multi-Conference on Industrial Engineering and Modern Technologies (FarEastCon), p. 1–4. IEEE (2018). DOI: https://doi.org/10.1109/FarEastCon.2018.8602562

12. Nguyen, T.X., Miura, N., Sone, A.: Analysis and control of compensation rope response in elevator system with timelyly length. In: 2017 11th Asian Control Conference (ASCC), p. 905–910 (2017). DOI: https://doi.org/10.1109/ASCC.2017.8287291
13. Shuangchang, F., Jie, C., Xiaoqing, C.: Analysis of the hidden danger for old elevator safety. In: 2020 3rd International Conference on Electron Device and Mechanical Engineering (ICEDME), p. 605–608. IEEE (2020). DOI: https://doi.org/10.1109/ICEDME50972.2020.00143
14. Zhang, H., Zhang, R., On, K., Liu, L.: Variable Universe Fuzzy Control of High-Speed Elevator Horizontal Vibration Based on Firefly Algorithm and Backpropagation Fuzzy Neural Network. IEEE Access **9**, 57020–57032 (2021). https://doi.org/10.1109/ACCESS.2021.3072648
15. Wang, J., Tang, S., X., Krstic, M.: Lateral Vibration Suppression of a Disturbed Mining Cable Elevator with Flexible Guideways. In: 2020 59th IEEE Conference on Decision and Control (CDC), p. 4436–4441. IEEE (2020). DOI: https://doi.org/10.1109/CDC42340.2020.9303756
16. Bonopera, M., Chang, K., Lee, Zheng-Kuan.: State-of-the-Art Review on Determining Prestress Losses in Prestressed Concrete Girders. Appl. Sci. (10), 72–57 (2020)
17. Jianqun, W., Shenghua, T., Zheng, H., Zhou, C., Zhu, M.: Flexural Behavior of a 30-Meter Full-Scale Simply Supported Prestressed Concrete Box Girder. Appl. Sci. **10**(9), 30–76 (2020)
18. Zou, J., Huang, Y., Feng, W., Chen, Y., Huang, Y.: Experimental study on flexural behavior of concrete T-beams strengthened with externally prestressed tendons. Math. Biosci. Eng. **16**(6), 6962–6974 (2019). https://doi.org/10.3934/mbe.2019349
19. Zhegulsky, V., Mironov, I., Lukashuk, O.: Design and calculation of crane metal structures. Ural Publishing House University, Russia (2019)
20. Tkachov, A., Tkachov, O., Sydorenko, I.: Improvement of the deformed state of flight beams of bridge cranes. Faculty of Architecture, Civil Engineering and Applied Arts **2**(4), 118–125 (2020)

Failure Probability of Ship Diesel Parts Under Operating Conditions

Gennady Ivanov and Pavlo Polyansky(✉)

Mykolayiv National Agrarian University, 9, G. Gongadze Street, Mykolayiv 54020, Ukraine
IvanovGO@MNAU.Edu.UA, PolyanskyPM@mnau.edu.ua

Abstract. The dependences are given to determine the probability of accidental failure for any transition, accidental failure to achieve maximum wear, and the density distribution of the probability of wear rate (cylinder bushings, main and connecting rod bearings). In the case of simultaneous action on the element (e.g., the cylinder bushing), the most common and severe factors that cause wear during operation (including during start-ups) and accidental failures. The other set of conditions or its partial case corresponds to other elements of the considered system. Calculation formulas are given for determining the average number of failure-free transitions during the standard service life, which can be replaced at any time from the start of operation of the parts to the time under investigation. The regularities are given, which allow determining the probability of emergency failure at any time and the probability that the whole period will not be an emergency failure and that the emergency failure will occur in the first transition.

Keywords: Diesel · Distribution · Random failures · Wear · Emergency failure · Two-dimensional law · Two-dimensional density · Energy efficiency

1 Introduction

Shipping is the essence of the international economy. Today, about ninety percent of world trade is carried out by sea through 50,000 merchant ships. These vessels carry various types of cargo and are staffed by more than a million sailors worldwide. Most of these vessels are started by main diesel engines, chereinafter referred to as main engines, due to their reliability and fuel efficiency.

The main engines are widely used in maritime transport as a direct connection to the propeller and as main or auxiliary electrical generator sets. The engine is the essential piece of equipment on the ship's platform, so the engine's reliability is a priority to optimize safety, life cycle costs, and the ship's energy.

However, numerous accidents occur due to the main engine's failure at sea. The main reason for this is an inadequate maintenance plan. To make an optimal maintenance plan, it is necessary to evaluate the reliability of the main engine. The problem of quality and reliability for maritime transport is significant, as it is related to maritime safety. To solve it, it is necessary to study the operating conditions of the main power plants and determine and predict the reliability of their work.

The primary purpose of the work is to study the reliability of the parts of the cylinder-piston group of the main engines of dry cargo ships.

V. Tonkonogyi et al. (Eds.): InterPartner 2022, LNME, pp. 414–423, 2023.
https://doi.org/10.1007/978-3-031-16651-8_39

2 Literature Review

A diesel turbo-piston engine is considered as a series chain of N elements (parts or assembly units), the reliability of which depends on the overall reliability of the engine [1]. Based on the operation experience, it is established that the reliability of system elements in the general case depends on accidental failures, significant wear out during operation, and other wearing during start-up [2].

This article [3] presents the two-dimensional distribution law of the system $(t_k, t_m)0 \leq k \leq m \leq n$ and its two-dimensional density.

The operation of the main engine parts when working on maneuvers is given in reference [4]. For refrigerated vessels of the «Priboi» series, during one docking (7 days), the average number of starts and reverses is 20. For quantitative assessment, it is assumed that one start from the cold state corresponds to the amount of wear for 5 h of running time. The amount of wear at one start from the hot state or reverse equals 1 h of running time. The total number of starts from the cold state belongs to 25%. The reverse is considered all hot, as the cold state is less than 4%. Then the average wear during one docking corresponds to the wear for 34.5 h of normal operation. Additional time proportional to maneuvers wear (during docking) will be 1070 h per year at 31 dockings during one year. The additional time is 46.5%, with an average annual running time of 2300 h.

SSTU ISO 9001: 2015 [5] introduces a quality management system which is a strategic decision of the organization that can help improve its overall effectiveness and provide a solid foundation of initiatives for sustainable development. There are potential benefits to the organization from implementing a quality management system based on this standard: a) the ability to continuously supply products and services that meet customer requirements, as well as legal and regulatory requirements, are applied; b) creating opportunities to increase customer satisfaction; c) taking into account the risks and opportunities associated with the environment and the goals of the organization; d) ability to demonstrate compliance with the established requirements for the quality management system.

Identification of car failures as a way of increasing their reliability is considered in the article [6] as random. In the operational enterprise's conditions, the classification of failures according to a source of occurrence is essential to increase the car's operational reliability. Today, most modern enterprises implement a quality management system to improve efficiency. From the point of view of the quality management system, given its constant improvement, it is necessary to identify problematic areas of operational reliability and make reasonable efforts to improve them.

Issues of information systems for monitoring the technical condition of cars are discussed in the paper [7]. In work [8], a general approach to the formation of models for assessing the technical condition of a car in operation, and in work [9], scientists present a research agent approach for monitoring the technical condition of vehicles.

An expedient method of creating models of materials and estimating the service life of cast iron cylinder heads under conditions of variable thermomechanical loading using computer engineering is presented in the article [10]. In addition, the effect of heat load and mechanical constraints on their service life is shown.

A study of the reliability of a high-speed radial marine diesel engine (M504B2) based on experimental data on engine malfunctions and the time required for repairing is presented in the article [11]. Operational reliability and failure rate were calculated based on operational data collected from the engine log. The mathematically calculated model of reliability of this engine showed a continuously increasing function of failure rates, and the Weibull distribution could reliably approximate the engine's reliability. A systematic approach to the analysis of the reliability of low-speed diesel engines with one shaft is presented in the article [12]. Available methods of analysis and the value of qualitative analysis are discussed, and a step-by-step approach to systematic qualitative analysis with examples is introduced.

The article [13] suggests a new approach to intelligent malfunction diagnosis for two-stroke diesel marine engines. Automated diagnostic systems are essential because they can run continuously in real-time, ensuring effective monitoring of engine operation. A fully automatic machine learning system for engine failure detection is presented. The experiments on an actual data set show that the suggested approach provides higher classification accuracy and low response time compared to a number of the most modern methods and is thus a suitable choice for real life.

The development of a 4-stroke high-speed marine diesel engine, which is used on military and civilian ships as the main engine, is described in the articles [14, 15]. The failure simulator is based on a one-dimensional thermodynamic model developed, corrected, and confirmed by experimental data from a real engine on a test bench. The novelty of this work is the applied methodology, which combines asset expertise knowledge, methodology, and failure modeling to obtain an accurate and reliable database for failure prediction, which is a key element of diesel engine failure.

The study [16] analyzes relevant data from various sources to determine the most appropriate failure model representing a particular component. The data collected and the model developed will be very useful for assessing the reliability of marine engines and planning maintenance activities on board of a ship. This can reduce the failure of marine engines, which will ultimately help reduce accidents in the shipping industry.

3 Research Methodology

A marine main diesel engine is considered as a series chain of N elements (parts or assembly units), the reliability of which depends on the overall reliability of the engine. The operating time, wear, and failure of each element comply with the distribution laws, the parameters of which are determined by the collected empirical information. To maintain reliability at a sufficiently high level at a certain time $t_1, t_2, \ldots, t_i$ preventive inspections and repairs are carried out. They are held in a dock and do not affect the running time.

Based on the operation experience, it is established that the reliability of system elements in the general case depends on accidental failures, major wear out during operation, and additional wearing during start-up.

Accidental failures of diesel engine components are a significant hazard during operation, as some parts (such as cylinder bushings and pistons) are usually replaced during repair. On the other hand, the preventive service does not eliminate accidental failures.

That is why the general problem of assessing the reliability of the diesel mathematical problem of assessing the reliability and durability, considering only random failures of its elements, is of the most significant practical importance.

The law of distribution of random failures of elements is expressed by the function of reliability-probability of trouble-free operation *P(t)* for the time from 0 to the moment *t*:

$$P(t) = \exp(-\alpha t), \tag{1}$$

where t is the time in thousand hours; α is the coefficient of the exponential law of distribution of docking time.

The speed of wear of the cylinder bushings during operation will be constant and independent of accidents, which is natural for the period of normal operation. The amount of wear ξ during one transition is a random variable ($\xi = \nu \cdot \tau$) with an exponential distribution law:

$$f(x) = \lambda \exp(-\lambda x). \tag{2}$$

Here $\lambda = \frac{\beta}{\nu}$, $x \geq 0$; β is the coefficient of the exponential law of distribution of transition time.

For the following decisions, it is necessary to have the value α, which refers to the parts we are interested in under different conditions. Note that the exponent of failure should be determined for the relative share of failing parts. If we have information on z_c ships, and the ship's engine has z_u parts of this type, their total number equals the multiplication of $z_c z_u$. The number of failures in any period of the statistical series is determined by the ratio $m_i/(z_c z_u)$. The final expression will be written as:

$$\exp(-\alpha t) = 1 - \frac{m_i}{z_c z_u}. \tag{3}$$

Hence at given values of t, we can find the value of α for the following cases:

1. Failure to replace α_3. It is possible to determine after what time all the originally delivered parts will fail, i.e., the term of their operation.
2. Failure to replace and restore α_e. Broken parts are replaced with new ones, so there will be mixed parts. This will allow you to calculate the number of failures of parts of this type for a given time (mixed outputs).

The values of α for the main engines of some types of vessels are given in Table 1.

Since failures are distributed according to the exponential law (1), the probability of failure for any period of normal operation can be written by the formula:

$$Q(t_1, t_2) = \exp(-\alpha t_1)\,[1 - \exp(-\alpha \tau)], \tag{4}$$

where $\tau = t_2 - t_1$– the duration of the *transition*.

Using formula (4), after performing a number of transformations, we obtain the probability of an emergency failure for the *i*-th transition:

$$Q_i = \frac{\alpha}{\beta}\left(\frac{\beta}{\alpha + \beta}\right)^i \; (i = 1,\ 2,\ 3,\ \ldots). \tag{5}$$

Table 1. Values of α for the main engines of some types of vessels.

Ships series	The value $of\,\alpha \cdot 10^{-2}$, 1 / thousand hours		
	Cylinder bushings	Pistons	Cylinder covers
«Simferopol»	2.08 / 3.38	2.08 / 2.40	-
«Murom»	2.80 / 3.38	2.70 / 3.04	3.13 / 3.46
«Lysychansk»	1.75 / 2.38	-	-
TR «Russian Island»	0.47 / 0.74	0.75 / 1.21	0.87 / 0.87

Note. Numeral – value α_3., denominator $-\alpha_e$.

Formula (5) gives the probability of accidental failure in any transition, i.e., the law of distribution of the moment of accidental failure (take the moments of transition numbers). Knowing the distribution law represented by formula (5), we determine the mathematical expectation and find the average number of faultless transitions:

$$M = \frac{\beta}{\alpha}. \tag{6}$$

Expression (6) is valid provided that no circumstances other than accidental failures disable the engine and the area of its normal operation is unlimited. It gives the average time of trouble-free operation in some idealized conditions. Let us call it "a theoretical average of the number of faultless transitions".

4 Results and Discussion

When the element reaches the limit wear "*b*", it is replaced with a new one. Therefore, accidental failure of an element can occur only in those periods that help to achieve the limit wear or in the transition when the limit wear is reached. The probability of occurring an accidental failure during the normal operation of the element (before its replacement due to extreme wear) is determined by the formula:

$$Q(b) = 1 - (\frac{\beta}{\alpha + \beta)}) \exp\left(-\frac{\alpha\lambda\beta}{\alpha + \beta}\right), \tag{7}$$

where λ is the value inverse of the average wear M_ξ per transition.

The probability of trouble-free operation of the element before its replacement at a given level "*b*" of wear limit is:

$$P(b) = \left(\frac{\beta}{\alpha + \beta}\right) \exp\left(-\frac{\alpha\lambda\beta}{\alpha + \beta}\right). \tag{8}$$

The average time of failure-free service, expressed in the number of transitions, is determined by mathematical expectations: $M\,\eta(b) = M\left[1 - \exp\left(\frac{-\alpha\cdot\gamma\cdot b}{\alpha+\beta}\right)\right]$. $M\,\eta(b) = M\left[1 - \exp\left(\frac{-\alpha\cdot\gamma\cdot b}{\alpha+\beta}\right)\right]$.

The density distribution of the probability of wear rate (cylinder bushings, main and connecting rod bearings) is described by the gamma distribution, which has the form:

$$f(c) = \frac{\gamma(\gamma c)^{k-1}}{\Gamma(k)} \exp(-\gamma c), \text{ at } 0 \le c \le \pm \propto . \quad (9)$$

Here c is the wear rate of the element, mm / (thousand hours);γ is the gamma distribution scale parameter;k is a parameter that characterizes the asymmetry and excess; $\Gamma(k)$ is the full gamma function. If k is an integer, then $\Gamma(k) = (k-1)!$.

Mathematical expectation: $M = \int_0^{\propto} cf(c)dc = \frac{\gamma}{\Gamma(k)} \int_0^{\propto} c(\gamma c)^{k-1} \cdot \exp(-\gamma c)dc.$

Marked $u = \gamma c$, we get:

$$M = \frac{1}{\gamma\Gamma(k)} \int_0^{\propto} u^{-k} e^{-u} du = \frac{k}{\lambda}. \quad (10)$$

Dispersion

$$D = \frac{\gamma}{\Gamma(k)} \int_0^{\propto} c^2(\gamma c)^{k-1} e^{-\gamma c} dc - M^2 = \frac{\Gamma(k+2)}{\gamma^2\Gamma(k)} - \left(\frac{k}{\gamma}\right)^2 = \frac{k}{\gamma^2}. \quad (11)$$

Solving together expressions (10) and (11), we obtain:

$$\gamma = \frac{M}{D}; \; k = \frac{M^2}{D}. \quad (12)$$

The values of M and D are found from the expressions:

$$M = \sum_1^n a_i \Delta c_i; \; D = \sum_1^n a_i \left(\Delta c_i - M^2\right). \quad (13)$$

Designation of values in formulas (10)-(13): M – mathematical expectation, mm/(thousand hours); D – dispersion, [mm/(thousand hours)]^2, γ – gamma distribution parameter, thousand hours/mm.

The moments considered above $t_i = \tau_1 + \tau_2 + \ldots + \tau_i$ form the simplest (Poisson) flow of random events with the parameter β_n. This is the flow of moments of arrival of the vessel at the dock (this, of course, does not take into account the docking time). Therefore, the number of N_m completed transitions over time T_H (normative service life), as a random variable, is subject to Poisson's law:

$$P_k(T_H) = P(N_m, k) = \frac{\exp -(\beta_n \cdot T_H)(\beta_n \cdot T_H)^k}{k!}, \; (k = 0, \; 1.2\ldots). \quad (14)$$

Let us denote η_m – the number of accident-free transitions for the normative period T_H. Obviously, $0 \le \eta_m \le N_m$. To establish the law of distribution of probabilities of the quantity, we η_m establish: $P_k(T_H) = P\{\eta_m = k\}, \; (k = 0, \; 1, \; 2\ldots)$.

First, we find the conditional distribution law:

$$P_k^{(n)}(T_H) = P\left(\frac{\eta_m = k}{N_m = n}\right), \ (0 \le k \le n). \tag{15}$$

The unconditional law of distribution of moments $t_i = \tau_1 + \tau_2 + \ldots + \tau_i$ was considered above – a gamma distribution, all components are τ_i independent and distributed by exponentially low [see formula (4)]. However, if it is known that during T_H time n events have occurred, then the moments $t_1 \le t_2 \le \ldots \le t_k \le \ldots \le t_n \le T_n$ receive a different distribution (as well as τ_1): $t_1 \le t_2 \le t_k \le \ldots \le t_n$. They form a variation series of considered independent values of a random variable evenly distributed over [0, T]. The distribution laws for t_k and the system's common (two-dimensional) distribution law (t_k, t_m) $0 \le k \le m \le n$ are known [3].

To obtain a two-dimensional law of distribution of the system (t_k, t_{k+1}), where $N_m = n$, we make the following transformations. According to [3], two-dimensional density is represented by the formula:

$$f(x,y) = \{\frac{n!}{(k-1)!(n-k-1)!}\Phi^{n-1}(x)[1-\Phi(y)]^{n-k-1} f(x)f(y) \ \text{для} \ 0 \le x \le y \le 1;$$

$$\{0 \ \text{для} \ x \succ y \quad x, y \succ T_н \ (n = 1, 2, \ldots, n-1, n).$$

$$\underset{t_0 t_1}{f}(x,y) = f(x) = n[1-\Phi(x)]^{n-1} \cdot f(x).$$

$$\text{Here} \quad \Phi(x) = \{\frac{x}{T} \ \text{для} \ 0 \le x \le T_н; \quad \{0 \ \text{для} \ x \prec 0, x \succ T_н; \qquad f(x) = \Phi'(x) = \frac{1}{T_н}, 0 \le x \le T_н. \tag{16}$$

(16)

If $0 \le k \le n-1$, it is the $P_k^{(n)}(T_H)$– probability that in the specified conditions. $(N_m = n)$ in the break (t_k, t_{k+1}), there will be an emergency failure.

The probability of emergency failure in any given interval (x, y) is calculated by formula (3) [3]:

$$Q(x, y) = \exp(-\alpha x) - \exp(-\alpha y). \tag{17}$$

Knowing the law of distribution of the system (t_k, t_{k+1}) in the form of (16) and the probability of emergency failure (17), we obtain the desired conditional probability of emergency failure for the $(k + 1)$-th transition. Let us denote this conditional probability by $Q_{k+1}^{(n)}(T_H)$. Applying the integral formula of the full probability of failures: $\{x \le t_k \prec x + dx; \{y \prec t_{k+1} \prec y + dy, \text{для} \ k = 1, 2, 3, \ldots$

$$Q_{k+1}^{(n)}(T_H) = \int_0^{\propto}\int f(x, y)[\exp(-dx)\exp(-dy)]dxdy = \frac{n!}{(k-1)!(n-k-1)!};$$

$$\int_0^{T_H} \exp(-dx)\left(\frac{x}{T}\right)^{k-1}\left[1 - \frac{x}{T}\right]\frac{dx}{T} - \frac{n!}{k!(n-k)!}; \tag{18}$$

$$\int_{0}^{T_H} \exp(-dy)\left(\frac{y}{T}\right)^k - \frac{y}{(1-T)^{n-k-1}}\frac{dy}{T}.$$

For $k = 0$, similarly, we obtain:

$$Q_1^{(n)}(T_H) = 1 - n\int_{0}^{T_H} \exp(-dx)\left(1 - \frac{x}{T}\right)^{n-1}\frac{dx}{T}. \tag{19}$$

For $k = n$, using formulas (18) and (19), we obtain for expression (15):

$$P_k^{(n)}(T_H) = P\left(\frac{\eta_m = n}{N_m = n}\right) = 1 - \sum_{i=0}^{n-1} \underset{i-1}{\overset{(n)}{Q}}(T_H) = n\int_{0}^{T_H} \exp(-dx)\left(\frac{x}{T}\right)^{n-1}\frac{dx}{T}. \tag{20}$$

After the transformations, we finally have:

$$\begin{aligned} P(\eta_{\mathrm{m}} = k) = {} & \frac{a^k}{(k-1)!}[k,\ (\alpha + \beta_n)T_H] - \frac{a^{k+1}}{(k+1)!}[k,\ (\alpha + \beta_n)T_H] + \\ & + \frac{a^{k+1}}{(k+1)!}[k,\ (\alpha + \beta_n)T_H]\exp[-(\alpha + \beta_n)T_H]. \end{aligned} \tag{21}$$

For $k = 1, 2, 3, \ldots; \alpha = \frac{\beta_n}{(\alpha + \beta_n)}$. Finally

$$P(\eta_m = k) = \frac{a^k(1-a)}{(k-1)!}[k,\ (\alpha + \beta_n)T_H] + a\frac{(\beta_n T_H)^k}{k!}\exp[-(\alpha + \beta_n)T_H]. \tag{22}$$

For $k = 0$ similarly, using (19):

$$P(\eta_m = 0) = \frac{\alpha}{\alpha + \beta_n}\exp[-(\alpha + \beta_n)T_H]. \tag{23}$$

Formulas (22) and (23) give the final solution to the problem. The first formula gives the probability that there will be no emergency failures for the entire period of T_H, and the second is that the emergency failure will occur in the first transition.

5 Conclusions

The elements that make up the ship's engine system under consideration operate under conditions specified by various distribution laws established during the statistical processing of information obtained from operational documents.

The dependences are obtained, which allow determining: $Q_i = \frac{\alpha}{\beta}\left(\frac{\beta}{\alpha+\beta}\right)^i$ $(i = 1, 2, 3, \ldots)$ – the probability of accidental failure for any transition, $Q(b) = 1 - \left(\frac{\beta}{\alpha+\beta}\right)\exp\left(-\frac{\alpha\lambda\beta}{\alpha+\beta}\right)$ – accidental failure before reaching the limit wear, and $P(T_H) =$

$\exp(-\alpha_\beta T_H)$ – the probability that there will be no accidental failure during the standard service life.

The most common and complicated case is simultaneous action on the element (for example, on the cylinder liner) of factors that cause wear during operation (including during start-ups) and accidental failures. Other system elements under consideration correspond to the same conditions or its partial case.

For the first time, the dependencies are obtained, which allows for determining: $Q(x, y) = \exp(-\alpha x) - \exp(-\alpha y)$ the probability of emergency failure at any time, as well as $P(\eta_m = k) = \frac{a^k(1-a)}{(k-1)!}[k, (\alpha + \beta_n)T_H] + a\frac{(\beta_n T_H)^k}{k!}\exp[-(\alpha + \beta_n)T_H]$ the probability that during the whole period will not be emergency failures and $P(\eta_m = 0) = \frac{\alpha}{\alpha+\beta_n}\exp[-(\alpha + \beta_n)T_H]$ that the emergency failure will occur in the first transition.

References

1. Dinkar, B.K., Mukhopadhyay, A.K., Chattopadhyaya, S., Sharma, S., Alam, F., Machado, J.: Statistical reliability assessment for small sample of failure data of dumper diesel engines based on power law process and maximum likelihood estimation. Appl. Sci. **11**, 5387 (2021). https://doi.org/10.3390/app11125387
2. Droździel, P., Ignaciuk, P., Kordos, P.: Research on wear of liners in diesel engines during start-ups. Transport and communications **9**(2), 5 (2021). https://doi.org/10.26552/tac.C.2021.2.2
3. Khare, V., Khare, C., Nema, S., Baredar, P.: Chapter 6 - Reliability Assessment Model. In: Khare, V., Khare, C., Nema, S., Baredar, P. (eds): Tidal Energy Systems, pp. 295–330. Elsevier (2019). https://doi.org/10.1016/B978-0-12-814881-5.00006-5
4. Alturki, W.: Marine diesel engine fixed and moving parts. Int. J. Eng. Res. Appl. **7**(11), 01–11 (2017). https://doi.org/10.9790/9622-0711060111
5. Chad, K.: Auditing Strategy for ISO 9001:2015. J. Qual. Particip. **39**(3), 25–28 (2016)
6. Chisa, S., Gambo, A.A., Kayode, V., Daniel, D.: Reliability analysis of car maintenance forecast and performance. American J. Eng. Res. **4**(7), 290–299 (2015)
7. Li, Z., Yan, X., Guo, Z., Zhang, Y., Yuan, C., Peng, Z.: Condition monitoring and fault diagnosis for marine diesel engines using information fusion techniques. Electronics Electrical Eng. **7**(123), 109–112 (2012). https://doi.org/10.5755/j01.eee.123.7.2387
8. Sujesha, G., Rameshb, S.: Modeling and control of diesel engines: a systematic review. Alex. Eng. J. **57**(4), 4033–4048 (2018). https://doi.org/10.1016/j.aej.2018.02.011
9. Zhang, P., et al.: Marine systems and equipment prognostics and health. a systematic review from health condition monitoring to maintenance strategy. Machines **10**, 72 (2022). https://doi.org/10.3390/machines10020072
10. Trampert, S., Gocmez, T., Pischinger, S.: Thermomechanical fatigue life prediction of cylinder heads in combustion engines. J. Eng. Gas Turbines Power **130**(1), 012806 (2008). https://doi.org/10.1115/1.2771251
11. Mihanovic, L., Karna, H., Matika, D.: Research, processing and analysis of exploitation reliability results of high-speed radial diesel engine. Engineering Review **41**(2), 136–150 (2021). https://doi.org/10.30765/ER.1580
12. Munir, A., Shah, H.: FPSO Propulsion Machinery Reliability - A Systematic Approach. Future Offshore Technology and Sustained Reliability. Westchase Marriott, Houston, Texas (2015)
13. Kowalski, J., Krawczyk, B., Woźniak, M.: Fault diagnosis of marine 4-stroke diesel engines using a one-vs-one extreme learning ensemble. Eng. Appl. Artif. Intell. **57**, 134–141 (2017). https://doi.org/10.1016/j.engappai.2016.10.015

14. Pagán Rubio, J.A., Vera-García, F., Hernandez Grau, J., Muñoz Cámara, J., Albaladejo Hernandez, D.: Marine diesel engine failure simulator based on thermodynamic model. Appl. Therm. Eng. **144**, 982–995 (2018). https://doi.org/10.1016/j.applthermaleng.2018.08.096
15. Vera García, F., Rubio, J.A.P., Grau, J.H., Hernández, D.A.: Improvements of a failure database for marine diesel engines using the RCM and simulations. Energies **1**(1), 104 (2019). https://doi.org/10.3390/en13010104
16. Anantharaman, M., Islam, R., Khan, F., Garaniya, V., Lewarn, B.: Data analysis to evaluate the reliability of a main engine. TransNav **13**(2), 403–407 (2019). https://doi.org/10.12716/1001.13.02.18

Twisting Deformation of Thin-Walled Metal-Composite Rods

Andrii Kondratiev[1] (✉), Igor Taranenko[2], Anton Tsaritsynskyi[2], and Tetyana Nabokina[2]

[1] O. M. Beketov National University of Urban Economy in Kharkiv, 17, Marshal Bazhanov Street, Kharkiv 61002, Ukraine
andrii.kondratiev@kname.edu.ua

[2] National Aerospace University "Kharkiv Aviation Institute", 17, Chkalova Street, Kharkiv 61070, Ukraine

Abstract. Thin-walled elements of structures made of composite materials are increasingly used in aircraft engineering, ship- and machine-building. The Damageability of these elements largely determines the possibility of further use of products containing them. Therefore, the adequate calculation of thin-walled element mechanical behavior is one of the urgent problems in the mechanics of deformable solids. The mathematical model is developed to describe the deformed state of the thin-walled rod with a non-uniform cross-section under arbitrary loading. The resolving equation is obtained to determine the angle of twist of the compound contour under the action of longitudinal forces. To confirm the validity of the developed model, a number of typical metal-metal and metal composite bars have been manufactured, for which experimental measurements of angular displacements in space relative to the selected frame of reference were taken. Results of calculation of the angles of rod twist show fairly good agreement with experimental values for angle bars with the composite reinforcing plate (the difference is 15...25%). Experimentally measured components of displacements match reasonably well with proposed analytical dependencies, and so the developed model can be recommended for practical implementation. Parametric and experimental studies allow us to formulate practical recommendations for forming a rational shape of bars with non-uniform sections.

Keywords: Non-uniform section · Angle bar · Compound contour · Product innovation

1 Introduction

The development of state-of-the-art technology is accompanied by extensive use of polymeric composite materials (PCM) [1, 2]. Owing to the high strength, stiffness, and low density of these materials, the weight efficiency of load-bearing structures is increased [3, 4]. They allow the required mechanical properties to form in specified directions [5]. In particular, PCM is useful for the elements of structures of various applications with high requirements for stiffness [6, 7]. Thin-walled compound rods

V. Tonkonogyi et al. (Eds.): InterPartner 2022, LNME, pp. 424–433, 2023.
https://doi.org/10.1007/978-3-031-16651-8_40

now represent one of the most promising areas of PCM application [8]. Such rods are manufactured by winding or laying unidirectional or woven tape at different angles to the axis onto the metal blank. Thin-walled rods are used as the elements of truss constructions, struts, spars, supporting beams, propellers of aircrafts and helicopters, and drive shafts [9, 10]. More extensive use of the compound thin-walled composite rods in various technology fields requires a thorough study of stress-strain behavior of this type of structure.

The research aim is to develop a mathematical model describing the deformed state of the composite thin-walled rod with a non-uniform cross-section under arbitrary loading.

2 Literature Review

At this time, refined analytical models, particularly the models constructed by the iterative method, are used to study the stress-strain behavior of thin-walled rods [11]. In these models, a distinction is made between coordinates of the initial and final rod states, taking into account the effect of angles of twist of cross sections on the magnitude and nature of the distribution of the internal forces [12]. Iterative methods are quite universal ones, but they are cumbersome and difficult for practical implementation at high refinement steps. At the same time, accurate solutions for metal composite rods are limited by choice of the real boundary conditions. However, they can serve as the basis for relatively simple but sufficiently accurate applied solutions to problems. For example, the problem in [13] is solved by directly integrating the system of elasticity equations under the condition of absolutely rigid contact of the rod layers. Relationships are obtained for all components of the stress-strain behavior. To consider the non-uniformity of the compound rod, piecewise constant functions of elastic characteristics are introduced into Hooke's law. The paper [14] is based on neglecting small terms making an insignificant contribution to the total energy of deformation of the composite rod. The original model that considers the rod layers' bending and twisting connection is proposed in [15]. The paper [16] deals with developing the asymptotically correct model of the thin-walled rod. The problem is solved in this paper with no additional conditions imposed due to the use of the mechanics of the genome structure. The paper [17] takes into account shear effects and interactions between elements of the compound rod. The theory of deformation of thin-walled rods with non-uniform properties along the contour is developed in [18]. The common disadvantage of the accurate analytical models lies in their limitations when considering the various types of loading and boundary conditions [19]. As a rule, recent studies of thin-walled rods are conducted by the finite element method [20, 21]. The problems are solved numerically in [22] using a stiffness matrix with polynomials of different degrees for rods of the channel section and I-section. A refined finite element model that reduces the problem dimension without compromising the calculation accuracy is proposed in [23]. The paper [24] pays special attention to the distortion of a thin-walled rod cross-section using Taylor's and Lagrange decomposition methods. The results show good agreement with the experimental data. The numerical methods have a common drawback of awkwardness and complexity for practical use at high refinement steps. The paper [25] deals with developing a new method that is much more efficient than the finite element method in terms of calculation accuracy. Here the complete 3D stress analysis problem is broken down into a two-dimensional problem applicable to the rod section and a one-dimensional problem applied along its length. The spectral method for studying thin-walled rods' stress-strain behavior is proposed in [26]. The basic systems satisfying both essential and natural boundary conditions have been developed. It allowed increasing the rate of convergence of the approximate solution. However, the high computational cost is a common disadvantage of these methods. The paper [27] deals with the experimental study of thin-walled rod behavior at a combined compression-bending-twisting load. Experimental results show the good ability of the elements to dissipate the energy at pure twisting and compression-bending-twisting loads. In contrast, the applied axial force improves the bearing capacity and stiffness. The drawback of this kind of experimental study is the use of expensive materials and energy-intensive process equipment. Furthermore, it is not possible to implement all theoretically possible structures of anisotropic PCM packs [28].

3 Research Methodology

We consider the thin-walled cylindrical rod of the open section, which bears the axial forces (normal forces P and shear forces T), transverse bending forces M and torques H (Fig. 1). In the general case, cross sections of the thin-walled rod are twisted because of occurrence of torque in them, determined by the joint action of the bending-twisting bimoment B, which is formed by the elementary force P_i, total torque L, consisting of the bending-twisting moment M_ω from axial shear forces and torque H from non-uniform distribution of shear stresses over the thickness of the rod wall [18]. Analysis of dependencies between the indicated force factors allows us to compose a general differential equation for angles of the rod twist:

$$EI_\omega \frac{d^4\theta}{dx^4} + GI_d \frac{d^2\theta}{dx^2} = m(x), \tag{1}$$

where $EI_\omega = \int_F \overline{E}_x(s)\omega^2(s)dF$ – mechanical sectorial moment of inertia of the section; ω, D – sectorial areas; $\overline{E}_x = \frac{E_x}{1-\mu_{zx}\mu_{xz}}$ – reduced modulus of elasticity; E_x, μ_{xz}, μ_{zx} – modulus of elasticity and Poisson's ratio of the rod material; $\theta(x)$ – angle of rod twist; x – longitudinal coordinate of the rod; $GI_d = \int_s \frac{\alpha_s}{3} G(s)\delta^3(s)sds$ – mechanical moment of inertia of the rod during its twist; α_s – empirical coefficient depending on the shape of rod cross-section; G – shear modulus of the section; δ – thickness of thin-walled section; $dF = \delta \cdot ds$ – differential of the cross-sectional area of the rod; F – cross-sectional area; $m(x)$ – linear external torque per unit length of the rod; relative to this case, this linear moment is equal to zero.

Linear coordinates of the areas of the thin-walled rod are determined in the system of mechanical major central axes of the section and measured from the mechanical center of gravity of the rod.

Equation (1) is a linear non-homogeneous differential equation with constant coefficients. The Eq. (1) is solved by replacing the variable $\frac{d^2\theta}{dx^2} = \frac{d\theta_1}{dx}$ to obtain a general solution to a homogeneous equation and applying the Cauchy rule to derive a particular

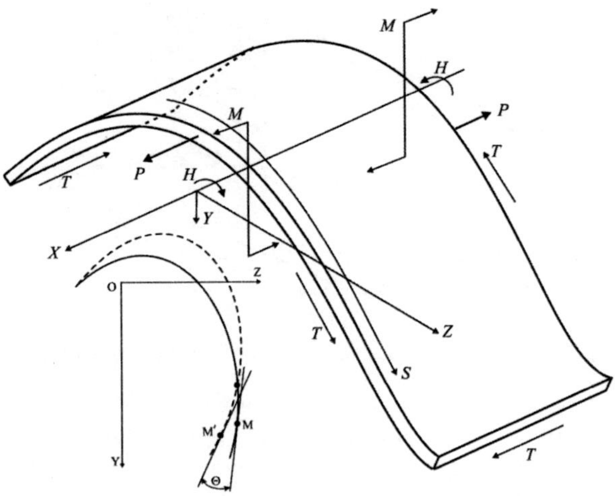

Fig. 1. Generalized view of the loaded thin-walled rod.

solution to the non-homogeneous equation. The general solution to such an equation is as follows:

$$\theta(x) = C_1 + C_2 x + C_3\, sh\frac{kx}{l} + C_4\, ch\frac{kx}{l} + \overline{\theta}(x), \tag{2}$$

where $\overline{\theta}(x)$ – particular solution to non-homogeneous equation; l – length of the rod; k – bending-twisting characteristic of the rod, $k = l\sqrt{\frac{GI_d}{EI_\omega}}$; C_1, C_2, C_3, C_4 – integration constants which are found from the boundary conditions.

If there is no external linear torque, then $\overline{\theta}(x) = 0$. Since the general differential equation of the angle of twist naturally includes the function $\theta(x)$ itself and its higher-order derivatives, then for the formation of boundary conditions it is possible to use the following parameters: θ– angle of rotation of an arbitrary section of the bar; θ'– changes in the angle of rotation of the bar section; $B = -EI_\omega \theta''_{xx}$– bending-twisting bimoment; $L = -EI_\omega \theta'''_{xxx} + GI_d \theta'_x$– total torque, the sum of bending-twisting moment M_ω from axial shear forces $\tau \cdot \delta$ and torque H from non-uniform distribution of shear stresses over the wall thickness.

Substituting the solution to the equation of twist into the boundary conditions, we obtain

$$\left.\begin{aligned}
\theta &= C_1 + C_2 x + C_3\, sh\frac{kx}{l} + C_4\, ch\frac{kx}{l};\\
\theta' &= C_2 + C_3 \frac{k}{l} ch\frac{kx}{l} + C_4 \frac{k}{l} sh\frac{kx}{l};\\
B &= -GI_d\left(C_3\, sh\frac{kx}{l} + C_4\, ch\frac{kx}{l}\right);\\
L &= GI_d\, C_2.
\end{aligned}\right\} \tag{3}$$

The position of the origin of coordinates (Fig. 2) can be chosen in the arbitrary section of the rod.

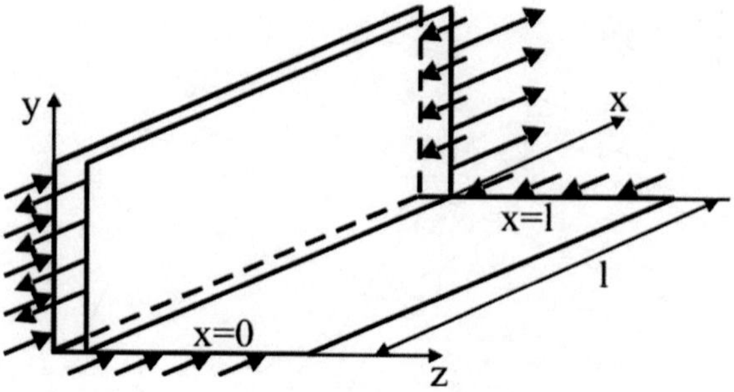

Fig. 2. Rod twisting due to the action of longitudinal forces.

Therefore, we assume that all geometric and static factors associated with the rod twisting under the sectorial laws have the known values. We determine two factors from each edge of the rod and denote these factors as θ_0, θ_0', B_0, L_0. . Let's assume that only these factors impact the rod under consideration. With no external effects, these factors are determined from the system (3) by substituting $x = 0$:

$$\left.\begin{aligned} \theta_0 &= \theta|_{x=0} = C_1 + C_4; \\ \theta_0' &= \theta'|_{x=0} = C_2 + C_3\frac{k}{l}; \\ B_0 &= B|_{x=0} = -C_4 GI_d; \\ L_0 &= L|_{x=0} = C_2 GI_d. \end{aligned}\right\} \quad (4)$$

When solving the system (4), we obtain:

$$C_1 = \theta_0 + \frac{B_0}{GI_d};\ C_2 = \frac{L_0}{GI_d};\ C_3 = \frac{l}{k}\theta_0' - \frac{l}{kGI_d}L_0;\ C_4 = -\frac{B_0}{GI_d}. \quad (5)$$

Substituting this solution into the general system for the boundary conditions (3), we get

$$\left.\begin{aligned} \theta &= \theta_0 + \frac{l}{k}\theta_0'\, sh\frac{kx}{l} - \frac{B_0}{GI_d}\left(ch\frac{kx}{l} - 1\right) + \frac{L_0}{GI_d}\left(x - \frac{l}{k}sh\frac{kx}{l}\right); \\ \theta' &= \theta_0'\, ch\frac{kx}{l} - \frac{k}{l}\frac{B_0}{GI_d}sh\frac{kx}{l} + \frac{L_0}{GI_d}\left(1 - ch\frac{kx}{l}\right); \\ B &= -\frac{l}{k}GI_d\,\theta_0'\, sh\frac{kx}{l} + B_0\, ch\frac{kx}{l} + \frac{l}{k}L_0 sh\frac{kx}{l}; \\ L &= L_0. \end{aligned}\right\} \quad (6)$$

Therefore, we obtain the system of general equations describing the twist angle of the compound thin-walled rod with a non-uniform cross-section under arbitrary loading.

When this system is used for thin-walled rod, to the ends of which longitudinal forces p_i are applied only, and torques M_ω and H are equal to zero, we can write the boundary conditions as follows:

$$\left.\begin{aligned} &if\ x = 0\ \ \theta_0 = \theta|_{x=0} = 0;\ \ B_0 = B|_{x=0} = -\int_s p_i \cdot \omega_i \cdot \\ &if\ x = l\ \ L_0 = L|_{x=l} = 0;\ \ B_0 = B|_{x=l} = -\int_s p_i \cdot \omega_i \cdot \end{aligned}\right\} \quad (7)$$

Substituting these boundary conditions into the system (6), we obtain the equation for the angle of twist of the rod along its length as a result of the action of longitudinal forces:

$$\theta(x) = \frac{\int_s p_i \omega_i}{GI_d}\left(ch\frac{kx}{l} - thk \cdot sh\frac{kx}{l} - 1\right) \tag{8}$$

4 Results and Discussion

For the experiments on studying the deformed state of long-length (900 mm) compound angle bars, we chose the combinations of titanium alloy – aluminum alloy (Fig. 3, a, b), aluminum alloy – CFRP element with the varying reinforcement structure (Fig. 3, c–e) and aluminum alloy – shortened CFRP plate (Fig. 3, f).

The physical and mechanical characteristics of alloys used in the experiments are presented in Table 1.

The aluminum section was bonded with the CFRP plate using epoxy adhesive. Curing (bonding) of the section and plate was carried out in the autoclave at the temperature of 165 ± 10 °C and overpressure of 0.08 ± 0.01 MPa during 3 – 4 h.

CFRP plates were manufactured by autoclave-vacuum molding based on the ELUR-0.08 unidirectional carbon tape and EDT 69N binder with the following characteristics: modulus of elasticity in the longitudinal direction $E_1 = 120 \pm 6$ GPa; modulus of elasticity in the transverse direction $E_2 = 10 \pm 2.6$ GPa; shear modulus $G_{12} = 6 \pm 2.3$ GPa; Poisson's ratio $\mu_{12} = 0.29 \pm 0.05$; coefficient of linear thermal expansion in the longitudinal direction $\alpha_1 = (-.5 \pm 0.08)\cdot 10^{-6}$ 1/K; in the transverse direction – $\alpha_2 = (28.0 \pm 1.9)\cdot 10^{-6}$ 1/K [29].

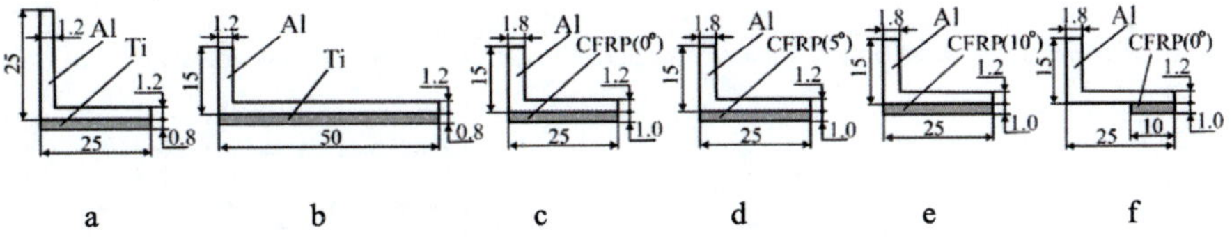

Fig. 3. Structures of compound bars under testing: a – aluminum equal angle bar with titanium plate; b – aluminum unequal angle bar with titanium plate; c – aluminum unequal angle bar with CFRP plate reinforced along the longitudinal axis of the bar; d – aluminum unequal angle bar with CFRP plate reinforced along the longitudinal axis of the bar at the angle of 5°; e – aluminum unequal angle bar with CFRP plate reinforced along the longitudinal axis of the bar at the angle of 10°; f – aluminum unequal angle bar with shortened CFRP plate reinforced along the longitudinal axis of the bar.

The angles of rod twist were measured on the working table of the milling machine using clamps and angular gage with the measurement accuracy of 0.1°. In measuring deviations of each point of the bar under study, the angle of rotation was measured 5 times. Then the values of mathematical expectation and variance were found.

Table 1. Physical and mechanical characteristics of alloys.

Alloy / Characteristics	Aluminum	Titanium
Modulus of elasticity, GPa	240±12.0	106±4.8
Poisson's ratio	0.3	0.32
Tensile strength, MPa	182±7.5	648±12
Coefficient of thermal expansion, 1/K	$(22\pm0.8)\cdot10^{-6}$	$(9.8\pm0.2)\cdot10^{-6}$

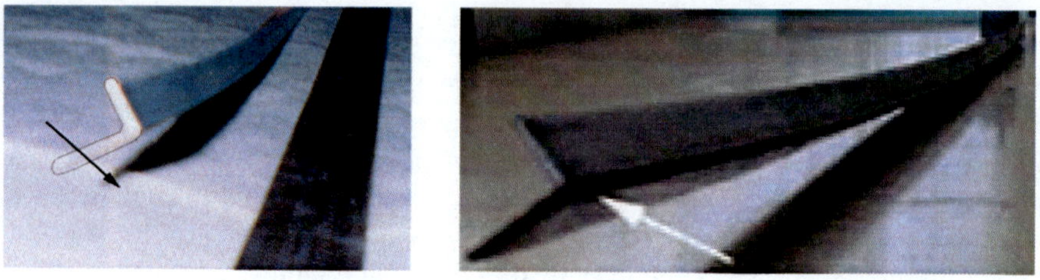

Fig. 4. Deformed state of the compound sections after molding.

The actual results of buckling deformation of the most twisted section of the rod with the CFRP plate laid at the angle of 10° are shown in Fig. 4.

In parallel with the experimental studies of the angle of twist of the section, we conducted parametric studies of the predicted angle of twist of such sections to confirm the developed method for designing the deformed state of the section. The results of comparing the experimental and theoretically predicted angles of twist of the compound metallic and metal composite sections are summarized in Table 2.

Table 2. Results of calculation and measurement of the deviations of the maximum angle of twist.

Bonded parts	Angle of twist θ, degrees	
	Estimated	Experimental
Aluminum angle bar and CFRP plate (reinforcement at the angle of 0°)	2.8	0.8
Aluminum angle bar and CFRP plate (reinforcement at the angle of 5°)	38.3	33.2
Aluminum angle bar and CFRP plate (reinforcement at the angle of 10°)	44.0	35.2
Aluminum angle bar and narrow (10 mm) CFRP plate (reinforcement at the angle of 0°)	4.5	0.2
Aluminum unequal angle bar and titanium alloy plate	2.3	0.2
Aluminum equal angle bar and titanium alloy plate	4.3	0

Based on the results of the analysis of experimental and parametric studies, we have made the following conclusions:

- results of calculation of the angles of rod twist show fairly good agreement with experimental values for angle bars with the composite reinforcing plate (the difference is 15…25%). The large absolute values of displacements explain it;
- since the absolute experimental displacements of aluminum angle bars with titanium plates are quite small, the error in determining angles in these cases is reasonably large. It is explained by the fact that the dial indicator with relatively low spring stiffness was chosen to measure deviations, but this stiffness also affects the measured value (causing the additional deviation of the bar). At a first approximation, the method for "transmission" measurement of deviations of the bar points was tested, and this method is to be improved in the future.

5 Conclusions

The mathematical model has been developed to describe the deformed state of the thin-walled rod with a non-uniform cross-section under arbitrary loading. To confirm the validity of the developed model, a number of typical metal-metal and metal composite bars have been manufactured, for which experimental measurements of angular displacements in space relative to the selected frame of reference were taken. Besides developing modified mathematical models novel approach to inner force factors determination is suggested, and practical recommendations for designer allowing to reduce the warping of the rod with inhomogeneous cross-section are developed.

References

1. Hsissou, R., Seghiri, R., Benzekri, Z., Hilali, M., Rafik, M., Elharfi, A.: Polymer composite materials: a comprehensive review. Compos. Struct. **262**, 113640 (2021). https://doi.org/10.1016/j.compstruct.2021.113640
2. Turpak, S.M., Taran, I.O., Fomin, O.V., Tretiak, O.O.: Logistic technology to deliver raw material for metallurgical production. Scientific Bulletin of National Mining University **1**, 162–169 (2018). https://doi.org/10.29202/nvngu/2018-1/3
3. Kombarov, V., Kryzhyvets, Y., Biletskyi, I., Tsegelnyk, Y., Aksonov, Y., Piddubna, L.: Numerical control of fiberglass pipe bends manufacturing. In: 2nd KhPI Week on Advanced Technology (KhPIWeek), pp. 357–362. IEEE, Kharkiv (2021). https://doi.org/10.1109/KhPIWeek53812.2021.9570068
4. Galych, I., Antoshchenkov, R., Antoshchenkov, V., Lukjanov, I., Diundik, S., Kis, O.: Estimating the dynamics of a machine-tractor assembly considering the effect of the supporting surface profile. Eastern-European J. Enterprise Technol. **1**(7(109)), 51–62 (2021). https://doi.org/10.15587/1729-4061.2021.225117
5. Bilym, P., Rogozin, A., Firsov, P., Zolotov, S.: Carbon fiber plastics for vehicles manufactured by resource-saving formation technology. IOP Conference Series: Materials Science and Eng. **1021**, 012034 (2021). https://doi.org/10.1088/1757-899X/1021/1/012034
6. Kovalov, A., Otrosh, Y., Rybka, E., Kovalevska, T., Togobytska, V., Rolin, I.: Treatment of determination method for strength characteristics of reinforcing steel by using thread cutting method after temperature influence. Mater. Sci. Forum **1006**, 179–184 (2020). https://doi.org/10.4028/www.scientific.net/MSF.1006.179

7. Slyvynskyi, V.I., Gajdachuk, V.E., Kyrychenko, V.V., Kondratjev, A.V.: Basic parameters' optimization concept for composite nose fairings of launchers. In: 62nd International Astronautical Congress, IAC 2011, vol. 9, pp. 5701–5710. Red Hook, NY, Curran, Cape Town (2011)
8. Bao, Y., Wang, B., He, Z., Kang, R., Guo, J.: Recent progress in flexible supporting technology for aerospace thin-walled parts: a review. Chin. J. Aeronaut. **35**, 10–26 (2021). https://doi.org/10.1016/j.cja.2021.01.026
9. Gavva, L.M., Firsanov, V.V.: Analytical review of account methods and experimental approaches to stress-strain state investigation of structurally-anisotropic aircraft panels made from composite materials. IOP Conference Series: Materials Science and Eng. **927**, 012067 (2020). https://doi.org/10.1088/1757-899x/927/1/012067
10. Smetankina, N., Kravchenko, I., Merculov, V., Ivchenko, D., Malykhina, A.: Modelling of bird strike on an aircraft glazing. In: Nechyporuk, M., Pavlikov, V., Kritskiy, D. (eds.) Integrated Computer Technologies in Mechanical Engineering. AISC, vol. 1113, pp. 289–297. Springer, Cham (2020). https://doi.org/10.1007/978-3-030-37618-5_25
11. Choi, S., Kim, Y.Y.: Higher-order Vlasov torsion theory for thin-walled box beams. Int. J. Mech. Sci. **195**, 106231 (2021). https://doi.org/10.1016/j.ijmecsci.2020.106231
12. Gorik, A.V., Koval'chuk, S.B.: Solution of a transverse plane bending problem of a laminated cantilever beam under the action of a normal uniform load. Strength Mater. **50**(3), 406–418 (2018). https://doi.org/10.1007/s11223-018-9984-7
13. Goryk, A.V., Koval'chuk, S.B.: Elasticity theory solution of the problem on plane bending of a narrow layered cantilever beam by loads at its free end. Mech. Compos. Mater. **54**(2), 179–190 (2018). https://doi.org/10.1007/s11029-018-9730-z
14. Harursampath, D., Harish, A.B., Hodges, D.H.: Model reduction in thin-walled open-section composite beams using variational asymptotic method. Part I: Theory. Thin-Walled Structures **117**, 356–366 (2017). https://doi.org/10.1016/j.tws.2017.03.018
15. Gunay, M.G., Timarci, T.: Stresses in thin-walled composite laminated box-beams with curvilinear fibers: Antisymmetric and symmetric fiber paths. Thin-Walled Structures **138**, 170–182 (2019). https://doi.org/10.1016/j.tws.2019.02.003
16. Deo, A., Yu, W.B.: Thin-walled composite beam cross-sectional analysis using the mechanics of structure genome. Thin-Walled Structures **152**, 106663 (2020). https://doi.org/10.1016/j.tws.2020.106663
17. Nemirovskii, Y.V., Mishchenko, A.V., Terletskii, R.F.: Dynamic analysis of composite rods under thermal and force loads. J. Math. Sci. **223**(1), 87–102 (2017). https://doi.org/10.1007/s10958-017-3340-5
18. Taranenko, I., Tsaritsynskyy, A., Nabokina, T., Kondratiev, A.: Bending-twisting deforming of thin-walled composite sections at thermal-mechanical loading. In: Nechyporuk, M., Pavlikov, V., Kritskiy, D. (eds): Integrated Computer Technologies in Mechanical Engineering 2021, ICTM 2021. Lecture Notes in Networks and Systems, vol. 367, pp. 375–387. Springer, Cham (2021). https://doi.org/10.1007/978-3-030-94259-5_33
19. Tusnin, A.: Finite element for calculation of structures made of thin-walled open profile rods. In: 2nd International Conference on Industrial Engineering, Icie-2016, vol. 150, pp. 1673–1679 (2016). https://doi.org/10.1016/j.proeng.2016.07.149
20. Orgiyan, A., Kobelev, V., Ivanov, V., Balaniuk, A., Aymen, A.: Ensuring the bending stiffness of pre-compressed cantilever boring bars during fine boring. In: Ivanov, V., Trojanowska, J., Pavlenko, I., Zajac, J., Peraković, D. (eds.) DSMIE 2020. LNME, pp. 315–324. Springer, Cham (2020). https://doi.org/10.1007/978-3-030-50794-7_31
21. Varbanets, R., et al.: Acoustic method for estimation of marine low-speed engine turbocharger parameters. J. Marine Science Eng. **9**(3), 321 (2021). https://doi.org/10.3390/jmse9030321

22. Choi, I.S., Jang, G.W., Choi, S., Shin, D., Kim, Y.Y.: Higher order analysis of thin-walled beams with axially varying quadrilateral cross sections. Comput. Struct. **179**, 127–139 (2017). https://doi.org/10.1016/j.compstruc.2016.10.025
23. Vojnic-Purcar, M., Prokic, A., Besevic, M.: A numerical model for laminated composite thin-walled members with openings considering shear lag effect. Eng. Struct. **185**, 392–399 (2019). https://doi.org/10.1016/j.engstruct.2019.01.142
24. Carrera, E., Kaleel, I., Petrolo, M.: Elastoplastic analysis of compact and thin-walled structures using classical and refined beam finite element models. Mech. Adv. Mater. Struct. **26**(3), 274–286 (2019)
25. Islam, A., Sheikh, A.H., Bennett, T., Thomsen, O.T.: An efficient model for laminated composite thin-walled beams of open or closed cross-section and with or without in-filled materials. Compos. Struct. **256**, 112998 (2021). https://doi.org/10.1016/j.compstruct.2020.112998
26. Tkachenko, D., Tsegelnyk, Y., Myntiuk, S., Myntiuk, V.: Spectral methods application in problems of the thin-walled structures deformation. J. Applied Computational Mechanics **8**(2), 641–654 (2022). https://doi.org/10.22055/JACM.2021.38346.3207
27. Wang, Y.H., et al.: Combined compression-bending-torsion behaviour of CFST columns confined by CFRP for marine structures. Compos. Struct. **242**, 112181 (2020). https://doi.org/10.1016/j.compstruct.2020.112181
28. Rimasauskiene, R., Jurenas, V., Radzienski, M., Rimasauskas, M., Ostachowicz, W.: Experimental analysis of active-passive vibration control on thin-walled composite beam. Compos. Struct. **223**, 110975 (2019). https://doi.org/10.1016/j.compstruct.2019.110975
29. Kondratiev, A., Píštěk, V., Smovziuk, L., Shevtsova, M., Fomina, A., Kučera, P:. Stress–strain behaviour of reparable composite panel with step–variable thickness. Polymers **13**(21), 3830 (2021). https://doi.org/10.3390/polym13213830

Dynamic Behavior of a Vibratory Plate Compactor Working on a Horizontal Elastic-Viscous-Plastic Surface

Vitaliy Korendiy(✉) and Oleksandr Kachur

Lviv Polytechnic National University, 12, S. Bandera Street, Lviv 79013, Ukraine
vitalii.m.korendii@lpnu.ua

Abstract. The paper is focused on the dynamic analysis of the double-mass oscillatory system of the vibratory plate compactor working on a horizontal surface modeled as an elastic-viscous-plastic medium. The main goal of the carried-out investigations is to study the vibratory plate compactor's kinematic and dynamic parameters under different operational conditions. The improved mathematical model of the plate compactor's double-mass vibration-driven locomotion system, taking into account the mechanical interaction between the compacting plate and the elastic-viscous-plastic surface being compacted, forms the major scientific novelty of the research. The numerical modeling results of the vibratory plate compactor operation are presented as the time dependencies of the oscillating bodies' kinematic parameters and the dynamic parameters describing the interaction between the compacting plate and the surface being compacted. In addition, the plastic deflection of the surface is considered, and the influence of the compactor's vibrations on the control (operating) handles is studied under different operational conditions. The obtained results have significant practical value and can be used by designers, technologists, and engineers while improving the existent vibratory compactors, developing new compacting technologies, and optimizing the compactors' operational parameters according to the technological requirements.

Keywords: Industrial growth · Vibration exciter · Unbalanced rotor · Dynamic model · Vibration isolator · Oscillatory system · Numerical modeling · Kinematic parameters

1 Introduction

Vibratory technologies are widely used for ramming, tamping, and compacting various materials: sand, grit, gravel, pebble, talus, asphalt, concrete, pavestone, sett, chippings, road metal, rubble, soil, etc. The mechanical and physical properties of a particular material define the specific conditions of its compacting. In addition, the final technological requirements set on the blankets and pavements also influence the working process of compactors [1]. The major problem occurring while performing ramming, tamping, and compacting operations with the help of the vibratory equipment consists in optimizing its

V. Tonkonogyi et al. (Eds.): InterPartner 2022, LNME, pp. 434–443, 2023.
https://doi.org/10.1007/978-3-031-16651-8_41

working conditions according to the technological requirements considering the physical and mechanical properties of a specific material [2]. Some aspects of the mentioned problem will also be studied in this paper.

In most cases, the vibratory compactors are equipped with inertial vibration exciters (unbalanced rotors). The operational conditions of such machines can be easily changed according to the technological requirements by controlling the exciter's forced frequency and the working member's speed (e.g., compacting plate) [3]. One of the vibratory compactors' main disadvantages is the vibrations' negative influence on the operator contacting the control handle. This paper will consider the double-mass design of the vibratory compactor, in which the control handle and the driving motor are sprung over the compacting plate, eliminating the negative influence of vibrations on the operator. While studying the dynamic behavior of the compactor, special attention must be paid to the self-propelling characteristics and compacting forces acting between the machine plate and the surface modeled as the elastic-viscous-plastic medium.

2 Literature Review

Studying the dynamic behavior of the vibratory compactors consists of two major subproblems: the vibrating plate interaction with the surface being compacted and the translational motion of the compactor under different operational conditions. The first subproblem has been thoroughly studied in numerous publications. The general characteristics of the paving compaction process are considered in [1], where the authors investigated a multi-degree of freedom nonlinear vibration system describing the interaction between the screed plate and the asphalt mixture. In [2], the technological process of surface soil compaction was investigated with the help of the inertial vibratory rammer. The paper [3] is devoted to the optimal selection method of the plate compactor weight and the analysis of its operation under different technological conditions. In [4], the authors developed specific software for optimizing the design parameters of the vibratory plate compactor according to technological requirements. The study on the impact force exerted by the tamping rammer upon the soil during the compaction process is presented in [5].

The paper [6] is devoted to modeling the dynamics of the interaction of the compactor (stacker) and the asphalt-concrete compound, considered an elastic-viscous body. In [7], the authors studied the operation of the three-mass vibratory system of the stacker (plate compactor). The sand compaction process performed with the help of a vibratory plate compactor under different technological conditions was investigated in [8]. The problems of improving the compaction performance and the paving quality of the vibratory screed systems were studied in [9]. The paper [10] considers the dynamic model of the vibratory tamper during its interaction with the asphalt compound. In [11], the dynamic characteristics of the improved vibro-impact eccentric-type compactor are studied. Considering the papers mentioned above, it can be stated that the problems of analyzing the influence of the compacting plate vibrations on the control (operating) handle haven't been thoroughly studied.

The second subproblem related to the translational motion of the vibratory systems under different operational conditions has been investigated in the following papers.

A vibratory locomotion system that consists of a main absolutely rigid body and two movable internal masses was investigated in [12]. A similar stick-slip locomotion system due to the dry friction forces was experimentally tested in [13]. The paper [14] considers the typical non-smooth dynamical behavior of the self-propelled vibratory systems based on the sliding bifurcation analysis. In [15], there is studied the dynamics of the vibro-impact systems acted by different types of constraints under various design and control parameters. The stick-slip motion of the self-propelled vibratory system with the internal mass performing periodic oscillations under dry friction is analyzed in [16]. A similar system moving in a resistive medium is investigated in [17]. The problems of optimizing the control strategies of the double-mass vibration-driven systems are considered in [18]. The paper [19] studies the locomotion principles of various multibody mobile robotic systems.

In [20], the authors analyzed the dynamics of the vibration-driven system consisting of a rigid body that slides along a rough horizontal plane and a mass point moving inside the body along a circular path. A similar system under the action of dry and viscous friction is considered in [21]. The paper [22] is dedicated to studying the effect of various types of friction on the dynamics of a self-propelled vibro-impact locomotion system. Similar research results extended for pure vibro-impact and vibration-driven systems are presented in [23]. The paper [24] is focused on analyzing the forced oscillations of a three-mass self-propelled vibro-impact locomotion system sliding along a horizontal surface characterized by dry friction. In [25], the authors presented the novel design of a vibration exciter with changeable eccentricity, which can be effectively used both in crank-type (eccentric-type) and inertial excitation mechanisms of mobile vibration-driven systems. Analyzing the papers mentioned above, it can be stated that the problems of optimizing the working conditions of the vibration-driven system according to the technological requirements and considering the mechanical and physical properties of the particular material being compacted haven't been thoroughly studied. Therefore, the scientific novelty of the present paper is focused on improving the mathematical model allowing for investigating the dynamic behavior of the vibratory plate compactor working on various elastic-viscous-plastic surfaces and for analyzing the influence of the compactor vibrations on the control (operating) handle.

3 Research Methodology

3.1 Design Peculiarities and Dynamic Diagram of the Compactor

The general design of the vibratory compactor is shown in Fig. 1. The compacting (ramming) plate (working member) 1 is in direct contact with the supporting surface being compacted due to the forced vibrations of the plate. To excite the vibrations of the compacting plate 1, there is used an unbalanced rotor 2 mounted on the plate. While performing the corresponding technological operations, the compactor is set into the translational motion by the pushing lever 6, on which the control handles 7 are fixed for applying the operator's pushing force. The shaft of the unbalanced rotor 2 is set into the rotary motion by the belt transmission 3 and the electric motor 4 fixed on the upper sprung (suspended) platform 8. The latter is connected with the lower unsprung platform (compacting-ramming plate) 1 with the help of the vibration isolators (absorbers) 9 eliminating the transmission of vibrations from the compacting plate to the upper sprung platform 8. The holding handles 5 are used for the manual carrying of the compactor. To ensure the additional vibration isolation of the pushing lever 6 and the control handles 7, there are used the rubber dumpers 10 fixed on the upper sprung (suspended) platform 8.

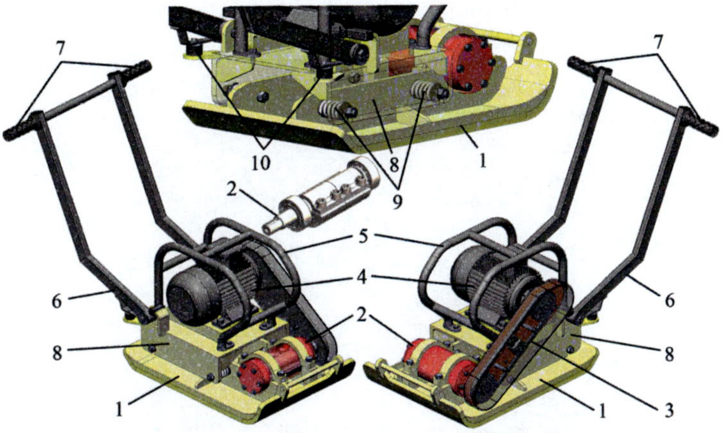

Fig. 1. General design of the studied vibratory compactor: 1 – compacting plate; 2 – vibration exciter (unbalanced rotor); 3 – belt transmission; 4 – electric motor; 5 – holding handles; 6 – pushing lever; 7 – control (operating) handles; 8 – upper sprung (shock-mounted) platform; 9 – vibration isolators (absorbers); 10 – rubber dampers.

The dynamic diagram of the double-mass self-propelled vibratory locomotion system of the considered vibratory plate compactor is presented in Fig. 2. The system is characterized by six uncontrollable degrees of freedom, and, therefore, its motion can be described by five generalized coordinates: x_A, y_A, x_{1B}, y_{1B}, φ, ψ. The coordinates x_A, y_A characterize the horizontal and vertical displacements of the compacting (ramming) plate working on the surface that is compacted; the angle φ describes the angular position of the compacting (ramming) plate relative to the horizontal surface; the coordinates x_{1B}, y_{1B}, ψ define the relative position of the sprung platform with respect to the compacting (ramming) plate (see Fig. 2). The degree of freedom describing the rotary motion of the vibration exciter is considered as a controlled one, i.e., the motion of the unbalanced rotor is performed at a constant rotational velocity ω. The mass center of the compacting plate is placed at point C (x_{1C}, y_{1C}); the mass center of the sprung platform is located at point B (x_{1B}, y_{1B}). The hinge D with the crank DE models the unbalanced rotor (vibration exciter) characterized by the eccentricity ρ and the unbalanced mass m_E. The springs with the stiffness k support the upper platform of the mass m_B on the lower unsprung platform. The mass of the compacting plate is equal to m_C.

In order to simulate the "compacting plate – supporting surface" interaction, the combined elastic-viscous-plastic medium model based on the Voigt and Bathelt soil models is used. The medium being compacted and interacting with the compacting plate is simulated as a rigid body characterized by the mass m_S supported by an elastic-viscous-plastic system (Fig. 2). The springs (k_{e1}, k_{e2}) and the absorbers (c_{s1}, c_{s2}) modeling the elastic and damping characteristics of the surface being compacted work regardless of the compacting force change. The springs (k_{p1}, k_{p2}) modeling the plastic characteristics of the surface work at the stage of the compacting force increasing (i.e., when the corresponding contact point attempts to move down). In the second stage, when the compacting force decreases and the contact point attempts to move up, the corresponding stiffness coefficients are equal to zero.

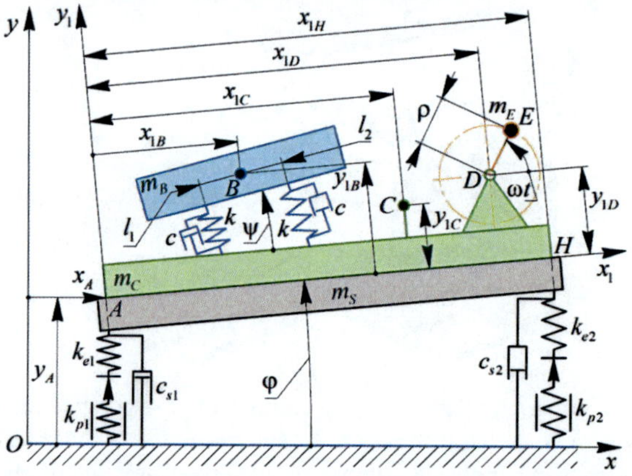

Fig. 2. Dynamic model of the vibratory plate compactor.

3.2 Mathematical Model Describing the Compactor Operation

The plane-parallel motion of the compactor's double-mass vibratory system is fully described by the system of six differential equations corresponding to the generalized coordinates $x_A, y_A, x_{1B}, y_{1B}, \varphi, \psi$. The degree of freedom that corresponds to the rotation of the unbalanced mass m_E is considered a controlled one: the crank rotates at the uniform rotational velocity ω. The Lagrange equations of the second kind (i.e., the Euler–Lagrange equations of motion) were used to derive the mathematical model describing the system motion. Due to the significantly large expressions obtained for each equation, let us consider the simplified system taking into account the following assumptions: the angular and the horizontal translational displacements of the upper suspended platform relative to the compacting plate are neglected ($\psi = 0, x_{1B} = \text{const}$); the compacting plate doesn't move in the horizontal direction ($x_A = 0$); the energy dissipation in the vibration isolators of the upper platform is negligible ($c = 0$). Therefore, the deduced system of three equations describing the compactor motion is following:

$$\begin{aligned}(m_B + m_C + m_E + m_S) \cdot \ddot{y}_A + (c_1 + c_2) \cdot \dot{y}_A + (k \cdot (1 + \cos(2 \cdot \varphi)) + k_1 + k_2) \cdot y_A \\ + c_2 \cdot x_{1H} \cdot \dot{\varphi} \cdot \cos\varphi + k \cdot (x_{1B} \cdot \sin(2 \cdot \varphi) - 2 \cdot y_{1B} \cdot \cos\varphi) \\ + k_2 \cdot x_{1H} \cdot \sin\varphi = m_E \cdot \rho \cdot \omega^2 \cdot \sin(\varphi + \omega \cdot t);\end{aligned} \tag{1}$$

$$m_B \cdot \ddot{y}_{1B} + (2 \cdot k - m_B \cdot \dot{\varphi}^2) \cdot y_{1B} + c_2 \cdot x_{1H} \cdot \cos\varphi \cdot (\dot{y}_A + x_{1H} \cdot \dot{\varphi} \cdot \cos\varphi) - 2 \cdot k \cdot (x_{1B} \cdot \sin\varphi + y_A \cdot \cos\varphi) = 0; \tag{2}$$

$$\left(\begin{array}{c} J_B + m_B \cdot (x_{1B}^2 + y_{1B}^2) + J_C + m_C \cdot (x_{1C}^2 + y_{1C}^2) + \\ + m_E \cdot ((x_{1D} + \rho \cdot \cos(\omega \cdot t))^2 + (y_{1D} + \rho \cdot \sin(\omega \cdot t))^2) \end{array} \right) \cdot \ddot{\varphi}$$
$$+ \left(\begin{array}{c} c_2 \cdot x_{1H}^2 \cdot \cos^2\varphi + 2 \cdot m_B \cdot y_{1B} \cdot \dot{y}_{1B} + \\ + 2 \cdot m_E \cdot \rho \cdot \omega \cdot (y_{1D} \cdot \cos(\omega \cdot t) - x_{1D} \cdot \sin(\omega \cdot t)) \end{array} \right) \cdot \dot{\varphi} + c_2 \cdot x_{1H} \cdot \cos\varphi \cdot \dot{y}_A$$
$$+ 2 \cdot k \cdot (x_{1B} \cdot \sin\varphi - y_{1B} + y_A \cdot \cos\varphi) \cdot (x_{1B} \cdot \cos\varphi - y_A \cdot \sin\varphi)$$
$$+ k_2 \cdot x_{1H} \cdot \cos\varphi \cdot (x_{1H} \cdot \sin\varphi + y_A) = m_E \cdot \rho \cdot \omega^2 \cdot x_{1D} \cdot \sin(\omega \cdot t), \tag{3}$$

where c_1, c_2 are the equivalent damping coefficients characterizing the viscosity properties of the surface being compacted; k is the stiffness factors of the springs supporting the compactor's upper platform; k_1, k_2 are the equivalent stiffness coefficients characterizing the elasticity properties of the surface being compacted; x_{1H}, x_{1D}, y_{1D} are the geometrical parameters of the compactor's double-mass vibratory system shown in Fig. 2; x_{1B} is the distance between the axis Ay_1 and the mass center of the upper sprung platform; x_{1C}, y_{1C} are the coordinates defining the position of the compacting plate's mass center in the x_1Ay_1 reference system; J_B, J_C are the moments of inertia of the upper (sprung) platform and the lower (unsprung) platform (compacting plate) about the axes intersecting the corresponding mass centers perpendicularly to the plane of the system's dynamic diagram; ρ, ω are the eccentricity and the rotational velocity of the unbalanced (eccentric) rotor exciting the oscillations of the vibratory compactor.

The elastic-viscous-plastic parameters of the surface can be described as follows:

$$k_i = \begin{cases} 0, \text{ if } \operatorname{sign}(\dot{y}_i) > 0, \ \ddot{y}_i > g; \\ (k_{ei} \cdot k_{pi})/(k_{ei} + k_{pi}), \text{ if } \operatorname{sign}(\dot{y}_i) \le 0; \\ k_{ei} \cdot (1 - k_{ei}/(k_{ei} + k_{pi})), \text{ if } \operatorname{sign}(\dot{y}_i) > 0, \ \ddot{y}_i \le g, \end{cases} \tag{4}$$

$$c_i = \begin{cases} 0, \text{ if } \operatorname{sign}(\dot{y}_i) > 0, \ \ddot{y}_i > g; \\ c_{si}, \text{ if } \operatorname{sign}(\dot{y}_i) \le 0; \\ c_{si} \cdot (1 - k_{ei}/(k_{ei} + k_{pi})), \text{ if } \operatorname{sign}(\dot{y}_i) > 0, \ \ddot{y}_i \le g, \end{cases} \tag{5}$$

where $i = 1, 2$ is the index number of the corresponding spring-damper element modeling the "compacting plate – surface" interaction; k_{ei}, k_{pi}, c_{si} are the equivalent elastic-plastic stiffness constants and the equivalent damping coefficients of the medium being compacted at the rear ($i = 1$) and front ($i = 2$) ends of the compacting plate; $y_1 = y_A$; $y_2 = y_H = y_A + x_{1H} \cdot \sin\varphi$; g is the free-fall (gravitational) acceleration.

4 Results and Discussion

4.1 Numerical Modeling of the Compactor Motion

To perform further calculations, the corresponding geometrical and inertial parameters are defined based on the compactor's 3D model designed in SolidWorks software: $m_B =$

41.28 kg, $m_C = 63.51$ kg, $m_E = 1.94$ kg, $\omega = 314$ rad/s, $\rho = 0.021$ m, $J_C = 2.7$ kg·m^2, $J_B = 1.62$ kg·m^2, $x_{1C} = 0.32$ m, $y_{1C} = 0.074$ m, $x_{1H} = 0.55$ m, $x_{1B} = 0.113$ m, $x_{1D} = 0.41$ m, $y_{1D} = 0.112$ m, $k = 1.25 \cdot 10^7$ N/m. The elastic-viscous-plastic characteristics of the compacted surface are chosen to be appropriate for the asphalt-concrete mixture: $m_S = 1.96$ kg, $c_{s1} = c_{s2} = 50$ N·s/m, $k_{e1} = k_{e2} = 1.4 \cdot 10^5$ N/m, $k_{p1} = k_{p2} = 8.5 \cdot 10^5$ N/m.

Figure 3 shows the compactor motion's numerical modeling results obtained with the Mathematica software's help. At the beginning of the compaction process, the oscillating masses reach the maximal displacements. The vibration amplitude of the compactor's rear end is about 3 mm. In contrast, the front end oscillates at the amplitude of 6 mm (Fig. 3a). After a definite period of time (Fig. 3b), the internal damping and plasticity of the medium being compacted cause the reduction of the amplitudes (to 0.5 mm and 1 mm, respectively) and the change of the initial position about which the oscillations are performed (the plastic deflection is about 3 mm).

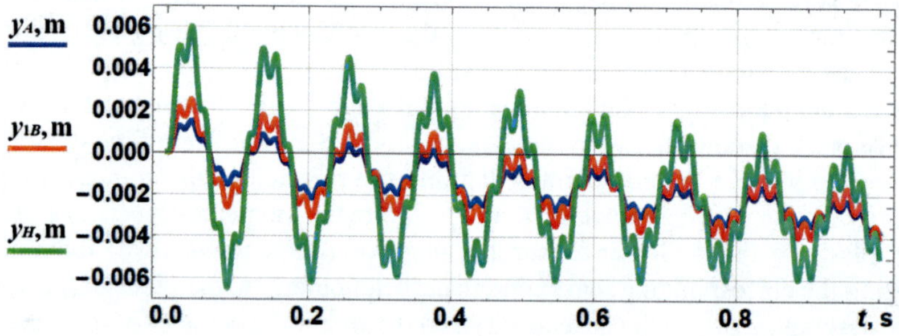

a

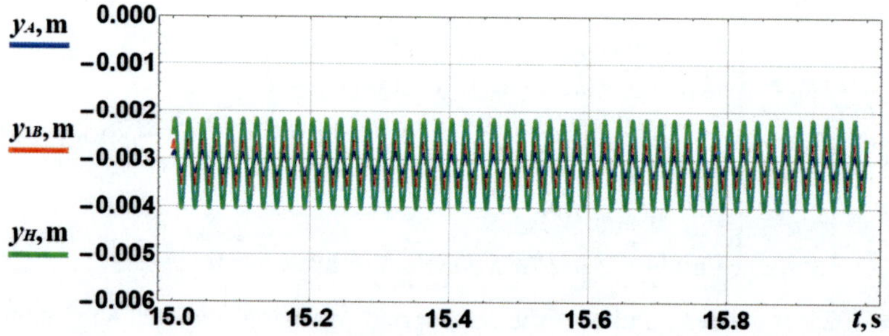

b

Fig. 3. Displacements of the compactor's representative points (Fig. 2) concerning their static equilibrium positions at the beginning (a) and the end (b) of the compaction process.

4.2 Analysis of the Compactor's Dynamic Characteristics

To analyze the action of the compacting (ramming) plate on the surface that is compacted, let us model the time dependence of the vertical (normal) force applied by the plate's front end to the surface (Fig. 4). The dynamic effect of the compaction force $F_{comp.}$

Depends on the compactor's front-end acceleration. Neglecting the inclination angle φ, the simplified expression of the compaction force can be derived as follows:

$$F_{comp.}(t) = (g - \ddot{y}_H) \cdot (m_B \cdot x_{1B} + m_C \cdot x_{1C} + m_E \cdot (x_{1D} + \rho \cdot \cos(\omega \cdot t)))/x_{1H}. \quad (6)$$

The initial value of the compaction force is equal to the reaction of the plate's front support in the state of rest ($t = 0$ s, $\ddot{y}_H = 0$ m/s^2): $F_{comp.}(0) = 461$ N. The maximal value exceeds 1000 N, whilst the smallest value is zero when the compactor's front end is under "jumping" conditions ($\ddot{y}_H > g$ and sign($\dot{y}_H$) > 0).

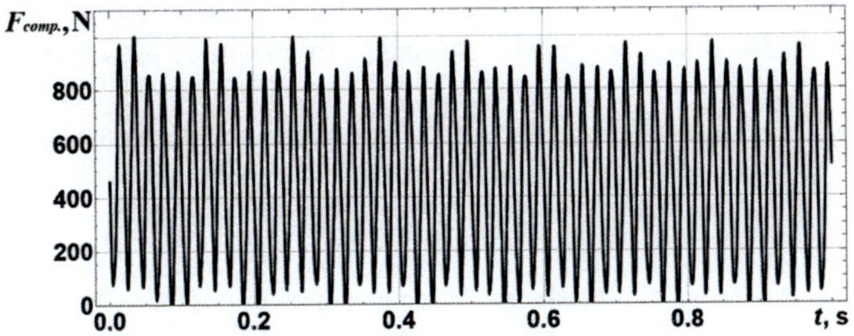

Fig. 4. The change of the compaction force in time.

The last stage of the carried-out investigations is devoted to analyzing the influence of the compacting plate vibrations on the control (operating) handle. As mentioned above, the control lever and handles are connected to the upper sprung platform and are used by the operator to push (pull) and to change the motion direction of the compactor. The absolute value of the force exerted on the operator through the control handles is equal to the product of the mass m_B of the upper platform and the acceleration $\ddot{y}_{1B}$: $F_{hand.} = m_B \cdot \ddot{y}_{1B}$. Considering the obtained results of the numerical modeling (Fig. 5), the maximal values of $F_{hand.}$ over 250 N are reached at $k = 1.25 \cdot 10^7$ N/m, while the smallest values of about 60 N – at $k = 10^6$ N/m.

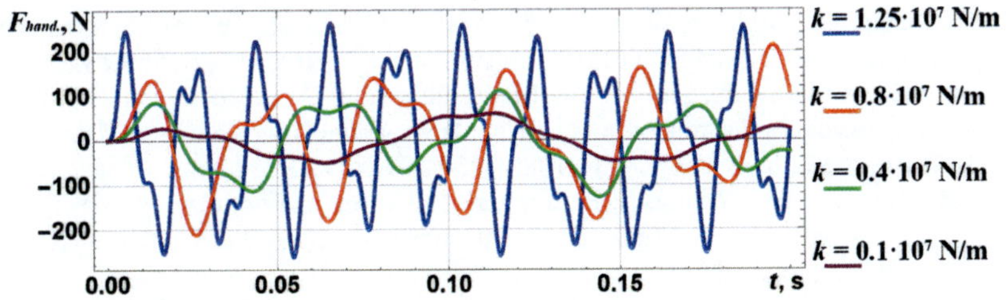

Fig. 5. Time dependencies of the forces acting on the control (operating) handles.

5 Conclusions

The design and operational peculiarities of the plate compactor (rammer) equipped with the inertial (centrifugal) vibration exciter (unbalanced rotor) are analyzed. The dynamic model of the compactor's oscillatory system is presented, and the corresponding equations describing the motions of the vibrating bodies are derived, taking into account the elastic-viscous-plastic properties of the compacted medium. Considering the compactor's front end, the numerical modeling results obtained in Mathematica software showed that the vibration amplitudes reach 6 mm at the beginning of the compaction process and 1 mm – at the end. The plastic deflection of the medium being compacted is about 3 mm. The largest value of the compaction force exceeds 1000 N.

Therefore, the present research's scientific novelty consists in improving the mathematical model of the plate compactor's vibration-driven locomotion system, taking into account the mechanical interaction between the compacting plate and the elastic-viscous-plastic surface compacted. The obtained results can be used by designers, technologists, and engineers while implementing new compacting technologies, improving and optimizing the existent vibratory compactors, rammers, screeds, etc.

References

1. Shi, Y., Liu, H., Wang, G.: Modeling of asphalt mixture-screed interaction: a nonlinear dynamic vibration model for improving paving density. Constr. Build. Mater. **311**, 125296 (2021). https://doi.org/10.1016/j.conbuildmat.2021.125296
2. Ivanchuk, Y.V.: Mathematical modeling for the technological process of surface soil compaction by the inertial vibratory rammer. Acad. J. Ser. Ind. Mach. Build. Civ. Eng. **1**(52), 21–29 (2019). https://doi.org/10.26906/znp.2019.52.1666
3. Morariu-Gligor, R.M.: Study of the influence of plate compactor weight on its functioning. Acta Technica Napoc. Ser. Appl. Math. Mech. Eng. **57**(2), 189–194 (2014)
4. Morariu-Gligor, R.M., Crişan, A.V., Şerdean, F.M.: Optimal design of an one-way plate compactor. Acta Technica Napoc. Ser. Appl. Math. Mech. Eng. **60**(4), 557–564 (2017)
5. Crişan, A.V., Morariu-Gligor, R.M.: A Study on the impact force in case of tamping rammers. Rom. J. Acoust. Vib. **16**(1), 78–83 (2019)
6. Prokopev, A.P., Nabizhanov, Z., Ivanchura, V.I., Emelyanov, R.T.: Modeling the non-destructive control of road surfaces compaction. J. Phys: Conf. Ser. **1399**(4), 044094 (2019). https://doi.org/10.1088/1742-6596/1399/4/044094
7. Emelyanov, R.T., Prokopev, A.P., Turysheva, E.S., Terekhova, I.I., Tkachenko, N.A.: Modeling of dynamic system "vibratory plate – soil" as an object quality control of compaction. J. Phys: Conf. Ser. **1399**(4), 044097 (2019). https://doi.org/10.1088/1742-6596/1399/4/044097
8. Sakhapov, R., Kadyrov, A., Gatiyatullin, M., Zemdikhanov, M.: Experimental studies of the working process of compaction of sand with a vibrating plate. IOP Conf. Ser. Mater. Sci. Eng. **890**, 012204 (2020). https://doi.org/10.1088/1757-899X/890/1/012204
9. Wang, P., Nguyen, V.L., Zhang, J.: Experimental research and optimal control of vibration screed system (VSS) based on fuzzy control. J. Vibroeng. **22**(6), 1415–1426 (2020). https://doi.org/10.21595/jve.2020.21560
10. Jiao, R., Nguyen, V., Zhang, J.: Analyzing the compacting energy and force distribution during the tamper-asphalt mixture interaction. J. Vibroeng. **23**(5), 1148–1159 (2021). https://doi.org/10.21595/jve.2021.21937
11. Metrikin, V.S., Igumnov, L.A., Nikiforova, I.V.: The dynamics of eccentric vibration mechanism (Part 1). J. Vibroeng. **19**(7), 4854–4865 (2017). https://doi.org/10.21595/jve.2017.18346
12. Zhan, X., Xu, J., Fang, H.: Planar locomotion of a vibration-driven system with two internal masses. Appl. Math. Model. **40**(2), 871–885 (2016). https://doi.org/10.1016/j.apm.2015.06.016

13. Du, Z., Fang, H., Zhan, X., Xu, J.: Experiments on vibration-driven stick-slip locomotion: a sliding bifurcation perspective. Mech. Syst. Signal Process. **105**, 261–275 (2018). https://doi.org/10.1016/j.ymssp.2017.12.001
14. Xu, J., Fang, H.: Improving performance: recent progress on vibration-driven locomotion systems. Nonlinear Dyn. **98**(4), 2651–2669 (2019). https://doi.org/10.1007/s11071-019-04982-y
15. Yan, Y., Liu, Y., Liao, M.: A comparative study of the vibro-impact capsule systems with one-sided and two-sided constraints. Nonlinear Dyn. **89**(2), 1063–1087 (2017). https://doi.org/10.1007/s11071-017-3500-7
16. Li, P., Jiang, Z.: Bifurcation analysis of stick-slip motion of the vibration-driven system with dry friction. Math. Probl. Eng. **2018**(3), 2305187 (2018). https://doi.org/10.1155/2018/2305187
17. Chernousko, F.L.: On the optimal motion of a body with an internal mass in a resistive medium. J. Vib. Control **14**(1–2), 197–208 (2008). https://doi.org/10.1177/1077546307079398
18. Bolotnik, N.N., Figurina, T.Yu., Chernous'ko, F.L.: Optimal control of the rectilinear motion of a two-body system in a resistive medium. J. Appl. Math. Mech. **76**(1), 1–14 (2012). https://doi.org/10.1016/j.jappmathmech.2012.03.001
19. Chernousko, F.L.: Locomotion of multibody robotic systems: dynamics and optimization. Theoret. Appl. Mech. **45**(1), 17–33 (2018). https://doi.org/10.2298/TAM171017001C
20. Bardin, B., Panev, A.: On dynamics of a rigid body moving on a horizontal plane by means of motion of an internal particle. Vibroeng. Procedia **8**, 135–141 (2016)
21. Bardin, B.S., Panev, A.S.: On the motion of a rigid body with an internal moving point mass on a horizontal plane. AIP Conf. Proc. **1959**(1), 030002 (2018). https://doi.org/10.1063/1.5034582
22. Nguyen, K.-T., La, N.-T., Ho, K.-T., Ngo, Q.-H., Chu, N.-H., Nguyen, V.-D.: The effect of friction on the vibro-impact locomotion system: modeling and dynamic response. Meccanica **56**(8), 2121–2137 (2021). https://doi.org/10.1007/s11012-021-01348-w
23. La, N.-T., Nguyen, T.-T., Nguyen, V.-D.: A comparative study on the two vibration driven locomotion systems in various friction levels. Vietnam J. Mech. **43**(2), 121–137 (2021). https://doi.org/10.15625/0866-7136/15662
24. Korendiy, V., Gursky, V., Kachur, O., Gurey, V., Havrylchenko, O., Kotsiumbas, O.: Mathematical modeling of forced oscillations of semidefinite vibro-impact system sliding along rough horizontal surface. Vibroeng. Procedia **39**, 164–169 (2021). https://doi.org/10.21595/vp.2021.22298
25. Lanets, O., Kachur, O., Korendiy, V., Lozynskyy, V.: Controllable crank mechanism for exciting oscillations of vibratory equipment. In: Ivanov, V., Pavlenko, I., Liaposhchenko, O., Machado, J., Edl, M. (eds.) DSMIE 2021. LNME, pp. 43–52. Springer, Cham (2021). https://doi.org/10.1007/978-3-030-77823-1_5

Check for updates

Analysis of CuZn5 Tube Buckling During Producing of the Crossover Bend for Metallurgical Unit

Volodymyr Kukhar[1](✉), Oleksandr Povazhnyi[1], and Oleksandr Grushko[2]

[1] Technical University "Metinvest Polytechnic" LLC, 71A, Sechenov Street, Mariupol 87524, Ukraine
kvv.mariupol@gmail.com

[2] Vinnytsia National Technical University, 95, Khmelnytsky Highway, Vinnytsia 21021, Ukraine

Abstract. Crossover bends in metallurgical units are used in pipelines for cooling systems and supplying liquids and gases under pressure. Because of the active placement environment, copper-based thick-walled tubular materials such as brass of the CuZn5 type are used. Due to the small need for crossover bends re- placing, their purchase is impractical, and low-tool methods should be considered for manufacturing. The expansion of the known method for bent axis parts obtained by buckling on the CuZn5 tubular blanks for the metallurgical unit's crossover bends production is proposed in this paper. The mathematical model for the stress-strain state determination at any point of a CuZn5 tubular blank under buckling has been developed. It is revealed that a Gaussian function is suited to describe the buckled tube shape for crossover bends. Collation of calculated and experimental results showed their satisfactory convergence: for change the blank axis length – 2.08%, for the deflection along the axis line – 15%, for the blank wall thickness – 6%. Dangerous zones have been identified where pipe wall folding can occur at a high upset reduction.

Keywords: Fitting · Brass · Buckling · Forming · Stress · Strain · Deflection · Product innovation

1 Introduction

Full crossover, bypass, and U-bends pipes of various designs are used as metallurgical equipment elements in industrial, automotive, and utility pipe systems for supplying liquids and gases. Materials, shape, production scale, requirements for geometry, and performance properties of the short bridge fitting or long crossover bend are considered when choosing a production method. Pipelines of equipment for transporting various substances under pressure, cooling systems for metallurgical units, and boilers are equipped with brass pipes and complete crossover bends (Fig. 1). Their use is justified in an active environment as corrosion-resistant parts [1, 2]. The manufacture of hollow bent parts is associated with complex die-tools and is justified only in the mass-scale production

V. Tonkonogyi et al. (Eds.): InterPartner 2022, LNME, pp. 444–454, 2023.
https://doi.org/10.1007/978-3-031-16651-8_42

of such bypasses. From this point of view, it is essential to develop low-cost dieless and low-tool methods for the bends production easily implemented in the immediate conditions of metallurgical shops.

2 Literature Review

The research regarding a bending process, which includes pushing a pipe through a channel of a flexible bending die located in the spherical bearing with a controlled trajectory of turns, is known [3, 4]. The disadvantage of this process includes the need for great force spent both for forming and overcoming friction in the die channel [5].

Moreover, the machine design is very complex, and its application is focused only on mass-scale production. As for the rotary draw bending [6], the implemented system of forces can lead to ovalization and destruction of the cross-section in the bending zone and longitudinal tube buckling. Therefore, it is necessary to carry out a complex preliminary adjustment of equipment and tools to ensure accurate centering of the feed device in the die hole [7]. Similar centering difficulties are encountered in producing pipes and hollow products using extrusion methods [8] that require uniform wall thickness and homogeneity of properties.

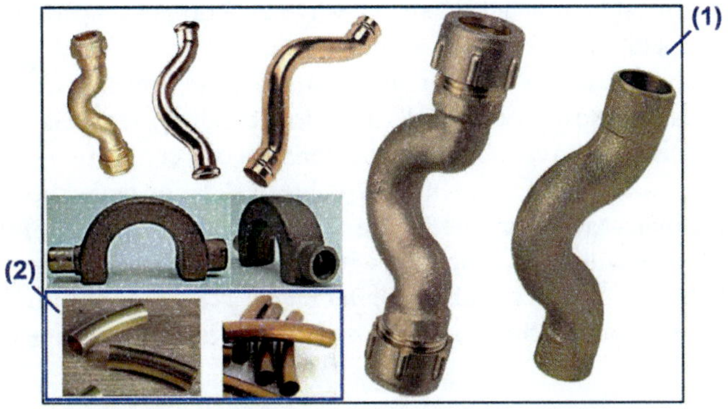

Fig. 1. Different designs of brass and Cu-based full crossover (1) and arc (2) bend pipes.

The work [9] showed production methods of hollow and tubular blanks using deep drawing, including a stepped (telescopic) punch. The resulting design of connection pipes with a variable diameter reduces hydraulic resistance in supply pipe systems. The large-sized pipes and hollow products of various cross-sections are produced by open die forging and dieless processes [10]. At the same time, the upsetting forging operation involves using only short blanks to exclude buckling during axial compression [11]. The paper [12] shows that the blanks' buckling phenomenon is turned into a positive solution for the case of dieless (impression-free) manufacturing of products and semi-finished with a bent axis. The cylindrical blanks with solid cross-sections were used here. Therefore, determining the application field for tube buckling in obtaining crossover bends shows promising scientific and practical interest. The initial information about the peculiarities of the tube deformation under critical axial compression is known due to the equipment operating experience [13]. It was revealed that the Euler buckling pattern deforms long (high) tubes and pipes under longitudinal compression [14]. It was found that the tube cross-section wrinkling during buckling is similar to one that occurs during pure [15] and compression bending [16]. In these cases, the stress-strain state is characterized by an increase in the size of the compression zone with a shift of the neutral line to the outer radius. This phenomenon has not been studied early for plastic buckling of tubular blanks.

The main difficulty is that most available studies aim to prevent pipe bending by predicting the critical force or strain for the tubular blank buckling [17]. In this case, only the pipe's geometric dimensions are considered. Zhen et al. [18] systematized and supplemented a significant amount of research to determine the critical strain for the beginning of pipe buckling. Studies for the supercritical region of pipe buckling, mainly aimed at ovalization research, are known [19]. They do not cover a comprehensive study of stress-strain state and form considering the plastic properties pattern changing for the blank material during buckling.

After considering the above operating conditions, the crossover bend pipes can be made of constructional, stainless, and low-carbon hardened steels [20, 21] and non-ferrous metals [22]. It should be noted that the strain-stress state and deformability of brass pipes are of most significant interest [23] (CuZn5 brass as a particular case). The mechanical and operational properties of pipe products are evaluated by standard tests [24]. And the study [25] shows the bending test application's results not only to identify strength properties but also to assess the deformation (bending), which is applicable to the tube buckling processes analysis. The rejection criteria of a connection pipe obtained by buckling include cracks, corrugations, unacceptable ovalization of the section, significant separation of the blank end from the flat die face, blank end crushing, folds, and local wall reduction. It was evident that buckling without distortion of the flow cross-section is possible only for relatively thick-walled tubular blanks. Thus, the existing technological limitations require a study of the tube stress-strain state to predict shape deformation.

3 Research Methodology

The pre-experiments showed that the implementation of the process requires the following conditions: $m_0 = L_0/D_0 = 3.3\ldots6.2$ and $S_0/D_0 \geq 0.17$, where L_0, D_0 and S_0 are the initial length (height), diameter, and thickness of the tube blank, respectively. The engineering strain (ε_x) was calculated as a relative decrease in height:

$$\varepsilon_x = (L_0 - H_f)/L_0), \tag{1}$$

where H_f – the final height after upsetting with buckling.

When analyzing the deformation mechanics, it was assumed that buckling applied the plane section hypothesis at the main stage of deformation. The following assumptions were used as limits: (i) the tube material is isotropic, hardening; (ii) the process is monotonous; (iii) the shell that is bending blank is considered to be momentless; (iv) corrugation, ovalization, crushing, and other geometric deviations were not taken into account, since they occur beyond the limiting upsetting degrees with buckling. The diagrams of CuZn5 stress-strain curves were formed as a result of tensile tests after annealing. The corresponding approximating dependence is as follows:

$$\sigma_i = \mathrm{A}e_i^n, \quad (2)$$

where σ_i – equivalent stress; e_i – equivalent strain; $A = 585$ MPa and $n = 0.196$ – the approximation coefficients for the stress-strain curve.

The balance of the elements of the outer (DE) and inner (CF) zones of the CDEF of the blank is considered (Fig. 2). Stresses σ_a and σ_θ of one sign: in the outer zone – tensile (Fig. 2b), in the inner zone – compressive (Fig. 2c). The elements have double curvature. The curvature in the cross section is determined by the radius of the middle surface of the pipe r, and in the meridional section – by the radius ρ of curvature of the axis, that is that is $R_1 = (\rho + r)$; $R_2 = (\rho - r)$.

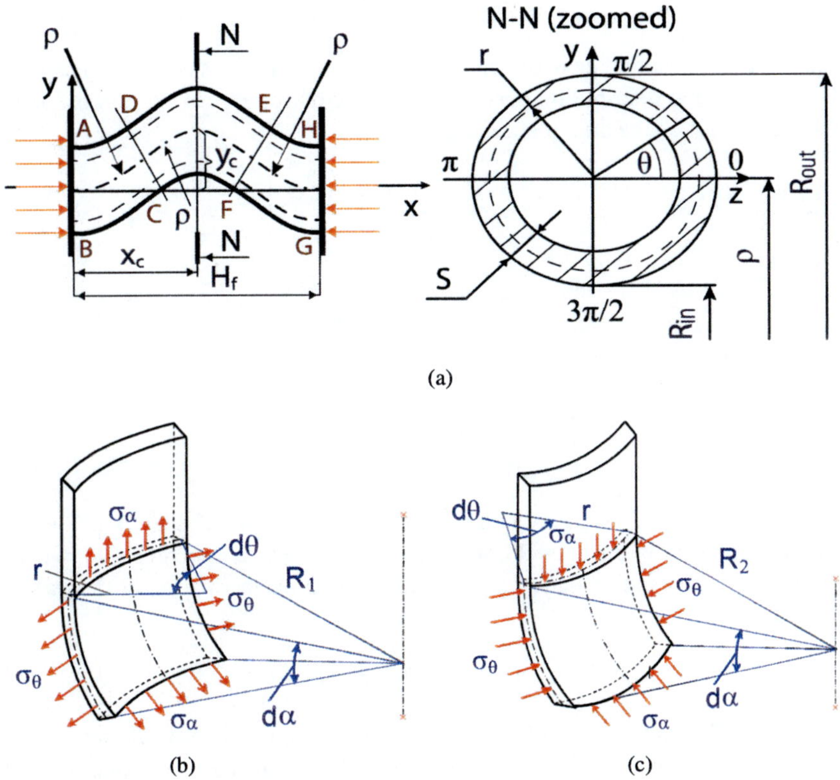

Fig. 2. The tube upset with buckling (a) and stresses in the outer (b) and internal (c) elements with double curvature: radius of curvature (ρ) and thickness (S) of the wall after deformation, internal R_{in} and outer R_{out} radii of curvature, $x_c = H_f/2$, deflection y_c along the curvature, circumferential radius r of the middle surface, solid angle θ.

The indicated radii are the principal radii of the shell. All the forces are projected onto the normal drawn to the middle surface of the element (Fig. 2b) and then calculated:

$$\sigma_a rSd\theta \sin d\alpha + \sigma_\theta R_1 Sd\alpha \sin d\theta = 0, \quad (3)$$

where S – the wall thickness of the deformed pipe, then:

$$\sigma_a/R_1 + \sigma_\theta/r = 0. \quad (4)$$

The action of the longitudinal compressive force in the cross-section of the blank causes the neutral layer to shift from the geometric axis of the shell by the amount of eccentricity $e = \rho\delta/(1 + \delta)$, here δ – the compressive deformation of the blank axis, which is hypothetically assumed to be uniform along the length. According to the plane section hypothesis, the longitudinal logarithmic deformations in a section with curvature (ρ + e) are calculated as follows:

$$e_a = \ln\frac{\rho + r\sin\theta}{\rho + e}. \quad (5)$$

The equations of circular arcs with a radius ρ or a function that is close in appearance to the Gaussian normal distribution curve can be selected for approximation of the blank axis that acquires a shape due to buckling:

$$y(x) = y_c \exp\left[-\left(\frac{x - x_c}{H_f w}\right)^2\right]. \quad (6)$$

Equation (6) was used for the solution, and the value $w = 0.319$ was obtained after processing the data of the preliminary experiments by the least squares method.

The curvature radius ρ at each point of the tube bent axis can be defined as

$$\frac{1}{\rho} = \frac{d^2y}{dx^2}/\left[1 + \left(\frac{dy}{dx}\right)^2\right]^{1.5}. \quad (7)$$

The parameter y_c (Fig. 2) is found under the condition that the change in the length L_f of the blank axis during upset buckling follows the pattern $L_f = L_0\ (1 - \delta)$, where the proposed empirical equation determines the compressive deformation of the axis in the first approximation:

$$\delta = k_\delta \cdot (1.716 - 0.263 \cdot m_0) \cdot \varepsilon_x. \quad (8)$$

While $k_\delta = 1\ldots2$, where values close to 1.0 should be applied for thick-walled blanks made of plastic materials by significant upsetting. Values close to 2.0 should be applied for thin-walled blanks made of less plastic materials at small ε_x. The following equation is used, applying the known parameters of the axial line function (6):

$$L_f = L_0(1 - \delta)\int_0^{H_f}\sqrt{1 + \left(\frac{dy}{dx}\right)^2}\,dx. \quad (9)$$

Solving Eqs. (9) for y_c, with taking into account Eq. (2):

$$e_a = \frac{3}{2}\frac{e_i}{\sigma_i}(\sigma_a - \sigma_0) = \frac{1}{2}\frac{e_i}{\sigma_i}(2\sigma_a - \sigma_\theta), \tag{10}$$

where σ_0 – the average normal stress.

Equivalent stress for the studied case:

$$\sigma_i = \sqrt{\sigma_a^2 + \sigma_\theta^2 - \sigma_a\sigma_\theta}. \tag{11}$$

Taking into account the power approximation (2) and Eqs. (4), (5), (10), and (11), after transformations, the following is obtained:

$$\sigma_a = Ae_\alpha^n 2^n\left(2 + \frac{r}{\rho + r\sin\theta}\right)^{-n}\left[1 + \frac{r}{\rho + r\sin\theta} + \left(\frac{r}{\rho + r\sin\theta}\right)^2\right]^{0.5n-0.5}. \tag{12}$$

The second principal stress is obtained from Eq. (4):

$$\sigma_\theta = -\sigma_a\frac{r}{\rho + r\sin\theta}. \tag{13}$$

The incompressibility condition $e_a + e_\theta + e_s = 0$ and equivalent plastic strain equation:

$$e_i = \frac{\sqrt{2}}{3}\sqrt{(e_\alpha - e_\theta)^2 + (e_\theta - e_s)^2 + (e_\alpha - e_s)^2} = \frac{\sqrt{2}}{3}\sqrt{e_\alpha^2 + e_\theta^2 + e_\alpha e_\theta}. \tag{14}$$

were used to calculate the rest components of the strain tensor.

From the power-law dependence (2) and expression (11), the equation transcendental to e_θ was compiled and solved in the Mathcad software:

$$A\left(\frac{\sqrt{2}}{3}\sqrt{e_\alpha^2 + e_\theta^2 + e_\alpha e_\theta}\right)^n - \sigma_i = 0. \tag{15}$$

The third strain component was calculated via incompressibility:

$$e_s = -e_\alpha - e_\theta. \tag{16}$$

The relative change in the pipe wall thickness was found as:

$$S/S_0 = \exp(e_s). \tag{17}$$

4 Results and Discussion

CuZn5 specimen with initial dimensions $D_0 = 14$ mm, $L_0 = 73$ mm, $S_0 = 2.5$ mm, i.e. $m_0 = 5.2$ and $S_0/D_0 = 0.18$, were upset to check the stress and strain fields calculations. The dimension is $H_f = 64$ mm, i.e. $_x = 0.137$ (Fig. 3). The measured length of the axis line is $L_f = 66$ mm, i.e. shortening of the blank axis is $\delta = 0.096$, and calculated according to the Eq. (8) is $\delta = 0.094$ (i.e. error is 2.08%).

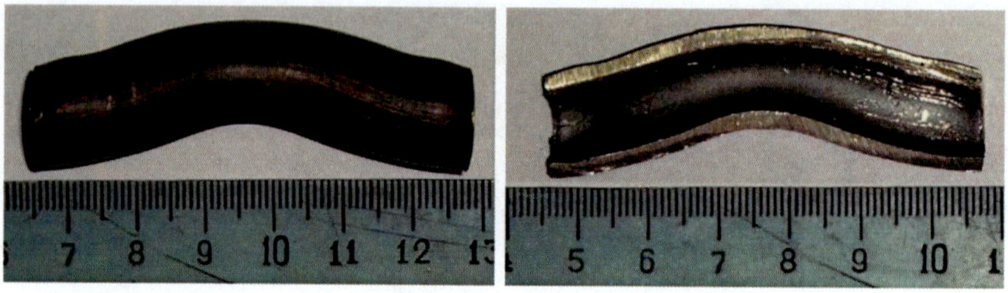

(a) (b)

Fig. 3. One-piece (a) and longitudinally sawn (b) CuZn5 tubular blank after buckling.

The values of the functions of the axial line $y(x)$, the principal radii $y(x) + r$, $y(x) - r$ and the neutral line $y(x) + e(x)$ of the blank were found (shown in Fig. 4) using expressions (6)–(9). It was found that the calculations adequately describe the shape change: the difference between the calculated value of the deflection $y_{c(calc)} = 9.78$ mm from the experimental value $y_{c(exp)} = 8.5$ mm was 15%.

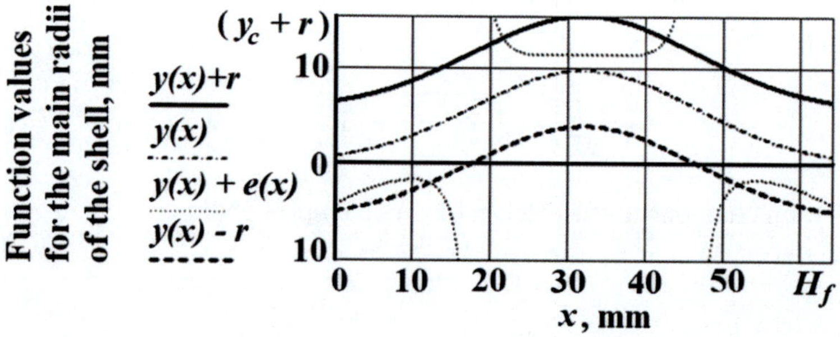

Fig. 4. Results of analytical constructions by Eqs. (6)–(9) for CuZn5 blank.

The developed mathematical apparatus makes it possible to determine the stress-strain state at any blank's point with coordinates (x, θ). In Fig. 5, the stress fields have been built along the height of the specimen for its outer ($\theta = \pi/2$) and inner ($\theta = 3\pi/2$) sides. The strain fields and the thickness change on the specimen's outer and inner sides are shown in Fig. 6 and Fig. 7, respectively.

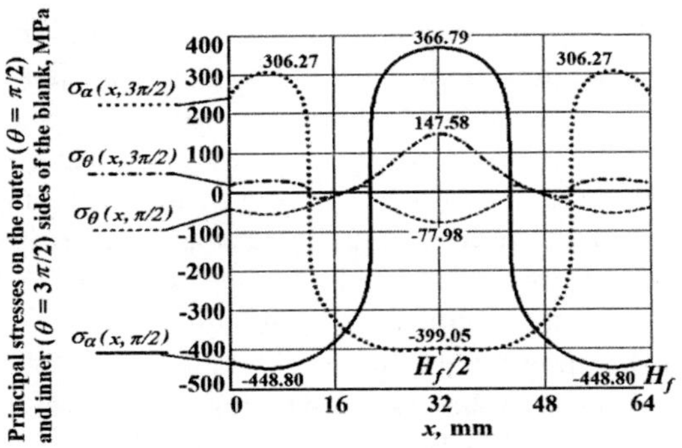

Fig. 5. Stress fields in upset buckled CuZn5 tubular blank.

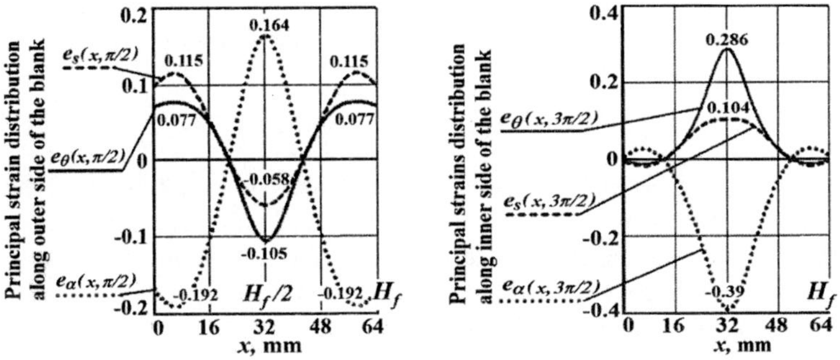

(a) (b)

Fig. 6. Strain fields in the outer (a) and inner (b) sides of buckled CuZn5 tubular blank.

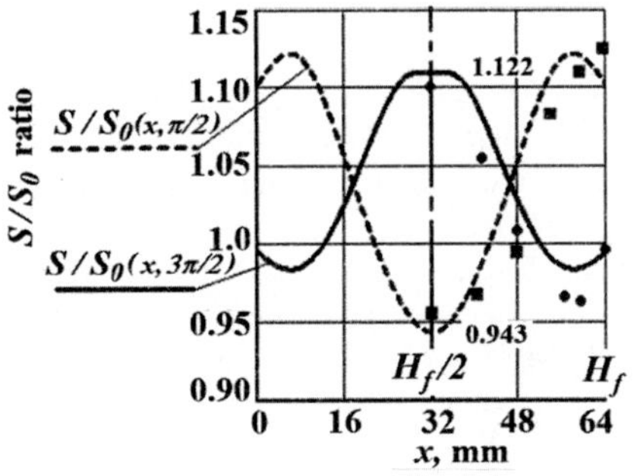

Fig. 7. Calculated and experimental S/S_0 ratio of buckled CuZn5 tubular blank.

The stress-strain state of the blank is inhomogeneous and in the region $x = L/2 = H_f/2$ is characterized by the maximum values of tensile stresses along the outer side and maximum compressive stresses along the inner side. The picture changed to the opposite when approaching the ends and near the coordinates $x \approx H_f/4$ and $x = 3\ H_f/4$ there are areas where only axial stresses act ($\sigma_\theta = 0$). Longitudinal and tangential strains take extreme values in the region $x = H_f/2$, and high absolute values in the inner zone and the outer zone in the frontal areas ($x \approx 0 \ldots H_f/8$ and $x \approx 7\ H_f/8 \ldots H_f$), can lead to the primary folding of the shell here during further upsetting.

The S/S_0 ratio of the upset tubular blank was calculated (Fig. 7) according to the Eq. (17). The results of the wall thickness S experimental measurements after sawing the specimen (Fig. 3b) were converted into ratio values and are shown in Fig. 7 only for the right side of the blank due to its symmetry. The most significant discrepancy between the experimental data and the calculated ones (towards the overestimation of the latter) is 6% for the outer regions with coordinates S/S_0 ($x = H_f/4$, $\theta = 3\pi/2$) and S/S_0 ($x = 3\ H_f/4$, $\theta = 3\pi/2$). Strains in the zones ($x = H_f/2$, $\theta = 3\pi/2$), ($x \approx 0 \ldots H_f/8$ and $x \approx 7\ H_f/8 \ldots H_f$, $\theta = \pi/2$) can lead to folding when the engineering strain ε_x exceeds a specific critical value that should be determined in prospect.

5 Conclusions

The tube buckling application has substantiated a new low-instrumental method for small-scale production of crossover bent for metallurgical units. The mathematical model for the stress-strain state determination at any point of a CuZn5 tubular blank, upset with buckling, has been developed based on the momentless theory and the shells theory. It is shown that a function close to the Gaussian normal distribution curve can be used with satisfactory accuracy to analyze the forming by buckling of full crossover bend pipes. Comparison of calculated and experimental results showed their satisfactory convergence: for the relative shortening (δ compressive deformation of the blank axis) – 2.08%, for the deflection along the axis line (y_c) – 15%, for the blank wall thickness (S) – 6%. It was found that strains in zones ($x = H_f/2$, $\theta = 3\pi/2$), ($x \approx 0\ldots H_f/8$ and $x \approx 7\ H_f/8\ldots H_f$, $\theta = \pi/2$) can lead to folding at large values of upsetting strain ε_x.

References

1. Callcut, V.: Introduction to Brasses (Part II). Copper Applications in Metallurgy of Copper & Copper Alloys (2000). https://www.copper.org/publications/newsletters/innovations/2000/01/brasses02.html. Accessed 27 Dec 2021
2. Kimstach, T.V., Uzlov, K.I., Repiakh, S.I., Solonenko, L.I.: Analysis of different environments influence on copper alloys corrosion resistance. Phys. Metall. Heat Treat. Met. **3**(94), 36–45 (2021). https://doi.org/10.30838/J.PMHTM.2413.010721.36.780
3. Guo, X., et al.: Simulation and experimental research of the free bending process of a spatial tube. J. Mater. Process. Technol. **255**, 137–149 (2018). https://doi.org/10.1016/j.jmatprotec.2017.11.062
4. Murata, M., Kuboki, T.: CNC tube forming method for manufacturing flexibly and 3-dimensionally bent tubes. In: Tekkaya, A.E., Homberg, W., Brosius, A. (eds.) 60 Excellent Inventions in Metal Forming, pp. 363–368. Springer, Heidelberg (2015). https://doi.org/10.1007/978-3-662-46312-3_56

5. Chen, H., et al.: Impact of bending dies with different friction forms on forming force and quality of tubes manufactured by free bending technology. Chin. J. Aeronaut. **34**(4), 253–264 (2021). https://doi.org/10.1016/j.cja.2020.08.029
6. Safdarian, R.: Investigation of tube fracture in the rotary draw bending process using experimental and numerical methods. Int.J. Mater. Form. **13**(4), 493–516 (2019). https://doi.org/10.1007/s12289-019-01484-5
7. Bostan, I., Mazuru, S., Casian, M.: Method of axial adjustment for precessional transmissions. MATEC Web Conf. **178**, 06024 (2018). https://doi.org/10.1051/matecconf/201817806024
8. Aliieva, L., Hrudkina, N., Aliiev, I., Zhbankov, I., Markov, O.: Effect of the tool geometry on the force mode of the combined radial-direct extrusion with compression. East. Eur. J. Enterp. Technol. **2**(1(104)), 15–22 (2020). https://doi.org/10.15587/1729-4061.2020.198433
9. Arhat, R., Puzyr, R., Shchetynin, V., Moroz, M.: The manufacture of cylindrical parts by drawing using a telescopic punch. In: Tonkonogyi, V., et al. (eds.) InterPartner 2020. LNME, pp. 363–372. Springer, Cham (2021). https://doi.org/10.1007/978-3-030-68014-5_36
10. Kukhar, V.V.: Producing of elongated forgings with sharpened end by rupture with local heating of the workpiece method. Metall. Min. Ind. **6**, 122–132 (2015)
11. Miyasaka, K.: Dieless tube-bending using high frequency induction heating. J. Japan Soc. Technol. Plast. **51**(591), 272–276 (2010). https://doi.org/10.9773/sosei.51.272
12. Kukhar, V., Burko, V., Prysiazhnyi, A., Balalayeva, E., Nahnibeda, M.: Development of alternative technology of dual forming of profiled workpiece obtained by buckling. East. Eur. J. Enterp. Technol. **3**(7(81)), 53–61 (2016). https://doi.org/10.15587/1729-4061.2016.72063
13. Florescu, V., Mocanu, S., Rece, L., Motounu, D.C., Gherghina, A., Burlacu, A.: Design contributions to the elaboration of new modeling schemes for the buckling assessment of hydraulic actuators. Metals **10**(9), 1143 (2020). https://doi.org/10.3390/met10091143
14. Ahn, K., Lim, I.-G., Yoon, J., Huh, H.: A simplified prediction method for the local buckling load of cylindrical tubes. Int. J. Precis. Eng. Manuf. **17**(9), 1149–1156 (2016). https://doi.org/10.1007/s12541-016-0139-0
15. Yudo, H., Yoshikawa, T.: Buckling phenomenon for imperfect pipe under pure bending. J. Mar. Sci. Technol. **20**(4), 703–710 (2015). https://doi.org/10.1007/s00773-015-0324-3
16. Jiang, Z.C., Qu, W.L.: Buckling analysis of the tube compression-bending member in elastic-plastic state with ANSYS. Adv. Mater. Res. **327**, 143–148 (2011). https://doi.org/10.4028/www.scientific.net/amr.327.143
17. Ji, L.K., et al.: An estimation of critical buckling strain for pipe subjected plastic bending. Cent. Eur. J. Eng. **4**(3), 326–333 (2014). https://doi.org/10.2478/s13531-013-0168-8
18. Zheng, M., Hu, J., Teng, H.P., Zhao, Y.: On the evaluation of plastic buckling of pipeline bending. Int. Rev. Appl. Sci. Eng. **8**(1), 25–35 (2017). https://doi.org/10.1556/1848.2017.8.1.5
19. Ji, L.K., et al.: Apparent strain of a pipe at plastic bending buckling state. J. Braz. Soc. Mech. Sci. Eng. **37**(6), 1811–1818 (2015). https://doi.org/10.1007/s40430-014-0302-4
20. Tarelnyk, V., et al.: New sulphiding method for steel and cast iron parts. IOP Conf. Ser. Mater. Sci. Eng. **233**, 012049 (2017). https://doi.org/10.1088/1757-899X/233/1/012049
21. Zurnadzhy, V.I., et al.: Mechanical properties of carbide-free lower bainite in complex-alloyed constructional steel: effect of bainitizing treatment parameters. Kovove Materialy **58**(2), 129–140 (2020). https://doi.org/10.4149/km_2020_2_129
22. Dragobetskii, V., Shapoval, A., Naumova, E., Shlyk, S., Mospan, D., Sikulskiy, V.: The technology of production of a copper-aluminum-copper composite to produce current lead buses of the high-voltage plants. In: 2017 Modern Electrical and Energy Systems, Kremenchuk, pp. 400–403. IEEE (2018). https://doi.org/10.1109/mees.2017.8248944
23. Smith, E., Wilding, M.J.: The effect of non-uniform bending deformation on the stability of circumferential growth of through-wall cracks in brass tubes. J. Mech. Phys. Solids **31**(3), 223–229 (1983). https://doi.org/10.1016/0022-5096(83)90023-6

24. Chen, M.T., Young, B.: Tensile tests of cold-formed stainless steel tubes. J. Struct. Eng. **146**(9), 04020165 (2020). https://doi.org/10.1061/(ASCE)ST.1943-541X.0002738
25. Nedelcu, D., Santo, L., Santos, A.G., Mazurchevici, S.P.: Mechanical behaviour evaluation of arboform material samples by bending deflection test. Materiale Plastice **52**(4), 423–426 (2015)

Stabilization of Natural Frequency Oscillation Equipment When Changing Its Weight

Victor Kurgan(✉), Ihor Sydorenko, Liubov Bovnegra, Andrii Pavlyshko, and Kateryna Kirkopulo

Odessa Polytechnic National University, 1, Shevchenko Avenue, Odessa 65044, Ukraine
kurgan@op.edu.ua

Abstract. Changes in mass in the operating cycle of technological equipment lead to changes in its natural frequency. A decreased mass of technological equipment increases the frequency of its free oscillations, the steady state of which is due to operating frequencies close to resonant. A series of passive vibration-isolating supports is considered, which solve the problem of "equal-frequency" vibration isolation at small oscillation amplitudes and high excitation frequencies. Based on the research, a vibration isolation system was developed, which ensured the stabilization of the natural frequency of oscillations in a certain range of amplitudes and low frequencies during the full technological cycle of the hopper-dryer of the SCM 200 technological machine. An elastic module of a passive vibration isolating device with mechanical feedback is proposed, and its geometrical parameters are determined. The control function of the elastic characteristic is calculated and presented in the form of a power polynomial. The possibility of using the cam mechanism as mechanical feedback has been tested. A new solution for vibration isolation of the SCM 200 technological machine is offered. The efficiency of vibration isolation systems was evaluated, which testifies to the greater efficiency of the created vibration isolation system against the background of its significant simplification.

Keywords: Vibration device · Nonlinear characteristics · Mechanical feedback · Actual stiffness · Multiplicity factor · Industrial growth

1 Introduction

The most dangerous operational mode of technological equipment, which leads to a sharp increase in dynamic load, is oscillating processes in the range closed to almost resonant or resonant frequencies. There is a large number of technological equipment, for which the technological cycle determines the operation in a wide range of excitation frequencies and the change in its own mass. This applies to lifting and transporting equipment, vehicles, and other machines, where the change in gross weight differs from the net weight several times and sometimes by order of magnitude. To prevent such an effect, a system was proposed for the first time to stabilize the natural frequency of oscillations in a certain range of amplitudes and frequencies [1]. It is advisable to develop a passive vibration isolation device with mechanical feedback.

V. Tonkonogyi et al. (Eds.): InterPartner 2022, LNME, pp. 455–464, 2023.
https://doi.org/10.1007/978-3-031-16651-8_43

2 Literature Review

Three groups of mechanical metastructures are considered: effective negative mass, effective negative stiffness, and negative Poison's ratio [2]. Experimental data show that the system's resonance frequency decreases when the friction force increases [3]. Design and simulation of the hysteresis of a new shape memory alloy (SMA) damper to mitigate structural vibrations [4]. The well-known systems with quasi-zero stiffness are observed, main advantages and disadvantages are discussed. The problem of determining the dynamic properties of large structures, in which the structure response is measured under external excitation, is associated with changes in the added mass and temperature [5]. Considered S&V ("shock and vibration"), corner supports are used to seal the transmitted sound vibration of various mechanisms and protect equipment from underwater explosions [6]. The vibration control of induction motors with sleeve bearings—mounted on soft steel frame foundations – using active motor foot mounts is analyzed [7]. Double isolation of rotating machines is considered, consisting of isolation of hull structures from machine vibrations and protection of machines during an earthquake to maintain their performance [8, 9]. Force-displacement loops under harmonic excitation with different frequencies and amplitude are tested and analyzed [10]. The effects of the piecewise stiffness and the gap on the dynamic response of the primary system and the NES are considered in forced vibration [11]. The design [12] of elastic coupling [13] with nonlinear mechanical feedback is considered [14]. Random oscillations of a vibration-resistant body on vibration dampers with straightened surfaces are studied [15]. It is proposed to use a corrective action block to compensate for the nonlinear effects of the feedback loop [17]. The analysis of existing methods and approaches for determining the special provisions of the mechanisms is carried out [18]. Separate issues of solving direct and inverse kinematics problems, determining the working zone and solving the problem of trajectory control of the mechanism during combined motion are considered [19]. The dynamic behavior of a platform-vibrator with shock is studied [20].

3 Research Methodology

If we consider the oscillations of the process equipment in one plane and estimate their intensity relative to the vertical axis with the common coordinate z, then the change in the mass of the equipment m leads to a corresponding change in the frequency of its free oscillations

$$\omega_{0z} = \frac{1}{2\pi}\sqrt{\frac{c_{ac}}{m}} \tag{1}$$

where c_{ac} – actual elastic support's stiffness of technological equipment or its suspension.

Expression (1) shows that the decrease in the mass of technological equipment leads to the frequency increase of its own oscillations, which can be dangerous for equipment whose steady state is due to operating frequencies close to resonant. That is why one of the important practical tasks of vibration isolation of technological machines is stabilizing the natural frequency of oscillations when changing their mass, called "equal frequency" vibration isolation.

The conclusions from work on vibration isolation of "equal frequency" indicate that the condition $\omega_{0z} = \text{const}$ can be ensured in the case when the actual stiffness of elastic supports or suspension of technological equipment c_{ac} is interconnected with its mass m, and there is a possibility of reproducing the necessary change $c_{ac} \sim m$. [9]. In works devoted to mathematical modeling of such systems, it is established that to stabilize the natural frequency at a certain level ω_{0z} – the main function of the actual stiffness of the elastic supports or suspension should be

$$c_{ac}(z) = F/e^{\frac{z}{a}} \tag{2}$$

where F – amplitude of external load; $a = g/\omega_{0z}^2$ – determined by design or technological considerations constant value; g – acceleration of gravity.

At present, the problem of "equal-frequency" vibration isolation is solved only at small oscillation amplitudes ($A \in [1; 15]$ mm) and high perturbation frequencies ($\omega_0 >$ 9 Hz). To do this, use serial passive vibration isolating devices in the form of vibration isolating supports type SV (Fig. 1a).

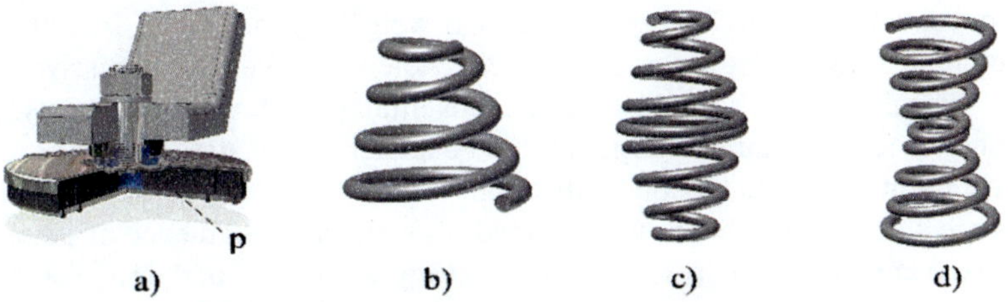

Fig. 1. Means of "equal frequency" vibration isolation.

The principle of operation of such passive vibration isolating device is based on the volumetric incompressibility of rubber, which is made from an elastic element of complex shape p [12]. According to the results of experimental studies conducted with vibration-insulating supports of the SV type, it is established that reproduction with their help with the accepted accuracy of the target elastic characteristic of actual stiffness corresponding to expression (2) is possible only in a certain range of compressive loads. For example, the vibration-insulating support SV – 31, type 2, can reproduce the target elastic characteristic with the accepted suitable accuracy $\Delta \pm 4$ Hz in the compressive load range, $F \in [630; 1010]$ N (Fig. 2).

The range of loads in which the passive vibration isolating device reproduces the target nonlinear elastic characteristic determines one of the essential indicators of its vibration-isolating properties – the multiplicity coefficient k_m. Therefore, analyzing the above graph of the elastic characteristics of the support SV, size 2, it was found that the coefficient of its multiplicity $k_m = 1{,}34$. Data analysis on other standard sizes of SV supports and supports that are structurally close to them allowed us to establish that their multiplicity coefficient can be $k_m \in [1{,}03; 1{,}5]$.

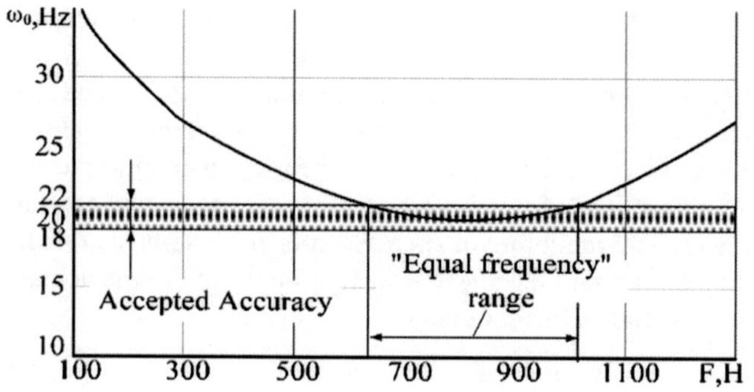

Fig. 2. The graph of the change in the natural frequency of oscillations SV supports of standard size 2 ($\omega_0 \in [18; 20]$ Hz).

The analysis terms of use of known technological equipment, for which the change of its mass is an integral part of the technological cycle, showed that passive vibration isolating devices with rubber elastic element having $k_m < 1{,}5$ allow to solve problems of "equal frequency" vibration isolation for the vast majority of metalworking machines. This is because metalworking machines, which usually operate at low amplitudes of high-frequency oscillations, rarely change their weight by more than 10…30% during the technological cycle. However, it should be noted that the use of passive vibration isolating devices with a rubber elastic element is limited by the presence of a significant disadvantage - reproducing the target elastic characteristic in static tests, they may not reproduce a similar characteristic in dynamic tests. This is due to the dependence of the dynamic stiffness of rubber on static load, as well as the influence of other factors, among which are dominated by temperature and aggressiveness of the environment [13].

The above shows that the multiplicity coefficient $k_m < 1{,}5$ is clearly too small to solve the "equal frequency" vibration isolation problem for lifting and transport equipment, forging and pressing equipment, vehicles, etc. If we consider that the characteristic feature of such equipment is operating at significant amplitudes of low-frequency oscillations – the use of passive vibration isolating devices with a rubber elastic element is not possible.

The analysis of works devoted to solving the problems of "equal frequency" vibration isolation at oscillation amplitudes $A > 40$ mm and low frequencies $\omega_{0z} < 2$ Hz using known passive vibration isolation devices with multiplicity $k_m > 1{,}5$ allowed to establish that almost all of them are built on non-cylindrical twisted springs (Fig. 1, b) and implement elastic characteristics close to the target with a multiplicity factor $k_m \in [1{,}5; 2{,}5]$. Currently, the Experimental Research Institute of Metal-Cutting Machines (Moscow) has developed and announced passive vibration isolating devices based on special non-cylindrical twisted springs (Fig. 1, types c and d), which provide a multiplicity factor of $k_m \in [3; 4]$ at amplitudes of oscillations $A \in [30; 180]$ mm. But given the high complexity and cost of technological preparation for production and the production of such springs - these passive vibration isolating devices are not widely used.

Thus, developing and implementing new passive vibration isolating devices in Ukraine that provide "equal frequency" vibration isolation with high multiplicity at significant amplitudes of low-frequency oscillations is an urgent scientific and practical task.

4 Results and Discussion

The development of methods for synthesizing, designing, and calculating passive vibration isolating devices with mechanical feedback [14] led to the creation of designs of fundamentally new passive vibration isolating devices with comprehensive functionality. Satisfactory tests of some samples of passive vibration isolating devices with mechanical feedback have determined the possibility of their industrial application in solving existing vibration isolation problems, including "equal frequency". As an example, based on the research, a vibration isolation system was developed that stabilized the natural frequency of oscillations in a certain range of amplitudes and low frequencies during the full technological cycle of the hopper-dryer of the SCM 200 process machine.

The hopper-dryer represents a design from thin-walled elements in which steam under pressure circulates. It is used as part of technological lines for processing post-alcohol bards into feed flour and food additives or raw meat into bone meal and is mass-produced by some Ukrainian enterprises (Fig. 3a).

At the beginning of the technological cycle, the gross weight of the hopper is 1500 kg, and at the end of the cycle, due to the evaporation of the liquid – 420 kg. Normal operation of loading and unloading devices of the hopper is due to the frequency of the oscillation generator $\omega_{0z} \in [0,7; 1,1]$ Hz, which is a conveying auger located inside the hopper. Based on the regulatory operating conditions in the range of pre-resonant frequencies ($\omega_{0z} < 0,8\omega$), vibration isolation is subject to requirements for stabilization of the natural frequency of the system $\omega_{0z} \in [0,4; 0,6]$ Hz in the range of amplitudes $A \in [10, 50]$ mm. When the hopper mass m changes 3.5 times during the technological cycle (i.e., the multiplicity of the vibration isolation system must be $k_m > 3.5$), the natural frequency of the system (1) must change 1.8 times. This fact leads to the operation of the system at resonant or close to resonant frequencies (Fig. 3b). This operation of the hopper is unacceptable, as it causes fluctuations beyond the strength and stability of the structure amplitudes $A > 80$ mm.

The SCM 200 hopper-dryer is usually installed on an open technological site, which causes a significant change in ambient temperatures, and the amplitude of its oscillations at a steady state is $A > 20$ mm. In such conditions, using passive vibration isolating devices with rubber elastic elements is impossible, and known passive vibration isolating devices based on non-cylindrical twisted springs have insufficient multiplicity.

That is why the vibration isolation of modern designs of the hopper-dryer, manufactured in series, is carried out comprehensively using an active vibration isolation system (AVS). It consists of six passive vibration isolation devices based on twisted cylindrical springs ($c_0 = 400$ N/mm), located under the frame of the hopper, and the automated control system – frequency control of the oscillation generator (ACS), which means the DC drive motor.

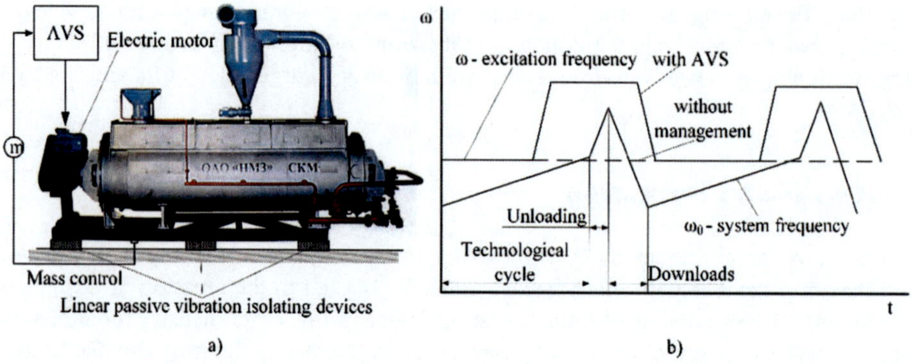

Fig. 3. Vibration isolation of SCM 200 in serial execution: and - an arrangement of elements of system of vibration isolation; b - frequency characteristics.

In practice, the vibration isolation system has a number of significant disadvantages, including impossible to start the drive with an empty hopper, due to the limited oscillations at close to resonant frequencies, so the drive is started when the hopper is full, which determines the starting mode at maximum loads; high cost of components, manufacture, and maintenance of AVS; multi-element and heterogeneity of the components of the AVS, which determine its lack of reliability.

Thus, based on these shortcomings, the problem was formulated to eliminate them and significantly simplify the existing vibration isolation system without loss of efficiency or with possible improvement.

Experience in installing and operating this equipment has shown that the implementation of the elastic connection between the hopper frame and the bearing surfaces of the assembly site leads to uneven loading of the passive vibration isolating device, making it very difficult to implement the calculated consolidated stiffness. The detected unevenness of the load is usually associated with a relatively low construction accuracy of the installation level when performing support platforms on open technological sites. Therefore, it was proposed to implement an elastic connection in the form of a hook directly between the hopper and the frame because the frame, due to the large contact surface, is less sensitive to errors in the reference level.

The basis of the proposed technical solution is an elastic module developed based on one of the synthesized and tested prototypes of a passive vibration isolating device with mechanical feedback [16]. The calculation of the basic geometric parameters of the elastic module and the strength of its elements are carried out provided that it is a statically defined system and the hopper-dryer SCM 200 is evenly filled with the maximum amount of raw materials. These conditions allowed calculating the required number of elastic modules taking into account the design features, which are the presence of two elastic elements in the industrial design instead of one in the prototype. The further design was reduced to the development of elastic suspension modules (Fig. 4a) and the completion of the serial frame of the hopper-dryer under the proposed suspension (Fig. 4b).

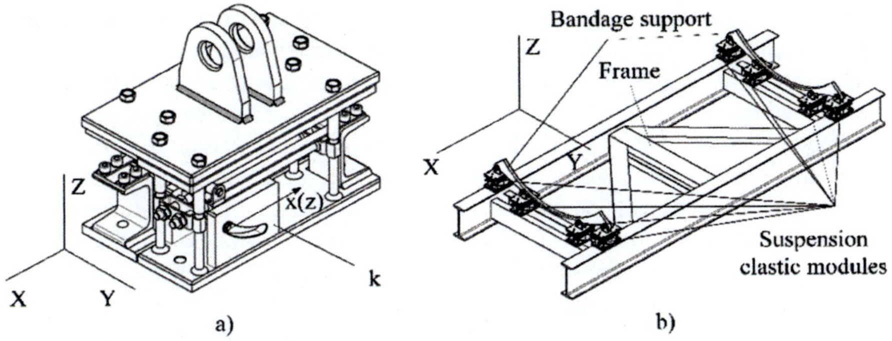

Fig. 4. Elements of SCM 200 vibration isolation system: a – elastic module; b – suspension.

The calculations of mechanical feedback of the industrial design used dependencies which describe the characteristics of the reducing elastic force of the prototype

$$F(z) = \frac{3EJ_z}{[l_{in} - x(z)]^3}, \tag{3}$$

where E – the modulus of elasticity of the second kind of material of the elastic element; J_z – the moment of inertia of the section of the elastic element; l_{in} – the initial distance from the point of application of the load on the elastic element to the place of its cantilever fixing in the device; x(z) – control function implemented by mechanical feedback in the form of horizontal movement of the point of application of the load to the elastic element.

Based on expressions (2) and (3), the required control function is calculated in the range $z \in [0; z_{max}]$. Given that the control function is implemented with sufficient accuracy by mechanical feedback in the form of a cam mechanism with a roller pusher k (Fig. 4, a), it was transformed into a step polynomial of the best approximation

$$x(z)|_1 = 4 \cdot 10^{-7}z^6 - 5 \cdot 10^{-5}z^5 + 0,0018z^4 - 0,258z^2 + 0,041z^2 + 1,7927z + 4 \cdot 10^{-8}. \tag{4}$$

The expediency of this transformation is because, in this form, the control function is a function of the coordinate marking of the cam profile, which was used in its manufacture using a CNC machine.

The magnitude and direction of the load acting on the components of the cam mechanism impose some restrictions on its geometric parameters. In this regard, the possibility of using the cam mechanism as mechanical feedback was tested. The constraint check for the synthesized cam profile was performed on the critical value of the pressure angle

$$\Theta_{cr} = \operatorname{arctg}(\frac{1}{f_{ac}}) - \varphi_{ac}, \tag{5}$$

where f_{ac} – actual coefficient of friction; φ_{ac} – the actual angle of friction of the roller on the axis of the pusher and between the roller and the cam.

Examination of the synthesized cam profile did not reveal any restrictions on its use as an element of mechanical feedback. Further design, manufacture and assembly of vibration isolation system, as well as its further testing and refinement of individual components allowed to offer the industry a new solution for vibration isolation technological machine SCM 200 (Fig. 5, a), which in comparison with serial is characterized by the following differences: the vibration isolation system is formed by means of eight elastic modules developed by the method of synthesis and calculations of passive vibration isolating devices with mechanical feedback, which form a suspension between the frame and the hopper; the characteristics of the elastic connection of the developed suspension determine the frequency of natural oscillations of the system $\omega_0 \in [0,43; 0,58]$ Hz and amplitude of oscillations [10, 80] mm in the whole range of changes in the mass of the hopper, which allowed to abandon the use of AVS (Fig. 5, b); instead of a DC electric motor, a cheaper asynchronous electric motor is used, which has significantly smaller dimensions and weight; during the installation of the suspension it is found that the accuracy of the location relative to the mounting level of the elastic modules can be significantly reduced, because the unevenness of their load does not affect the actual stiffness of "equal frequency" passive vibration isolating devices; the evaluation of the effectiveness of vibration isolation systems indicates the greater efficiency of the created vibration isolation system against the background of its significant simplification (Table 1).

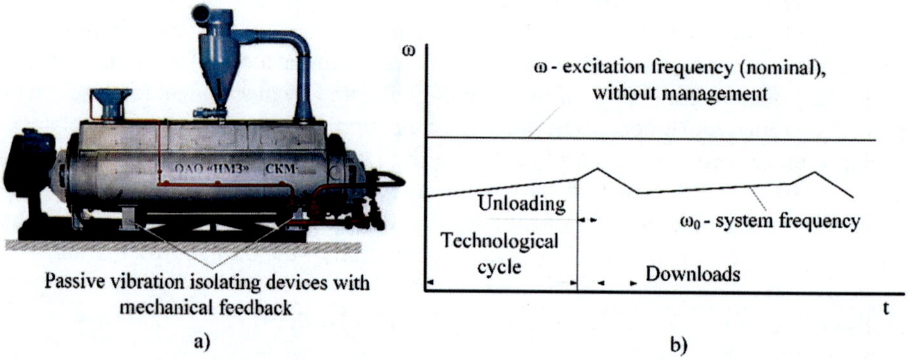

Fig. 5. Vibration isolation of SCM 200 by passive vibration isolating devices with mechanical feedback: a – location of passive vibration isolating devices with mechanical feedback; b – frequency characteristics of work processes.

Table 1. Evaluation of SCM 200 vibration isolation system efficiency.

Performance indicators	Conditional designations	Vibration isolation system	
		Serial	Using PVID with MF
1. Vibration isolation	k_R	0,87	0,89
2. Dynamics	k_X	0,64	0,72
3. Multiplicities	k_m	4	4
4. Failure time (hours)	L_h	1800	2500

5 Conclusions

Practical testing of theoretical positions was implemented, and the efficiency of one of the synthesized structures of passive vibration isolating devices with mechanical feedback was evaluated by testing it in industrial conditions, which showed an increase in vibration efficiency by 2.3%, dynamism by 7.5% and resource by 40%. The presence in the structures of passive vibration isolating devices with mechanical feedback of typical parts, elements, and components, each of them has separately tested methods of calculation and design, reduces the task of synthesis of such devices to common use, and in some cases to combine known techniques.

Thus, the positive result of industrial use of passive vibration isolating devices with mechanical feedback in creating vibration isolation systems of technological equipment indicates the possibility of widespread use of this device in various industries.

References

1. Cveticanin, L.: On mechanical metastructures applied in vibration suppression—review. In: Altenbach, H., Amabili, M., Mikhlin, Y.V. (eds.) Nonlinear Mechanics of Complex Structures. ASM, vol. 157, pp. 3–17. Springer, Cham (2021). https://doi.org/10.1007/978-3-030-75890-5_1
2. Ngo, Q.-H., Ho, K.-T., Nguyen, K.-T.: Experimentally investigating the resonance of the vibration of two masses one spring system under different friction conditions. In: Tien Khiem, N., Van Lien, T., Xuan Hung, N. (eds.) Modern Mechanics and Applications. LNME, pp. 63–71. Springer, Singapore (2022). https://doi.org/10.1007/978-981-16-3239-6_5
3. Bui, D.Q., Nguyen, H.Q., Hoang, V.L., Mai, D.D.: Design and hysteresis modeling of a new damper featuring shape memory alloy actuator and wedge mechanism. In: Tien Khiem, N., Van Lien, T., Xuan Hung, N. (eds.) Modern Mechanics and Applications. LNME, pp. 125–136. Springer, Singapore (2022). https://doi.org/10.1007/978-981-16-3239-6_10
4. Zotov, A., Valeev, A.: Vibration isolating and impact protecting systems with quasi-zero stiffness providing wide operating area. In: Radionov, A.A., Kravchenko, O.A., Guzeev, V.I., Rozhdestvenskiy, Y.V. (eds.) ICIE 2019. LNME, pp. 299–307. Springer, Cham (2020). https://doi.org/10.1007/978-3-030-22041-9_34
5. Wynne, Z., Kanellopoulos, G., Koutsomarkos, V., Law, A., Stratford, T., Reynolds, T.P.S.: Mass and temperature changes in operational modal analysis. In: Rainieri, C., Fabbrocino, G., Caterino, N., Ceroni, F., Notarangelo, M.A. (eds.) CSHM 2021. LNCE, vol. 156, pp. 69–81. Springer, Cham (2021). https://doi.org/10.1007/978-3-030-74258-4_4
6. Battula, S.K., Raju, P.R.M., Ratnam, C.: Shock and vibration prevention using angular mounts with different types of oil-based elastomers. In: Vijay Sekar, K.S., Gupta, M., Arockiarajan, A. (eds.) Advances in Manufacturing Processes. LNME, pp. 619–629. Springer, Singapore (2019). https://doi.org/10.1007/978-981-13-1724-8_57
7. Werner, U.: Vibration control of soft mounted induction motors with sleeve bearings using active motor foot mounts: a theoretical analysis. Arch. Appl. Mech. **88**(9), 1657–1682 (2018). https://doi.org/10.1007/s00419-018-1393-7
8. Rana, R., Soong, T.: Control of seismic and operational vibrations of rotating machines using semi-active mounts. Earthq. Eng. Eng. Vib. **3**, 85–100 (2004). https://doi.org/10.1007/BF02668854
9. Yorish, Y.N.: Vibrometry. Mashgiz, Moscow (1969)

10. Shen, X., Li, J., Guan, D., Zhang, C., Shen, H.: Experimental test on the dynamic damping performance of energy harvesting shock absorbers with overrun clutch under open circuit condition. In: Jing, X., Ding, H., Wang, J. (eds.) ICANDVC 2021. LNEE, vol. 799, pp. 16–28. Springer, Singapore (2022). https://doi.org/10.1007/978-981-16-5912-6_2
11. Geng, X.-F., Ding, H., Mao, X.-Y., Chen, L.-Q.: Research on a limited NES with forced vibration. In: Jing, X., Ding, H., Wang, J. (eds.) ICANDVC 2021. LNEE, vol. 799, pp. 113–126. Springer, Singapore (2022). https://doi.org/10.1007/978-981-16-5912-6_9
12. Rivin, E.I.: Vibration isolators and installation systems for equipment with automatic control. Machine Tool Industry, Moscow (1971)
13. Rivin, E.I.: Machine drive dynamics. Mechanical Engineering, Moscow (1976)
14. Kurgan, V., Sydorenko, I., Prokopovich, I., Yeputatov, Y., Levynskyi, O.: Synthesis of elastic characteristics based on nonlinear elastic coupling. In: Tonkonogyi, V., et al. (eds.) InterPartner 2020. LNME, pp. 166–175. Springer, Cham (2021). https://doi.org/10.1007/978-3-030-68014-5_17
15. Bissembayev, K., Iskakov, Z., Sagadinova, A.: Investigation of random vibrations of a rigid body on vibration dampers with straightened surfaces. In: Khang, N.V., Hoang, N.Q., Ceccarelli, M. (eds.) ASIAN MMS 2021. MMS, vol. 113, pp. 805–813. Springer, Cham (2022). https://doi.org/10.1007/978-3-030-91892-7_77
16. Sydorenko, I., Gutyrya, S., Atmazhov, S.: Nonlinear dynamic demper with a mechanical feed-back. In: Proceedings of Odessa Polytechnic University, vol. 1(33)–2(34), pp. 28−32 (2010)
17. Jatsun, S., Malchikov, A., Yatsun, A., Saveleva, E.: Studying of copying control system with nonlinear measurer. In: Ronzhin, A., Shishlakov, V. (eds.) Electromechanics and Robotics. SIST, vol. 232, pp. 13–23. Springer, Singapore (2022). https://doi.org/10.1007/978-981-16-2814-6_2
18. Pashchenko, V., Romanov, A., Chaikin, M., Zakharov, V., Pashchenko, V., Romanov, A.: Determination of special positions for solving the problem of joint-relative manipulation mechanisms kinematic control. In: Ronzhin, A., Shishlakov, V. (eds.) Electromechanics and Robotics. SIST, vol. 232, pp. 25–36. Springer, Singapore (2022). https://doi.org/10.1007/978-981-16-2814-6_3
19. Orekhov, S., Zaychikov, N., Petrukhin, K., Tsepurkin, A., Tsepurkin, N.: Kinematic modeling in study of manipulative mechanism of combined movement. In: Ronzhin, A., Shishlakov, V. (eds.) Electromechanics and Robotics. SIST, vol. 232, pp. 37–47. Springer, Singapore (2022). https://doi.org/10.1007/978-981-16-2814-6_4
20. Bazhenov, V., Pogorelova, O., Postnikova, T.: Crisis-induced intermittency and other nonlinear dynamics phenomena in vibro-impact system with soft impact. In: Altenbach, H., Amabili, M., Mikhlin, Y.V. (eds.) Nonlinear Mechanics of Complex Structures. ASM, vol. 157, pp. 185–203. Springer, Cham (2021). https://doi.org/10.1007/978-3-030-75890-5_11

Wave Propagation Speed Analysis in Polyurethane Foams

Olena Mikulich(✉)

Lutsk National Technical University, 75, Lvivska Street, Lutsk 43018, Ukraine
shyprao@gmail.com

Abstract. The article is devoted to developing an analytical approach to studying wave propagation speeds in structurally inhomogeneous materials. Studies have been conducted on polymer foam materials with closed cells. The apparatus of the micropolar theory of elasticity, the Cosserat continuum, was used to analyze the wave propagation speeds change in foam materials. Using this approach allowed the account of the influence of shear-rotation deformations. Using the known analytical approaches of wave mechanics, the defining equation for the case of a two-dimensional problem was obtained. The value of the characteristic frequency was determined, and the dispersion properties of shear-rotation waves in foam materials were analyzed. High-frequency waves in structurally inhomogeneous materials are difficult to detect because the wavelength must be much longer than the length of the microstructure to avoid scattering and attenuation of waves. Moreover, only the dispersion ratio of one shear-rotation wave is available when the frequency is below the limit value. Therefore, experimental studies for such materials are quite complex. The method of numerical modelling based on the proposed analytical approaches is convenient and effective for such types of research.

Keywords: Polyurethane foam · Cosserat elasticity · Shear-rotation wave · R&D investment

1 Introduction

In modern construction, there is a significant increase in the use of polymer foams for installation and insulation. To expand the scope of these materials, which have low density and good vibration-absorbing and heat-retaining characteristics, it is necessary to develop approaches to the analysis and study of the mechanical behavior of foam materials under different types of loads.

For the analysis of the stress-deformation state of foam materials in the literature are widely used refined models of the continuous mechanics, which account for the influence of heterogeneity of the microstructure of the material. In particular, the micropolar theory of elasticity - Cosserat elasticity, within which the influence of shear-rotation deformations of microparticles of the medium can be accounted for, has gained wide popularity.

V. Tonkonogyi et al. (Eds.): InterPartner 2022, LNME, pp. 465–472, 2023.
https://doi.org/10.1007/978-3-031-16651-8_44

Using such models makes it possible to account for the influence of structural inhomogeneity of foam materials on their mechanical behavior under the action of different types of external excitations. Also, applying such approaches allows obtaining analytical or analytical-numerical solutions for some problems, making it possible to study the influence of material microstructure based on the analysis of shear-rotational deformations on the wave propagation speed of such materials.

A lot of known papers on the study of the effect of wave loading on foam materials are experimental. Based on the experimental data, obtained models can predict specific types of foams' mechanical behavior and estimate their damping rate.

However, it is rather difficult to generalize the models obtained based on the experiment of specific foam materials to investigate an arbitrary foam media. Also, in the case of overlapping waves of different frequencies, experimental studies do not provide an opportunity to have a clear picture of the influence of the intensity of each of the waves.

Therefore, an approach based on a model of Cosserat mechanics is more general. Such an approach will make it possible to obtain formulas that will provide an opportunity to determine the corresponding speeds at different frequencies and classify them.

2 Literature Review

The wide use of foam materials in production leads to great interest in researching the properties of these materials. A large number of such studies are experimental or based on numerical simulation.

The wave method and finite element analysis are used in [1] to examine open-cell foams' elastic wave propagation phenomena. The wave propagation characteristics analysis of nanobeams made of nanoporous metal foams is investigated in [2]. Experimental exploring of the ability of polyurethane foam to mitigate intense and short-duration stress waves is described in [3]. The compaction stress level is used as an indicator of the mitigation capability of the foams. The effect of cell size, connectivity, and relative pore diameter on sound absorption of polyurethane foam by numerical simulation analysis is performed in [4]. In the study [5], experiments and simulations perform the interaction of solitary waves in a granular crystal with a plastically compressible and rate-sensitive test medium in the form of rigid polyurethane foam. The experimental results of impact hammer tests conducted on sand and composite sand-polyurethane samples at different confining pressures are shown in [6].

Analytical methods for studying wave effects in foams are based on applying refined models of solid medium mechanics. In the literature, an approach based on the use of motion equations of the micropolar elasticity is used for obtaining equations for determining the speeds of wave propagation within the Cosserat continuum [7, 8]:

$$\sigma_{ji,j} + X_i = \rho \ddot{u}_i, \tag{1}$$

$$\in_{kji} \sigma_{ij} + \mu_{jk,j} + Y_k = J \ddot{\phi}_k, \tag{2}$$

where σ_{ji} is the force stress, μ_{ji} is the couple stress, ρ is the material density, X = $\{X_i\}$ is the mass forces vector, Y = $\{Y_i\}$ is the couple forces vector, J is the inertia of unit volume rotation, $\in_{klm}$ is the permutation symbol, u = $\{u_i\}$ is the displacement vector, ϕ = $\{\phi_k\}$ is the rotation vector. Functions u and ϕ are continuous functions.

Here and further, the Einstein summation convention is used. A comma at subscript denotes differentiation for a coordinate indexed after the comma, i.e. $u_{j,i} = \partial_i u_j = \frac{\partial u_j}{\partial x_i}$. Under the condition of plane deformation indices vary from 1 to 2, and $k = 3$.

For the stress determining in the research [8], Nowacki's formulas are used:

$$\begin{aligned}\sigma_{ji} &= (\mu + \alpha)\gamma_{ji} + (\mu - \alpha)\gamma_{ij} + \lambda\gamma_{rr}\delta_{ij},\\ \mu_{ji} &= (\gamma + \varepsilon)\kappa_{ji} + (\gamma - \varepsilon)\kappa_{ij} + \beta\kappa_{rr}\delta_{ij},\end{aligned} \tag{3}$$

where λ, μ are Lame parameters in the framework of classical elasticity; α, β, γ, κ are the elastic constant required to describe an isotropic constrained Cosserat elastic solid, $\gamma_{ij} = u_{i,j} - \in_{kji}\phi_k$ is the asymmetric deformation tensor, $\kappa_{ij} = \phi_{i,j}$ is the torsion bending tensor. Along with the elastic characteristics of the material in the framework of the classical theory of elasticity λ and μ, elastic characteristics in the framework of the micropolar elasticity α, β, γ and κ are used.

R.S. Lakes proposed the method for determining the elastic characteristics of foam materials within the Cosserat continuum [9]. In this case, such elastic characteristics as Young's modulus or the shear modulus are determined by accounting for the microstructure of the material [9, 10]:

$$E = \frac{(2\mu + \kappa)(3\lambda + 2\mu + \kappa)}{2\lambda + 2\mu + \kappa}, \quad G = \mu + \frac{\kappa}{2}, \tag{4}$$

where κ is elastic constant in Cosserat elasticity, Lame parameters λ and μ are obtained in the framework of classical elasticity.

R.S. Lakes used an approach based on the use of Eringen's formulas [11] to determine the force and couple stresses [8–10, 12]:

$$\begin{aligned}\sigma_{ji} &= 2G\varepsilon_{ji} + \lambda\varepsilon_{rr}\delta_{ji} + \kappa \in_{jik} (\omega_k - \phi_k),\\ \mu_{ji} &= \alpha\phi_{r.r}\delta_{ji} + \beta\phi_{j,i} + \gamma\phi_{i,j},\end{aligned} \tag{5}$$

where $\varepsilon_{ij} = (u_{i,j} - u_{j,i})/2$ is the small strain, $\omega_k = (\in_{kji} u_{i,j})/2$ is the macrorotations. The value of Lame parameter λ is the same for the case of Cosserat and classical elasticity. Other elastic constants of materials α, β, γ, κ and G are elastic constants, which described an isotropic constrained Cosserat elastic solid.

R.S. Lakes [9, 10] determined the values of elastic characteristics for different types of closed and open cells foam. All these elastic constants were calculated by accounting for the material microstructure in the framework of Cosserat elasticity.

Therefore, we apply an approach based on using Eringen's formulation [11] formulas for the force and moment stresses for determining the speed of wave propagation in foam materials. This approach is more convenient and practical for practical application, as the techniques proposed in [8] allow analyzing and studying microstructure effects for different foam types.

3 Research Methodology

Carry out similar transformations to [8] to obtain the motion equations of the Cosserat continuum. The dynamic motion equations in displacements and microrotations we obtain by substituting the force and couple stresses formulas (5) into the motion equation of Cosserat elasticity (1)–(2). After a few simplifications, we get the motion equation in vector form in the absence of the mass and couple forces vector:

$$\begin{aligned}&(\lambda + 2G) grad\ div\mathbf{u} - \left(G + \frac{\kappa}{2}\right) rot\ rot\ \mathbf{u} + \kappa rot\phi = \rho\ddot{\mathbf{u}},\\ &(\alpha + \beta + \gamma) grad\ div\ \phi - \gamma rot\ rot\phi + \kappa rot\mathbf{u} - 2\kappa\phi = J\ddot{\phi}.\end{aligned} \tag{6}$$

Under the condition of plane deformation in complex form the displacement vector can be written as $\mathbf{u} = u_1 + i\,u_2$; the macrorotation vector is $\omega_3 = \frac{1}{2}(\partial_1 u_2 - \partial_2 u_1)$; the microrotation vector is $\boldsymbol{\phi} = \phi_3$. The motion equation for a plane problem can be written as:

$$\begin{aligned}&(\lambda + G)\partial_1\theta + G\Delta u_1 - \kappa\partial_2(\omega_3 - \phi_3) = \rho\ddot{u}_1,\\ &(\lambda + G)\partial_2\theta + G\Delta u_2 + \kappa\partial_1(\omega_3 - \phi_3) = \rho\ddot{u}_2,\\ &2\kappa(\omega_3 - \phi_3) + \gamma\Delta\phi_3 = J\ddot{\phi}_3,\end{aligned} \tag{7}$$

where $\theta = \partial_1 u_1 + \partial_2 u_2$.

Representing the vectors of displacements $\mathbf{u}$ and microrotations $\boldsymbol{\phi}$ in the form of the sum of scalar and vector potentials [7, 8]:

$$\mathbf{u} = grad\ \Phi + rot\Psi; \quad \phi = grad\ \Sigma + rot\mathrm{N}; \tag{8}$$

and substituting them into the motion equations Cosserat continuum (6), we obtain the following wave equations:

$$\begin{aligned}&\ddot{\Phi} - c_1^2\Delta\Phi = 0;\\ &\ddot{\Psi} - (c_2^2 + c_3^2)\Delta\Psi + 2c_3^2 rot\Psi = 0;\\ &\ddot{\Sigma} - (c_4^2 + c_5^2)\Delta\Sigma - \omega_*^2\Sigma = 0;\\ &\ddot{\mathrm{N}} - c_5^2\Delta\mathrm{N} - \omega_*^2\mathrm{N} = 0,\end{aligned} \tag{9}$$

where wave speeds $c_1,\ c_2,\ c_3, c_4$ and the value of the characteristic frequency ω_* in the framework of Cosserat elasticity are determined:

$$c_1^2 = \frac{\lambda + 2G}{\rho};\ c_2^2 = \frac{G}{\rho};\ c_3^2 = \frac{\kappa}{2\rho};\ c_4^2 = \frac{\alpha + \beta}{J};\ c_5^2 = \frac{\gamma}{J};\ \omega_*^2 = \frac{2\kappa}{J}. \tag{10}$$

Given the values of the expansion and shear wave speeds within the classical elasticity [8], the expressions for the speeds of the Cosserat continuum in (10) can be written as:

$$c_1^2 = c_\ell^2 + c_3^2; \quad c_2^2 = c_\tau^2 + c_3^2.$$

Analysis of the first wave Eqs. (9) shows that the expansion waves, as in the classical theory of elasticity, do not have dispersion properties. The third equation from (9) describes the waves of longitudinal rotation, while the characteristic speed c_4 does not coincide with their phase speed. However, as the frequency increases, the phase speed goes to c_4:

$$v_f = \sqrt{\frac{(\alpha + \beta + \gamma)\omega^2}{J\omega^2 - 2\kappa}} \to c_4, \quad \omega \to \infty. \tag{11}$$

The second and fourth equations in (9) show the shear waves (whose velocities c_2, c_3 are determined from (10)) and the transverse rotation (whose speed c_5) form shear-rotation waves that have dispersion properties.

To determine the speed of shear-rotation waves in the Cosserat continuum, we present the potentials Φ, Ψ, Σ, N in the form:

$$\begin{aligned} \Phi &= A \exp(-i\omega t + ikx_1); \quad \Psi = B \exp(-i\omega t + ikx_1); \\ \Sigma &= C \exp(-i\omega t + ikx_1); \quad \mathrm{N} = D \exp(-i\omega t + ikx_1). \end{aligned} \tag{12}$$

The isotropic Cosserat continuum produces two dispersion shear-rotation waves for the plane case. To determine them, we substitute the dependencies (12) in Eq. (7), accounting for the representations for displacements and microrotations (8). We obtain the characteristic equation of the form:

$$-\frac{c_5^2\ell_r^2}{4N}k^4 + \left(c_2^2 - \omega^2\frac{\ell_r^2}{2N} + \omega^2\ell_c^2\right)k^2 + \omega^2\left(1 - \left(\frac{\omega}{\omega_*}\right)^2\right) = 0, \tag{13}$$

where $\ell_r = \sqrt{\frac{J}{\rho}}$ and $\ell_C = \sqrt{\frac{\gamma}{2\kappa}}$ are the radius of inertia of the elementary volume of a homogeneous and centrally symmetric element and the scale parameter of the Cosserat continuum [8], $N = \sqrt{\frac{\kappa}{2G+\kappa}}$ is coupling number [9, 10, 12].

Investigating Eq. (13), we can see that for $\omega < \sqrt{\frac{2\kappa}{J}}$ the characteristic equation has one positive and one negative root for k^2, and for $\omega \geq \sqrt{\frac{2\kappa}{J}}$ both roots of k^2 the characteristic equation are positive.

Therefore, for low-frequency ranges there is one dispersion branch, and for $\omega \geq \sqrt{\frac{2\kappa}{J}}$ dispersion branches are two. This suggests that the phase speed of shear-rotation waves at low frequencies is close to the speed of shear waves in the classical theory of elasticity: $V_f \approx c_\tau$. For high frequencies, the asymptotic solution of Eq. (13) gives two values of the phase velocity: $V_f \approx c_2$ and $V_f \approx c_4$.

For polymeric foam materials, studies are conducted in the low-frequency ranges [13]. Therefore, there is one dispersion branch. In this case, for the considered types of materials, the relationship between the speeds in the Cosserat continuum and the speeds in the classical theory of elasticity can be written:

$$c_1 > c_\ell > c_2 > c_\tau > c_3 > c_4 >_5 .$$

For the low-frequency bands, two real roots are obtained from characteristic Eq. (12), one positive and the other negative. The speeds of shear-rotation waves in the Cosserat continuum are determined:

$$V_I^2 = \frac{c_2^2 + c_3^2}{\frac{c_2^2}{2\ell_C^2\omega^2} - \aleph - 1 + \sqrt{\left(\frac{c_2^2}{2\ell_C^2\omega^2} - \aleph - 1\right)^2 + \left(\frac{1}{\omega^2} - \frac{1}{\omega_*^2}\right)\frac{\aleph}{\ell_C^2}}};$$

$$V_{II}^2 = -\frac{c_2^2 + c_3^2}{\aleph + 1 - \frac{c_2^2}{2\ell_C^2\omega^2} + \sqrt{\left(\frac{c_2^2}{2\ell_C^2\omega^2} - \aleph - 1\right)^2 + \left(\frac{1}{\omega^2} - \frac{1}{\omega_*^2}\right)\frac{\aleph}{\ell_C^2}}}, \qquad (14)$$

where $\aleph = \frac{c_2^2 + c_3^2}{c_5^2}$ determines the relationship between shear and rotation speeds in the Cosserat elasticity.

Analyzing the shear-shear wave speeds based on dependencies (14), it is seen that when $\ell_C \to 0$ (when the influence of the microstructure of the material is so small that it can be neglected) the characteristic frequency goes to infinity: $\omega_* \to \infty$, while the shear wave speed c_2 in the Cosserat continuum tends to the according to speed c_τ within the classical elasticity. The speed of the rotation wave tends to zero:$c_4 \to 0$. That is, the speed of one of the waves V_I tends to the speed of the shear waves c_τ, and the other with an imaginary wave number $i\omega/V_{II}$ disappears. That is, the results of the calculations coincide with the classical theory of elasticity for isotropic materials in the case when the influence of microstructure can be neglected.

4 Results and Discussion

Let's investigate the change in shear-rotation speeds in polyurethane foams WF51, WF110, and WF300 with a change in frequency ω using the results of experimental studies [9]. Here, the values of such elastic characteristics within the Cosserat continuum as shear modulus G, dimensional characteristics in bending ℓ_b and torsion ℓ_t, and coupling number N were determined. The corresponding values of elastic characteristics are given in Table 1.

The values of the characteristic frequencies for polyurethane foams are: $\omega_* = 6.3058 \cdot 10^5\,\text{s}^{-1}$ (WF300), $\omega_* = 3.506 \cdot 10^5\,\text{s}^{-1}$ (WF110), $\omega_* = 4.051 \cdot 10^5\,\text{s}^{-1}$ (WF51).

Since the shear-rotation speeds have features when approaching the frequency of external load to the value of the characteristic frequency ($\omega = \omega_*$), the numerical analysis is performed for the case of the frequency range, which is 60% of the limit value of the characteristic frequency. Figure 1 shows the dependence of the change in the speed V_I and the phase speed of V_f with a change in frequency for three types of polyurethane foam: curves 1 and 2 correspond to the case of WF 300, curves 3 and 4 correspond to the case of WF 110, curves 5 and 6 correspond to the case of WF 51.

Table 1. Values of polyurethane foam characteristic [9].

	WF300	WF110	WF51
Shear modulus G *(MPa)*	285	104	65
Characteristic length, bending ℓ_b *(mm)*	0.77	0.33	0.55
Characteristic length, torsion ℓ_t *(mm)*	0.8	0.62	0.54
Coupling number N^2	0.04	0.04	0.01
Cell size *(mm)*	0.65	0.5	0.4
Density ρ *(kg/m³)*	380	340	60

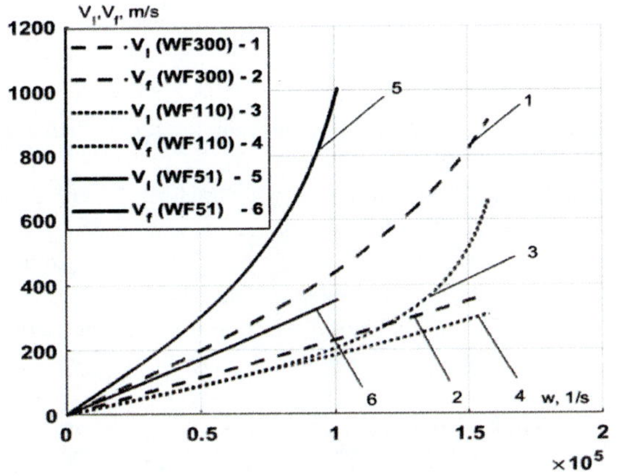

Fig. 1. Dependence of shear-rotation speeds on frequency.

Figure 1 shows that in the case where the frequency ω does not exceed 30% of the cut-off characteristic frequency ω_*, the dependence of the shear-rotation speed V_I on frequency ω is linear. With a further increase in frequency, this dependence degenerates into nonlinear. In addition, numerical calculations have shown that the behavior of the shear-rotation wave speed V_I is asymptotic when the frequency ω approaches to the value ω_*.

Numerical calculations confirm the results of formula (11) analytical studies regarding the difference between the phase speed V_f and the speed V_I.

5 Conclusions

The dispersion properties characteristic of shear-rotation waves plays an important role in determining the parameters of the Cooserat continuum and in predicting the behavior of foam materials under different types of loads. High-frequency waves in such structurally inhomogeneous materials are difficult to detect because the wavelength must be much longer than the length of the microstructure to avoid scattering and attenuation of the waves. In addition, only the dispersion ratio of one shear-rotation wave is available when the frequency is below the limit value. Therefore, experimental studies for such materials are quite complex. It is for these types of research is convenient and effective is the method of numerical modeling based on the approaches proposed in work, which can be used to assess the effectiveness of foam materials for vibration and noise protection of structures and in determining the optimal characteristics of the foam material with predetermined properties.

Acknowledgment. The work performed within the state grants of applied research № 0122U001064 "Methodology of predicting mechanical behavior and optimizing the effective characteristics of foam and porous materials". The team of authors expresses their sincere gratitude to the Ministry of Education and Science of Ukraine.

References

1. Bayat, A., Gaitanaros, S.: Elastic wave propagation in open-cell foams. J. Appl. Mech. **86**(5), 051008 (2019)
2. Wangab, Ya., Lianga, Ch.: Wave propagation characteristics in nanoporous metal foam nanobeams. Results Phys. **12**, 287–297 (2019)
3. Pradel, P., Malaise, F., de Rességuier, C., et al.: Stress wave propagation and mitigation in two polymeric foams. AIP Conf. Proc. **1979**, 110015 (2018)
4. Chen, S., Lei, S., Zhu, J., Zhang, T.: The influence of microstructure on sound absorption of polyurethane foams through numerical simulation. Macromol. Theory Simul. **30**(5), 2000075 (2021)
5. Schiffer, A., Leeb, D., Kimb, E., Kimc, T.: Interaction of highly nonlinear solitary waves with rigid polyurethane foams. Int. J. Solids Struct. **152–152**, 39–50 (2018)
6. Placido, M., Gatto, A., Montrasio, L., Zavatto, L.: Experimental analysis and theoretical modelling of polyurethane effects on 1D wave propagation through sand-polyurethane specimens. J. Earthq. Eng. **9**, 1961933 (2021)
7. Erofeev, V.I.: Wave Processes in Solids with Microstructure. World Scientific, Singapore (2003)
8. Sulym, H., Mikulich, O., Shvabyuk, V.: Investigation of the dynamic stress state of foam media in Cosserat elasticity. Mech. Mech. Eng. **22**(3), 739–749 (2018)
9. Lakes, R.S.: physical meaning of elastic constants in cosserat, void, and microstretch elasticity. J. Mech. Mater. Struct. **11**(3), 217–229 (2016)
10. Lakes, R.: Softening of Cosserat sensitivity in a foam: warp effects. Int. J. Mech. Sci. **192**, 106125 (2021)
11. Hassanpour, S., Hepple, G.: Micropolar elasticity theory: a survey of linear isotropic equations, representative notations, and experimental investigations. Math. Mech. Solids, 1–19 (2015)
12. Rueger, Z., Lakes, R.S.: Cosserat elasticity of negative Poisson's ratio foam: experiment. Smart Mater. Struct. **25**, 1–8 (2016)
13. Ogam, E., Fellah, Z., Fellah, M., Depollier, C.: Theoretical and experimental study of micropolar elastic materials using acoustic waves in air. J. Sound Vib. **510**, 116298 (2021)

A Method for Calculating the Strength Performance of Cast Parts

Olga Ponomarenko[1], Nataliia Yevtushenko[1](✉), Tetiana Berlizieva[1], Igor Grimzin[2], and Tatiana Lysenko[3]

[1] National Technical University "Kharkiv Polytechnic Institute", 2, Kyrpychova Street, Kharkiv 61002, Ukraine
natalya0899@ukr.net

[2] Research and Production Center "European Engineering Technologies", 101, Velyka Panasivska Street, Kharkiv 61017, Ukraine

[3] Odessa Polytechnic National University, 1, Shevchenko Avenue, Odessa 65044, Ukraine

Abstract. Based on the probabilistic approach and the statistical theory of fatigue fracture, a method has been developed for calculating the strength, reliability, and resource of cast parts operating under conditions of variable loads of a random nature. It allows linking the endurance limit of a part with its geometric parameters, physical and mechanical properties of the alloy, and statistical characteristics of the load on the part. The above technique was used to calculate the strength reliability of a cast part of a tractor hinge body. Relationships were determined between the values of operational reliability and bearing capacity of the cast part under various operating conditions of the tractor. The bearing capacity of the studied part, its average value, and the coefficient of variation were established. The safety factor of the cast part was determined, and it was found that the design of the studied part is well adapted to the specific conditions of its production. The work showed ways to reduce a part's mass by increasing its dimensions' geometric accuracy. Their practical use makes it possible to reduce the weight of the casting by 15–25% while maintaining the reserve of its bearing capacity.

Keywords: Reliability · Bearing capacity · Physical properties · Mechanical properties · Dimensional accuracy · Industrial innovation

1 Introduction

Cast parts for engineering applications hold a dominant place in the total volume of casting production [1]. Heavy-duty castings' main requirements are a high service life level and performance with minimum metal consumption [2]. The rated features are determined at the design stage [3], developed in its implementation [4], and emerge during the operation [5, 6].

The strength of cast parts is understood as their ability to resist deformation or fracture under the action of applied loads. When designing parts, the designer uses reference data on the material used for the part and calculates its geometric dimensions in order to so

V. Tonkonogyi et al. (Eds.): InterPartner 2022, LNME, pp. 473–481, 2023.
https://doi.org/10.1007/978-3-031-16651-8_45

that the product can withstand the expected design loads. Safety factors are introduced to reduce the risk of breakdowns.

The margin of safety considers the spread of the mechanical properties of the material, inaccurate knowledge of the acting loads and stresses, and the deviation in the geometry of the parts associated with the production conditions of cast parts. The safety factor is always greater than one, usually $n = 1.3$–2.5. However, an increase in the safety margin more than the required value leads to an increase in the dimensions of the part [7], which is not only economically unprofitable, but sometimes it is simply unacceptable (for example, in aircraft structures).

Therefore, the main trend in improving the quality of castings in industrially developed countries is to reduce their weight [8] and improve their use properties [9]. Therefore, the design and manufacture of cast parts of minimum metal consumption with a given level of service life and reliability for the operation of parts in specific conditions is a relevant scientific [10] and practical task [11].

Until recently, the main criterion for calculating cast parts for strength and reliability was the safety factor of the part [10]. Its significant drawback was revealed during the long-term use of the safety factor concept. It consists of the absence of an adequate assessment of the situation of an actual assessment of the reliability of the part [12]. For instance, a case of the disruption of a cast rolling mill bed under the action of alternating loads has been described, although the bed's working sections were designed based on the use of a twenty-fold safety factor [13].

2 Literature Review

In the first third of the 20th century, a probabilistic understanding of the nature of the safety factor was developed [14]. In the late 1950s, the time factor was introduced into the strength design model [15]. Combined with a probabilistic approach, this led to the creation of the concept of failures of mechanical structures [16] as events consisting of outbreaks of a random process of loading a part beyond the permissible level due to the bearing capacity of the part [17].

According to this concept, a statistical theory of fatigue failure of machine parts and structures has been developed and widely used in practice [18, 19], which allows linking the endurance limit of a part of an arbitrary profile with its geometric parameters [20, 21], metal properties and statistical characteristics of the load on the parts [22, 23].

Therefore, developing a method for calculating the main system characteristics of a cast part is a prerequisite for solving problems associated with the design of high-quality castings [24].

The study aims to develop a method for calculating the strength performance and service life of cast parts operating under alternating loads of a random nature based on the probabilistic approach and the statistical theory of fatigue failure.

3 Research Methodology

The implementation of the method includes the following stages.

The loading characteristics of the section of a cast part are determined. The distribution functions of the geometrical dimensions of the loaded section, the stress concentration coefficient (α), and the perimeter of the working section of a cast part (L) are found, and the relative stress gradient is calculated

$$G = \frac{1}{\sigma_{max}} \left| \frac{\sigma}{dx} \right|_{x=a} \tag{1}$$

where G is the relative stress gradient; a is the coordinates of the point at which the stresses σ take their maximum value.

The data on load strain measurement are processed on a section of length l_δ, which is the length of the load block, expressed in kilometers of run, hours of operation or the number of process cycles. Strain measurement results are processed and presented in the form of histograms.

The parameters are introduced that characterize the physical and mechanical properties of the alloy used and their probabilistic characteristics:

$\left(\sigma_{bp}(l), PP(l),\ (l = 1, PP)\right)$ is the histogram of the tensile strength of the alloy;

N_0, S are the fatigue curve parameter's mean value and standard deviation.

The Monte Carlo method determines the amplitudes of the loads

$$\sigma_{ak} = \frac{1}{F} \sum\nolimits_k A_k \cdot Q_k \tag{2}$$

where σ_{ak} is the stress amplitude of the k-th loading cycle; F is the cross-sectional area; A_k is the transfer function from the external load Q_k to the part.

The load block is determined by the histogram as follows:

$$\left(\frac{\sigma_{ai}}{\sigma_{a,max}}, t_i\right) \quad (i = 1, 2, 3, \ldots, r) \tag{3}$$

where $\sigma_{a,max}$ is the maximum stress in the block; t_i is the relative number of repetition cycles of the i-th stress amplitude in the load block.

For the case under consideration, the value of the coefficient K considers all factors' impact on the fatigue resistance of the part.

The Monte Carlo method determines the mean values and histograms of the parameters m, ν_δ, and S [5];

m is the parameter of stress distribution, characteristics of the material that determine its sensitivity to stress concentration;

ν_δ is the scale factor of the root-mean-square deviation of the similarity criterion S.

The mean value and the histogram of the endurance limit of the part are calculated according to the formula:

$$\sigma_{-1D} = \frac{\sigma_{-1}}{2\alpha}\left(1 + \theta^{-\nu_\sigma} 10^{U_p S}\right), \tag{4}$$

where α is the theoretical stress concentration cooefficient;

σ_{-1} is the endurance limit of a standard laboratory sample;

U_p is the quantile of the normal distribution corresponding to the probability p;

θ is the elative criterion for fatigue failure, determined by the formula:

$$\theta = \frac{\left(\frac{L}{G}\right)}{\left(\frac{L_0}{G_0}\right)} = \frac{L}{(88.3 \cdot G)} \tag{5}$$

where $\frac{L_0}{G_0}$ – the value of the similarity criterion for a smooth cylindrical specimen.

The value is calculated as follows

$$\varepsilon = \sum_i \frac{\sigma_{ai}}{\sigma_{a,max}} \cdot t_i \tag{6}$$

for all stages of the load block.

The sum with respect to the durabilities α_p is calculated based on the corrected linear hypothesis of the summation of fatigue damages according to the formula:

$$\alpha_p = \frac{\sigma_{a,max} \cdot \varepsilon - 0,5 \cdot \sigma_{-1D}}{\sigma_{a,max} - 0,5 \cdot \sigma_{-1D}} \tag{7}$$

If the value calculated by this formula is less than 0.1, then $\alpha_p = 0.1$ is taken.

The mean value and the histogram of the number of load blocks λ before the failure of the part are calculated as follows:

$$\lambda = \frac{\mathrm{a_p} \cdot \sigma_{-1D}^{m} \cdot N_0}{\sigma_{a,max}^{m} \cdot \nu \cdot \sigma \cdot \sum_i \left(\frac{\sigma_{ai}}{\sigma_{a,max}}\right)^m \cdot t_i} \tag{8}$$

and the corresponding service life (MTBF) is calculated according to the formula:

$$L = \lambda \cdot l_\delta \tag{9}$$

where l_δ is the length of the load block.

Based on the found MTBF histogram, the distribution function of the strength reliability of the cast part is determined as follows.

$$F(PN) = Prob(PN < L_k)$$

where PN is the strength reliability of the part under consideration;

L_k is the k-th value of the MTBF of the part.

4 Results and Discussion

According to the results of the above calculations, the distribution functions of the endurance limit of the part, service life, and strength reliability are printed out.

The above technique was used to calculate the strength performance of the cast part of the hinge body of the T-150K tractor. The casting material is steel 45L according to GOST 977–88. Its most loaded place is the lower eye ring; geometrically, it represents the intersection of the I-beam with the cylinder.

To determine the dispersion of the geometric dimensions of the casting in production conditions, statistical studies were conducted. Based on the processing of the obtained data and the use of the simulation modeling method, the dispersion of the cross-sectional area of the part was studied. It was found that at a rated value of 34 cm^2, the cross-sectional area fluctuates in the range of 26.1 to 41.3 cm^2. The range of accuracy of castings made in raw sand molds is within 6…14 accuracy classes according to GOST 26645–85, which is a relatively good indicator.

The key issue for determining the strength performance of a cast part is to identify the relationship between the values of operational reliability and bearing capacity values. This is because the failure of a part occurs when the load taken by the part exceeds its bearing capacity. The relationship between the load and the bearing capacity of the part in question is shown in Fig. 1. The loads experienced by the hinge body are shown in the graph as two separate sections. The left side of the graph shows the normal operating loads that occur during normal operation of the tractor in plowing with forwarding movement in a straight line, with a trailer and a semi-trailer in various modes. The range of loads, in this case, is from 0.3–3.5 tons to 23.6 tons. Shows the bearing capacity of the hinge body in non-standard situations, for example, the load when hanging (raising the plow) when turning the tractor.

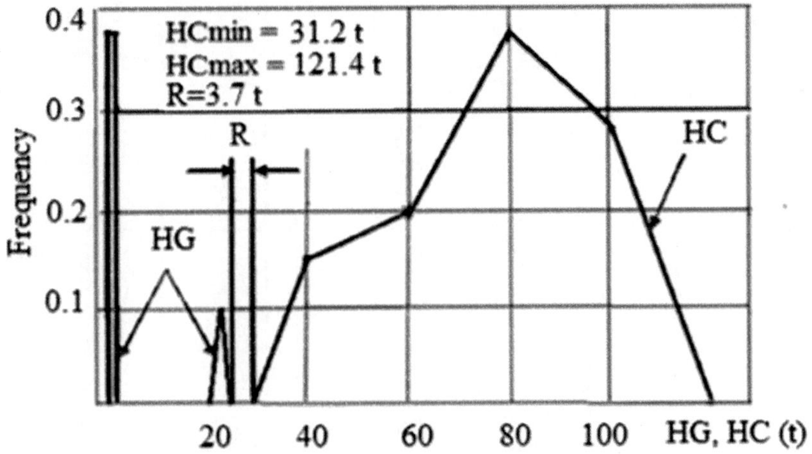

Fig. 1. Polygons of the load and the bearing capacity of the part.

The value of the bearing capacity was determined by numerical methods based on the quantitative account of the following factors: scattering of the value of the endurance limit of a part of a given design; variations in the mechanical properties of cast steel in the conditions under study; fluctuations in the cross-sectional area of the part under consideration. According to the analysis of the numerical study results, it was determined that the value of the bearing capacity of the part under study (HC) is in a range of 31.2 to 121.4 t, its mean value is 76.3 t, and the coefficient of variation is 0.32.

For the part under consideration, between the maximum load and the minimum bearing capacity of the part under the given manufacturing conditions, there is a certain interval *R*, the value of which is the bearing capacity reserves, its value, in this case, is 3.7 tons. In this case, the determined safety factor equal to the ratio of the average bearing capacity to the calculated load is 3.1.

The given data show that the design of the part under study is well adapted to the specific conditions of its production.

In this regard, it is of great interest to study the possibility of reducing the part's mass by improving its dimensional accuracy in the manufacture of casting. This possibility is because, with improved geometric accuracy, the actual minimum size of the area of the cross-section of the part can be guaranteed at lower nominal values of the dimensions of this cross-section.

The study of the effect of the geometric accuracy of the dimensions of the part under study on its strength reliability and the possibility of reducing the mass has resulted in the following. An increase in the casting accuracy class from 11 to 7 increases the value of the bearing capacity margin from 3.7 to 9.8 t. Also, with a decrease in accuracy to Class 12, the value of *R* becomes negative, and the probability of failure appears; the value of which for a given part under the studied conditions is 0.009.

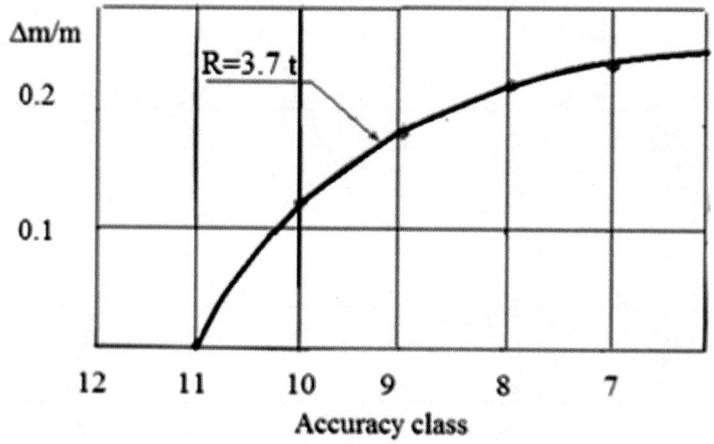

Fig. 2. Possible decrease in the part's mass depending on its accuracy.

The study of the possibility of reducing the mass of a part by increasing the accuracy while maintaining its bearing capacity at the existing level (R = 3.7) has showed that with an increase in the accuracy class from 11 to 10, the mass of the part can be reduced by 7%. With a further increase in accuracy to Class 7, the maximum achieved class for this casting method, it is possible to reduce the mass of the casting by 25% while maintaining the existing level of reliability (Fig. 2).

The efficiency of such calculations in the future can be increased through the use of applied integrated computer systems designed to simulate casting processes and calculate castings' thermal and stress-strain state.

To do this, it is necessary to create a 3D model of the part using CAD packages of various levels that work without conflict in the PDM system environment. The 3D model is the basis for various structural calculations with operational loads in the future. Then, on its basis, a 3D model of the casting is created. To obtain a high-quality casting without defects, it is necessary to use computer-integrated design packages to simulate the process of filling the mold with a melt, cooling the casting, and detecting the locations of internal defects.

Getting quality casting is possible using new methods and software products such as SolidWorks and LVMFlow. This approach to design allows you to simultaneously link such parameters as the optimal design of the part and its manufacture's manufacturability and lay these parameters at the pre-production stage while reducing the time for their design.

Thus, with the help of CAD/CAE programs, a three-dimensional casting model was created, and a foundry technology for casting it was worked out, considering the mold's uniform filling with metal and obtaining a high-quality casting. To simulate casting processes, but today it is most expedient to use the LVMFlow software package for computer simulation through finite-difference calculation algorithms. When using finite-difference numerical models, special attention should be paid to the correct choice of initial and boundary conditions and the creation of a grid.

For engineering modeling of the thermal and stress-strain state of cast parts, the use of the ANSYS complex, which has high interaction with existing CAD and CAE systems, which, in turn, has full interaction with Workbench Products and classic ANSYS, as well as a relatively large number of mathematical solutions, will allow quickly and efficiently perform calculations in Workbench Products.

5 Conclusions

Based on the probabilistic approach and the statistical theory of fatigue fracture, a method has been developed for calculating the strength, reliability, and resource of cast parts operating under conditions of variable loads of a random nature. It allows linking the endurance limit of a part with its geometric parameters, physical and mechanical properties of the alloy, and statistical characteristics of the load on the part.

The relationship between the values of operational reliability and its bearing capacity was determined to determine the strength reliability of the cast part. The value of the bearing capacity was determined by numerical methods based on the quantitative account of the following factors: dispersion of the value of the endurance limit of a part of a given design; variations in the mechanical properties of cast steel under the conditions under study; fluctuations in the cross-sectional area of the part under consideration.

It was found that the value of the bearing capacity of the part under study is in the range of 31.2–121.4 t, its average value is 76.3 t, and the coefficient of variation is 0.32.

The safety factor of the cast part was determined, equal to the ratio of the average bearing capacity to the design load. For the investigated part, its value is 3.1. The given data show that the design of the studied part is well adapted to the specific conditions of its production.

The work showed ways to reduce the part's mass by increasing its dimensions' geometric accuracy. Their practical use makes it possible to reduce the weight of the casting by 15–25% while maintaining the reserve of its bearing capacity.

The efficiency of such calculations in the future can be increased through the use of applied integrated computer systems designed to simulate casting processes and calculate castings' thermal and stress-strain states.

References

1. Hovorun, T.P., Berladir, K.V., Pererva, V.I., Rudenko, S.G., Martynov, A.I.: Modern materials for automotive industry. J. Eng. Sci. **4**(2), F8–F18 (2017). https://doi.org/10.21272/jes.2017.4(2).f8
2. Ponomarenko, O.I., Lysenko, T.V., Stanovskiy, A.L., Shinsky, O.I.: Management of foundry systems and processes. NTU "KhPI", Kharkov (2012). (in Ukrainian)
3. Berladir, K., Hovorun, T., Gusak, O., Reshetniak, Y., Khudaybergenov, D.: Influence of modifiers-ligatures on the properties of cast aluminum alloy AK5M2 for the automotive industry. In: Ivanov, V., Trojanowska, J., Pavlenko, I., Zajac, J., Peraković, D. (eds.) DSMIE 2020. LNME, pp. 473–482. Springer, Cham (2020). https://doi.org/10.1007/978-3-030-50794-7_46
4. Ponomarenko, O., Grimzin, I., Yevtushenko, N., Lysenko, T., Marynenko, D.: Advanced technologies of manufacturing readily removable cores for obtaining high-quality castings. In: Ivanov, V., Trojanowska, J., Pavlenko, I., Zajac, J., Peraković, D. (eds.) DSMIE 2021. LNME, pp. 565–574. Springer, Cham (2021). https://doi.org/10.1007/978-3-030-77719-7_56
5. Tokarev, A.O., Mironenko, N.G.: Failures of machine parts. Analysis of causes, technical diagnostics and prevention. Infra-engineering, Moscow (2020). (in Russian)
6. Zaitsev, A.V., Glotov, V.A., Maslov, E.B.: Calculation of elements and joints of metal structures of machines. Direct Media, Moscow-Berlin (2019). (in Russian)
7. Karpus, V.E., Ivanov, V.A.: Locating accuracy of shafts in V-blocks. Russ. Eng. Res. **32**(2), 144–150 (2012). https://doi.org/10.3103/S1068798X1202013X
8. Kolobov, A.B.: Strength reliability and durability of machine parts and structures. Infra-engineering, Moscow-Vologda (2020)
9. Kuznetsov, S.M., Kuznetsova, K.S.: Justification of the reliability of machinery and equipment. Direct Media, Moscow-Berlin (2020). (in Russian)
10. Getman, A.A.: Criteria for assessing the reliability of cast parts. Foundry **1**, 34–35 (2000)
11. Grebenik, V.M., Gordienko, V.M., Tsapko, V.K.: Improving the reliability of metallurgical equipment. Metallurgy, Moscow (1988).(in Russian)
12. Zheldubovskiy, A.V. Pogrebnyak, K.S., Regulsky, M.N., Serditov, A.G.: To the assessment of the safety factor of machine parts subjected to asymmetric loading, taking into account the stress concentration. Bull. NTTU "KPI" **3**(75), 42–46 (2015)
13. Yusha, V.L., Busarov, S.S., Aistov, I.P., Titov, D.S., Vansovich, K.A.: Influence of wall thickness and properties of structural materials on the discharge temperature and strength characteristics of slow-speed long-stroke stages. AIP Conf. Proc. **1876**, 020040 (2017). https://doi.org/10.1063/1.4998860
14. Kuzin, O., Kusyj, J., Topilnytskyy, V.: Influence of technological heredity on reliability parameters products. Technol. Audit Prod. Reserv. **1**(21), 15–21 (2015). https://doi.org/10.15587/2312-8372.2015.37678
15. Tovo, R., Lazzarin, P., Berto, F., Cova, M., Maggiolini, E.: Experimental investigation of the multiaxial fatigue strength of ductile cast iron. Theoret. Appl. Fract. Mech. **73**, 60–67 (2014). https://doi.org/10.1016/j.tafmec.2014.07.003

16. Fragassa, C., Minak, G., Pavlovic, A.: Tribological aspects of cast iron investigated via fracture toughness. Tribol. Ind. **1**, 1–10 (2016)
17. Raabe, D., Tasan, C.C., Olivetti, E.A.: Strategies for improving the sustainability of structural metals. Nature **575**, 64–74 (2019). https://doi.org/10.1038/s41586-019-1702-5
18. Merzliakov, I., Pavlenko, I., Chekh, O., Sharapov, S., Ivanov, V.: Mathematical modeling of operating process and technological features for designing the vortex type liquid-vapor jet apparatus. In: Ivanov, V., et al. (eds.) DSMIE 2019. LNME, pp. 613–622. Springer, Cham (2020). https://doi.org/10.1007/978-3-030-22365-6_61
19. Liaposhchenko, O., Pavlenko, I., Monkova, K., Demianenko, M, Starynskyi, O.: Numerical simulation of aeroelastic interaction between gas-liquid flow and deformable elements in modular separation devices. In: Ivanov, V., et al. (eds.) DSMIE 2019. LNME, pp. 765–774. Springer, Cham (2020). https://doi.org/10.1007/978-3-030-22365-6_76
20. Chigarev, V.V., Kovalenko, I.V.: Investigation of the process of fatigue failure of elements and parts, ship auxiliary mechanisms and structures in the event of various stress concentrators. Bull. Azov State Univ. **34**, 80–85 (2017)
21. Hovorun, T., et al.: Improvement of the physical and mechanical properties of the cutting tool by applying wear-resistant coatings based on Ti, Al, Si, and N. J. Eng. Sci. **8**(2), C13–C23 (2021). https://doi.org/10.21272/jes.2021.8(2).c3
22. Krizhevich, G.B.: Methodology for calculating the ultimate fatigue strength of structures of marine technology in low-temperature conditions. Proc. Krylov State Sci. Cent. **2**(388), 41–54 (2019)
23. Bondar, V.S., Abashev, D.R., Petrov, V.K.: Some features of predicting the resource of materials and structures under cyclic loading. Vesnik Perm Natl. Res. Polytech. Univ. **1**, 18–26 (2019)
24. Kostyk, K., et al.: Simulation of diffusion processes in chemical and thermal processing of machine parts. Processes **9**(4), 698 (2021). https://doi.org/10.3390/pr9040698

Lyapunov Function-Based Approach to Estimate Attractors for a Dynamical System with the Polynomial Right Side

Volodymyr Puzyrov[1,2], Nataliya Losyeva[2,3], Nina Savchenko[4(✉)], Oksana Nikolaieva[5], and Olga Chashechnikova[6]

[1] Universitat Politècnica de Catalunya, 31, Carrer de Jordi Girona, 08034 Barcelona, Spain
[2] Nizhyn Gogol State University, 2, Grafska Street, Nizhyn 16600, Ukraine
[3] Universitat de Barcelona, 585, Gran Via de les Corts Catalanes, 08007 Barcelona, Spain
[4] National Aerospace University "KhAI", 17, Chkalova Street, Kharkiv 61070, Ukraine
n.savchenko@khai.edu
[5] National University of Food Technologies, 68, Volodymyrska Street, Kyiv 01033, Ukraine
[6] Sumy State Pedagogical University, 87, Romenska Street, Sumy 40002, Ukraine

Abstract. Stability analysis is an essential part of the study of the behavior of a dynamic system. Typically, this analysis includes finding the stationary points or limit cycles, determining their stability or instability, and identifying the regions of attraction (RoAs) of attractors. There are several classical methods for obtaining RoAs estimates, which may be divided into Lyapunov and non-Lyapunov methods; at the same time, due to the limitations of existing methods, the identification of a complex RoAs boundary is practically impossible, and it also leads to a high computational cost. The existing methods are quite effective for systems of the second and third orders. However, an increase in the dimension of the system or uncertain mechanical parameters leads to an exponential increase in the required calculations. In this regard, it is essential to design relatively simple algorithms in terms of the number of necessary operations and, at the same time, give acceptable from a practical point of view estimates of RoAs. The present paper deals with the problem of obtaining estimates of the domains of attraction and stability for a nonlinear dynamical system with a polynomial right-hand side. It is based on a particular procedure of polynomial Lyapunov function construction. As an example, this procedure is applied to estimate the domain of attraction for the mechanical system of two coupled oscillators.

Keywords: Region of attraction · Lyapunov function · System with polynomial right-hand side · Iterative procedure · Mechanical system · Industrial innovation

1 Introduction

The stability of a dynamical system to disturbances is a common requirement for the stable operation of most technical systems. The stability analysis consists of finding the attractors of a system, which can be fixed points, limit cycles or toruses, determining the stability nature of the attractors, and identifying the region of attraction (RoA)

V. Tonkonogyi et al. (Eds.): InterPartner 2022, LNME, pp. 482–494, 2023.
https://doi.org/10.1007/978-3-031-16651-8_46

of each attractor. RoA is defined as a set of initial conditions for which the system approaches a specific attractor. Over the last decades, much effort has been made to develop different approaches to find the boundary of RoA. Such approaches can be categorized into non-Lyapunov-based methods (NLB) and Lyapunov function-based methods (LFB). LFB methods have a strong mathematical fundamental and are applied in various forms, including the "direct" construction of LF according to a specific rule, the sum of squares programming (SoS), methods that use both simulation and SoS programming, and procedures using the moment theory. In classical problems of stability of motion, an essential role is played by the estimation of the attraction domain of the zero solution. This estimate guarantees the value of the initial perturbation, at which the perturbed solution remains in a small neighborhood of the beginning and tends to zero with increasing time. In those cases when it is possible to construct the LF providing the asymptotic stability of the zero solution, this function allows one to obtain this kind of estimate. For systems of high dimension ($n \geq 3$) or with uncertain parameters, the LF is often sought in the class of positive definite quadratic forms, based on the condition that the constructed form is the Lyapunov function (QLF) for the corresponding linearized system. In addition, the question may be raised about the construction of the QLF with some given properties, determined by the problem's features. In particular, in nonlinear dynamic problems, when the researcher is interested in not only qualitative but also quantitative characteristics of the system, there is a need for restrictions on the first derivative of the Lyapunov function along the trajectories of the system.

The research aims to suggest a method of constructing LF which allows obtaining an adequate estimate of the region of attraction for dynamical systems in case $n > 3$. In this regard, we developed the approach proposed earlier in [1] and applied it to the fourth order system. It is shown that computations needed are not very laborious (for instance, grid search or sampling methods are much more computationally costly).

2 Literature Review

Generally, the approaches described in the literature may be divided into NLB and LFB methods. Most methods of the NLB group are based on simulations [for example, 2, 3]. The work in [4] is based on topological approaches like the viability theory. Genesio et al. [5] used the trajectory reversal method, in which the trajectories of the systems are used to find a part of the RoA. In [6], the problem of increasing the estimate of the attraction domain was discussed. The improved estimate was obtained as a discrete set of points due to backward integration of the points lying on the original estimate. The approach was used to estimate the attraction domain of the vertical equilibrium of a pendulum stabilized by a linear controller. In paper [7], a combined approach was proposed in which, based on SOS programming and trajectory reversal, an initial RA estimate was established. Then, a state-dependent edge impulse was applied to obtain an extended RA estimate. The drawback of simulation-based methods is their dependence on the initial conditions, i.e., low predictability.

Significant research has been carried out to develop LFB methods. During the last two decades, various techniques have been developed. Among them, there are methods based on LF-based approaches, including methods based on linear matrix inequalities (LMI)

[8, 9], relatively novel SoS programming [10, 11], genetic and evolutionary algorithms [12, 13], sampling methods [14, 15], and others [16, 17]. In [18], an approach is proposed to estimate the RoA for a class of nonpolynomial systems. In this case, the Chebyshev approximations are used, which are more efficient than the Taylor approximation since they give a more accurate estimate. Estimating RoA for systems with uncertain parameters was discussed in [10, 19]. The papers considered various approaches to estimating RoA for the power systems [20, 21]. This line of research is developing in various directions. Some authors tend to improve the volume of the RoA estimation [6, 7], others wish to reduce the computation efforts [14, 22], and third intend to simplify the mathematical model involved in the computation process [17, 23]. Each technique has advantages and drawbacks, and no "gold" rule exists for LF choice.

3 Research Methodology

The class of systems investigated in this paper is given by the following state space differential equation

$$\dot{\boldsymbol{x}} = f(\boldsymbol{x}),\ \boldsymbol{x}(0) = \boldsymbol{x}_0 \tag{1}$$

where $\boldsymbol{x}$ represents the state vector of the system, $f : \Omega \to \mathbb{R}^n$ is locally Lipschitz function, and $\Omega \subset \mathbb{R}^n$ is a domain containing the origin. We assume that $f(0) = 0$, i.e., the origin is the stationary point of the system (1), and the matrix

$$\boldsymbol{A} = \frac{\partial f}{\partial \boldsymbol{x}}|_{x=0} \tag{2}$$

is a Hurwitz's matrix (it has all eigenvalues in the left open half plane of the complex plane).

Let

$$V = \boldsymbol{x}^T\boldsymbol{K}\boldsymbol{x}, \boldsymbol{K} = \boldsymbol{K}^T \in \mathbb{R}^{n\times n}, \tag{3}$$

where $\boldsymbol{K}$ is a positive definite matrix. The hyper-surfaces described by equality

$$\dot{V}(x) = 0(x \neq 0) \tag{4}$$

define the boundary of the region of negative definiteness of $\dot{V}$ in which we seek the guaranteed estimation for Ω_c.

In the case of LF is taken in the class of quadratic forms, the resulting estimate for RoA is the interior of the n-dimensional ellipsoid. Our goal is to find such a QFL for which the volume of the corresponding ellipsoid will be the largest.

Definition [21]. Let $\phi(t, x)$ be the solution of system (1) that starts at the initial state x_0 at time $t = 0$. The region of attraction of the origin is denoted by R_a is defined by

$$R_a = \{x \in D : \phi(t, x) \to 0\ as\ t \to \infty\}. \tag{5}$$

For most nonlinear systems, the exact determination of region (2) is not possible. Therefore, the task is to obtain an estimate

$$S \subset R_a, 0 \in S, \tag{6}$$

so that S can be represented in a relatively simple form. Such an estimate can be obtained using the Lyapunov function for system (1).

Consider a locally positive definite function $V(x)$ whose derivative $\dot{V}(x)$ is locally negative in a neighborhood of origin. We shall use the following statement.

Theorem [24]. Let S is a compact set, and the following conditions hold:

$$\exists c : 0 < V(\boldsymbol{x}) \leq c, \forall \boldsymbol{x} \in S(\boldsymbol{x} \neq 0) \textit{ and } \forall \boldsymbol{x} \in \partial S : V(\boldsymbol{x}) = c; \tag{7}$$

$$\forall \boldsymbol{x} \in S : \dot{V}(\boldsymbol{x}) < 0(\boldsymbol{x} \neq 0), \tag{8}$$

then $S \subset R_a$.

Note that the set $\boldsymbol{x} : V(\boldsymbol{x}) \leq c$ may not be compact, it is enough only that condition (4) is satisfied in the considered region S.

If all the eigenvalues of the system linearized in a neighborhood of the equilibrium are located in the left half-plane, the equilibrium is asymptotically stable regardless of the nonlinear expansion terms of the right-hand side. In terms of the direct Lyapunov method, it is sufficient to use quadratic LF. However, when estimating the region of attraction, the influence of nonlinear terms is significant. To obtain an effective estimate, it is necessary to add terms of a higher order to the LF $V(\boldsymbol{x})$ that can improve the properties of the derivative dV/dt.

Technically, the proposed approach is an iterative process for obtaining lower bounds for a DA to maximize the volume of the corresponding n-dimensional ellipsoid. Namely, some initial QLF candidate V_0 is chosen, and its improvement goes on throw the algorithm described below. We consider two different ways of constructing V_0. However, the subsequent calculations are carried out according to one scheme.

Let LF $V(x)$ be chosen, and c_{max} is the most considerable value which satisfies the Theorem. Thus, for any $c < c_{max}$ system $V(\boldsymbol{x}) = c, \dot{V}(\boldsymbol{x}) = 0$ is inconsistent. Geometrically it means that corresponding hyper-surfaces do not intersect in state space. This means that above-mentioned surfaces have a common tangent h-plane at $\boldsymbol{x}_0$ which leads to relations

$$\Phi_j = \frac{\partial V}{\partial x_j}\frac{\partial \dot{V}}{\partial x_s} - \frac{\partial V}{\partial x_s}\frac{\partial \dot{V}}{\partial x_j} = 0, j, s = 1, \ldots, k+m, j \neq s. \tag{9}$$

1. Calculate the expression for the derivative of LF.
2. We solve (as a rule, numerically) the system of equations.
3. Choose the appropriate solution. Such will be the real set $x_1^{(0)}, \ldots, x_n^{(0)}$ to which the minimum value of $c_0 = V_0(\boldsymbol{x}^{(0)}, \boldsymbol{\alpha}^{(0)})$ corresponds. Note that conditions (9) characterize both the internal contact (the region $0 \leq V_0 \leq c_0$ belongs to the region $\dot{V}_0 \leq 0$) and the external contact (in the part of the region S the requirement $\dot{V}_0 \leq 0$ is violated). Therefore, to check the correctness of the choice of $\boldsymbol{x}^{(0)}$, one can calculate $\dot{V}$ at a point on the surface $V_0(\boldsymbol{x}) = c_0$ close to the point of contact.
4. We perturb the "parameters of influence" $k_{jl} = k_{jl}^{(0)} + \varepsilon k_{jl}^{(1)}$, where ε is a small parameter. We are looking for a solution to system (9) in the form

$$\boldsymbol{x} = \boldsymbol{x}^{(0)} + \varepsilon \boldsymbol{x}^{(1)} + \boldsymbol{O}\left(\varepsilon^2\right). \tag{10}$$

The unknown component $\boldsymbol{x}^{(1)}$ of this solution may be found in the form $\boldsymbol{x}^{(1)} = \boldsymbol{B}(k_{jl}^{(1)})$, where $\boldsymbol{B}$ is some column vector which elements are known linear combinations of perturbations $k_{jl}^{(1)}$.

5. Substituting $\boldsymbol{x}^{(1)}$ into QLF, we have now the perturbation in the value of c in the following view

$$\delta c = c_1 - c_0 = \varepsilon\Big[\left(\boldsymbol{x}^{(0)}\right)^{\tau}\left(\boldsymbol{K}_0\boldsymbol{B} + \boldsymbol{K}^{(1)}\boldsymbol{x}^{(0)}\right) + \boldsymbol{B}^{\tau}\boldsymbol{K}_0\boldsymbol{x}^{(0)}\Big] = \langle \boldsymbol{P}, \boldsymbol{k}\rangle, \tag{11}$$

$$\boldsymbol{k} = (k_{11}, k_{12}, \ldots, k_{nn}).$$

The upper subscript "τ" means the transposition, angle brackets mean the scalar product, and $\boldsymbol{P}$ is some numerical vector determined by the nonlinear part of equations (9) and the choice of V_0. The described algorithm is the modification of the approach proposed in the paper [1].

Case A. Intuition-based choice of V_0. This way is widespread in the literature and usually consists in choosing as QLF candidate the sum of the squares of the phase variables. The reason is that in many (relatively simple) examples, this provides an acceptable initial estimate for DA. On the other hand, in more complex cases, when the nonlinear part of the right-hand side of the system (1) is not structured, it is almost impossible to make a meaningful choice since the analysis of the condition $\dot{V} < 0$ is complicated to implement. Under the conditions of such uncertainty, a good choice of V_0 is a matter of chance, when the choice of the simplest admissible option is justified. The advantage of such a choice of V_0 The simplicity of the initial processing cycle (solving a system of equations (9)) can be considered. A disadvantage is the need to verify that $\dot{V}_{lin}$ is the negative definite form, and the function V remains positive definite during the following iterations.

Case B. This choice of V_0 is oriented on the distribution of the eigenvalues of matrix A. Let this matrix has m pairs of complex conjugate eigenvalues $-\sigma_j \pm i\omega_j (j = \overline{1, m})$ and s real ones $-\sigma_{m+1}, \ldots, -\sigma_{m+s} (n = 2m + s)$, and its canonical Jordan form is diagonal matrix. Let $\beta_j \left(j = \overline{1, n}\right)$ are normalized eigenvectors of the matrix **A**. Then, the initial QLF candidate is taken in the form

$$V_0(x_1, \ldots, x_n, \alpha_1, \ldots, \alpha_{m+s}) = \sum_{j=1}^{m} \alpha_j z_j \bar{z}_j + \sum_{j=1}^{s} \alpha_{j+m} u_j^2, \begin{pmatrix} z \\ u \end{pmatrix} = \boldsymbol{T}^{-1}\boldsymbol{x}, \tag{12}$$

where the transformation matrix $\boldsymbol{T}$ is composed of columns of β_j.

4 Results and Discussion

Let us first consider two examples to demonstrate the proposed approach's effectiveness.

Example 1. Let's consider the second order system (system (18) from the article [7])

$$\dot{x}_1 = x_2, \ \dot{x}_2 = -\sin(x_1) - 0.3x_2. \tag{13}$$

As the initial LF candidate, we take the quadratic form

$$V^{(2)} = \left(\left(a^2 + d\right)x_1^2 + 2ax_1x_2 + x_2^2\right)/2,$$

where a, d are still unknown parameters ($a > 0$).

Since the eigenvalues of the Jacobian are complex $-0.1500 \pm 0.9887i$, then we take the initial approximation of the QLF in the form

$$V_0^{(2)} = \alpha_1 z_1 \bar{z}_1 = 0.255754(x_1^2 + 0.3x_1x_2 + x_2^2).$$

Thus, $a_0 = 0.15, d_0 = 0.9775$. Applying the procedure described in Sect. 3 after the first iteration, we find $a_1 = a_0 + 0.1 = 0.25, d_1 = d_0 - 0.015 = 0.9625$, after the second iteration $a = a_1 + 0 = 0.25, d = d_1 - 0.1 = 0.8625$. The estimates obtained are shown in Fig. 1a.

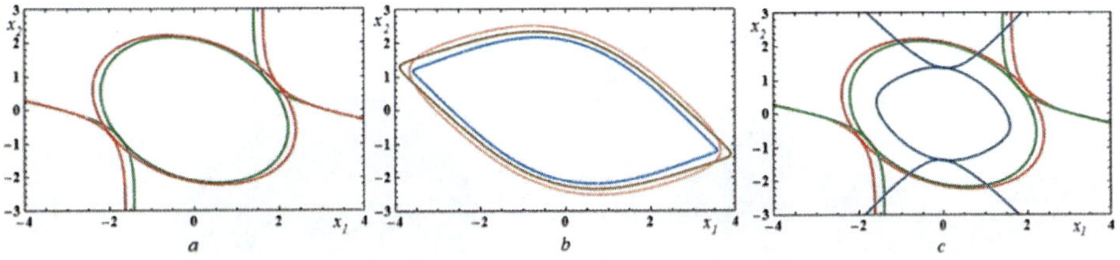

Fig. 1. Regions of estimation: a) $a = 0.25$, red color: $d = 0.8$; green color:$d = 0.9$; b) pink: $g = -0.3$; brown: $g = -0.1$; blue: $g = 0$; c) blue color - the estimation from [7].

If we supplement $V^{(2)}$ with a fourth-order form

$$V^{(4)} = k_{04}x_2^4 + k_{13}x_1x_2^3 + k_{22}x_1^2x_2^2 + k_{31}x_1^3x_2 + k_{40}x_1^4,$$

where $k_{40} = -0.0076 - 0.04g$, $k_{04} = -0.053g + 0.024$, $k_{22} = 2.54k_{04} + 0.01175g$, $k_{31} = 0.042 - 0.0625g$, $k_{13} = 1.2k_{04} + 0.0025g$,

then we obtain an improvement in the RoA estimates (Fig. 1b). Notably, even the estimates obtained using the QFL are better than the estimates obtained in work [7] based on SOS programming and using the sixth order LF (Fig. 1c)

$$\tilde{V} = (2x_1^6 + 6x_1^5x_2 + 10x_1^4x_2^2 + 2x_1^3x_2^3 + 20x_1^2x_2^4 - 6x_1x_2^5 + 6x_2^6 + 0.6x_1^4$$
$$+20x_1^3x_2 + 30x_1^2x_2^2 + 30x_1x_2^3 + 5x_2^4 + 100x_1^2 + 20x_1x_2 + 100x_2^2) \cdot 10^{-3}.$$

Example 2.

$$\dot{x}_1 = -x_1 + 3x_2 + 2x_3 + x_2x_3,$$

$$\dot{x}_2 = 3x_1 - x_2 - x_3 + x_1x_3, \tag{14}$$

$$\dot{x}_3 = -2x_1 + x_2 - x_3 + x_1x_2.$$

This system considers an example from work [25]. The following quadratic form was proposed as an LF candidate:

$$V^{(2)} = x_1^2 + 2x_1x_2 + x_1x_3 + 2x_2^2 + 2x_2x_3 + 2x_3^2.$$

According to our procedure, the following function is taken

$$V = 4\left(x_1^2 + 4x_1x_2 + 3x_1x_3 + 4x_2^2 + 12x_2x_3 + 9x_3^2\right)a + 5\left(13x_1^2 + 10x_2^2 + 5x_3^2 - 4x_1x_2 - 6x_1x_3 - 12x_2x_3\right).$$

where a > 0 is a free parameter.

As can be seen from Fig. 2, even one iteration with $a = 1$ is enough to get a significantly broader estimate for RoA.

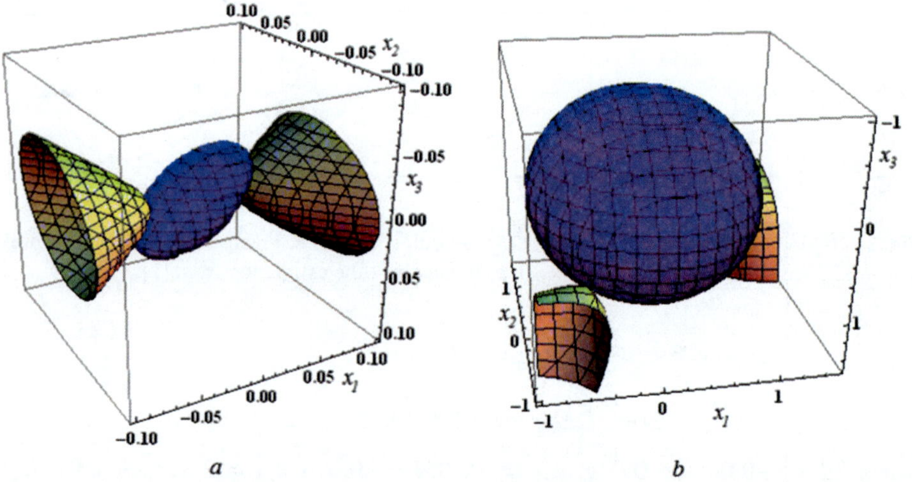

Fig. 2. Estimation of RoA for system (14): a – based on $V^{(2)}$; b – based on V.

Now let us consider the mechanical system, which consists of two coupled oscillators and is schematically presented in Fig. 3. Here x_1, x_2 are the displacements of the masses m_1, m_2 respectively, c_2 characterizes the viscous damping of the coupling between oscillators, and k_1, k_2^{lin} and k_2^{nonlin} represent the stiffnesses of the springs. The motion equations are

$$m_1\ddot{x}_1 + c_2(\dot{x}_1 - \dot{x}_2) + k_1x_1 + k_2^{lin}(x_1 - x_2) - k_2^{nonlin}(x_1 - x_2)^3 = 0,$$

$$m_2\ddot{x}_2 + c_2(\dot{x}_2 - \dot{x}_1) + k_2^{lin}(x_2 - x_1) - k_2^{nonlin}(x_2 - x_1)^3 = 0. \quad (15)$$

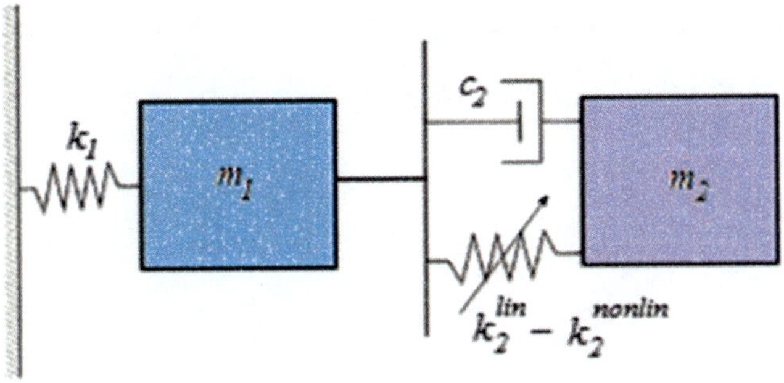

Fig. 3. Mechanical system.

Introducing the dimensionless parameters and time by formulas

$$\mu = \frac{m_2}{m_1}, \omega_1 = \sqrt{\frac{k_1}{m_1}}, \omega_2 = \sqrt{\frac{k_2^{lin}}{m_2}}, \kappa = \frac{\omega_2^2}{\omega_1^2}, h = \frac{c_2\omega_1}{2\mu^2 k_1}, \tag{16}$$

$$x_j = \sqrt{\frac{m_2}{k_2^{nonlin}}}\,\tilde{x_j}\ (j = 1, 2), \tau = \omega t,$$

we can rewrite the system (15) in the following form

$$\tilde{x}_1'' + 2\mu h\left(\tilde{x}_1' - \tilde{x}_2'\right) + (1 + \mu\kappa)\tilde{x}_1 - \mu\kappa\tilde{x}_2 - \mu(\tilde{x}_1 - \tilde{x}_2)^3 = 0, \tag{17}$$

$$\tilde{x}_2'' - 2h\left(\tilde{x}_1' - \tilde{x}_2'\right) - \kappa(\tilde{x}_1 - \tilde{x}_2) + (\tilde{x}_1 - \tilde{x}_2)^3 = 0.$$

For convenience, the symbol "~" is omitted below.

The matrix of the linear part of the system (17) written in normal form has the following form

$$\begin{pmatrix} 0 & 0 & 1 & 0 \\ 0 & 0 & 0 & 1 \\ -1-\mu\kappa & -\mu\kappa & -2\mu h & 2\mu h \\ \kappa & -\kappa & 2h & -2h \end{pmatrix} \tag{18}$$

To get an estimate for the domain of attraction for the system (18), we use the procedure described in the previous section, which relies on the canonical Jordan form of a matrix (18). This matrix is needed to compose the Lyapunov function. To illustrate the procedure, let us take the following parameter values: $\mu = \frac{1}{4}, \kappa = \frac{2}{3}, h = \frac{1}{3}$.

The corresponding transformation matrix had the following view

$$\begin{pmatrix} 1 & 1 & -\frac{1}{8}(1 - i\sqrt{15}) & -\frac{1}{8}(1 + i\sqrt{15}) \\ \frac{1}{2}(3 - i\sqrt{23}) & \frac{1}{2}(3 + i\sqrt{23}) & 1 & 1 \\ \frac{1}{6}(-1 + i\sqrt{23}) & -\frac{1}{6}(1 + i\sqrt{23}) & -\frac{1}{16}(7 + i\sqrt{15}) & \frac{1}{16}(-7 + i\sqrt{15}) \\ \frac{1}{3}(5 + i\sqrt{23}) & \frac{1}{3}(5 - i\sqrt{23}) & -\frac{1}{4}(1 - i\sqrt{23}) & -\frac{1}{4}(1 + i\sqrt{23}) \end{pmatrix},$$

According to the procedure described in Sect. 3, we take the candidate of LF:

$$V(x_1, x_2, x_3, x_4) = z_1\bar{z}_1 + \alpha z_2\bar{z}_2(x_3 = \dot{x}_1, x_4 = \dot{x}_2), \tag{19}$$

where

$$z_1\bar{z}_1 = \frac{3}{92}(12x_1^2 + 2x_2^2 + 8x_3^2 + 4x_4^2 - 2x_1x_2 - 4x_1x_3 - 13x_1x_4 + 8x_2x_3 + 3x_2x_4 + 6x_3x_4),$$

$$z_2\bar{z}_2 = \frac{2}{15}(20x_1^2 + 2x_2^2 + 12x_3^2 + 3x_4^2 - 10x_1x_2 - 15x_1x_4 + 6x_2x_3 + 3x_2x_4 - 3x_3x_4).$$

$$\begin{aligned} V' = &-[\left(\frac{3}{23} + \frac{4}{3}\alpha\right)x_1^2 + \left(\frac{1}{46} + \frac{2}{15}\alpha\right)x_2^2 + \left(\frac{2}{23} + \frac{4}{5}\alpha\right)x_3^2 + \left(\frac{1}{23} + \frac{1}{5}\alpha\right)x_4^2 \\ &-\left(\frac{1}{46} + \frac{2}{3}\alpha\right)x_1x_2 - \frac{1}{23}x_1x_3 - \left(\frac{13}{92} + \alpha\right)x_1x_4 + \left(\frac{2}{23} + \frac{2}{5}\alpha\right)x_2x_3 \\ &+\left(\frac{3}{92} + \frac{1}{5}\alpha\right)x_2x_4 + \left(\frac{3}{46} - \frac{1}{5}\alpha\right)x_3x_4 - \left(\frac{9}{23} + 2\alpha\right)x_1^4 \\ &-\left(\frac{3}{92} + \frac{1}{5}\alpha\right)x_2^4 + \left(\frac{111}{46} + \frac{31}{5}\alpha\right)x_1^3x_2 + 3\left(\frac{1}{46} - \frac{2}{5}\alpha\right)x_1^3x_3 \\ &+\frac{3}{2}\left(\frac{13}{92} + \frac{3}{5}\alpha\right)x_1^3x_4 - 3\left(\frac{39}{92} + \frac{11}{5}\alpha\right)x_1^2x_2^2 - 9\left(\frac{1}{46} - \frac{2}{5}\alpha\right)x_1^2x_2x_3 \\ &-\frac{9}{2}\left(\frac{13}{92} + \frac{3}{5}\alpha\right)x_1^2x_2x_4 + \left(\frac{45}{92} + \frac{13}{5}\alpha\right)x_1x_2^3 + 9\left(\frac{1}{46} - \frac{2}{5}\alpha\right)x_1x_2^2x_3 \\ &+\frac{9}{2}\left(\frac{13}{92} + \frac{3}{5}\alpha\right)x_1x_2^2x_4 - 3\left(\frac{1}{46} - \frac{2}{5}\alpha\right)x_2^3x_3 - \frac{3}{2}\left(\frac{13}{92} + \frac{3}{5}\alpha\right)x_2^3x_4. \end{aligned}$$

Further, we solve the nonlinear system (9) supplemented by equation $V'(x_1, x_2, x_3, x_4) = 0$. This system has 38 nontrivial solutions, mainly complex. The only pair of real solutions is $x_1^{(0)} \approx 0.068, x_2^{(0)} \approx -0.31, x_3^{(0)} \approx 0.11, x_4^{(0)} \approx 0.13$ (another solution has the opposite signs). As a result, we find

$$c_0 = V\left(\boldsymbol{x}^{(0)}\right) \approx 0.155 \cdot 10^{-2}.$$

To complete step 3 of the procedure, we check the condition of the "inner" tangency of the hyper-surfaces $V(x_1, x_2, x_3, x_4) = c_0$ and $V'(x_1, x_2, x_3, x_4) = 0$. To do this, we fix three coordinates in the neighborhood of the point $M_0\left(\boldsymbol{x}^{(0)}\right)$, for instance $x_1 = 0.07, x_2 = -0.31, x_3 = 0.11$. Then we find two values of x_4 from equation $V = c_0$. These values are 0.2321 and 0.1379. Calculating the value of V' we have

$$V'(M_1) \approx -0.0378 < 0, \quad V'(M_2) \approx -0.00016 < 0.$$

To do the fourth step, we put $\alpha = 1 + \alpha_1, x_j = x_j^{(0)} + \alpha_1 x_j^{(1)} \left(j = \overline{1,4}\right)$ and take the linear parts of expansions of functions V, Φ_{js} on parameter α_1. This leads to the following linear system

$$0.5185x_1^{(1)} - 0.1360x_2^{(1)} - 0.0040x_3^{(1)} - 0.2230x_4^{(1)} = 0.0082\alpha^{(1)},$$

$$-0.9701x_1^{(1)} + 2.2368x_2^{(1)} + 5.2018x_3^{(1)} + 1.1597x_4^{(1)} = -0.00057\alpha^{(1)},$$

$$-4.8643x_1^{(1)} - 1.0261x_2^{(1)} - 2.5388x_3^{(1)} + 2.3268x_4^{(1)} = 0.2775\alpha^{(1)},$$

$$4.5591x_1^{(1)} - 1.6020x_2^{(1)} - 2.8772x_3^{(1)} - 3.5203x_4^{(1)} = -0.1328\alpha^1.$$

The solution of this system is

$$x_1^{(1)} = -0.048\alpha^{(1)}, x_2^{(1)} = -0.27\alpha^{(1)}, x_3^{(1)} = 0.11\alpha^{(1)}, x_4^{(1)} = 0.013\alpha^{(1)}.$$

Substituting the found values into a formula for the volume of 4-D ellipsoid $V = c$,

$$Volume = \frac{c^2}{\sqrt{\det(hessian(V))}}$$

we have $Volume|_{\alpha=1} \approx (0.3818 + 1.7556\alpha_1) \cdot 10^{-3}$.

As coefficient at the α_1 is positive, then increasing the value of α will lead to increasing the value of the volume of ellipsoid $V = c$. Thus, now we take $\alpha = 2$ and repeat the calculation procedure.

In the same way as before, we find

$$x_1 = 0.0538 - 0.05890\alpha_1, x_2 = -0.3508 - 0.1084\alpha_1,$$

$$x_3 = 0.1292 + 0.0365\alpha_1, x_4 = 0.1324 - 0.0027\alpha_1,$$

$$Volume|_{\alpha=2} \approx (0.6937 + 0.7518\alpha_1) \cdot 10^{-3}.$$

For the subsequent iterations, we have

$$Volume|_{\alpha=3} \approx (0.7938 + 0.1246\alpha_1) \cdot 10^{-3},$$

$$Volume|_{\alpha=4} \approx (0.7853 - 0.1511\alpha_1) \cdot 10^{-3}.$$

Thus, the optimal value of α belongs to the interval $(3, 4)$. After a couple of other iterations, we find that the best approximation is obtained when $\alpha \approx 3.39$ and is equal to $0.8007 \cdot 10^{-3}$ (see Fig. 4 and Table 1).

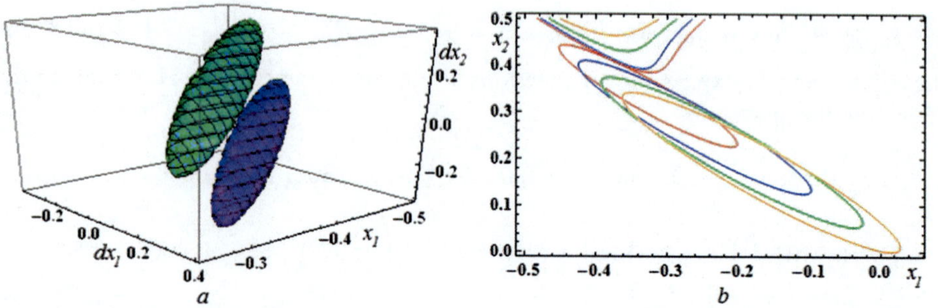

Fig. 4. The boundaries of domains $0 < V < c, V < 0$: a) the intersection with hypersurface $x_1 - x_2 = 0.4$; b) red line: $dx_1 = 0.13, dx_2 = -0.1257$; blue line: $dx_1 = 0.1, dx_2 = -0.0966$; green line: $dx_1 = 0.08, dx_2 = -0.0773$; orange line: $dx_1 = 0.06, dx_2 = -0.058$.

Table 1. The values of α, c and volumes.

α	c	Volume, 10^{-3}	α	c	Volume, 10^{-3}
1	0.0475	0.382	3.4	0.1267	0.7992
2	0.9006	0.6912	3.6	0.1302	0.797
2.8	0.114	0.7854	4	0.1363	0.7854
3	0.1187	0.7942	4.4	0.1413	0.7665
3.2	0.1229	0.7986	4.8	0.1454	0.7226

5 Conclusions

We modify the method proposed in [1] for constructing the polynomial LF to obtain the RoA estimation for 2-DoF nonlinear mechanical system. In state space, these estimations have the shape of stretched ellipsoids. The algorithm showed its usefulness, and the objective of the investigation was achieved. Notably, there are very few examples for systems of fourth order in literature, and it is impossible to compare our estimates with other results.

Simultaneously, the system studied in [6] is close enough to our case, and the shape of estimates in this work is the same ([6], Fig. 6).

Improvements of the proposed algorithms and study of the influence of nonlinear components of stiffness onto the shape and volume of RoA estimation will be considered further.

References

1. Awrejcewicz, J., Bilichenko, D., Cheib, A.K., Losyeva, N., Puzyrov, V.: Estimating the region of attraction based on a polynomial Lyapunov function. Appl. Math. Model. **90**, 1143–1152 (2021). https://doi.org/10.1016/j.apm.2020.10.010
2. Sliwa, I., Grygiel, K.: Periodic orbits, basins of attraction and chaotic beats in two coupled Kerr oscillators. Nonlin. Dyn. **67**(1), 755–765 (2012)

3. de Freitas, M.S., Viana, R.L., Grebogi, C.: Basins of attraction of periodic oscillations in suspension bridges. Nonlin. Dyn. **37**(3), 207–226 (2004)
4. Cruck, E., Moitie, R., Seube, N.: Estimation of basins of attraction for uncertain systems with affine and Lipschitz dynamics. Dyn. Control **11**(3), 211–227 (2001)
5. Genesio, R., Tartaglia, M., Vicino, A.: On the estimation of asymptotic stability regions: state of the art and new proposals. IEEE Trans. Autom. Control **30**(8), 747–755 (1985)
6. Kant, N., Chowdhury, D., Mukherjee, R., Khalil, H.K.: An algorithm for enlarging the region of attraction using trajectory reversing. In: 2017 American Control Conference (ACC), pp. 4171–4176 (2017)
7. Li, Y., Li, C., He, Z., Shen, Z.: Estimating and enlarging the region of attraction of multi-equilibrium points system by state-dependent edge impulses. Nonlin. Dyn. **103**(3), 2421–2436 (2021). https://doi.org/10.1007/s11071-021-06259-9
8. Chesi, G., Garulli, A., Tesi, A., Vicino, A.: LMI-based computation of optimal quadratic Lyapunov functions for odd polynomial systems. Int. J. Robust Nonlin. Control **1**(15), 35–49 (2005)
9. Topcu, U., Packard, A.K., Seiler, P.: Local stability analysis using simulations and sum-of-squares programming. Automatica **44**, 2669–2675 (2008)
10. Chesi, G.: Domain of Attraction. Springer, London (2011)
11. Tan, W., Packard, A.: Stability region analysis using polynomial and composite polynomial Lyapunov functions and sum-of-squares programming. IEEE Trans. Autom. Control **53**(2), 565–571 (2008)
12. Grosman, B., Lewin, D.R.: Automatic generation of Lyapunov functions using genetic programming. IFAC Proc. **38**(1), 75–80 (2005)
13. McGough, J.S., Christianson, A.W., Hoover, R.C.: Symbolic computation of Lyapunov functions using evolutionary algorithms. In: Proceedings of the 12th IASTED International Conference, vol. 15, pp. 508–515 (2010)
14. Najafi, E., Babuška, R., Lopes, G.: A fast sampling method for estimating the domain of attraction. Nonlin. Dyn. **86**(2), 823–834 (2016)
15. Bobiti, R., Lazar, M.: Automated sampling-based stability verification and DOA estimation for nonlinear systems. IEEE Trans. Autom. Control **63**(11), 3659–3674 (2018)
16. Henrion, D., Korda, M.: Convex computation of the region of attraction of polynomial control systems. IEEE Trans. Autom. Control **2**(59), 297–312 (2014)
17. Khodadadi, L., Samadi, B., Khaloozadeh, H.: Estimation of region of attraction for polynomial nonlinear systems: a numerical method. ISA Trans. **53**, 25–32 (2014)
18. Han, D., Panagou, D.: Chebyshev approximation and higher order derivatives of Lyapunov functions for estimating the domain of attraction. In: 2017 IEEE 56th Annual Conference on Decision and Control (CDC), pp. 1181–1186 (2017)
19. Anghel, M., Milano, F., Papachristodoulou, A.: Algorithmic construction of Lyapunov functions for power system stability analysis. IEEE Trans. Circuits Syst. I. Regul. Pap. **60**(9), 2533–2546 (2013)
20. Izumi, S., Somekawa, H., Xin, X., Yamasaki, T.: Estimation of regions of attraction of power systems by using sum of squares programming. Electric. Eng. **100**(4), 2205–2216 (2018). https://doi.org/10.1007/s00202-018-0690-z
21. Khalil, H.: Nonlinear Systems, 3rd edn. Prentice Hall, New Jersey (2002)
22. Ji, Z., Wu, W., Feng, Y., Zhang, G.: Constructing the Lyapunov function through solving positive dimensional polynomial system. J. App. Math. **2013**, 859578 (2013). https://doi.org/10.1155/2013/859578
23. Wu, M., Yang, Z., Lin, W.: Domain-of-attraction estimation for uncertain non-polynomial systems. Commun. Nonlin. Sci. Numer. Simulat. **19**, 3044–3052 (2014)
24. Rouche, N., Habets, P., Laloy, M.: Stability Theory by Liapunovs Direct Method. Springer, New York (1977)

25. She, Z., Xia, B., Xiao, R., Zheng, Z.: A semi-algebraic approach for asymptotic stability analysis. Nonlin. Anal. Hybrid Syst. **3**(4), 588–596 (2009). https://doi.org/10.1016/j.nahs.2009.04.010

Contact of a Ball Piston with a Running Track in a Hydrovolumetric Transmission Regarding the Elastic Properties of the Material

Mykola Tkachuk(✉), Andrey Grabovskiy, Mykola Tkachuk, Iryna Hrechka, and Hanna Tkachuk

National Technical University Kharkiv Polytechnic Institute, 2, Kyrpychova Street, Kharkiv 61002, Ukraine
m.tkachuk@tmm-sapr.org

Abstract. Radial hydrovolumetric transmissions are commonly used in modern heavy-tracked vehicles. Some of these hydrovolumetric transmissions have ball pistons. Such pistons perform two types of motion: translation along the circular trajectory and rolling over the running track. The maximal stresses are concentrated around the contact spot between the interacting ball piston and the cylindrical running track. The magnitude of the contact pressure largely dictates the strength of the pistons. The small initial gap determines the contact spot's shape and the contact pressure distribution. The elastic properties of the ball piston are another crucial factor considered in this paper. The choice of materials suitable for this critical component of the hydrovolumetric drive is limited by the strength and stress concentration. The maximal contact pressure and equivalent stress can be interpolated from the parametric analysis data for an arbitrary combination of material elastic properties.

Keywords: Contact interaction · Contact pressure · Material properties · R&D investment

1 Introduction

Heavy wheeled and tracked vehicles tend to be equipped with continuously variable transmissions. These vehicles include quarry dump trucks, tractors, road machinery, various armored combat vehicles, and many others. Radial hydrovolumetric transmissions are used besides more traditional technical solutions such as electric transmissions, axial hydraulic drives, or hydrodynamic drives. They have certain advantages concerning dimensional limitations and high specific power requirements. The particular design studied in this paper is a radial hydrovolumetric transmission with ball pistons GOP-900 [1].

The working capacity of this type of transmission depends highly on the contact interaction of the ball pistons with the stator ring and the local stress-strain state.

V. Tonkonogyi et al. (Eds.): InterPartner 2022, LNME, pp. 495–505, 2023.
https://doi.org/10.1007/978-3-031-16651-8_47

Two major factors determine the stress-strain state of the ball pistons. Firstly, that is the complex shape of the running track on the surface of the stator ring. It is recommended to make it geometrically close to the ball pistons, which means that the initial gap is very small in one of the directions. Consequently, traditional models of contact, such as the well-known Hertzian theory [2], are not applicable in this case. Secondly, the deformations depend on the mechanical properties of the contacting bodies. In particular, the possible changes in the pistons' material can be considered. Once they are manufactured from a material with different elastic constants, the shape of the contact spot and the contact pressure distribution will be different. Hence a more favorable combination of stress magnitude and the material strength limit may be achieved. The paper delivers an analysis of this problem.

2 Literature Review

Modern hydraulic transmissions are based on various design solutions. The radial hydrovolumetric transmission GOP-900 [1] shown in Fig. 1 has a straightforward working principle. The hydraulic liquid is pumped from a radial cylinder at high pressure under the action of the cyclic translational motion of the ball cylinders. This motion is resulted from the rotor rotation with the cylinder block, which forces the rolling of the ball pistons over the stator surface along a circular trajectory. Due to the misalignment of the rotor and the stator axes, the radial distance of the ball pistons is reciprocating. The stroke of the piston is twice as large as the misalignment distance. The hydropump part of the transmission is set with the fixed maximal misalignment, while the axial distance between the rotor and stator is variable in the hydromotor. This design requires the ball pistons to transmit extreme forces in contact with the stator. This is the major challenge to the strength of these crucial transmission components [3].

Other problems are also typical for hydraulic machines. The oscillations of the interaxial distance due to the manufacturing error of the working surfaces are studied in works [4]. These intolerances negatively affect the output characteristics of orbital hydromotors [5]. As justified, the limits for rotor surfaces deviation need to be kept in order to avoid oscillations of the radial gap and to maintain the stable output characteristics of the orbital hydraulic motor. The work [6] is considers rotary hydraulic machines with improved output characteristics. A multi-criteria optimization is employed in order to design mechatronic systems with the desired output characteristics. An experimental installation for testing high-torque hydraulic motors is developed in [7]. A methodology for testing hydromechanical efficiency of planetary hydraulic motors is proposed.

A planetary motor dynamics computational model is proposed in [8]. The simulations predict unsteady torque on the hydraulic motor shaft and pressure in the discharge line of the mechatronic system. These results are used to modify the distribution system of the upgraded hydraulic motor.

Various computational models and analyses have been developed for contact problems of complex shaped machine parts. The most general theoretical framework is provided by variation inequalities [9]. A unilateral contact can be consistently formulated as a convex optimization problem for the unknown displacement field subjected to the inequality constraints [10]. The alternative approach is to take the contact pressure as the primary variable [11]. Thus integral boundary equations in a strong form of Kalker's variational principle [11] as a weak form can be applied to solve this class of nonlinear mechanical problems [12]. Since the unknown contact pressure distribution is defined on the boundary or even more narrowly on the contact interface (instead of the bulk of the interacting bodies), the problem's dimensionality is reduced by one [13].

The displacement-based formulations are generally treated using the finite element methods [14]. The implementation of contact is available in most commercial finite element analysis packages [15]. On the contrary, the boundary element methods are more suitable for the traction-based approach. Both approaches have their advantages and shortcomings. Due to the reduced dimensionality, the boundary element method requires a much smaller model size.

On the other hand, there are fewer boundary element implementations available for the researchers compared to the finite elements. In the previous works [16] we have utilized both proposed approaches. When required, the newly developed implementations of the computational models were extended to the peculiar features of the studied objects, which includes surface roughness [17], adhesion [18], intermediate contact layers with linear [3] and nonlinear stiffness [19], presence of gas or liquid on the interface [20]. The present work extends the research described in [3] by accounting for variable elastic parameters of one of the contacting bodies.

3 Research Methodology

The research objective is to evaluate the contact interaction of a ball piston with a running track on a stator ring depending on the material properties of the first body. Following the work [3], a finite element model of the two contacting elements of a hydrovolumetric drive GOP-900 was built.

The Young's modulus $E \equiv p_1$ and Poisson's coefficient $\nu \equiv p_2$ are varied in the range: $p_1 \in [2 \cdot 10^{10}; 2 \cdot 10^{12}]$ Pa; $p_2 \in [0.1; 0.48]$. A basic steel alloy with the elastic parameters $E^0 \equiv p_1{}^0 = 2 \cdot 10^{11}$ Pa; $\nu \equiv p_2{}^0 = 0.3$ is considered as a reference material.

We introduce dimensionless coefficients

$$\alpha_1 = \log\left(p_1/p_1^0\right);\ \alpha_2 = \log\left(p_2/p_2^0\right), \tag{1}$$

that vary in the range

$$\alpha_1 \in [-1;\ 1];\ \alpha_2 \in \left[\log(1/3),\ \log 1.6\right]. \tag{2}$$

Contact interaction is characterized by the contact pressure distribution q, Pa in the contact interface area S_c. The maximal obtained value of the contact traction is controlled in the parametric analysis

$$q_{\max} = q_{\max}(p_1, p_2) = q_{\max}(\alpha_1,\ \alpha_2). \tag{3}$$

In order to estimate the strength, the maximal level of von Mises equivalent stress is considered as well

$$\sigma_{\max} = \sigma_{\max}(p_1, p_2) = \sigma_{\max}(\alpha_1,\ \alpha_2). \tag{4}$$

The reference values given as

$$q_{\max}^0 = q_{\max}\left(p_1^0, p_2^0\right);\ \sigma_{\max}^0 = \sigma_{\max}\left(p_1^0, p_2^0\right), \tag{5}$$

and the dimensionless ratio characteristics can be introduced

$$q^\wedge = q^\wedge(p_1, p_2) = q^\wedge(\alpha_1, \alpha_2) = q_{\max}(p_1, p_2)/q_{\max}^0, \tag{6}$$

$$\sigma^\wedge = \sigma^\wedge(p_1, p_2) = \sigma^\wedge(\alpha_1, \alpha_2) = \sigma_{\max}(p_1, p_2)/\sigma_{\max}^0. \tag{7}$$

These relations represent the local loading of the material in terms of the contact tractions on the surface and the equivalent stresses inside the ball piston of hydrovolumetric transmission. Either the entire distributions of q i σ; their maximal values $q_{\max}$ i $\sigma_{\max}$; or the relative dimensionless factors $q^\wedge$ i $\sigma^\wedge$ may be used in strength criteria. Varying the parameters p_1 i p_2 (or their dimensionless counterparts α_1 i α_2) one can examine possible material choices and their effect on the local stress-strain state. Ultimately, these data can be used to assess the potential designs concerning the strength of the ball pistons as the most critical components of hydrovolumetric transmission.

4 Results and Discussion

4.1 Finite Element Model

The model of the hydrovolumetric transmission GOP-900 (Fig. 1) used in this study was built in the works [1, 3]. Geometrical parameters (Fig. 2) are equivalent to those in the original work [3]: Rp = 0,03175 m is the piston radius, Rsp = 0,128 m is the radius of the piston center circular trajectory, Rst = 0,15975 m is the radius of the stator ring, Rrot = 0,145 m is the radius of the rotor housing, pressing force is P = 100 kN. The finite element model limited to a quarter segment due to symmetry is shown in Fig. 3.

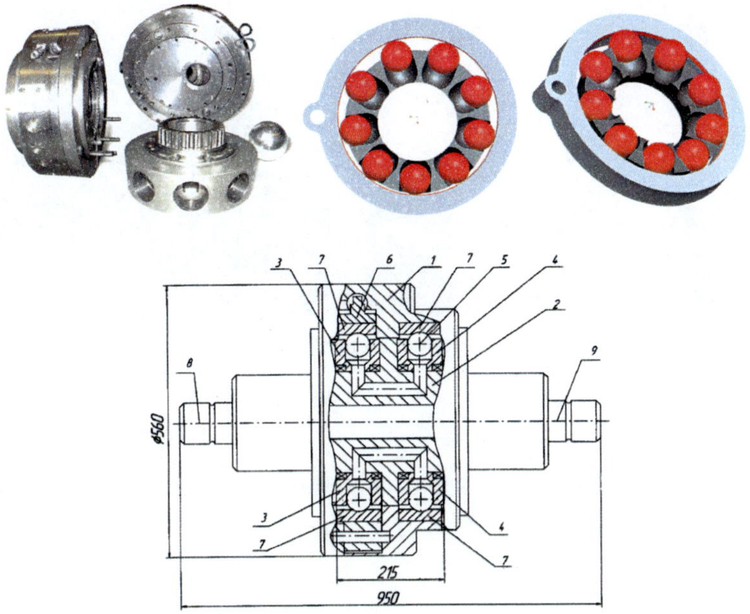

Fig. 1. The hydrovolumetric transmission GOP-900 with ball pistons: 1 – housing; 2 – block of pin distributors; 3 – pump cylinder block (rotor); 4 – cylinder block of the hydraulic motor (rotor); 5 – ball-piston; 6 – pump stator; 7 – a running track on the pump and the hydraulic motor; 8 and 9 – input and output shafts of hydraulic transmission [1, 3].

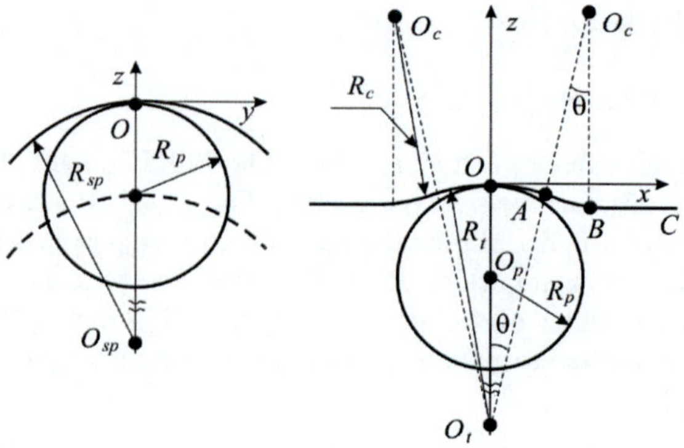

Fig. 2. Geometric parameters of the ball piston and the running track [3].

Fig. 3. Geometric and finite-element models of the two-body system "ball piston - running track".

4.2 Research Results

The contact pressure distributions q for the sample design points with the specified parameters α1, α2 are given in Table 1.

Table 1. Contact pressure distributions q (MPa) on the interface between the ball piston and the running track for the sample design points with parameters α_1, α_2.

α_1 / α_2	−1	0	1
log (1/3)	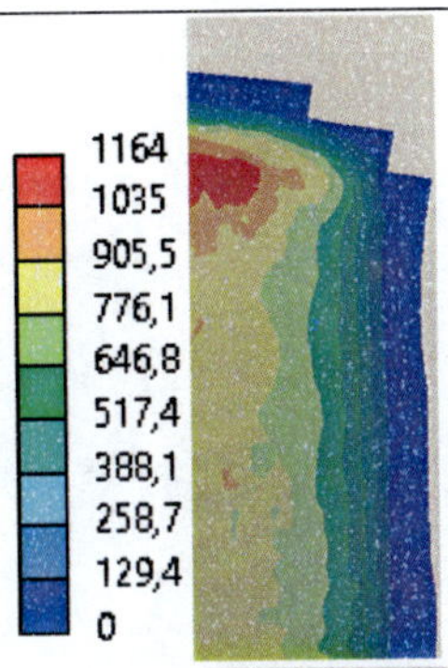	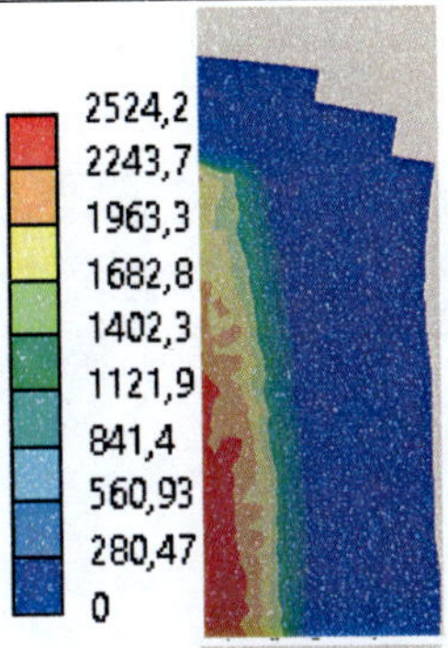	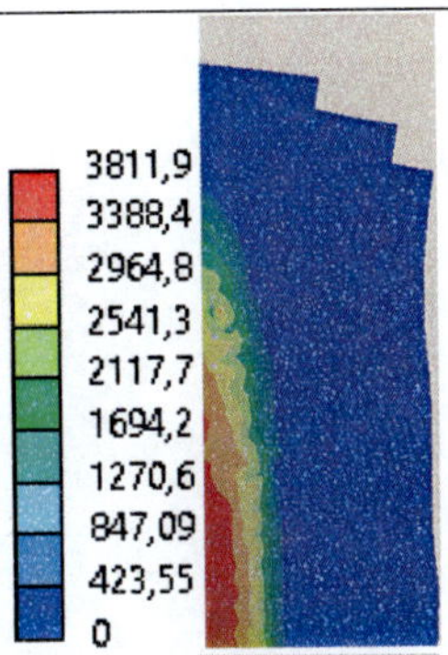
0	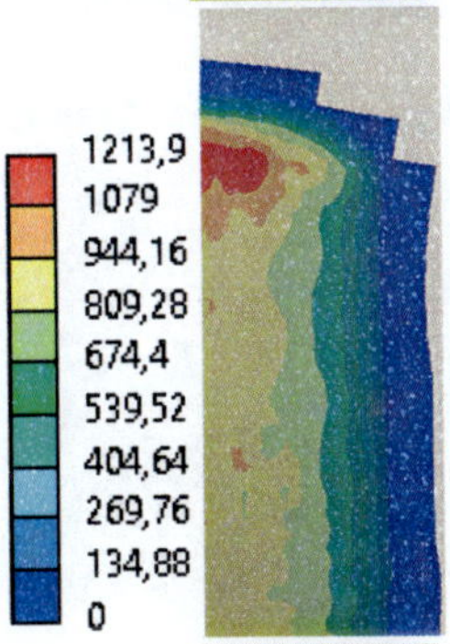	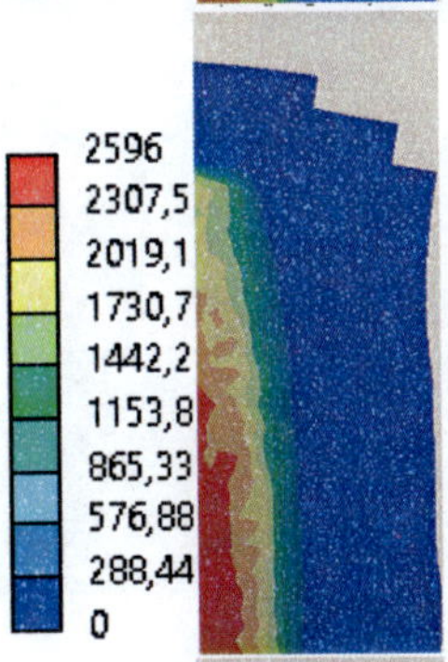	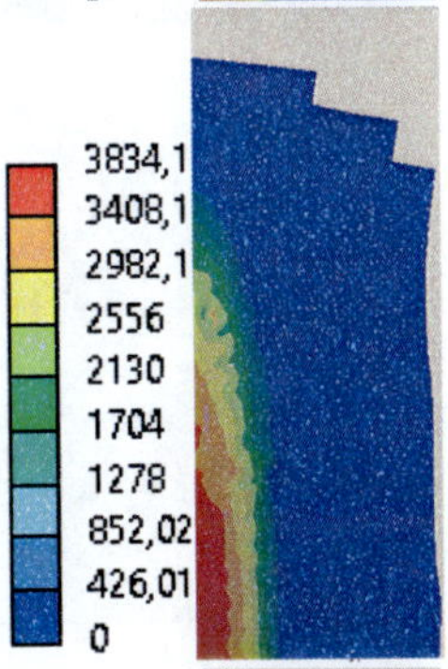
log 1.6	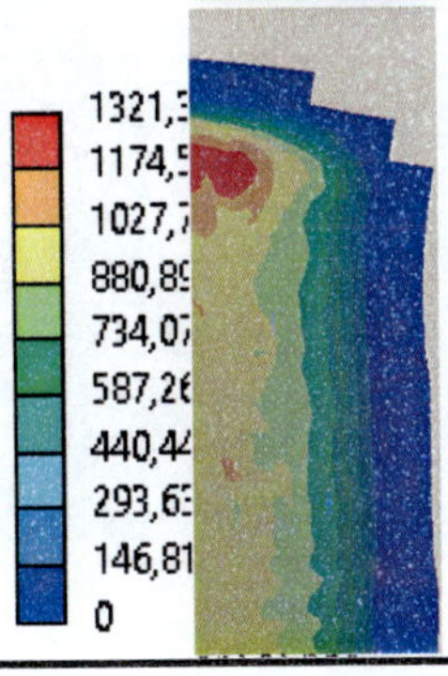	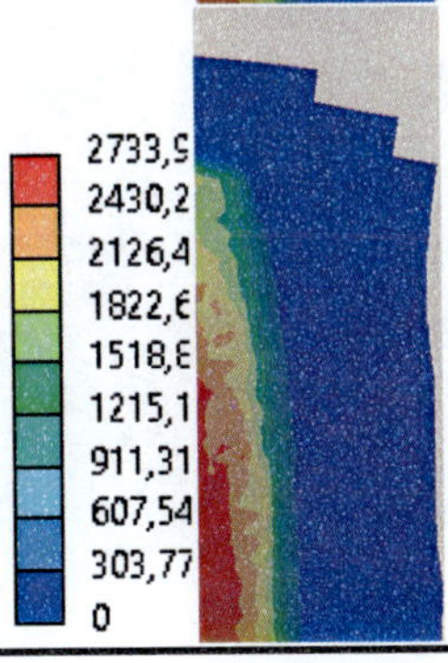	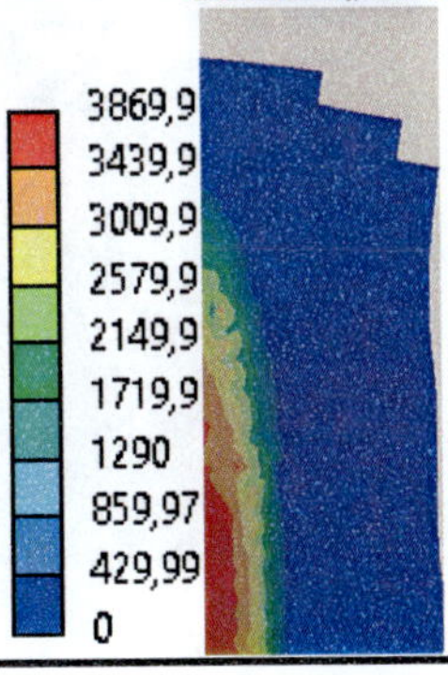

A similar case study of the von Mises stress σ is shown in Table 2.

Table 2. Equivalent stress distributions σ (MPa) in the ball piston for the sample design points with parameters α_1, α_2.

α_1 / α_2	−1	0	1
log (1/3)	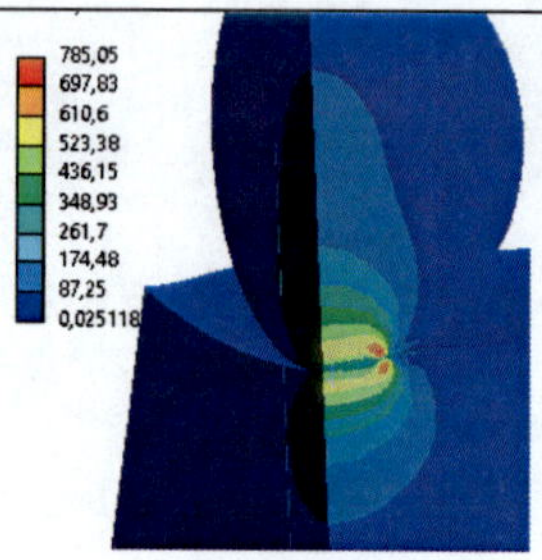	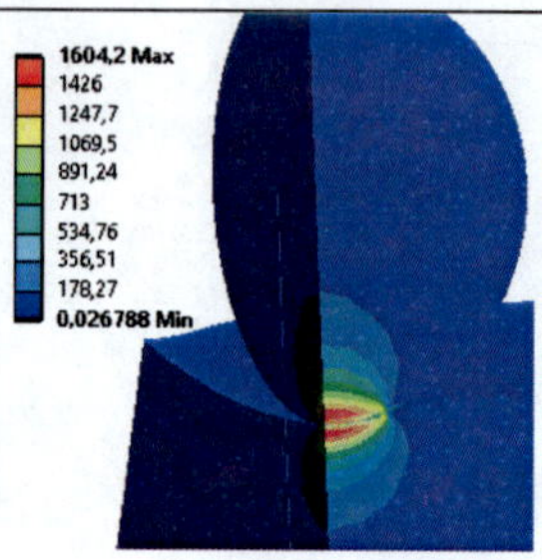	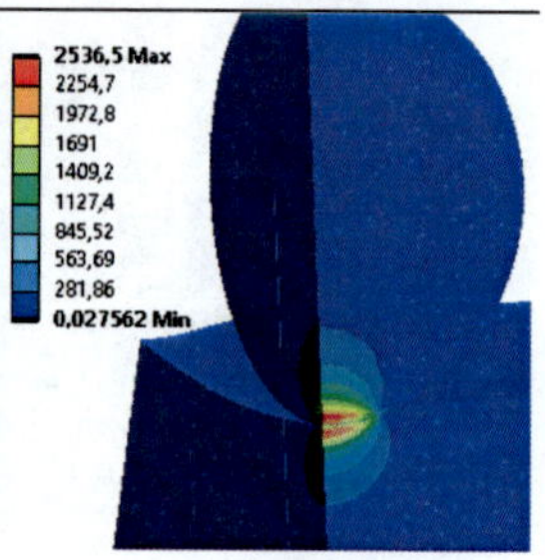
0	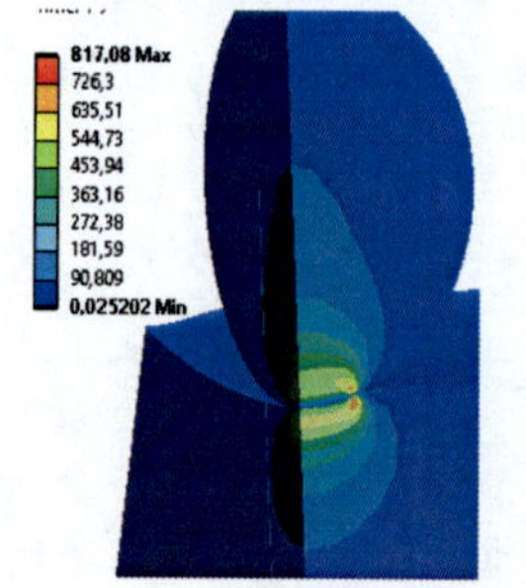	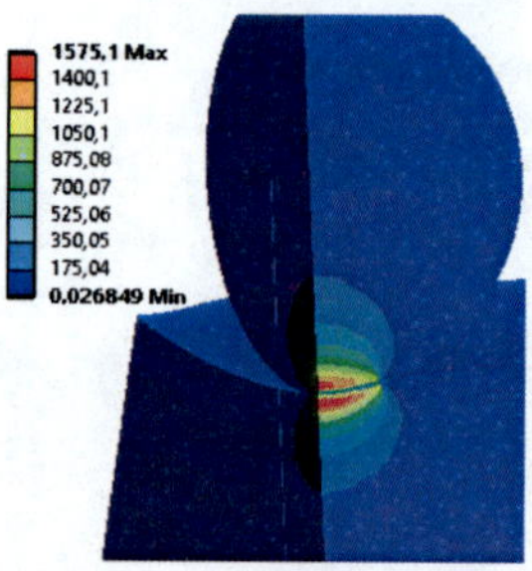	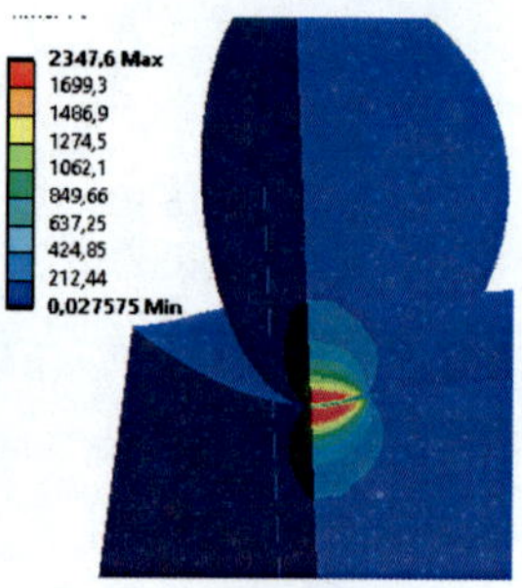
log 1.6	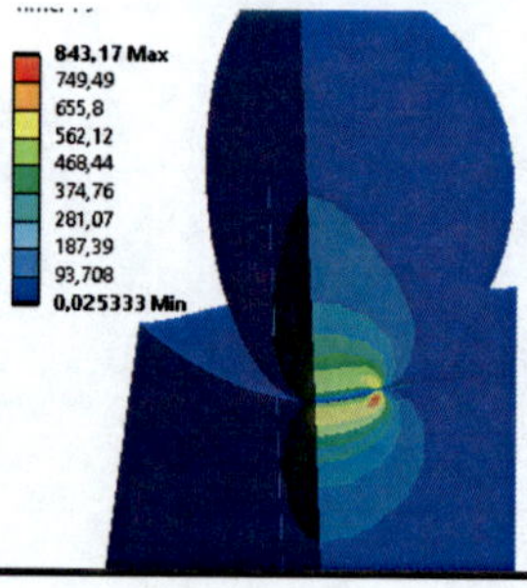	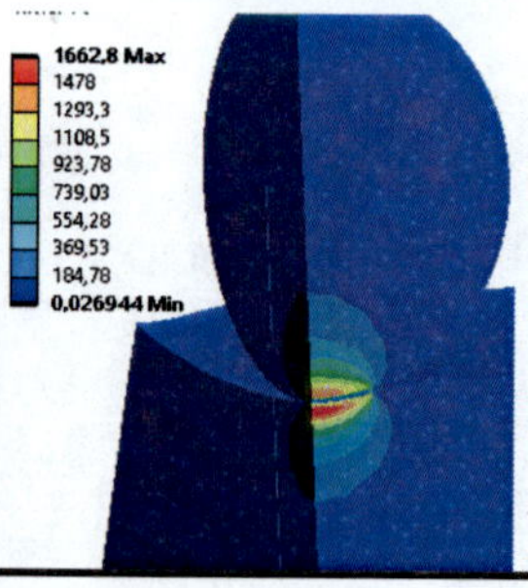	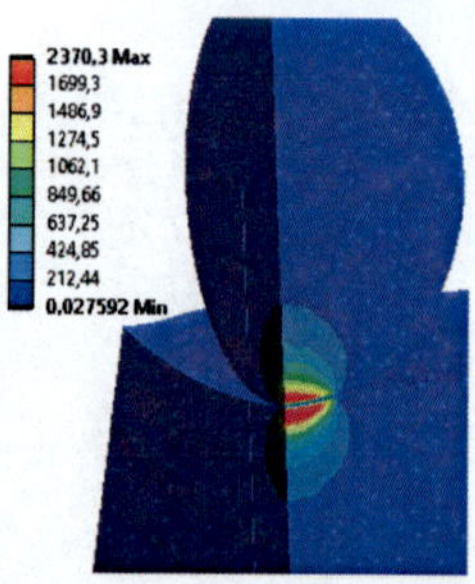

The parameter study results in the whole range of the variable elastic constants are given by dimensionless relations q^(α1, α2) and σ^(α1, α2) of the relative maximal contact pressure and the relative maximal von Mises stress in Fig. 4 and Fig. 5, correspondingly.

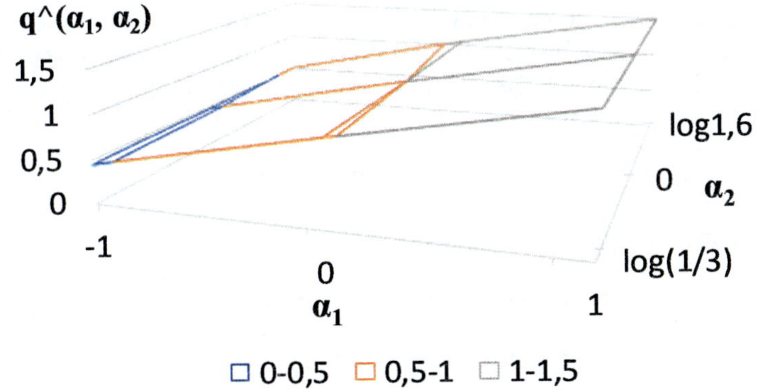

Fig. 4. Relative maximal contact pressure $q\hat{}(\alpha_1, \alpha_2)$.

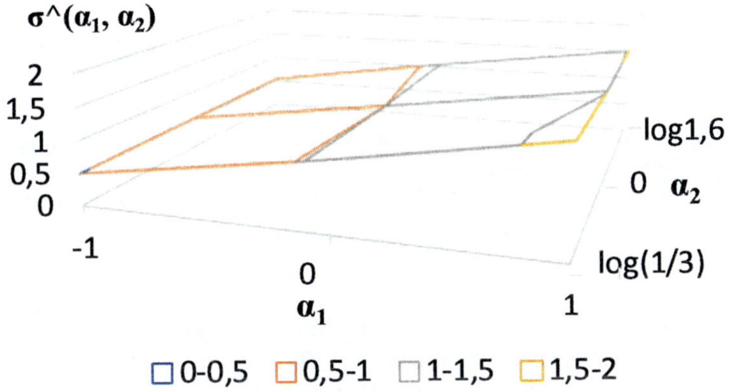

Fig. 5. Relative maximal von Mises stress $\sigma\hat{}(\alpha_1, \alpha_2)$.

4.3 Results Analysis

The parameter study displayed the strong relation of the local stresses on the elastic properties of the contacting bodies. The lower the stiffness of one of the body materials, the wider the contact spot. It extends first in the transverse direction keeping the nearly elliptic shape. Once the contact reaches the transition part of the running track profile, the contact spot takes a dumbbell shape. Meanwhile, the maximum contact pressure shifts from the central point of the initial contact to the side parts of the running track. The Poisson's coefficient has a much less profound effect on the contact pressure distribution q and the shape of the contact area S_C.

The characteristics $q\hat{}(\alpha_1, \alpha_2)$ and $\sigma\hat{}(\alpha_1, \alpha_2)$ represent a quantitative change in the stress-strain state. These values can be used to evaluate the material choice for the ball piston concerning its strength. The strength criteria can be formulated as a programming problem in terms of either the relative maximal contact pressure

$$q^\wedge(\alpha_1, \alpha_2) \to \min; \; q^\wedge(\alpha_1, \alpha_2) \leq [q^\wedge], \tag{8}$$

or the relative maximal von Mises stress

$$\sigma^\wedge(\alpha_1, \alpha_2) \to \min; \; \sigma^\wedge(\alpha_1, \alpha_2) \leq [\sigma^\wedge], \tag{9}$$

where $[q\hat{}]$, $[\sigma\hat{}]$ are the limiting values.

5 Conclusions

The performed parameter study provides new valuable insights for the design of radial hydrovolumetric transmissions. As outlined, the material choice of the ball pistons and the stator ring are crucial for the structure. The obtained data can be interpolated for an arbitrary practical combination of material properties considered for application. Hence, the performed study provides effective tools for choosing future design solutions considering the predicted local stressed state of the two considered critical components.

Variations of Young's modulus E and the Poisson's ratio ν tend to have a quantitative effect on the contact pressure distribution between the ball piston and the running track on the stator of the hydrovulumetric transmission GOP-900. The character of the contact pressure distribution, the contact shape, and the maximal level of the contact traction change with varied stiffness of the material. Higher deformations in a softer material provide better compliance between the bodies. The contact expands over a larger area of the interface surface. It extends up to the transitional rounded edges of the running track. The maximum of the contact pressure shifts there as well. Compared to the 3.9 GPa for the worst case with the contact pressure localized in the center of the elliptic contact spot, there is almost a 3 times reduction to 1.2 GPa in the case of more flexible material.

The elastic material properties on the equivalent stress distribution σ and its maximum level σ_{max} have qualitatively and quantitatively similar effects, mainly when analyzed in dimensionless relative quantities. In equivalent stress, the same threefold ratio is reduced from 2.5 GPa to 0.8 GPa. These characteristics are crucial for assessing the mechanical strength of the ball piston.

The developed models and the performed analysis resulted in a representative data set in the broad range of the design parameters. The variation of Young's modulus and Poisson's ratio in the analysis spans most structural materials, including metal alloys and ceramics. The proposed quantitative strength criteria can drive the material choice.

References

1. Avrunin, G., Kabanenko, S., Khavil, V.: Volumetric hydraulic transmission with ball pistons GOP-900: characteristics and technical level. Mech. Mech. Eng. **1**, 14–21 (2004)

2. Johnson, K.L.: Contact Mechanics. Cambridge University Press (1985)
3. Tkachuk, M., Grabovskiy, A., Tkachuk, M., Hrechka, I., Sierykov, V.: Contact interaction of a ball piston and a running track in a hydrovolumetric transmission. In: Ivanov, V., Pavlenko, I., Liaposhchenko, O., Machado, J., Edl, M. (eds.) DSMIE 2021. LNME, pp. 195–203. Springer, Cham (2021). https://doi.org/10.1007/978-3-030-77823-1_20
4. Panchenko, A., Voloshina, A., Milaeva, I., Panchenko, I., Titova, O.: The influence of the form error after rotor manufacturing on the output characteristics of an orbital hydraulic motor. Int. J. Eng. Technol. **7**(4.3), 1–5 (2018). https://doi.org/10.14419/ijet.v7i4.3.19542
5. Panchenko, A., Voloshina, A., Luzan, P., Panchenko, I., Volkov, S.: Kinematics of motion of rotors of an orbital hydraulic machine. IOP Conf. Ser.: Mater. Sci. Eng. **1021**, 012045 (2021)
6. Panchenko, A., Voloshina, A., Titova, O., Panchenko, I., Caldare, A.: Design of hydraulic mechatronic systems with specified output characteristics. In: Ivanov, V., Pavlenko, I., Liaposhchenko, O., Machado, J., Edl, M. (eds.) DSMIE 2020. LNME, pp. 42–51. Springer, Cham (2020). https://doi.org/10.1007/978-3-030-50491-5_5
7. Voloshina, A., Panchenko, A., Titova, O., Pashchenko, V., Zasiadko, A.: Experimental studies of a throughput of the distribution systems of planetary hydraulic motors. IOP Conf. Ser.: Mater. Sci. Eng. **1021**, 012054 (2021)
8. Voloshina, A., Panchenko, A., Titova, O., Panchenko, I.: Changes in the dynamics of the output characteristics of mechatronic systems with planetary hydraulic motors. J. Phys.: Conf. Ser. **1741**, 012045 (2021)
9. Kikuchi, N., Oden, J.T.: Contact problems in elasticity: a study of variational inequalities and finite element methods (Studies in Applied and Numerical Mathematics, Series Number 8). Society for Industrial and Applied Mathematics, Philadelphia (1986)
10. Hlaváček, I., Haslinger, J., Nečas, J., Lovíšek, J.: Solution of Variational Inequalities in Mechanics. Springer, New York (1988)
11. Kalker, J.J.: Variational principles of contact elastostatics. Inst. Math. Appl. **20**, 199–221 (1977)
12. Li, Q., Pohrt, R., Lyashenko, I.A., Popov, V.L.: Boundary element method for nonadhesive and adhesive contacts of a coated elastic half-space. Proc. Inst. Mech. Eng. Part J: J. Eng. Tribol. **234**(1), 73–83 (2018). https://doi.org/10.1177/1350650119854250
13. Vollebregt, E.A.H.: 100-fold speed-up of the normal contact problem and other recent developments in "CONTACT". In: Proceedings of the 9th International Conference on Contact Mechanics and Wear of Rail/Wheel Systems, vol. 96, pp. 201–209 (2012)
14. Zienkiewicz, O.C., Taylor, R.L., Zhu, J.Z.: The Finite Element Method: Its Basis and Fundamentals, 7th edn. Butterworth-Heinemann, Oxford (2013)
15. Lee, H.-H.: Finite Element Simulations with ANSYS Workbench (2020)
16. Tkachuk, M.M., Grabovskiy, A., Tkachuk, M. A., Saverska, M., Hrechka I.: A semi-analytical method for analys of contact interaction between structural elements along aligned surfaces. Eastern-Eur. J. Enterp. Technol. **1/7**(103), 16–25 (2020). https://doi.org/10.15587/1729-4061.2020.193985
17. Tiwari, A., Persson, B.N.J.: Cylinder-flat contact mechanics with surface roughness. Tribol. Lett. **69**(1), 1–7 (2021)
18. Li, Q., Pohrt, R., Lyashenko, I.A., Popov, V.L.: Boundary element method for nonadhesive and adhesive contacts of a coated elastic half-space. Proc. Inst. Mech. Eng. Part J: J. Eng. Tribol. **234**(1), 73–83 (2020)
19. Martynyak, R.M., Prokopyshyn, I.A., Prokopyshyn, I.I.: Contact of elastic bodies with nonlinear Winkler surface layers. J. Math. Sci. **205**, 535–553 (2015)
20. Kozachok, O., Martynyak, R.: Contact problem for wavy surfaces in the presence of an incompressible liquid and a gas in interface gaps. Math. Mech. Solids **24**(11), 3381–3393 (2019)

Sliding of a Particle on the Horizontal Plane Under Oscillating and Rotary Movements

Tatiana Volina[1,3](✉), Serhii Pylypaka[1], Vitaliy Babka[1], Olha Zalevska[2], and Alla Rebrii[3]

[1] National University of Life and Environmental Sciences of Ukraine, 15, Heroyiv Oborony Street, Kyiv 03041, Ukraine
t.n.zaharova@ukr.net

[2] National Technical University of Ukraine "Igor Sikorsky Kyiv Polytechnic Institute", 37, Prosp. Peremohy, Kyiv 03056, Ukraine

[3] Sumy National Agrarian University, 160, Kondratieva Street, Sumy 40021, Ukraine

Abstract. Numerous engineering tasks, which are related to the interaction of the working bodies of machines with particles of technological material, require analytical dependencies of the particle movement on a rough moving plane. The reciprocating oscillations of the horizontal plane and the translational oscillations, when all points of the plane describe circles, are thoroughly investigated. In these cases of oscillations, there is no rotation of the plane. However, multitudinous details of machines and mechanisms carry out such movement. Against this background, the relative movement of a particle on a rough horizontal plane that performs complex oscillations is considered in the article. The plane moves in a circle with a constant angular velocity relative to the circle's center and simultaneously rotates with the same angular velocity in the opposite direction. Differential equations of the particle sliding are compiled and solved by numerical methods. The trajectories of the relative movement of the particle on the plane are constructed. A partial case of oscillations is considered when the lengths of the crank and the slider are equal to zero. The obtained dependencies significantly expand the theory of particle movement on the surface. In addition, it can be applied to the geometric designing of crank-type mechanisms, in which the length of the slider is equal to the length of the crank.

Keywords: Particle movement · Moving plane · Surface movement · Analytical dependencies · Trajectory · Process innovation

1 Introduction

There has been a long-standing interest in analytical dependencies of particle movement on different surfaces, which are widely used for constructing machines' working bodies and mechanisms. For example, rational parameters of a screw feeder [1], winnowing machines [2], vortex-type apparatuses [3] and jet devices [4], and modular separation devices [5] can be justified by means of geometrical methods. Moreover, a great deal is being written and said about geometric modeling methods. Thus, geometric modeling of

V. Tonkonogyi et al. (Eds.): InterPartner 2022, LNME, pp. 506–514, 2023.
https://doi.org/10.1007/978-3-031-16651-8_48

varieties with given differential characteristics is presented in [6], geometric modeling of torse surfaces – in [7], rational designing [8], and solid modeling of geometric objects in point calculus – in [9, 10]. It can be seen that objects' geometric design methods are quite diverse. Besides, particle trajectories depend not only on the surface's shape but also on its movement.

2 Literature Review

It is well-known that the rough plane, which performs oscillating motion, is a working element of many machines and mechanisms for screening and separating bulk technological material [11]. Importantly, reciprocating oscillations of the horizontal plane are quite common. The particle of material, in this case, slides similarly to the oscillations of the plane but with a smaller amplitude [12]. Besides, in the case of translational oscillations, when all points of the plane describe circles. Particles of material after motion stabilization slide on the plane and describe circles of smaller radius. There is no rotation of the plane in both described cases of oscillations. The plane oscillations with rotational motion affect the sliding trajectory of the particles in some way. We predict that the analytical dependencies of such movement allow expanding the theory of particle movement on the surface significantly.

In this article, differential equations of particle sliding are compiled and solved by numerical methods. The trajectories of the relative movement of the particle on the plane are constructed. A partial case of oscillations is considered when the lengths of the crank and the slider are equal to zero. The obtained results can be applied to the increase of reliability and durability of working bodies of machines – so, to the solving of the task, which is solving nowadays much more complicated, for example, by the implementation of new methods of details manufacturing [13, 14] or surfaces improvement [15]. Besides, analytical dependencies of particle movement can be applied in real production in different areas: hydrodynamics [16], building [17, 18], separation of mixtures [19, 20], parts manufacturing [21, 22], and field operations [23].

M.Ye. Zhukovskiy was the first to solve the problem of the motion of a material particle on a plane that performs circular oscillating motion in the geometric interpretation. I.I. Blechmann generalized and extended it to cases of elliptical oscillations. P.M. Vasylenko composed the differential equations of motion of a particle in projections on the axes of a movable coordinate system, which is rigidly tied to an oscillating plane, and I.I. Blechman – in projections on the axes of a fixed coordinate system. Some of our recent research works are conducted on a similar problem. In [24], particle motion over the edge of an inclined plane that performs axial movement was investigated.

Moreover, the complex movement of a point on a plane with the predefined plane displacement was researched in [25]. To the best of our knowledge, there are no results in the literature regarding the particle movement on a rough horizontal plane, a point that describes a circle regarding a stationary horizontal plane, and the rough plane itself rotates around this moving point. Against this background, the present article investigates the laws of motion of material particles on a plane that performs such movement.

3 Research Methodology

The analytical dependencies of particle movement on a rough horizontal plane that performs complex oscillations are considered using differential geometry methods. The calculations and visualization were realized using Mathematica, MatLab, and AutoCAD software products. The relative and absolute trajectories of particles sliding on the plane were found. The main mathematical equations are given below.

In Fig. 1, the moving plane μ is represented by a rectangle highlighted by thickened segments. It is attributed to the rectangular system *Auv*. The movement of the moving plane μ will be carried out regarding the fixed coordinate system *Oxy*.

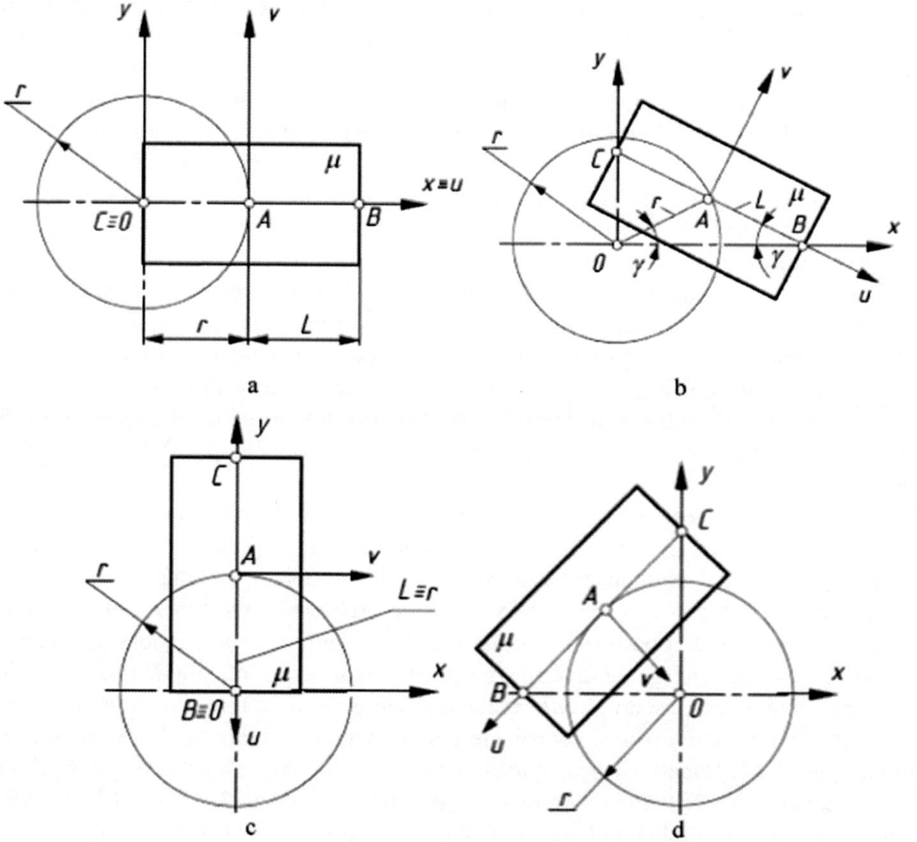

Fig. 1. Graphical illustrations for moving of a plane μ: a) angle $\gamma\ = 0°$; b) $\gamma = 30°$; c) $\gamma = 90°$; d) $\gamma = 135°$.

At the beginning of the movement of the plane μ, the axes *Au* and *Ox* coincide, and the axes *Oy* and *Av* are parallel and offset by the value *r* of the radius of the circle (Fig. 1, a), on which the origin of the moving plane μ moves.

Let the segment $r = OA$ be a crank and the segment $L = AB$ – a slider. They are equal, i.e., have the same length. A fixed plane μ is attached to the slider, so they move together. If the crank *OA* is rotated at the angle γ (Fig. 1, b), the slider *AB* also rotates at the angle γ. It takes place because, according to the initial conditions, point *B* moves along the axis *Ox* and $OA = AB$, so the triangle *OAB* is isosceles, and the base angles are equal. It is important to emphasize that the crank *OA* rotates counterclockwise and the slider *AB* – clockwise.

If the angle γ is an independent variable, the parametrical equations of the circle (the set of positions of the point *A* of the crank *AB*) are:

$$x_A = r\cos\gamma; \qquad y_A = r\sin\gamma. \tag{1}$$

The slider *AB* and the movable system *Auv* rotate at an angle $(-\gamma)$ respectively to the fixed coordinate system *Oxy*. Rotation of the movable system regarding the fixed system:

$$x_{uv} = u\cos(-\gamma) - v\sin(-\gamma); \qquad y_{uv} = u\sin(-\gamma) + v\cos(-\gamma). \tag{2}$$

Let us assume that the crank *OA* rotates with a constant angular velocity ω, so $\gamma = \omega t$, where t is time. With this in mind, the parametrical equations of the positions of the points of the moving plane μ, given by the coordinates u, v, in the projections on the fixed plane *Oxy* can be obtained by adding two motions (1) and (2):

$$x = (r + u)\cos\omega t + v\sin\omega t; \qquad y = (r - u)\sin\omega t + v\cos\omega t. \tag{3}$$

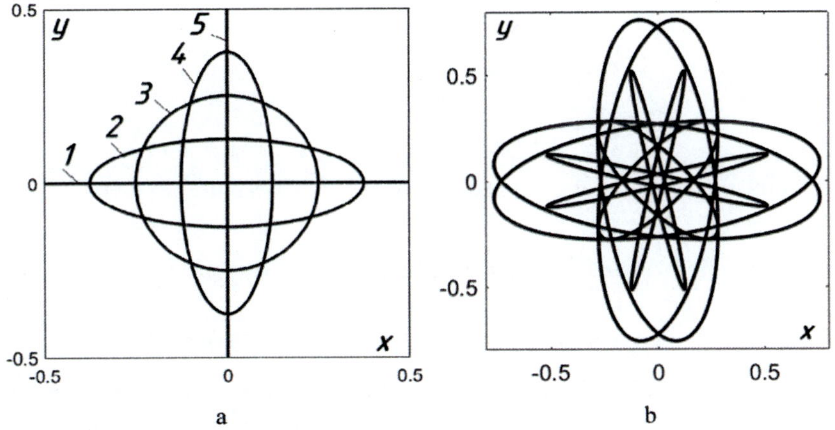

Fig. 2. Absolute trajectories of points in *Oxy* system that are fixed regarding the movable coordinate system *Auv*: a) trajectories of points located on the segment *AB* (at $v = 0$): 1 – point *B* ($u = 0.25$), 3 – point *A* ($u = 0$), 5 – point *C* ($u = -0.25$), and others ($u = \pm 0.125$); b) trajectories of points located on the plane μ at the previous values of the coordinate u and the values of the coordinate $v = \pm 0.125$.

Point *B* in the movable system has coordinates: $u = AB = r$, $v = 0$. By substituting these coordinates in Eq. (3) one can obtain: $x = 0$, i.e., point *B* slides along the axis *Ox*. Point *C* of the slider has coordinates: $u = -r$, $v = 0$. Therefore, it slides on the axis

Oy. Point A describes a circle. All other points describe ellipses. In Fig. 2, the absolute trajectories of individual points of the moving plane μ at $r = 0.25$ are constructed.

The trajectories in Fig. 2 are constructed for fixed points of a moving plane. However, when a particle hits the plane μ, it begins to slide on it, so the coordinates u and v are variable and time-dependent t, that is $u = u(t)$ and $v = v(t)$. These dependencies will describe the line in the movable system Auv, along which the particle slides (the relative trajectory). Hence, it is essential to find the unknown dependencies $u = u(t)$ and $v = v(t)$. For this, let us compose the equation of motion in the form $m\overline{w} = \overline{F}$, where m is the mass of the particle, $\overline{w}$ is the vector of absolute acceleration, $\overline{F}$ is the resulting vector of forces applied to the particle. Such forces are the force of gravity mg ($g = 9.81$ m/s^2), balanced by the plane's reaction force N, and the friction force F = fN = fmg, directed in the opposite side of the particle sliding (f is the coefficient of friction). The force of gravity mg and the reaction force N act in the vertical direction and are balanced, so only one friction force $F = fmg$ acts in the plane. The vector of its action is directed along with the tangent to the relative trajectory in the opposite direction of the relative velocity. The last one is determined by the first derivatives $\dot{u} = \dot{u}(t)$ and $\dot{v} = \dot{v}(t)$. The absolute value of the relative velocity is the geometric sum of the derivatives: $V_r = \sqrt{\dot{u}^2 + \dot{v}^2}$. The projections of the unit vector of relative velocity in the system Auv are:

$$\left\{\frac{\dot{u}}{\sqrt{\dot{u}^2 + \dot{v}^2}}; \frac{\dot{v}}{\sqrt{\dot{u}^2 + \dot{v}^2}}\right\}. \tag{4}$$

To find the absolute acceleration, it is necessary to differentiate the absolute trajectory (3), bearing in mind that $u = u(t)$ and $v = v(t)$. After the first differentiation, the components of absolute velocity can be obtained:

$$\dot{x} = (\dot{v} - \omega u - \omega r)\sin\omega t + (\dot{u} + \omega v)\cos\omega t;$$

$$\dot{y} = (\dot{v} - \omega u + \omega r)\cos\omega t - (\dot{u} + \omega v)\sin\omega t. \tag{5}$$

By differentiating expressions (5) the components of absolute acceleration can be received:

$$\ddot{x} = [\ddot{v} - \omega(2\dot{u} + \omega v)]\, sin\,\omega t + [\ddot{u} + \omega(2\dot{v} - \omega u - \omega r)]\, cos\,\omega t;$$

$$\ddot{y} = [\ddot{v} - \omega(2\dot{u} + \omega v)]\, cos\,\omega t - [\ddot{u} + \omega(2\dot{v} - \omega u + \omega r)]\, sin\,\omega t. \tag{6}$$

The vector equation $m\overline{w} = \overline{F}$ should be decomposed in projections on the axes of the fixed coordinate system. But the unit vector (4) of relative velocity is found in the system Auv without considering its rotation. To bring this vector in line with the fixed coordinate system Oxy, it must also be rotated at the angle $(-\gamma)$ by formulas (2):

$$\left\{\frac{\dot{u}\cos\omega t + \dot{v}\sin\omega t}{\sqrt{\dot{u}^2 + \dot{v}^2}}; \frac{\dot{v}\cos\omega t - \dot{u}\sin\omega t}{\sqrt{\dot{u}^2 + \dot{v}^2}}\right\}. \tag{7}$$

Taking in mind that the friction force $F = fmg$ is directed in the opposite side from the vector (7) the differential equations take form:

$$m\ddot{x} = -fmg\frac{\dot{u}\cos\omega t + \dot{v}\sin\omega t}{\sqrt{\dot{u}^2 + \dot{v}^2}};$$

$$m\ddot{y} = -fmg\frac{\dot{v}\cos\omega t - \dot{u}\sin\omega t}{\sqrt{\dot{u}^2 + \dot{v}^2}}. \quad (8)$$

After substituting in (8) the expressions of absolute acceleration (6) and solving it regarding the second derivatives:

$$\ddot{u} = -\frac{fg\dot{u}}{\sqrt{\dot{u}^2 + \dot{v}^2}} + \omega(\omega r\cos 2\omega t - 2\dot{v} + \omega u);$$

$$\ddot{v} = -\frac{fg\dot{v}}{\sqrt{\dot{u}^2 + \dot{v}^2}} + \omega(\omega r\sin 2\omega t + 2\dot{u} + \omega v). \quad (9)$$

Numerical methods allow for solving this system of differential equations. The found dependencies $u = u(t)$ and $v = v(t)$ in the form of graphs represent the trajectory of the relative movement of the particle, namely, it is the trace of its sliding on the plane μ. Parametrical Eqs. (3) and these dependencies allow the construction of the trajectory of absolute motion.

4 Results and Discussion

As a result, in Fig. 3, the relative trajectory of the particle sliding on the plane μ is indicated with the number 1. The absolute trajectory of its movement regarding the fixed coordinate system *Oxy* is indicated with the number 2. The particle hit the plane μ at points *B*, *A*, and *C* (Fig. 1, a). Over time, the particle's motion became predictable: relative – along the spiral and absolute – along the curve that intersects the spiral turns at approximately the same angle.

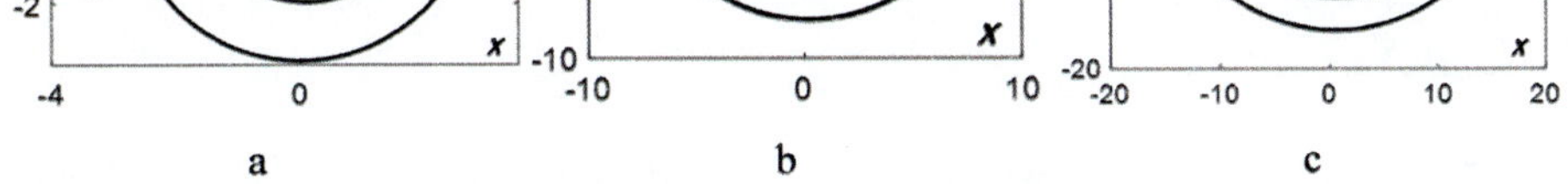

Fig. 3. Relative – 1 and absolute – 2 trajectories of the particle movement at $r = 0.25$, $w = 15$ and $v = 0$: a) $u = 0.25$; b) $u = 0$; c) $u = -0.25$.

In Fig. 4, a, b, the particle was falling on the plane μ on both sides of the point *B* with the coordinate value $v = \pm 0.1$, and in Fig. 4, c – at $r = 0$, that is at the rotational motion of the plane.

Understandable that if a particle gets onto a rough horizontal plane, which is moving while remaining horizontal, it begins to slide on it. The shape of the sliding trajectories

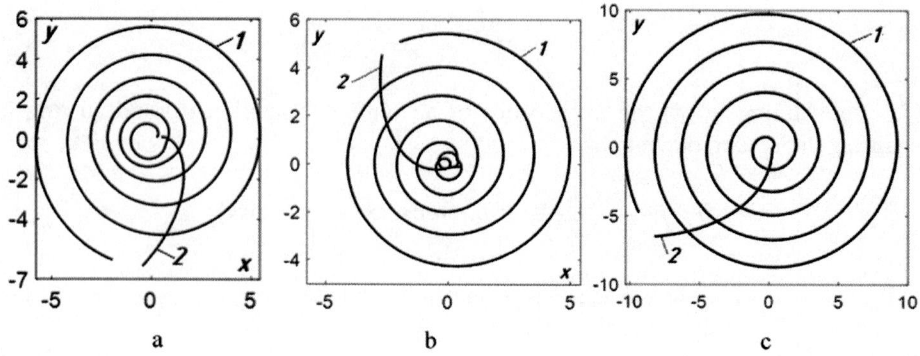

Fig. 4. Relative – 1 and absolute – 2 trajectories of the particle movement at $w = 15$ and $u = 0.25$: a) $u = r = 0.25$, $v = 0.1$; b) $u = r = 0.25$, $v = -0.1$; c) $r = 0$, $v = 0.1$.

depends on the nature of the plane's motion. Thus, the movement of a particle on a rough horizontal plane that performs complex oscillations is investigated in the study. The plane moves in a circle with a constant angular velocity relative to the circle's center and simultaneously rotates with the same angular velocity in the opposite direction. Differential equations of the particle sliding are compiled and solved by numerical methods. The trajectories of the relative movement of the particle on the plane are constructed. Without any doubt, the obtained dependencies allow expanding the theory of particle movement on the surface significantly.

During translational movement, when all points of the plane describe the same curves (for example, circles), the sliding trajectories of the particle are similar to these curves. When a plane rotates, and all its points describe concentric circles, the trajectory of the relative movement of the particle is a spiral with a regular shape. During combining both motions, when the points of the plane describe ellipses and their partial case – a circle or a line, the relative movement of the particle at the initial stage was somewhat chaotic because the spiral was partially distorted. But during rotational motion, the relative movement trajectory (spiral) has the correct shape.

The obtained results can also be applied to the geometric designing of crank-type mechanisms, in which the slider's length is equal to the length of the crank.

5 Conclusions

Numerous engineering tasks, which are related to the interaction of the working bodies of machines with particles of technological material, require analytical dependencies of the particle movement on a rough moving plane. Trajectories of particle movement depend not only on the surface's shape but also on the surface's movement. The present research investigates the laws of motion of material particles on a plane, which performs complex oscillations. It must be borne in mind that this study was only conducted on a material particle (an object with rest-mass and an observable position in space but with no geometrical extension, being confined to a single point). Further research is needed to change the shape of the surface. The relative and absolute (regarding the fixed coordinate system) trajectories of the particle sliding on the plane at r = 0.25, w = 15, v = 0, and different values of the variable u (0.25, 0, −0.25) were received. Over time, the particle's motion became predictable: relative – along the spiral and absolute – along the curve that intersects the spiral turns at approximately the same angle. Moreover, the relative and absolute trajectories of the particle movement at w = 15, u = r = 0.25, with the coordinate value v = ±0.1 and at r = 0, are within the plane's rotational motion.

References

1. Baranovsky, V., Hevko, R., Dzyura, V., Klendii, O., Klendii, M., Romanovsky, R.: Justification of rational parameters of a pneumoconveyor screw feeder. INMATEH Agricult. Eng. **54**(1), 15–24 (2018)
2. Shrestha, K., Parajuli, P., Baral, B., Shrestha, B.: Mathematical modeling, simulation and analysis of rice grain movement for design and fabrication of low-cost winnowing machine. J. Mech. Eng. Res. **9**(1), 1–14 (2017). https://doi.org/10.5897/JMER2016.0403
3. Merzliakov, I., Pavlenko, I., Chekh, O., Sharapov, S., Ivanov, V.: Mathematical modeling of operating process and technological features for designing the vortex type liquid-vapor jet apparatus. In: Ivanov, V., et al. (eds.) DSMIE 2019. LNME, pp. 613–622. Springer, Cham (2020). https://doi.org/10.1007/978-3-030-22365-6_61
4. Sharapov, S., Husiev, D., Krmela, J.: Experimental stand for studying the working process in a liquid-vapor jet device with replaceable diffuser parts. J. Eng. Sci. **9**(1), F21–F26 (2022). https://doi.org/10.21272/jes.2022.9(1).f4
5. Liaposhchenko, O., Pavlenko, I., Monkova, K., Demianenko, M, Starynskyi, O.: Numerical simulation of aeroelastic interaction between gas-liquid flow and deformable elements in modular separation devices. In: Ivanov, V., et al. (eds.) DSMIE 2019. LNME, pp. 765–774. Springer, Cham (2020). https://doi.org/10.1007/978-3-030-22365-6_76
6. Konopatskiy, E., Bezditnyi, A., Shevchuk, O.: Modeling geometric varieties with given differential characteristics and its application. In: Proceedings of the 30th International Conference on Computer Graphics and Machine Vision (GraphiCon 2020), Part 2, pp. 31-1–31-8 (2020). https://doi.org/10.51130/graphicon-2020-2-4-31
7. Konopatskiy, E., Bezditnyi, A., Litvinov, A.: Geometric modeling of torse surfaces in BN-calculus. IoP Conf. Ser. J. Phys. Conf. Ser. **1791**, 012050 (2021). https://doi.org/10.1088/1742-6596/1791/1/012050
8. Karpus, V.E., Ivanov, V.A.: Choice of the optimal configuration of modular reusable fixtures. Russ. Eng. Res. **32**(3), 213–219 (2012). https://doi.org/10.3103/S1068798X12030124
9. Konopatskiy, E., Bezditnyi, A.: Solid modeling of geometric objects in point calculus. In: CEUR Workshop Proceedings, vol. 3027, pp. 666–672 (2021)
10. Pavlenko, I., Liaposhchenko, A., Ochowiak, M., Demyanenko, M.: Solving the stationary hydroaeroelasticity problem for dynamic deflection elements of separation devices. Vib. Phys. Syst. **29**, 2018026 (2018)
11. Ahmed, T., Younes, M., Wu, L., Hincke, M.: A survey of recent patents in engineering technology for the screening, separation and processing of eggshell. Front. Bioeng. Biotechnol. **9**, 677559 (2021). https://doi.org/10.3389/fbioe.2021.677559
12. Nakata, S., Kayahara, K., Yamamoto, H., Skrobanska, P.: Reciprocating motion of a self-propelled rotor induced by forced halt and release operations. J. Phys. Chem. C **122**(6) (2018). https://doi.org/10.1021/acs.jpcc.7b12089
13. Tarel'nik, V., Martsinkovskii, V., Zhukov, A.: Increase in the reliability and durability of metal impulse end seals 1. Chem. Pet. Eng. **53**(1–2), 114–120 (2017). https://doi.org/10.1007/s10556-017-0305-y

14. Tarel'nik, V., Martsinkovskii, V., Zhukov, A.: Increase in the reliability and durability of metal impulse seals 3. Chem. Pet. Eng. **53**(5–6), 385–389 (2017). https://doi.org/10.1007/s10556-017-0351-5
15. Tarelnyk, V., et al.: New method for strengthening surfaces of heat treated steel parts. IOP Conf. Ser. Mat. Sci. Eng. **233**, 012048 (2017). https://doi.org/10.1088/1757-899x/233/1/012048
16. Fesenko, A., Yevsiukova, F., Basova, Y., Ivanova, M., Ivanov, V.: Prospects of using hydrodynamic cavitation for enhancement of efficiency of fluid working medium preparation technologies. Periodica Polytechnica Mech. Eng. **62**(4), 269–276 (2018). https://doi.org/10.3311/PPme.11877
17. Konopatskiy, E., Bezditnyi, A.: Application of mixed geometric interpolants for modeling the strength characteristics of steel fiber concrete. IoP Conf. Ser. J. Phys. Conf. Ser. **1546**, 012037 (2020). https://doi.org/10.1088/1742-6596/1546/1/012037
18. Konopatskiy, E., Bumaga, A., Bezditnyi, A.: Geometric approach to finding the best possible solutions based on composition optimization of the mixed aggregate of fine-grained concrete. IoP Conf. Ser. Mater. Sci. Eng. Conf. Ser. **962**, 032031 (2020). https://doi.org/10.1088/1757-899X/962/3/032031
19. Bulgakov, V., et al.: Theory of motion of grain mixture particle in the process of aspiration separation. Agron. Res. **18**(2), 1177–1188 (2020). https://doi.org/10.15159/AR.20.069
20. Bulgakov, V., et al.: Theory of grain mixture particle motion during aspiration separation. Agron. Res. **18**(1), 18–37 (2020). https://doi.org/10.15159/ar.20.057
21. Karpus, V.E., Ivanov, V.A.: Locating accuracy of shafts in V-blocks. Russ. Eng. Res. **32**(2), 144–150 (2012). https://doi.org/10.3103/S1068798X1202013X
22. Kostyk, K., et al.: Simulation of diffusion processes in chemical and thermal processing of machine parts. Processes **9**(4), 698 (2021). https://doi.org/10.3390/pr9040698
23. Kobets, A., Ponomarenko, N., Kobets, O., Tesliuk, H., Kharytonov, M., Yaropud, V.: Study of fertilizer spreader centrifugal type under field conditions. INMATEH Agricult. Eng. **57**(1), 253–260 (2019)
24. Pylypaka, S., Klendii, M., Nesvidomin, V., Trokhaniak, V.: Particle motion over the edge of an inclined plane that performs axial movement in a vertical limiting cylinder. Acta Polytechnica **59**(1), 67–76 (2019). https://doi.org/10.14311/AP.2019.59.0067
25. Pylypaka, S., Volina, T., Nesvidomin, V., Pavlov, A., Dranovska, S.: The possibility to apply the Frenet trihedron and formulas for the complex movement of a point on a plane with the predefined plane displacement. East. Eur. J. Enterp. Technol. **3**(7(111)), 45–50 (2021)

Process Engineering

Temperature Distribution in Parts of the Vehicle Disk Brake

Gustav Gudz[1], Ihor Zakhara[2], Tetyana Voitsikhovska[2], Vasyl Vytvytskyi[2](✉), and Liubomyr Ropyak[2]

[1] Lviv Polytechnic National University, 12, Bandera Street, Lviv 79013, Ukraine
[2] Ivano-Frankivsk National Technical University of Oil and Gas, 15, Karpatska Street, Ivano-Frankivsk 76019, Ukraine
vytvytskyi.v.s@gmail.com

Abstract. The paper analyzes brake elements' materials and coatings and considers temperature distributions in friction elements. The authors used the finite element method to study brake temperature mode due to the complexity of heating transfer analytical solution and the wide scope of simulation applications. Simulation models were developed and studied for ventilated and non-ventilated discs of an A-172 bus brake to study temperature distribution and simulate friction heating/cooling of open brakes. The authors conducted a type I test of 20 simulations of 60 s duration to obtain the frictional elements' temperature distribution. It was found that quasi-steady maximum surface temperature is reached after the 10th braking cycle for the ventilated brakes and until the 14-16th braking cycle for the non-ventilated ones. Experimental tests of friction element temperatures were carried out, and results were compared with simulated data. Deviations between simulated data and experimental ones did not exceed 5%. The hygienic assessment of air pollution by brake wear products within type I tests showed that chrysotile asbestos fiber concentration (grade 1, carcinogen) in the laboratory working zone was 0.07 fibers/cm^3 at the friction zone temperature of 180 °C. This concentration does not exceed the maximum permissible concentration of 0.1 fibers/cm^3, regulated by DIRECTIVE 2009/148/EC.

Keywords: Coating · Polymer brake pad · Type-1 test · 3D-modeling · Computer simulation model · Mechatronic system · Hygienic evaluation · Wear products

1 Introduction

Modern vehicle development is characterized by continuous improvement to increase vehicle productivity and performance and make the environment friendly.

Vehicle's high-speed movement is impossible without a braking system with high braking efficiency and optimal braking performance in order to provide vehicle stability and handling. Therefore, the abovementioned options define traffic safety and the full usage of rated speed capability.

Vehicle reliability is one of the main requirements of the vehicle braking system, so it is essential to gain its operating mode and energy loading data to develop the braking

V. Tonkonogyi et al. (Eds.): InterPartner 2022, LNME, pp. 517–529, 2023.
https://doi.org/10.1007/978-3-031-16651-8_49

system with sufficiently stable initial characteristics in conditions of high-energy load (Regulation No. 13 of the UN/ECE). Therefore, developing new combined approaches to technical diagnostics and thermal analysis of brake friction pair elements is an issue at stake.

2 Literature Review

The systems approach is widely used in the study of machines and vehicles. It comprehensively describes the full range of external and internal connections and the functionality of the brake system of vehicles, hoisting machines and more.

When designing brakes, researchers focus on the rational choice of materials and coatings, the study of temperature, force load, and wear tests [1], including considering tribocurrents [2] and tribocorrosion [3]. The following materials are used to make brakes: steel, cast iron, and composite materials. Brake discs are usually made of cast or rolled billet, and in recent years 3D printing is applied.

Considerable attention is paid to ensuring the accuracy and quality of machining [4], considering the technological heredity [5] to ensure the functioning of the product during the life cycle [6].

Composite materials and coatings are used to increase the durability of the brake. The studies Radchenko et al. [7] and Levchuk et al. [8] present theoretical approaches to the development of new composite materials and reinforcement technologies. The authors Shihab et al. [9] and Melnick et al. [10] developed a method of thermodynamic prediction of the phase composition of high-entropy alloys of transition metals for parts of friction units. Tungsten-free cermets with Cu-Ni-Mn bonds have been proposed to increase the durability of friction pairs [11, 12].

For disks made of deformed aluminum alloys, plasma electrolytic oxidation [13] is used. The working surfaces of steel brake discs are strengthened by laser treatment [14], and coatings are formed in different ways: vacuum-arc multilayer [15], chemical and thermal processing [16], a combined electron-beam method [17], and combined electrospark deposition [18]. The use of other types of coatings to strengthen brake parts is not always effective.

For example, coatings obtained by electrochemical chromium plating of steel parts in a flowing electrolyte [19] and electrodeposited from an ionic liquid based on choline chloride [20] have high resistance to wear, oxidation, and corrosion, heat-resistant and heat-resistant. They are characterized by the absence of burrs on the friction surfaces.

However, these coatings are not used to strengthen the brake discs due to the low coefficient of friction paired with the pad. The use of welded wear-resistant composite coatings is promising [21, 22].

For the safe operation of vehicles, it is necessary to study brake parts' stress-strain and temperature conditions. The studies present an analytical calculation of the stress-strain state of layered structures during the application of forces perpendicular to the surface of the layers [23] and parallel to them [24]. Analysis of the limit state of elements with cracks, considering the contact of the crack edges, is considered in the article [25]. The authors [26, 27] developed a direct method for studying heat transfer and calculating nonstationary temperature fields in multilayer bodies with imperfect thermal contact.

The temperature effects of the interaction of elastic bodies under friction contact and local heating conditions are considered, particularly in [28], including considering the action of the cooling medium [29, 30].

In [31], the long-term thermal heating of a steel brake disc with zirconium oxide coating on its working surface was investigated. To determine the temperature distribution in the coating and the disk, the authors formulated the problem by considering the temperature properties of the material related to the temperature within braking and established the distribution of temperature and temperature stresses in the disk and its coating.

Computer modeling is used to study the dynamics of complex parts [32]. Pryhorovska et al. [33] investigated the force interaction of metal elements of friction pairs during their interaction with a non-metallic rock by the finite element method but did not consider temperature factors. Researchers [34] performed numerical simulations of the frictional contact of the disc brake and established the temperature distribution in solid and ventilated brake discs to predict and determine the critical temperatures of the friction vapor elements. The article [35] presents friction disk/pad simulation results considering the effects of thermomechanical connection.

The authors [36, 37] provide a comparative analysis of temperature fields in braking discs. The study [38, 39] proposed finite element models for determining the parameters of flows. However, the calculations were performed separately for different parts, introducing the heat distribution coefficient. Friction materials with asbestos are widely used in brake pads. High temperatures arise in the "paddle- metal disk" contact area, which causes pad destruction and its wear, carcinogenic and mutagenic substance emissions to air. Therefore, in order to define maximum permissible concentrations of materials emitted within frictions, it is necessary to determine the safe temperature operation modes of these materials and the composition of products emitted during wear [40].

Usually, the works mentioned above investigate the permanent friction contact of parts, but in the brakes, there is a disconnection of the friction pair parts, so the previously obtained results of theoretical studies are complicated to apply to the design and improvement of the brakes. In addition, theoretical techniques have a limited application for brakes with ventilated discs. This way, there is a need to develop new combined theoretical and experimental approaches for technical diagnostics and thermal analysis of open and closed brakes for studying wear product composition and hygienic assessment of their environmental impact.

The study aims to compare the experimental and FEM simulated friction element temperature state of the ventilated and non-ventilated disk brakes during their cyclic tests and hygienic evaluation of wear products.

The following tasks were set to achieve the aim defined: to develop a simulation model for disk brake heating/cooling studying with consideration of their open/close state; to study temperature distribution into friction elements according to the developed model; to compare experimental and simulated results of disk brake friction within cyclic testing of the vehicle; to make a hygienic assessment of air pollution by disk brake wear products.

3 Research Methodology

3.1 Studied Brake Unit, Materials of Its Friction Elements

The authors carried out bench and performance tests for technical diagnostic of the brake of LAZ A-172 bus disc brakes manufactured by the Lviv Bus Plant, Ukraine (Fig. 1).

Fig. 1. General view of the disk-pad brake: 1 – ventilated disk; 2 – brake caliper.

The brake disk was made of gray cast iron MH 19 according to State Standard GOST 1412–85 (analogue of cast iron Gh 190) with physical parameters E = 1105 MPa, σ = 250 MPa (Table 1).

Table 1. Chemical composition and properties of disk's material.

Chemical composition, mass %						Mechanical and physical properties				
C	Si	Mn	P	S	Fe	ρ, kg/m^3	σ_B, MPa	k·10^6, 1/deg	λ, W/(m·K)	α, W/(m^{2}K)
3.3	2.0	0.8	up 0.2	up 0.12	others	7200	250	10	50	50

The friction pads were made of MKV–50A according to Branch Stand-ard OST 1 90115–74 with physical parameters E = 1105 MPa, σ = 34 MPa (see Table 2).

The friction coefficient is in the range of 0.3–0.5. The outer diameter of the disks is 430 mm. The thickness of the ventilated disk is 45 mm, unventilated – 30 mm. Width of a friction belt of 90 mm.

Table 2. Chemical composition and properties of pad's material.

Chemical composition, mass %							Mechanical and physical properties				
C	SiC	Cu	$FeSO_4$	B_4C	Asbestos	Fe	ρ, kg/m^3	HB	k 10^6, 1/deg	λ, W/(m·K)	α, $W/(m^2K)$
8	5	10	5	5	2.5–4	64	5000	900	12.67	23.0	75

3.2 Bench and Performance Test Methods

Tests were carried out for hot brakes. The authors used the type I test method to heat the studied brakes. The preliminary test stage of 20 sequential braking cycles at speeds ranging from 60 km/h to 30 km/h was carried out for M3-category vehicles. According to the type I test method, the total cycle duration was 60 s (Regulation No 13 of the UN/ECE). Each experiment was repeated three times in order to provide a stable value of the studied brake options. A mechatronic system with thermocouples allows measuring the temperature in the friction zone. Chrysotile asbestos fibers concentration in the air was defined according to phase-contrast optical microscopy, recommended by WHO.

3.3 Simulation Model Development Method

The developed 3D model of the brake element is presented in Fig. 2 with specified friction element thermophysical characteristics and initial and boundary conditions. Ansys® Academic Research was used for modeling.

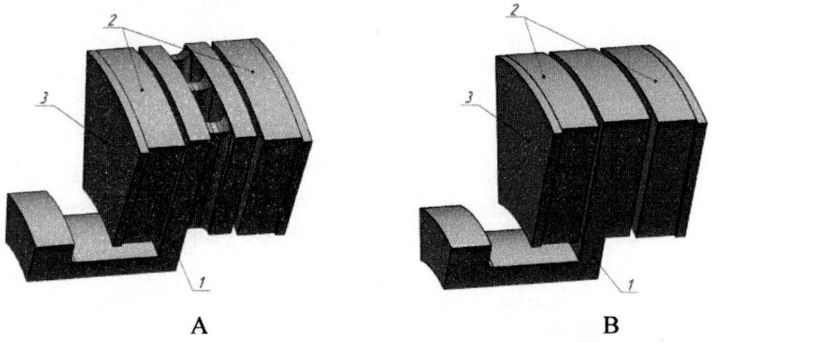

Fig. 2. Sector of the developed disk-pad brake model with ventilated (A) and unventilated disk (B): 1 – disk; 2 – pads; 3 – overlays.

In the simulation, we considered the process of nonstationary heat transfer during the braking-cooling cycle was considered, where the braking period alternates with the cooling period. The finite-element model of the solid disk contained $7.4 \cdot 10^4$ elements with a total number of $1.4 \cdot 10^5$ nodes, and the ventilated disk had $9.1 \cdot 10^4$ elements with $1.7 \cdot 10^5$ nodes. The disk was fixed, and the friction pad could only move normally to the working surface of the disk.

At the initial moment, we set the heat flow on the working surfaces of friction pair parts, referring to the braking power. There was the perfect thermal contact on the coupling surface of the disc and pad. Convective heat exchange was carried out on free surfaces (heat flow through the surface is proportional to the temperature difference), and radiation heat exchange occurred on working surfaces. In the next braking cycle, the initial temperature was taken as the temperature reached at the end of the previous cycle.

4 Results and Discussion

The preliminary stage of the type I test consists of 20 heating/cooling cycles. Heat transfer coefficients α for surfaces of open brake elements were specified, according to Tables 1 and 2, and the friction surface's heat transfer coefficient was adjusted, considering the friction element overlapping coefficient of 0.35 for the considered friction pair.

Figure 3 a, c, e shows simulated temperature distribution in the brake friction elements across the ventilation channel section and Fig. 3 b, d, f – across the section of the ventilation channel edge.

Figure 3 a, b shows that friction surface temperature is 110 °C at the 1st braking cycle, while the brake disk and pad temperatures do not exceed 45 °C. This is because the brake elements interact for a short time.

Figure 3 c, d shows the temperature distribution in the brake disk at the end of the cooling of the 19th cycle of the type I test. There was a re-distribution of heat transfer in the disk, and its maximum temperature was 81 °C. Two symmetrical pads heated significantly less than the disk, and their temperatures ranged from 20 °C (pad side) to 45 °C (friction belt).

Figure 3 e, f shows the temperature field of the closed brake disk within friction element interaction before the 20th I-type test cycle starts. Figure 3 e, f shows pad-disk head re-distribution and heat transfer to disk's flange. This way, the maximum temperature on the friction belt reached 162 °C, disk's flange warmed up from 30 °C to 60 °C, and pad overlays were heated to temperatures ranging from 60 °C to 120 °C.

In order to compare simulation results, temperature distributions for different cycles of braking for the non-ventilated disk were obtained (Fig. 4) in the same way.

This way, the developed computer model studied brake temperature fields for all subsequent braking cycles.

Figure 5 shows the simulated temperature range for A-172 bus disc brake heating and cooling for the preliminary stage of type I tests. Plot analysis of experimental and simulated data showed that quasi-stationary temperature is observed after the 10-th test cycle for the ventilated brakes and practically at the end of the tests for the non-ventilated brakes.

The proposed and applied simulation algorithm was compared with the following experimental data to provide the assumptions' accuracy and reliability. Simulation results (Fig. 5) showed that the temperature mode of the ventilated brake disk at the end of type I tests is lower by 9.3–10.7% than for the non-ventilated ones.

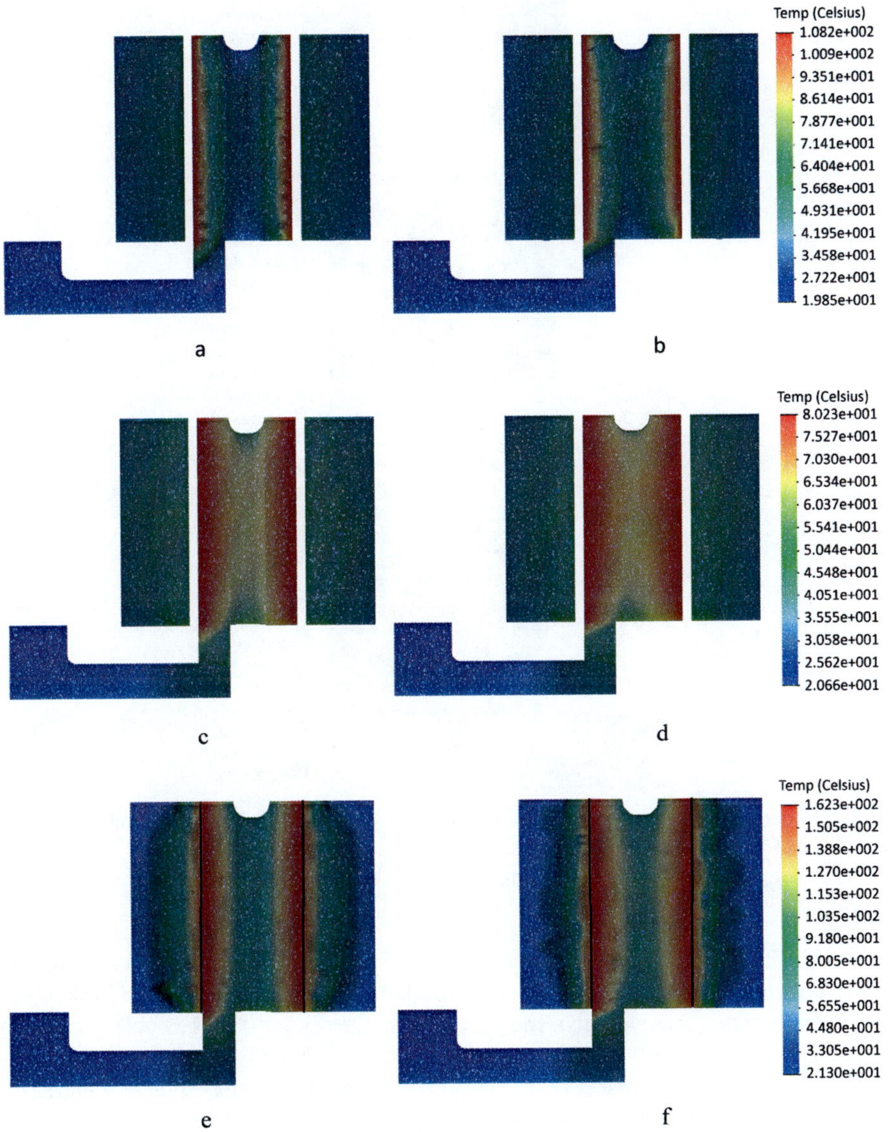

Fig. 3. Temperature distribution in the friction element of A-172 bus front brake with ventilated disk within type I tests: a, b – at the end of the 1st braking cycle; c, d – at the end of the 19th cooling cycle; e, f – at the beginning of the 20th braking cycle.

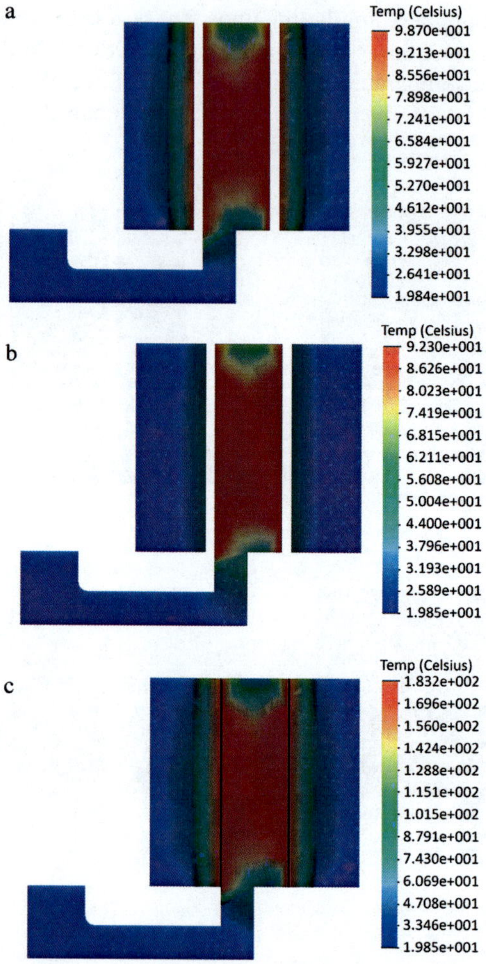

Fig. 4. Temperature distribution in the friction element of A-172 bus front brake with non-ventilated disk during type I tests: a – at the end of the 1st braking cycle; b – at the end of the 19th cooling cycle; c – at the beginning of the 20th braking cycle.

Besides, it is worth mentioning that the temperature of the ventilated brake disks is higher until the 5–6 test cycles than the temperature for the non-ventilated brake disks. The reason for this is less weight of the ventilated brake discs. After the cycles mentioned above, an 11–12% lag between ventilated and non-ventilated brake disk temperature is observed at the end of the test (Fig. 5). This is due to heat exchange intensity increasing due to the presence of ventilation channels.

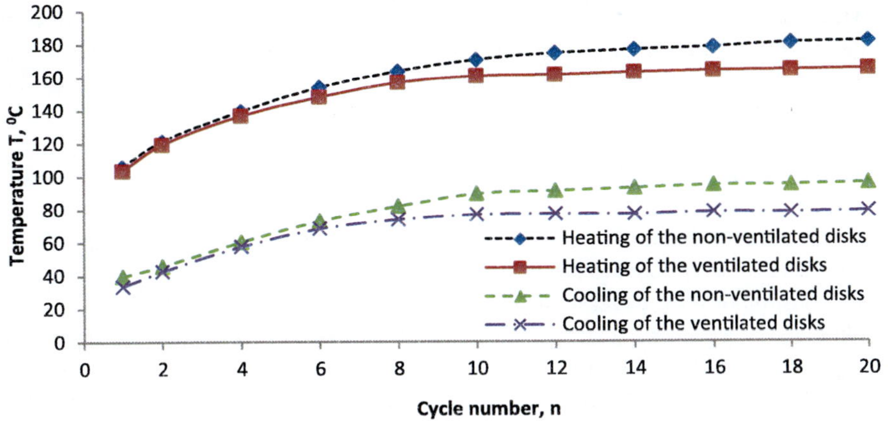

Fig. 5. Heating and cooling maximum temperature ranges for the ventilated and non-ventilated brake discs of the bus A-172 during type I tests.

Table 3 shows the simulation and experimental maximal surface temperatures of the A-172 bus brake with ventilated disks at the beginning (1-st cycle) and end (20th cycle) of the previous stage of type I tests.

Table 3 shows that experimental and simulated data deviation does not exceed 5%.

Experimental studies showed that friction zone temperature reached 180 °C for 20 braking cycles of bench testing.

Chrysotile asbestos fibers concentration in the air was defined to provide a hygienic assessment of brake pad's wear product impact on environmental pollution within bench testing.

Table 3. Simulation and experimental maximal surface temperatures of the A-172 bus brake with ventilated disks at the beginning and end of the previous stage of type I tests.

Braking cycle	Disk temperature, °C		Pad temperature, °C	
	Simulation	Experiment	Simulation	Experiment
1st cycle	$\frac{108}{35}$	$\frac{106}{34}$	$\frac{108}{25}$	$\frac{106}{23}$
20th cycle	$\frac{162}{78}$	$\frac{161}{76}$	$\frac{162}{42}$	$\frac{161}{41}$

Note: numerator – for heating; denominator – for cooling.

Chrysotile asbestos fiber concentration (grade 1, carcinogen) in the laboratory working zone was 0.07 fibers/cm^3 at the friction zone temperature of 180 °C. This concentration does not exceed the maximum permissible concentration of 0.1 fiber/cm^3, regulated by directive DIRECTIVE 2009/148/EC.

Due to the complex mathematical description of heat exchange, the analytical solutions for the heat equations were obtained only for simple parts with uniform physical parameters [26, 41], which do not include the brake friction elements.

Numerical simulation results made it possible to determine the temperature distribution quickly in friction parts for their direct contact and the open brakes when brake element geometric options change.

Experimental temperatures of friction elements are somewhat higher in their magnitudes due to assuming constant environmental temperature within the simulation, although the experimental brake friction parts are air flushed and get additional cooling within bus movement.

As far as experimental and simulated result deviation does not exceed 5%, it is possible to select friction materials and rational geometry options for brakes, considering energy consumption for different test conditions, when designing brakes.

Thus, the research showed the relevance of the developed disk brake simulations within brake designing and part testing, simplifying and reducing brake design costs.

In addition, it should be noted that Regulation No. 13 does not specify the maximum permissible concentration of the chrysotile asbestos fibers in the air within brake testing for different brake types, although DIRECTIVE 2009/148/EC regulates the maximum permissible concentration.

5 Conclusions

In order to simplify the technical diagnostics, an A-172 bus disc brake spatial model was developed and studied. It allows FEM simulation of heating and cooling brake elements within type I tests.

The simulation showed that quasi-steady maximum surface temperature is reached after the 10th braking cycle for the ventilated brakes and until the 14–16-th braking cycle for the non-ventilated ones.

Experimental tests of friction element temperatures were carried out, and results were compared with simulated data. Deviations between simulated data and experimental ones did not exceed 5%. The experimental temperatures of friction elements are somewhat higher due to assuming constant environmental temperature within simulations even though the experimental friction parts are air flushed within bus movement and get additional cooling.

The hygienic assessment of air pollution by brake wear products within type I tests showed that chrysotile asbestos fiber concentration (grade 1, carcinogen) in the laboratory working zone was 0.07 fibers/cm^3 at the friction zone temperature of 180 °C. This concentration does not exceed the maximum permissible concentration of 0.1 fibers/cm^3, regulated by directive DIRECTIVE 2009/148/EC.

Acknowledgments. The authors are grateful to the Ministry of Science and Education of Ukraine for the grant to implement the project D 8-21-P (RK 0121U109591) and D2-22-P (RK 0122U002082). The team of authors express their gratitude to the reviewers for valuable recommendations that have been taken into account to improve significantly the quality of this paper.

References

1. Bembenek, M., Krawczyk, J., Pańcikiewicz, K.: The wear on roller press rollers made of 20Cr4/1.7027 steel under conditions of copper concentrate briquetting. Materials **13**(24), 5782 (2020). https://doi.org/10.3390/ma13245782
2. Kryshtopa, S.I., Prun'ko, I.B., Dolishnii, B.V., Panchuk, M.V., Bogatchuk, I.M., Mel'nyk, V.M.: Regularities of wear of metal-polymer friction couples under the influence of tribocurrents. Mater. Sci. **55**(2), 193–200 (2019). https://doi.org/10.1007/s11003-019-00288-x
3. Vynar, V.A., Pokhmurs'kyi, V.I., Zin', I.M., Vasyliv, K.B., Khlopyk, O.P.: Determination of the mechanism of tribocorrosion of D16T alloy according to the electrode potential. Mater. Sci. **53**(5), 717–723 (2018). https://doi.org/10.1007/s11003-018-0128-9
4. Bazaluk, O., Velychkovych, A., Ropyak, L., Pashechko, M., Pryhorovska, T., Lozynskyi, V.: Influence of heavy weight drill pipe material and drill bit manufacturing errors on stress state of steel blades. Energies **14**(14), 4198 (2021). https://doi.org/10.3390/en14144198
5. Kusyi, Y.M., Kuk, A.M.: Investigation of the technological damageability of castings at the stage of design and technological preparation of the machine life cycle. J. Phys. Conf. Ser. **1426**(1), 012034 (2020). https://doi.org/10.1088/1742-6596/1426/1/012034
6. Kopei, V.B., Onysko, O.R., Panchuk, V.G.: Principles of development of product lifecycle management system for threaded connections based on the Python programming language. J. Phys. Conf. Ser. **1426**(1), 012033 (2020). https://doi.org/10.1088/1742-6596/1426/1/012033
7. Radchenko, T.M., Gatsenko, O.S., Lizunov, V.V., Tatarenko, V.A.: Martensitic α''-Fe16N2-type phase of non-stoichiometric composition: current status of research and microscopic statistical-thermodynamic model. Prog. Phys. Metals **21**(4), 580–618 (2020). https://doi.org/10.15407/ufm.21.04.580
8. Levchuk, K.H., Radchenko, T.M., Tatarenko, V.A.: High-temperature entropy effects in the tetragonality of the ordering interstitial-substitutional solution based on the body-centred tetragonal metal. Metallofizika i Noveishie Tekhnologii **43**(1), 1–26 (2021). https://doi.org/10.15407/mfint.43.01.0001
9. Shihab, T., Prysyazhnyuk, P., Semyanyk, I., Anrusyshyn, R., Ivanov, O., Troshchuk, L.: Thermodynamic approach to the development and selection of hardfacing materials in energy industry. Manag. Syst. Prod. Eng. **28**(2), 84–89 (2020). https://doi.org/10.2478/mspe-2020-0013
10. Melnick, O.B., Soolshenko, V.K., Levchuk, K.H.: Thermodynamic prediction of phase composition of transition metals high-entropy alloys. Metallofizika i Noveishie Tekhnologii **42**(10), 1387–1400 (2020). https://doi.org/10.15407/mfint.42.10.1387
11. Shihab, T.A., Shlapak, L.S., Namer, N.S., Prysyazhnyuk, P.M., Ivanov, O.O., Burda, M.J.: Increasing of durability of mechanical seals of oil and gas centrifugal pumps using tungsten-free cermet with Cu-Ni-Mn binder. J. Phys. Conf. Ser. **1741**(1), 012031 (2021). https://doi.org/10.1088/1742-6596/1741/1/012031
12. Student, M.M., et al.: Influence of the porosity of a plasma-electrolytic coating on the corrosion resistance of D16 alloy. Mater. Sci. **54**(6), 899–906 (2019). https://doi.org/10.1007/s11003-019-00278-z
13. Duriagina, Z., Kulyk, V., Kovbasiuk, T., Vasyliv, B., Kostryzhev, A.: Synthesis of functional surface layers on stainless steels by laser alloying. Metals **11**(3), 434, 1–19 (2021). https://doi.org/10.3390/met11030434
14. Beresnev, V.M., et al.: Effect of the deposition parameters on the phase–structure state, hardness, and tribological characteristics of Mo2N/CrN vacuum–arc multilayer coatings. J. Superhard Mater. **38**(2), 114–122 (2016). https://doi.org/10.3103/S1063457616020052

15. Kostyk, K., et al.: Simulation of diffusion processes in chemical and thermal processing of machine parts. Processes **9**(4), 698 (2021). https://doi.org/10.3390/pr9040698
16. Antonyuk, V.S., Bondarenko, Y.Y., Bilokin', S.O., Andrienko, V.O., Bondarenko, M.O.: Research of microhardness of thin ceramic coatings formed by combined electron-beam method on dielectric materials. J. Nano Electron. Phys. **11**(6), 06024 (2019). https://doi.org/10.21272/jnep.11(6).06024
17. Tarelnyk, V., Konoplianchenko, I., Tarelnyk, N., Kozachenko, A.: Modeling technological parameters for producing combined electrospark deposition coatings. In: Materials Science Forum 968MSF, pp. 131–142 (2019). https://doi.org/10.4028/www.scientific.net/MSF.968.131
18. Bazaluk, O., Dubei, O., Ropyak, L., Shovkoplias, M., Pryhorovska, T., Lozynskyi, V.: Strategy of compatible use of jet and plunger pump with chrome parts in oil well. Energies **15**(1), 83 (2022). https://doi.org/10.3390/en15010083
19. Protsenko, V.S., Bobrova, L.S., Baskevich, A.S., Korniy, S.A., Danilov, F.I.: Electrodeposition of chromium coatings from a choline chloride based ionic liquid with the addition of water. J. Chem. Technol. Metall. **53**(5), 906–915 (2018)
20. Ivanov, O., Prysyazhnyuk, P., Lutsak, D., Matviienkiv, O., Aulin, V.: Improvement of abrasion resistance of production equipment wear parts by hardfacing with flux-cored wires containing boron carbide/metal powder reaction mixtures. Manag. Syst. Prod. Eng. **28**(3), 178–183 (2020). https://doi.org/10.2478/mspe-2020-0026
21. Trembach, B., et al.: Effect of exothermic addition (CuO-Al) on the structure, mechanical properties and abrasive wear resistance of the deposited metal during self-shielded flux-cored arc welding. Tribol. Ind. **43**(3), 452–464 (2021). https://doi.org/10.24874/ti.1104.05.21.07
22. Ropyak, L.Ya., Makoviichuk, M.V., Shatskyi, I.P., Pritula, I.M., Gryn, L.O., Belyakovskyi, V.O.: Stressed state of laminated interference-absorption filter under local loading. Funct. Mater. **27**(3), 638–642 (2020). https://doi.org/10.15407/fm27.03.638
23. Velychkovych, A., Ropyak, L., Dubei, O.: Strength analysis of a two-layer PETF-concrete column with allowance for contact interaction between layers. Adv. Mater. Sci. Eng. **2021**, 4517657 (2021). https://doi.org/10.1155/2021/4517657
24. Shatskyi, I., Makoviichuk, M., Perepichka, V., Dalyak, T.: Effect of cracks closure in plates and shells under combined tension and bending. In: 23rd International Conference Engineering Mechanics 2017, pp. 866−869 (2017)
25. Tatsiy, R.M., Pazen, O.Yu., Vovk, S.Ya., Kharyshyn, D.V.: Direct method of studying heat exchange in multilayered bodies of basic geometric forms with imperfect heat contact. Naukovyi Visnyk Natsionalnoho Hirnychoho Universytetu **2021**(1), 60–67 (2021). https://doi.org/10.33271/nvngu/2021-1/060
26. Tatsii, R.M., Stasyuk, M.F., Pazen, O.Y.: Direct method of calculating nonstationary temperature fields in bodies of basic geometric shapes. J. Eng. Phys. Thermophys. **94**(2), 298–310 (2021). https://doi.org/10.1007/s10891-021-02302-z
27. Malanchuk, N.I., Slobodyan, B.S., Martynyak, R.M.: Friction sliding of elastic bodies in the presence of subsurface inclusions. Mater. Sci. **52**(6), 819–826 (2017). https://doi.org/10.1007/s11003-017-0026-6
28. Stepanov, M., Ivanova, L., Litovchenko, P., Ivanova, M., Basova, Y.: Model of thermal state of the system of application of coolant in grinding machine. In: Ivanov, V., et al. (eds.) DSMIE 2018. LNME, pp. 156–165. Springer, Cham (2019). https://doi.org/10.1007/978-3-319-93587-4_17
29. Vdovin, A., Gustafsson, M., Sebben, S.: A coupled approach for vehicle brake cooling performance simulations. Int. J. Therm. Sci. **132**, 257–266 (2018). https://doi.org/10.1016/j.ijthermalsci.2018.05.016

30. Yevtushenko, A., Kuciej, M., Och, E.: Influence of thermal sensitivity of the materials on temperature and thermal stresses of the brake disc with thermal barrier coating. Int. Commun. Heat Mass Transf. **87**, 288–294 (2017). https://doi.org/10.1016/j.icheatmasstransfer.2017.07.021
31. Kopei, V.B., Onysko, O.R., Panchuk, V.G.: Component-oriented acausal modeling of the dynamical systems in Python language on the example of the model of the sucker rod string. PeerJ. Comput. Sci. **2019**(10), e227 (2019). https://doi.org/10.7717/peerj-cs.227
32. Karpus, V.E., Ivanov, V.A.: Choice of the optimal configuration of modular reusable fixtures. Russ. Eng. Res. **32**(3), 213–219 (2012). https://doi.org/10.3103/S1068798X12030124
33. Pryhorovska, T.A., Chaplinskiy, S.S.: Finite element modeling of rock mass cutting by cutters for PDC drill bits. Neftyanoe Khozyaystvo Oil Ind. **1**, 38–41 (2018). https://doi.org/10.24887/0028-2448-2017-1-38-41
34. Belhocine, A., Abdullah, O.I.: A thermomechanical model for the analysis of disc brake using the finite element method in frictional contact. J. Therm. Stresses **43**(3), 305–320 (2020). https://doi.org/10.1080/01495739.2019.1683482
35. Belhocine, A., Abdullah, O.I.: Modeling and simulation of frictional disc/pad interface considering the effects of thermo-mechanical coupling. World J. Eng. **17**(6), 761–784 (2020). https://doi.org/10.1108/WJE-04-2020-0124
36. Yevtushenko, A., Kuciej, M., Grzes, P., Wasilewski, P.: Comparative analysis of temperature fields in railway solid and ventilated brake discs. Materials **14**(24), 7804 (2021). https://doi.org/10.3390/ma14247804
37. Towoju, O.A.: Braking pattern impact on brake fade in an automobile brake system. J. Eng. Sci. **6**(2), E11–E16 (2019). https://doi.org/10.21272/jes.2019.6(2).e2
38. Yevtushenko, A.A., Kuciej, M., Grzes, P., Wasilewski, P.: Temperature in the railway disc brake at a repetitive short-term mode of braking. Int. Commun. Heat Mass Transf. **84**, 102–109 (2017). https://doi.org/10.1016/j.icheatmasstransfer.2017.04.007
39. Merzliakov, I., Pavlenko, I., Chekh, O., Sharapov, S., Ivanov, V.: Mathematical modeling of operating process and technological features for designing the vortex type liquid-vapor jet apparatus. In: Ivanov, V., et al. (eds.) DSMIE 2019. LNME, pp. 613–622. Springer, Cham (2020). https://doi.org/10.1007/978-3-030-22365-6_61
40. Hovorun, T.P., et al.: Physical-mechanical properties and structural-phase state of nanostructured wear-resistant coatings based on nitrides of refractory metals Ti and Zr. Funct. Mater. **26**(3), 548–555 (2019). https://doi.org/10.15407/fm26.03.548
41. Stojanovic, N., Glisovic, J., Abdullah, O.I., Belhocine, A., Grujic, I.: Particle formation due to brake wear, influence on the people health and measures for their reduction: a review. Environ. Sci. Pollut. Res. **29**(7), 9606–9625 (2021). https://doi.org/10.1007/s11356-021-17907-3

Numerical Modeling of Point Defect Formation Processes During the Nuclear Power Plants Operation

Vladislav Opyatyuk[1], Igor Kozlov[1](✉), Kostiantyn Karchev[1], and Raul Turmanidze[2]

[1] Odessa Polytechnic National University, 1, Shevchenko Avenue, Odessa 65044, Ukraine
kozlov_i.l_@ukr.net
[2] Georgian Technical University, 77, Kostava Street, 0175 Tbilisi, Georgia

Abstract. This paper considers the defects' influence on the nuclear power reactor operation when the uranium fuel environment. The following tasks were solved: a referential sources review on the current state issue and considering various types of defects that can occur during the reactor operation. Numerical modeling of these processes with the process itself visualization is carried out. To create a numerical simulation algorithm, a number of theoretical assumptions were used, namely spatial isotropy, ergodic hypothesis, the superposition principle, and the dense packaging concept. It was also assumed that when the uranium-235 nucleus interacts with a neutron, a certain energy amount is released, with two fission fragments forming. According to the model, the debris scattering energy equals the released energy evenly distributed between those fragments. Further, these fragments were transformed by β-decay into isotopes of xenon (tellurium, iodine) and gadolinium (samarium, europium). The model considers the first three possible stages of decay. The visual analysis of possible cracks obtained at that simulation result and photographs of cracks resulting from the nuclear fuel operation allows the conclusion that the elaborated algorithm is successful and operational.

Keywords: Nuclear fuel · Point defect · Fuel rods · Spatial isotropy · Stresses structure · Energy efficiency

1 Introduction

One of the reasons for the decrease in nuclear power plants' installed capacity coefficient is the formation of defects in reactors' heat-releasing elements (relevant for WWER [1], PWR, and others [2]). Therefore, it is essential to identify the main regularities of fuel assemblies' state and behavior depending on the damaging nature and operation duration.

The purpose of this study was to investigate the processes of defect occurrence depending on the processes taking place in the fuel during reactor operation.

The article provides statistics of modern defects directly related to the issue of revising the requirements of the standard mode.

V. Tonkonogyi et al. (Eds.): InterPartner 2022, LNME, pp. 530–539, 2023.
https://doi.org/10.1007/978-3-031-16651-8_50

2 Literature Review

In this section, the processes of those defects' appearance will be considered, depending on the processes in the fuel during the reactor operation.

The statistics of defects occurring at the contemporary stage and directly related to the issue of the revision of requirements for standard modes are given.

In most cases, the reason for the relatively small values of the installed power factor (IPF) is the static occurrence of damage to fuel assemblies due to the defect formation in the fuel rod shells [3, 4].

During planned fuel refueling, a reduction in the time of cladding tightness control (CLC) in the presence of a fuel assembly with damaged fuel elements in the fuel load is needed [5, 6] to prevent of unplanned shutdown due to the separation fragments' excessive activity in the coolant beyond the specified operational limit.

Based on the results in the field of irradiated fuel assemblies post-reactor surveys [7, 8], as well as taking into account the materials of the study of defects in the early stages of nuclear energy industry development [9, 10], it should be noted the main mechanisms of damage to fuel rods are due to the: radiation increase [10]; thermomechanical interaction between fuel and shell [3]; radiation and thermal creep [8]; deflection of fuel rods (associated with thermomechanical interaction in the beam) [7]; radiation decreases plasticity [9].

In turn, the change in fuel rod diameter and length during operation is due to various effects. So, for example, under the influence of excessive coolant pressure immediately after the operation starts, the fuel rod's diameter decreases. Figure 1 shows the change in the WWER – 1000 fuel element shell size during fuel burnout [11].

As the burnout process increases, the size reduction rate drops to zero. After that, the diameter increases. Also, in parallel with the process of the shell diameter reduction, there is an increase in the fuel pellet diameter.

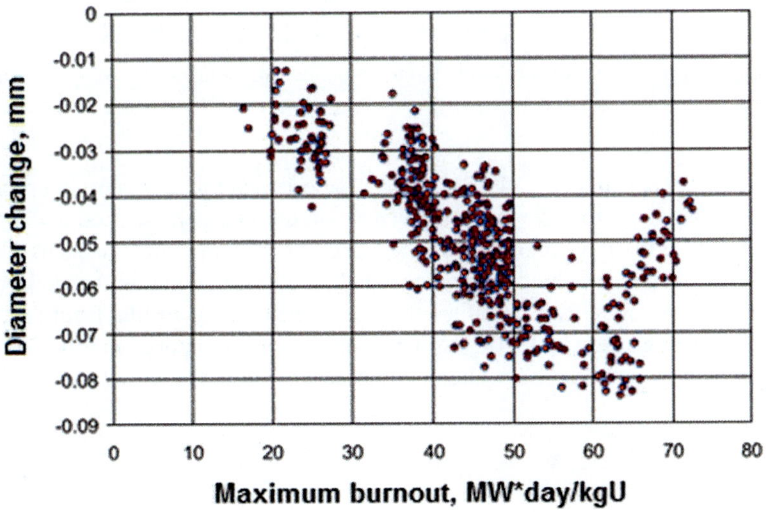

Fig. 1. Dependence of the change in the WWER - 1000 fuel element shell size during fuel burnout.

The graphs below show that the burnout diameter's minimum values are ~60 MW·-day/kg U [12, 13].

Figure 2 shows the dependence of the fuel core outer diameter and the shell inner diameter on the amount of fuel burnout.

The red and blue lines' intersection corresponds to the contact between the fuel and the shell during burnouts of ~48 MW·day/kg U for fuel rods WWER – 1000 [14, 15].

Figure 3 shows images of fuel tablets' sections in the shell before the start of operation. Before the start of the operation, both axial and radial cracks in the fuel pellets are visible.

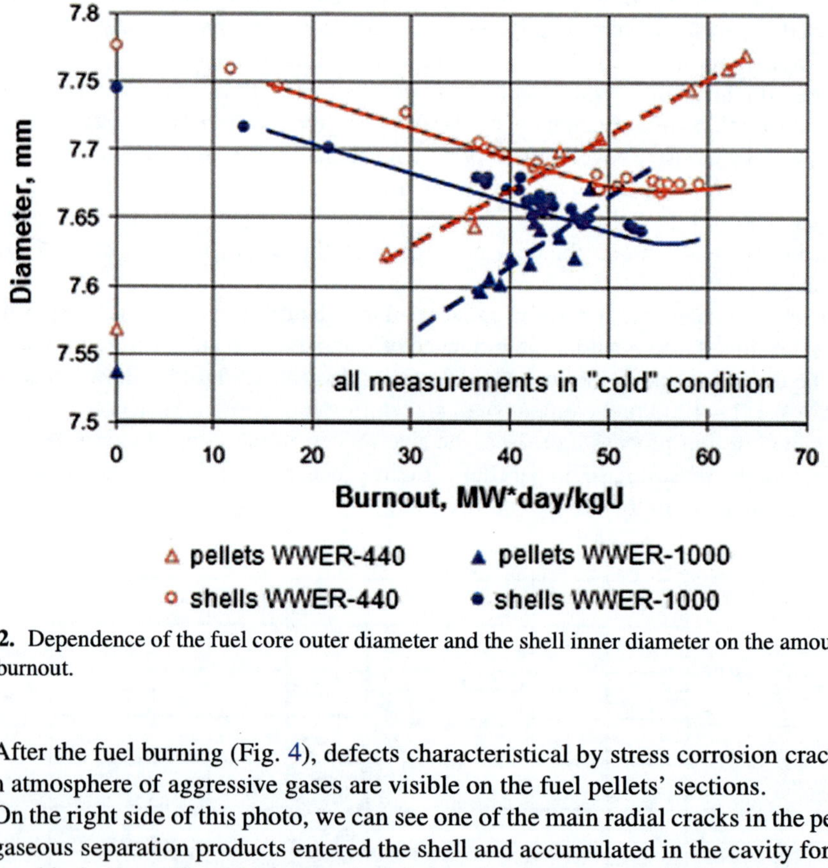

Fig. 2. Dependence of the fuel core outer diameter and the shell inner diameter on the amount of fuel burnout.

After the fuel burning (Fig. 4), defects characteristical by stress corrosion cracking in an atmosphere of aggressive gases are visible on the fuel pellets' sections.

On the right side of this photo, we can see one of the main radial cracks in the pellet: the gaseous separation products entered the shell and accumulated in the cavity formed by the chip in the pellet. The initial shell defect development results from an increase in the shell's tensile stress under the influence of aggressive separation products. As the result, a crack was formed with sequential depressurization of the fuel element.

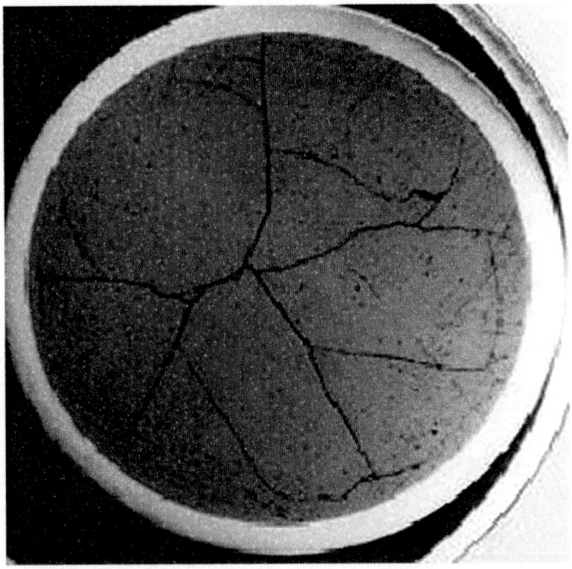

Fig. 3. Fuel pellet cross-section at the start of operation.

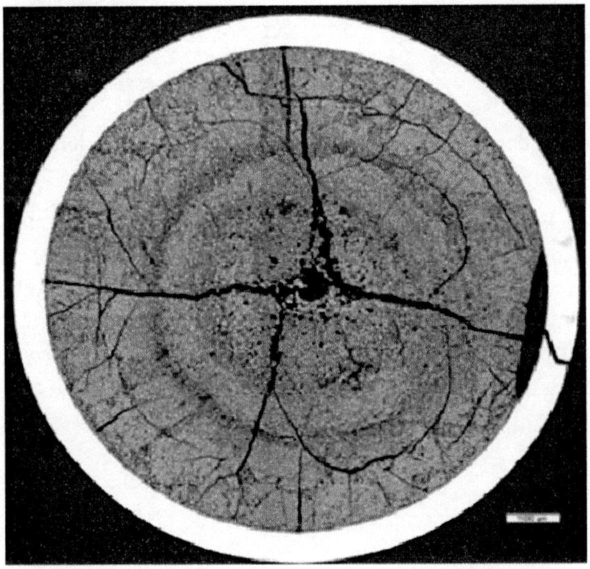

Fig. 4. Fuel pellet cross-section after the fuel burning.

3 Research Methodology

As follows from the review, the process of defects in fuel formation during the nuclear power plant operation bears a multiparametric, sequentially parallel, and multistage character. To create a numerical simulation algorithm, we will use several theoretical assumptions: spatial isotropy, ergodic hypothesis, superposition principle and the concept of dense packing.

Let's also assume that when the uranium-235 nucleus interacts with a neutron, 2 fission fragments (tellurium and samarium) are formed, and an energy of 210 meV is released [16, 17]. According to the model, the distribution of the fragments scattering energy is equal to the released energy and is evenly distributed between them. Further, these fragments are transformed by β-decay into isotopes of xenon (tellurium, iodine) and gadolinium (samarium, europium). According to the referred data, there can be up to 40 such stages of decay. The first three possible stages are considered in the model. Let's represent this process using a linear scheme (Fig. 5):

U^{235} + n^{1}_{0}

Stage 1 ↓ ↓

Te + Sm

Stage 2 / ↓ \ / ↓ \

Te J Xe Sm Eu Gd

Stage 3 / ↓ \ / ↓ \ / ↓ \ / ↓ \ / ↓ \ / ↓ \

Fig. 5. The supposed decay scheme of uranium 235 nucleus after interaction with a neutron.

Since the reactor fuel pellet consists of a mixture of uranium 238 powdered oxides with inclusions of uranium 235, all nuclear fission products for the fuel pellet structure embody point defects. These defects, such as inclusion and embedding, lead to the accumulation of stresses in the fuel pellet structure and further form cracks. The dynamics of point defects formation are described by a system of Eqs. (1–6):

$$\frac{dC}{dt} = \frac{u}{V}(C_0 - C) - ze^{\frac{-E}{RT}}C, \tag{1}$$

$$\Phi(\mathrm{E}) = \Phi_1(\mathrm{E}) + \Phi_2(\mathrm{E}) \tag{2}$$

$$\Phi_1(E) = \frac{\int_E^{\infty}\frac{\Sigma_{el}(E')}{\Sigma_t(E')}Q(E')\left\{exp\left[\int_{E'}^{\infty}\frac{-\Sigma a(E'')dE''}{|\xi_{el}|\Sigma_t(E'')\left[E''+\frac{1}{A}x\frac{3}{2}kT\right]}\right]\right\}dE'}{\left[E+\frac{1}{A}\cdot\frac{3}{2}kT\right]\Sigma_t(E)|\xi_{el}|} + \frac{\frac{\Sigma_{el}(E)}{\Sigma_t(E)}Q(E)}{\Sigma_t(E)} \tag{3}$$

$$\Phi_2(E) = \frac{\int_E^{\infty}\frac{\Sigma_{in}(E')}{\Sigma_t(E')}Q(E')\left\{exp\left[\int_{E'}^{\infty}\frac{-\Sigma a(E'')dE''}{|\xi_{in}|\Sigma_t(E'')\left[E''+\frac{1}{A}x\frac{3}{2}kT\right]}\right]\right\}dE'}{\left[E+\frac{1}{A}\cdot\frac{3}{2}kT\right]\Sigma_t(E)|\xi_{in}|} + \frac{\frac{\Sigma_{in}(E)}{\Sigma_t(E)}Q(E)}{\Sigma_t(E)} \tag{4}$$

$$\frac{\partial C_V(r,t)}{\partial t} + \omega\nabla\vec{J}_V(\vec{r},t) = -\alpha C_V(\vec{r},t)C_I(\vec{r},t) \tag{5}$$

$$\frac{\partial C_I(r,t)}{\partial t} + \omega\nabla\vec{J}_I(\vec{r},t) = -\alpha C_V(\vec{r},t)C_I(\vec{r},t) \quad (6)$$

where $C_V(r,t)$ and $C_I(r,t)$ – concentrations of vacancies and interstitial atoms, respectively; $J_V(r,t)$ – densities of interstitial atoms and vacancies flows; $\boldsymbol{\omega} \sim \boldsymbol{a}^3$ – atomic volume; $\boldsymbol{a}$ – lattice constant; $\boldsymbol{\gamma}$ - coefficient of fuel distribution mutual recombination; $Q(E)$ – is the number of generated neutrons with energy E per unit volume of time; Φ_1 – the expression for the flux density of elastically scattered neutrons slowing, [18]; Φ_2 – the expression for the non-elastically scattered slowing neutrons flux density;

$z = f = f(E)$ – the rate of generation of particles in the reaction [19]; C_0 and C – are the initial and current concentration of the fuel component; $\boldsymbol{\varepsilon}$ – is the activation energy, T is the reactor temperature; $\boldsymbol{\alpha}$ – is the heat transfer coefficient; $\boldsymbol{u}$ – is the nuclear fuel burning wave velocity.

4 Results and Discussion

Considering the spatial isotropy (the distribution fragments are scattered uniformly in all directions), in a two-dimensional approximation, the picture of point defects localization after the linear scheme first stage implementation is represented as (Fig. 6).

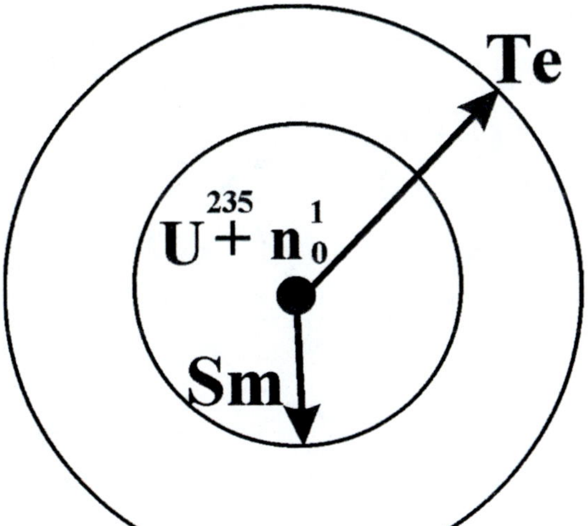

Fig. 6. 2-D picture of point defects localization after implementing the first stage of the uranium 235 nucleus linear decay scheme.

The lighter fission fragments (Te), according to the model, scatter further than the uranium-235 with the neutron interaction site; the heavier ones are located closer (Sm).

The exact model representations apply to the effects of secondary and tertiary decays of fission products. It is assumed that the areas of these decays-produced defects localization will be located in the dense packing of some small chains, and these small chains'

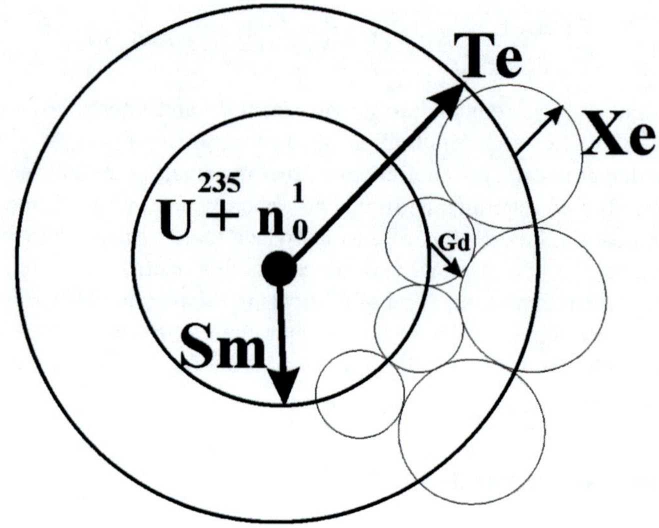

Fig. 7. 2-D picture of point defects localization after implementing the second stage of the uranium 235 nucleus linear decay scheme.

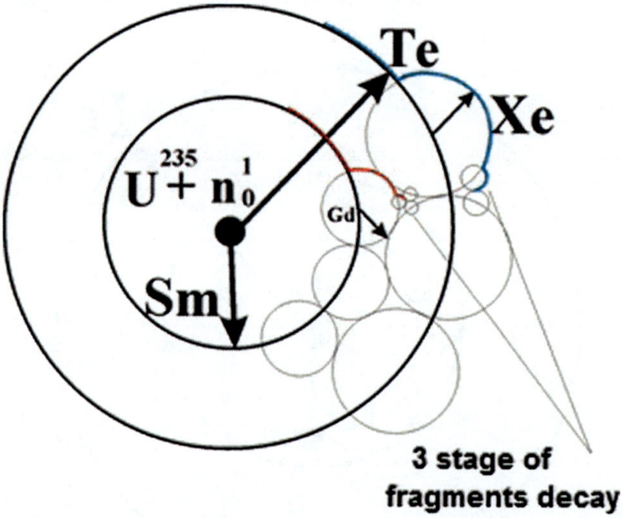

Fig. 8. 2-D picture of point defects localization after the implementation of the uranium 235 nucleus linear decay scheme in the second and third stages.

centers are located on the chains from the first (Fig. 7) and second (Fig. 8) stages of the linear decay scheme:

Figure 8 shows the envelope lines resulting from the simulation (red and blue). Along these lines, within the model framework, stress accumulation processes in the fuel pellet are supposed to occur, leading to cracking.

Based on the criterion for achieving the goals and objectives defined in this study introduction, it is necessary to compare the simulation results with actual photographic materials of defects obtained from the nuclear fuel operation process. These comparison results are shown in Figs. 9 and 10.

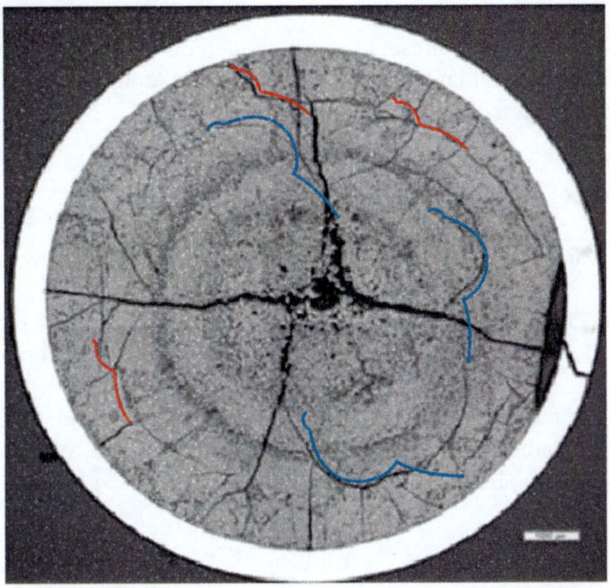

Fig. 9. Photo of fuel pellet defects after the fuel burning session + results of numerical simulation according to the proposed model.

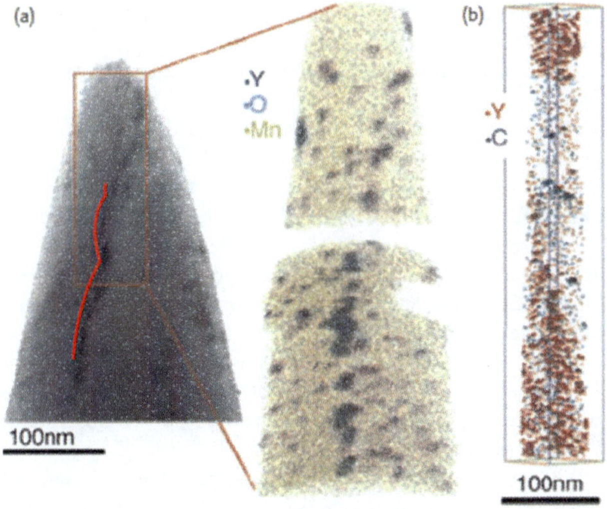

Fig. 10. a) ODS Eurfer97steel photo taken with an electron microscope by the Tea and ART methods, showing the presence of large particles along the grain boundary (material kindly provided by Dr. Lindau, FZK); b) three-dimensional reconstruction of the ODS Fe-12 alloy with weight.% Cr showing non-uniform particles distribution (material provided by Professor Pareja, Carlos III University, Madrid) + numerical simulation results based on the proposed model (red line).

The visual similarity of the envelope lines obtained from modeling and photographs of cracks obtained from the nuclear fuel operation allows us to conclude that the developed algorithm is successful.

In the three-dimensional representation, point defects are located between the "bubble" spheres in the spherical layer. This idea has not been considered in this study. Still, further research is expected.

5 Conclusions

Considered are various types of defects that formation is possible during the reactor operation.

An algorithm has been developed for numerical simulation of the point defects formation, considering the multistage process of fission fragments formation in reactor fuel.

The analysis of the pattern of point defects localization in numerical modeling and the pattern of these defects formation during the operation of reactor fuel pellets showed visual similarity, which allows us to conclude that the proposed method is correct.

References

1. Perepelkin, S.O., Markov, D.V., Polenok, V.S., et al.: The results of post-reactor studies of leaky WWER fuel rods. In: Proceedings of the FSUE "SSC RF NIIAR", Dimitrovgrad, vol. 4, pp. 12–21 (2007). (in Russian)
2. Review of Fuel Failures in Water Cooled Reactors. IAEA Nuclear Energy Series (NF-T-2.1). International Atomic Energy Agency, Vienna (2010)
3. Sokolova, I.D.: Experience of fuel operation in PWR reactors. Nuclear Eng. Abroad **6**, 3–11 (2010). (in Russian)
4. Review of Fuel Failures in Water Cooled Reactors. Technical Repost Series. International Atomic Energy Agency, Vienna (1998)
5. Proceedings of the 1997 International Topical Meeting on Light Water Reactor Fuel Performance, Portland, Oregon (1997)
6. Burukin, A.V., Markov, D.V., Borisov, K.V., et al.: Results of studies of WWER—1000 fuel rods operability after tests in stationary mode with increased power and surface boiling. In: Bulgarian - Russian Scientific and Technical seminar on the experience of operation and introduction of new generation WWER—100 fuel, Nessebar, Bulgaria (2008)
7. Polenok, V.S., Pavlov, S.V., Smirnov, A.V., et al.: Studies on the problem concerned with the VVER 1000 fuel assembly bending during operation. In: Proceedings of the Vth Interindustry Conference on Reactor Material Science, vol. 1, pp. 47–58. JSC SSC RIAR Publ., Dimitrovgrad (1998). (in Russian)
8. Markov, D.V., Polenok, V.S.: Change of tension between fuel rods and cells of spacer grids during operation of serial fuel assemblies WWER – 1000. In: The 9th International Conference on Nuclear Power Plants, Dimitrovgrad-Ulyanovsk, pp. 212–217 (1998). (in Russian)
9. Markov, D.V., Pavlov, S.V., Novoselov, A.E.: VVER and RBMK fuel of a new generation: results of post-reactor studies, substantiation of reliability and performance. In: IX Russian Conference on Reactor Materials Science, pp. 24–46. JSC "SSC RIAR", Dimitrovgrad (2009). (in Russian)

10. Pavlov, S.V.: Key results of VVER-1000 fuel assemblies post-irradiation examinations. In: 10th International Conference on VVER Fuel Performance, Modeling and Experimental Support, Sandanski, Bulgaria, pp. 213–227 (2013)
11. Pavlov, S.V.: Changes in the bending stiffness of VVER-1000 fuel assemblies during operation. Izvestiya vuzov. Nucl. Energy **3**, 42–52 (2006). https://doi.org/10.26583/npe.2016.3.05
12. Smirnov, V.P., Markov, D.V., Smirnov, A.V., et al.: WWER fuel: results of post irradiation examination. In: Proceedings of the 2005 Water Reactor Fuel Performance Meeting, pp. 217–226. AESJ, Kyoto, Japan (2005)
13. Markov, D.V., Smirnov, V.P., Smirnov, A.V., et al.: WWER fuel; results of post irradiation examination. In: European Nuclear Conference, Versailes, France, p. 0009 (2005)
14. Markov, D.V., Polenok, V.S., Smirnov, A.V., et al.: WWER fuel: results of post-reactor studies. In: The Ukrainian-Russian Scientific Practical Seminar on the Experience of Operation and Introduction of New Generation WWER Fuel Khmelnitsky NPP, Ukraine (2005). (in Russian)
15. Smirnov, V.P., Markov, D.V., Polenok, V.S., et al.: Results of post-reactor studies of WWER fuel assemblies with high fuel burnout. In: The 5th International Conference "Safety, Efficiency and Economics of Nuclear Energy". FSUE "Concern ROSenergoatom", Moscow (2006). (in Russian)
16. Smirnov, V.P., Markov, D.V., Smirnov, A.V., et al.: WWER fuel: result of post irradiation examination. In: Fontevraud 6th International Symposium on Contribution of Materials Investigations to Improve the Safety and Performance of LWRs, A092-T09. Fontevraud, France (2006)
17. Smirnov, V.P., Markov, D.V., Polenok, V.S., et al.: The state of WWER fuel assemblies with high fuel burnout. In: IAEA Technical Committee "Post-Reactor Studies in the Hot Chambers of Fuel Assemblies of Water Reactors and their Inspection in the Holding Pools, Buenos Aires, Argentina (2006)
18. Smirnov, A.V., Markov, D.V., Smirnov, V.P., et al.: Results of post-reactor studies of the structural elements of the WWER fuel assemblies made of E110 and E635 alloys. In: The 6th International Conference "Modern Problems of Nuclear Physics - 2006", Tashkent, Uzbekistan (2006). (in Russian)
19. Opiatiuk, V.V., Kozlov, I.L., Skalozubov, V.I., Ostapenko, I.A.: Study of parametric interactions in the nuclear reactor control with feedback. ENERGETIKA. Proc. CIS High. Educ. Inst. Power Eng. Assoc. **64**(6), 517–524 (2021). https://doi.org/10.21122/1029-7448-2021-64-6-517-524

The Changes in the Output Parameters of Planetary Hydraulic Machines with the Increase in the Gap Between Their Rotors

Anatolii Panchenko[1], Angela Voloshina[1](✉), Shahriyor S. Sadullozoda[2], Igor Panchenko[1], and Viacheslav Mitin[1]

[1] Dmytro Motornyi Tavria State Agrotechnological University, 18, B. Khmelnitsky Avenue, Melitopol 72310, Ukraine
voloshinaa2012@gmail.com

[2] Tajik Technical University named after academician M. Osimi, 10, Academicians Rajabov's Avenue, 734042 Dushanbe, Tajikistan

Abstract. The current research was devoted to improving self-propelled vehicles energy efficiency in agriculture mainly through the hydraulic drive of active working bodies' output characteristics. The hydraulic drive with planetary-type hydraulic machines is widely used on agricultural equipment. The previous research revealed that during the operation of planetary hydraulic machines, the gap between the rotors often exceeded its critical value, accompanied by non-standard changes in its output parameters. It was predicted that the solution to that problem would make it possible to stabilize the output characteristics of planetary hydraulic machines. For this purpose, the physical model of the mutual movement of the rotors (considering a gap between them) and the diagram of their movement were developed. The mathematical dependences of the change in the angular velocity of the orbital hydraulic motor rotors were obtained. The nature of the change in the output parameters confirmed the results of the current theoretical studies, the decrease in the angular velocity and volumetric and overall efficiency.

Keywords: Rotor movement · Leakage flow rate · Critical clearance · Machinery energy efficiency · Volumetric efficiency

1 Introduction

Increasing the energy efficiency of self-propelled vehicles is inextricably linked with one of the areas identified by the SDGs, the development of the agricultural (national) economy. To drive active working bodies [1] and running systems [2] of agricultural, construction, municipal, logging, road, and other self-propelled equipment, as a gearless hydraulic drives of mechatronic systems, planetary-type hydraulic machines are widely used [3]. A distinctive design feature of planetary hydraulic machines is the presence of external and internal rotors with a special (hypocycloidal) tooth contour [4]. In the

V. Tonkonogyi et al. (Eds.): InterPartner 2022, LNME, pp. 540–551, 2023.
https://doi.org/10.1007/978-3-031-16651-8_51

process of operation, the centers of the teeth of the rotors can change their relative position. It has been established [4] that the rotors can move for a planetary hydraulic motor due to the lack of a constant kinematic connection, occupying a position where "self-sealing" occurs. Other movements of the rotors of the planetary hydraulic motor, caused by an increase in the gap between them, are accompanied by non-standard changes in its output parameters, for example, a decrease in the angular velocity of rotation of the planetary hydraulic motor shaft with a constant volumetric efficiency [4].

The results of the first phase of the research, which was devoted to determining the effect of the gap between the rotors on the change in the output parameters of planetary hydraulic machines, were presented in [4]. At that stage, the limitation of the research was that the changes in the output characteristics were considered at values of the diametrical clearance up to its critical value. During the operation of planetary hydraulic machines, the gap between the rotors often exceeds its critical value, so the next stage of the investigation appeared to be relevant.

In this regard, the second stage of our research was devoted to the issue of the influence of the gap between the rotors (when it is higher than the critical one) on the change in the output parameters of planetary hydraulic machines. The solution to this problem will ensure the stabilization of the output characteristics of hydraulic drives of self-propelled machines to enable the increase of their energy efficiency.

2 Literature Review

The solution was based on the results obtained before. We considered that the effect of pumped Bingham liquids on the performance of swirl chamber superchargers had been studied [5], and the characteristics of the vortex flow were determined [6]. The modeling of the flow of the working fluid in the flow parts was carried out by solving the RANS equations using the SST turbulence model for vortex-chamber [7], and labyrinth-screw [8] pumps, a linear model of the shaft rotation frequency of an axial piston hydraulic motor was obtained depending on the gas content of the working fluid [9]. After considering these fundamentals, it was revealed that the studies of the effect of gaps between rotors on the output characteristics of rotary hydraulic machines had not been carried out.

The theoretical analysis of the internal rotors of epitrochoidal, hypotrochoidal [10], original rotary [11], and hypogerotor [12] hydraulic machines was carried out, and the design of the gear contour of a gerotor machine with epitrochoidal and hypotrochoidal gearing was presented in [13]. The theory of gearing is proposed for constructing a mathematical description of a gerotor hydraulic machine with internal cycloidal gearing of rotors with a difference of one tooth [14]. The analytical model of surface wear [15], and its hydrodynamic effect [16] with the related mathematical model [17] allowed to describe the geometry of the geared contour of the rotor in a parametric form. The influence of the rotors' gear contour's geometric and kinematic parameters was studied and presented in [18], and the forces and moments acting on them were considered in [19]. The results of the displacement analysis were given in [20]. A comparison of analytical and numerical studies in the rotating elements of hydraulic machines were described in [21, 22]. The optimal gerotor design for their reduction was proposed in [23], and the method of factorial experiment – in [24, 25]. The multicriteria optimization of the geometry of a gerotor cogwheel [26] using a genetic algorithm considering the kinematic and actual flow pulsations [27] and using evolutionary strategy algorithms [28] using the ANSYS-CFX [29, 30] and GEROLAB [31] packages were presented. A numerical simulation of gerotor-type hydraulic machines was carried out. Its results were confirmed experimentally [32]. The geometric parameters of fluid distribution systems and the initial conditions for dynamic processes in hydraulic drives and their elements were also substantiated [33]. Experimental studies of the throughput distribution systems of planetary hydraulic motors [34] have been carried out, and substantiated the parameters of multicriteria optimization of hydraulic drive control elements [1]. At the same time the issues of modeling the operation of planetary hydraulic machines with a change in the gap between the rotors were not considered.

Technology for manufacturing a cycloidal pump with a negative punch clearance [35] has been proposed, the geometry of conjugated rotor profiles has been developed [36], and the engagement [37] of a cycloidal pump has been calculated and simulated. A parametric calculation of the rotor [38] using a cycloid and a circular arc curve [39] is carried out, the geometry is described [40] and an optimal tooth profile is proposed [41], the mechanism of stress and deformation occurrence in cycloidal gears [42] is considered. A mathematical model of engagement with gaps between the profiles of the teeth [43] has been proposed, the effect of the gap between the teeth on the pressure and film thickness of gear [44] and trochoidal [45] pumps has been investigated. On the other hand, the effect of changing the gap between the rotors of hydraulic machines [46, 47] on changing its output characteristics has not been investigated.

Overall, it can be argued that practically no attention was paid to the issue of modeling the operating conditions of a planetary hydraulic machine with a change in the gap between the rotors. Therefore, the study of the effect of the clearance value exceeding its critical value between the rotors of planetary hydraulic machines on the change in the output parameters of these hydraulic machines is an urgent task. The solution to this problem makes it possible to stabilize the output characteristics of planetary hydraulic machines operating as part of mechatronic systems of hydraulic drives of self-propelled equipment to increase its energy efficiency, which is extremely important for the development of agriculture as the Sustainable Development Goals indicate it.

3 Research Methodology

To study the influence of the gap between the rotors (when it is higher than the critical one) on the change in the output parameters of planetary hydraulic machines, it is necessary:

- to develop a physical model and a mathematical apparatus that describes the dependence of the change in the steady-state angular velocity of movement of the planetary hydraulic motor rotors when there is a gap;
- to justify the movement of the rotors of hydraulic machines of planetary type with a gap value exceeding the critical;
- to investigate the influence of the gap between the rotors (when higher than the critical one) on the change in the output parameters of planetary hydraulic machines.

The studies carried out in the first stage [4] found that the gap between the rotors of the planetary hydraulic motor causes additional movements of the rotors without rotation. This statement allows us to develop a physical model of the movement of the rotor (Fig. 1). Using the developed physical model (Fig. 1), explaining the principle of operation of the orbital hydraulic motor, the movement of the inner rotor 1 was demonstrated, which, under the action of the resulting pressure force of the working fluid p, rolls without sliding along the "infinite" sweep of the outer rotor 2. In the absence of a diametral gap, theoretical angular velocity ω_T of the inner rotor 1 (Fig. 1, curve 4) will be equal to [4].

$$\omega_T = \frac{\pi \cdot Q}{30 \cdot V_0}, \tag{1}$$

where Q is the actual flow rate of the working fluid; V_0 is the volume of the planetary hydraulic machine.

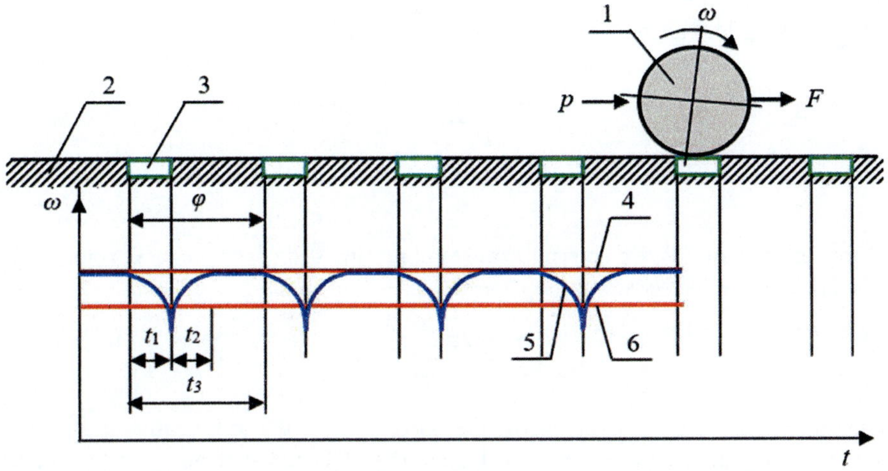

Fig. 1. Physical model of the mutual movement of rotors when there is a gap between them: 1 – inner rotor; 2 – "infinite" sweep of the outer rotor; 3 – "ascent" (deceleration) zone; 4 – theoretical angular velocity; 5 – real speed; 6 – an average speed of movement; t_1 – the "ascent" time; t_2 – the acceleration time; t_3 – the total turning time.

On the physical model (Fig. 1), the inner rotor 1, moving in the ascent zone 3, moves without rotation, translationally (slides). The "ascent" time t_1 required to move the inner rotor 1 (Fig. 1) in the "ascent" zone 3 (length of the sliding section) is determined by the size of the diametrical gap. During the "ascent" t_1, the angular velocity ω_p of rotation of the inner rotor 1 slows down. After the "self-sealing" of the rotors, during the rotation during the time t_2 of the inner rotor 1, in the beginning, its acceleration is observed during the time t_3. Therefore, the real value of the steady-state angular velocity ω_p, considering the translational displacements (slip) of the inner rotor 1, can be depicted by curve 5, and the average value of the parameter ω_p - by curve 6 (Fig. 1).

We will take the following assumptions when determining the average value of the angular velocity parameter ωp (Fig. 1, curve 6). All sections of the "ascent" zone 3 are equal to each other and are determined by the size of the diametrical gap G, and the "ascent" time t_1 (deceleration) of the inner rotor 1 is its acceleration time t_2.

Since during the "ascent" the inner rotor 1 moves translationally by the value of the diametrical gap G, then, considering the geometric parameters of the inner rotor 1, the "ascent" time t_1 can be determined as [4]

$$t_1 = \frac{G \cdot D_{e_1} \cdot b_1}{4Q \cdot \cos\delta}, \quad (2)$$

where D_{e1} and b_1 are the diameter of the location of the center of the teeth and the width of the inner rotor 1, respectively; cos δ is the angle of the "ascent" normal.

From this, it follows that with an increase in the diametrical clearance G (with wear), the "ascent" time t_1 increases.

The angle of rotation φ of the inner rotor 1, which is carried out during the time t3, depends on the number of its teeth z1 and is equal to $\varphi = 2 \cdot \pi / z1$.

Then, taking into account the assumptions, we have:

$$\omega_T = \frac{2\pi \cdot / z_1}{t_3 - 2t_1}, \quad (3)$$

$$\omega_p = \frac{2\pi \cdot / z_1}{t_3}, \quad (4)$$

From expression (3), we determine the total turning time t_3 of the inner rotor, which characterizes the steady-state turning speed.

$$t_3 = \frac{2\pi}{\omega_T \cdot z_1} - 2t_1 \quad (5)$$

Investigations of the kinematics of the movement of the inner rotor of an orbital hydraulic motor in the presence of a diametral gap G made it possible to obtain a mathematical apparatus describing the dependence of the change in the steady-state angular velocity ω_p, which is determined by a set of expressions (2), (3), (4) and (5).

Analysis [4] of the kinematics of movement of the inner rotor (Fig. 2, a) showed that the absence of a "rigid" kinematic connection between the rotors, in the presence of a diametric gap G, allows the inner rotor 1 to occupy a position with tangency at points B and C. In this position and there is a "self-sealing" of the rotors, despite the presence of a diametrical gap G at point D. When the value of the diametral gap G is equal to its critical value Glim (G = Glim), the teeth of the inner rotor 1 simultaneously touch the teeth 3 of the outer rotor 2 at three points B, C, and E (Fig. 2, a). Further increase in the gap between the rotors G is due to the wear of their toothed surface. When the value of the diametrical clearance G is greater than the critical G_{lim} ($G > G_{lim}$), then the movement of the inner rotor 1 under the action of the pressure force of the working fluid p (when it "rises") leads to the formation of a gap G at point C when the teeth contact in the point B and E (Fig. 2, b).

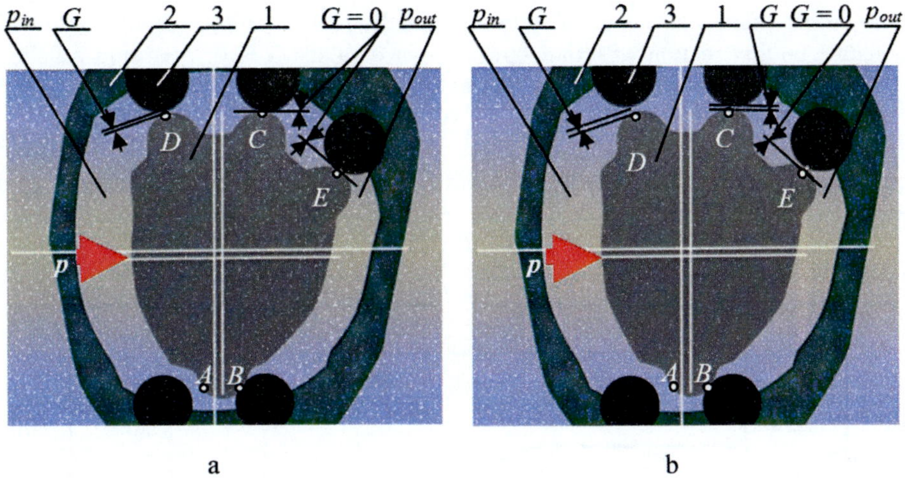

Fig. 2. The scheme of the movement of the rotors of hydraulic machines of planetary type with an increase in the gap between them: a – with $G = G_{lim}$; b – for $G > G_{lim}$; p_{in} – high-pressure chamber; p_{out} – low-pressure chamber; 1 – inner rotor; 2 – outer rotor; 3 – tooth (roller).

The presence of a gap G at point C (Fig. 2, b) indicates that when the diametral gap G is greater than the critical G_{lim} ($G > G_{lim}$), there is no "self-sealing" between the chambers of high p_{in} and low p_{out} pressures. The absence of the "self-sealing" effect means that a leakage rate has appeared between the pressure chambers p_{in} and the discharge chamber p_{out}, and, therefore, the characteristics of the orbital hydraulic motor with a value of the diametrical clearance $G > G_{lim}$ will change as in a standard positive displacement hydraulic machine.

4 Results and Discussion

The developed physical model of the mutual movement of the rotors with an increase in the gap between them (Fig. 1), reasonable schemes for the movement of the rotors of planetary-type hydraulic machines (Fig. 2), and the developed mathematical apparatus allowed the second phase of the research to be carried out as a continuation of the work [4]. In order to compare the obtained data with previous studies [4], all the initial conditions, restrictions, known mathematical apparatus, and hydraulic machines are accepted and described in [4].

The results of modeling the movement of the rotors of a planetary hydraulic machine obtained using the VisSim dynamic simulation system allow us to combine them with previous studies [4] and obtain a complete dependence of the change in the output parameters of planetary hydraulic machines on the gap between the rotors (Fig. 3).

The analysis of the presented dependences of the output parameters of planetary hydraulic machines on the change in the gap G between the rotors shows (Fig. 3) that the critical value of the gap was $G_{lim} = 0.4$ mm. When changing the diametral clearance G, depending on the degree of wear of the toothed surfaces of the rotors, the change in the output characteristics of the orbital hydraulic motor could be divided into two sections. The first section with diametral clearance values G = 0… 0.4 mm was less than critical G_{lim} ($G < G_{lim}$), and the second one, with diametral clearance values G = 0.4… 0.8 mm, was larger than critical G_{lim} ($G > G_{lim}$).

All considered dependences (Fig. 3, curves 2–5, and 7) of changes in the output parameters of planetary-type hydraulic machines on the gap G between the rotors, presented in the first section (at $G < G_{lim}$) were parallel to the abscissa axis. This testifies to the "self-sealing" of the orbital hydraulic motor rotors, which do not have a "rigid" center-to-center distance [4].

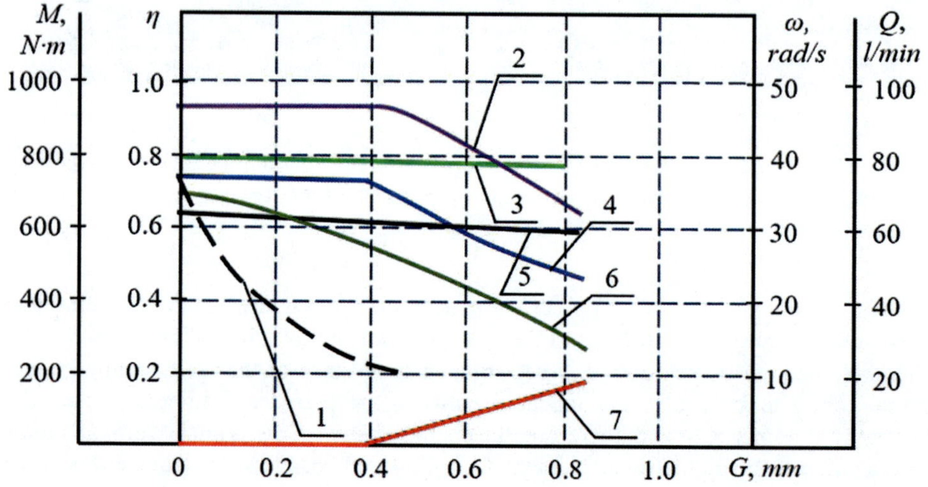

Fig. 3. The dependences of the change in the output parameters of planetary type hydraulic machines on the gap between the rotors [4]: 1 – the overall efficiency of the axial piston hydraulic motor; 2 – volumetric efficiency; 3 – hydromechanical efficiency; 4 – overall efficiency; 5 – torque; 6 – angular velocity; 7 – leakage rate.

A comparison of the dependences of the change in the overall efficiency (Fig. 3, curve 1) of the axial piston motor and the overall efficiency of the orbital hydraulic motor (Fig. 3, curve 4) indicates a significant difference in their change. This difference is explained by the fact that a standard rotary hydraulic machine with a diametral clearance G = 0… 0.4 mm (first section) has large leaks in the working chambers. In this regard, the volumetric efficiency, and hence the overall efficiency of the axial piston hydraulic motor, decreases to 0.2, which makes its further operation impossible.

Noteworthy is the sharp decrease in the angular velocity from 34 to 27 rad/s (Fig. 3, curve 6) caused by additional displacements of the inner rotor of the orbital hydraulic motor in the area under consideration. A more detailed analysis of the change in the above dependencies in the first section is presented in earlier studies [4].

The nature of the change in the output characteristics of the orbital hydraulic motor (Fig. 3), presented in the second section (at $G > G_{lim}$) confirms the studies of the kinematics of the movement of the inner rotor and explains the non-standard changes in the considered characteristics. When the diametral clearance G = 0.4… 0.8 mm, the change in the overall efficiency of the axial piston motor (Fig. 3, curve 1) was not considered due to the inoperability of the hydraulic machine in this range of clearances between the functional elements.

All other dependencies (Fig. 3, curves 2–7) of changes in the output characteristics from the diametrical gap, presented in the second section (with $G > G_{lim}$) can be divided into two groups. The first group is dependencies, the nature of the change of which is a continuation of the corresponding dependencies noted in the first section ($G < G_{lim}$) in previous studies [4]. This group (Fig. 3) includes the dependences of changes in hydromechanical efficiency (curve 3), torque M_{tor} (curve 5), and angular speed ω (curve 6), the nature of the change of which has remained unchanged. In the range of variation of the diametral gap G = 0… 0.8 mm, the decrease in the hydromechanical efficiency is 3% (0.8… 0.78), the torque is 5% (620… 595 N·m), and the angular velocity ω is 56% (34… 15 rad/s).

The second group consists of dependencies. The nature of the change sharply differs from the corresponding dependencies presented in the first section ($G < G_{lim}$). This group includes (Fig. 3) the dependences of changes in volumetric efficiency (curve 2), overall efficiency (curve 4), and leakage rate Qr (curve 7). In the range of variation of the diametral gap G = 0.4–0.8 mm, the decrease in volumetric efficiency is 18% (0.94…0.68), and the overall efficiency is 35% (0.73… 0.48), and the increase in leakage rate Q_r from 0 to 17 l/min. These changes confirm the studies of the kinematics of the inner rotor movement and are explained by the presence of leakage rate Q_r (Fig. 3, curve 7) through the formed gap G at the point C of contact of the rotor teeth (Fig. 2, b). The presence of leakage flow rate Q_r (Fig. 3, curve 7), between the chambers of the high pin and low-pressure chamber (Fig. 2), was a factor that reduced volumetric (Fig. 3, curve 2), and as a result, the overall efficiency (Fig. 3, curve 4) of the investigated planetary hydraulic motor with a change in the gap between the rotors in the range G = 0.4…0.8 mm.

The studies make it possible to predict changes in the output characteristics of orbital hydraulic motors and hydraulic drives of self-propelled equipment in general, both at the development and design stages.

5 Conclusions

The studies of the physical model of the mutual displacement of the rotors of a planetary hydraulic machine made it possible to obtain a mathematical apparatus that describes the dependence of the change in the angular velocity on the gap between the rotors of this hydraulic machine. This will make it possible to determine the quantitative values of the diametrical clearance during the operation of the orbital hydraulic motor.

By substantiating the kinematics of the movement of the inner rotor, it was found that, depending on the value of the diametrical gap, it can occupy two positions: less than the critical one, at which “self-sealing” of the rotors occurs, and more than the critical one, at which there is no “self-sealing”. This made it possible to develop an algorithm for modeling the kinematics of the rotor motion using the VisSim dynamic modeling system.

The studies carried out have established that at diametral clearance values of 0…0.4 mm (first position), all considered dependences of the change in the output characteristics of the orbital hydraulic motor are parallel to the abscissa axis, which indicates the "self-sealing" of the rotors. It was found that additional movements of the inner rotor cause a significant decrease in the angular velocity from 34 to 27 rad/s.

When the diametral clearance values are 0.4…0.8 mm (second position), all the obtained dependences of changes in output characteristics were presented in two groups: the first is the dependences of changes in hydromechanical efficiency, torque, and angular velocity, the nature of which remains unchanged; the second is the dependences characterizing a decrease in volumetric efficiency by 18%, overall efficiency by 35% and an increase in the leakage rate from 0 to 17 l/min, which is explained by the presence of leakage rates between the high and low-pressure chambers.

The studies make it possible to predict changes in the output characteristics of orbital hydraulic motors and hydraulic drives of self-propelled equipment in general, both at the development and design stages.

References

1. Panchenko, A., Voloshina, A., Titova, O., Panchenko, I., Caldare, A.: Design of hydraulic mechatronic systems with specified output characteristics. In: Ivanov, V., Pavlenko, I., Liaposhchenko, O., Machado, J., Edl, M. (eds.) DSMIE 2020. LNME, pp. 42–51. Springer, Cham (2020). https://doi.org/10.1007/978-3-030-50491-5_5
2. Nadykto, V., Kyurchev, V., Chaplinskyi, A., Ayubov, A.: Ways to increase the traction efficiency of modular draft device. IOP Conf. Ser Mater. Sci. Eng. **1021**(1), 012043 (2021)
3. Askarhodjaev, T., Rustamov, K., Aymatova, F., Husenova, G.: Justification of the hydraulic system parameters of the excavation body of a multi-purpose road construction vehicle based on the TTZ tractor. JCR **5**(7), 229–234 (2020). https://doi.org/10.31838/jcr.07.05.40
4. Panchenko, A., Voloshina, A., Luzan, P., Panchenko, I.: Volkov, S: kinematics of motion of rotors of an orbital hydraulic machine. IOP Conf. Ser Mater. Sci. Eng. **1021**, 012054 (2021)
5. Rogovyi, A., Korohodskyi, V., Medvediev, Y.: Influence of Bingham fluid viscosity on energy performances of a vortex chamber pump. Energy **218**, 119432 (2021). https://doi.org/10.1016/j.energy.2020.119432
6. Rogovyi, A., Khovanskyi, S., Hrechka, I., Gaydamaka, A.: Studies of the swirling submerged flow through a confuser. In: Ivanov, V., Pavlenko, I., Liaposhchenko, O., Machado, J., Edl, M. (eds.) DSMIE 2020. LNME, pp. 85–94. Springer, Cham (2020). https://doi.org/10.1007/978-3-030-50491-5_9
7. Rogovyi, A., Korohodskyi, V., Khovanskyi, S., Hrechka, I., Medvediev, Y.: Optimal design of vortex chamber pump. J. Phys: Conf. Ser. **1741**, 012018 (2021). https://doi.org/10.1088/1742-6596/1741/1/012018
8. Andrenko, P., Rogovyi, A., Hrechka, I., Khovanskyi, S., Svynarenko, M.: Characteristics improvement of labyrinth screw pump using design modification in screw. J. Phys: Conf. Ser. **1741**, 012024 (2021). https://doi.org/10.1088/1742-6596/1741/1/012024
9. Andrenko, P., Rogovyi, A., Hrechka, I., Khovanskyi, S., Svynarenko, M.: The influence of the gas content in the working fluid on parameters of the the hydraulic motor's axial piston. In: Ivanov, V., Pavlenko, I., Liaposhchenko, O., Machado, J., Edl, M. (eds.) DSMIE 2021. LNME, pp. 97–106. Springer, Cham (2021). https://doi.org/10.1007/978-3-030-77823-1_10
10. Bonandrini, G., Mimmi, G., Rottenbacher, C.: Theoretical analysis of internal epitrochoidal and hypotrochoidal machines. Proc. Inst. Mech. Eng. **223**, 1469–1480 (2009). https://doi.org/10.1243/09544062JMES1163

11. Bonandrini, G., Mimmi, G., Rottenbacher, C.: Theoretical Analysis of an Original Rotary Machine. J. Mech. Des. **132**(2), 024501 (2010). https://doi.org/10.1115/1.4000698
12. Kwon, S.-M., Kang, H.S., Shin, J.-H.: Rotor profile design in a hypogerotor pump. J. Mech. Sci. Technol **23**, 3459–3470 (2010). https://doi.org/10.1007/s12206-009-1007-y
13. Hsieh, C.-F., Hwang, Y.-W.: Geometric design for a gerotor pump with high area efficiency. J. Mech. Des. **129**, 1269–1277 (2007). https://doi.org/10.1115/1.2779887
14. Hwang, Y.-W., Hsieh, C.-F.: Geometric design using hypotrochoid and nonundercutting conditions for an internal cycloidal gear. J. Mech. Des. **129**(4), 413–420 (2007). https://doi.org/10.1115/1.2437806
15. Kostyk, K., et al.: Simulation of diffusion processes in chemical and thermal processing of machine parts. Processes **9**(4), 698 (2021). https://doi.org/10.3390/pr9040698
16. Kwon, S.-M., Kim, C.-H., Shin, J.-H.: Analytical wear model of a gerotor pump without hydrodynamic effect. J. Adv. Mech. Des. Syst. Manuf. **2**, 230–237 (2008). https://doi.org/10.1299/jamdsm.2.230
17. Bonandrini, G., Mimmi, G., Rottenbacher, C.: Design and simulation of meshing of a particular internal rotary pump. Mech. Mach. Theory **49**, 104–116 (2012). https://doi.org/10.1016/j.mechmachtheory.2011.11.001
18. Ivanović, L., Mackić, T., Stojanović, B.: Analysis of the instantaneous friction coefficient of the trochoidal gear pair. J. Balk. Tribol. Assoc. **22**, 281–293 (2016)
19. Ivanović, L., Devedžić, G., Mirić, N., Ćuković, S.: Analysis of forces and moments in gerotor pumps. Proc. Inst. Mech. Eng. **224**, 2257–2269 (2010). https://doi.org/10.1243/09544062JMES2041
20. Yang, D.C.H., Yan, J., Tong, S.-H.: Flowrate formulation and displacement analyses for deviation function-based gerotor pumps. Proc. Inst. Mech. Eng. **225**, 480–487 (2011). https://doi.org/10.1243/09544062JMES2203
21. Ivanović, L., Rakić, B., Stojanović, B., Matejić, M.: Comparative analysis of analytical and numerical calculations of contact stresses at rotational elements of gerotor pumps. Appl. Eng. Lett. **1**, 1–7 (2016)
22. Merzliakov, I., Pavlenko, I., Chekh, O., Sharapov, S., Ivanov, V.: Mathematical modeling of operating process and technological features for designing the vortex type liquid-vapor jet apparatus. In: Ivanov, V., et al. (eds.) DSMIE 2019. LNME, pp. 613–622. Springer, Cham (2020). https://doi.org/10.1007/978-3-030-22365-6_61
23. Kwak, H.S., Li, S.H., Kim, C.: Optimal design of the gerotor (2-ellipses) for reducing maximum contact stress. J. Mech. Sci. Technol. **30**(12), 5595–5603 (2016). https://doi.org/10.1007/s12206-016-1128-z
24. Ivanović, L.T., Veličković, S.N., Stojanović, B.Ž., Kandeva, M., Jakimovska, K.: The selection of optimal parameters of gerotor pump by application of factorial experimental design. FME Trans **45**, 159–164 (2017)
25. Fesenko, A., Yevsiukova, F., Basova, Y., Ivanova, M., Ivanov, V.: Prospects of using hydrodynamic cavitation for enhancement of efficiency of fluid working medium preparation technologies. Periodica Polytechnica Mechanical Engineering **62**(4), 269–276 (2018). https://doi.org/10.3311/PPme.11877
26. Robison, A., Vacca, A.: Multi-objective optimization of circular-toothed gerotors for kinematics and wear by genetic algorithm. Mech. Mach. Theory **128**, 150–168 (2018). https://doi.org/10.1016/j.mechmachtheory.2018.05.011
27. Robinson, A., Vacca, A.: Multi-Objective Optimization of Gerotor Port Design by Genetic Algorithm with Considerations on Kinematic vs. Actual Flow Ripple. SAE Technical Papers 2019–01–0827 (2019). https://doi.org/10.4271/2019-01-0827
28. De Martin, A., Jacazio, G., Sorli, M.: Optimization of gerotor pumps with asymmetric profiles through evolutionary strategy algorithm. Machines 7, 17 (2019). https://doi.org/10.3390/machines7010017

29. Bae, J., Kwak, H., San, S., Kim, C.: Design and CFD analysis of gerotor with multiple profiles (ellipse–involute–ellipse type and 3-ellipses type) using rotation and translation algorithm. Proc. IMechE Part **230**, 804–823 (2016). https://doi.org/10.1177/0954406215583888
30. Liaposhchenko, O., Pavlenko, I., Monkova, K., Demianenko, M, Starynskyi, O.: Numerical simulation of aeroelastic interaction between gas-liquid flow and deformable elements in modular separation devices. In: Ivanov, V., et al. (eds.) DSMIE 2019. LNME, pp. 765–774. Springer, Cham (2020). https://doi.org/10.1007/978-3-030-22365-6_76
31. Gamez-Montero, P.J., Castilla, R., Mujal, R., Khamashta, M., Codina, E.: GEROLAB package system: Innovative tool to design a trochoidal-gear pump. J. Mech. Des. **131**, 074502 (2009). https://doi.org/10.1115/1.3125889
32. Pellegri, M., Vacca, M., Frosina, E., Buono, D., Senatore, A.: Numerical analysis and experimental validation of Gerotor pumps: A comparison between a lumped parameter and a computational fluid dynamics-based approach. Proc. IMechE. Part C (2016). https://doi.org/10.1177/0954406216666874
33. Voloshina, A., Panchenko, A., Titova, O., Panchenko, I.: Changes in the dynamics of the output characteristics of mechatronic systems with planetary hydraulic motors. J. Phys: Conf. Ser. **1741**, 012045 (2021). https://doi.org/10.1088/1742-6596/1741/1/012045
34. Voloshina, A., Panchenko, A., Titova, O., Pashchenko, V., Zasiadko, A.: Experimental studies of a throughput of the distribution systems of planetary hydraulic motors. IOP Conf. Ser. Mater, Sci. Eng. **1021**, 012054 (2021)
35. Chen, T.-T., et al.: High-precision technology with negative punch clearance for the manufacturing of a cycloid pump. Int. J. Adv. Manuf. Technol. **95**(1–4), 1179–1183 (2017). https://doi.org/10.1007/s00170-017-1312-z
36. Vecchiato, D., Demenego, A., Argyris, J., Litvin, F.L.: Geometry of a cycloidal pump. Comput. Methods Appl. Mech. Eng. **190**, 2309–2330 (2001). https://doi.org/10.1016/S0045-7825(00)00236-X
37. Demenego, A., Vecchiato, D., Litvin, F.L., Nervegna, N., Mancó, S.: Design and simulation of meshing of a cycloidal pump. Mech. Mach. Theory **37**, 311–332 (2002). https://doi.org/10.1016/S0094-114X(01)00074-X
38. Zhang, E., Chen, M., Deng, B.: Parametric Design on Internal Gear of Cycloid Gear Pump with NX10.0. IOP Conf. Ser. Mater. Sci. Eng. **250**, 1–5 (2017). https://doi.org/10.1088/1757-899X/250/1/012068
39. Choi, T.H., Kim, M.S., Lee, G.S., Jung, S.Y., Bae, J.H., Kim, C.: Design of rotor for internal gear pump using cycloid and circular-arc curves. J. Mech. Des. **134**, 011005 (2012). https://doi.org/10.1115/1.4004423
40. Paffoni, B.: Pressure and film thickness in a trochoidal hydrostatic gear pump. Proc. Inst. Mech. Eng. **217**, 179–187 (2003). https://doi.org/10.1243/095441003769700744
41. Biernacki, K.: Selection of the optimum tooth profile for plastic cycloidal gears. Proc. Inst. Mech. Eng. **228**, 3395–3404 (2014). https://doi.org/10.1177/0954406214531408
42. Biernacki, K., Stryczek, J.: Analysis of stress and deformation in plastic gears used in gerotor pumps. J. Strain Anal. Eng. Des. **45**, 465–479 (2010). https://doi.org/10.1243/03093247JSA630
43. Ivanović, L., Devedžić, G., Ćuković S., Mirić, N.: Modeling of the Meshing of Trochoidal Profiles with Clearances. J. Mech. Des. **134**(4), 041003 (2012). https://doi.org/10.1115/1.4005621
44. Paffoni, B., Progri, R., Gras, R.: Teeth clearance effects upon pressure and film thickness in a trochoidal hydrostatic gear pump. Proc. Inst. Mech. Eng. **218**, 247–256 (2004). https://doi.org/10.1243/0954410041872799
45. Gamez-Montero, P.J., Garcia-Vilchez, M., Raush, G., Freire, J., Codina, E.: Teeth clearance and relief grooves effects in a Trochoidal-Gear Pump using new modules of GeroLAB. J. Mech. Des. **134**, 054502 (2012). https://doi.org/10.1115/1.4006440

46. Pavlenko, I.V., Simonovskiy, V.I., Demianenko, M.M.: Dynamic analysis of centrifugal machines rotors supported on ball bearings by combined application of 3D and beam finite element models. IOP Conf. Ser. Mater. Sci. Eng. **233**(1), 012053 (2017). https://doi.org/10.1088/1757-899X/233/1/012053
47. Onysko, O., Karabegović, I., Dašić, P., Penderetskyi, M., Melnyk, O.: The stress state of compact mechatronic satellites of a cycloidal reducer. J. Eng. Sci. **8**(2), D12–D17 (2021). https://doi.org/10.21272/jes.2021.8(2).d3

Check for updates

Improvement of Vortex Chamber Supercharger Performances Using Slotted Rectangular Channel

Andrii Rogovyi[1](✉), Artem Neskorozhenyi[2], Sergey Krasnikov[2], Irina Tynyanova[1], and Serhii Khovanskyi[3]

[1] National Technical University "Kharkiv Polytechnic Institute", 2, Kyrpychova Street, Kharkiv 61002, Ukraine
asrogovoy@ukr.net
[2] Kharkiv National Automobile and Highway University, 25, Yaroslava Mudrogo Street, Kharkiv 61002, Ukraine
[3] Sumy State University, 2, Rymskogo-Korsakova Street, Sumy 40007, Ukraine

Abstract. Abrasive wear, the influence of shock loads, temperature, and chemical aggressiveness of the pumped liquids lead to a decrease in the reliability and durability of pumping equipment when working on heterogeneous mediums that are difficult to pump. The implementation of the idea of using the centrifugal force effect on a solid particle in a vortex chamber has led to the vortex chamber supercharger creation that has high pumping efficiency of solid abrasive mediums. However, these blowers have a significant drawback – using a technological drainage channel leads to pumped medium losses. Using a slotted outlet nozzle in the tangential blower outlet channel made it possible to reduce and even nullify such losses of solid particles. The study consisted of three stages: flow mathematical modeling inside the vortex chamber supercharger; experimental study of the magnitude of losses in the drainage channel; determination of the mathematical model adequacy and further search for effective ways to reduce losses. It was found that, in contrast to the cylindrical outlet channel, where minimal losses are observed for a density of 2000...3000 kg/m^3, for a cylindrical outlet channel, losses decrease with increasing the pumped medium density.

Keywords: Energy efficiency · Vortex chamber supercharger · Experiment · Numerical simulation · Drainage channel · Granular material losses

1 Introduction

The wear issues with mechanical moving parts when pumping abrasive solids is very acute in many industries [1, 2]. Using classical vane and positive displacement pumps [3, 4] for pumping mediums containing solid particles significantly decreases pump efficiency [5, 6]. In addition, frequent seals wear requires stopping production from replacing worn seals, which leads to economic losses [7, 8]. The combination of abrasive wear and the shock loads influence, liquid temperature, and chemical aggressiveness lead

V. Tonkonogyi et al. (Eds.): InterPartner 2022, LNME, pp. 552–561, 2023.
https://doi.org/10.1007/978-3-031-16651-8_52

to a decrease in the reliability and durability of pumping equipment when operating on heterogeneous mediums that are difficult to pump [9, 10].

It is possible to improve the reliability and durability of pumps and blowers using jet pumps [11]. They have very high reliability and uptime but low energy efficiency. The jet ejector efficiency can be improved using the hydrodynamic effects of swirling flows in the vortex chamber [12]. So, creating a new vortex chamber supercharger can improve the energy efficiency of jet pumps and the efficiency of pumping abrasive solids mediums.

Thus, this paper aims to investigate the pumping parameters and properties of the pumped medium, which will reduce the bulk losses to zero by using a slotted tangential channel in the vortex chamber supercharger.

2 Literature Review

Jet pumps with a vortex chamber are usually used for closed volumes extraction and rarely for pumping homogeneous mediums due to low efficiency [13]. Implementing the centrifugal force effect on a solid particle in a vortex chamber requires an additional tangential channel creation for the mixed flow outlet at the vortex chamber periphery [14, 15]. Such jet superchargers, called vortex chamber superchargers, have high pumping efficiency rates of solid abrasive particles. The efficiency of pumping solids with air is twice that of classic straight jet pumps and ejectors.

Vortex chamber superchargers have a significant drawback – using a technological drainage channel leads to pumped medium losses. These losses depend on the flow kinematic parameters, swirl inside the vortex chamber, and the chamber and channels' geometric dimensions. In paper [14], the blower energy characteristics are shown, but drainage flow rate dependence is not mentioned on other modes and geometric parameters. In most cases, researchers pay special attention to the equipment's energy efficiency [16, 17]. However, loss reduction solves the issue of energy efficiency indirectly by improving the parameters of the energy-efficient workflow.

Using the same pump geometry, it is possible to implement two working processes with and without losses of the pumped medium [11, 14]. A working process without a drainage channel will be less energy efficient due to the reduction in energy parameters of the pumped medium at the device outlet.

In previous articles, the authors considered the losses of the pumped fluid in the drainage channel but did not consider the ways to assess these losses and the factors that affect them. Using a slotted outlet nozzle in the pump tangential channel makes it possible to reduce losses, but they do not reduce to zero, and, to date, the influence of operating parameters on the losses has not been revealed.

In order to minimize losses in the drainage channel, it is necessary to evaluate the influence of all possible factors on the particle trajectories, which is difficult to achieve with the help of experimental studies. A modern approach based on CFD calculation techniques [18], i.e., using solid-phase motion simulation [19, 20] makes it possible to understand what steps should be taken to minimize losses of solid particles in the supercharger.

Thus, it is topical to evaluate the losses of the pumped flow in the drainage channel and identify the factors influencing them. Minimizing the losses of pumped flows can lead to a broader distribution of vortex chamber superchargers in industry and the creation of autonomous installations based on renewable energy [21, 22]. It is also possible to use them in conditions of limited overall dimensions, such as transport vehicles [23, 24].

3 Research Methodology

The study consisted of three stages: flow mathematical modeling inside the vortex chamber supercharger; experimental study of the vortex chamber supercharger; determination of the mathematical model adequacy and further search for effective ways to reduce losses of pumped bulk.

Mathematical modeling was carried out in the OpenFoam software package based on the solution of the Reynolds-averaged Navier-Stokes (RANS) equations using the SST turbulence model and its rotation-curvature correction. To determine the trajectories of the solid particles, the superposition method was used to determine the kinematic parameters of a solid particle after determining the velocity and pressure fields of a homogeneous liquid medium (One-Way Coupled). The simulation was carried out based on solving stationary equations.

There devices that have passed metrological verification were used in experimental investigations: pressure gauges (relative error no more than 1%), stopwatch (relative error no more than 1% for determining the mass flow rate of a solid medium by volumetric method), scales (relative error no more than 1%). Air mass flow rates in all channels were indirectly determined using Venturi flowmeters.

4 Results and Discussion

4.1 Vortex Chamber Supercharger

The vortex chamber supercharger scheme, its 3D model, and mesh are presented in Fig. 1.

Centrifugal force in the vortex chamber supercharger is similar to its use in a centrifugal pump. In the vortex chamber supercharger, the role of the centrifugal pump impeller is performed by the primary flow, which is supplied into the tangential inlet channel. This flow swirls in the vortex chamber and leaves it through the technological axial drainage channel. Due to the primary flow swirling, hydrodynamic effects arise, such as gauge pressure on the periphery of the vortex chamber and vacuum on the axis. Their use allows energy transfer to the passive flow entering the vortex chamber through the axial inlet channel. The pumped flow is sucked into the vortex chamber due to vacuum, then moves along spiral trajectories to the vortex chamber periphery and the tangential outlet channel. One of the main advantages of such blowers is that they are straightforward to manufacture and operate [25, 26], like other jet technology devices [27, 28]. In this work, both in theoretical and experimental studies, the supercharger with the following geometric dimensions was used: the diameter of the vortex chamber is 50 mm, the height is 10 mm, the diameter of the tangential supply channel is 8 mm, the diameter of the outlet tangential channel is 6 mm, the diameter of the axial inlet channel is 5 mm, the diameter of the axial drainage channel is 10 mm.

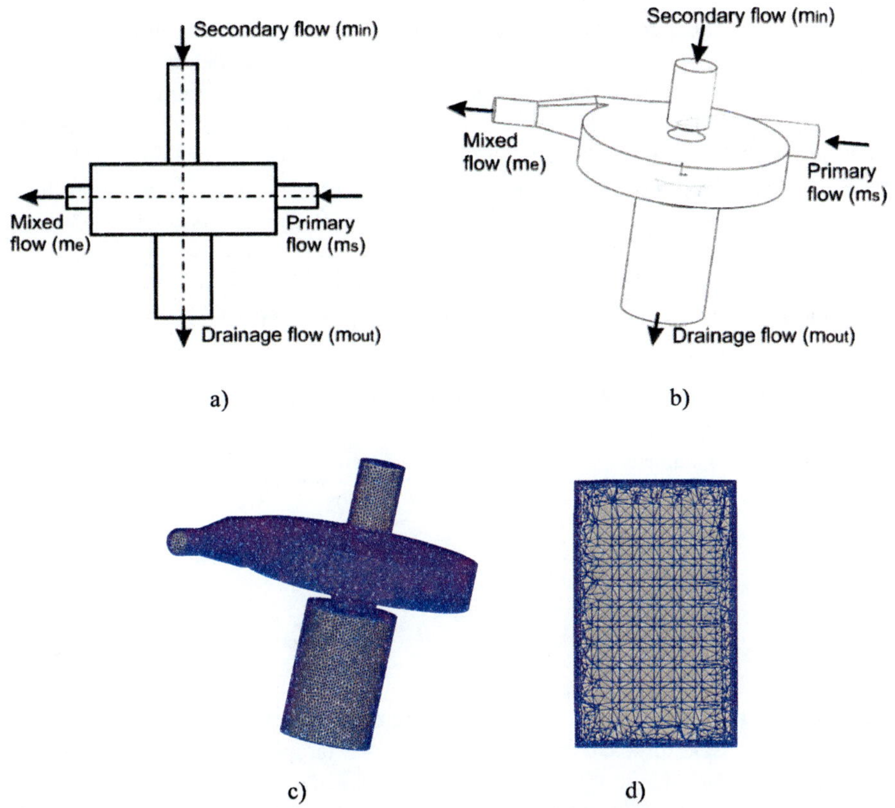

Fig. 1. Geometrical parameters of the investigated vortex chamber supercharger: a) design scheme; b) solid model; c) mesh; d) mesh in the outlet slotted tangential channel.

Mathematical modeling was carried out based on solving the Reynolds equations using the SST turbulence model and rotation-curvature corrections. The model equations are given in [11, 29]. Modeling of the solid particle trajectories was carried out based on solving the basic dynamics equation of a solid particle, taking into account the following forces [30]: drag force, gravity force, Saffman lift force, virtual mass force, and pressure gradient force. Calculating solid particle trajectories was carried out after finding the velocity and pressure fields of the main gas flow in the supercharger, which significantly reduces the calculation time (One-Way Coupled) with sufficient accuracy for estimating the solid particles' mass flow rates. The simulation was carried out in a stationary setting and ended after reaching the residuals of the main equations up to values of 10^{-5}. In addition, the constancy of the flow rate in the supercharger channels was controlled, i.e., the calculation ended after reaching constant flow rates [31].

Grid partitions were created in such a way as to provide y^+ values less than 3. In general, three meshes were used with a number of elements: 3, 7, and 15 million elements [32, 33]. On the basis of sensitivity analysis, it was found that if the number of elements is more than 7 million, the results do not depend on the grid. Checking the sensitivity, as well as comparison with experimental data, was carried out on the basis of the integral flow parameters comparison: the flow rate at the device outlet, the mass flow rate of solid particles in the drainage channel. To check the mathematical model used adequacy, an assessment was made of the correctness of calculating the solid medium flow rate in the drainage channel. The additional complexity of such verification is added by correctly estimating the solid particle size distribution in the experiment. For this, analysis was carried out by sifting particles through a system of sieves. Based on the sieving, the set of solid particles was determined to satisfy the Rosin–Rammler distribution [34] with the parameters given in Table 1. The particles were chosen in such a way that their shape was as spherical as possible. The same distribution was set as the initial conditions for calculating the trajectories of solid material in the supercharger.

Table 1. Particle diameter distribution.

Parameter	Details
Total Flow Rate, kg/s	0.008
Minimum diameter, m	$3 \cdot 10^{-6}$
Maximum diameter, m	0.001
Mean diameter, m	$5 \cdot 10^{-5}$

Simulations made it possible to determine the main parameters that affect the magnitude of solid particle losses: dynamic pressure in the tangential supply channel (swirl number in the vortex chamber), particle diameter and density. As a result of varying these parameters, it was possible to completely eliminate the loss of granular medium in the drainage channel. It was determined that in order to reduce losses to zero, the particle size must be greater than a certain diameter (Table 2).

Table 2. Minimum particle size for no loss in the drainage channel.

Dynamic pressure, kPa	Solid particle density, kg/m^3		
	700	1400	2000
5	$3 \cdot 10^{-5}$ m	$4 \cdot 10^{5}$ m	–
7.5	$2 \cdot 10^{-5}$ m	$2.2 \cdot 10^{5}$ m	–
10	$5 \cdot 10^{-6}$ m	$2 \cdot 10^{5}$ m	–
16	$7 \cdot 10^{6}$ m	$1.8 \cdot 10^{5}$ m	–
21	$7 \cdot 10^{6}$ m	$9 \cdot 10^{6}$ m	$3 \cdot 10^{5}$ m

This result is facilitated, first of all, by the use of a rectangular slotted receiving channel. Using a cylindrical receiving channel did not lead to the complete disappearance of the pumped medium mass flow rate in the drainage channel. The calculation results are summarized in Table 2. It shows that it was possible to get rid of losses only for densities less than 2000 kg/m^3. The negative result for higher densities can be explained by the fact that the particles become harder, do not have time to acquire a flow velocity before they enter the drainage channel, and cannot return to the vortex chamber during the movement.

4.2 Experimental Results

Experimental studies were carried out on an installation that allows determining the main integral energy characteristics of the supercharger (Fig. 2).

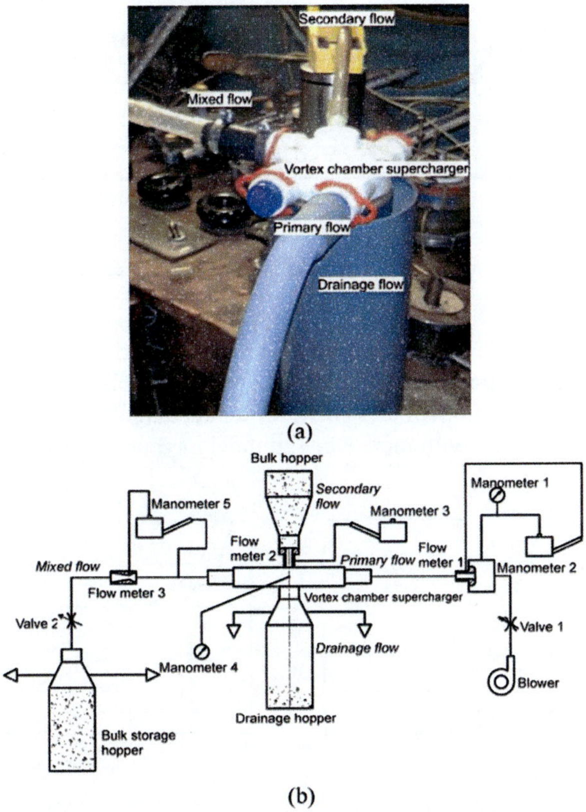

Fig. 2. Schematic diagram of experimental setup: (a) laboratory setup; (b) experiment system.

The comparison results of the calculation and experiment are shown in Fig. 3. Here, the rectangular slotted outlet channel can significantly reduce the flow rate of solid particles in the drainage channel. In addition, we can assume that the mathematical model adequately describes the flow inside the supercharger. Some discrepancies between the calculation and experimental results of the relative flow rate in the drainage channel is explained by the fact that the mathematical model is general. Moreover, the SST model calculates the vacuum near the axis with an error, leading to the underestimated flow rate. This problem manifests less with an increased swirl number (dynamic pressure at the supply channel) in the vortex chamber. In addition, the calculation error can influence the error in determining the particle diameters in the experiment, as well as their nonsphericity.

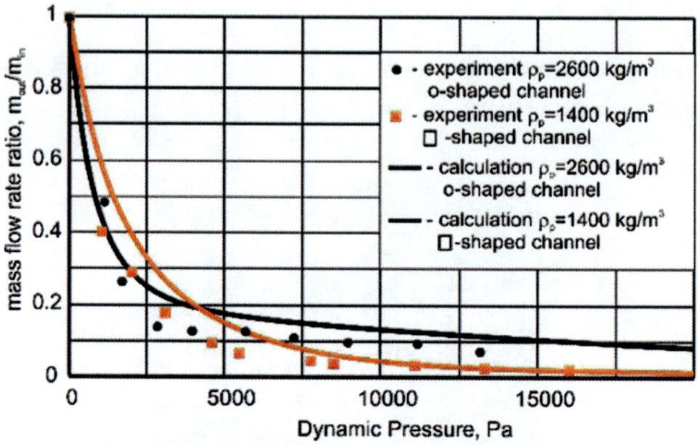

Fig. 3. Comparison of calculation results with experiment.

Figure 4 shows that the drainage channel's relative flow rate depends on the working medium's density. In contrast to the cylindrical outlet channel, where minimal losses are observed for a density of 2000–3000 kg/m^3, for the rectangular outlet channel, the solid medium flow rate decrease with increasing pumped medium density.

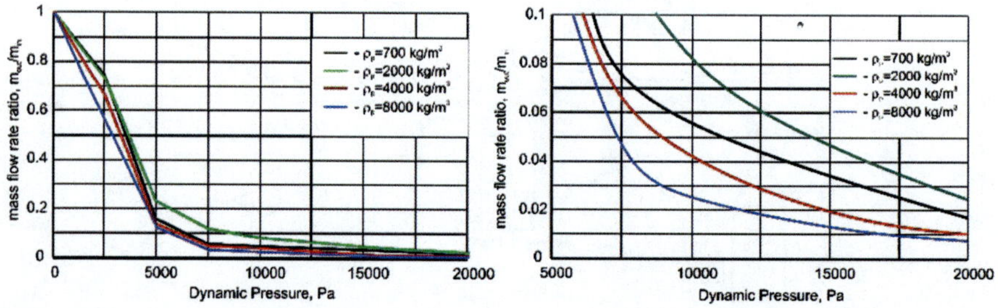

Fig. 4. Influence of the pumped medium density on the flow rate in the drainage channel.

Curves of the density influence depend quite strongly on the distribution of particles by mass, and it is necessary to carry out further studies on the specific particle sizes 8 influence drainage flow rates. Also, in this study, the influence of turbulent fluctuations in the main fluid is not considered, which can make it possible to obtain more accurate modeling results based on averaging and probabilistic estimation since the particle size is small enough that the turbulence parameters [35, 36] can influence the particle trajectories.

5 Conclusion

Based on experimental and numerical studies of two-phase medium flows in a vortex chamber supercharger, the losses of the pumped medium in the technological drainage channel of the supercharger are estimated.

For the first time, the use of a slotted (rectangular) outlet channel made it possible, at appropriate values of the diameter of the solid particles, their density, as well as the dynamic pressure at the inlet (the swirl number in the vortex chamber), completely get rid of the granular medium losses in the drainage channel. The scientific novelty lies in that earlier in the vortex chamber supercharger, in which the working process with a drainage channel is implemented, it was impossible to avoid losses of the pumped medium, ranging from 5 to 10%. Using the parameters described in this paper allows the vortex chamber supercharger without any losses of pumping medium.

In contrast to the cylindrical outlet channel, where minimal losses are observed for a density of 2000–3000 kg/m^3, for the rectangular outlet channel, losses decrease with an increase in the pumped medium density.

References

1. Noon, A.A., Kim, M.H.: Erosion wear on centrifugal pump casing due to slurry flow. Wear **364**, 103–111 (2016). https://doi.org/10.1016/j.wear.2016.07.005
2. Panchenko, A., Voloshina, A., Titova, O., Panchenko, I., Caldare, A.: Design of hydraulic mechatronic systems with specified output characteristics. In: Ivanov, V., Pavlenko, I., Liaposhchenko, O., Machado, J., Edl, M. (eds.) DSMIE 2020. LNME, pp. 42–51. Springer, Cham (2020). https://doi.org/10.1007/978-3-030-50491-5_5
3. Voloshina, A., Panchenko, A., Boltyansky, O., Zasiadko, A., Verkholantseva, V.: Improvement of the angular arrangement of distribution system windows when designing planetary hydraulic machines. In: Tonkonogyi, V., Ivanov, V., Trojanowska, J., Oborskyi, G., Pavlenko, I. (eds.) InterPartner 2021. LNME, pp. 53–63. Springer, Cham (2022). https://doi.org/10.1007/978-3-030-91327-4_6
4. Panchenko, A., Voloshina, A., Luzan, P., Panchenko, I., Volkov, S.: Kinematics of motion of rotors of an orbital hydraulic machine. IOP Conf. Ser.: Mater. Sci. Eng. 1021, 012045 (2021)
5. Andrenko, P., Rogovyi, A., Hrechka, I., Khovanskyi, S., Svynarenko, M.: Characteristics improvement of labyrinth screw pump using design modification in screw. J. Phys.: Conf. Ser. 1741, 012024 (2021)
6. Voloshina, A., Panchenko, A., Titova, O., Pashchenko, V., Zasiadko, A.: Experimental studies of a throughput of the distribution systems of planetary hydraulic motors. IOP Conf. Ser.: Mater. Sci. Eng. 1021, 012054. (2021)
7. Sokolov, V., Porkuian, O., Krol, O., Stepanova, O.: Design calculation of automatic rotary motion electrohydraulic drive for technological equipment. In: Ivanov, V., Trojanowska, J., Pavlenko, I., Zajac, J., Peraković, D. (eds.) DSMIE 2021. LNME, pp. 133–142. Springer, Cham (2021). https://doi.org/10.1007/978-3-030-77719-7_14
8. Tkachuk, M.M., Grabovskiy, A., Tkachuk, M.A., Shut, O.: Computational–experimental evaluation of stiffness response in elastic supports of rotor systems. In: Altenbach, H., Amabili, M., Mikhlin, Y.V. (eds.) Nonlinear Mechanics of Complex Structures. ASM, vol. 157, pp. 353–366. Springer, Cham (2021). https://doi.org/10.1007/978-3-030-75890-5_20
9. Pavlenko, I.: Static and dynamic analysis of the closing rotor balancing device of the multistage centrifugal pump. Appl. Mech. Mater. **630**, 248–254 (2014). https://doi.org/10.4028/www.scientific.net/AMM.630.248

10. Krol, O., Porkuian, O., Sokolov, V., Tsankov, P.: Vibration stability of spindle nodes in the zone of tool equipment optimal parameters. Comptes rendus de l'Acade'mie bulgare des Sciences **72**(11), 1546–1556 (2019). https://doi.org/10.7546/CRABS.2019.11.12
11. Rogovyi, A., Korohodskyi, V., Medvediev, Y.: Influence of Bingham fluid viscosity on energy performances of a vortex chamber pump. Energy **218**, 119432 (2021). https://doi.org/10.1016/j.energy.2020.119432
12. Merzliakov, I., Pavlenko, I., Chekh, O., Sharapov, S., Ivanov, V.: Mathematical modeling of operating process and technological features for designing the vortex type liquid-vapor jet apparatus. In: Ivanov, V., et al. (eds.) DSMIE 2019. LNME, pp. 613–622. Springer, Cham (2020). https://doi.org/10.1007/978-3-030-22365-6_61
13. Evdokimov, O.A., Piralishvili, S.A., Veretennikov, S.V., Guryanov, A.I.: CFD simulation of a vortex ejector for use in vacuum applications. J. Phys. Conf. Ser. **1128**, 012127 (2018)
14. Rogovyi, A., Korohodskyi, V., Khovanskyi, S., Hrechka, I., Medvediev, Y.: Optimal design of vortex chamber pump. J. Phys. Conf. Ser. **1741**, 012018 (2021)
15. Kondus, V.Y., Gusak, O.G., Yevtushenko, J.V.: Investigation of the operating process of a high-pressure centrifugal pump with taking into account of improvement the process of fluid flowing in its flowing part. J. Phys. Conf. Ser. **1741**, 012012 (2021)
16. Patlins, A., Hnatov, A., Arhun, S.C., Bogdan, D., Dzyubenko, O.: Development of an energy generating platform for converting kinetic energy into electrical energy using the kinematic synthesis of a three-stage multiplier. In: Transport Means 2019: Proceedings of the 23rd International Scientific Conference. Part 1, pp. 403–408 (2019)
17. Arhun, S., Migal, V., Hnatov, A., Hnatova, H., Ulyanets, O.: System approach to the evaluation of the traction electric motor quality. EAI Endorsed Trans. Energy Web **7**(27), 1–9 (2020). https://doi.org/10.4108/eai.13-7-2018.162733
18. Fesenko, A., Yevsiukova, F., Basova, Y., Ivanova, M., Ivanov, V.: Prospects of using hydrodynamic cavitation for enhancement of efficiency of fluid working medium preparation technologies. Periodica Polytech. Mech. Eng. **62**(4), 269–276 (2018). https://doi.org/10.3311/PPme.11877
19. Hasečić, A., Imamović, J., Bikić, S., Džaferović, E.: Investigation of the Contamination Influence on the Parameters of Gas Flow Through Multihole Orifice Flowmeter. IEEE Trans. Instrum. Meas. **70**, 1–8 (2021). https://doi.org/10.1109/TIM.2021.3063198
20. Berberović, E., Bikić, S.: Computational Study of Flow and Heat Transfer Characteristics of EG-Si3N4 Nanofluid in Laminar Flow in a Pipe in Forced Convection Regime. Energies **13**(1), 74 (2020). https://doi.org/10.3390/en13010074
21. Zaitsev, R.V., Khrypunov, G.S., Veselova, N.V., Kirichenko, M.V., Kharchenko, M.M., Zaitseva, L.V.: The cadmium telluride thin films for flexible solar cell received by magnetron dispersion method. J. Nano- Electron. Phys. **9**(3), 03015 (2017). https://doi.org/10.21272/jnep.9(3).03015
22. Minakova, K.A., Zaitsev, R.V.: Improving the solar collector base model for PVT system. J. Nano- Electron. Phys. **12**(4), 04028 (2020). https://doi.org/10.21272/jnep.12(4).04028
23. Voloshina, A., Panchenko, A., Titova, O., Panchenko, I.: Changes in the dynamics of the output characteristics of mechatronic systems with planetary hydraulic motors. J. Phys. Conf. Ser. **1741**, 012045 (2021)
24. Voloshina, A., Panchenko, A., Titova, O., Milaeva, I., Pastushenko, A.: Prediction of changes in the output characteristics of the planetary hydraulic motor. In: Tonkonogyi, V., et al. (eds.) InterPartner 2020. LNME, pp. 744–754. Springer, Cham (2021). https://doi.org/10.1007/978-3-030-68014-5_72
25. Krol, O., Sokolov, V.: Research of toothed belt transmission with arched teeth. Diagnostyka **21**(4), 15–22 (2020). https://doi.org/10.29354/diag/127193

26. Krol, O., Sokolov, V.: Research of modified gear drive for multioperational machine with increased load capacity. Diagnostyka **21**(3), 87–93 (2020). https://doi.org/10.29354/diag/126026
27. Torregrosa, A.J., Broatch, A., Quintero, P., Redondo, Á., Morin, F.: Assessment of the fluid-dynamic and acoustic behaviour of a swirl static mixer for after-treatment systems. Appl. Acoust. **186**, 108446 (2022). https://doi.org/10.1016/j.apacoust.2021.108446
28. Evdokimov, O.A., Piralishvili, S.A., Veretennikov, S.V., Elkes, A.A.: Experimental study of cleaning aircraft GTE fuel injectors using a vortex ejector. J. Phys. Conf. Ser. **925**, 012027 (2017)
29. Smirnov, P.E., Menter, F.R.: Sensitization of the SST turbulence model to rotation and curvature by applying the Spalart-Shur correction term. J. Turbomach. **131**(4), 041010 (2009). https://doi.org/10.1115/1.3070573
30. Sommerfeld, M., Sgrott, O.L., Jr., Taborda, M.A., Koullapis, P., Bauer, K., Kassinos, S.: Analysis of flow field and turbulence predictions in a lung model applying RANS and implications for particle deposition. Eur. J. Pharm. Sci. **166**, 105959 (2021). https://doi.org/10.1016/j.ejps.2021.105959
31. Rezvaya, K., Krupa, E., Drankovskiy, V., Makarov, V.: Optimization of the water passage of a pump-turbine based on a numerical study of its hydrodynamic characteristics. In: 2019 IEEE 2nd Ukraine Conference on Electrical and Computer Engineering (UKRCON), pp. 460–463. IEEE (2019). https://doi.org/10.1109/UKRCON.2019.8879889
32. Brazhenko, V., Mochalin, I.: Numerical simulation and experimental tests of the filter with a rotating cylindrical perforated filter element. Proc. Inst. Mech. Eng. C J. Mech. Eng. Sci. **235**(12), 2180–2191 (2021). https://doi.org/10.1177/0954406220950346
33. Rezvaya, K., Krupa, E., Shudryk, A., Drankovskiy, V., Makarov, V.: Solving the hydrodynamical tasks using CFD programs. In: 2018 IEEE 3rd International Conference on Intelligent Energy and Power Systems (IEPS), pp. 205–209. IEEE (2018). https://doi.org/10.1109/IEPS.2018.8559548
34. Riha, Z., Zelenak, M., Soucek, K., Hlavacek, A.: Flow field analysis inside and at the outlet of the abrasive head. Materials **14**(14), 3919 (2021). https://doi.org/10.3390/ma14143919
35. Mandryka, A., Majid, A.P., Ratushnyi, O., Kulikov, O., Sukhostavets, D.: Ways for improvement of reverse axial pumps. Journal of Engineering Sciences **9**(1), D14–D19 (2022). https://doi.org/10.21272/jes.2022.9(1).d3
36. Gusak, O., Cherkashenko, M., Potetenko, O., Hasiuk, A., Rezvaya, K.: Improvement of operating processes of high-head tubular horizontal hydraulic turbines. In: Ivanov, V., Pavlenko, I., Liaposhchenko, O., Machado, J., Edl, M. (eds.) DSMIE 2021. LNME, pp. 126–136. Springer, Cham (2021). https://doi.org/10.1007/978-3-030-77823-1_13

Failure Analysis of Refractories in Rotary Kilns

Valerii Scherbyna[1(✉)], Aleksandr Gondlyakh[1], Aleksandr Sokolskiy[1], Yaroslav Shilovich[1], and Nataliia Bulavina[2]

[1] National Technical University of Ukraine "Igor Sikorsky Kyiv Polytechnic Institute", 37, Prospect Peremohy, Kyiv 03056, Ukraine
xpsm@ukr.net

[2] Modern Art Research Institute of the National Academy of Arts of Ukraine, 18-D, Ye. Konovaltsa Street, Kyiv 01133, Ukraine

Abstract. This efficiency of rotary kilns largely depends on the rational use of structural elements that ensure heat apparatuses' safe, long-lasting, and reliable operation. This is especially relevant for refractory lining, in which fractures or chipping occur during operation due to high temperatures and pressures. To study the destruction of lining refractories, a calculation method has been developed that makes it possible to analyze the formation of destruction (splitting) in the lining of a rotary kiln when metal plates are installed between refractories. The mathematical model considers the nonlinear dependence of the physical and mechanical characteristics of refractories and plates on temperature. Various fracture criteria for brittle materials were used for credible analysis. The work of the lining of a rotary kiln 4.0 × 60 m has been studied. The possibility of destruction of refractories installed in different parts of the furnace unit with thickness change of the lining due to wear during operation of the furnace is considered. As a result of a numerical experiment, it was found that destruction occurs only at the initial stage of the furnace operation in the refractory zone, which is located at a distance of 20–100 mm from a more heated surface. Operating experience in rotary kilns confirms this. A decrease in the thickness of the lining decreases the zone width and the failure criterion with a simultaneous offset to the work surface. The article also suggests methods for preventing the destruction of refractory.

Keywords: Lining · Strength · Thermal stresses · Split · Brittleness · Energy saving · Industrial growth

1 Introduction

Rotating drum-type thermal units, such as rotary kilns, are widely used in oil refining, petrochemical, chemical, and gas industries [1, 2]. However, they are most widely used in the building materials industry [3], the main equipment for producing cement clinker, expanded clay, lime, perlite, and other materials [4].

From the point of view of technical implementation, rotary kilns are complex units, both physical and chemical reactors, and combustion chambers. Simultaneously, most such equipment is operated under high temperatures, pressures, aggressive environments,

V. Tonkonogyi et al. (Eds.): InterPartner 2022, LNME, pp. 562–573, 2023.
https://doi.org/10.1007/978-3-031-16651-8_53

and increased dustiness, which requires special measures to improve their reliability, safety, and durability. Their application largely depends on using efficient and rational structural elements, which should ensure the long-term and reliable operation of the kiln unit [5].

At the same time, an essential factor is the reliability and stability of the lining, which provides the geometric shapes of the working chamber, the duration of operation, and the technical and economic indicators of the process of obtaining the final product [6]. The analysis shows that increasing the life of the refractory lining contributes to the extension of its service life, which is an essential factor for significant energy savings. Therefore, this work devoted to the improvement of refractories is highly relevant, as it contributes to the intensification of the furnace operation and makes it possible to implement more intensive modes of use of heat equipment.

The work aims to develop a methodology and perform calculations to determine the possibility of fractures or chips in the refractories of the lining of rotating heat aggregates and determine methods for their prevention.

2 Literature Review

Under thermal action on the lining of a kiln in refractories, such physical phenomena as thermal expansion, plastic deformation, and fracture (splitting) occur [7], which violate the durability of the lining, and the duration of operation, which can lead to an emergency [8].

Cracking of the lining is the surface cracking phenomenon on the refractories and the lining as a whole. It leads to the destruction of the surface layer and its delamination. In the literature, two types of spalls are mainly considered: spalling of thin layers of products with a thickness of up to 1–3 cm (delamination) and spalling of pieces with a thickness of 5–8 cm [9]. The following types of impact explain their appearance [10]:

- thermal – a consequence of internal stresses caused by the difference in thermal expansion of the lining zones during thermal shocks;
- mechanical – the formation of cracks under the action of mechanical forces;
- structural changes in the refractory's chemical composition and physical properties under the action of high temperatures.

However, among the listed reasons causing chipping of refractories, such an essential factor as destruction due to the contact of the refractory with metal plates is not included, which is especially important for large rotary kilns. In this case, the reason for their formation may be increased thermal stresses caused by the difference in the coefficients of thermal expansion of the refractory and metal plates.

To make the refractory masonry monolithic, metal plates are placed between individual refractory bricks, which are used as a binder material [11]. Therefore, of particular interest is the study of stresses in the lining and refractory to determine the conditions under which the fracture and spalling of bricks occur, which will make it possible to eliminate the causes of destruction.

Simulation of this process will make it possible to select the appropriate laying method, lining thickness, quantity, width, placement order, and type of plates, which will ensure the refractory's durability and the lining's unreliability.

3 Research Methodology

The formation of fracture zones (splitting) in the refractory is caused by the fact that, depending on the temperature change, the metal plate can be partially in an elastic, plastic, and molten state.

The refractory can move freely in the area where it partially passes into the molten state, which does not cause significant stresses and thermal deformations in the circumferential direction.

This phenomenon is shown in Fig. 1. The refractory in this area is in a more accessible state and not in a "compressed" state.

As a result, compressive and rupture deformations arise in different planes, which cause bending stresses, resulting in cracks of separation or shearing (Fig. 2).

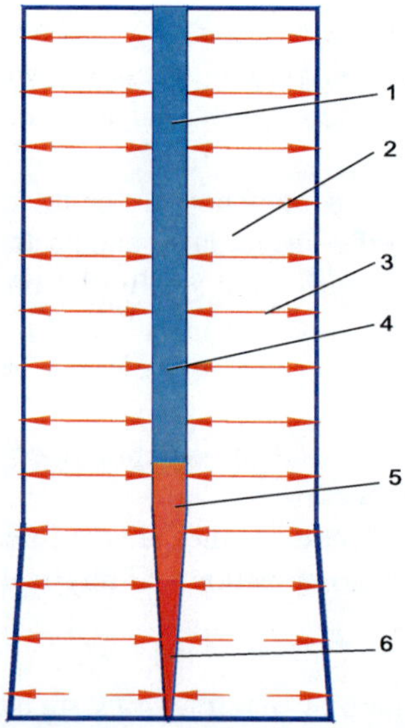

Fig. 1. The action of on the plate: 1 - metal plate; 2 - refractory; 3 - stresses; 4 – elastic zone; 5 - plastic zone; 6 - melt zone.

Fig. 2. Splitting of refractories [12].

When computer modeling the operation of a rotary kiln, we consider that, from the point of view of the basic provisions of structural mechanics, the kiln body can be represented as a thin-walled cylindrical shell mounted on several supports and lined with refractory bricks inside.

The supporting elements are bandages. Near one of the furnace supports, the driving gear wheel of the drive is attached to the body. The main loads are the own weight from the furnace body, lining, bandages, and drive gear [13]. A much smaller part is the weight of the processed material, so it is usually not considered in the calculations. The considered rotary kiln with dimensions of 4.0 × 60 m of the Kryvyi Rih cement plant is shown in Fig. 3.

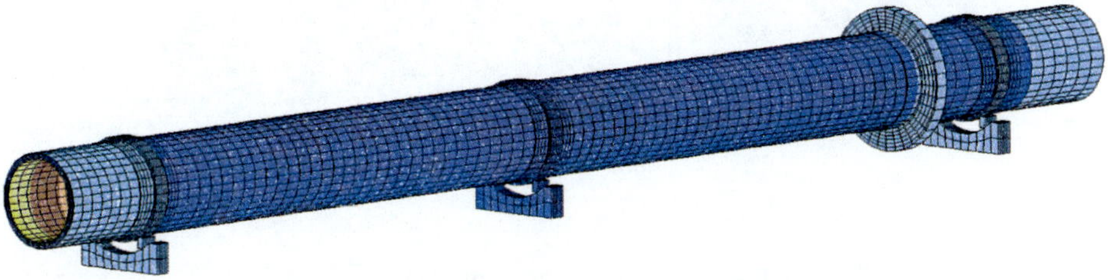

Fig. 3. Calculation scheme of a rotary kiln.

The calculations also consider the kiln as a supporting structure and a thermal aggregate. The strength of the kiln is greatly influenced by the forces that arise under the action of uneven temperature fields in the working areas of the thermal unit and cause corresponding thermomechanical stresses in the kiln body and refractories. The calculation is made for a stationary thermal regime. The dependence [14] sets the gas flow temperature and varies from 950 °C to 1750 °C along with the furnace length in the sintering zone (Fig. 4).

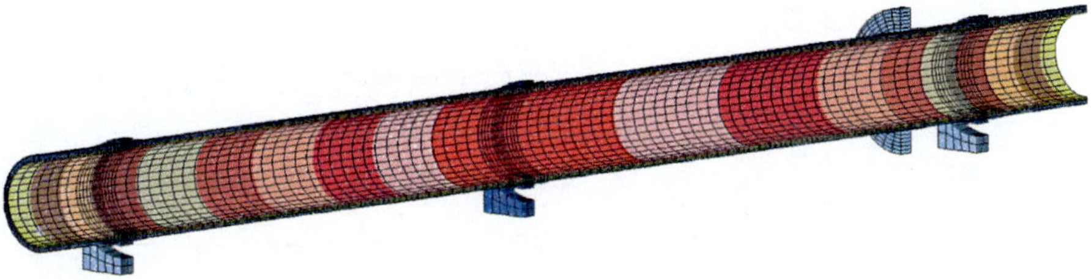

Fig. 4. Gas flow temperature.

The lining is modeled with chromium-magnesite refractories 230 × 150 × 60 mm in size and metal plates 230 × 150 × 2 mm, which are placed between the refractories, conditionally shown in Fig. 5. The calculations consider refractories with a compressive strength of 45 MPa and tensile strength of 15 MPa [15, 16]. The heat transfer coefficient to the environment was determined from the known empirical dependence $\alpha_{oc} = 3.5 + 0.062\,T_c$, and the gas flow to the lining was determined by considering the flow of gases, temperature, and dimensions [10].

To adequately display the physics of processes, the paper considers the transition of plates at high temperatures from the elastic to the plastic and molten state by considering the dependence of physical and mechanical characteristics on temperature.

Therefore, the dependences of the elastic modulus (Fig. 6) and the thermal expansion coefficient on temperature (Fig. 7) were used in the model.

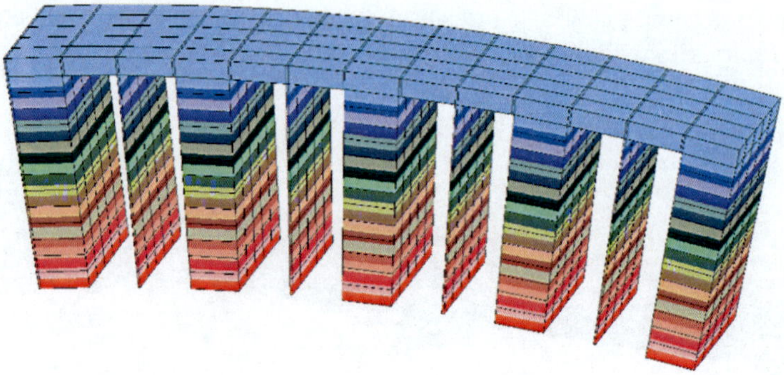

Fig. 5. Plate layout with temperature distribution (simplified).

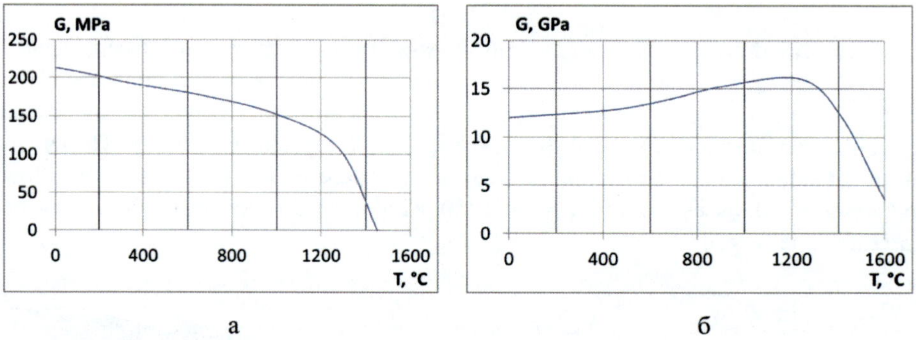

Fig. 6. Dependence of the modulus of elasticity on temperature: a – kiln body, plates; b – lining.

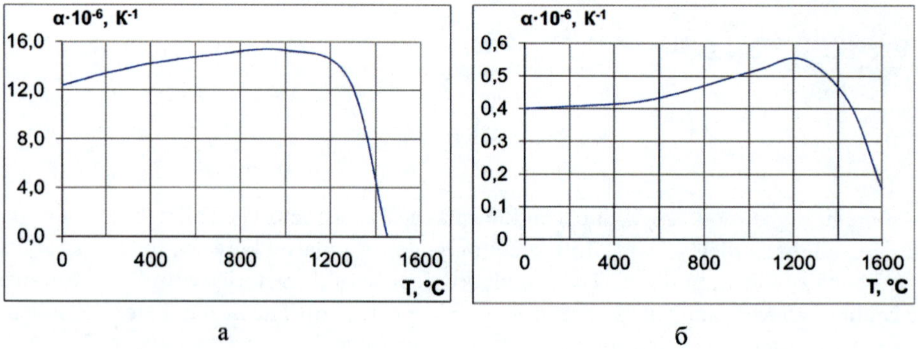

Fig. 7. Dependence of the linear expansion coefficient on temperature: a – kiln body, plates; b – lining.

When determining the strength criteria for refractories, we assume that this is a brittle body in a comprehensive deformed state. To date, a sufficiently large number of general theories are designed to simulate the destruction of brittle bodies [17]. However, in the general case, they are not universal and accurately describe some materials' destruction process, while for others that differ in structure, they give erroneous results. Therefore, we consider and analyze the results obtained using several strength theories.

One of the theories of strength that give adequate results for brick products is the Balandin's theory, which is based on experiments for a brittle material under uniaxial tension and compression [18]. The formula that reduces a complex stress state (for the case when $\sigma 1 > \sigma 2 = \sigma 3$) to a uniaxial state is:

$$\sigma_{ci} = \frac{(1-\psi)(\sigma_1 + 2\sigma_3) - \sqrt{(1-\psi)^2(\sigma_1 + 2\sigma_3)^2 + 4\psi(\sigma_1 - \sigma_3)^2}}{2\psi} \leq R_c \quad (1)$$

$$K_S = \frac{\left(\frac{\sigma_1}{R_c} - \frac{\sigma_3}{R_c}\right)^2}{\left(\frac{\sigma_1}{R_c} - 2\frac{\sigma_3}{R_c}\right)} \leq 1$$

where σ_{ci}– stress equivalent to uniaxial stress state, MPa; $\sigma_1, \sigma_2, \sigma_3$ – principal stresses, MPa; $\psi = \frac{R_p}{R_c}$ – fragility index; R_c – ultimate strength in uniaxial compression, MPa; R_p – uniaxial tensile strength, MPa; R_τ – shear strength, MPa. The values used in the calculations $R_c = 40$–55 Мпа, $R_p = Rc/3$.

According to (1), the strength of a brittle material, such as refractory, is estimated by three main indicators: R_c, R_p и R_τ. This circumstance and the accuracy of the results explain this theory's widespread use. The previous makes it possible to single out the uniaxial compressive strength as the main strength characteristic in describing fracture.

Also, for brittle materials, the Hoek-Brown and Parchevsky–Shashenko criteria [17] are widely used, in which it is assumed that the brittle material under consideration contains structural defects in the form of cracks. The strength criterion proposed by Hoek and Brown is as follows:

$$\sigma_1 = \sigma_3 + R_c\left(m\frac{\sigma_3}{R_c} + s\right)^{0,5}, \quad K_S = \left(\frac{\sigma_1}{R_c} - \frac{\sigma_3}{R_c}\right)^2 - m\frac{\sigma_3}{R_c} \leq 1 \quad (2)$$

where s is a constant based on body type; m is a constant based on triaxial load test or tabulated data.

The analytical expression of the Parchevsky–Shashenko strength theory, obtained from the same premises as the Balandin theory, has the form:

$$\sigma_{3KB} = \frac{(\psi - 1)(\sigma_1 + \sigma_3) + \sqrt{(1-\psi)^2(\sigma_1 + \sigma_3)^2 + 4\psi(\sigma_1 - \sigma_3)^2}}{2\psi} \leq R_c \quad (3)$$

$$K_S = \frac{\left(\frac{\sigma_1}{R_c} - \frac{\sigma_3}{R_c}\right)^2}{\left(\frac{\sigma_1}{R_c} - \frac{\sigma_3}{R_c}\right)} \leq 1,$$

According to these theories, fracture zones appear in refractories in places where the strength criterion exceeds 1.0.

The process of modeling the heat-stressed state of structures is carried out in two stages. In the first stage, the problem of heat conduction is solved. Since the system of equations is nonlinear, an iterative algorithm is used. The second stage consists of forming the stiffness matrix and the vector of equivalent nodal thermal and forces loads, considering the obtained temperature field and the nonlinearity of the physical and mechanical characteristics for steel and refractory. After solving it, based on the obtained values of nodal displacements, the deformations and stresses in the elements of the finite element model are calculated, and the overall stress-strain state of the structure is analyzed.

To solve the problem of calculating the temperature fields and the stress-strain state, the VESNA software package was used, developed at the Department of HPSM Igor Sikorsky Kyiv Polytechnic Institute. Igor Sikorsky. The specified software package simulates hydrodynamic and thermal processes and performs strength calculations using FEM. The system allows the study of linear and nonlinear deformation processes of spatial combined systems made of isotropic and anisotropic materials [19, 20].

4 Results and Discussion

The temperature of the inner surface of the lining along the length of the furnace varies from 850 °C to 1450 °C. The external temperature of the furnace body and the cold section of the refractory varies from 190 °C in the zones of cooling and exothermic reactions to 275 °C in the sintering zone. There is also a slight decrease in the temperature in the area of installation of bandages and gears. This fact is explained by more intensive cooling by the external environment.

Below, for comparison, we consider the calculation of the kiln structure's stress-strain state without considering the dependencies of the elastic modulus and the coefficient of linear expansion on temperature.

Figure 8 shows the values of equivalent stresses without considering the expansion. The temperature of the open surface of the refractory is 1300 °C.

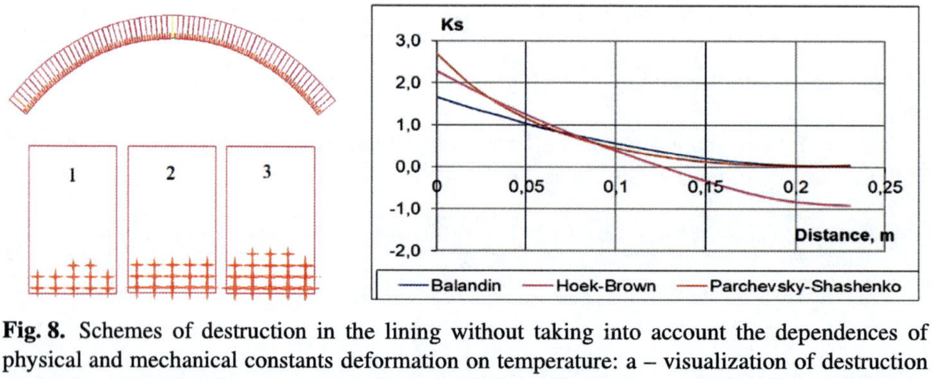

Fig. 8. Schemes of destruction in the lining without taking into account the dependences of physical and mechanical constants deformation on temperature: a – visualization of destruction zones; b – strength assessment according to different criteria; 1 – according to the Balandin criterion; 2 – according to the Hoek–Brown criterion; 3 – according to the Parchevsky–Shashenko criterion.

This and the following graphs show the results for strength evaluation according to three strength criteria: Balandin, Hoeck–Brown, and Parchevsky–Shashenko.

As seen from the above data, the highest equivalent stresses occur directly on the open working surface of the lining, where destruction occurs, and chipping and cracks in the refractory do not occur. However, this phenomenon is not typical for refractories of rotary kilns in practical use. When considering the dependencies of physical and mechanical characteristics on temperature, more physically substantiated results were obtained and confirmed in practice.

Figure 9 shows the stresses in refractories, considering the temperature dependence of the modulus of elasticity and linear expansion coefficient. In this case, the temperature of the open surface of the refractory corresponds to 1300 °C. Figure 9a shows schemы of refractory destruction. Figure 9b shows the strength of refractories with respect to length.

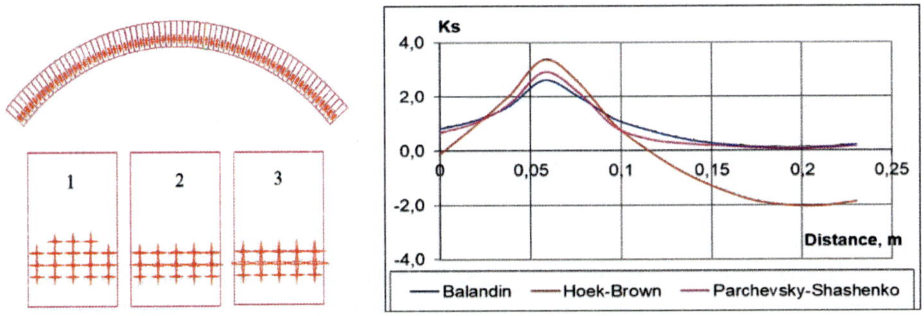

Fig. 9. Schemes of destruction in the lining, taking into account the dependences of physical and mechanical characteristics on temperature: a – visualization of destruction schemes; b – strength assessment according to different criteria; 1 – according to the Balandin criterion; 2 – according to the Hoek-Brown criterion; 3 – according to the Parchevsky-Shashenko criterion.

The graphs show a pronounced extremum, at which the failure criterion Ks exceeds the value of 1.0, indicating the possible destruction of refractories in this area. Moreover, the values in which the equivalent stresses exceed 1.0, which indicates the possibility of cracks, do not differ significantly for various criteria.

As follows from the given data, the zone of possible destruction and the most considerable stresses occur at a certain distance from the surface of the refractory, in contrast to the destruction scheme without taking into account the dependencies of physical and mechanical characteristics on temperature shown in Fig. 8.

From Fig. 9, it can be seen that the maximum is at a distance of 60 mm, and the zone in which the value exceeds 1.0 is at a distance of 20–100 mm from the "hot" refractory surface. In other areas, the strength criterion does not exceed 1.0. In the contact area near the furnace body, the value of Ks is approximately 0.1–0.2. This means that destruction in these places is less likely. This pattern is confirmed by literature data indicating that cracks occur no closer than 50–60 mm from the working surface and practical experience in operating rotary kilns [12, 21].

Considering that the temperature changes along the length of the furnace, let us consider the possibility of destruction of refractories installed in different parts of the kiln aggregate. Figure 10 shows the strength of the refractory obtained by the cross-section according to the Balandin criterion. Refractories with a temperature of 1100 °C on the "hot" surface and 260 °C in the contact zone near the metal case were studied. Refractories with temperatures 1200–292 °C, 1300–318 °C, and 1450–342 °C, respectively, were also analyzed.

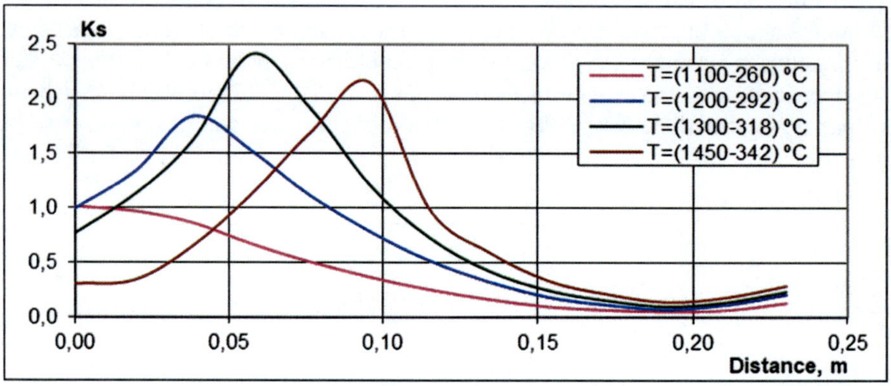

Fig. 10. Evaluation of the strength of refractories with a change in surface temperature.

An increase in the temperature of the "hot" surface of the refractory contributes to the formation of a chipped area. In this case, moving zones of maximum stresses from the working surface of the refractory are observed. A decrease in temperature decreases the plot size. Therefore, destruction does not occur at a temperature of 1100 °C and below.

According to the calculation, at a temperature of 1450 °C, the formation of a chip is possible in the area of 55–115 mm from the "hot" surface of the refractory, where the strength criterion exceeds 1.0. The highest values are located at a distance of 90–100 mm. At T = 1300 °C, destruction is possible in 20–100 mm. The maximum destruction is in the area of 65 mm. At T = 1200 °C in the area of 5–80 mm, and maximum – 40 mm. At T = 1100 °C and below, the stresses do not exceed the tensile strength.

During the operation of a rotary kiln, the lining refractories wear out. Usually, its thickness decreases from the initial 230 mm to 80 mm. It is valuable to simulate the possibility of refractory destruction when its length changes to 230, 180, 150, 120, and 80 mm. The calculation results are shown in Fig. 11.

It can be seen from the given data that with a decrease in the thickness of the lining, the probability of refractory splitting decreases, and the maximum values are 2.41 for refractory 230 mm, 2.06–180 mm, 1.7–150 mm, 0.94–12 mm. These values are explained by a decrease in the size of the plastic and molten zones, which affects the overall stress-strain state of the refractory. Stresses at the point of contact of the refractory with the kiln body increase monotonously, without exceeding the allowable ones, and amount to 0.36–230 mm, 0.78–80 mm. At the same time, for a refractory with a size of 80 mm, the stress at the contact area with the furnace body exceeds the stress peak inside the refractory.

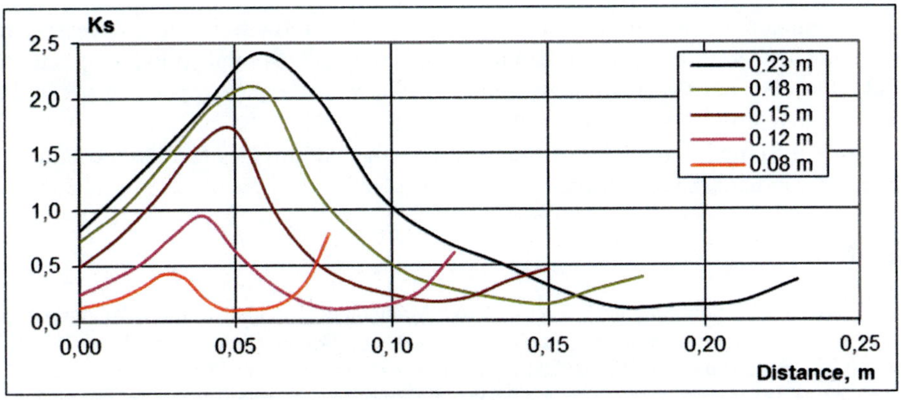

Fig. 11. Evaluation of strength when changing the thickness of the lining.

In furnaces, chipping occurs at the initial stage of operation, that is, in less worn refractories. Notably, with a decrease in the refractory's height, the zone's width and the magnitude of the highest stress decrease. In this case, the zone moves closer to the working surface. This indicates that worn or shortened refractories are less susceptible to crack initiation and are less likely to occur. When choosing the geometric formats of refractories for the lining of rotary kilns, these data should be considered.

5 Conclusions

The proposed method, taking into account the dependencies of physical and mechanical characteristics on temperature and using the criteria for the strength of brittle materials, makes it possible to carry out more accurate calculations when simulating the operation of lining refractories, to avoid their destruction and the formation of the chipping effect.

As a result of the calculations, it was found that:

- it is necessary to model both refractories and metal plates, which are made in the form of spacers, in the calculation scheme;
- an adequate solution to the problem of modeling refractory failure is possible only by taking into account the dependence of the elastic modulus and linear expansion coefficient on the temperature;
- destruction occurs at the initial stage of furnace operation in a zone located at a distance of 20–100 mm from the most heated surface. In this case, the maximum stresses occur at a distance of 60 mm, which is confirmed by practical experience;
- with a decrease in the thickness of the lining due to wear, the refractory failure probability decreases, and the maximum values of the failure criterion Ks is 2.41 for a refractory of 230 mm, 2.06–180 mm, 1.7–150 mm, 0.94–120 mm.

The data obtained should be considered when choosing the geometric dimensions of refractories for the lining of rotary kilns.

To eliminate the possibility of damage (splitting) of refractories, it is necessary to control the temperature regime of the lining, especially in the initial period of operation. In addition, instead of standard metal plates, use corrugated ones or plates made in the form of a wire grid. The specified technical solution reduces the magnitude of thermal stresses in the circumferential direction of the lining and provides favorable conditions for the operation of the refractory.

References

1. Sharikov, Y.V., Sharikov, F.Y., Krylov, K.A.: Mathematical model of optimum control for petroleum coke production in a rotary tube kiln. Theor. Found. Chem. Eng. **55**(4), 711–719 (2021). https://doi.org/10.1134/S0040579521030192
2. Liang, R., Zhang, Z., Jin, Z., Yu, X., Zhang, S.: Engineering application of indirect heating rotary kiln in oil shale pyrolysis treatment. Chin. J. Environ. Eng. **15**(9), 3029–3034 (2021). https://doi.org/10.12030/j.cjee.202104216
3. Kurdowski, W., Jelito, E.: Rotary kilns in current cement industry. Cement, Wapno, Beton **25**, 127–136 (2020). https://doi.org/10.32047/CWB.2020.25.2.5
4. Mungyeko Bisulandu, B.-J. R., Marias, F.: Modeling of the thermochemical conversion of biomass in cement rotary Kiln. Waste Biomass Valoriz. **12**(2), 1005–1024 (2020). https://doi.org/10.1007/s12649-020-01001-9
5. Sekisov, A.N., Serga, G.V., Gura, D.A., Vyrodova, I.G., Danko, V.P.: Bases of increasing operational characteristics of the equipment for cement production. Civil Eng. Architect. **9**(5), 1498–1505 (2021). https://doi.org/10.13189/cea.2021.090521
6. Kashcheev, I.D.: The use of refractories in the lining of rotary cement Kilns1. Refract. Ind. Ceram **56**(5), 483–485 (2016). https://doi.org/10.1007/s11148-016-9873-1
7. Szczerba, J., Śnieżek, E., Antonovič, V.: Evolution of refractory materials for rotary cement Kiln sintering zone. Refract. Ind. Ceram. **58**(4), 426–433 (2017). https://doi.org/10.1007/s11148-017-0123-y
8. Asadi, F., André, D., Emam, S., Doumalin, P., Huger, M.: Numerical modelling of the quasi-brittle behaviour of refractory ceramics by considering microcracks effect. J. Eur. Ceram. Soc. **42**(3), 1149–1161 (2022). https://doi.org/10.1016/j.jeurceramsoc.2021.11.016
9. Ramanenka, D., Stjernberg, J., Jonsén, P.: FEM investigation of global mechanisms affecting brick lining stability in a rotary kiln in cold state. Eng. Fail. Anal. **59**, 554–569 (2016). https://doi.org/10.1016/j.engfailanal.2015.10.023
10. Ramanenka, D., Gustafsson, G., Jonsén, P.: Influence of heating and cooling rate on the stress state of the brick lining in a rotary kiln using finite element simulations. Eng. Fail. Anal. **105**, 98–109 (2019). https://doi.org/10.1016/j.engfailanal.2019.06.031
11. Andreev, K., Sinnema, S., Rekik, A., Blond, E., Gasser, A.: Effects of dry joints on compressive behaviour of refractory linings. In: International Forum on Science and Education of Refractories, Wuhan, China, pp. 63−66 (2012)
12. Shubin, V.I.: The effect of temperature on the lining of rotary cement kilns. Refract. Ind. Ceram. **42**(5–6), 216–221 (2001). https://doi.org/10.1023/A:1012346718095
13. Herz, F.: Process modeling of direct-fired rotary kilns for the assessment of the thermal stress of the refractory material. Keram. Z. **70**(1–2), 26–34 (2018)
14. Singh, A.P., Ghoshdastidar, P.S.: Computer simulation of heat transfer in alumina and cement rotary kilns. J. Thermal Sci. Eng. Appl. **14**(3), 031001 (2022). https://doi.org/10.1115/1.4051376
15. Dai, Y., Gruber, D., Harmuth, H.: Determination of the fracture behaviour of MgO-refractories using multi-cycle wedge splitting test and digital image correlation. J. Eur. Ceram. Soc. **37**(15), 5035–5043 (2017). https://doi.org/10.1016/j.jeurceramsoc.2017.07.015

16. Oliveira, R.L.G., Rodrigues, J.P.C., Pereira, J.M., Lourenço, P.B., Ulrich Marschall, H.: Normal and tangential behaviour of dry joints in refractory masonry. Eng. Struct. **243**, 112600 (2021). https://doi.org/10.1016/j.engstruct.2021.112600
17. Shashenko, O., Kovrov, O., Rakishev, B.: Failure criteria for structurally heterogeneous materials. Mining Miner. Deposits **10**(3), 84–89 (2016). https://doi.org/10.15407/mining10.03.084
18. Hoek, E., Carranza-Torres, C., Corkum, B.: Hoek-Brown criterion – 2002 edition. In: Proceedings of the NARMS-TAC Conference, Toronto, vol. 1, pp. 267–273 (2002)
19. Gondlyakh, A., Chemeris, A., Kolosov, A., Sokolskiy, A., Antonyuk, S.: Simulation of delamination processes of multilayer mechanical engineering structures. In: Tonkonogyi, V., et al. (eds.) InterPartner 2020. LNME, pp. 129–138. Springer, Cham (2021). https://doi.org/10.1007/978-3-030-68014-5_13
20. Gondliakh, O., Krytskyi, V., Onopriienko, V., Chemerys, A., Krytska, N.: Computer analysis of thermomechanical state of sealing steel lining for containment of NPPs with VVER-1000/V-320 in emergencies. Nucl. Radiat. Saf. **4**(76), 28–39 (2017)
21. Qian, F., Duan, X., Yang, W., Liu, G., Li, H.: Research progress of magnesia chrome refractories and their application in greenization for high temperature furnace. Cailiao Daobao/Mater. Rep. **33**(12), 3882–3891 (2019). https://doi.org/10.11896/cldb.18110166

The Efficiency of Convective Heat Exchange at the Airflow of Metal Friction Elements of Brakes

Vasiliy Skripnik[1], Oleksandr Vudvud[2](✉), Dmitry Zhuravlev[1], Sergiy Nikipchuk[3], and Tetiana Danulyak[4]

[1] Ivano-Frankivsk National Technical University of Oil and Gas, 15, Karpatska Street, Ivano-Frankivsk 76019, Ukraine

[2] Odessa National Polytechnic University, 1, Shevchenko Avenue, Odesa 65044, Ukraine
o.m.vudvud@op.edu.ua

[3] Lviv Polytechnic National University, 12, Stepan Bandera Street, Lviv 79000, Ukraine

[4] Drohobych Applied College of Oil and Gas, 57, Hrushewskoho Street, Drohobych 82100, Ukraine

Abstract. In the materials of the article, the following issues were examined: border heat layer and its role in the efficiency of convective heat exchange; interaction of dynamic border layer with heat layer in processes of the airflow around surfaces; heat transfer from metal brake friction elements. It has been established that the «longevity» of the boundary dynamic and thermal layers of the surrounding air-heated surfaces of metallic friction elements at the open brake friction pairs are different. What unites them is that the boundary heat layer «sticks» to the metal surface, and the dynamic boundary layer «sticks» to the bottom of the main flow of the surrounding air. The frequency of application of the brake is the main factor in the break-up of the boundary heat layer of the air, which is a kind of insulator for convective heat exchange. The rotation speed of metallic friction elements during injections and open brake friction pairs determines the laminar or turbulent mode of washing with air streams of their surfaces. The dependencies of the total coefficient of resistance (C) on the Reynolds number (Re) during the longitudinal washing by the flow of air with a variable angle of attack (45–125°), the outer surfaces of the rim of the pulley, and the brake drum are determined, By separating zones of constant and significantly decreasing values of coefficient C. The method of calculating heat transfer coefficients which are «closing» in estimating heat transfer coefficients, has been developed.

Keywords: Friction element · Braking device · Thermal layer · Convection · Heat exchange · Energy efficiency

1 Introduction

Research into the thermal regimes of metallic friction elements has raised new problems in the thermal conductivity theory, which forms convective and radiation fields [1]. From the perspective of heat physics, the friction unit of different brakes is a system of

V. Tonkonogyi et al. (Eds.): InterPartner 2022, LNME, pp. 574–583, 2023.
https://doi.org/10.1007/978-3-031-16651-8_54

many bodies with energy sources and effluents that are complex in space and time [2]. Note that complex body systems are found in braking [3] and other areas of mechanical engineering. Therefore, researchers are growing interested in the theory of heat exchange in general and, in particular, in convection in the airflow of metal brake friction elements [4]. The main issues of the article are the boundary heat layer and its role in the efficiency of convective heat exchange; the interaction of the dynamic boundary layer with the thermal layer in airflow processes; heat transfer from metal friction elements of brakes.

The work aims to evaluate the efficiency of convective forced air cooling of metallic friction elements in relation to boundary dynamics and heat layers surrounding their air flows.

2 Literature Review

Heat exchange processes are essential and often decisive in high-stress heat modes of brake friction pairs on which their effectiveness depends.

In work [5], the heat balance of the pulley rim of the strip block type brake of the drilling winch was investigated. At the same time, the matt and polished surfaces of the pulley were considered for the radiative heat exchange and the convective heat exchange of the surface area thereof, surfaces washed by air streams. The pulley was regarded both as a moving and as a stationary object.

Work [6] is devoted entirely to thermal conductivity in a solid field, gases, and liquids. Electron, phonon, and «photon» thermal conductivity are distinguished in the solid body.

Thermal radiation of the solid body is given attention in work [7]. Of interest are heat transfer issues by radiation: in the case of specified surface-volume temperature fields, when the transport equation is «closed by» the heat conductivity equation.

Work [8] illuminates the technology's fundamentals of radiation and complex heat exchange. Convective heat exchange in a homogeneous medium (heat transfer) is the subject of work [9]. It distinguishes the thermal and dynamic boundary layers resulting from the air washing of metal brake friction components.

Analysis of literature sources [10] has shown that the nature of the formation and separation of the boundary heat layer, which merges with the dynamic boundary layer during longitudinal washing of the outer surfaces of the pulley rim and drum by air flows and assessing their effect on convective heat transfer, it is necessary to establish the following:

- directly on the surfaces of metallic friction elements, the border heat layer shields them and thereby reduces the efficiency of forced air cooling at open brake friction pairs;
- the frequency and time of application of the brake are the main factors in the break-up of the thermal boundary layer, which is an insulator for heat exchange surfaces;
- the speed of rotation of metallic friction elements during injections and open friction pairs of brakes determines the laminar or turbulent mode of washing with air streams of their surfaces; The source of artificial turbulence is the rough external surface of the brake drum rim;

- the dependence of the total coefficient of resistance (C) on the Reynolds number (Re) when the air is washed longitudinally on the outer surfaces of the pulley rim and the brake drum, emitting zones of constant and significantly decreasing values of the C coefficient;
- the regularity of changes in airspeed and pressure gradients at different values when the dynamic boundary layer interacts with the thermal boundary layer on the outer surfaces of the pulley and brake drum;
- there is no uniform method for evaluating heat transfer from the surfaces of pulley and drum rims and solid and self-ventilated discs; based on the data obtained, a calculation of heat transfer through metal friction elements of brakes is carried out.

3 Research Methodology

3.1 Border Heat Layer and Its Role in Heat Exchange Efficiency

As a concrete example, consider the longitudinal airflow of the round cylinders, which are the rotating working and non-working surfaces of the pulley rims and drum.

The interaction between rim surfaces and the air surrounding them is expressed in the formation of viscous forces under which a thermal boundary layer is formed [11].

The ratio between conductive heat transfer intensity of conductivity and convection changes within the boundary layer is considered. Directly at the surface, the speed of movement of air is zero, and the transfer of heat is entirely due to the effect of thermal conductivity. Therefore, the specific heat flow is determined. As the distance from the working surface of the pulley rim increases, the relative role of convective transfer increases, and this form of heat redistribution becomes crucial at the outer edge of the thermal boundary layer. For the boundary heat layer, it is generally characteristic that within its boundary, the heat transfer processes caused by thermal conductivity and convection are commensurate in intensity. Exo- or endothermic reactions do not occur in the thermal boundary layer.

The non-circular air flow around the non-working surface of the brake drum rim does not disturb its thermal boundary layer (Fig. 1a). In contrast, vortices' circular flow of upper and lower air streams increases and decreases speed, creating a positive speed gradient (Fig. 1b). The latter also destroys the boundary heat layer on the working surface of the brake pulley rim [12].

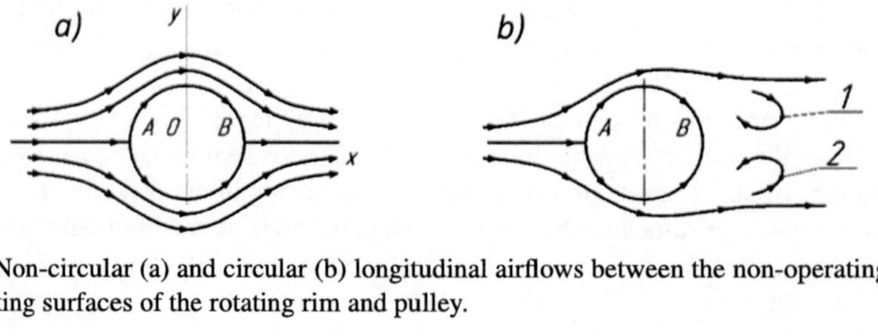

Fig. 1. Non-circular (a) and circular (b) longitudinal airflows between the non-operating air and the working surfaces of the rotating rim and pulley.

Let us now consider the picture of the flow around a cylinder under conditions of fully developed separation (Fig. 1). At the bifurcation point of flow A, a thermal boundary layer begins to form, and the thickness gradually increases. However, the body coverage angle is limited since, at some point, the separation of the boundary layer from the surface must occur. All quantitative features of this picture are the change in the thickness of the heating layer $\frac{\delta}{R}$ in the angle θ, measured from the bifurcation point, the separation point location, the velocity profile in the heating boundary layer, and the nature of its change along the flow.

The stability of the flow pattern around the body is manifested in the invariability of their interaction intensity. A quantitative characteristic of the dynamic interaction of a body with the flow around it is a special kind of quantity C. It is constructed similarly to the coefficient of hydraulic resistance ζ and is determined by the equation

$$F = C\frac{\rho U_0^2}{2}A \tag{1}$$

where F is the resistance force; A is the transparent frontal area.

The name of the coefficient of resistance is retained for the proportionality factor C. It can be represented as a sum of the coefficients of friction resistance and resistance to a specific pressure. The resistance coefficient is formed by assigning the resistance force to the transparent frontal area. However, the coefficient of friction determined by the equation in which the friction is related to the surface area washed by air is also widely applied. The resistance coefficient is a function of the Reynolds criterion. Consideration of the graph of this function, presented in logarithmic anamorphosis (Fig. 2), leads to the following conclusions.

Areas of very small values Re (i.e., the region of continuous flow around the cylinder) correspond to a straight line inclined to the abscissa axis at an angle of 45°. Consequently, within this area, the law of inverse proportionality is valid

$$C\,Re = \text{const or } \frac{F}{A\rho U_0^2}\frac{\rho U_0 d}{\mu} = \frac{F}{A}\frac{d}{\mu U_0} = \text{const}, \tag{2}$$

where ρ, μ are the air density and its dynamic viscosity.

The pattern of longitudinal viscous flow is presented as follows. The actual process conditions consistent with the above symmetry scheme are only at very low Reynolds values. At Re values of the order of 10°, this simple flow form begins to break down under the influence of the separation of the thermal boundary layer. Initially, this fact, weakly expressed, increases with increasing Re and eventually leads to a complete restructuring of the flow.

In the graph (Fig. 2) of the resistance coefficient of the streamlined longitudinal cylinder, the self-similarity interval is replaced by an area of extreme decrease of coefficient C.

The explanation of this unexpected effect should be sought in the interaction between the boundary heat layer and the dynamic boundary layer, which is the base of the main flow, which must occur on the surface of their partition. First, this interaction consists of the exchange of momentum, i.e., in a process that, to one degree or another, neutralizes the effect of energy dissipation on the flow in the thermal boundary layer.

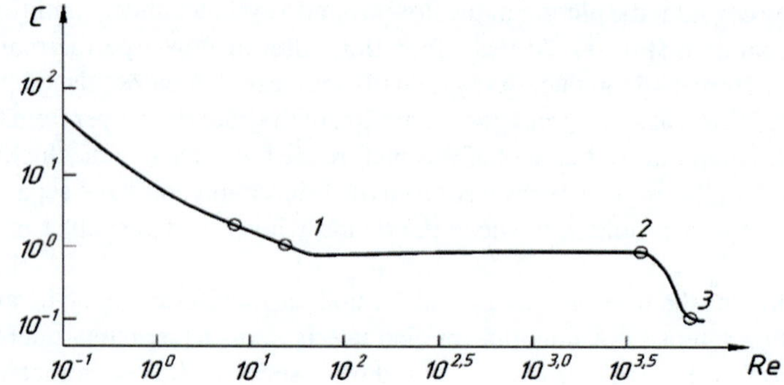

Fig. 2. Dependence of the total coefficient of resistance (C) on the Reynolds number (Re) in the longitudinal air flux on the pulley and brake drum's outer surfaces: Sects. 1 and 2 - areas of constant and significantly decreasing coefficient C.

However, as long as the flow in the dynamic layer is laminar, the intensity of this process is very low. The resulting recovery of the kinetic energy in the boundary layer is sufficient to move the air into the backpressure area only under flow conditions around bodies of slight curvature, bodies of a well-streamlined shape (since in this case, the backpressure increases rather slowly). But this effect is too insignificant for effect on the moment of separation of the flow from the surface (i.e., on the angle of coverage of the body by the flow) in the case of a body with a poorly streamlined shape. The position changes significantly if the flow in the dynamic boundary layer becomes turbulent. The intensity of the exchange rate increases considerably.

Figure 2 shows that for the coefficient of resistance (C) in the function from Re, the flow adjustment process is reflected in the form of a Sect. 2 of a gradual but very significant decrease in C. The considered effect is known as the flow crisis (although the flow adjustment occurs gradually, it does not create the idea of a "crisis").

3.2 Interaction of the Boundary Dynamic Layer with the Heat Layer in Air-Flow Processes

Under normal airflow conditions in the thermal boundary layer, the longitudinal component of the velocity u monotonously increases over the normal y from zero on the surface to the value U at the outer boundary of the layer. With that, the velocity distribution curve has a characteristic shape, in which its continuous increase, starting from the surface, and a smooth transition from a slow flow in the boundary heat layer to a flow in the external flow are reflected, corresponding to the boundary dynamic layer (Fig. 3a). The latter is part of the main airflow, which has a more significant density gradient and stops the boundary heat layer, which, as it were, sticks to the surfaces of the pulley and drum when the brake friction pairs are open. In this case, the velocity distribution along the entire length corresponds to a positive derivative $\frac{\partial u}{\partial y}$, that gradually decreases as it moves away from the surface and tends to zero when approaching the outer boundary of the heating layer. Accordingly, the second derivative $\frac{\partial^2 u}{\partial y^2}$ remains negative over the entire thickness of the boundary heat layer [13].

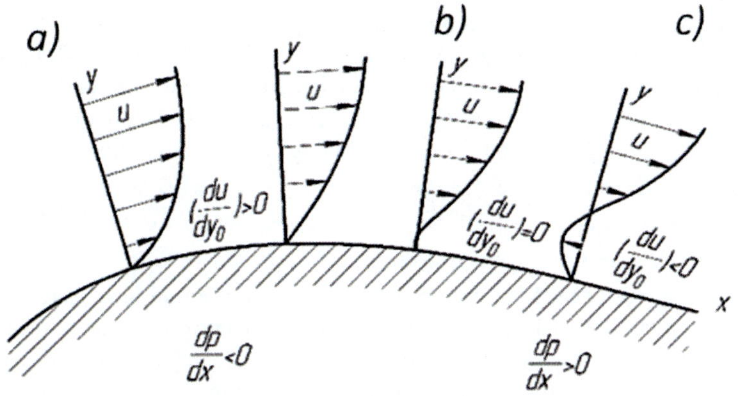

Fig. 3. Regularities of changes in the gradients of speed and air pressure at their various values during the interaction of a dynamic layer with a heat layer on the outer surfaces of the pulley rim and brake drum.

Differently, the kinematic situation develops in the flow by separating the boundary heat layer because of part of the air of the dynamic boundary layer. As a result of its heating (i.e., by expansion), it moves in the opposite direction. Therefore, the derivative $\frac{\partial u}{\partial y}$ directly at the surface (and within a specific interval, the value of y) should be negative (Fig. 3c). The velocity distribution curve near the surface has an entirely different form than in the conditions of a continuous flow, and only when approaching the outer edge of the boundary heat layer (due to with the requirement of a smooth transition to the conditions of the external flow) gets the usual shape, which is characterized by a positive first derivative tending to zero $\left(\frac{\partial u}{\partial y}\right) \geq 0$, and a negative second derivative $\left(\frac{\partial^2 u}{\partial y^2}\right) < 0$. In the section that separates the zones of continuous flow and flows with separation, a velocity profile should be established (Fig. 3b), which is a curve limiting for curves of both the first (Fig. 3a) and the second (Fig. 3c) type. On this curve, the reverse flow region should be tapered to a point located on the surface (at the base of the profile - point A, Fig. 3b). At this point, the derivative $\frac{\partial u}{\partial y} = 0$. Therefore, the tangent to the profile coincides with the y-axis. The condition

$$\left(\frac{\partial u}{\partial y}\right)_{y=0} = 0 \tag{3}$$

is characteristic of the boundary between forward and reverse flows. Therefore, the separation point location depends on shear phenomena at the boundary of the thermal and dynamic layers.

A more detailed understanding of the curve and the proximity to the surface depending on the process conditions can be obtained based on the following consideration. With $y = 0$, due to $u = \upsilon = 0$ should be

$$\left(\frac{\partial^2 u}{\partial y^2}\right)_{y=0} = \frac{1}{\mu}\frac{dp}{dx} \tag{4}$$

Consequently, the velocity's 2nd derivative sign along the normal direction at the surface is entirely determined by the change in the specific pressure along the flow and falls with the sign of the specific pressure gradient. Therefore, under conditions of accelerated motion, which is associated with a drop in specific pressure, the second derivative is already negative from the beginning. As it was found that the requirement $\frac{\partial^2 u}{\partial y^2} < 0$ for the entire thickness of the boundary heat layer is a necessary prerequisite for the possibility of forming a profile of the first type corresponding to a normal continuous flow. In contrast, in slow motion with the recovery of specific pressure should be $\frac{\partial^2 u}{\partial y^2} > 0$.

Thus, the analysis of the kinematic pattern of the longitudinal flow around the outer surface of the pulley rim (with conclusions obtained on an entirely different basis) shows that the separation of the boundary heat layer from the body surface can occur only under flow conditions with a positive gradient of specific pressure.

3.3 Heat Transfer from Metal Friction Elements of the Brakes

Let us dwell on the features of heat transfer (K) through a solid brake disc. The heat flux that penetrates the body of the disk is equal to:

$$q = K(T_1 - T_2) \tag{5}$$

where T1, T2 are surface temperatures of the friction belts of the disk, °C.

At the same time, according to Fourier Law, the heat flux is expressed by the dependence:

$$q = \frac{\lambda}{\delta} \nabla T \tag{6}$$

where ∇T is the operator of the temperature gradient over the thickness of the disk.

The increase in the heat flow after the subsequent cyclic braking of the vehicle is written in the form:

$$\delta' q = \frac{\lambda}{\delta} \delta' (\nabla T) - \left(\delta' \frac{\lambda}{\delta} \right) \nabla T \tag{7}$$

where δ' is the symbol of variation, in the particular case when $\lambda = \text{const}$, we obtain dependence (7).

To describe the processes of heat transfer, the concept of "heat transfer coefficient of a rotating metal friction element" is introduced [14]

$$\alpha_{\Sigma}(\tau) = \frac{1}{A} \int_A \alpha(\tau, x, y, z) dA \tag{8}$$

Subsequently, using the empirical formula $\alpha = 7.14\upsilon^{0.78}$ (υ is the speed of movement of the washing medium related to the cooled surface of the metal friction element), and after performing a number of transformations, we obtained dependence of the form

$$\alpha_{\Sigma}(n) = \frac{1{,}23\upsilon^{0{,}78}}{A} \int_A 0{,}5D^{0{,}78} dA \tag{9}$$

where n is the rotational speed of the metal friction element.

Having accepted that

$$\frac{1{,}23\upsilon^{0{,}78}}{A}\int_A 0{,}5D^{0{,}78}dA = g \tag{10}$$

wherein the coefficient g is entirely determined by the features of the geometry of the metal friction element.

Formula (10) is written in the following form

$$\alpha_{\Sigma}(n) = gn^{0{,}78} \tag{11}$$

After obtaining experimental data concerning the energy loading of the metal friction elements of braking devices and establishing the relationship between the criteria included in Eq. (11), and under the test condition α_2/K, less and more than one was obtained [15]

$$\alpha_2 = 0{,}75K\left(\frac{\alpha_1 d}{\lambda_2}\right)^{0{,}5}\left(\frac{\upsilon d}{D}\right)^{0.25}\left(\frac{\upsilon\rho c_p}{\lambda_2}\right)^{0.1} \tag{12}$$

The criterion Eq. (12) is a regulator of heat exchange processes using the parameters $\alpha_{2,}$ and K. K increases rapidly with increasing α_1 until α_1 and α_2 become approximately equal. With a further increase in α_1, the growth of K slows down and then practically stops. Thus, at $\alpha_1 \ll \alpha_2$, to increase K, it is necessary to increase α_1, which is equivalent to a decrease in the heat transfer of thermal resistances $1/\alpha_2$. After reaching the equality $\alpha_1 \approx \alpha_2$, heat transfer coefficients can be increased to intensify heat transfer.

4 Results and Discussion

The calculations based on dependence (11) are shown for various metal friction elements of braking devices.

From the analysis of the graphical dependencies (Fig. 4a, b, c), it follows that the heat transfer coefficient for various metal friction elements grows due to the increase: in matte and polished areas; friction radii, and rotational frequency, as well as washing surfaces with moist air and water in wet weather.

Discussion of the results on the nature of the formation and separation of the boundary heat layer, which merges with the dynamic boundary layer during longitudinal washing of heated outer surfaces of the pulley rim and drum by air flows and assessing their effect on convective heat transfer, it was possible to establish the following:

- the understanding of the phenomena mechanism, it is helpful to alter some of the ideas we have, based on the consideration of a detailed kinematic picture, the boundary heat layer that occurs directly at the surfaces of the metal friction elements shields them and thereby reduces the efficiency of forced air cooling with open friction pairs of the brake;
- the frequency of switching on the brake is the main factor in the destruction of the boundary heat layer of the air, which is a kind of insulator;

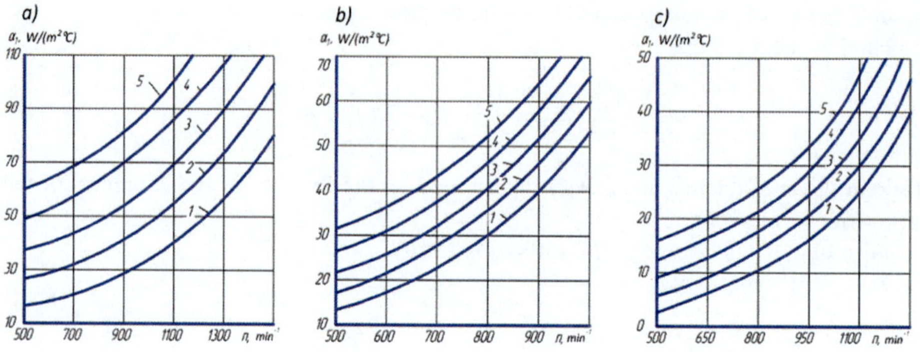

Fig. 4. a, b, c. Regularities of the change in the heat transfer coefficients α_i of metal friction elements from the rotation frequency n for brake discs (a), pulleys (b), and drums (c).

- rotation frequency of metal friction elements when braking and open friction pairs of brakes determines the laminar or turbulent mode of washing their surfaces with air flows; the source of artificial turbulence is the roughness of the outer surface of the brake drum rim;
- the dependence of the total resistance coefficient (C) on the Reynolds number (Re) during longitudinal air flow of the outer surfaces of the pulley rim and brake drum, highlighting the zones of constant and significantly decreasing values of the C coefficient;
- the regularities of changes in the gradients of air velocity and pressure at their various values during the interaction of the dynamic with a thermal boundary layer on the outer surfaces of the pulley rim and brake drum.

5 Conclusions

During the study, an assessment was made of the effectiveness of convective heat transfer when air flows around the elements of the brake.

It has been proven that the thermal boundary layer on the surface of open brake elements creates a shielding effect that reduces the cooling efficiency.

It has been established and confirmed by calculated dependencies that the "survivability" of the shielding layer depends on the rotational speed of the brake elements under the condition of a turbulent flow near the surface of the brake drum or disk, as well as the rotational speed.

In the study, graphical dependencies were obtained that allow us to trace the patterns of change in the heat transfer coefficients αi of metal friction elements on the rotational speed n for brake discs, pulleys, and drums.

The criterion Eq. (12) has been obtained, which makes it possible to establish changes in the heat transfer coefficients K depending on the heat transfer coefficients from the outer α_1 and inner α_2 surfaces of the rotating brake elements.

References

1. Xu, H., Xing, Z., Wang, F., Cheng, Z.: Review on heat conduction, heat convection, thermal radiation and phase change heat transfer of nanofluids in porous media: fundamentals and applications. Chem. Eng. Sci. **195**, 462–483 (2018). https://doi.org/10.1016/j.ces.2018.09.045
2. Ciavarella, M., Johansson, L., Afferante, L., Klarbring, A., Barber, J.: Interaction of thermal contact resistance and frictional heating in thermoelastic instability. Int. J. Solids Struct. **40**, 5583–5597 (2003). https://doi.org/10.1016/S0020-7683(03)00313-5
3. Nosko, A., Tarasiuk, W., Sharifullin, I., Safronov, E.: Tribotechnical and ecological evaluation of friction pairs of brake devices in lifting and transport machines. Friction Wear **41**, 347–353 (2020). https://doi.org/10.3103/S106836662004008X
4. Javadov, M., Volchenko, D., Skrypnyk, V., Volchenko, N., Vudvud, A.: Physical methods for evaluating the load of friction pairs of braking devices (Part I). Herald Azerbaijan Eng. Acad. **13**(2), 58–68 (2021). https://doi.org/10.52171/2076-0515_2021_13_02_58_68
5. Dzhanakhmedov, A., Volchenko, D.: Design and Verification Calculation of Frictional Units of Tape-Shoe Brakes of Draw Works. Apostrophe, Baku (2016)
6. Han, J., Wright, L.: Analytical Heat Transfer, 2nd edn. CRC Press, Boca Raton (2022). https://doi.org/10.1201/9781003164487
7. Al-Maghalseh, M., Mahkamov, K.: Methods of heat transfer intensification in PCM thermal storage systems. Renew. Sustain. Energy Rev. **92**, 62–94 (2018). https://doi.org/10.1016/j.rser.2018.04.064
8. Howell, J., Mengüç, M., Daun, K., Siegel, R.: Thermal Radiation Heat Transfer, 7th edn. CRC Press, Boca Raton (2020). https://doi.org/10.1201/9780429327308
9. Özişik, M., Orlande, H., Colaço, M., Cotta, R.: Finite Difference Methods in Heat Transfer, 2nd edn. CRC Press, Boca Raton (2017). https://doi.org/10.1201/9781315121475
10. Volchenko, N., et al.: Features of the estimation of the intensity of heat exchange in self-ventilated disk-shoe brakes of vehicles. East.-Eur. J. Enterp. Technol. **5**(97), 47–53 (2019). https://doi.org/10.15587/1729-4061.2019.154712
11. Volchenko, N., Volchenko, D., Polyakov, P., Krasin, P., Fedotov, E., Evchenko, A.: Pulse-contact frictional interaction of microprotrusions of friction pairs of brake devices. IOP Conf. Ser. Mater. Sci. Eng. **560**, 012194 (2019). https://doi.org/10.1088/1757-899X/560/1/012194
12. Lee, J., Ramamurthi, K.: Fundamentals of Thermodynamics, 1st edn. CRC Press, Boca Raton (2022). https://doi.org/10.1201/9781003224044
13. Belyakov, N., Nosko, A.: Heat frictional contact of semi-bounded solids. Motorization Power Ind. Agric. **10**(A), 83–91 (2008)
14. Belyakov, N., Nosko, A.: Non-ideal Thermal Contact of Bodies During Friction. Librokom (2010)
15. Zhang, S., et al.: Simulation study on friction and wear law of brake pad in high-power disc brake. Math. Probl. Eng. **2019**, 6250694 (2019). https://doi.org/10.1155/2019/6250694

Non-uniform Nanocapillary Fluid Cooling of the Drawworks' Band-Shoe Brake Friction Couples

Dmytry Volchenko[1], Vasiliy Skripnik[1], Dmytry Zhuravlev[1](✉), Yaroslav Savchyn[2], and Mykhailo Savchyn[2]

[1] Ivano-Frankivsk National Technical University of Oil and Gas, 15, Karpatska Street, Ivano-Frankivsk 76019, Ukraine
dmytro.2103@ukr.net

[2] Drohobych Applied College of Oil and Gas, 57, Mykhailo Hrushevsky Street, Drohobych 82100, Ukraine

Abstract. The following issues are considered in the materials of the article: capillary-porous bodies formed from nanoparticles; design and operation of a nanocapillary liquid cooling system for friction pairs of a band-shoe brake; experimental studies and discussion results. According to the principles of non-equilibrium thermodynamics, the processes of heat generation in friction pairs and its removal from the inner surface of the brake pulley rim are studied. For this purpose, briquettes made of nanoparticles and forming capillary structures are used. These briquettes rest in metal frames and are installed in three rows in wide circular grooves on the inner surface of the rim of the pulley to which the fluid chamber is attached. In addition, the interaction of the inner surface of the pulley rim with the nanocapillary structure in the briquettes is described. Experimental studies of cooling are also given, and the efficiency of the brake pulley is established.

Keywords: Band-shoe brake · Friction couples · Brake pulley rim · Nanocappillary cooling · Nanoparticle briquettes · Fluid · R&D investment

1 Introduction

Suspensions based on solid phase nanoparticles are called nanofluids [1]. The thermal conductivity of suspensions with a low concentration of solid phase particles can be described by Maxwell's theory [2]. The theory is based on a number of assumptions: the concentration of solid phase particles is small (the distance between the particles significantly exceeds their size); particles are immobile in a fluid; particles have a spherical shape, and the equations of conductive heat transfer are valid to describe the heat transfer process. In the case of using particles of the nanometer range, shrouded in a polymer film, which is dipoles, when the nanofluid moves in cooling systems, a transformation of charges occurs, leading to the formation of electronic and ionic zones. The driving force in nanofluids is jumps of various kinds of potentials.

V. Tonkonogyi et al. (Eds.): InterPartner 2022, LNME, pp. 584–593, 2023.
https://doi.org/10.1007/978-3-031-16651-8_55

The large scatter of experimental data is associated with a number of objective reasons: the method of nanoparticle synthesis, the distribution function of the nanoparticle size, the nanofluid production technology, as well as the method of measuring thermal conductivity and integrating the results.

It is necessary to solve the problem of using nanoparticles in the briquettes form that make up the nanocapillary structure in the system of forced fluid cooling of a band-shoe brake friction couples and evaluate its efficiency.

This work aims to substantiate the uneven nanocapillary fluid cooling of the brake pulley rim to equalize its energy load.

2 Literature Review

In paper [1] related to the study of heat transfer in nanofluids, it is shown that the thermal conductivity of suspensions of ultrafine oxides of aluminum, silicon, and titanium in water at a volume concentration of the order of several percent exceeds the thermal conductivity of a pure fluid by tens of percent.

The experiment results with nanoparticles of various sizes show that the thermal conductivity of a fluid based on larger particles is well described using Maxwell's theory [3].

First, the data obtained with theoretical models were constructed to describe the thermal conductivity of coarsely dispersed suspensions. The first such model was created by Maxwell [2], who obtained the relation between the thermal conductivity coefficient of the suspension λ and the base fluid λ_0.

The analysis of the effect of nanoparticles size on a nanofluid's thermal conductivity coefficient (λ) shows that the coefficient λ increases with nanoparticles size increase [4].

The modeling of nanofluid's thermal conductivity coefficient is presented in [5] as a function of a different nanoparticle mass. The authors found that λ of nanofluids at fixed sizes and concentrations of nanoparticles increases with an increase in mass [6]. The dependence of the increase in nanofluid's thermal conductivity coefficient on the nanoparticles' mass simultaneously means the same dependence on the density of particles of the same size [7].

The nanofluids' thermal conductivity coefficient is also influenced by the shape of nanoparticles, which can be spherical, cylindrical, prismatic, flat, and elliptical. Thus, in [8, 9], the thermal conductivity of nanofluids with ZnO nanoparticles having prismatic and spherical shapes was experimentally investigated at various volume concentrations of nanoparticles in the range from 0.05 to 5.0% [10]. It was found that the λ coefficient of nanofluids with zinc oxide nanoparticles increased by 12% and 18%, respectively, for the spherical and prismatic shape of nanoparticles at $\varphi = 5.0\%$, compared to the λ coefficient of the base fluid – water [11].

However, many experimental data obtained to date have a wide scatter and often contradict each other. Some data indicate an abnormal increase in nanofluids' thermal conductivity compared to the theory. However, in the course of joint research carried out by organizations from different countries, no abnormal increase in thermal conductivity at low concentrations of nanoparticles was found [12].

The fluid (water) cooling of band-shoe brake's friction couples was investigated on a new type of composite brake pulley with a band-shoe brake chamber in test-bed conditions, and the cooling efficiency was 10–14% according to [13].

Modes of motion and changes in nanofluid and steam parameters along the pulley rim inner wall length were studied. The presence of the following zones was found from the fixed edge of the rim to its free edge [14].

The efficiency of air-nanofluid cooling of a band-shoe brake friction couples in test-bed conditions of the material is 16–18% [15].

However, the considered forced cooling systems of friction couples of a band-shoe brake have the following disadvantages: there was sedimentation (falling) of nanoparticles under the influence of gravitational field and centrifugal forces settling to the bottom of the pulley chamber, which reduced the forced cooling efficiency; forced cooling contributes to an uneven distribution of volumetric temperatures from the fixed edge of the pulley rim to the free one due to the complex conductive heat transfer between the rim, its flanges, and the fastening protrusion.

3 Research Methodology

3.1 Capillary-Porous Bodies Formed from Nanoparticles

The pore distribution curve along its radius is the main structural characteristic of capillary-porous bodies formed from nanoparticles. The integral pore distribution curve characterizes the change in the relative pore volume V* (the ratio of pore volume to body volume) along the capillary radius r. The curve V* = f(r) starts from a certain value r_{min} (the minimum radius of the capillary pore) and crosses the axis of the volume V^* at the value $r = r_{max}$.

The total pore volume per body unit volume equals the porosity of the body and is determined by the ratio

$$\prod_{v} = V^*_{max} = \int_{r_{min}}^{r_{max}} \frac{dV^*}{dr} dr = \int_{r_{min}}^{r_{max}} f_v(r)dr. \tag{1}$$

The maximum fluid content ω_{max} will be equal:

$$\omega_{max} = V^*_{max} \rho_l = \prod_V \rho_l, \tag{2}$$

here ρ_l – fluid density.

The specific fluid content or its relative concentration (the mass of fluid in a porous body, referred to as the mass unit of an absolutely dry body) is equal to:

$$u = \frac{1}{\rho_s} \omega = \frac{\rho_l}{\rho_s} \int_{r_{min}}^{r} f_v(r)dr, \tag{3}$$

here ρ_s – density of an absolutely dry porous body.

Let the cross-sectional area **of** the body be 1 cm^2, part of this area is occupied by capillaries (doors), and the rest is the substance of the body itself. If all capillaries are filled with fluid, then the resistivity is:

$$u = \frac{\int_0^L \rho_l A_l dL}{\rho_s \cdot 1 \cdot L}, \tag{4}$$

where A_l – the area occupied by the fluid, that is, the area of all holes filled with fluid and located on 1 sm^2 of the body's cross-section; L – the depth (thickness) of the layer.

Let us denote the number of reduced cylindrical capillaries in a given section, having a radius from r to $r + dr$, through dn_s, then the surface porosity (clearance) is equal to:

$$\prod_s = A_l = \int_{r_{\min}}^{r_{\max}} \pi r^2 \frac{dn_s}{dr} dr, \tag{5}$$

here n_s – the number of capillaries per unit cross-sectional area of the body.

If A_l does not depend on L, that means the surface porosity is the same in any section of the body, which is equivalent to the equality of the surface and volume porosity of the body ($\Pi_V = \Pi_S$), then we can write:

$$u = \frac{\rho_l}{\rho_s} A_l = \frac{\rho_l}{\rho_s} \int_{r_{\min}}^{r_{\max}} \pi r^2 \frac{dn_s}{dr} dr. \tag{6}$$

For many, especially those rotating with a period π, the body surface porosity changes along the body coordinates. Therefore, along with the differential volumetric characteristic of pores $f_V(r)$, a differential surface characteristic of pores $f_S(r)$ is introduced and equals to:

$$f_s(r) = \frac{dA_l}{dr} = \pi r^2 \frac{dn_s}{dr}. \tag{7}$$

The secondary dehumidification and absorption curves form a hysteresis loop *RA* with all subsequent reference points lying within this loop. The primary dehumidification sweep curves, which begin on the absorption curve *A*, either meet at the intersection of the curve (Fig. 1b) or converge on the secondary dehumidification curve in the area close to the intersection. The behavior of the absorbent sweep curves is similar. Any point within the hysteresis loop can be obtained in many ways.

In the process of intense fluid flow through a porous body, in addition to capillary and gravitational forces, inertial forces are of great importance, which can be estimated by the value of the Reynolds criterion (Re), criteria of Bond (Bo), and Weber (We)

$$\mathrm{Re} = \frac{\rho \upsilon L}{\eta}; \; Bo = \frac{\rho g L^2}{\sigma}; \; We = \frac{\rho \upsilon^2 L}{\sigma}, \tag{8}$$

here υ - fluid velocity; L - typical dimensions.

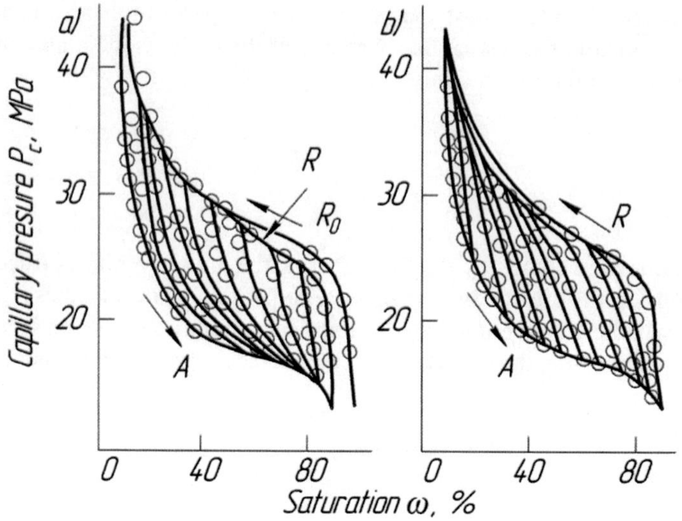

Fig. 1. Capillary pressure hysteresis for microcapillaries with a diameter of 48 μm (a) and 60 μm (b).

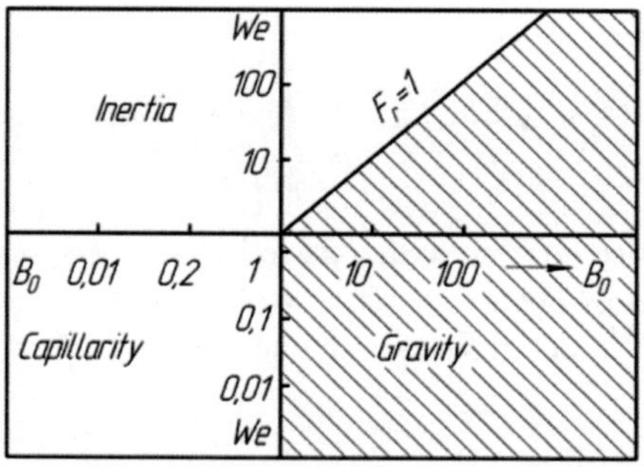

Fig. 2. Schematic relation of inertial, gravitational, and capillary forces during fluid flow through a porous body.

Combining the We and Bo criteria, one can obtain the well-known Froude criterion:

$$Fr = \frac{\upsilon}{\sqrt{gL}}, \tag{9}$$

which characterizes the ratio of inertial forces to gravity forces.

There are modes when the dominant are: 1) inertial forces; 2) capillary forces; 3) gravitational. At small We and Bo numbers, the motion of the fluid is controlled by capillary forces (Fig. 2), and the effect of gravity can be neglected. At small numbers Bo, the criterion We determines the decisive role of capillary and gravitational forces. The gravity effects can also be neglected for large values of the criterion Fr.

3.2 Design and Operation of a Nanocapillary Fluid Cooling System for Brake Friction Couples

The principle of capillary-fluid cooling of a band-shoe brake friction couples is based on the following effects: convective, vortex, radiant, conductive, and evaporative condensation in capillary structures in briquettes. Let's dwell on a brief analysis of cooling types that have been considered.

Let us turn directly to the design features of the nano-capillary fluid cooling system for the tribocoupling of the drawworks' band-shoe brake (Fig. 3).

A band-shoe brake with a nanocapillary and fluid cooling system operates as follows.

When pressing the brake control lever 16, the brake band 13 is tightened, and the working surfaces 10 of the polymer linings 11 interact with the working surface 8 of the brake pulley 6, which contributes to the generation of heat on their surfaces. In this case, a significant part of the heat is absorbed by the pulleys 6, a thermal energy accumulator.

The first case is illustrated in Fig. 3 c, when fluid 24 does not wash the polished non-working surface 9 of the brake pulley rim 6 and its briquettes, and a gap has been formed between their surfaces.

The second case is shown in Fig. 3c, when fluid 24 is on the polished non-working surface 9 of the brake pulley rim 6 and in its briquettes. In this case, convective-conductive heat transfer is substantial when the layers of fluid 24 interact with the polished non-working surface 9 of the pulley rim 6 with weak radiant heat transfer.

From Fig. 3c, it follows that the thermal state of the brake pulley parts located at different poles in the vertical plane is not the same due to changes in the thermodynamic parameters of the fluid and the washing air, which contributes to its gradients change, and as a consequence, the intensification of conductive, convective and radiative heat transfer in the proposed nanocapillary fluid cooling system.

Thus, when operating in the modes of brake pulley rotation or frictional interaction of band-shoe brake friction couples, the following types of heat transfer take place:

- in the first mode - convective air and fluid, conductive, as well as radiant with a working and non-working surface (polished) of the pulley rim; as well as evaporative-condensation heat exchange with nanocapillary structures of briquettes;
- in the second mode - convective, air and fluid, conductive, as well as radiant with a polished non-working surface of the brake pulley rim; in this case, the nanocapillary structure in briquettes is saturated with fluid (Fig. 3d).

4 Results and Discussion

The fluid movement structure in the cooling chamber cavity depends on its shape and location in the heated pulley rim space.

Based on Table 1, let's write down the ratio of the form

$$\frac{\lambda_1}{\lambda_2} = \frac{A_1}{A_2}; \tag{10}$$

$$\frac{\lambda_2}{\lambda_3} = \frac{A_2}{A_3}; \tag{11}$$

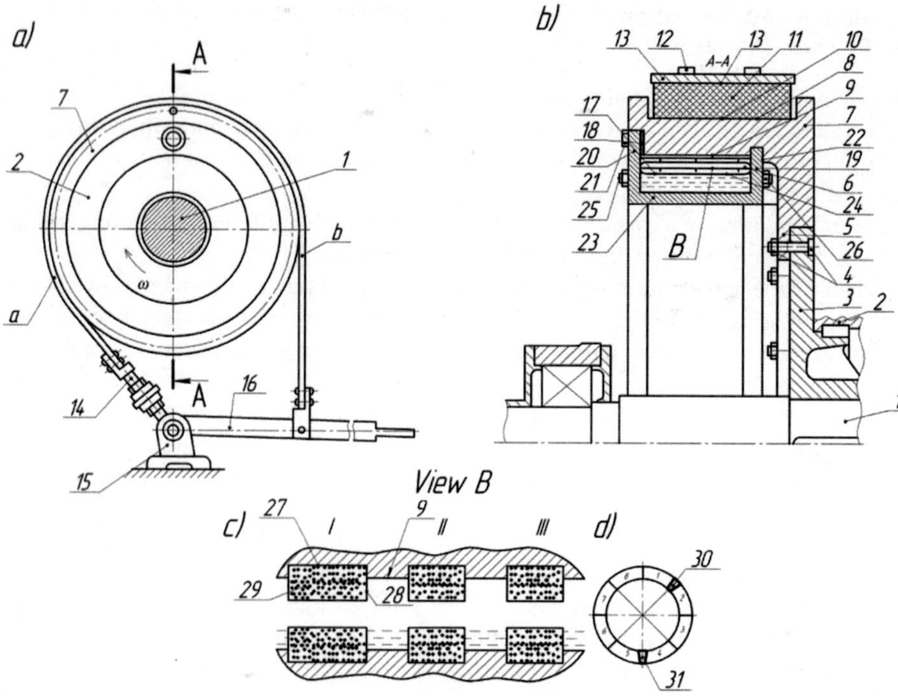

Fig. 3. Band-shoe brake with a forced nanocapillary fluid cooling system: a - general view; b - section along A-A on a; c - longitudinal section of cooling devices (view B on b) located in the upper and lower parts of the pulley rim; Fig. 1 g - diffusers and confusers formed between the ends of stationary briquettes during rotation of the pulley rim.

$$\frac{\lambda_1}{\lambda_3} = \frac{A_1}{A_3}; \tag{12}$$

The discrepancy between the values in relation (10) is 13.0%, in (11) only 5.0% and in (12) 6.0%, and the average is 8.0%. This suggests that the materials for nanoparticles in briquettes are selected correctly. This is evidenced by the ratio for powder nanoparticles for aluminum (Al) - $\frac{0{,}748}{0{,}9}$; copper (Cu) - $\frac{0{,}8}{0{,}95}$; silicon carbide (SiC) - $\frac{0{,}63}{0{,}7}$ of thermal conductivity coefficients, W/(m·°C) [nanoparticles in a fluid are in the numerator, in the denominator - nanoparticles in sector briquettes washed by a fluid.

Experimental studies of friction couples "steel 35KHNL - FK-24A" of band-shoe brake model, the pulley rim of which was equipped with a 200 dm^3 fluid chamber, and on the polished inner surface of the rim, there were installed sector briquettes with a gap into annular grooves of various widths (eight pieces). The weight of each type of briquette was: I - 80 g; II - 50 g; III - 25 g. Simultaneously, the weight of the sheet copper frame was equal: I - 55.4 g; II - 28.7 g; III - 13.4 g. While the weight of the perforated frame was twice lighter. Table 2 shows the experimental studies' results.

Table 1. The selection of nanoparticle materials for sector briquettes by thermal conductivity coefficients and their cross-sectional area.

Nanoparticle selection options	Thermal conductivity coefficient λ, W/(m·°C): I – Al; II – Cu; III – SiC	Segment number of briquettes		
		I	II	III
		Relation between thermal conductivity coefficients, λ, W/(m·°C)		
		I – II	II - III	I - III
		$\lambda_1/\lambda_2 = 1{,}7$	$\lambda_2/\lambda_3 = 2{,}0$	$\lambda_1/\lambda_3 = 3{,}3$
		Relation between the cross-sectional areas of briquettes, sm^2		
		$A_1/A_2 = 1{,}9$	$A_2/A_3 = 2{,}1$	$A_1/A_3 = 3{,}5 \ldots 4{,}1$

Table 2. Experimental data of serial pulley rim energy loading (in the numerator) and with its nanocapillary fluid cooling.

Segment number of briquettes		I	II	III
Temperatures, °C	Unit loading, MPa	$\frac{0{,}75^*}{0{,}7}$	$\frac{0{,}85}{0{,}75}$	$\frac{0{,}95}{0{,}8}$
	Superficial	$\frac{320{,}0^*}{300{,}0}$	$\frac{340{,}0}{320{,}0}$	$\frac{360{,}0}{340{,}0}$
	Volume	$\frac{100{,}0^*}{20{,}0}$	$\frac{110{,}0^*}{85{,}0}$	$\frac{120{,}0^*}{90{,}0}$

*Note: the regularities of changes in energy loading parameters follow a linear law (from the fixed edge of the rim to the free one, III - I).

Analysis of non-uniform forced cooling by local heat exchangers (sector briquettes) of the brake pulley rim made it possible to establish the following: a 60% increase in thermal conductivity coefficients λ_1, λ_2, λ_3, from 0.748; 0.8 and 0.63, W/(m·°C) up to 0.9; 0.95 and 0.7 W/(m·°C) respectively; a change in the nanofluid parameters in sector briquettes was achieved at a heat flux density $q = 2 \cdot 10^2 \ldots 2 \cdot 10^4$ W/m^2, an average bulk temperature $t_o = 85$ °C of the pulley rim at an average linear speed from 2.0 to 6.0 m/s rotation; the heat transfer coefficient was 100… 350 W/(m^2·°C) from the polished inner surface of the pulley rim; the heat transfer coefficient was 75… 300 W/(m^2·°C) through the multilayer structure.

Theoretical and experimental studies of non-uniform nanocapillary liquid cooling of friction pairs of a drawworks band-shoe brake made it possible to state the following: according to the principles of non-equilibrium thermodynamics, the processes of heat generation in friction pairs and the local nature of its removal from the inner surface of the brake pulley rim were studied; means providing local heat removal are briquettes made of nanoparticles from various materials and forming capillary structures resting in perforated metal frames; the latter is installed in three rows of different widths (circular grooves on the inner surface of the pulley rim, to which the fluid chamber is attached); it has been established that the thermal conductivity of nanoparticles in briquettes with different capillary structures does not go beyond a certain limit level with an increase in their concentration; thermal conductivity coefficient of nanoparticles in briquettes is some integral property of this non-standard two-phase system; experimental studies of the forced local cooling system showed that in briquettes with nanoparticles from various materials and capillary structures realized in them, an increase in the thermal conductivity coefficient by an average of 60% was achieved; under a given loading mode of a model band-shoe brake, fluctuations in specific loads along the width of its rim amounted to 0.7–0.8 MPa, surface (300–340 °C) and bulk (80–90 °C) temperatures, which is approximately 35% in on average less than in a water-cooled serial rim.

5 Conclusions

During the research, an assessment was made of non-uniform nanocapillary liquid cooling of friction pairs of a drawworks band-shoe brake to equalize its energy load. This, in turn, contributes to the stabilization of the operational parameters of the friction pairs of the brakes and, as a result, a decrease in their wear.

Based on the calculated dependences, the capillary pressure hysteresis for microcapillaries with a diameter of 48 μm and 60 μm was constructed from which it follows that for many materials, the minimum saturation (wetting phase volume corresponding to high capillary pressures) is the same for the initial and secondary drying curves.

It is proved that the Reynolds, Bond, and Weber criteria can be used to evaluate inertial forces, which, in addition to capillary and gravitational forces, occur when a fluid flows through a porous body.

Based on theoretical and experimental studies, the design of the friction unit of a band-shoe brake with a forced system of nanocapillary liquid cooling has been developed, which makes it possible to increase the braking efficiency by 15%.

Experimental data have shown that the heat transfer coefficient in the nanocapillary structure system can be increased by varying the particle size, i.e., their area of interaction with the liquid or by its transformation into vapor.

The results of theoretical and experimental studies can be used in design bureaus, and factories, and the further development of friction pairs of brake friction units.

References

1. Nosko, A., Tarasiuk, W., Sharifullin, I., Safronov, E.: Tribotechnical and ecological evaluation of friction pairs of brake devices in lifting and transport machines. Frict. Wear **41**, 347–353 (2020). https://doi.org/10.3103/S106836662004008X
2. Al-Maghalseh, M., Mahkamov, K.: Methods of heat transfer intensification in PCM thermal storage systems. Renew. Sustain. Energy Rev. **92**, 62–94 (2018). https://doi.org/10.1016/j.rser.2018.04.064
3. Xu, H., Xing, Z., Wang, F., Cheng, Z.: Review on heat conduction, heat convection, thermal radiation and phase change heat transfer of nanofluids in porous media: fundamentals and applications. Chem. Eng. Sci. **195**, 462–483 (2018). https://doi.org/10.1016/j.ces.2018.09.045
4. Balitskii, A., et al.: Hydrogen containing nanofluids in the spark engine's cylinder head cooling system. Energies **15**, 59 (2022). https://doi.org/10.3390/en15010059

5. Ciavarella, M., Johansson, L., Afferante, L., Klarbring, A., Barber, J.: Interaction of thermal contact resistance and frictional heating in thermoelastic instability. Int. J. Solids Struct. **40**, 5583–5597 (2003). https://doi.org/10.1016/S0020-7683(03)00313-5
6. Özişik, M., Orlande, H., Colaço, M., Cotta, R.: Finite Difference Methods in Heat Transfer, 2nd edn. CRC Press, Boca Raton (2017). https://doi.org/10.1201/9781315121475
7. Rudyak, V.Y., Minakov, A.V., Pryazhnikov, M.I.: Thermophysical properties of nanofluids and similarity criteria. Lett. J. Tech. Phys. **42**(24), 9–16 (2016)
8. Choi, E.S., et al.: Enchancement of thermal and ekectrical properties of carbon nanotube polymer composites by magnetic field processing. J. Appl. Phys. **94**(9), 6034–6039 (2003)
9. Zhang, S., et al.: Simulation study on friction and wear law of brake pad in high-power disc brake. Math. Probl. Eng. **2019**, 1–15 (2019). https://doi.org/10.1155/2019/6250694
10. Lee, J., Ramamurthi, K.: Fundamentals of Thermodynamics, 1st edn. CRC Press, Boca Raton (2022). https://doi.org/10.1201/9781003224044
11. Ropyak, L.Y., Makoviichuk, M.V., Shatskyi, I.P., Prytula, I.M., Gryn, L.O., Belyakovskyi, V.O.: Stressed state of laminated interference-absorption filter under local loading. Funct. Mater. **27**(3), 638–642 (2020)
12. Malanchuk, N., Martynyak, R., Monastyrskyy, B.: Thermally induced local slip of contacting solids in vicinity of surface groove. Int. J. Solids Struct. **48**(11–12), 1791–1797 (2011)
13. Belyakov, N., Nosko, A.: Non-ideal thermal contact of bodies during friction. Book House "Librokom" (2010)
14. Howell, J., Mengüç, M., Daun, K., Siegel, R.: Thermal Radiation Heat Transfer, 7nd edn. CRC Press, Boca Raton (2020). https://doi.org/10.1201/9780429327308
15. Han, J., Wright, L.: Analytical Heat Transfer, 2nd edn. CRC Press, Boca Raton (2022). https://doi.org/10.1201/9781003164487

Author Index

V. Tonkonogyi et al. (Eds.): InterPartner 2022, LNME, pp. 595–597, 2023.
https://doi.org/10.1007/978-3-031-16651-8